Fish Reproductive Biology

Dedication

Bernard Megrey, known as Bern among his family and friends, passed away peacefully at his home on 1 October 2010, at the age of 60. He had just returned from the ICES Annual Science Conference in Nantes, France.

Bern was one of the driving forces when the idea of this book was born and virtually baptised during a late night/early morning at the ICES Annual Science Conference in Aalborg, Denmark during fall 1995. In the years following, the idea was further developed and finally turned into this book and the first edition was published in 2009. Bern's background was perfect for the book. During his doctoral research he started working for the NOAA, Alaska Fishery Science Center in Seattle, USA. He was assigned to the Groundfish Assessment Group, preparing the first stock assessment for the newly discovered walleye pollock fishery in the Gulf of Alaska. His work enabled a team of scientists to provide timely forecasts of abundance and biomass to the North Pacific Fishery Management Council. In 1987, he started to work in the Fisheries Oceanography Coordinated Investigations (FOCI) programme where his charge was to develop recruitment prediction models and oversee the analytical personnel. During his tenure in the FOCI programme, Bern continued to work on recruitment prediction, but also broadened his focus from single species to ecosystems. The Gulf of Alaska recruitment prediction model that he developed for walleye pollock is one of the few models that incorporate both environmental and biological data in predictions based on an underlying mechanistic model. Over the years he put a lot of effort into organisations such as ICES, PICES and AFS, both as a working group member and preparing theme sessions at annual conferences, resulting in several books and special issues of journals, where he acted as both co-editor and author. He served for six years (2001–2007) on the editorial board of the *ICES Journal of Marine Science*. Bern had an enormous working capacity and during his career he worked together with colleagues from many regions of the world. He had the gift of being positive, encouraging colleagues and enjoying discussing new ideas and technologies that could advance science.

As friends and colleagues, we are very grateful for the time we had together with Bern. He made our lives richer, influenced our way of thinking, and we are proud of what we were able to achieve together. Bern was too young when he passed away, he had so many ideas he wanted to work on and regrettably never will be able to pursue. We miss you!

Tore Jakobsen
Michael J. Fogarty
Erlend Moksness

Fish Reproductive Biology

IMPLICATIONS FOR ASSESSMENT AND MANAGEMENT

SECOND EDITION

Edited by

Tore Jakobsen
Institute of Marine Research,
Bergen, Norway

Michael J. Fogarty
Northeast Fisheries Science Center,
National Marine Fisheries Service, Woods Hole, Massachusetts, USA

Bernard A. Megrey
Deceased

Erlend Moksness
Institute of Marine Research,
Flødevigen Marine Research Station, Arendal, Norway

WILEY Blackwell

This edition first published 2016 © 2016 by John Wiley & Sons Ltd
First edition published 2009 © 2009 by Blackwell Publishing Ltd

Registered office: John Wiley & Sons, Ltd, The Atrium, Southern Gate, Chichester, West Sussex, PO19 8SQ, UK

Editorial offices: 9600 Garsington Road, Oxford, OX4 2DQ, UK
The Atrium, Southern Gate, Chichester, West Sussex, PO19 8SQ, UK
111 River Street, Hoboken, NJ 07030-5774, USA

For details of our global editorial offices, for customer services and for information about how to apply for permission to reuse the copyright material in this book please see our website at www.wiley.com/wiley-blackwell.

Library of Congress Cataloging-in-Publication Data

Names: Jakobsen, Tore, editor.
Title: Fish reproductive biology : implications for assessment and management / edited by Tore Jakobsen and three others.
Description: Second edition. | Chichester, West Sussex, U.K. ; Hoboken, NJ : Wiley-Blackwell, 2016. | Includes bibliographical references and index.
Identifiers: LCCN 2015033704 | ISBN 9781118752746 (hardback)
Subjects: LCSH: Fish stock assessment. | Fishes–Reproduction. | Recruitment (Population biology) | Fishery management. | BISAC: TECHNOLOGY & ENGINEERING / Fisheries & Aquaculture.
Classification: LCC SH329.F56 F57 2016 | DDC 333.95/611–dc23 LC record available at http://lccn.loc.gov/2015033704

A catalogue record for this book is available from the British Library.

Wiley also publishes its books in a variety of electronic formats. Some content that appears in print may not be available in electronic books.

Cover image: GettyImages-155307256/Lee Chin Yong

Set in 10/12.5pt Times by Aptara Inc., New Delhi, India

1 2016

Contents

Part III Incorporation of Reproductive Biology and Recruitment Considerations into Management Advice and Strategies

Contributors

Gavin A. Begg, South Australian Research and Development Institute, 2 Hamra Avenue West Beach, South Australia 5024, Australia

Steven X. Cadrin, University of Massachusetts, School for Marine Science and Technology, 200 Mill Road, Fairhaven MA 02719, USA

Kevern L. Cochrane, Department of Ichthyology and Fisheries Science, Rhodes University, P.O. Box 94, Grahamstown 6140, South Africa

Natal'ya G. Emel'yanova, Moscow State University, Faculty of Biology, Department of Ichthyology, Moscow 119992, Russia

Michael J. Fogarty, Northeast Fisheries Science Center, National Marine Fisheries Service, 166 Water Street, Woods Hole, MA 02543, USA

Edward D. Houde, University of Maryland Center for Environmental Science, Chesapeake Biological Laboratory, P.O. Box 38, Solomons, MD 20688, USA

Tore Jakobsen, Institute of Marine Research, P.O. Box 1870 Nordnes, 5817 Bergen, Norway

Olav S. Kjesbu, Institute of Marine Research (IMR) and Hjort Centre for Marine Ecosystem Dynamics, P.O. Box 1870 Nordnes, NO-5817 Bergen, Norway; Centre for Ecological and Evolutionary Synthesis (CEES), Department of Biosciences, University of Oslo, P.O. Box 1066 Blindern, NO-0316 Oslo, Norway

Nancy C. H. Lo, NOAA/NMFS Southwest Fisheries Science Center, 8604 La Jolla Shores Drive, La Jolla, California, CA 92037, USA

Lise Marty, Centre for Ocean Life, Technical University of Denmark, National Institute of Aquatic Resources, Kavalergården 6, 2920 Charlottenlund, Denmark

C. Tara Marshall, University of Aberdeen, School of Biological Sciences, Zoology Building, Tillydrone Avenue, Aberdeen AB24 2TZ, UK

Bernard A. Megrey, Deceased.

Erlend Moksness, Institute of Marine Research, Flødevigen Marine Research Station, 4817 His, Norway

Carl M. O'Brien, Centre for Environment, Fisheries & Aquaculture Science, Lowestoft Laboratory, Pakefield Road, Lowestoft, Suffolk NR33 0HT, UK

Loretta O'Brien, Northeast Fisheries Science Center, National Marine Fisheries Service, 166 Water Street, Woods Hole, MA 02543, USA

Dimitri A. Pavlov, Moscow State University, Faculty of Biology, Department of Ichthyology, Moscow 119992, Russia

John. G. Pope, NRC (Europe) Ltd, The Old Rectory, Staithe Road, Burgh St Peter, Norfolk NR34 0BT, UK

Marie-Joëlle Rochet, IFREMER, B.P. 21105, 44311 Nantes Cedex 03, France

Paul E. Smith, NOAA/NMFS Southwest Fisheries Science Center, 8604 La Jolla Shores Drive, La Jolla, California, CA 92037, USA

Motomitsu Takahashi, Seikai National Fisheries Research Institute, Fisheries Research Agency, 1551-8, Taira-machi, Nagasaki-shi, Nagasaki, 851-2213, Japan

Preface

Six years have elapsed since the first edition of this book was published. The book was intended to reach a fairly wide scientific readership, attempting to link different scientific disciplines under a common theme, and we see the request for an update as a sign that the book has at least to some extent has succeeded in achieving this. The Second Edition contains an update of nine of the eleven chapters (the exceptions being Chapters 5 and 11), and considering the changes in professional and private circumstances that will have occurred among the 17 authors of the first edition during a six-year interval, we are pleased and grateful to get this response. The update primarily adds new developments in the scientific fields described, but we believe there also is an improvement of the overall quality of the text. We have decided to keep our Introduction unchanged, since this broad description of the chapters is not significantly affected by the updates.

The Editors

Acronyms

ACF	Actual fecundity
ACFM	Advisory Committee for Fisheries Management
AF	Absolute Fecundity
AFRB	Applied Fish Reproductive Biology
AOGCM	Atmosphere-Ocean General Circulation Model
ASAP	Age-Structured Assessment Program
CalCOFI	California Cooperative Oceanic Fisheries Investigations
CalVET	CalCOFI Vertical Egg Tows
CANSAR-TAM	Catch-at-age ANalysis for SARdine - Two Area Model
CEFAS	Center for the Environment, Fisheries and Aquaculture Science
COADS	Comprehensive Ocean Atmosphere Dataset
CPUE	Catch-Per-Unit-Effort
CUFES	Continued Underwater Fish Egg Sampler
CV	Coefficients of Variation
DARDNI	Department of Agriculture and Rural Development - Northern Ireland
DBOBL	Double-oblique Deep Bongo
Defra	Department for Environment, Food and Rural Affairs
DEPM	Daily Egg Production Method
DFO	Department of Fisheries and Oceans (Canada)
DNA	Deoxyribonucleic Acid
DO	Dissolved Oxygen
Dy	Yolk Diameter
ENSO	El Nino Southern Oscillation
EPM	Egg Production Methods
ERSEM	European Regional Seas Ecosystem Model
ESOHS	Evolutionary Stable Optimal Harvesting Strategies
ESS	Evolutionarily Stable Strategy
ETM	Estuarine Turbidity Maximum Zone
EU	European Union
FAO	Food and Agriculture Organization (United Nations)
FF	Final fecundity
FRS	The Fisheries Research Services Marine Laboratory
FSB	Female-only Spawning Biomass
GAM	Generalised Additive Model
GCMs	General Circulation Model

GIS	Geographic Information System
HAMSOM	Hamburg Shelf Ocean Model
IBM	Individual-based Model
IBTS	International Bottom Trawl Survey
ICCAT	International Commission for Conservation of Atlantic Tunas
ICES	International Council for Exploration of the Sea
ICNAF	International Commission for the Northwest Atlantic Fisheries
ICPBMs	Individual-based, Coupled Bio-Physical model
IF	Individual fecundity
IMR	Instantaneous Mortality Rate
IRF	Individual Relative Fecundity
LIDAR	Light Detection and Ranging
LSRP	Long-term Stock Reproductive Potential
MPA	Marine Protected Area
MRM	Marine Resource Management
MSP	Marine Spatial Planning
MSVPA	Multi-Species Virtual Population Analysis
MSY	Maximum Sustainable Yield
NAFO	Northwest Atlantic Fisheries Organization (NAFO)
NAO	North Atlantic Oscillation
NOAA	National Oceanic and Atmospheric Administration (USA)
NRC	National Research Council (USA)
OECD	Organization for Economic Cooperation and Development
PDO	Pacific Decadal Oscillation
PF	Population fecundity
PGC	Primordial Germ Cells
PGEGGS	Planning Group on (North Sea Cod and Plaice) Egg Surveys
PNR	Point-of-No-Return
POF	Post-ovulatory Follicles
POM	Princeton Ocean Model
RA	Risk Assessment
RCM	Regional Climate Models
RF	Relative Fecundity
RFP	Relative Potential Fecundity
RFP,S	Relative Somatic Potential Fecundity
RFP,W	Relative Body Weight Potential Fecundity
RV	Reproductive Value
SEERAD	Scottish Executive Environment and Rural Affairs Department
SF	Species fecundity
SNAC	Statistical emulator of the North-east Atlantic Circulation
SRP	Stock Reproductive Potential
SS2	Stock Synthesis 2
SSB	Spawning Stock Biomass
SSB/R	Spawning Stock Biomass per Recruit
SST	Sea Surface Temperature

STECF	Scientific, Technical and Economic Committee for Fisheries (EU)
STEREO	Stock Effects on Recruitment Relationships
TAC	Total Allowable Catch
TEP	Total Egg Production
TL	Total Length (Average Body Length)
UEA	University of East Anglia
VPA	Virtual Population Analyses
YOY	Young of Year

Introduction

Tore Jakobsen, Michael J. Fogarty, Bernard A. Megrey and Erlend Moksness

As long as humans have exploited marine fish resources, fluctuations in availability and yields have been experienced. Nearly a century ago Hjort (1914, 1926) linked variation in yield to variability in recruitment, and today scientists still identify recruitment as a major driving force in stock fluctuations. This is reflected in the research focus on reproductive dynamics and recruitment over the past 30 years (see: http://www2.ncsu.edu/elhs/elhspubs.html) and in recent strategic plans by the International Council for the Exploration of the Sea (ICES) (Anon. 2001). Most international research programmes focusing on reproductive biology and recruitment of marine fishes do so to improve the understanding of the underlying processes controlling survival and growth during the early life stages. Few are aimed primarily at directly linking these processes to the assessment and management of harvested stocks, although progress in this area is evident. As exploited fish stocks decline, the demand for information on recruitment dynamics and for better prediction of recruitment typically increases. Furthermore, there is an increasing awareness of the importance of understanding these mechanisms for fisheries management.

The realisation that recruitment processes are of fundamental importance to the prosecution and management of fisheries has resulted in concerted efforts to monitor recruitment and to understand the factors controlling variability of marine fish populations. These efforts provide an unparalleled opportunity to study processes regulating fish populations and to understand and predict the impacts of harvesting on living marine resources. An extremely valuable compendium of estimates of adult biomass and recruitment has been assembled for exploited marine resources throughout the world's oceans (http://www.mscs.dal.ca/~myers/welcome.html), offering opportunities to examine patterns of recruitment variability, compensatory dynamics, and the current status of these stocks. The economic importance of fishes and their societal and cultural relevance provide powerful incentives for large-scale, sustained studies of their dynamics. Few other taxonomic groups – terrestrial or aquatic – offer such rich databases for examination of these processes as those available for fishes.

Fish Reproductive Biology: Implications for Assessment and Management, Second Edition.
Edited by Tore Jakobsen, Michael J. Fogarty, Bernard A. Megrey and Erlend Moksness.

Scope and organisation of the book

The overall goal of this book is to provide a picture of the present use of information on fish reproductive biology in assessment and management and its potential for improving management of these resources. We have divided this book into three main parts. The first part sets the stage by focusing on recruitment processes, reproductive biology and the effects of fishing on exploited marine fishes. Here, we describe the critical role of recruitment in replenishing an exploited population, the importance of fundamental reproductive dynamics in this process, and how natural and anthropogenic forcing factors affect recruitment and sustainability. The second part explores the fundamental elements for any evaluation of fish population dynamics. These encompass issues related to identifying populations and stock units, estimation methods for obtaining abundance and demographic information at different life history stages, and the data requirements for more refined estimates of reproductive output and dynamics for inclusion in assessment and management. The final part describes both the current approach to management and ways in which a detailed understanding of reproductive processes can inform new approaches to management. Contributions to each of these parts are described in greater detail below. We also provide below references to key texts for further reading which complement the material presented in the individual chapters.

Biology, population dynamics and recruitment

Consideration of the form of the relationship between the reproductive output of the population and the resulting recruitment lies at the very heart of any understanding of how a fish population will respond to sustained perturbations such as fishing (Chambers & Trippel 1997). This issue has been a focus of fisheries research for the past half century. Any such representation depends on an understanding of the life-cycle dynamics of the population from the production of viable eggs, through the gauntlet of processes that affect the probability of survival, to the age or size at recruitment. If we are to predict the likely effects of fishing on an exploited population we require conceptual and analytical models of this process. In Chapter 1, these considerations are used to introduce the principal themes recurring throughout this book, including how a refined understanding of reproductive dynamics influences our perception of the status of the population, the relationship between the adult population and recruitment, and the choice of effective management strategies. The earliest recruitment models were cast in terms of total egg production. However, the general lack of time series of information on fecundity at the time necessitated the use of a proxy for this quantity – usually the total adult biomass of the population (Beverton & Holt 1957). The focus of these early efforts therefore was on the compensatory mechanisms that shape the relationship between spawning stock biomass and recruitment. There is now accumulating evidence that the spawning stock biomass alone is not always an adequate measure of the spawning potential of a fish stock. Maternal factors such as fecundity and egg quality are known to be affected by growth, condition, body size and spawning class. Furthermore, sex ratios of many populations change with increasing fishing pressure in combination with selective harvesting practices. Collectively, these considerations argue for a return to the origins of recruitment theory with its recognition of the importance of the actual reproductive output of the population. Translated into a management setting, we see that ignoring the effective reproductive output of a population and status of the adult population in some cases can lead to an overly optimistic view of the condition of the population with attendant risks to sustainability.

The importance of gaining a detailed understanding of reproductive processes of fish in the context of recruitment studies has long been appreciated (Potts & Wootton 1984). Chapter 2 provides essential background on the reproductive biology of fishes with considerations spanning cellular development of primordial germ cells, individual fecundity, reproductive strategies, ontogenetic development and factors affecting the quality of reproductive products. An understanding of the reproductive processes of fish at the cellular level is ultimately necessary to correctly determine and interpret the potential reproductive output of a population. An understanding of reproductive strategies and mating systems is no less important for some species. Semelparous life histories, in which adults spawn once and die, are notable among species such as capelin and Pacific salmon, while iteroparity involving multiple reproductive opportunities throughout the lifespan is common in most other fish taxa. Most fishes maintain separate sexes throughout the lifespan, but various forms of sequential hermaphroditism are also known among a number of important exploited species. Discrete seasonal patterns of spawning are common in temperate and boreal systems and are often linked to seasonal primary and secondary production cycles. In contrast, many tropical and subtropical species spawn throughout the year.

Factors underlying the characteristically large variation in recruitment of fishes, broadly classified into trophodynamic and physical/environmental components, are described in Chapter 3. The striking prevalence of highly variable recruitment patterns in marine fishes and the recognised underlying importance of stabilising mechanisms has been called the stock–recruitment paradox (Rothschild 1986). Trophodynamic considerations such as prey availability during the pre-recruit stages and the risk of predation on the pre-recruits strongly influence survival. Physical processes such as turbulence can directly affect the probability of prey capture, while other physical factors such as temperature affect activity levels and metabolic requirements. The role of transport, retention and loss has been linked to variation in survival during the early life stages of fish. Survival depends on successful transport to and/or retention within favourable habitats. Different early life stages exhibit different vulnerabilities to these environmentally driven events. Although recruitment variability obviously is linked to processes in the local environment, there is evidence that these processes are driven by large-scale environmental variations. Thus, major climate and oceanic events have been shown to have effects on fish populations over a wide area. Fish populations respond to biotic and abiotic environmental forcing on both short (high-frequency) and long (low-frequency) time scales. The high levels of interannual variation in recruitment characteristically observed in fish stocks reflect high frequency forcing, while long-term regime shifts in environmental factors are followed by changes in overall recruitment levels. High- and low-frequency changes in recruitment hold very different implications for the development and evaluation of management strategies. In the former case, stochasticity in recruitment should be taken into account in making short-term tactical management decisions. In the latter, adjustments of biological reference points used in management may be necessary to accommodate persistent shifts in productivity.

In Chapter 4, the direct and indirect effects of fishing on abundance and demographic structure of fish populations are described. Among the direct effects are a reduction in biomass of the adult population and truncation of the age structure. Changes in age composition, sex ratio, age or size at maturation, and other demographic characteristics may in turn be critical for recruitment. The potential indirect effects include the impact of fishing activities on the structure of ecological communities affecting the prey and predators of the species of interest, disruption of habitat, and so on (Hall 1999). These effects also have important implications for recruitment. An understanding of the mechanisms by which exploited populations can potentially compensate for changes in abundance or population structure induced by harvesting is crucial. Many life-history traits of fishes have been assumed to be plastic, responding to environmental change. Currently, important efforts are under way in an attempt to separate environmental effects from potential evolutionary change induced by artificial selection due to fishing. Well-documented changes in the age or size of maturation under size-selective harvesting for a number of fish species have been examined in both laboratory and field studies. The main concern is that fishing could lead to a loss in genetic diversity and thereby produce non-reversible, or very slowly reversible, changes in the fish populations. Hence, rebuilding stocks that have collapsed can, as experience has shown, be a very slow process, and this means that overfishing poses a larger risk than previously expected.

Information critical to successful assessment and management: methods and data

The rationale and methods employed in scientific surveys of pre-recruit stages of fish are described in Chapter 5. Plankton surveys have been used to measure egg and larval abundance as well as other components of planktonic communities, including zooplankton species that are both predators and prey of fish larvae (Gunderson 1993). Stage-specific estimates of egg abundance are routinely used for some species to back-calculate the abundance of spawners, based on knowledge of fecundity and estimates of egg mortality rates. Larval abundance estimates have also been used for this purpose, and in some cases for making recruitment predictions. Estimates of juvenile abundance derived from net-based sampling, direct visual observation (e.g., in coral reef systems) and other approaches are used to provide forecasts of recruitment to the fishery. Mortality rates during the early life stages can be estimated based on serial sampling of successive life stages. Mortality estimates and their variability provide crucial information on expected recruitment variability and the probability distribution of recruitment. This can also provide important insights into the timing of critical population events, such as where in the life cycle density dependence is important or where the highest interannual variability in mortality occurs. The overall spatial scales on which sampling of the early life stages is conducted, and the volume filtered by the sampling gear in relation to small-scale patchiness of the organisms, are important factors in the calculation of abundance indices for the early life stages. Consideration of small-scale distribution patterns is increasingly possible with new optical and acoustic sampling tools.

In Chapter 6, the critically important issue of defining population or stock units is addressed. Often, stocks used as units in management are defined more from practical considerations, such as the spatial resolution of catch data or national borders, than biological considerations (Cadrin *et al.* 2004). This is clearly neither defensible nor desirable considering the importance of knowing the true dynamics of exploited populations in management, and scientific advice will attempt to address biological stocks whenever there is adequate data and the stock identity is known. Stock identification is complicated by the fact that fish stocks rarely are completely isolated from each other. Mixing may occur at all life stages, and in some cases individuals may transfer from one stock to another. Stocks are normally most clearly separated during the spawning periods when the fish tend to aggregate, and it may then be possible to map the distribution of their eggs and larvae. If the distribution of later stages in the life cycle is also known, stock identity may not be a problem. However, there may be mixing of stocks even on the spawning grounds, and recruits originating from different spawning grounds may produce mixed catches when they enter the fishery. Furthermore, adult fishes are usually distributed over a wide geographical area between spawning periods, and mixing of individuals from different spawning populations on the fishing grounds is not uncommon. Such mixing of stocks requires mapping of the population structure to define the unit of analysis, and a number of methods are being applied. The tools available to identify populations include the examination of meristic characters, morphometric analysis, infestation rate of various parasites, analysis of nuclear and/or mitochondrial DNA, fatty acid profiles, otolith microstructure and otolith microchemistry. The methods are quantitative, but may give somewhat diverging results and the overall evaluation tends to be qualitative. Recently, consideration of metapopulation structure of fishes and the potential management implications have been explored with particular reference to issues such as the placement of marine protected areas.

As described in Chapter 7, fish stock assessments serve as a vehicle for synthesis of diverse information on stock status, and prediction of the probable outcomes of alternative management outcomes (Quinn & Deriso 1999). The main purpose of stock assessment is to provide fisheries managers with the information needed to make effective management decisions. Fisheries management requires a determination of the current state of a stock, for example whether the exploitation rate is above a sustainable level or the biomass is low compared to earlier years. In addition, predictions of catch and biomass are needed for managers to know the most likely future effects of alternative management actions. Stock assessment is highly dependent on the available data, and a number of different classes of model have been developed to meet different needs. The analyses depend on an evaluation of information derived from the fishery (catches, discards, fishing effort, age or size composition of the catch, etc.) and from fishery-independent sources, mainly research surveys. The simplest models do not include estimates of spawning stock biomass (SSB) and recruitment may be assumed to be constant, whereas more complex models typically include annual estimates of both. Recruitment is related to the abundance of the adult population, although the form of the stock–recruitment relationship may be obscure. It is typically masked by environmental influences and often apparent only when the stock has been driven to low levels. Considerations related to data availability have meant that the reproductive output of fish populations has traditionally been measured in terms of adult biomass as a proxy for total egg production. However, the fundamental models used to estimate population size by size or age classes in traditional stock assessments provide an important framework for extension to more refined estimates of reproductive output as information accrues on changes in sex ratios over time, female condition, and fecundity for an increasing number of species. Similarly, the models used to frame management advice can also be modified to incorporate more detailed considerations of reproductive biology.

Consideration of reproductive strategies and tactics and estimation of reproductive potential at the cellular and organismal levels are described in Chapter 8. Reproductive strategies encompass the range of expression of reproductive traits over the full spectrum of environmental conditions (Potts & Wootton 1984). Reproductive tactics refer to the manifestation of specific reproductive traits under particular environmental conditions. This distinction sets the stage for the consideration of factors affecting the regulation of fecundity in marine fishes. The majority of marine fish species are highly fecund and produce a 'superabundance' of eggs. In these species the parental energetic investment per individual egg is relatively low and mortality during the pre-recruit stages is very high. Some species (notably the elasmobranchs), however, produce relatively few young per spawning event, some exhibit parental care, and others are ovoviviparous. The expression of factors such as fecundity and egg size under different environmental conditions is, of course, a critical element of stock reproductive potential. Fundamental reproductive characteristics such as whether a species exhibits determinate or indeterminate spawning have important implications for our ability to measure fecundity at the individual level. In turn, this affects our ability to estimate total egg production of a population. Furthermore, emerging evidence suggests that a clear distinction between determinate and indeterminate fecundity for some species in some circumstances may not always be possible. Finally, the transition from estimates of potential egg production to realised egg production, including consideration of atresia and other mechanisms of downregulation of fecundity, is critically important in estimating the reproductive dynamics of a population.

Incorporation of reproductive biology and recruitment considerations into management advice and strategies

The forms of biological advice on management of fish stocks currently given on both international and national levels are described in Chapter 9. This advice is traditionally framed in terms of benchmarks related to fishing mortality rates or biomass levels relative to defined 'optimum' or 'risk' levels (Charles 2001). These benchmarks are called 'biological reference points'. Management advice is mostly given only for the short term and often concerns a total allowable catch (TAC) for the next year, while national, regional and fleet quotas are decided by political processes. The advice may have the form of a clear recommendation of a TAC, or may present options within a biologically acceptable range of catch levels, describing the short-term effects of each option. Increasingly, however, advice is given for a management strategy which may aim at rebuilding the stock or stabilising catch and biomass levels over a specified time frame. It is now strongly recommended that management advice be based on the 'Precautionary Approach'. The underlying philosophy is to avoid a reduction of SSB to levels where recruitment will be impaired. The SSB and the fishing mortality rate, both of which have a defined set of reference points, are the most important elements of the advice. The basis for estimating biological reference points ranges from simple production models, models that consider only the effects of fishing on a cohort of fish (yield per recruit models), to full age-structured models that explicitly account for the stock–recruitment relationship. In all cases, an appropriate measure of the actual reproductive output of the population is critical. During the past decade, emphasis has been placed on limit reference points, serving as warning signs of overfishing and stock declines. If the problem of overexploitation can be overcome, target reference points aimed at optimising yield or economic returns will assume greater importance in management.

Chapter 10 explores new approaches to management, grounded in detailed information on environmental influences on recruitment, the oceanographic setting, reproductive biology, ecological interactions and spatial dynamics. These points are crystallised in a detailed case study of cod population dynamics around the British Isles. This perspective is clearly in keeping with the move toward a more holistic ecosystem approach to management of fishery resources which has been increasingly advocated around the world (Jennings *et al.* 2001). Many of the concepts raised in previous chapters are highlighted, and new dimensions considered. The importance of incorporating these more detailed biological and ecological considerations is made clear in this case study. The development of spatially explicit simulation models incorporating information on patterns of spawning aggregation, advective transport of eggs and larvae, larval settlement, vital rates of juvenile and adult cod, and exploitation patterns as in this example, provides a powerful tool for synthesis, integration and prediction.

Chapter 11 concludes this book with a compelling argument for the need to move towards the use of total egg production and consideration of demographic characteristics in our evaluation of stock reproductive condition. Egg viability can be related to the age and reproductive history of the female. Truncating the age composition towards younger spawners can have a disproportionate effect on recruitment that is not reflected in simple measures of the adult population such as SSB. Changes in sex ratio in response to harvesting in species with dimorphic growth can be very important in estimating the actual reproductive output of the population compared with estimates based on total adult biomass. Furthermore, these changes can alter mating systems and other aspects of behaviour in some fishes with direct effects on spawning and recruitment.

Although constraints on the availability of time series of fecundity estimates have hindered progress, these limitations are beginning to ease. In the interim, recognition of the broader availability of sex ratio information over time has allowed estimation of female spawning biomass for an increasing number of stocks as a stepping stone to enhanced consideration of reproductive dynamics. In other cases, it has been possible to employ other measures based on female energetic reserves as a proxy for effective egg production. Inclusion of other factors such as the age diversity of female spawners has also proven useful in some circumstances in improving the predictability of recruitment.

Summary

A full appreciation of reproductive dynamics is critical for assessing the impacts of harvesting on fish populations and in devising appropriate management strategies. Attempts to ascertain limits to exploitation and defining optimal harvesting strategies have typically been based on proxies of reproductive potential of stocks – most notably simple measures of the biomass of the adult population. However, we need measures of the actual reproductive capacity and output of the population. This will entail an understanding of reproduction biology, behaviour and demographic characteristics of the population to provide adequate measures of reproductive capacity. We further need to understand the factors that affect the survival through the early life stages before recruitment to the fishery. The confluence of factors affecting egg condition and environmental effects on survival is critical in this regard.

Accounting for these factors in management will place renewed emphasis on demographic and other characteristics of the stock. Attention to the age and size structure of the population, sex ratio, and so on, will lead to new ways of measuring the reproductive capacity and replace simpler measures such as total spawning biomass. Management tools to specifically address these issues will also require a shift from simple considerations of TAC to measures that are designed both to limit the catch and to control its demographic composition. This will entail strategies such as the use of marine protected areas to protect segments of the population, and the development of more selective fishing gears. Consideration of factors such as preserving multiple reproductive opportunities for individual females will become increasingly important. We anticipate a shift towards increased emphasis on long-term management strategies from the current focus on short- and, in some cases, medium-term management. A full understanding of the stock–recruitment relationship will be essential in this endeavour.

There is an emerging international acceptance of the need for a holistic ecosystem approach to management for marine systems with the objective of preserving ecosystem structure and function, biological diversity and habitat. The ecosystem approach will involve consideration of the cumulative impacts of human activities in the sea and an evaluation of trade-offs among potentially competing uses of the marine environment. (Fogarty & McCarthy, 2014) Within this broader context, however, regulation of individual ocean use sectors will remain important. Fisheries exert a dominant influence in many marine ecosystems. It will remain necessary to determine the status of individual stocks and to predict the effect of alternative management actions on these stocks, and on the ecosystem as a whole. The advances in understanding reproductive dynamics and recruitment outlined in this book serve as a benchmark against which to measure future progress in meeting the goal of incorporating greater biological and ecological realism in management of fishery resources within this broader context.

References

Anon. (2001) The ICES Strategic Plan. International Council for the Exploration of the Sea, Copenhagen, 12 pp. (http://www.ices.dk/iceswork/strategic%20plan-final.pdf)

Beverton, R.J. & Holt, S.J. (1957) *On the Dynamics of Exploited Fish Populations.* Fisheries Investigations Series II, **19**. Ministry of Agriculture, Fisheries and Food, London. 533 pp.

Cadrin, S.X., Friedland, K.D. & Waldman, J.R. (Eds) (2004) *Stock Identification Methods: Applications in Fishery Science.* Elsevier, Amsterdam.

Chambers, R.C. & Trippel E.A. (1997) *Early Life History and Recruitment in Fish Populations.* Chapman & Hall, London.

Charles, A. (2001) *Sustainable Fishery Systems.* Blackwell Science, Oxford.

Fogarty, M.J. & McCarthy, J.M. (2014). Marine Ecosystem Based Management. Vol. 16. The Sea. Harvard University Press. Cambridge.

Gunderson, D.R. (1993) *Surveys of Fishery Resources.* J. Wiley & Sons, New York.

Hall, S.J. (1999) *The Effects of Fishing on Marine Ecosystems and Communities.* Blackwell Science, Oxford.

Hjort, J. (1914) Fluctuations in the great fisheries of northern Europe viewed in the light of biological research. *Rapports et Procès-verbaux des Réunions, Conseil International pour l'Exploration de la Mer*, **20**, 1–228.

Hjort, J. (1926) Fluctuations in the year classes of important food fishes. *Journal du Conseil International pour l'Exploration de la Mer*, **1**, 1–38.

Jennings, S., Kaiser, M.J. & Reynolds, J.D. (2001) *Marine Fisheries Ecology.* Blackwell Science, Oxford.

Potts, G.W. & Wootton, R.J. (1984) *Fish Reproduction: Strategies and Tactics.* Academic Press, London.

Quinn, T.J., II & Deriso, R.B. (1999) *Quantitative Fish Dynamics.* Oxford University Press, Oxford.

Rothschild, B.J. (1986) *Dynamics of Marine Fish Populations.* Harvard University Press, Cambridge, MA.

Part I

Biology, Population Dynamics, and Recruitment

Chapter 1
Recruitment in Marine Fish Populations

Michael J. Fogarty and Loretta O'Brien

1.1 Introduction

The production of viable eggs by a population provides the raw material for recruitment (the number of young ultimately surviving to a specified age or life stage). Recruitment processes in the sea reflect the interplay of external forcing mechanisms such as physical drivers in the environment that affect demographic rates, and stabilizing mechanisms exhibited by the population. Many marine populations fluctuate widely in space and time (Fogarty *et al.* 1991). These dramatic changes are attributable to fluctuations in biotic and abiotic factors affecting growth and/or mortality rates during the early life history (Fogarty 1993a). Potentially countering these sources of variability are internal regulatory mechanisms that can compensate for population changes. Considerable attention has been devoted to the development of recruitment models embodying different types of compensatory processes operating during the pre-recruit phase of the life history (see Rothschild 1986, Hilborn & Walters 1992, Quinn & Deriso 1999 and Walters & Martell 2004 for reviews). In contrast, the issue of compensatory changes in factors such as fecundity, adult growth, and maturation affecting reproductive output has received less attention in modeling recruitment dynamics (but see Ware 1980, Jones 1989, Rothschild & Fogarty 1989, 1998). We argue that a complete model of population regulation of marine fishes must allow for the possibility of compensatory processes operating during both the early life history and the adult stages, and that a refined understanding of reproductive processes as described in the contributions to this book is essential in the quest to understand recruitment of marine fishes. In particular, integrating our emerging understanding of maternal effects on reproductive success of fish, as documented in this volume, into management models is essential.

In this chapter, we attempt to set the stage for several themes found throughout this volume—factors controlling the effective reproductive output of the population, the fate of fertilized eggs and larvae, and the implications for assessment and management of exploited marine species. In subsequent chapters these issues are explored in greater individual detail. An understanding of recruitment processes is essential if we are to predict the probable response

Fish Reproductive Biology: Implications for Assessment and Management, Second Edition.
Edited by Tore Jakobsen, Michael J. Fogarty, Bernard A. Megrey and Erlend Moksness.

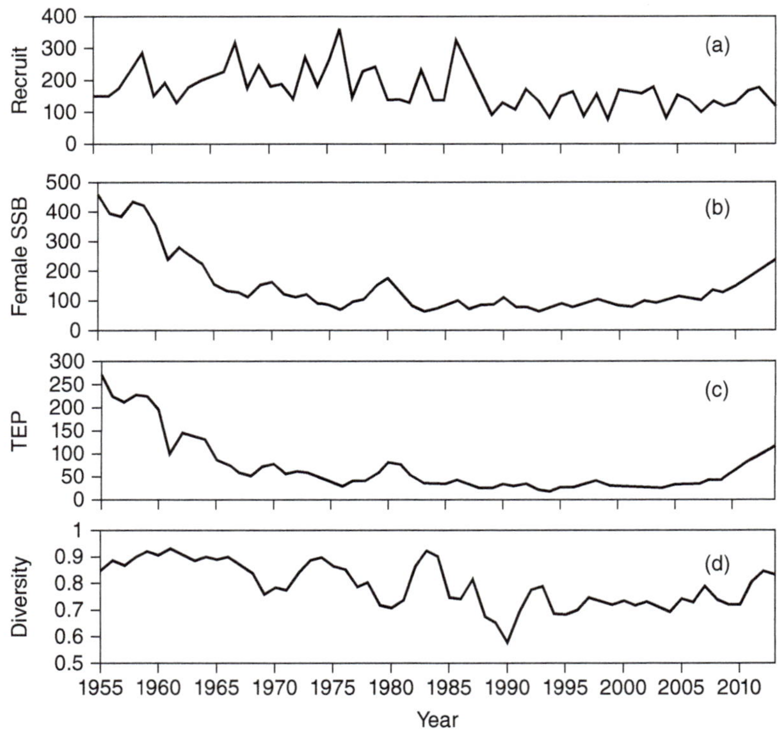

Figure 1.1 Time series of estimates of (a) recruitment (millions of 3-year-old fish), (b) spawning stock biomass (thousand Mt), (c) total viable egg production (trillions), and (d) age diversity of spawners (Shannon–Weiner index) for Icelandic cod. Based on assessment data from ICES (2014) and fecundity relationships from Martinsdottir & Begg (2002).

of a population to exploitation and to proposed management actions. These predictions require an analytical framework. Here, we trace the theoretical developments relating recruitment to the adult population to provide such a framework. Our interest centers on exploring the consequences of different recruitment mechanisms, demonstrating how these processes can be modeled, and illustrating their importance for stability and resilience of the population. In a variable environment, sustainable exploitation is possible only if the population exhibits some form of compensation in response to variation in population size at some stage in the life history. The general issue of the role of compensation in population dynamics is therefore of both theoretical and practical importance. Correctly accounting for the effective reproductive output of the population, including the consideration of factors such as maternal effects on egg and larval viability, the age composition of the adult population, female condition, and how these are affected by population density or abundance, is critical in understanding the form of the relationship between recruitment and egg production and how the population will respond to exploitation.

An illustration of the magnitude of change in these population components is provided by trajectories of recruitment and adult biomass over the past five decades for Icelandic cod, an economically and ecologically important fish population (Figure 1.1a,b). Attempts have now been made to refine estimates of reproductive output by reconstructing the total egg production by the female population (Figure 1.1c) and to understand how factors such as the age diversity of the spawning stock (Figure 1.1d) affect recruitment success. Estimates of each of these quantities are becoming increasingly available for more marine fish populations (e.g. Marteinsdottir & Thorarinsson 1998, Trippel 1999, Marteinsdottir & Begg 2002, Marshall *et al.* 1998, 2003, Morgan *et al.* 2011, Cervino *et al.* 2013, Macchi *et al.* 2013). We will return to the relationship between recruitment and spawning stock biomass (SSB) or total egg production for Icelandic cod in Section 1.2.4 to further explore these issues, and in Section 1.8.1 we address the issue of whether consideration of the age diversity of the adult population improves the predictability of recruitment for this population (Marteinsdottir & Thorarinsson 1998).

In the following, we describe several models incorporating factors affecting survivorship from the egg stage to recruitment. These include competition for limiting resources, cannibalism, and the interaction of compensatory growth and size-dependent mortality. Our initial treatment will focus on deterministic processes for a single pre-recruit stage. We then broaden our development to encompass consideration of compensatory processes operating during the post-recruit phase of the life history, the stability properties of these models, multistage life history patterns, the implications of maternal effects, and the effects of environmental and demographic stochasticity. Throughout, the implications of these factors for management of exploited populations is of primary interest.

1.2 Recruitment theory

Consider the life cycle diagram depicted in Figure 1.2. For the population to persist, a sufficient number of progeny must, on average, survive to replace the parental stock. For the purposes of illustration, we show several stanzas including egg, larval, juvenile and adult stages. The eggs produced by the different adult stages can, in principle, exhibit different viabilities and have different probabilities of successful transition to the larval stage. For the purposes of this simple illustration we do not trace the effect of the size or age of the adult females beyond the egg stage, but we extend this treatment to later stages as well in a subsequent section. The transitions between stages represent the probability of surviving and growing into the next stage during a specified time interval. Note that the population becomes vulnerable to exploitation following the first juvenile stage in this example. In the following, we use the

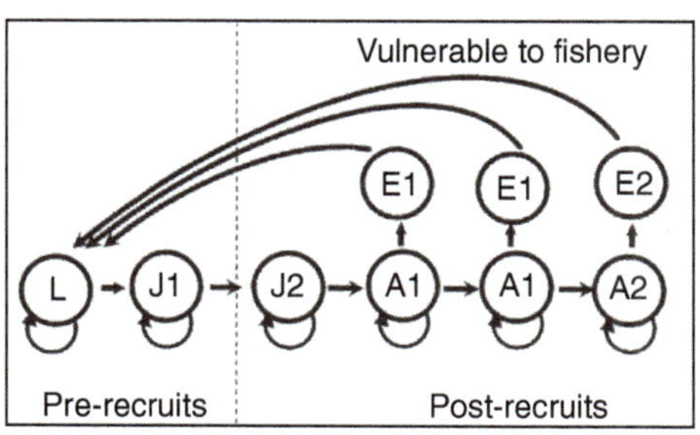

Figure 1.2 Life cycle diagram including egg, larval, juvenile and adult stages. Eggs produced by adults of different ages can have different viabilities.

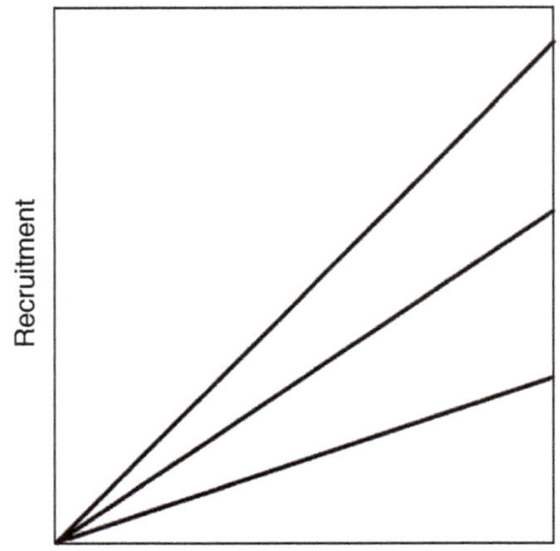

Figure 1.3 Density-independent model relating recruitment and egg production for three levels of the density-independent mortality rate.

size or age at first harvest as the demarcation point for recruitment. The life cycle is completed with the production of eggs by the adult component of the population. The fishery reduces the probability of survival in the late juvenile and the adult stages with important consequences for the overall reproductive output of the population. The number and quality of eggs produced by different segments (age or size classes) of the adult female population varies in relation to spawning history, condition, and other factors – a central theme of many contributions found in this volume (see Chapters 2, 4, and 8) with potentially important management implications (see Chapters 9, 10, and 11). In this section, we address the issue of the form of the relationship between the production of viable eggs and recruitment in which survivorship from hatching to recruitment does not depend on factors such as maternal age or reproductive experience. Here, viability is taken to represent an amalgam of fertilization success and hatchability.

To model this process, we begin with the simple observation that, for a closed population, the number of individuals in a cohort can only decline over time. A cohort is defined as the number of individuals hatched in a specified period (spawning season, year, etc.). In the very simplest case where no compensation occurs, the number of recruits (R) is given by the product of the proportion surviving (S) from the egg to the recruit stage and the initial number in the cohort (the number of viable eggs – designated E):

$$R = S\mathrm{E} \tag{1}$$

This gives a simple linear relationship between egg production and recruitment with slope equal to the survival fraction (Figure 1.3). We can think of E as the sum of egg production by each female spawner age class multiplied by an age-specific viability coefficient (see Section 1.3.1). For a closed population, the relationship goes through the origin. Being able to correctly identify the members of the population and their spatial domain is of course a critical prerequisite for defining this relationship (see Chapter 7). For metapopulation structures with interchange among populations, the relationship may not pass through the origin (e.g., for a sink population receiving a subsidy from a source population; see also Section 1.2.2).

In subsequent sections we will expand the density-independent case to include compensatory processes resulting in nonlinear relationships between the number of viable eggs produced and recruitment, random variation in vital rates, and other factors. For now, we will focus on the underpinnings of the simple density-independent model. We will consider this to be our null recruitment model. Note that a straight line with zero slope is not an appropriate null model in this context – it implies that recruits can be produced when the egg production has been reduced to zero. Adopting such a null model would entail high risk to the population (see Fogarty *et al.* 1992, 1996).

The null model can be derived from first principles by describing the rate of change of a cohort:

$$\frac{dN}{dt} = -\mu N \tag{2}$$

where N is the number in the cohort and μ is the instantaneous rate of mortality during the pre-recruit phase. This model of course captures the idea that the number in the cohort can only decline over time (in this case, at a constant rate). Separating variables we have:

$$\int_{N=E}^{R} \frac{dN}{N} = -\mu \int_{t=0}^{tr} dt \tag{3}$$

where E is again the initial number in the cohort (the number of viable eggs produced), and R is the number surviving to the age of recruitment (t_r). The solution to this simple model is given by:

$$R = \mathrm{E}\, e^{-\mu t} \tag{4}$$

where for simplicity we have set $t = t_r – t_o$ and where $e^{-\mu t}$ is the survival fraction (S; cf., Equation (1)).

1.2.1 Compensatory and overcompensatory models

The null recruitment model implies that there are no constraints on the number of recruits produced for a given level of egg production, leading to unrealistic predictions of unrestrained population growth (see Chapter 7). We can readily extend the density-independent recruitment model to incorporate various types of compensatory processes affecting growth and survival during the pre-recruit phase. Because the density-independent model cannot account for limitations in recruitment that emerge as a result of competition for limiting resources (food, space, etc.) or factors such as cannibalism known to be important in many marine populations, we need to expand our consideration of underlying recruitment mechanisms. (For a lucid verbal description of the underpinnings of the classical stock-recruitment models embodying these mechanisms, see Chapter 7.) These considerations lead to nonlinear models with important implications for the stability of the population. In the following, we will refer to a compensatory recruitment model as one in which the per capita rate of change of a cohort is a decreasing linear function of the number in the cohort during the pre-recruit period. In contrast, an overcompensatory model is one in which the per capita rate of change is a decreasing linear function of the initial number in the cohort or a related metric (e.g. the number of adults exerting cannibalistic controls on recruits). The principal focus of this book is in incorporating increased biological realism into our measures of reproductive output of the population. We are no less interested in incorporating biological realism in the development of recruitment models. We view recruitment models not simply as heuristic guides to the shape of the egg production–recruitment relationship but as the elaboration of testable biological hypotheses concerning different compensatory mechanisms.

1.2.1.1 Intracohort competition

In situations where members of the cohort compete for critical resources (food, space, etc.), density-dependent mortality may be critically important. The simple null model can be extended to account for a linear increase in mortality with increasing cohort density by making the substitution $\mu = (\mu_o + \mu_1 N)$. Our model for the rate of decay of the cohort can then be expressed as:

$$\frac{dN}{dt} = -(\mu_o + \mu_1 N)N \tag{5}$$

where μ_o is the instantaneous rate of density-independent mortality and μ_1 is the coefficient of density-dependent mortality (Beverton & Holt 1957). Note that this model simply indicates that the per capita rate of change of cohort size (dN/Ndt) declines linearly with increasing N.

$$R = \left[\frac{1}{\mathrm{E}} e^{\mu_o t} + \frac{\mu_1}{\mu_o}(e^{\mu_o t} - 1)\right]^{-1} \tag{6}$$

The solution is given by:

which can be simplified to:

$$R = \left[\frac{\alpha}{\mathrm{E}} + \beta\right]^{-1} \tag{7}$$

where $\alpha = \exp(\mu_o t)$ and $\beta = ((\mu_1/\mu_o)(\exp(\mu_o t)–1))$. For this model, recruitment initially increases rapidly with increasing egg production and then approaches an asymptote (Figure 1.4). We further note that intracohort cannibalism could also result in a model of this general form.

In this chapter, we will refer to this asymptotic form as a compensatory recruitment model and will distinguish it from "overcompensatory" models in which recruitment actually declines at higher levels of egg production (see next section), although some authors define these terms differently. Rothschild & Fogarty (1998) describe generalized models in which the per capita rate of change as a function of cohort size is not limited to the linear case as in the model above.

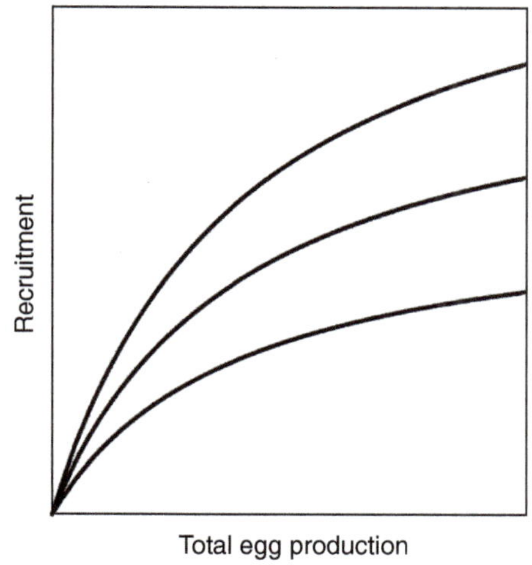

Figure 1.4 Beverton–Holt-type model relating recruitment and egg production for three levels of the parameter α.

1.2.1.2 *Cannibalism by adults*

Cannibalism has been shown to be an important population regulatory mechanism in many marine fish populations (Dominey & Blumer 1984). In many cases, the adults are the principal predators of earlier life stages. To represent intraspecific predation by adults on pre-recruits, we can let $\mu = (\mu_o + \mu_2 P)$, and the model for the decay of the cohort now can be specified:

$$\frac{dN}{dt} = -(\mu_o + \mu_2 P)N \tag{8}$$

where μ_2 is the coefficient of "stock-dependent" mortality (Harris 1975), and P is a measure of the cannibalistic component of the adult population. Here, the per capita rate of change declines linearly with the adult population size metric. Note that some segments of the adult population may contribute more to cannibalism, and the index of the adult population used can and should reflect this fact where available. The solution is:

$$R = \mathrm{E}\, e^{-(\mu_o + \mu_2 P)t} \tag{9}$$

and in this form, we require information on both total egg production (E) and the relevant index of adult population size. For some applications we are ultimately interested in a bivariate model relating recruitment to total egg production. This requires a substitution of the index of population size by one for total egg production in the model. Later in this chapter, the potentially complex relationship between egg production and population size is explored in the context of these models. For the moment we will consider only the simplest case where egg production is related to the measure of cannibalistic adult population size by a constant of proportionality (ω) to illustrate the translation to a bivariate form. Letting $\kappa = \exp(-\mu_o t)$ and $\delta = \mu_2 t/\omega$, the model can be written as:

$$R = \kappa \mathrm{E}\, e^{-\delta \mathrm{E}} \tag{10}$$

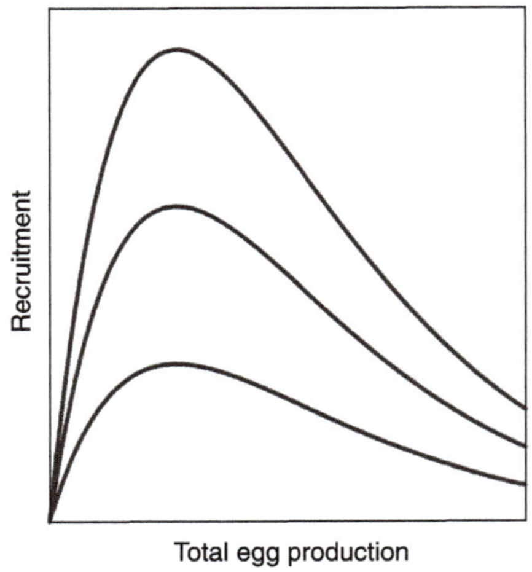

Figure 1.5 Ricker-type model relating recruitment and egg production for three levels of the slope at the origin parameter.

This overcompensatory model produces a characteristically dome-shaped relationship between recruitment and egg production (Figure 1.5). We note that the model implicitly assumes random encounter between the progeny and the adult predators. If the early life stages are aggregated and the encounter probabilities are nonrandom, the degree of curvature of the relationship decreases (i.e., becomes less convex; see Ricker 1954).

Ricker (1954) also noted that in instances where there is a delayed response by a predator to the initial number in the cohort, an overcompensatory response may be generated. In this case, our specification of the model for the rate of change of the cohort would directly include a term for the number of eggs produced, generating a model identical in form to Equation (10) but with a different interpretation of the parameter in the exponent.

1.2.1.3 Size-dependent processes

Compensatory recruitment models based on size-specific mortality rates have also been developed to reflect the interaction of compensatory growth and mortality rates. If smaller individuals are more vulnerable to predation, then density-dependent factors that affect the time required to grow through a "window of vulnerability" to predation will have a direct effect on recruitment (see Chapter 3 for an overview). In particular, size can have critical effects on vulnerability when the ratio of predator to prey size is relatively low (Miller *et al.* 1988). Accordingly, density-related effects on growth can have potentially important implications for survival rates even if mortality itself is independent of density. Maternal influences can result in differential growth rates of larvae, and the incorporation of this factor can have important implications for recruitment dynamics (see Section 1.3.2).

Beverton & Holt (1957) first illustrated this concept in a derivation of a two-stage pre-recruit life history model. The pre-recruits were subjected to differing levels of mortality during the two stages. Beverton & Holt (1957) considered the case where the time required to grow from the first to the second stage was inversely proportional to the food supply and directly proportional to the initial number in the cohort, and showed that such a formulation resulted in an overcompensatory stock–recruitment relationship.

It is possible to directly model growth processes and their interaction with mortality during the pre-recruit stage. Consider a model for individual growth in weight:

$$\frac{dW}{dt} = G(W) \tag{11}$$

where $G(W)$ is a compensatory function for individual growth. If the mortality rate is size-dependent, then we have:

$$\frac{dN}{dt} = -\mu(W)N \tag{12}$$

and the rate of change of cohort size with respect to weight (size) is therefore:

$$\frac{dN}{dW} = -\frac{\mu(W)}{G(W)}N \tag{13}$$

The solution to this model is:

$$R = N(W_1) = N(W_o)\,e^{-\int \frac{\mu(W)}{G(W)}dW} \tag{14}$$

where $N(W_1)$ is the number in the population surviving to weight (size) W_1, which we will take to be the size at recruitment. This model has been discussed by Werner & Gilliam (1984). Without further specification of the functions $\mu(W)$ and $G(W)$, it is not possible to determine the functional form of this size-based recruitment function. However, if the growth rate is taken to be dependent on the cohort size and the mortality rate to be density-independent, then the recruitment function will generally be compensatory. If instead, the growth rate is taken to be dependent on the *initial* number in the cohort, then the recruitment function will be overcompensatory (Ricker-type) (see Rothschild & Fogarty 1998).

Shepherd & Cushing (1980) assumed that $G = G^*/(1 + N/K)$, where G^* is the maximum growth rate, N is cohort size, and K is a constant related to the abundance of food. It is further assumed that the mortality rate μ is independent of density. When $N = K$, the growth rate is exactly one half of the maximum rate. Separating variables, we can then write the model as:

$$\frac{dW}{W} = -\frac{G^*}{\mu}\frac{dN}{[(1+\frac{N}{K})N]} \tag{15}$$

and the solution is:

$$log_e\left(\frac{W_1}{W_o}\right) = -\frac{G^*}{\mu}\,log_e\left[\frac{(K+\mathrm{E})\,N_1}{(K+N_1)\mathrm{E}}\right] \tag{16}$$

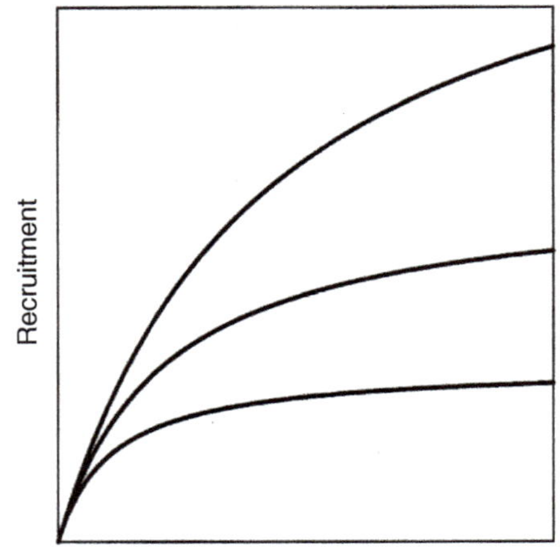

Figure 1.6 Cushing–Shepherd-type model relating recruitment and egg production for three levels of the density-dependent parameter *K*.

where again, the initial number in the cohort (E) emerges as the lower limit to integration on the right-hand side of Equation (16). Exponentiating and letting $A = exp\{-(\mu/G^*)ln(W_1/W_0)\}$, the model becomes (after further rearranging terms):

$$R = N(W_1) = \frac{AE}{1 + (1 - A)E/K} \tag{17}$$

which describes an asymptotic relationship between total egg production and recruitment (here, the number surviving to some specified weight class ($R = N(W_1)$; see Figure 1.6).

These examples should suffice to show that many different mechanisms can underlie recruitment dynamics and that, in some cases, very different mechanisms can give rise to similarly shaped recruitment curves. Therefore, it will not generally be possible to understand the important regulatory mechanisms operating in the population based on information on egg production and the resulting recruitment alone. However, an understanding of the underlying biological mechanisms can guide the choice of appropriate recruitment models, an issue of considerable importance in the face of the characteristically high levels of recruitment variability exhibited by many marine populations which tends to obscure the underlying relationship (see Section 1.8).

1.2.2 Depensatory processes and the Allee effect

The preceding sections have focused on compensatory and overcompensatory mechanisms. For closed populations, these processes generally lead to stable non-zero equilibrium points (see Section 1.4), although for the case of overcompensatory models, quite complex dynamics can emerge (Ricker 1954), including chaos. *Depensatory* mechanisms of various types are also potentially of interest and can lead to multiple equilibria. Depensatory recruitment dynamics occur when the per capita rate of change of recruitment increases over some range

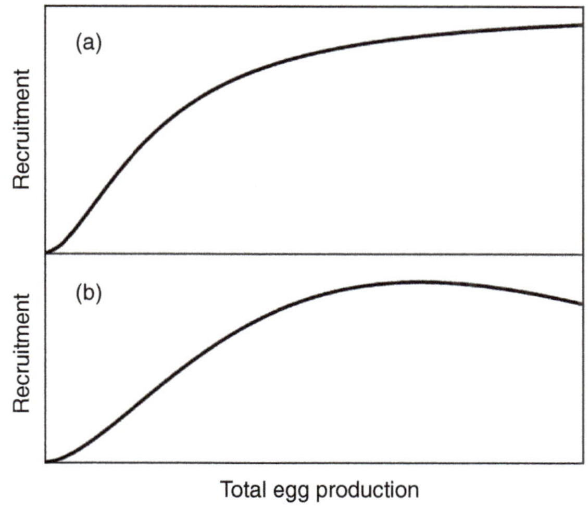

Figure 1.7 Models allowing for depensatory effects based on generalizations of (a) Beverton–Holt-type and (b) Ricker-type models relating recruitment and egg production.

of population or cohort size rather than declining monotonically as in compensatory and overcompensatory models. For such a system, we observe an inflection in the relationship between egg production and recruitment, and this characteristic can lead to multiple equilibrium points for the population (see Section 1.4). For the case of "critical depensation" a lower unstable equilibrium point exists, and if the effective egg production by the population is driven below some threshold level, a sudden population collapse is predicted.

Depensation can occur under a number of mechanisms, including when fertilization success is low at low population densities or there is a reduced probability of finding a mate. More broadly, when fitness or population growth is enhanced in the presence of conspecifics over some range of population size we have a so-called "Allee effect." (For a description of the array of behavioral and ecological mechanisms that can lead to this effect, see Stephens *et al.* 1999.) Among the mechanisms of direct interest in this chapter are effects related to fluctuations in the sex ratio at low population sizes (Stephens *et al.* 1999) which affect fertilization success.

The Beverton–Holt model can be generalized to allow for depensation as follows:

$$R = \left[\frac{\alpha}{\mathrm{E}^{\gamma}} + \beta\right]^{-1} \tag{18}$$

where γ is a "shape" parameter and all other terms are defined as before (when $\gamma > 1$, depensatory dynamics occur; see Figure 1.7a). Similarly for a generalized Ricker model, we can write:

$$R = \kappa \mathrm{E}^{\gamma}\, e^{-\delta \mathrm{E}} \tag{19}$$

where, for economy of notation, γ again represents the shape parameter (Figure 1.7b). Attempts to discern widespread evidence for depensatory dynamics in exploited fish populations have so far provided relatively few direct examples (Myers *et al.* 1995), but a lack of information at very low population levels may be responsible, in part, for this result. Marshall *et al.* (2006) did find that the relationship between recruitment and SSB for Northeast Arctic cod was depensatory when the analysis focused on more recent years (since 1980), although estimates based on female spawning biomass and total egg production did not indicate depensatory dynamics.

Frank & Brickman (2000) considered a Ricker-type model incorporating a specific form of Allee effect in which no recruitment at all occurs below a threshold population level. Frank & Brickman (2000) further considered a system comprising a number of spatially defined substocks (i), each of which is subject to the Allee effect. Reframing this model in our notation and expressing in terms of egg production levels, we have:

$$R_i = \kappa_i(\mathrm{E}_i - E_o)e^{-\delta(\mathrm{E}_i - E_o)} \tag{20}$$

where E_o is the threshold level of egg production below which no recruitment occurs, and the subscript i indicates an individual substock. Frank & Brickman (2000) showed that if Allee effects are important and managers either ignore or are unaware of the substock structure, the Allee effect may be masked and lead to risk-prone decisions concerning appropriate harvest levels. This example reinforces the importance of both understanding the true population structure (see Chapter 6) and the nature of population regulatory mechanisms.

1.2.3 Maternal effects on pre-recruit survivorship

In the models described above, maternal effects were restricted to the development of "viable" eggs defined by fertilization and hatching success. Shelton *et al.* (2012) provided an approach to specifying differential survival of progeny of spawners of different ages to effectively generate recruitment curves for different aged spawners. Although they developed the method for an overcompensatory recruitment model, the approach can in principle be applied to a broader array of recruitment models. Adapting our earlier notation, the model can be written as:

$$\frac{dN_i}{dt} = -(\mu_{o.i} + \mu_{2,i}\, P)N_i \tag{21}$$

where the subscript i denotes progeny (N) or mortality processes (μ) associated with spawners of age i and P is the biomass of fish contributing to density-dependent mortality. The solution is:

$$R_i = \mathrm{E}_i\, e^{-(\mu_{o,i} + \mu_{2,i}\, P)t} \tag{22}$$

and the total recruitment is then the sum over all age-classes of spawners:

$$R = \sum\nolimits_{i-0}^{I} \mathrm{E}_i\, e^{-(\mu_{o,i} + \mu_{2,i}\, P_D)t} \tag{23}$$

Shelton *et al.* (2012) allowed the mortality terms to be simple linear functions of maternal age.

1.2.4 *Total egg production or spawning stock biomass: does it matter?*

We have framed our analysis of recruitment dynamics in terms of total viable egg production by the population and factors affecting growth and survival during the pre-recruit period. Because estimates of total egg production were not widely available at the time, the earliest recruitment models were recast in terms of SSB. Both, Ricker (1954) and Beverton & Holt (1957) assumed a simple proportional relationship between egg production and adult biomass, and used the latter as a proxy for the former (Chapter 11). Rothschild & Fogarty (1989) noted that the assumption of proportionality may be questionable, and Marshall (Chapter 11) showed that other implicit assumptions such as a constant sex ratio and mean fecundity are not generally valid. As noted by Marshall (see Chapter 11), the use of spawning biomass as a proxy for total egg production remains the standard today, and will likely remain so until refined estimates of reproductive output are more widely available.

Estimates of recruitment and adult population size are available for many species using well-established stock assessment methods (see Chapter 7), and these provide an important foundation for our analysis of recruitment dynamics. Although fecundity estimates are routinely made for a relatively few but growing number of populations, rapid measurement techniques have been developed that promise to increase the availability of this type of information (Chapter 11). With the diversity of reproductive patterns in marine fishes, and the range of reproductive strategies and tactics represented, obtaining a proper accounting of fecundity and reproductive output is no trivial matter (see Chapters 2 and 8), but important progress is now quite evident. Given reliable estimates of fecundity in concert with age-specific estimates of sex ratios and abundance, it is possible to derive estimates of total egg production. Alternatively, for some populations, egg abundance can be measured directly at sea and corrected to provide estimates of viable egg production (Chapter 5). Given the documented changes in sex ratios, female condition, and other factors over time (Trippel, 1999, Marshall *et al.* 1998, 1999, 2000, 2003, 2006), there is ample justification for broader application of estimates of total egg production in recruitment studies (see Chapter11).

Relationships between recruitment and adult biomass and between recruitment and total egg production for Icelandic cod are illustrated in Figure 1.8. The high recruitment variability that is common to many marine fishes is clearly evident in both representations (see Section 1.8). Cod are cannibalistic, and we accordingly fit Ricker-type models to these data. For this population, a recruitment model based on egg production explains somewhat more of the variability in recruitment than does one based on SSB. We show in Section 1.8.1 that further improvements in the fit of the model are obtained by also considering the age diversity of spawners.

In addition, the modeled relationship between recruitment and egg production reveals subtle differences that are important in understanding how a population will respond to exploitation when compared to a model based on SSB. In particular for Icelandic cod, the slope of the recruitment curve at the origin is steeper for the recruitment–egg production relationship (Figure 1.9). Relatively small differences in the slope of the recruitment curve at the origin can have important implications for inferences concerning the resilience of a population to high levels of exploitation. Later in this chapter, we will explore how these considerations shape our view of the resilience of a population to harvesting and the ways in which a refined

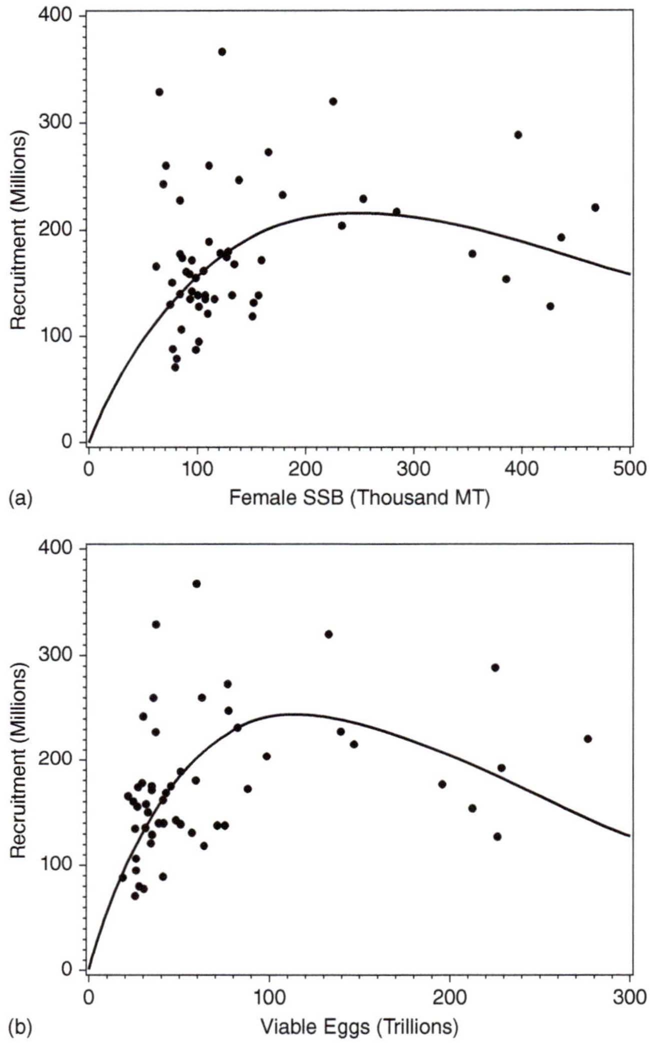

Figure 1.8 Relationship between (a) recruitment and female spawning stock biomass and (b) recruitment and total viable egg production for Icelandic cod.

understanding of the reproductive output of a population can help in setting appropriate management objectives.

1.3 Completing the life cycle

The previous sections have dealt strictly with processes during the pre-recruit phase, operating on the initial number of viable eggs produced by the population. To complete our consideration of the life cycle dynamics of a cohort, we next examine models of the reproductive output of the adult population. Many of the topics covered in this book of course deal with this issue in detail.

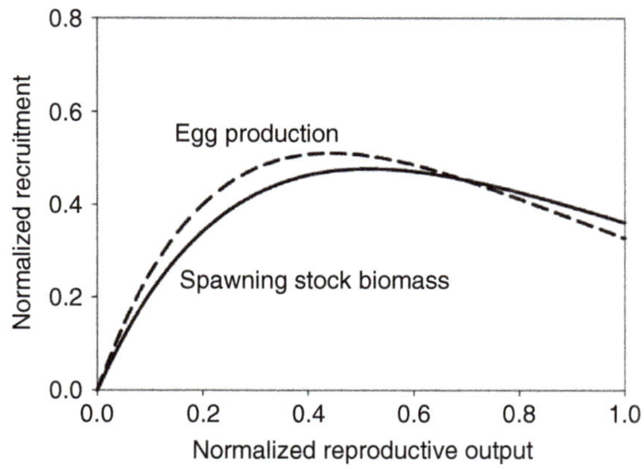

Figure 1.9 Fitted Ricker models for normalized recruitment and reproductive output using total egg production (solid line) and spawning stock biomass (dashed line) for Icelandic cod.

1.3.1 Viable egg production

The number of individuals of a cohort alive at each successive age following recruitment is simply the product of the survival rates over the age classes considered and the number of recruits (taken as the starting point for this phase of the life history):

$$N_{a+1} = R \prod_{a=a_r}^{a_{\max}} \exp^{-(M_a + p_a F)} \tag{24}$$

where M_a is the natural mortality rate at age a, p_a is the proportion vulnerable to the fishery at age a, F is the instantaneous rate of fishing mortality, $a = a_r$ is the age at recruitment, and a_{max} is the maximum age. The expected number of viable eggs produced by a cohort over its lifespan can be expressed:

$$\mathrm{E} = \sum_{a=a_r}^{a_{max}} v_a m_a f_a s_a N_a \tag{25}$$

where v_a is the relative viability of eggs produced by females at age a (expressed as a proportion), m_a is the proportion of mature females, f_a is the fecundity, s_a is the sex ratio, and N_a is the number in the population at age a (see also Rothschild & Fogarty 1998; Chapter 11). If we normalize these results for the initial number of recruits, we define the egg production-per-recruit (EPR) – a quantity of interest in a number of analyses presented later in this chapter. We can further define a percent spawning potential ratio (SPR), expressed as the ratio of the EPR at each level of fishing mortality divided by the EPR at $F = 0$ (see Brooks 2013 for a recent exposition). It is possible to define the SPR resulting in “maximum excess recruitment” (SPR_{MER}) – the level of recruitment beyond that needed for replacement of the population. This quantity is a function of the slope of the recruitment curve at the origin and the EPR at $F = 0$ (reviewed in Brooks 2013). Chapter 7 describes the conceptual foundation for this

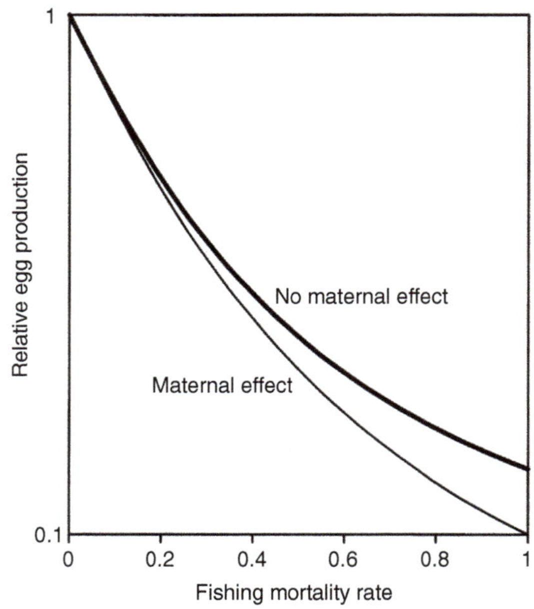

Figure 1.10 Normalized egg production per recruit (proportion of maximum) as a function of fishing mortality assuming no maternal age effects (thick line) and a maternal age effect on egg viability (thin line). Normalized egg production on a logarithmic (base 10) scale.

approach in terms of spawning biomass as currently applied to most fish stocks (see also Chapter 9). In principle, each of these parameters can be expressed as functions of some measure of population size to reflect compensatory processes operating during the post-recruit phase (see below). We have explicitly allowed for differences in the viability of eggs produced by females of different ages or size classes. Larger females may produce larger eggs with higher energetic reserves (see Chapters 2 and 11); developmental success may also be higher for the progeny of these individuals (Chapters 2 and 8). In principle, it may also be possible to make the age-specific viability term time-dependent and account for interannual variation in the condition of female spawners. For example, information on total lipid energy in liver tissue has successfully been used as a proxy for egg production (Marshall *et al.* 1999, 2000; Chapter 9), and this can be used to develop a time-varying index of egg viability.

With increasing levels of fishing mortality, the expected lifetime reproductive potential of the cohort decreases nearly exponentially (Figure 1.10). Note that if eggs produced by older females have a higher hatching success, the decline is sharper than for the case of no age-specific differences in egg viability. This holds important implications for understanding the stability and resilience of populations to harvesting pressure (see Section 1.7).

It is also possible to modify this expression to reflect not only the age of spawners but also their reproductive history (Murawski *et al.* 2001, Scott *et al.* 2006). If hatching success is a function of the previous number of reproductive events experienced by a female, we have:

$$\mathrm{E} = \sum_{a=a_r}^{a_{max}} \sum_{j}^{n} m_a f_a s_a p_{a_j} h_j N_a \tag{26}$$

where $p_{a,j}$ is the proportion of females of age a spawning for the jth time, h_j is the hatching success for a female experiencing her jth spawning event, and all other terms are defined as before.

1.3.2 Larval viability and survival

Some empirical studies have indicated that egg hatching success and egg survival *per se* may not depend on the age of the spawners, but that larval viability does increase with egg size and energetic reserves, which in turn are functions of the age and, possibly, the reproductive history of the female. If we incorporate larval survival rates, the output then would be the number of larvae surviving to some specified age or size (see Murawski *et al.* 2001, O'Farrell & Botsford 2006, Spencer *et al.* 2007). We then have:

$$L = \sum_{a=a_r}^{a_{max}} m_a f_a s_a \ell_a N_a \tag{27}$$

where L is the number of larvae surviving to a specified point in time (e.g., settlement), l_a is the proportion of larvae from females of age a surviving to this point, and all other terms are defined as before. Murawski *et al.* (2001) provide an expression for the case where the reproductive history of the female is also considered.

1.3.3 Sex ratios

As noted by Marshall (Chapter 11), it has often been assumed that the sex ratio remains constant over time in analyses that attempt to substitute spawning biomass as a proxy for total egg production. Larkin (1977) had earlier called attention to the potential pitfalls of making this assumption (see Chapter 9). For species exhibiting sexual dimorphism in growth, the vulnerability to size-selective fishing gear differs by sex. Changing rates of fishing mortality in turn result in systematic changes in the sex ratio. Sex-specific differences in natural mortality and longevity may also contribute to this effect. Given an equal sex ratio at hatching, the expected proportion of females in a cohort as a function of age, fishing mortality, and natural mortality can be expressed:

$$s_a = \left[\frac{e^{-(M_{a,f}+p_{a,f}F)}}{e^{-(M_{a,m}+p_{a,m}F)}} + 1 \right]^{-1} \tag{28}$$

where the subscripts m and f indicate males and females. For species in which the females grow more rapidly and reach larger sizes (e.g., many flatfish species), the ratio of females to males will decline with increasing age as the fishing mortality rate increases. An illustration is provided in Figure 1.11 for a hypothetical long-lived population. Conversely, if males grow more quickly, the ratio of females to males will increase (see also Chapter 11). At low population levels in particular, distortions in the sex ratio of this type can lead to adverse effects on fertilization success. This can exacerbate chance variations in sex ratios at low population sizes, and lead to Allee effects by affecting the probability of finding a mate or other mechanisms.

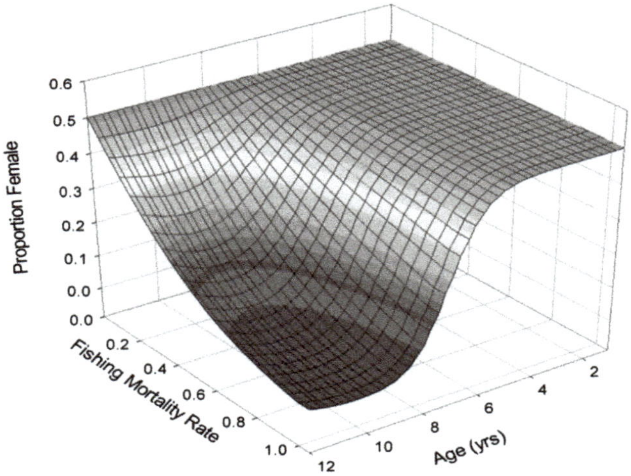

Figure 1.11 Proportion of females in a population as a function of age and fishing mortality when females exhibit faster growth and males and females experience identical natural mortality rates. The sex ratio at birth is assumed to be 1:1. For a color version of this figure, please see Plate 1 in the color plate section.

1.3.4 Effects on genetic structure

We have thus far focused on ecological processes and dynamics. It must be appreciated that exploitation also may potentially affect the genetic structure of populations with important consequences for sustainability and reversibility of the effects of fishing (see Chapter 4 for a detailed overview). Life history theory predicts that increased adult mortality will select for earlier maturation (Gadgil & Bossert 1970), and that increased juvenile mortality will select for later maturation (Reznick *et al.* 1990). Direct evidence that size-selective removals from a population can contribute to rapid evolutionary change has been examined in small-bodied fishes amenable to experimental manipulation and/or field observation (Conover & Munch 2002, Reznick *et al.* 1997). Reznick *et al.* (1997) demonstrated that size-selective predation in natural populations of guppies resulted in a significant evolution of life history traits of age and size at maturity (Reznick *et al.* 1990, 1997). Experiments in size-selective harvesting over four generations of Atlantic silverside resulted in the evolution in egg size, larval growth, and other life history traits (Conover & Munch 2002).

Attempts to determine potential evolutionary effects of fishing on natural populations of marine fishes have focused on estimating "reaction norms" (see Chapter 4). A reaction norm is derived by measuring the phenotypic expression of one genotype when exposed to different environmental conditions. Although reaction norms are generally determined experimentally, the need to understand the possible genetic impact of fishing on natural populations has led to an emphasis on the development of probabilistic reaction norms in wild populations. These studies have focused in particular on maturation, and have attempted to disentangle environmental effects on phenotypic characteristics from genetic effects attributable to harvesting (see Marshall & Browman 2007 and contributions within). Evidence for changes in maturation attributable to selective fishing effects have now been reported for a number of marine fishes based on probabilistic maturation reaction norms (Dieckmann & Heino 2007), and work continues on further attempts to separate environmental from fishing effects (Marshall & McAdam 2008).

1.3.5 *Compensation during the post-recruitment phase*

If no compensatory processes operate during the post-recruit phase of the life history affecting fecundity, maturation schedules, and so on, then the relationship between viable egg production and recruitment is linear and the slope of the relationship is a function of the age-specific fishing and natural mortality rates. We next turn to the case where compensatory response in fecundity and maturation schedules is important.

1.3.5.1 Compensatory fecundity

Fecundity is potentially affected by changes in abundance (see Ware 1980, Rothschild 1986, Rothschild & Fogarty 1989, Cushing 1995). Although direct estimates of fecundity as a function of population size are comparatively rare, there is substantial information on changes in the body size of fish as a function of abundance. The fecundity of marine fish is generally a linear function of body weight, and we can infer changes in mean fecundity with changes in body size if we can evaluate tradeoffs between allocations of energy for growth and reproductive output.

Ware (1980) and Rothschild & Fogarty (1989) considered the case where the total population fecundity is a nonlinear function of the spawning biomass (S):

$$\mathrm{E} = dS\,e^{-gS} \tag{29}$$

where d and g are model parameters which incorporate terms for sex ratio and mean fecundity. This model arises when the mean fecundity per unit biomass decays exponentially with increasing population biomass. Ware (1980) combined this result for stock-dependent fecundity with a density-dependent mortality structure to derive his energetically based stock–recruitment model. Rothschild & Fogarty (1989) showed that this relationship, when combined with a density-independent mortality function, results in a Ricker-type stock–recruitment function.

A simpler power function may be appropriate in some instances to describe the relationship between total egg production and spawning biomass:

$$\mathrm{E} = kS^{h} \tag{30}$$

where k and h are model parameters which again incorporate terms for sex ratio and mean fecundity. For the case when mean fecundity declines geometrically with increasing stock biomass, we obtain a compensatory relationship between total egg production and stock size (in this case, $h < 1.0$). However, Marshall *et al.* (1998) found that for Northeast Arctic cod during the period 1985–96, total egg production *increased* with spawning biomass ($h = 1.286$; Figure 1.12), implying a depensatory relationship over the range of available observations (see also Section 1.2.2).

1.3.5.2 Maturation schedules

Shifts in the age or size at maturation with changes in abundance have been documented for a number of exploited fish populations (see reviews in Rothschild 1986, Cushing 1995). If

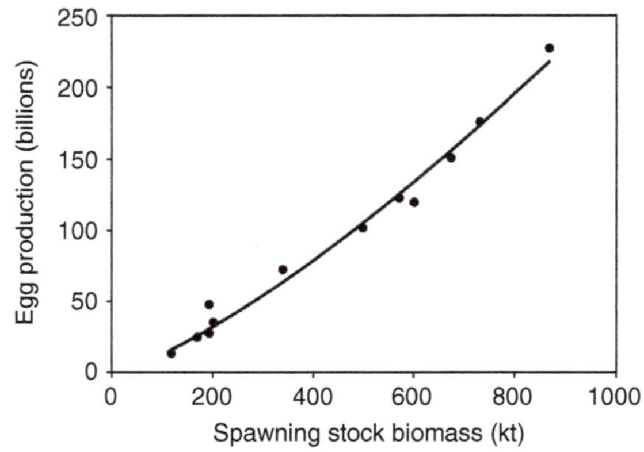

Figure 1.12 Relationship between spawning stock biomass and total egg production for Northeast Arctic cod for the period 1985–1996 (after Marshall *et al.* 1998).

the maturation schedule is affected by population size (reflecting density-dependent effects on energy available for growth and maturation), the proportion of mature females for the *i*th age or size class can be described by the logistic function:

$$m_i = \frac{1}{1 + e^{a - b_i + cN'}} \tag{31}$$

where a, b, and c are coefficients, i represents age or size, and N' is a measure of population size (e.g., total abundance, adult abundance, abundance of the *i*th size or age class, etc.). We have focused on density effects in this context because they ultimately have direct consequences for the shape of the recruitment–egg production curve. We can, in principle, also extend this approach to consider the effects of other factors affecting reproduction schedules (e.g., food supply) that reflect broader ecosystem changes.

Density-dependent maturation results in a nonlinear relationship between the number of recruits and the lifetime egg production of the cohort. We show the expected form of the relationship between egg production and recruitment in Figure 1.13 for the case where

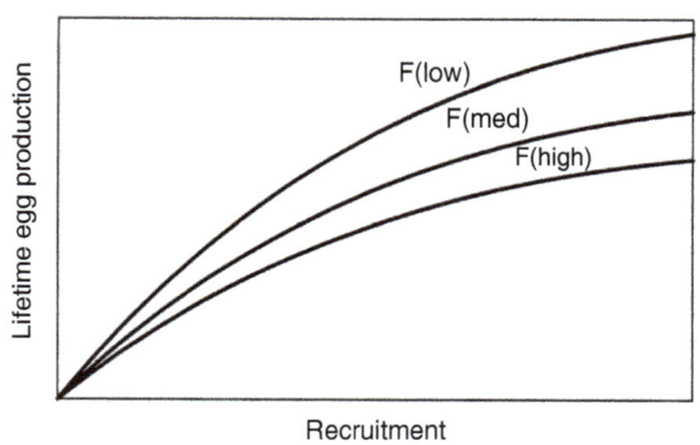

Figure 1.13 Lifetime egg production as a function of recruitment for a model incorporating density-dependent maturation at three levels of fishing mortality.

maturation follows the logistic maturation model with explicit consideration of abundance effects (Rothschild & Fogarty 1998). Harvesting alters the lifetime reproductive output by affecting the number of reproductive opportunities; accordingly, we provide results in Figure 1.13 for several levels of fishing mortality.

1.4 Stability properties

We next examine the stability and resilience of the population to sustained perturbations such as exploitation. The previous sections have illustrated relationships between egg production and recruitment and between recruitment (Section 1.2), and lifetime expected egg production (Section 1.3). We can combine these representations of the two major stanzas of the life history to examine stability points. First, we return to the relationship between egg production and recruitment and use an overcompensatory relationship to represent this life history stanza (Figure 1.14a). Next, if compensatory mechanisms are not important in the post-recruit phase of the life history, we can represent the relationship between recruitment and lifetime egg production as a family of straight lines for different levels of fishing mortality (Figure 1.14b). Essentially, for any level of fishing mortality we have a single value of egg production per recruit (refer to Figure 1.10) which specifies the slope of these lines. Next, can we overlay these relationships (Figure1.14a and b) on the same graph (this will involve exchanging the

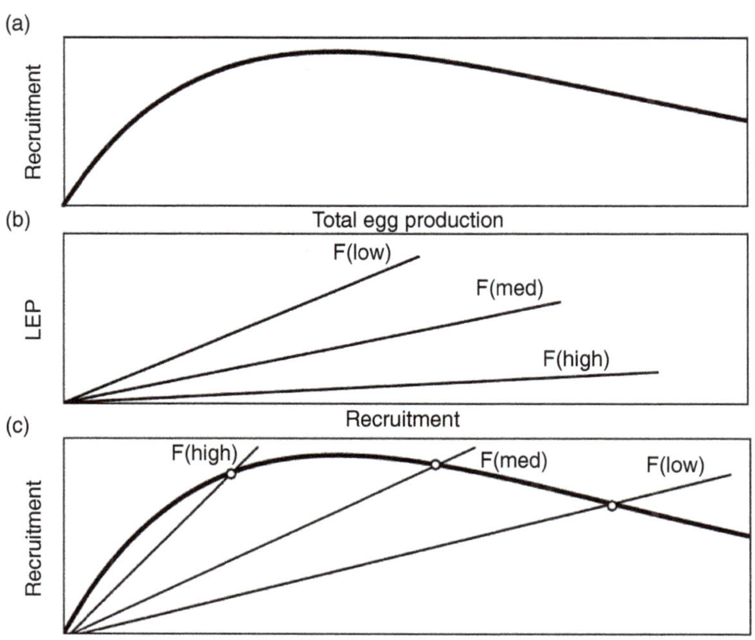

Figure 1.14 The relationship between (a) recruitment and total egg production, (b) lifetime egg production (LEP) and recruitment for three levels of fishing mortality (low, medium, and high) assuming no compensation in post-recruitment processes, and (c) superposition of panels (a) and (b) to illustrate intersection points representing stable equilibria.

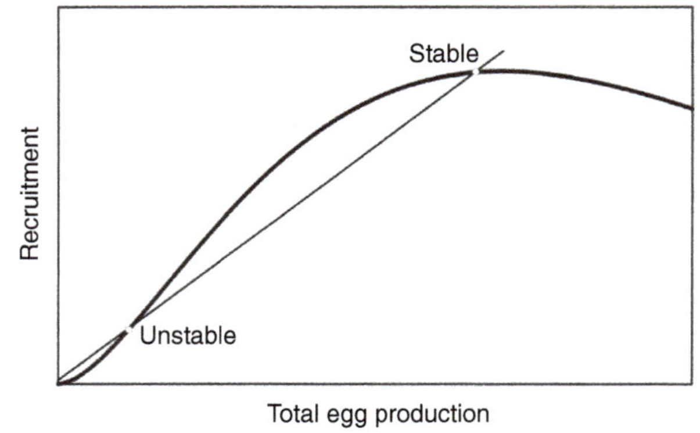

Figure 1.15 Stable and unstable equilibrium points for a depensatory model at one level of fishing mortality assuming no compensation in post-recruitment processes.

axes for the recruitment–lifetime egg production relationship so that the relationships can be superimposed; Figure 1.14c).

The points where the relationships for egg production–recruitment and recruitment–lifetime egg production now intersect represent equilibrium points. Note that, as the fishing mortality rate continues to increase, we eventually reach a level where there is no intersection point and a stock collapse is predicted. It follows that the steeper the recruitment curve at the origin, the more resilient the population will be to exploitation. However, because consideration of maternal effects may not only affect the slope of the recruitment curve at the origin but also the egg production per recruit curve, we must consider the interplay of these two factors in determining the stability properties and resilience of the population. With this type of information, we can estimate the levels of fishing mortality rate that would result in a high risk of population collapse and employ a precautionary approach to ensure that these levels are not approached.

These basic principles hold when we also observe compensatory processes during the post-recruit phase. Now, we have nonlinear relationships between recruitment and lifetime egg production, but as long as we have an intersection point between the curves an equilibrium point exists (see Rothschild & Fogarty 1998 for a graphical illustration).

As noted earlier, for the case of critical depensation, we have the possibility of multiple equilibria. This is illustrated in Figure 1.15 for the case where the post-recruit dynamics are density-independent and for one level of fishing mortality. Note that in this case, the upper intersection point at higher egg production gives a stable equilibrium point, while the lower one represents an unstable equilibrium. If the total egg production is driven below the lower level, a sudden population collapse is predicted.

1.5 Multistage models

In the preceding sections, we have collapsed the life history into two principal stanzas: pre- and post-recruitment. We can represent the life history with finer resolution within each of these stanzas. We have seen that it is important to account for relationships between the adult female population and viable egg production, between the egg and larval stages, from larvae

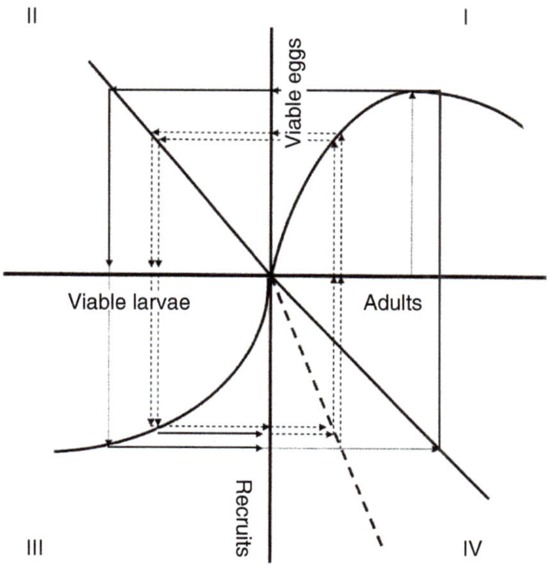

Figure 1.16 Paulik diagram for a four-stage life history pattern with nonlinear dynamics in two quadrants. Two levels of fishing mortality are represented in Quadrant IV. Arrows trace the trajectories of the population over several generations under the lower and higher fishing mortality rates.

to recruits, and from recruits to adults. An illustration of such a system is provided in Figure 1.16 (see also Rothschild & Fogarty 1998). In effect, this approach allows us to combine consideration of the factors explored in Sections 1.3 and 1.4 into a single framework. This graphical representation (or Paulik diagram; Paulik 1973, Rothschild 1986) allows a ready visualization of the implications of linear and nonlinear transitions between life stages for the stability properties of the population. In this example, the relationship between the adult female population and egg production is taken to be overcompensatory (Quadrant I), the transition between eggs and the larval settlement stage is linear (no density-dependence; Quadrant II), and the relationship between larvae and recruits (Quadrant III) is compensatory. In this example, the relationship between recruitment and the adult stage is linear. This relationship of course varies with the exploitation rate and we have shown results for two levels of fishing mortality, with the dashed line in Quadrant IV representing higher exploitation. For the lower exploitation rate we trace the transitions between life stages for two generations from an arbitrary starting point (see thin lines). Note that the compensatory processes result in a stabilization after two generations from the starting conditions in this example. When we increase the fishing mortality rate in the exploitation module (Quadrant IV; see dashed line), we find that tracing the trajectories between successive life stages results ultimately in convergence to a lower adult population level, but one that still results in a stable population.

In principle, any of these quadrants or modules can involve compensatory, overcompensatory, or depensatory processes. Some interesting mechanisms affecting the relationship between larval settlement and recruitment (Quadrant III) have been explored by Walters & Juanes (1993) and Walters & Martell (2004). Walters & Juanes (1993) evaluate the tradeoffs between individual growth and predation risk associated with the foraging behavior of juvenile fish. The system comprises spatial refuges and nearby foraging areas. The survivorship during the juvenile phase (S_j) can then be expressed:

$$S_j = e^{-(M_o+M_1T_f)} \tag{32}$$

where M_o is the instantaneous mortality rate due to all sources other than predation, M_1 is the instantaneous predation risk per unit time, and T_f is the time spent foraging. Walters & Juanes (1993) define fitness as the product of the survival rates in the juvenile and adult stages and the mean fecundity, and show that for a simple but robust relationship between fecundity and foraging time, the optimum time spent foraging (R_{opt}) is:

$$R_{opt} = R_o + 1/M_1 \tag{33}$$

where R_o is the minimum foraging time required to survive and reproduce after leaving the juvenile refuge area. If the optimum time spent foraging is directly proportional to the larval settlement (L_s), then the recruitment relationship will be overcompensatory:

$$R = L_s e^{-(M_o+M_1{'}L_s)} \tag{34}$$

where $M_1{'}$ is the product of the predation risk coefficient and the constant of proportionality between R_{opt} and L_s (Walters & Juanes 1993). If, instead, the foraging time of the cohort changes continuously with abundance of the cohort, a compensatory (asymptotic) recruitment curve results.

Walters & Martell (2004) considered a different scenario in which juveniles compete for a limited number of shelter sites. In this model, the juvenile population is partitioned into two groups: one group which is in a dispersal state, and another group comprising individuals who have located shelter. The rate of change for the dispersal component is:

$$\frac{dN_d}{dt} = -M_dN_d - aN_d(k - N_s) \tag{35}$$

where M_d is the mortality rate for dispersers, a is the search rate for dispersing individuals, k is the number of shelter sites, and N_s is the number of individuals in shelters. The model for individuals having found shelter is:

$$\frac{dN_s}{dt} = -M_sN_s + aN_d(k - N_s) \tag{36}$$

Although this system of equations does not appear to have an analytical solution, Walters & Martell (2004) showed that the numerical solution indicates an asymptotic recruitment model. Interestingly, the solution reveals a more abrupt transition to the asymptotic state than the Beverton–Holt model, and is more similar to the results obtained with the so-called "hockey-stick" representation of Barrowman & Myers (2000), in which recruitment increases linearly from the origin until a threshold level of reproductive output is reached and recruitment remains constant thereafter (see Marshall *et al.* 2006 for an application to Northeast Arctic cod).

1.6 Yield and sustainable harvesting

In general, information from a stock–recruitment model can be combined with information from a yield- and spawning-per-recruit analysis to estimate total equilibrium yield (see Chapter 7 for an overview and verbal description). Here, we illustrate this process when viable egg production is the metric used to represent reproductive output. In the following, we will consider just the case where the post-recruitment processes do not exhibit any form of compensation. To begin, note that we can solve the recruitment model in terms of total egg production. For example, returning to our earlier specification for the Ricker model (see Equation (10)), we can write:

$$\log_e \left[\frac{R}{\mathrm{E}}\right] = \log_e \kappa - \delta \mathrm{E}. \tag{37}$$

Solving for total egg production we have:

$$\mathrm{E} = \frac{1}{\delta} \log_e \left[\kappa \left(\frac{\mathrm{E}}{R}\right)\right]. \tag{38}$$

Notice that the expression inside the brackets includes (E/R), or egg production-per-recruit. Given estimates of the parameters of the Ricker model and an egg production-per-recruit analysis, we can substitute estimates of E/R for different levels of fishing mortality $(\mathrm{E}/R)_F$ to determine the total egg production for each of these fishing mortality rates. This general approach can of course be followed using other recruitment models. Once the total spawning biomass corresponding to a particular level of fishing mortality is determined, the corresponding recruitment can be obtained by the simple identity:

$$R_F = \frac{\mathrm{E}_F}{(\mathrm{E}/R)_F} \tag{39}$$

and in essence, we have simply provided the analytical framework for the graphical analysis presented in Figure 1.14.

We can now obtain the predicted equilibrium yield for each level of fishing mortality by combining the yield per recruit at each level of fishing mortality with this predicted recruitment level to obtain an estimate of the total yield at each level of fishing mortality:

$$Y_F = (Y/R)_F R_F \tag{40}$$

An illustration of this approach is provided in the next section to evaluate the implications of factors such as maternal effects on viable egg production. Specification of these yield curves allows us to determine key biological reference points, such as the maximum sustainable yield (MSY) and the fishing mortality rate resulting in MSY (F_{MSY}). The fishing mortality rate resulting in a predicted collapse of the population is designated F_{crash}.

1.7 Implications of maternal effects

When maternal effects on egg viability are important, our *perceptions* of the relationship between egg production and recruitment (the pre-recruit phase) and the relationship between recruitment and lifetime egg production (the post-recruit phase) are altered. In particular for observed levels of recruitment, if the viable egg production is lower than the nominal total egg production because of fishing effects on the age-structure of the population, the slope of the recruitment curve at the origin may be steeper than our perception if we ignore or are unaware of this effect. This can occur because the observed levels of recruitment are actually derived from lower levels of effective reproductive output relative to our perception if we use egg production uncorrected for maternal effects. This seemingly paradoxical result also of course holds when spawning biomass is used as the index of reproduction when maternal effects on viable egg production are in fact important (as shown for Icelandic cod in Figure 1.9).

Countering this effect is the fact that the lifetime reproductive output of viable eggs by a cohort will be lower than perceived if maternal effects are important and we do not take them into account (see Figure 1.10). If we ignore or are unaware of maternal effects, the consequences of fishing on the population will then depend on the interplay of these two factors – the potential underestimation of the slope of the recruitment curve at the origin, and the overestimation of the lifetime reproductive output of viable eggs.

Recall the development in Figure 1.14 showing equilibrium points and the limiting level of fishing mortality beyond which the risk of stock collapse is high. Suppose now we have a "perceived" relationship between egg production and recruitment in which maternal effects are not recognized (see Figure 1.17a, thin line) and an "actual" relationship with a steeper slope at the origin (Figure 1.17a, bold line) with a proper accounting of realized egg production levels. Next, consider the case where we calculate the lifetime egg production per recruit for a particular level of fishing mortality but are unaware of important maternal effects (Figure 1.17b, thin line; labeled F(s)) and contrast this with two cases of maternal effects on viable egg production at the same level of fishing mortality as for line F(s), one of which reflects a stronger maternal effect (labeled F(m2) in Figure 1.17b). Note that "stronger" here refers to the case where there is a larger differential in egg or larval viabilities with age. We now overlay panels (a) and (b) as before. For the case where we do not account for maternal effects either in the egg production–recruitment relationship or in the calculation of lifetime egg production, we do have an intersection point and we would predict that this level of fishing mortality would be sustainable under our (mis)perceived view of the dynamics of this population. Now consider the case where we correctly portray maternal effects both in both the egg production–recruitment relationship and in the calculation of lifetime viable egg production. Under the more moderate maternal effect on lifetime egg production (labeled F(m1)), we still predict a stable equilibrium point. In effect, for this hypothetical case the steeper slope of the recruitment curve at the origin for the "actual" case is sufficient to offset the lower level of lifetime egg production at this level of fishing mortality when maternal effects are in fact important. However, with the second case representing a stronger maternal effect on lifetime viable egg production, we no longer have an intersection point with the egg production–recruitment curve and we predict a stock collapse at this level of fishing mortality. In this case, our prediction of a sustainable fishery when we did not properly account for maternal effects would place the population at risk.

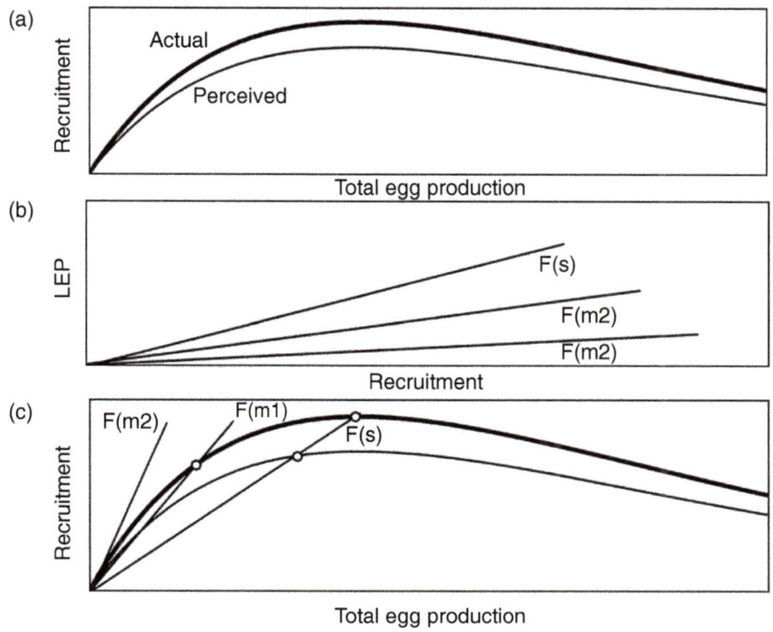

Figure 1.17 (a) Recruitment curves for the case of maternal age effects ("actual"; thick line) and when maternal effects are important but are unrecognized ("perceived"; thin line); (b) estimated lifetime egg production (LEP) as a function of recruitment at the same level of fishing mortality assuming the "standard" model of no maternal effect (F(s)), a "moderate" maternal effect (F(m1) and a "strong" maternal effect (F(m2)); (c) superposition of these relationships to determine equilibrium points.

Emerging evidence suggests that different outcomes are predicted for different species/populations. Lower overall resilience to high exploitation for different cod stocks using different metrics for maternal effects have been reported (Marshall *et al.* 2000, 2006, Murawski *et al.* 2001, Martinsdottir & Begg 2002). In contrast, Cervino *et al.* (2013) found that an increase in the slope of the R–E curve relative to the R–S curve more than offset the decreased EPR_F curve relative to SPR_F, resulting in a higher estimate of F_{MSY}. Spencer *et al.* (2007) and Spencer & Dorn (2013) each reported a lower resilience to high exploitation rates for Pacific ocean perch when viable larvae was used as the index of reproduction relative to either SSB or total larval production, although the F_{MSY} levels were similar for each metric of reproductive activity. O'Farrell & Botsford (2006) found that if maternal effects were only important for a restricted range of younger age groups of rockfish species on the west coast of the United States, little effect on the perceived resilience of the population to harvesting would be evident. However, if maternal effects were important for a broader range of ages in the adult population, the discrepancy between the perceived and actual resilience of the population would be greater, possibly resulting in a higher risk to the population.

Using the approach described in Section 1.6, we can construct yield curves corresponding to the hypothetical cases described above. The "normalized" yield predicted for the case of no maternal effects is depicted in Figure 1.18 (thick line). We contrast the normalized yield as a function of fishing mortality for the case where maternal effects on viable egg production are important (Figure 1.18, thin line). In this instance, the population is less resilient at high

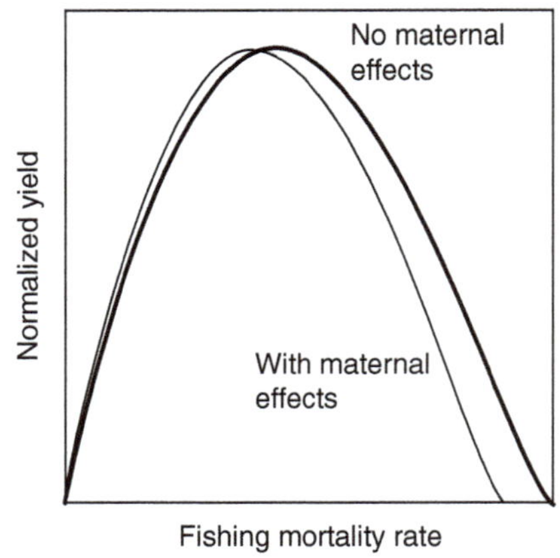

Figure 1.18 Normalized yield as a function of fishing mortality for the case of no maternal age effects (thick line) and maternal age effects on viable egg production (thin line).

levels of fishing mortality when maternal effects on lifetime viable egg production is more than sufficiently strong to offset the steeper slope of the recruitment curve at the origin.

1.8 Recruitment variability

The models described above do not consider exogenous environmental effects (either biotic or abiotic) on recruitment, and nor do they explicitly account for factors such as the age diversity of spawners that may serve to dampen the effects of fluctuations in the environment. Yet, as noted earlier, recruitment is extremely variable, largely as a result of the effects of exogenous forcing factors. In the context of the major theme areas of this book, we are interested in questions such as whether a proper accounting of the relationship between reproductive output and recruitment improves the predictability in recruitment (see Chapter 11), and how recruitment variability is related to egg production (see below). Houde (see Chapter 3) describes the many sources of variability in growth and mortality during the early life stages and their implications for recruitment variability. In the following, we explore two general approaches to this problem. In the first approach, additional factors are explicitly represented in the recruitment model in an attempt to partition the variance in recruitment into definable sources. In the second approach, recruitment is treated as a stochastic process as a result of random variation in mortality rates during the pre-recruit stage (environmental stochasticity) or, for small populations, chance variation in the number of deaths in a given time interval (demographic stochasticity).

1.8.1 Multidimensional recruitment models

The models described earlier treat recruitment solely as a function of egg production; other aspects of the biotic and abiotic environment are not explicitly considered. However, it is appreciated that recruitment processes are highly dimensional (Rothschild 1986, Fogarty *et al.* 1991). To the extent that specific environmental factors affecting recruitment can be identified and quantified, these should be incorporated into recruitment–stock formulations (see Chapters 3, 9, and 10). This approach has received considerable attention as an extension to traditional

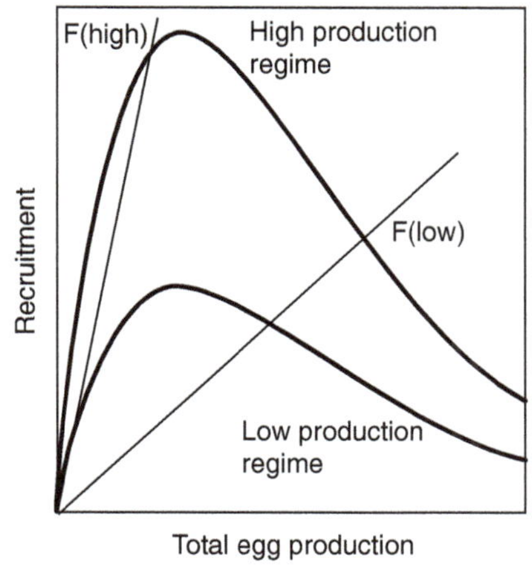

Figure 1.19 Relationship between recruitment and egg production under two environmental regimes and two levels of fishing mortality, demonstrating the interaction between harvesting and changes in productivity states on stability of the population.

recruitment models (see Hilborn & Walters 1992 for an overview and caveats). Marshall *et al.* (2000) demonstrated the utility of developing recruitment models incorporating both maternal effects and temperature for Northeast Arctic cod.

Consider a simple extension of the Ricker model to account for an additional physical or biological environmental variable:

$$R = \kappa \mathrm{E}\, e^{-\delta \mathrm{E} + \upsilon X} \tag{41}$$

where υ is the coefficient for the environmental factor X and all other terms are defined as before. We now have a three-dimensional surface rather than a bivariate egg production–recruitment plane. The effect of projecting multidimensional data into an artificially reduced two-dimensional system, as in the classical egg–production recruitment models, will then show a potentially highly variable representation as a result of this compression (Marshall *et al.* 2000). Pope (see Chapter 7) shows that incorporating temperature into the stock–recruitment relationship for Northeast Arctic cod substantially improves the fit of the model. For examples of recruitment models including biotic (multispecies) interactions, see Chapter 10. Cochrane (see Chapter 9) reviews additional applications involving physical variables. Note that this approach can accommodate multiple environmental variables.

We can also think of this multidimensional system as a family of recruitment curves for given levels of the biotic and abiotic variables. Consider the egg production–recruitment relationship under two environmental regimes representing low and high productivity states (Figure 1.19). We can see immediately that, although a relatively low fishing mortality rate may be sustainable under either the low- or high-productivity regime (there are equilibrium points under both environmental states when the exploitation rate is low), the combination of low productivity and high fishing mortality can result in a stock collapse at an exploitation rate that is sustainable when productivity is high (Figure 1.19). We therefore need to be concerned about the interaction between environmental change and fishing pressure and not ascribe changes to either fishing or the environment alone (Fogarty *et al.* 1991).

It is also possible to consider additional types of explanatory variables in these recruitment models. In keeping with the focus of this book on identifying critical factors affecting spawning success, we considered the effect of including an index of the age diversity of spawners in an extended Ricker model for Icelandic cod. Marteinsdottir & Thorainsson (1998) had previously demonstrated an improvement in the fit of a recruitment model when age-diversity of spawners was incorporated in the model for this population. Age diversity was measured using the well-known Shannon–Weiner diversity index (Figure 1.1d). It has been hypothesized that higher levels of age-diversity in the spawning stock increases the spawning window in time, and may serve as an important bet-hedging strategy for reproduction in a variable environment (see Chapter 11) by spreading reproductive effort over a range of environmental conditions throughout the spawning season. Conversely, truncation of the age distribution can reduce the length of the spawning period by up to a month and a shift in the peak spawning period (Scott *et al.* 1999).

In our updated analysis using a linearization of Equation (41), we found a significant improvement in the fit of the model. The adjusted coefficient of determination increased from 0.50 to 0.68 when the age diversity of spawners was included. Other analyses, including consideration of the age-diversity of spawners, have provided mixed results; for example, O'Brien *et al.* (2003) found that including an index of age diversity along with measures of bottom temperature and the spatial distribution of spawned eggs improved the fit of a model of egg survival rates for Georges Bank cod. Morgan *et al.* (2007), however, found no consistent improvement in model fit when adding consideration of the age composition of the spawning stock for three Canadian cod populations and one American plaice population.

1.8.2 Environmental stochasticity

We have thus far provided an overview of key issues in pre-recruit processes in a deterministic setting to focus attention on fundamental demographic principles. However, as described above, the high dimensionality of the marine environment virtually assures that the full complexities of pre-recruit processes cannot be represented in simple models. Variation in physical forcing mechanisms and predator and prey fields translates into variation in survival and/or growth rates during the pre-recruit phase (Houde 1987, 1989; see Chapter 3). Clearly, when these mechanisms have been identified, they should be incorporated into recruitment models (see Section 1.6.1 and discussions in Chapters 9 and 10). In this section, we describe an alternative approach for the case where key physical and biological environmental variables have not been identified (or the case where substantial random variation remains after key variables have been incorporated into recruitment models). We will illustrate the development of stochastic recruitment models for the density-independent case to explore the expected form of recruitment variability when the pre-recruit mortality rate is not constant but instead is a random variable (see Fogarty 1993a for an overview of stochastic models for other functional forms). We will also consider questions such as whether the magnitude of egg production affects the expected level of variability in recruitment.

First, consider the case where the density-independent mortality term is assumed to be a normally distributed random variable. The assumption of normally distributed mortality rates can be justified under the Central Limit Theorem (see Fogarty 1993b). If the mortality coefficient varies randomly during the pre-recruit phase, then the overall mortality can be viewed as the accumulated sum of random variables. For the case of independent mortality rates and

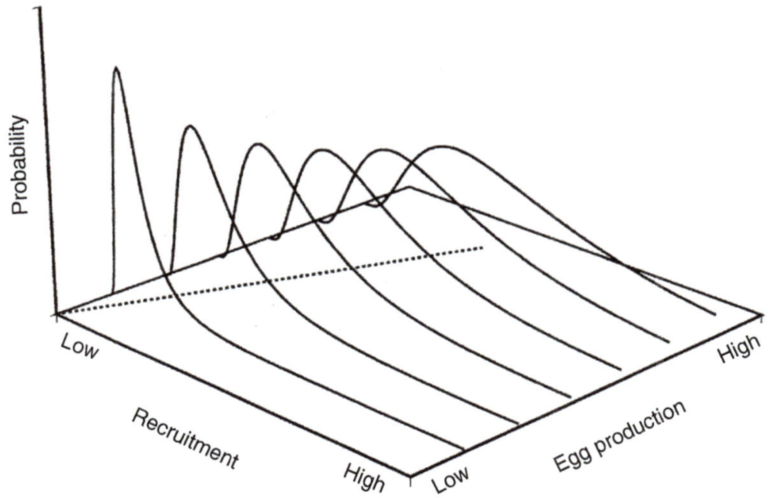

Figure 1.20 Representation of the conditional probability distributions of recruitment for the density-independent models (after Fogarty *et al.* 1991).

for a relatively large number of such intervals, the overall mortality rate will be normally distributed. However, the Central Limit Theorem holds under much more general conditions for non-independent stationary processes (see Fogarty 1993b for a review). In the following, we will illustrate some results using the null (density-independent) model, although extension to the various compensatory recruitment models considered earlier can be readily made (Fogarty 1993a).

Under these assumptions, the conditional probability density function of recruitment for the density-independent model is lognormal:

$$p(R\,|\,\mathrm{E}) = \frac{R^{-1}}{\sqrt{2\pi}\,\sigma_\mu\,t}\exp\frac{[-(\log_e(R/\mathrm{E}) + \bar{\mu}t)^2]}{2\,\sigma_\mu^2\,t^2} \tag{42}$$

where μ is the mean density independent mortality rate and $\sigma_\mu{}^2$ is its variance. The shape of the distribution for several levels of initial cohort size is illustrated in Figure 1.20. The mean is:

$$\bar{R} = \mathrm{E}\,e^{(-\bar{\mu}+\sigma_\mu^2/2)t} \tag{43}$$

and its variance is:

$$V(R) = \mathrm{E}^2\,e^{(-2\bar{\mu}+\sigma_\mu^2)t^2}\,[e^{\sigma_\mu^2 t^2} - 1] \tag{44}$$

The mean recruitment level is higher for the stochastic model than for the corresponding deterministic case by the factor $\exp[(\sigma_\mu{}^2/2)t^2]$. The modal recruitment is, however, lower for the stochastic case than the deterministic recruitment level. Notice that the expression for the variance of recruitment conditioned on egg production is a function of the square of the total egg production. We would therefore expect that populations with higher levels of egg production would exhibit higher absolute variability in recruitment (although the coefficient of variation of recruitment is independent of the initial number of eggs; see below).

Several general consequences emerge from the properties of the lognormal recruitment distributions. First, we expect low to moderate recruitment levels most of the time, with occasional strong year classes (represented by the strong right-hand tails of the distributions). Second, a relatively small variation in mortality rates translates into a large variation in recruitment. To see this, note that the coefficient of variation (CV) for the model described above can be expressed:

$$CV(R) = (e^{\sigma_\mu^2 t^2} - 1)^{1/2} \tag{45}$$

The CV therefore depends on the variance of the density-independent mortality rate, but not on the number of eggs in this model. The instantaneous rate of density-independent mortality is typically on the order of $\mu = 10.0$. Even a relatively low level of variability in the mortality rate translates into a high level of recruitment variability. For example, when $\mu = 10.0$, a CV as low as 10% in mortality would result in over a 100% CV in recruitment. It is clear that small changes in mortality can result in large changes in recruitment. Indeed, given the potential range of variation, it is surprising that fish populations are not more variable than actually observed (Gulland 1982; Chapter 3). This observation suggests that mechanisms exist to dampen variability (see also Chapter 3). Fogarty (1993a) noted that density-dependent mechanisms can serve to reduce variability in this context. These general results also suggest that even with a more detailed accounting of effective reproductive output, we should still expect high levels of variability in recruitment for many marine species because of events following the production of viable eggs by the population. It is therefore perhaps not surprising that equivocal results have been obtained for the proportion of variation explained by recruitment models in comparisons using SSB as the indicator of reproductive output with those using total egg production (see Brunel 2010, Morgan *et al.* 2011; Chapter 11).

Attempts to relate recruitment variability to fecundity have provided conflicting results (Rothschild & DiNardo 1987, Mertz & Myers 1996, Rickman *et al.* 2000). Rothschild & DiNardo (1987) found such a relationship for anadromous fish but not for marine fish. Mertz & Myers (1996) were also unable to detect an empirical relationship between fecundity and recruitment variability. In contrast, Rickman *et al.* (2000) demonstrated a relationship between fecundity and recruitment variability for marine fish after accounting for phylogenetic relationships in an analysis of 52 fish stocks.

1.8.3 Demographic stochasticity

The model for environmental stochasticity described above is appropriate for the case where mortality rates vary during the course of the pre-recruit period. We can also consider a model for the number of individuals in a cohort based on time-invariant survival probabilities where the cohort size can assume only discrete values (see Fogarty 1993b). Models of this type are particularly appropriate for species with relatively low population sizes and substantial energetic investment in the progeny by the parents such that variability in mortality is relatively low.

For example, application to certain threatened elasmobranch populations might be appropriate in this context. In the following, it is assumed that the probabilities for each individual are independent and constant through time. The probability of an individual dying in an interval of time Δt is $\mu\Delta t + o(\Delta t)$, where $o(\Delta t)$ is a quantity of smaller order than Δt. The probability of obtaining exactly N individuals at time $t+\Delta t$ is given by:

$$P_N(t + \Delta t) = P_{N+1}(t)\mu(N + 1)\Delta t + P_N(t)(1 - \mu N\Delta t) \tag{46}$$

where the first term on the right-hand side is the probability of exactly one death in a small interval of time Δt, and the second term is the probability of no deaths in the interval. The interval is defined to be sufficiently small that the probability of two or more events is negligible. These probabilities are taken to be independent, and therefore additive. The model can be expressed as:

$$\frac{P_N(t + \Delta t) - P_N(t)}{\Delta t} = \mu(N + 1)\,P_{N+1}(t) - \mu(N)\,P_N(t) \tag{47}$$

and taking the limit as $\Delta t \to 0$ and solving yields the probability density function (PDF) for recruitment. The PDF is now binomial:

$$P(R|\mathrm{E}) = \binom{E}{R}(e^{-\mu t})^R(1 - e^{-\mu t})^{\mathrm{E}-R} \tag{48}$$

with mean:

$$\bar{R} = \mathrm{E}\,e^{-\mu t} \tag{49}$$

and variance:

$$V(R) = \mathrm{E}\,e^{-\mu t}[1 - e^{-\mu t}] \tag{50}$$

The coefficient of variation is given by:

$$CV(R) = \left[\frac{1 - e^{-\mu t}}{\mathrm{E}\,e^{-\mu t}}\right]^{1/2} \tag{51}$$

Although the exact shape of the probability distribution is dependent on the level of mortality, in general it is much more symmetrical than for the case of the environmental stochasticity (it also of course is a discrete rather than continuous distribution). Note that, in contrast to the model of environmental stochasticity, the mean is identical to the deterministic level and the coefficient of variation does depend on E; in this case, the CV *decreases* with increasing E and the results converge to the deterministic case.

Shelton (1992) constructed a simulation model in which the fate of individual eggs and patches of eggs was traced. For the latter case, individuals in a patch all shared the same fate. Shelton further considered the important case where individuals or patches could exhibit different survival probabilities. Fogarty (1993b) showed that Shelton's (1992) simulations could be cast as a pure death stochastic process of the form described above. In Shelton's (1992) simulations the mean recruitment can be written:

$$\bar{R} = n\mathrm{E}^{*}S \tag{52}$$

where n is the number of patches, E* is the number of eggs in the group, and S is the survival fraction ($S = e^{-\mu t}$); the total egg production is of course $\mathrm{E} = n\mathrm{E}^{*}$. The variance is now:

$$V(R) = n\mathrm{E}^{*2}S(1 - S) \tag{53}$$

and the coefficient of variation is:

$$CV(R) = \left[\frac{\{nS(1 - S)\}^{1/2}}{nS}\right] \tag{54}$$

Notice that the CV does not depend on the number of eggs in a patch in this case. The significance of this formulation is that it is now possible to consider the fate of groups of eggs that could, for example, represent reproductive events for individual females (each comprising a patch) or eggs derived from more than one female in a given location forming a patch. In his simulations, Shelton contrasted results for individual eggs (E = 1) with patches of eggs (E > 1). Shelton (1992) equated the former with bet-hedging strategies in which the risk is spread by placing a large number of eggs in space and time with effectively independent survival probabilities.

Shelton's (1992) extension to the case of different survival probabilities for individual eggs or groups of eggs further allows consideration of factors such as maternal effects or the effects of spatial heterogeneity in survival rates. The mean recruitment is now:

$$(\bar{R}) = \mathrm{E}^{*}\sum_{i=1}^{n} S_i \tag{55}$$

where the subscript i indexes the individual or group. The variance is given by:

$$V(R) = \mathrm{E}^{*2}\sum_{i=1}^{n} S_i(1 - S_i) \tag{56}$$

Interestingly, the variance is actually lower relative to the case of identical survival rates. The coefficient of variation for this case is:

$$CV(R) = \left[\frac{\left\{\sum_{i=1}^{n} S_i(1 - S_i)\right\}^{1/2}}{\sum_{i=1}^{n} S_i}\right] \tag{57}$$

and we see that the CV depends only on the survival rates for individuals or patches.

It is therefore possible to incorporate some of the considerations related to maternal effects and other factors in these simple models to understand what we might expect for patterns of variability in recruitment. The models can be extended to consider more complicated cases where the survival within a patch is not a completely dependent all-or-nothing response structure (Fogarty 1993b). It is also possible to include environmental stochasticity in these models (Fogarty 1993b).

The simple consideration of patch dynamics in the above representation is a stepping stone to increased realism in modeling. O'Brien (see Chapter 10) provides an illustration of a much more detailed spatially explicit model for cod around the British Isles, where careful attention is paid to physical processes, ecological interactions and demographic structure. Scott *et al.* (2006) have developed a model with full representation of size and age structure, female condition and seasonal dynamics to estimate reproductive output.

1.9 Summary

The renewed emphasis on recruitment processes as a function of total egg production of the population rather than SSB represents a return to the origins of recruitment theory. The contributions to this book explore the current state of knowledge and the importance of returning to these roots. Information is now accruing that will allow a broader utilization of information on the relationship between total egg production and recruitment in fishery management, although a concerted effort is needed to collect the necessary information for many more populations. Although timed series of fecundity estimates required to estimate total egg production remain relatively rare, the development of rapid estimation methods promises to revolutionize the acquisition of this vital information. In general, it is currently possible to examine the relationship between total egg production and recruitment only for restricted time periods and for selected species. To reconstruct longer time series, it has sometimes been possible to use predictive relationships derived from the shorter series and/or proxies, and to expand these to the full-time series (see Marshall *et al.* 1998, 2006; Chapter 11). If information on fecundity is not available, it will be valuable to account for changes in sex ratios over time to develop estimates of female spawning biomass in place of total SSB (see Chapter 11).

An understanding of how a fish population will respond to harvesting requires not only an accurate accounting of its effective reproductive output but also an understanding of the relative importance of compensatory mechanisms operating at different points in the life cycle. Under some circumstances, the choice of different measures of reproductive output (total spawning biomass, female spawning biomass, total egg production) results in different views of the types of regulatory mechanisms that are operative (Morgan *et al.* 2009). Marshall *et al.* (2006) showed that this choice results in models for Northeast Arctic Cod which indicate depensatory dynamics when SSB is used, while overcompensatory models are indicated when female SSB or total egg production is employed.

Even relatively low levels of variability in growth and mortality can translate into high levels of recruitment variability. This variability in recruitment should be viewed as an integral part of the life history strategies of many marine species. Occasional strong year classes (represented by the tails of the distributions depicted in Figure 1.20) can be important in maintaining the population, and this so-called "storage" effect (Chesson 1984) may permit population persistence in a variable environment. It must be appreciated that harvesting can interfere with this mechanism by truncating the age distribution of females and reducing the number of lifetime reproductive opportunities (Fogarty 1993a, Longhurst 2002, Beamish *et al.* 2006). Coupled with information showing the importance of maternal effects, and that the age-diversity of spawners is a critical attribute of many populations, the implications of overharvesting can be more severe than previously appreciated.

Evidence from a growing number of species currently suggests that maternal effects are important in determining egg and larval viability, and that the age, condition, and/or reproductive history of an individual female may be critically important. The consequences of these maternal effects can depend greatly on their relative importance for our perception of the relationship between recruitment and the adult population (particularly the slope of the recruitment curve at the origin), and their implications for the effective lifetime reproductive output of an individual female. The interplay of these two factors determines whether the resilience of the population to exploitation will be wrongly estimated if maternal effects are ignored or unappreciated.

These considerations can in turn inform the choice of tactical management tools (for an overview of management tools and their application in a conventional fishery management setting, see Chapter 9). Although management strategies such as reductions in fishing effort and fishing mortality result in an increased probability of survival to older and larger sizes, other methods can be tailored specifically to enhance the population of older individuals. For example, the use of marine protected areas has been shown to be effective in rebuilding the age and size structure of fish populations within reserves for species characterized by low to moderate mobility. The potential buildup of larger, older individuals within the reserve can then serve as a source for the replenishment of adjacent areas through the export of eggs and larvae. If larger, older females produce eggs and/or larvae with higher survival probabilities, the utility of the protected area will be enhanced and this should be accounted for in evaluations of the efficacy of the reserve as a fishery management tool. The elimination of size-selective harvesting in at least parts of the population area may also help to reduce adverse selective pressure on the genetic composition of exploited species (Law 2007).

For harvesting methods that result in the capture of fish that can be released alive (e.g., hook and line, certain traps), it is possible to consider the use of maximum size limits in which the taking of larger individuals is prohibited, or "slot" limits in which both minimum and maximum size limits are employed. Of course, in the face of high exploitation rates the number of individuals that reach the upper size limit can be sharply reduced, and slot or maximum size limits must be used in conjunction with controls on overall fishing mortality to be effective.

These considerations all point to the utility of developing a refined understanding of reproductive dynamics in order to enhance the resilience of harvested populations to exploitation through the management choices we make. By quantifying the effect of factors such as maternal effects on egg and larval survival, we can develop more realistic models of the dynamics of exploited populations and devise more effective fishery management strategies. The following chapters provide important insights into current the state of knowledge and directions for future research in reproductive dynamics that will substantially advance this critically important goal.

References

Barrowman, N.J. & Myers, R.A. (2000) Still more spawner–recruitment curves: the hockey stick and its generalizations. *Canadian Journal of Fisheries and Aquatic Sciences*, **57**, 665–76.

Beamish, R.J., McFarlane, G.A. & Benson, A. (2006) Longevity overfishing. *Progress in Oceanography*, **68**, 289–302.

Beverton, R.J. & Holt, S.J. (1957) *On the dynamics of exploited fish populations*. Fisheries Investigations Series 2, 19. Ministry of Agriculture, Fisheries and Food, London. 533 pp.

Brooks, E.N. (2013) Effects of variable reproductive potential on reference points for fisheries management. *Fisheries Research*, **138**, 152–8.

Brunel, T. (2010) Age-structure-dependent recruitment: a meta-analysis applied to Northeast Atlantic fish stocks. *ICES Journal of Marine Science*, **67**, 1921–30.

Cervino, S., Domínguez-Perez, R., Jardim, E., Mehault, S., Piñeiro, C. & Saborido-Rey, F. (2013) Impact of egg production and stock structure on MSY reference points and its management implications for southern hake (*Merluccius merluccius*). *Fisheries Research*, **138**, 168–78.

Chesson, P.L. (1984) The storage effect in stochastic population models. *Lecture Notes in Biomathematics*, **54**, 76–89.

Conover, D.O. & Munch, S.B. (2002) Sustaining fisheries yields over evolutionary time scales. *Science*, **297**, 94–6.

Cushing, D.H. (1995) Population production and regulation in the sea. Cambridge University Press, Cambridge. 354 pp.

Dieckmann, U & Heino, M. (2007) Probabilistic maturation reaction norms: their history, strengths, and limitations. *Marine Ecology Progress Series*, **335**, 253–69.

Dominey, W.J. & Blumer, L.S. (1984) Cannibalism of early life stages in fishes. In: G. Hausfater & S. Blatter Hardy (Eds). pp 43–64. *Infanticide: Comparative & Evolutionary Perspectives*. Aldine, New York.

Fogarty, M.J. (1993a) Recruitment in randomly varying environments. *ICES Journal of Marine Science*, **50**, 247–50.

Fogarty, M.J. (1993b) Recruitment distributions revisited. *Canadian Journal of Fisheries and Aquatic Sciences*, **50**, 2723–8.

Fogarty, M.J., Mayo, R.K., O'Brien, L. & Rosenberg, A.A. (1996) Assessing uncertainty and risk in exploited marine populations. *Reliability Engineering and System Safety*, **54**, 183–95.

Fogarty, M.J., Rosenberg, A.R. & Sissenwine, M.P. (1992) Fisheries risk assessment – sources of uncertainty; a case study of Georges Bank haddock. *Environmental Science and Technology*, **26**, 440–7.

Fogarty, M.J., Sissenwine, M.P. & Cohen, E.B. (1991) Recruitment variability and the dynamics of exploited marine populations. *Trends in Ecology and Evolution*, **6**, 241–6.

Frank, K.T. & Brickman, D. (2000) Allee effects and compensatory population dynamics within a stock complex. *Canadian Journal of Fisheries and Aquatic Sciences*, **57**, 513–7.

Gadgil, M. & Bossert, W.H. (1970) Life historical consequences of natural selection. *American Naturalist*, **104**, 1–24.

Gulland, J.A. (1982) Why do fish numbers vary? *Journal of Theoretical Biology*, **97**, 69–75.

Harris, J.G.K. (1975) The effect of density-dependent mortality on the shape of the stock and recruitment curve. *Journal du Conseil International pour l'Exploration de la Mer*, **36**, 144–9.

Hilborn, R. & Walters, C.J. (1992) *Quantitative Fisheries Stock Assessment*. Chapman & Hall, New York.

Houde, E.D. (1987) Fish early life dynamics and recruitment variability. *American Fisheries Society Symposium*, **2**, 17–29.

Houde, E.D. (1989) Subtleties & episodes in the early life of fishes. *Journal of Fish Biology*, **35** (Supplement A), 29–38.

ICES. 2014. Report of the North-Western Working Group (NWWG) 24 April–1May 2014. ICES HQ, Copenhagen Denmark. ICES CM 2014/ACOM-07 Ref. ACOM 915 p.

Jones, R. (1989) Towards a general theory of population regulation in marine teleosts. *Journal du Conseil International pour l'Exploration de la Mer*, **45**, 176-89.

Larkin, P.A. (1977) An epitaph for the concept of maximum sustainable yield. *Transactions of the American Fisheries Society*, **106**, 1–11.

Law, R. (2007) Fisheries-induced evolution: present status and future directions. *Marine Ecology Progress Series*, **335**, 271–7.

Longhurst, A. (2002) Murphy's law revisited: longevity as a factor in recruitment to fish populations. *Fisheries Research*, **56**, 125–31.

Macchi, G.J., Leonarduzzi, E., Diaz, M.V., Renzi, M. & Rodrigues, K. (2013) Maternal effects on fecundity and egg quality of the Patagonian stock of Argentine Hake (*Merluccius hubbsi*). *Fishery Bulletin*, **111**, 325–36.

Marshall, C.T. & Browman, H.I. (2007) Introduction. *Marine Ecology Progress Series*, **335**, 249–51.

Marshall, C.T. & McAdam, B.J. (2008) Integrated perspectives on genetic and environmental effects on maturation can reduce potential for errors of inference. *Marine Ecology Progress Series*, **335**, 301–10.

Marshall, C.T., Kjesbu, O.S., Yaragina, N.A., Solemdal, P. & Ulltang, Ø. (1998) Is spawner biomass a sensitive measure of the reproduction & recruitment potential of Northeast Arctic cod? *Canadian Journal of Fisheries and Aquatic Sciences*, **55**, 1766–83.

Marshall, C.T., Needle, C.L., Thorsen, A., Kjesbu, O.S., & Yaragina, N.A. (2006) Systematic bias in estimates of reproductive potential of an Atlantic cod (*Gadus morhua*) stock: implications for stock-recruit theory and management. *Canadian Journal of Fisheries and Aquatic Sciences*, **63**, 980–94.

Marshall, C.T., O'Brien, L., Tomkiewicz, J., Marteinsdóttir, G., Morgan, M.J., Saborido-Rey, F., Köster, F.W., Blanchard, J.L., Secor, D.H., Kraus, F., Wright, P., Mukhina, N.V. & Björnsson, H. (2003) Developing alternative indices of reproductive potential for use in fisheries management: case studies for stocks spanning an information gradient. *Journal of Northwest Atlantic Fisheries Science*, **33**, 161–90.

Marshall, C.T., Yaragina, N.A., Ådlandsvik, B. & Dolgov, A.V. (2000) Reconstructing the stock/recruit relationship for Northeast Arctic cod using a bioenergetic index of reproductive potential. *Canadian Journal of Fisheries and Aquatic Sciences*, **57**, 2433–42.

Marshall, C.T., Yaragina, N.A., Lambert, Y. & Kjesbu, O.S. (1999) Total lipid energy as a proxy for total egg production by fish stocks. *Nature*, **402**, 288–90.

Marteinsdottir, G. & Begg, G. (2002) Essential relationship in incorporating the influence of age, size, and condition on variables required for the estimation of reproductive potential in Atlantic cod (*Gadus morhua* L.). *Marine Ecology Progress Series*, **235**, 235–56.

Marteinsdottir, G. & Thorarinsson, K. (1998) Improving the stock–recruitment relationship in Icelandic cod (*Gadus morhua* L.) by including age diversity of spawners. *Canadian Journal of Fisheries and Aquatic Sciences*, **55**, 1372–7.

Mertz, G. & Myers, R.A. (1996) Influence of fecundity on recruitment variability of marine fish. *Canadian Journal of Fisheries and Aquatic Sciences*, **53**, 1618–25.

Miller, T.J., Crowder, L.B., Rice, J.A. & Marschall, E.A. (1988) Larval size and recruitment in fishes: Toward a conceptual framework. *Canadian Journal of Fisheries and Aquatic Sciences*, **45**, 1657–70.

Morgan, M.J., Perez-Rodriquez, A. & Saborido-Rey, F. (2011) Does increased information about reproductive potential result in better prediction of recruitment? *Canadian Journal of Fisheries and Aquatic Sciences*, **68**, 1361–8.

Morgan, M.J., Shelton, P.A. & Brattey, J. (2007) Age composition of the spawning stock does not always influence recruitment. *Journal of Northwest Atlantic Fisheries Science*, **38**, 1–12.

Murawski, S.A., Rago, P.J. & Trippel, E.A. (2001) Impacts of demographic variation in spawning characteristics on reference points for fishery management. *ICES Journal of Marine Science*, **58**, 1002–14.

Myers, R.A., Barrowman, N.J., Hutchings, J.A. & Rosenberg, A.A. (1995) Population dynamics of exploited fish stocks at low population levels. *Science*, **269**, 1106–8.

O'Brien, L., Rago, P.J., Berrien, P. & Lough, R.G. (2003) Incorporating early-life history parameters in the estimation of the stock–recruit relationship of Georges Bank Atlantic cod (*Gadus morhua*). *Journal of Northwest Atlantic Fisheries Science*, **33**, 191–205.

O'Farrell, M.R. & Botsford, L.W. (2006) The fisheries management implications of maternal-age-dependent larval survival. *Canadian Journal of Fisheries and Aquatic Sciences*, **63**, 2249–58.

Paulik, G.J. (1973) Studies of the possible form of the stock–recruitment curve. *Rapport et Procès-verbaux des Réunions, Conseil International pour l'Exploration de la Mer*, **164**, 302–15.

Quinn, T.J. II & Deriso, R.B. (1999) *Quantitative Fish Dynamics*. Oxford University Press, New York. 542 pp.

Reznick, D.A., Bryga, H. & EndlerJ.A. (1990) Experimentally induced life-history evolution in a natural population. *Nature*, **346**, 357–9.

Reznick, D. N., Shaw, F. H., Rodd, F. H. & Shaw, R. G. (1997) Evaluation of the rate of evolution in natural populations of guppies (*Poecilia reticulata*). *Science*, **275**, 1934–7.

Ricker, W.J. (1954) Stock and recruitment. *Journal of the Fisheries Research Board of Canada*, **11**, 559–623.

Rickman, S.J., Dulvy, N.K., Jennings, S. & Reynolds, J.D. (2000) Recruitment variation related to fecundity in marine fishes. *Canadian Journal of Fisheries and Aquatic Sciences*, **57**, 116–24.

Rothschild, B.J. (1986) *Dynamics of marine fish populations*. Harvard University Press, Cambridge. 277 pp.

Rothschild, B.J. & DiNardo, G.T. (1987) Comparison of recruitment variability & life history data among marine and anadromous fishes. *American Fisheries Society Symposium*, **1**, 531–46.

Rothschild, B.J. & Fogarty, M.J. (1989) Spawning stock biomass as a source of error in recruitment–stock relationships. *Journal du Conseil International pour l'Exploration de la Mer*, **45**, 131–5.

Rothschild, B.J. & Fogarty, M.J. (1998) Recruitment and the population dynamics process. In: A. Robinson & K. Brink (Eds) *The Sea*. Vol. 10. pp. 293–325. John Wiley, New York.

Scott, B., Marteinsdottir, G. & Wright, P. (1999) The potential effects of maternal factors on spawning stock–recruitment relationships under varying fisheries pressure. *Canadian Journal of Fisheries and Aquatic Sciences*, **56**, 1882–90.

Scott, B.E., Marteinsdottir, G., Begg, G.A., Wright, P.J. & Kjesbu, O.S. (2006) Effects of population size/age structure, condition and temporal dynamics of spawning on reproductive output in Atlantic cod (*Gadus morhua*). *Ecological Modelling*, **191**, 383–415.

Shelton, P.A. (1992) The shape of recruitment distributions. *Canadian Journal of Fisheries and Aquatic Sciences*, **49**, 1734–61.

Shelton, A., Munch, S.B., Keith, D. & Mangel, M. (2012) Maternal age, fecundity, egg quality, and recruitment: linking stock structure to recruitment using an age-structured Ricker model. *Canadian Journal of Fisheries and Aquatic Sciences*, **69**, 1631–41.

Shepherd, J.G. & Cushing, D.H. (1980) A mechanism for density-dependent survival of larval fish as the basis for a stock–recruitment relationship. *Journal du Conseil International pour l'Exploration de la Mer*, **39**, 160–7.

Spencer, P., Hanselman, D. & Dorn, M. (2007) The effect of maternal age at spawning on estimation of F(msy) for Alaska Pacific Ocean Perch. In: J. Heifetz, J. DiCosimo, A.J. Gharrett, M.S. Love, V.M. O'Connell & R.D. Stanley (Eds) *Biology, Assessment, and Management of North Pacific Rockfishes*. pp. 513–33. Alaska Sea Grant College Program AK-SG-07-01.

Spencer, P.D. & Dorn, M.W. (2013) Incorporation of weight-specific relative fecundity and maternal effects in larval survival into stock assessments. *Fisheries Research*, **138**, 159–67.

Stephens, P.A., Sutherland, W.J. & Freckleton, R.P. (1999) What is the Allee effect? *Oikos*, **87**, 185–90.

Trippel, E. (1999) Estimation of stock reproductive potential: history and challenges for Canadian Atlantic gadoid stock assessments. *Journal of Northwest Atlantic Fisheries Science*, **25**, 61–81.

Walters, C.J. & Juanes, F. (1993) Recruitment limitation as a consequence of natural selection for use of restricted feeding habitats and predation risk taking by juvenile fishes. *Canadian Journal of Fisheries and Aquatic Sciences*, **50**, 2058–70.

Walters, C.J. & Martell, S.D. (2004) *Fisheries Ecology & Management*. Princeton University Press, Princeton. 448 pp.

Ware, D.M. (1980) Bioenergetics of stock and recruitment. *Canadian Journal of Fisheries and Aquatic Sciences*, **37**, 1012–24.

Werner, E.E. & Gilliam, J.F. (1984) The ontogenetic niche and species interactions in size-structured populations. *Annual Reviews in Ecology and Systematics*, **15**, 393–425.

Chapter 2
Reproductive Dynamics

Dimitri A. Pavlov and Natal'ya G. Emel'yanova

2.1 Introduction

Fishes are not only the most diverse vertebrate taxon in terms of species numbers, but also in terms of the diversity of life styles and morphologies. All sorts of mating systems are represented, from spawning in large groups to strict monogamy. Parental behaviour is also highly variable, with no-care the most common pattern. Most commercial species lack parental care, but even species with a sex life traditionally considered as simple may have complex mating dynamics (Amundsen 2003).

The reproductive styles and population dynamics are closely connected with the types of gametogenesis, gonadal differentiation, and mechanisms of determination of final fecundity. In this chapter, a brief description of these events is followed by an analysis of reproductive strategies and types of early ontogeny that have implications for sustainable exploitation of fish stocks.

2.2 Determination of final fecundity in fishes with different life styles

2.2.1 *Development of primordial germ cells*

Fecundity is adapted to maintaining optimal population abundance in a variable environment. As in other groups of animals, the level of fecundity in fishes is determined in the early ontogeny and depends on cell differentiation. Primary cell differentiation of the embryo is based on ooplasmic segregation of the mature oocyte. In many organisms, the oocyte contains specialised granular cytoplasm that is often found in the vegetal region, and cells developing later around this material remain undifferentiated and physically separated from most early developmental events (Devlin & Nagahama 2002). Two modes of differentiation of primordial germ cells (PGCs) in fishes are recognised. *Epigenesis* describes the induction of PGCs from pluripotent cells by signals from surrounding somatic tissues. In contrast, PGCs in many fish species are specified cell-autonomously by maternally derived molecules known as 'germ

Fish Reproductive Biology: Implications for Assessment and Management, Second Edition.
Edited by Tore Jakobsen, Michael J. Fogarty, Bernard A. Megrey and Erlend Moksness.

plasm', a complex of substances determining the development of gamete-forming cells and preventing the cells from somatic differentiation (Aizenshtadt 1984, Johnson *et al.* 2011). Special cells (blastomeres) containing germ plasm subsequently develop into PGCs, which ultimately form gamete-forming cells within the gonad. The cytoplasmic substances responsible for inducing PGC development have not been fully defined as yet, but appear to be specialised maternally deposited RNAs encoded by maternal genes involved in PGC formation. Genes involved in germ-cell development have been identified in different fish species, one of which is a maternal mRNA encoding a DNA helicase (vasa gene) and a homologous vasa-like gene (Yoon *et al.* 1997, Herpin *et al.* 2007).

As in other vertebrates, the sex cells are formed outside of the presumptive anlages of the gonads. In different fish species, the first PGCs are registered at different developmental stages, but always during the embryonic period before hatching. The migration of PGCs to the area of the presumptive anlages of the gonads, as a rule, is completed after hatching (Makeyeva *et al.* 1988, Kobayashi *et al.* 2003, Extavour 2007, Chilke 2012), though in several species this may occur before hatching (Persov 1975).

2.2.2 Gonadal differentiation

Formation of the presumptive gonads usually begins following the completion of migrations of PGCs into the area of gonadal anlages. In this area, the peritoneal epithelium – which covers all internal organs – develops into small nodes or folders. The number of cells of the node increases as a result of mitotic divisions, and these somatic cells surround the PGCs. In addition to mitosis, the cells of the peritoneal epithelium can migrate into the area of the presumptive gonad, and, thus, the somatic zone of the gonad is increased.

The initial number of PGCs located inside of the presumptive gonad is usually small, at fewer than 50 (Lebrun *et al.* 1982, Makeyeva *et al.* 1988). Following the completion of migration, the PGCs begin to cleave so as to form the whole pool of sex cells of the organism. In the majority of fishes this event occurs during the larval and juvenile periods, with the exclusion of salmonids (*Salmo* and *Oncorhynchus*), where such processes are observed during the embryonic period (Persov 1975, Fujimoto *et al.* 2010).

2.2.2.1 Initial development

Mitotic cleavage of the PGCs leads to the formation of uniform gonia of the first order. A part of the gonia remains in the resting condition forming the reserve fund, but other gonia undergo mitotic cleavage. The reproduction of the gonia is very intensive such that gonia of subsequent orders are formed, and their number is substantially increased. At this stage, the sex of the fish cannot be determined. This period of gonadal development is termed 'indifferent', and its duration varies in different groups of fishes and, to a certain degree, it determines the age of sexual maturity. For example, the duration of the indifferent period can reach several years in sturgeons and from one to several months in the majority of teleost fishes (Moiseeva *et al.* 1988, Selyukov 2012). This duration depends on the environmental conditions, in particular the water temperature. In sterlet *Acipenser ruthenus* (L.), the indifferent period reaches approximately 2 and 7 months at 24 and 13 °C, respectively (Akhundov 2000). The period is completed with the onset of gonadal sexualisation.

2.2.2.2 Anatomical and cytological changes

Anatomical and cytological differentiation can be seen during the development of indifferent gonads into ovaries and testes. Anatomical differentiation means that the sexual differences are determined by the gonadal structure, while the morphology of sex cells (represented by gonia) in prospective male and female gonads is similar. At cytological differentiation, the sexual differences are connected with the features of sex cells, but the gonadal morphology can be similar in both sexes.

In the majority of fish species, anatomical differentiation occurs before cytological differentiation. Exceptions to this rule are known, however, such as salmonids of the genus *Oncorhynchus*, European whitefish *Coregonus lavaretus* (L.), round goby *Neogobius melanostomus* (Pallas) and White Sea herring *Clupea pallasii marisalbi* Valenciennes (Persov 1975, Moiseeva 1983, Zelenkov 1990). The timing of gonadal differentiation in fish is highly variable, even between closely related species (Devlin & Nagahama 2002, Chiasson *et al.* 2008). In salmonids of the genus *Salmo*, muksun *Coregonus muksun* (Pallas) and some cyprinids, anatomical differentiation begins approximately 2 months after hatching, and onset of cytological differentiation in females is registered 1–3 months later. In the males, the temporal interval between the two modes can reach one year (Persov 1975, Makeyeva, 1992, Selyukov 2012). In several species, cytological gonadal differentiation occurs before hatching.

During the process of cytological differentiation, oocytes at the prophase of the first meiotic division (with morphology different from that in the gonia) appear in females, but the paternal sex cells are still represented by gonia. The development of sex cells in males normally begins later than in females. However, the development of oocytes is arrested at the prophase of the first meiotic division, while sex cells at the meiotic phase develop continuously in males up to the terminal stages (formation of spermatozoa).

In gonochoristic species possessing purely ovarian or testicular tissues, two main types of gonadal differentiation can be seen (Figure 2.1). In the majority of species, gonadal development proceeds from an indifferent gonad directly to the ovary or testis. These species are called 'primary gonochorists' or 'differentiated gonochorists'. In other species, all individuals initially possess unisexual gonads, after which resorption of the cells of one sex and development (from the gonia) of the cells of the other sex are registered. Consequently, the sex cells of both sexes are present in the gonads during a certain developmental interval. Such a feature is termed 'juvenile hermaphroditism'. More often, the gonads develop directly into females and indirectly (throughout the initial female phase) into males; this is termed 'juvenile protogyny' (Persov 1975, Maack & Seoner 2003). The opposite mode ('juvenile protandry') is an initial development of testes with their subsequent transformation (in a part of the individuals) into ovaries. Species with juvenile hermaphroditism are referred to as 'undifferentiated gonochorists'.

In several fish species, the sex ratio can be changed under the influence of environmental conditions, mainly temperature. For example, in Argentine atherinid *Odontesthes argentinensis* (Valenciennes), a lower water temperature during the embryonic development leads to an increased number of females. Conversely, in honmoroko *Gnathopogon caerulescens* (Sauvage), a higher incubation temperature (34°C) leads to the appearance of only females in the progeny (Fujioka 2001). Similar 'environmental sex determination' mechanisms are observed in several species of tilapia and in the cyprinodontids.

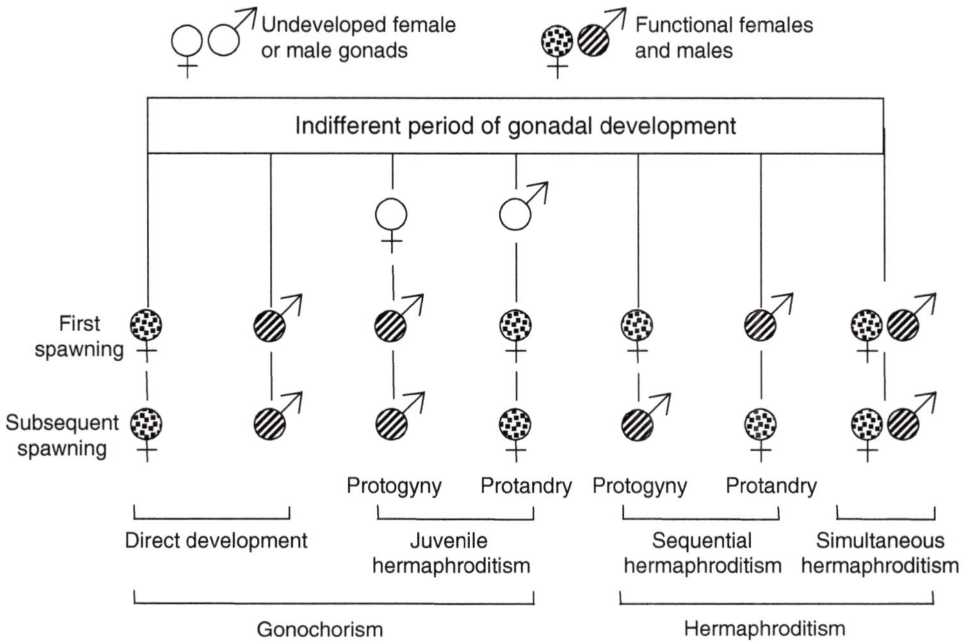

Figure 2.1 Types of sex differentiation in fishes.

2.2.3 Oocyte resorption as a mechanism for determination of final fecundity

Fecundity is an adaptation to life in a certain environment. The formation of fecundity begins in early ontogeny after gonadal differentiation. The reserve fund of the sex cells determining potential fecundity is formed by mitotic cleavage of the oogonia and the subsequent development of oocytes. In sexually mature females, potential fecundity is transformed into final fecundity. The values of final fecundity are always lower than those of potential fecundity.

Substantial differences in the formation of potential fecundity are observed between the monocyclic species (which spawn once and then die) and polycyclic species. In monocyclic species, a reserve fund of sex cells is absent, and all gonia undergo subsequent development. The maximal number of previtellogenic oocytes (which determine potential fecundity) appears at the early stages of juvenile development. Potential fecundity depends heavily on the feeding conditions. In polycyclic species, potential fecundity is established at the start of each spawning season and it is determined by the number of oocytes at the start of the period of vitellogenesis. The reserve fund of sex cells is represented by oogonia, and previtellogenic oocytes are always present.

The decrease in the number of sex cells, leading to the formation of final fecundity, is a result of resorptive processes ('downregulation'; see Chapter 8). In monocyclic species, the resorption of sex cells at all phases of their development is registered over the entire life up to spawning. In polycyclic species, final fecundity is determined by oocyte resorption during the periods of vitellogenesis and maturation, and occasionally, at the end of previtellogenesis (Ivankov 2001).

The resorption of unreleased oocytes is registered in the ovaries of spent females. Resorption is also found among the sex cells of senescent fishes with a reduced function of the reproductive system, as well as in hermaphrodites during sex reversion. The processes of oocyte resorption are common in fishes subjected to unfavourable conditions, and can be caused by stress, an inadequate temperature regimen and photoperiod, unfavourable feeding conditions, and water quality. Sometimes, a whole generation of oocytes filled with yolk and designed for release in the coming spawning season may be resorbed. In hatcheries, a decreasing egg quality caused by the initial processes of oocyte resorption is usually observed at the end of the breeding season (Szabo *et al.* 2000, Pavlov *et al.* 2004a, Grier *et al.* 2008, Muñoz *et al.* 2010).

Oocyte resorption occurs mainly under the action of follicular cells (as registered for vitellogenic oocytes) and also histiocytes, leucocytes and macrophage-like cells (these are more often found during atresia of previtellogenic oocytes). An initial follicular resorption is associated with the activation of lysosomes of follicular cells, the main organelles for the utilisation of oocyte components (Makeyeva 1992, Linares-Casenave *et al.* 2002, Durkina 2006). At the same time, decreased levels of 17β-estradiol, testosterone and vitellogenin are registered in the blood serum (Linares-Casenave *et al.* 2002). Occasionally (as in gadoid fish), the oocytes are encapsulated in cystic structures that might remain in the ovary for several months, altering the female reproductive potential during future spawning seasons (Domínguez-Petit *et al.* 2011).

The duration of oocyte resorption is shorter in fishes with spring or summer spawning than in those which spawn in autumn or winter. The duration is also connected to the number of oocytes subjected to resorption. If the number is small, resorption is comparatively rapid and is terminated during two or three weeks in summer, but in fishes with synchronous oocyte development that release eggs once in the spawning season the total resorption can last for several months or even over the entire year, sometimes causing the next spawning season to be omitted. In fishes with asynchronous oocyte development and multiple spawning, the resorption and development of a new oocyte generation can occur simultaneously (Koshelev 1984, Emel'yanova & Pavlov 2012), and mass resorption can lead to a retarded development of the next oocyte generations. Partial oocyte resorption represents a special mechanism for self-regulation of the physiological status of the female, leading to increased metabolism and maintenance of the physiological condition essential for normal reproduction.

The atretic bodies developing after the termination of resorption can remain in the gonads of fishes of the temperate zone for at least one year (Makeyeva 1992). The number of these bodies is an indicator of the spawning conditions in the previous year.

Thus, the resorptive processes in the ovaries of fish can determine: (1) the number of oocytes to be released in the coming spawning season; (2) the developmental rate of the next oocyte generations; and (3) the rate of subsequent sex cycles.

2.2.4 Individual and population fecundity

The main definitions related to fecundity terms are as follows:

- *Fecundity of fish (F)* as a general term. The total number of mature oocytes released by a female. In viviparous fishes, it is the total number of embryos or larvae released into water.
- *Individual fecundity (IF)*. The total number of mature oocytes released by a female during the spawning season. In fact, it is the value of absolute fecundity (AF) of a female, which

increases with larger body weight of the fish and then decreases with ageing. However, the latter process is not observed practically in natural populations that are subjected to a strong anthropogenic pressure. Different methods of assessing individual oocyte production in species with determinate and indeterminate fecundity (see Section 2.3.1) can be found in the reviews of Murua *et al.* (2003) and Witthames *et al.* (2009). In particular, the abundance of a specific type of oocyte (defined by morphological criteria) per unit volume can be measured using the stereological method and image-analysing system. This method is based on a principle which states that the fractional volume of a component is proportional to its cross-sectional area. In fish species with indeterminate fecundity, computing annual reproductive output requires estimates of batch fecundity and the annual number of batches, which in turn depends on the spawning frequency and duration of the spawning season. It should be noted, that batch fecundity, spawning duration and spawning frequency changes with age (for a review, see Fitzhugh *et al.* 2012).

- *Actual fecundity (ACF)*. The total number of oocytes obtained from a female during artificial reproduction. The values of ACF are always lower than the values of IF, especially in batch-spawning species.
- *Individual relative fecundity (IRF)* or '*Relative fecundity*' *(RF)*. The total number of mature oocytes released by a female during the spawning season per gram body weight of the gutted fish. Less often, the IRF is expressed per unit body length.
- *Potential fecundity (PTF)*. The total number of oocytes in the ovaries of the female representing the reserve fund. This fund determines the total number of eggs released by a female during the current spawning season. In monocyclic species, it is the number of oocytes of protoplasmic growth; in polycyclic species, it is the number of oocytes in the beginning of trophoplasmic growth. Potential fecundity can be assessed based on the numbers of sex cells at histological sections of ovaries (Murua *et al.* 2003). In addition, potential fecundity can be estimated using a solid displacement pipette to remove quantitative oocyte samples and a power equation that links follicles per gram ovary to the mean vitellogenic follicular diameter, the autodiametric method (Witthames *et al.* 2009; see also Chapter 8):

$$\mathrm{Ln} = F_{ow} = a \times \ln D_f + b \tag{1}$$

where F_{ow} is fecundity per gram of ovary, D_f is mean follicular diameter and a and b are equation constants. Potential fecundity is calculated based on ovarian weight and F_{ow}.
- *Final fecundity (FF)*. The total number of mature oocytes in the ovary before spawning.
- *Species fecundity (SF)*. The total number of mature oocytes released by females from all populations of the species over their reproductive life. Opinions on the definition of the term differ, however. According to another definition, this term should mean the total number of oocytes released by all females of the species during the spawning season.

A formula of Ioganzen (1955) for the determination of species fecundity is as follows:

$$SF = \sqrt[PJ]{N \times X} \tag{2}$$

where N is the number of oocytes released during the spawning event, X is the number of spawning events over the whole life of the fish, P is the period between the spawning events, and J is age at sexual maturation. However, an assessment of population fecundity is more informative and important for practical purposes.

- *Population fecundity (PF).* The total number of mature oocytes released by all females of the spawning stock during the spawning season. Based on the another definition, the population fecundity is equal to the average individual fecundity multiplied by the average number of spawning events of the females from the population over their lives. A formula of Serebryakov (1990) to determine population fecundity is as follows:

$$PF = \sum N_x \overline{F}_x \quad (3)$$

where N_x is the number of females aged x, and $\overline{F}_x$ is the mean fecundity per female aged x, or the age-specific fecundity.

Based on the value of PF, the dynamics of reproductive potential in each population can be followed and a comparison of these potentials in different populations is possible. Each new generation reaching the spawning condition will differ from other generations in certain biological characteristics, including age, growth rate, body length and weight, sex ratio, AF, IF, and RF.

Formation of the year class depends much on the survival conditions during initial periods of life (embryonic, larval or juvenile). For example, the survival index (to the age when the generation enters the spawning part of the population) ranges from 0.8 to 25 individuals per million eggs (ipm) in Northeast Arctic cod *Gadus morhua* L., from 0.1 to 451 ipm in Norwegian spring spawning herring *Clupea harengus* L., and from 0.3 to 38 ipm in Northeast Arctic haddock *Melanogrammus aeglefinus* (L.) (Bondarenko *et al.* 2003). Based on the retrospective analysis of the monitoring conducted over many years, the levels of PF providing different recruitment levels at certain ecological conditions during early ontogeny can be calculated (Serebryakov 1990):

  - The 'safe level' guarantees the production of a strong year class under average survival conditions for eggs and larvae.
  - The 'minimum acceptable level' provides a generation characterised by the average abundance in 'average' ecological conditions but low abundance during unfavourable conditions.
  - The 'critical level' provides a poor generation with a low abundance even in favourable ecological conditions that can lead to irreversible changes in the population.

Each level is connected with a certain biomass and abundance of the spawning stock. Taking these basic characteristics into account is considered necessary to obtain the goal of sustainable fisheries, but how this is done depends on the management objectives.

2.3 Reproductive strategies

2.3.1 Types of oogenesis and spawning

In fishes, oogenesis includes four periods: (I) division; (II) previtellogenesis; (III) vitellogenesis; and (IV) maturation. Each period is characterised by a certain morphophysiological condition of the sex cells (Figure 2.2).

During the division period, a fund of sex cells is formed as a result of the mitotic cleavage of oogonia. The modes of oocyte size distribution in the ovary and egg release are shown in Figure 2.3. In monocyclic fishes, all oocytes develop synchronously, especially prior to

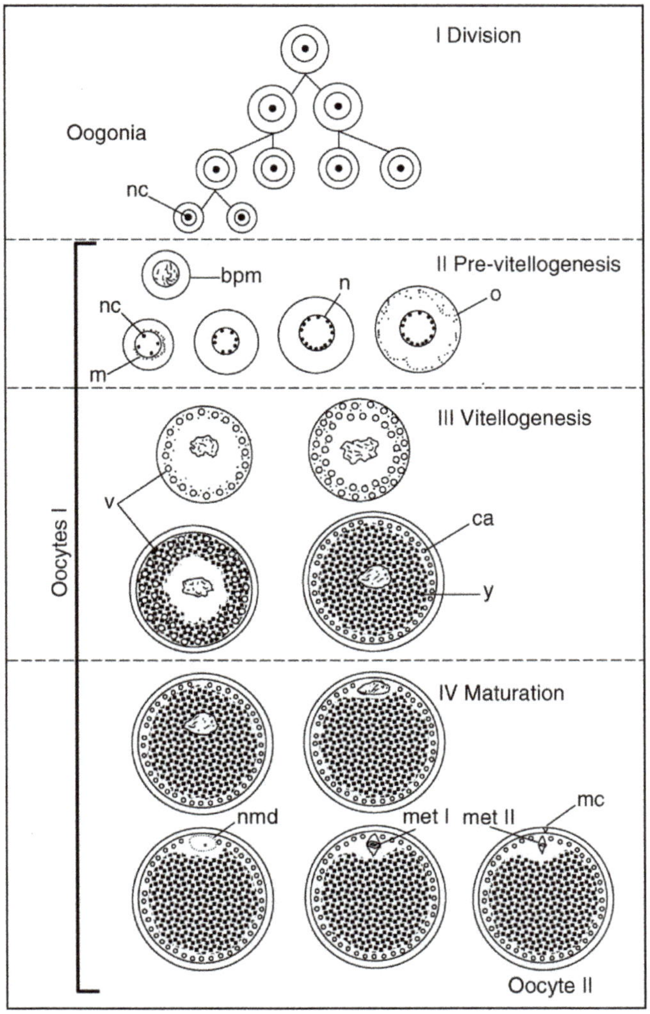

Figure 2.2 Scheme of oogenesis in cyprinid fishes. BPM, oocyte in the beginning of prophase of the first meiotic division; CA, cortical alveoli; M, mitochondria; MC, micropyle; Met I, metaphase of the first meiotic division; Met II, metaphase of the second meiotic division; N, nucleus; NC, nucleolus; NMD, nuclear membrane disintegration; O, organelles; Oocytes I, oocytes of the first order; Oocyte II, oocyte of the second order; V, vacuoles; Y, yolk granules. Reproduced from Makeyeva 1992; modified, with kind permission of Moscow University.

spawning, and sex cells of the reserve fund are absent. In polycyclic species, only a part of these cells develop to the terminal stages. Based on the features of the separation (from the reserve fund of previtellogenic oocytes) of the oocyte generation for current spawning, two types of oogenesis can be recognised (Götting 1961, Oven 2004). The discontinuous type of oogenesis is characterised by a clear separation of the oocyte generation designed for the spawning. Such a separation is absent when the oogenesis is continuous: the recruitment of vitellogenic oocytes from the reserve fund occurs continuously.

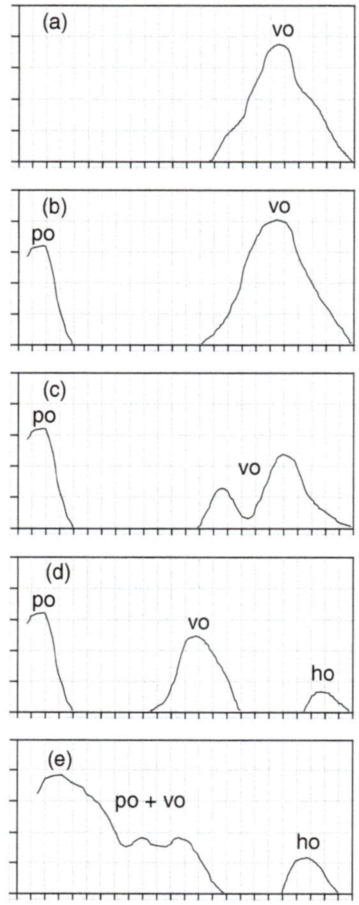

Figure 2.3 Oocyte size frequency for different types of oogenesis and spawning. (a) Monocyclic fishes, synchronous oocyte development; (b–e) polycyclic fishes. (b) Discontinuous type of oogenesis, synchronous development of vitellogenic oocytes, release of eggs once in the spawning season (group-synchronous oocyte development). (c) Discontinuous type of oogenesis, asynchronous development of vitellogenic oocytes, release of two to three egg batches in the spawning season (group-synchronous oocyte development). (d) Discontinuous type of oogenesis, synchronous development of vitellogenic oocytes, multiple egg release (group-synchronous oocyte development). (e) Continuous type of oogenesis, asynchronous development of oocytes, multiple egg release. HO, hydrated oocytes; PO, previtellogenic (reserved) oocytes; VO, vitellogenic oocytes. Hypothetical examples, no values on axes. Reproduced from Murua & Saborido-Rey 2003; modified with kind permission of NAFO.

The difference between the two main types of oogenesis has great importance for the assessment of fecundity. In fishes with discontinuous oogenesis, the last oocyte generation to be released in the current spawning season is represented by vitellogenic oocytes, and the fecundity can be determined easily ('determinate fecundity'). In fishes with continuous oogenesis, however, an exact assessment of fecundity is impossible due to the permanent recruitment of new oocytes from the reserve fund ('indeterminate' fecundity).

Discontinuous oogenesis includes the synchronous and asynchronous development of vitellogenic oocytes and three types of egg release: once in the spawning season, in a small number of batches, and in many batches (multiple egg release). Synchronous development means that all oocytes have a similar developmental state to the end of vitellogenesis. However, those eggs released once are usually deposited in small sub-batches during a short time period (up to several days). This pattern is mostly found in fishes of high latitude or in temperate zones. If the eggs are released in many batches, the spawning season will be protracted for up to two months or more. Such a mode of spawning is mostly observed in fishes of temperate zones and lower latitudes (Lisovenko 2000, Oven 2004). In the asynchronous development of oocytes, two or three groups of vitellogenic cells are separated in the ovary during the period of vitellogenesis, and are then released in the same number of batches; hence, the spawning season is protracted (normally by 2–3 months). The latter mode of oocyte development is also found in fishes of temperate zones and lower latitudes.

Continuous oogenesis means that the oocytes develop asynchronously over the entire reproductive period, and transition between the sex cells of the smallest and largest diameters is smooth. Just prior to release into the water, a group of oocytes is subjected to a substantial hydration. The spawning season is usually very protracted, reaching several months or even an entire year. Such a mode of spawning is usual in marine fishes of low latitudes (Oven 2004, Emel'yanova & Pavlov 2012, Emel'yanova *et al.* 2013). The continuous type of oogenesis associated with multiple spawning represents a strategy for investment of larger numbers of smaller eggs when resources are patchy on a relatively large spatial scale (Winemiller & Rose 1993, Oven 2004). This mode of spawning is possible if the environmental conditions are comparatively stable over a prolonged period of the year.

2.3.2 *Types of spermatogenesis and duration of spawning*

Four separate periods are identified in the development of male sex cells (spermatogenesis): (I) division; (II) growth; (III) maturation; and (IV) formation of spermatozoa or spermiogenesis (Figure 2.4). The first period begins with the mitotic divisions of primary spermatogonia that have originated from PGCs. Some of the spermatogonia do not undergo cleavage, and these cells will be the largest in the testes. Other spermatogonia divide several times, decreasing in size and transforming into the spermatogonia of subsequent orders. In fishes, the number of subsequent cleavages of spermatogonia can be up to 14, and is larger than in other vertebrates (Ruzen-Range 1980). Following the mitotic series with full cytotomy, the cells begin to divide with incomplete separation of the daughter cells from each other. As a result, clones of spermatogonia are represented by cells joined by cytoplasmic bridges, forming a syncytium. The clones are surrounded by somatic, follicular or bearing cells, known as Sertoli cells, while the sex cells are surrounded by follicular cells that form the cysts. Based on the syncytial structure, the sex cells inside each cyst are characterised by a uniform developmental stage.

The last mitotic division leads to the formation of spermatogonia of type B, characterised by a premeiotic duplication of DNA. These enter into the prophase of the first meiotic division, leading to the appearance of spermatocytes of the first order and indicating onset of the next developmental (growth) period. During the subsequent period (maturation), the first meiotic division is completed, forming haploid spermatocytes of the second order; the spermatids appear after the second meiotic division. During the period of spermiogenesis, a spermatozoon forms from each spermatid, and at the same time both the cytoplasmic bridges and the cysts break down.

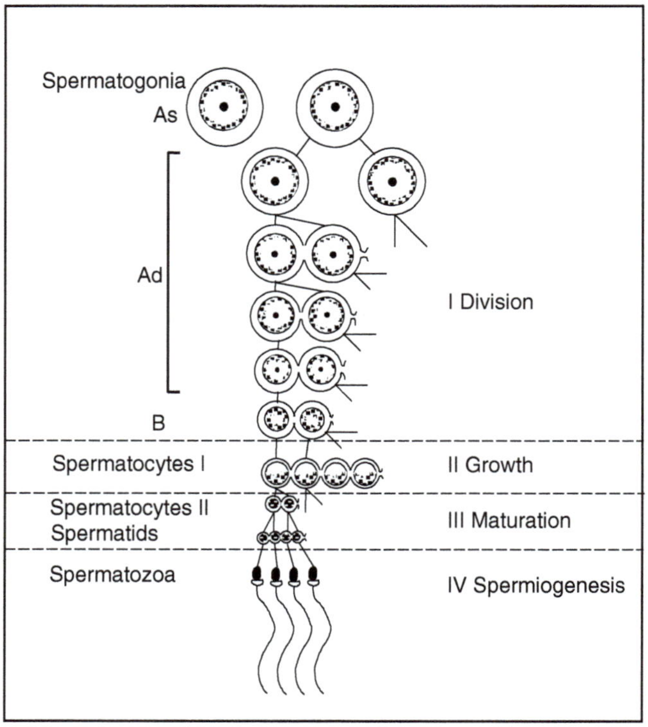

Figure 2.4 Scheme of spermatogenesis in fishes. AS, primary stem spermatogonia of the type A; AD, differentiated spermatogonia of the type A; B, spermatogonia of the type B. Reproduced from Makeyeva 1992; modified, with kind permission of Moscow University.

Short-term spawning (ranging from several hours in fishes showing parental care to several days in non-guarding fishes) is observed in species that reproduce once in the season. The males are characterised by a clear alternation of the maturity stages, a total completion of spermatogenesis in the prespawning period, an absence of new sex-cell generations undergoing meiosis, a fast release of sperm, and transition to the 'spent' condition. The formation of semen fluid occurs in the testicular canals due to secretion of the follicular cells. A new wave of spermatogenesis (as the sex cells enter meiosis) begins after the completion of resorption.

A protracted spawning season (ranging from one month to one year) in males is usual when females have a multiple release of eggs. This spawning type is characterised by the following features: (1) an additional wave of spermatogenesis; (2) continuous spermatogenesis where the males never reach the 'spent' condition; or (3) a final maturation of sex cells in a part of the cysts and a protracted maturation of spermatozoa in other cysts over the entire spawning season. The formation of semen fluid occurs mainly in the follicular cells of the spermiduct.

Two types of spermatogenesis are known in teleost fishes: cystic (when all spermatogenesis occurs inside of the cysts); and semi-cystic (when spermatogenesis can be partly observed outside of the cysts). In the latter type, spermatids are released from the cysts, and spermiogenesis (i.e., the development of spermatozoa from spermatids) is registered in the testicular canals outside of the cysts (Mattei *et al.* 1993). This type of spermatogenesis leads to an asynchronous development of sex cells and a protracted spawning period of males. In cystic spermatogenesis, follicular or Sertoli cells might be more efficient in supporting germ cell development (Schulz *at al.* 2010).

Two types of spatial arrangement have been described for the teleost testes, based on the distribution of spermatogonia in the germinal compartment (Parenti & Grier 2004). In the first type (restricted spermatogonial distribution), the primary spermatogonia are located in distal regions of the germinal compartment. This arrangement is found in the higher teleosts, such as the orders Atheriniformes, Cyprinodontiformes, and Beloniformes. In the second type (unrestricted spermatogonial distribution), the spermatogonia are spread along the germinal compartment throughout the testis. The unrestricted distribution of spermatogonia is considered a more primitive pattern that is found in less-derived taxonomic groups, such as the orders Cypriniformes, Characiformes, and Salmoniformes.

2.3.3 Special types of reproduction

2.3.3.1 Hermaphroditism

The majority of fish species are bisexual (gonochoristic). Hermaphroditism is a derived and polyphyletic condition in fishes that is documented in about 2% of all extant teleost species scattered across more than 20 taxonomic families from nine orders (Avise & Mank 2009). The majority of hermaphrodite species is distributed in seas of low latitudes. Hermaphrodites are rare in fresh water.

Based on the features of development and function of sex cells, the hermaphrodites are classified in two groups, namely sequential (ontogenetic); and simultaneous (Figure 2.5). The so-called ‘potential’ hermaphrodites can be referred to the former group.

In the gonads of sequential hermaphrodites, the function of the ovarian and testicular tissues is not synchronous. In the majority of representatives from this group, the younger fish function as females as the male zone of the gonad is weakly developed and remains in an inactive condition. Following spawning events, the ovarian tissue is reduced and resorption of the oocytes occurs while the testicular zone develops intensively and simultaneously. Thus, the fishes of younger age groups are represented by females, and older fish are functional males. The development of gonads according to this mode is termed *protogyny* (Kroon & Liley 2000). Protogynous hermaphrodites are often represented by harem species, usually with one male and several females; subsequently, if the male dies the dominant female will undergo sex inversion and transits to a male. The protogynous hermaphrodites are found in such families as Labridae, Scaridae, Pomacantidae, Gobiidae, Lethrinidae, Serranidae, Synbranchidae, and Muraenidae.

The primary development of males – termed *protandry* – occurs less often. In this case, the younger fishes function as males while the older individuals reproduce as females. Examples of protandrous hermaphrodites are found in the families Pomacentridae, Latidae, Sparidae, and Clupeidae (Micale *et al.* 2002, Abou-Seedo *et al.* 2003).

Bi-directional sex reversion has also been observed (Kroon *et al.* 2003). The sex allocation patterns in several fish species, which belong to simultaneous hermaphrodites, have a sequential component, as found in representatives of the family Gobiidae (St Mary 1997). Several groups of fish have different degrees of development of sex cells of ovarian and testicular

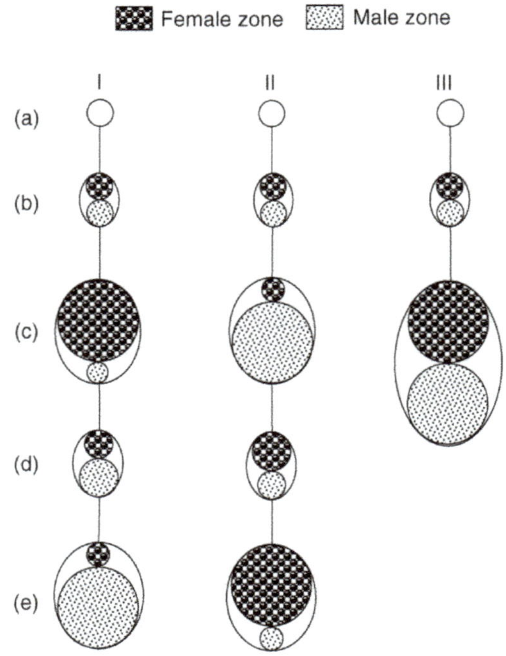

Figure 2.5 Types of hermaphroditism in fishes. I, sequential hermaphroditism, protogyny; II, sequential hermaphroditism, protandry; III, simultaneous hermaphroditism. (a) Indifferent gonad; (b) Immature ovarian and testicular zones; (c) Mature pre-spawning gonad: (I) functional female, (II) functional male, (III) functional female and male zones; (d) Gonad after spawning: (I) regression of female zone and development of male zone, (II) regression of male zone and development of female zone; (e) Mature pre-spawning gonad: (I) functional male, (II) functional female.

zones, and the individuals are capable of sex inversion in both directions to achieve maximal reproductive success.

In some fish species the gonads develop according to protogyny, but clearly separated testicular zones are absent. Rather, the testes are formed by the development of sex cells of the reserve fund (gonia), and this 'potential' hermaphroditism can be referred to the sequential mode. The sexual dimorphism in hermaphrodites is often well expressed, and this can lead to mistaken sex identification.

In simultaneous hermaphrodites, the ovarian and testicular zones of the gonad develop synchronously, and have independent ducts. Both, female and male sex cells can function during the spawning period. This group includes many species of the family Serranidae (Bubley & Pashuk 2010), and also several representatives of the families Cyprinodontidae (Reinboth 1980) and Gobiidae (St Mary 1997). In Nature, self-fertilisation, as a rule, is absent: the individual changes its role, alternatively spawning as a female or as a male. Only two species of the genus *kryptolebias* (Rivulidae) are able to self-fertilisation that is unique among vertebrates (Avise & Tatarenkov 2015). The gonad of sand perch *Diplectrum formosum* (L.) (Serranidae) contains ovarian and testicular tissue separated by a thin basement membrane, with no means of internal self-fertilisation. There are also trends indicating that testicular tissue is more likely to be in a spawning condition in the presence of hydrated oocytes within the accessory structure. This may indicate use of the accessory structure as a storage site for hydrated oocytes until a mate can be located (Bubley & Pashuk 2010).

A precise classification of a hermaphrodite is not always possible. For example, individuals developing as pure females or pure males (as in bisexual gonochorists) can be found among fishes of the same population undergoing sex inversion; such species are termed 'diandrous' hermaphrodites. Conversely, hermaphrodites where all individuals are capable of sex inversion are termed 'monandrous' hermaphrodites. The environmental conditions can have a direct effect on sex reversion in hermaphroditic fishes. At a high level of natural mortality, sex inversion occurs at lower body size than at a low level. Consequently, the sizes of females and males at the transitional condition may overlap substantially (Platten *et al.* 2002), and the reproductive strategy of such species is directed to a very fast response to the environmental conditions, aiming at a maximum reproductive success. The onset of sexual inversion in individuals is often determined by the sex ratio and a presence of fishes of a certain sex among older age groups (Kroon *et al.* 2003). Therefore, a change of a sex ratio can be a response to changing fishing mortality.

2.3.3.2 Gynogenesis

The special modes of sexual reproduction in animals represent substantial modifications of the meiotic and fertilisation processes. Gynogenesis, sometimes referred to 'pseudogamy' or 'pseudofertilisation', is a form of unisexual reproduction whereby females produce offspring with all their inheritance from the female parent. Only certain Amphibia, Reptilia and Pisces are known among the Vertebrata to reproduce by a gynogenetic mechanism in which the sperm of a related species is used to trigger embryonic development without any contribution from the paternal genome. With this rare type of sexual reproduction, insemination (i.e., contact between gametes) is necessary but the nuclear apparatus of the spermatozoon entering the egg cytoplasm is inactivated. The chromosomes of the spermatozoon are then eliminated, and subsequent development is controlled by the mother genotype (Figure 2.6). Thus, an essential part of fertilisation – karyogamy (fusion of the pronuclei) – is excluded (Cherfas 1987). Hence, two sequential mutations leading to genetic inactivation of the spermatozoon and a resumption of diploid number of chromosomes are required for the appearance of natural gynogenesis

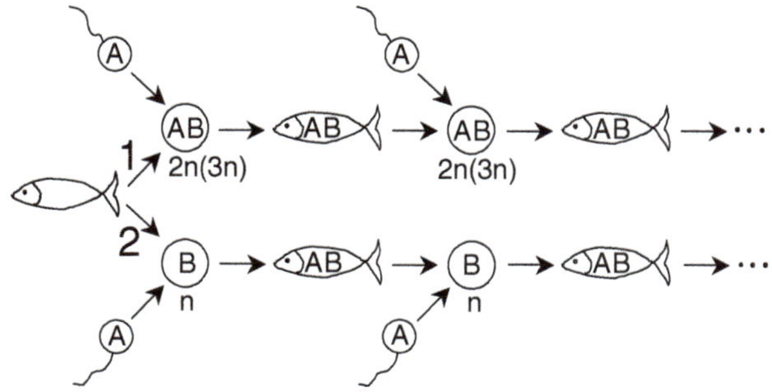

Figure 2.6 Scheme of (1) gynogenesis and (2) hybridogenesis in fishes. A and B are the genomes of different species. Reproduced from Katasonov & Gomel'skii 1991; modified, with kind permission of Agropromizdat, Moscow.

(Cherfas & Emel'yanova 1986). Gynogenetic populations include exclusively females, and natural gynogenesis is reported in several representatives of such families as Poeciliidae, Atherinidae, Cyprinidae, and Cobitidae.

The general features of the gynogenetic forms are an absence of reduction in chromosome number and a genetic homogeneity inside of the gynogenetic clonal lines. The sources of genetic variation are the mutations and clones evolving *de novo* (Vasil'ev 1985, Cherfas & Emel'yanova 1986). An occurrence of gynogenetic forms leads to the origin of well-adapted stable genetic systems. In spined loaches of the genus *Cobitis* from the diploid–polyploid complex, the larvae of triploid gynogenetic forms are larger than those of diploid bisexual forms, and are more resistant to parasites (Pavlov *et al.* 2004b).

2.3.3.3 Hybridogenesis

Hybridogenesis is also a very rare reproductive mode, which has been described only in fishes of the genus *Poeciliopsis.* This mode differs from gynogenesis in the occurrence of a true fertilisation, but the paternal chromosomes in the sex cells of the progeny are eliminated, and the phenotype expression is matroclinous (Figure 2.6). The hybridogenous forms are represented exclusively by females, which spawn with males of related species (Schultz 1977). The genetic variation is restricted. As in gynogenesis, transition to all-female populations leads to an increasing reproductive rate and, as a consequence, to a wider geographical distribution.

2.3.3.4 Androgenesis

Androgenesis is a developmental process that facilitates the inheritance of an exclusively paternal genome. Natural androgenetic populations are *absent*, but androgenesis has been induced in the laboratory in many fish species. The androgenesis involves two steps: (1) the elimination or inactivation of maternal chromosomes; and (2) monospermic or dispermic activation of embryonic development by the haploid or diploid gamete. Haploid monospermic activation requires the restoration of diploidy by suppression of the first mitotic cleavage (Kirankumar & Pandian 2004, Grunina & Recoubratsky 2005). Androgenesis serves as a method for obtaining high-inbred males without hormonal sex inversion, and is important for the conservation of rare genomes in threatened fish species.

2.3.3.5 Parthenogenesis

Parthenogenesis involves egg development without insemination and fertilisation, but the parthenogenetic development of eggs with a total exclusion of spermatozoa is impossible; hence, only so-called 'pseudoparthenogenesis' can be observed in fishes. In fishes that lay their eggs in compact clutches, any unfertilised eggs can survive for several days to several months, and these represent an adaptation to increased survival of the clutch; however, they will die at the time of hatching in fertilised eggs.

2.3.4 *Classification of reproductive styles*

Fish reproductive patterns represent the key element in the sustainable management of fish stocks, with reproductive styles being classified according to spawning tactics and ecological niches for development. In 1949, the Russian scientist Sergei Kryzhanovskii was the first to propose a classification for some freshwater fishes, separating them into five ecological groups based on the spawning substrates. The most comprehensive classification of reproductive styles, which included up to 36 subdivisions (guilds) was created by Eugene Balon (1985, 1990), and was based on the ideas of Kryzhanovskii. A modified scheme is shown in Table 2.1. In this scheme, the guilds of obligate lecithotrophic live bearers and matrotrophic live bearers are joined because maternal nutrient transfer to developing embryos is observed, even in those species referred previously to the former guild (Wourms 1991). In addition, the title of the guild C.2.1 (facultative internal bearers) has been changed to use the term 'zygoparity', as proposed by Wourms (1991). This refers to the oviparous reproductive mode in which internally fertilised ova are retained within the female reproductive tract for short periods.

Owing to the extremely large diversity and plasticity of life styles, new reproductive guilds can be added, though some species cannot be referred to a certain guild. For example, the eggs of moustache sculpin *Triglops murrayi* Günther are attached to the tubes of polychaetes (Pavlov *et al.* 1992), and apparently represent a special guild of non-guarding open substratum spawners. Sympatric lacustrine forms of Arctic charr *Salvelinus alpinus* (L.) can spawn on silty substratum and rocks (ethological section A1) or they may hide their eggs in gravel (ethological section A2) (Pavlov & Osinov 2008). The substratum type is not important for Chinese sleeper *Perccottus glenii* Dybowski; the eggs are deposited on submerged leafs, grass stems, sticks and even plastic bags, and guarded by the male (Voskoboinikova & Pavlov 2006).

The adaptations of reproductive styles will determine the features of adult ecology, migration, and distribution, while the concept of reproductive guilds reflects evolutionary lines to a certain extent. Both, the succession of groups within the ethological sections and the succession of guilds in each ecological group, represent a trend from a lifestyle characterised by small unprotected eggs and high fecundity to a life style with larger eggs, lower fecundity, and a more complex protection of eggs and offspring. These and some other characteristics are applied by Balon (1985, 1990) to describe two types of ontogeny ('indirect' and 'direct') and relative alternative life strategies ('altricial' and 'precocial'). These life strategies can be observed at a level of species or larger taxonomic groups, as well as within a population or even among the progeny of a female.

2.3.5 *Viviparity: evolution, diversity and importance*

Viviparous fishes are known from three large groups: Chondrichthyes, Teleostei, and Actinistia (Coelacanths). From over 25 000 teleost fish species, approximately 500 species are viviparous, and one-half of these species is represented by mainly freshwater fishes from three families of the order Cyprinodontiformes (Anablepidae, Goodeidae and Poeciliidae) and from the family Hemiramphidae of the order Atheriniformes. The largest number of viviparous fishes (approximately 200 species) is registered within the family Poeciliidae. Based on generally accepted opinion, oviparity is a primary mode of reproduction, while viviparity is regarded as the secondary phylogenetic adaptation characterised by selective advantages. However, the

Table 2.1 Classification of reproductive styles (guilds) in fish. Modified after Blaxter (1988) and Balon (1990).

Ethological section	Ecological group	Reproductive guilds
A Non-guarders		
1	Open and substratum spawners	1 Pelagic spawners 2 Rock and gravel spawners with pelagic larvae 3 Rock and gravel spawners with benthic larvae 4 Non-obligatory plant spawners 5 Obligatory plant spawners 6 Sand spawners 7 Terrestrial spawners, damp conditions
2	Brood hiders	1 Beach spawners, damp conditions 2 Annual spawners, eggs estivate 3 Rock and gravel spawners 4 Cave spawners 5 Spawners in live invertebrates
B Guarders		
1	Substratum spawners	1 Pelagic spawners, at surface of hypoxic waters 2 Above-water spawners; male splashes around 3 Rock spawners 4 Plant spawners
2	Nest spawners	1 Froth nesters 2 Miscellaneous substratum and materials nesters 3 Rock and gravel nesters 4 Glue-making nesters 5 Plant material nesters 6 Sand nesters 7 Hole nesters 8 Anemone nesters; at base of host
C Bearers		
1	External bearers	1 Transfer brooders; eggs carried before deposition 2 Auxiliary brooders; adhesive eggs carried on skin under fins, etc. 3 Mouth brooders 4 Gill-chamber brooders 5 Pouch brooders
2	Internal bearers	1 Zygoparous internal bearers; occasional internal fertilisation of normally oviparous fish or normal retention of internally fertilised eggs within the female's body and releasing them at early developmental stages 2 Lecithotrophic or matrotrophic live bearers including adelophages (one or a few eggs developing at the expense of other eggs or embryos) 3 Viviparous trophoderms; nutrition partially or entirely from female via 'placental structures'

transition to viviparity has not been realised in the majority of fish species due to complexities connected with structural and functional transformations required for respiration and nutrition of developing eggs in the female's ovary.

In fish evolution, viviparity evolved independently at least 29 times in different evolutionary lineages from the Palaeozoic and Mesozoic to comparatively recent times (Blackburn 2005). Evolutionary changes connected with the transition to viviparity remain mostly unclear, although intermediate stages between oviparity and viviparity can be followed in several groups, including representatives of the suborder Scorpaenoidei (Scorpaeniformes) (see the review of Pavlov & Emel'yanova 2013). In particular, the transition from the spawning of free pelagic eggs (Congiopodidae and Synanceiidae) to release externally inseminated pelagic eggs in groups embedded into gelatinous matrix (Scorpaenidae), to release internally inseminated pelagic eggs (at early developmental stages) in groups embedded into gelatinous matrix (Sebastidae: *Sebastolobus* and *Helicolenus*) and to viviparity (Sebastidae: *Sebastes* and *Sebasticus*), can be followed.

The scorpaenoid fishes (as mentioned above) possess a different evolutionary history that is reflected in their ovarian structure which, in turn, is not in accordance with the present taxonomic position of the species. Viviparous representatives of the family Sebastidae are characterised by a location of ovigerous lamellae at the periphery of the ovary that is usual for the majority of teleost fishes (Figure 2.7a). Nevertheless, oviparous but internally inseminated fishes of the genera *Sebastolobus* and *Helicolenus* from the same family, as well as fishes from the family Scorpaenidae with external insemination, are characterised by ovarian structure with a central location of the stroma (Koya & Muñoz 2007, Muñoz 2010, Pavlov & Emel'yanova 2010), that is important for the production of a hollow gelatinous egg clutch (Figure 2.7b). In the representatives of the latter family, a hypertrophied urinary

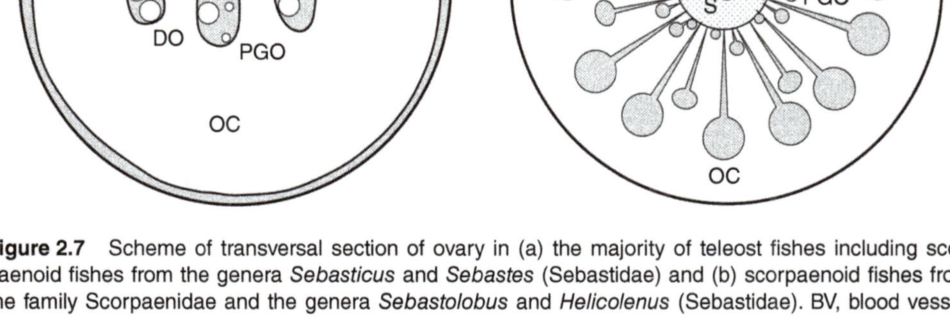

Figure 2.7 Scheme of transversal section of ovary in (a) the majority of teleost fishes including scorpaenoid fishes from the genera *Sebasticus* and *Sebastes* (Sebastidae) and (b) scorpaenoid fishes from the family Scorpaenidae and the genera *Sebastolobus* and *Helicolenus* (Sebastidae). BV, blood vessel; DO, derived oocyte; OC, ovarian cavity (parovarian and entovarian in the majority of teleost fishes and exoovarian in scorpaenoid fishes); OL, ovigerous lamella; OW, ovarian wall; P, peduncle; PGO, primary growth oocyte; S, stroma.

bladder is used for the storage of mature sperm (Pavlov & Emel'yanova 2013). Evolutionary changes associated with the appearance of this unique ovarian morphology, as well as with formation of the additional function of the urinary bladder, represent a subject for future investigations.

Internal insemination connected mainly with an increasing efficiency of fertilisation and not accompanied by viviparity evolved independently in several groups of fishes. In particular, in the suborder Cottoidei (Scorpaeniformes), only the endemics of Lake Baikal, namely big Baikal oilfish *Comephorus baicalensis* (Pallas) and little Baikal oilfish *C. dybowski* Korotneff from the family Comephoridae, are viviparous. In some species from three other families of the suborder (Cottidae, Hemitripteridae and Agonidae), a special mode of fertilisation including copulation, 'internal gametic association' and subsequent external fertilisation is registered (Abe & Munehara 2009). An occurrence of internal insemination and the release of fertilised eggs before cleavage (i.e., zygoparity) have been described in the species from the families Anarhichadidae and Zoarcidae (Perciformes: Zoarcoidei) (Johannessen *et al.* 1993, Pavlov 1994, 2007, Yao and Crim 1995), and several species from the genus *Zoarces* are viviparous. However, the reproductive strategies of fishes from the suborders Scorpaenoidei and Zoarcoidei are substantially different. The evolution of viviparity in scorpaenoid fishes was not associated with any substantial increase in egg size, leading to a prolonged consumption of yolk by the embryo and a considerable decrease in fecundity (i.e., adaptations common for the majority of viviparous fishes), and the *r* selection strategy (see Section 2.4.2.2) was retained. A high fecundity (sometimes exceeding five million eggs), as well as a long life span (tens of years in many viviparous species of the genus *Sebastes*), increases the probability of appearance of strong year classes under favourable environmental conditions. In the representatives of the suborder Zoarcoidei, the reproductive strategy is directed to a maximum survival of progeny by means of increases in egg size and parental investment (see Section 2.4.2.1).

An extremely high level of specialisation is observed in several other viviparous taxa. In fishes from the family Embiotocidae, the male's anal fin develops a complex system of appendages and tubular structures used for sperm transfer. Dense aggregations of sperm are introduced into the female reproductive tract via spermatophore packages, and the sperm are maintained within pockets in the ovarian epithelium for about six months before ovulation and fertilisation. Embryos are retained within the ovarian compartments, where they absorb nutritive materials and oxygen from ovarian fluid through modifications that occur in the developing embryos and ovary. The progeny are born in an almost adult condition (Liu & Avise 2011). Even more complex adaptations to viviparity with placental analogues are found in sharks.

In teleost marine fishes, viviparity at low latitudes is not usual, and a possible explanation of this phenomenon is as follows. In the conditions of not-pronounced seasons of the year and substantial spatial variation of abiotic and biotic factors (in particular, food resources for larvae), continuous oogenesis and multiple spawning are most advantageous. In this mode of reproduction, a probability of survival of at least a part of the progeny is increased, and a prolonged and frequent pregnancy accompanied by release of a small number of embryos would be disadvantageous. Viviparity associated with the production of numerous progeny has a selective advantage under conditions of temperate climate with distinct seasons of the year, with more predictable and stable conditions for the feeding of larvae and juveniles.

2.3.6 *Types of spawning stocks*

In the majority of fish species, a spawning population consists of fish reaching sexual maturity for the first time, repeat-spawners, and old fish that are unable to reproduce, either partly or totally. The relative abundance of the groups is, to a certain degree, species-specific, but the situation can change depending on environmental conditions. Based on an analysis of the structure of reproductive parts of populations, three types of spawning populations have been revealed (Monastyrskii 1953):

(1) The repeat-spawners and old fish are absent, and all mature fish (monocyclic species) die after the first spawning. The age structure of mature fish can differ. In some gobies (*Aphya*, *Benthophilus*), all fish attain sexual maturity at a certain age, but generations of Pacific salmon may reach sexual maturity at different ages. In chum salmon *Oncorhynchus keta* (Walbaum), the spawning stock consists of fish at the age from 2+ to 7+ years, but in pink salmon *O. gorbuscha* (Walbaum), it is represented by only two age groups (1+ and 2+). Thus, a stock of pink salmon will almost totally disappear under the influence of two subsequent unfavourable years, but the stock of chum salmon is more robust.

(2) The spawning population includes both first-spawning fish (*recruitment part*) and repeat-spawners. Old fish unable to reproduce are sometimes present. Repeat-spawners and old fish represent the *residual part* of the spawning population. In the spawning stock, the recruitment part is larger than the residual part. This type is usual in Caspian shad *Alosa kessleri* (Grimm), Atlantic salmon *Salmo salar* L., and many other fish species.

(3) The proportion of the residual part is larger than that of the recruitment part. The period of sexual maturation is prolonged, and the age composition of the residual part is complex. This part includes fish that have spawned every year and as well as fish which have omitted one or several spawning seasons. In some species (e.g., sturgeons), all individuals do not spawn annually. A large proportion of old fish unable to spawn is usual for the spawning populations of this type.

Whilst the types of spawning populations are mainly species-specific, the borders between them are not stable (Nikolsky 1974). For example, in some populations of Atlantic salmon all fish can die after the first spawning, and thus the spawning population can be referred to as type 1 instead of type 2. In a species, the type of spawning population can also be different if the ratio between the recruitment and residual parts changes in various years.

The spawning stock of fishes with prolonged life cycles consists of a large number of age groups, in both the recruitment and residual parts. For example, the spawning stock of cod with a low level of exploitation consists of seven to ten age groups. In this situation, the annual recruitment represents a small proportion of the spawning population, and the total abundance of the stock remains relatively stable. In populations with a large life span of individuals and a complex age structure, only a small proportion of the spawning stock can be exploited by the fishery. Early-maturing fish with a simple population structure and a small number of age groups are adapted to great fluctuations of their abundance, and a larger part of the spawning population can be removed.

2.4 Egg structure and features of early ontogeny in fishes with different reproductive strategies

The variability of fish eggs is associated mainly with their external structure, including size, shape, and coloration. In the majority of fishes, the eggs possess a spherical form but they can be elongated or enclosed in the egg capsule (as in oviparous sharks and skates). The coloration of eggs of teleost fishes can be absent (in many pelagophils), but otherwise can vary from greenish to yellow, orange, and cherry colours.

2.4.1 Classification of eggs

The eggs of all fishes have a polar distribution of yolk and cytoplasm, and belong to the telolecithal type. The concentration of yolk at the vegetal pole of the egg is different in species from various taxonomic groups. The eggs also have different features of yolk distribution in the cytoplasm and, based on this characteristic, two egg subtypes are known: (i) eggs with a nonseparated yolk; and (ii) eggs with a separated yolk (Makeyeva 1992) (Figure 2.8).

The patterns of egg cleavage and development are different in the two subtypes. In the eggs of the first subtype, the yolk is distributed among the cytoplasm as granules or conglomerates.

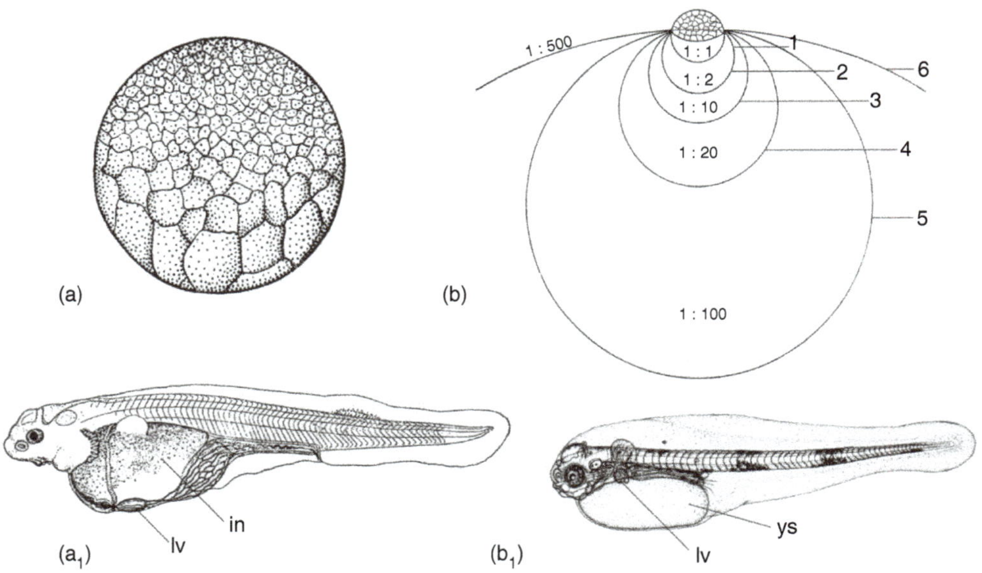

Figure 2.8 Two subtypes of development in telolecithal eggs. Subtype with non-separated yolk: (a) total (holoblastic) cleavage in starred sturgeon *Acipenser stellatus* Pallas; (a$_1$) free embryo of kaluga *Huso dauricus* (Georgi). Subtype with separated yolk: (b) various blastodisc to yolk ratios in mature eggs of different species of teleost fishes (1, *Gobius niger* L.; 2, *Osmerus eperlanus* (L.); 3, *Coregonus lavaretus* (L.); 4, *Thymallus thymallus* (L.); 5, a representative of the subfamily Salmoninae; 6, a representative of sharks); (b$_1$) free embryo of *Gadus morhua* L. IN, intestine; LV, liver; YS, yolk sack. Drawings from Soin (1981), reproduced with kind permission of 'Interperiodica'.

These eggs undergo full (holoblastic) cleavage; that is, the whole egg goes through cell cleavage after fertilisation. The eggs of this subtype are common for lower bony fishes spawned in fresh water: Dipnoi (lungfishes, e.g., *Neoceratodus*), Chondrostei (sturgeons), and Holostei (*Amia* and *Lepisosteus*).

All teleosts, most of the sharks (Chondrichtyes), and the coelacanth *Latimeria chalumnae* (Smith) produce eggs of the second subtype, where the yolk is separated from the cytoplasm. As only the cytoplasm is subjected to cleavage, this cleavage is termed meroblastic or discoidal.

The spatial separation of the yolk and cytoplasm leads to the possibility of a substantial variation of yolk content in the egg that represents a main source for the early ontogeny variation in teleost fishes (see Section 2.4.2).

Egg activation represents a complex of changes, including release of the developmental block of meiosis at the metaphase of the second meiotic division, consecutive breakdown of the cortical alveoli from the animal to the vegetal pole of the egg, and formation of the perivitelline space between the thin cytoplasmic layer surrounding the yolk and the internal surface of the egg envelope (Figure 2.9). In teleost fishes, egg activation is induced by the penetration of a spermatozoon (in the majority of marine fish), by contact with water (in the majority of freshwater fishes and salmonids), or even by mechanical stimulation (Ginzburg 1968). After this initial activation, the cytoplasm begins to concentrate towards the animal pole, forming the cytoplasmic disc, which is ready for the first cell cleavage. The second meiotic division is completed, and the female pronucleus appears. The spermatozoon head then transforms into the male pronucleus, after which the two pronuclei fuse, causing a resumption of the diploid chromosome number. Owing to the water uptake, the egg swells and the egg envelope hardens while, at the same time, the external layer of the egg envelope – the chorion – possesses a high adhesiveness to the substrate. In pelagic eggs and demersal eggs of some species (e.g., salmonids), the chorion is reduced or absent. In several groups of fishes, hardening of the egg envelope is accompanied by an adhesion of the egg envelope to neighbouring eggs of the clutch, but not to other substrates.

Three egg types can be recognised based on the cytoplasm volume to yolk volume ratio (at the stage before the first cleavage of the cytoplasmic disc): polyplasmatic (>40%); mesoplasmatic (25–40%); and oligoplasmatic (<25%). The eggs of the first type are known in many cyprinids and several perches, the eggs of the second type are registered in the representatives of whitefishes, loaches and silurids, and the eggs of the third type are usual in pikes and salmonids.

Sars (1869) was the first to discover that fishes may have pelagic eggs floating freely in seawater. As was found later, the majority of marine fishes release pelagic eggs that are fertilised externally. In general, pelagic eggs are small, ranging from about 0.6 to 4.0 mm in diameter, with a mean of 1 mm (Kendall *et al.* 1984) and a median of 1.1 mm (Chambers & Leggett 1996), and most of them float in the upper surface layers. The eggs of several fish species sink in stagnant water, although they can float under the influence of water currents or develop above a more dense water mass with higher salinity (see, e.g. Haug *et al.* 1984).

Many coastal and freshwater fishes, as well as some species of the open sea, lay demersal eggs, which are generally larger than pelagic eggs. The demersal eggs are often adhesive, can be laid in some sort of nest or cluster, and are often protected by a male, a female or both parents.

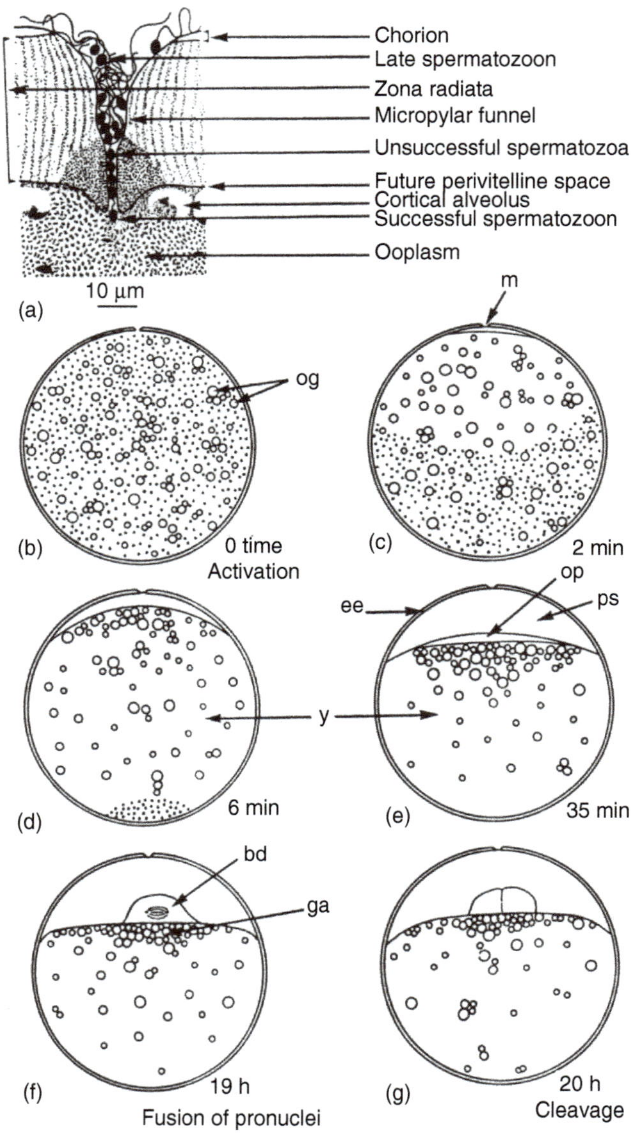

Figure 2.9 Schematic representation of the beginning of ontogeny in salmonid fishes at 4.4 °C. (a) Insemination shown in a cross-section through the egg envelopes, adjacent to the micropyle, consisting of a zona radiata and chorion, and with competing spermatozoa; (b–g) Fertilisation process in its broadest sense, in side views. (b) Activation by water; (c) Discharge of the cortical alveoli at the animal pole causing formation of the perivitelline space between the oocyte envelope and egg envelope; (d) Cortical reaction near completion; (e) Concentration of the cytoplasm at the animal pole; (f) Fusion of male and female pronuclei inside of the formed cytoplasmic disc signifying fertilisation in a strict sense; (g) First cleavage. Dots, cortical alveoli; BD, blastodisc; EE, egg envelopes; GA, globule aggregation; M, micropyle; OG, oil globules (droplets); OP, ooplasm; PS, perivitelline space; Y, yolk. Figure redrawn from Balon (1990), with kind permission.

2.4.2 *Main features of early ontogeny*

2.4.2.1 *Types of ontogeny and relative duration of ontogenetic intervals*

A larva is a transitory form of life, which often inhabits an entirely different niche than the adult form. The larval period is considered to start from the transition of the organism to exogenous feeding (Kryzhanovskii 1949, Soin 1968, Balon 1985, 1990, Peňáz 2001). Thus, the subperiod of development of the free embryo until the beginning of external feeding belongs to the embryonic period. (According to another opinion, e.g. Kamler (1992, 2002), the larval period begins from hatching.) Larvae may have a different body shape than the adults, and are characterised by temporary larval organs and tissues such as an unpaired finfold with respiratory blood vessels and a dorsal sinus. Spines and filamentous appendages to obtain buoyancy may be present.

The process of metamorphosis or remodelling of the organism often involves substantial changes in larval morphology. For example, the transition from pelagic to demersal life in flatfishes is associated with a flattening of the body and the migration of one eye to the opposite side of the head, which becomes the upper side of the body. The body shape of European eel *Anguilla anguilla* (L.) changes from leaf-like to definitive eel-like during a prolonged migration from the Sargasso Sea to the coast of Europe. According to Rass (1948), the states of prelarvae and finformed larvae are not observed in the life cycles of several Arctic forms; in some species these states occur inside the egg envelope, and thus the metamorphosis is absent or weakly expressed. It is important to note, that the term 'metamorphosis' is not clearly defined. Peňáz (2001) suggests applying this term exclusively for the interval from the onset of exogenous feeding to the beginning of the juvenile period.

According to Balon (1985, 1990, 2002), *indirect* ontogeny is characterised by a well-expressed metamorphosis and a prolonged larval period. This ontogenetic type is usual for the majority of fishes with pelagic eggs. *Direct* ontogeny means that the larval period is eliminated: the organism transits directly from the embryonic to the juvenile period. The direct type is found in many viviparous species. According to Peňáz (2001), real direct ontogeny with a total elimination of the larval period can be observed only in a small number of oviparous fishes. *Transitory* ontogeny is characterised by the beginning of exogenous feeding at a comparatively advanced morphological state of the organism. The larval period is short and remains as a 'larval vestige'. The latter type is registered in salmonids, as well as in several marine and freshwater fishes with demersal eggs. Larvae of some representatives of the suborder Zoarcoidei with different types of ontogeny are shown in Figure 2.10. In the ocean pout *Zoarces americanus* (Bloch & Schneider), an increased survival of progeny is achieved by means of a substantial increase in egg diameter (up to 8.5 mm) and very high level of embryonisation (the organism develops inside the egg envelope up to the juvenile state and hatches at 33–41 mm total length, TL) that is associated with intensive protection of the clutch by the female. The same result (release of the young at the juvenile state at 33–45 mm TL) is attained by means of a matrotrotrophic form of viviparity in eelpout *Zoarces viviparus* (L.). In representatives of the suborder Scorpaenoidei, viviparity leads to the release of larvae of a substantially larger size and in a more derived morphological condition in comparison to oviparous species (Figure 2.11). However, the type of ontogeny can be defined as indirect both in oviparous and viviparous forms.

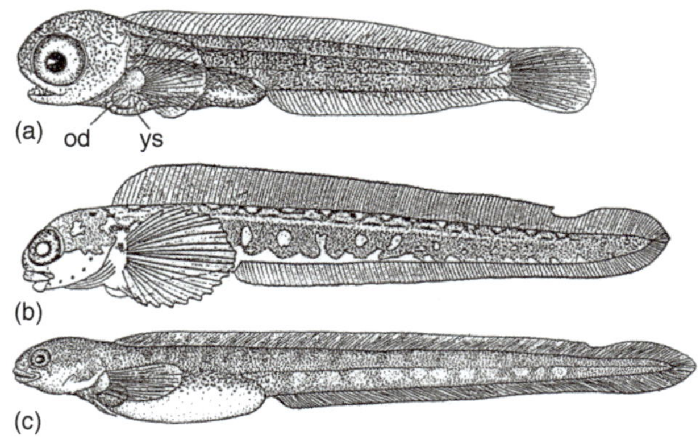

Figure 2.10 Larvae of some species from the suborder Zoarcoidei with different types of ontogeny at the beginning of exogenous feeding. (a) Transitory ontogeny: common wolffish *Anarhichas lupus* L., TL = 21.2 mm; average egg diameter before swelling 5.5 mm; OD, oil (lipid) droplet; YS, yolk sac. Reproduced from Pavlov & Moksness (1994), with kind permission of Springer Science and Business Media; (b) Direct ontogeny: ocean pout *Macrozoarces* (= *Zoarces*) *americanus* (Bloch & Schneider), 40.4 mm TL; average egg diameter before swelling 8.5 mm. Reproduced from Methven & Brown (1991), with kind permission of NRC Research Press; (c) Direct ontogeny: viviparous fish, eelpout *Zoarces viviparus* (L.), 44.4 mm TL; average egg diameter before swelling 3.0 mm. Reproduced from Pavlov (1999), with kind permission of Springer Science and Business Media.

For comparison of the relative duration of the main ontogenetic intervals in fishes from different systematic groups, exclusively morphological criteria have been applied (Pavlov 1999). The key events in the ontogeny of fishes are the transitions to the larval and juvenile states. The beginning of the larval state is the morphological condition when the organism has the ability to feed orally, while the beginning of the juvenile state is the morphological condition when the organism no longer has larval characters and most adult characters have appeared. Onset of the larval state suggests a similar degree of development of the digestive system, but represents various degrees of development of other organs in different species, while the majority of the organ systems reach a definitive structure at the juvenile state. Figure 2.12 shows that the relative duration of the larval period (which is usually characterised by a high mortality level) is variable in different fish species, and this period is eliminated in the eelpout. In the majority of fish species, transitions to the larval and juvenile states coincide with the start of the larval and juvenile periods, although in several fish species such a coincidence is absent.

2.4.2.2 Relationship between egg parameters and life history

A general evolutionary trend in reproduction among vertebrates is the production of a smaller number of larger offspring. The increase in the egg size is accompanied by a higher degree of parental care (Sargent *et al.* 1987). The origin of each species is connected with a trade-off between the production of a larger number of offspring of smaller size or a smaller number of offspring of larger size of the progeny of each female. An analysis of data on absolute fecundity

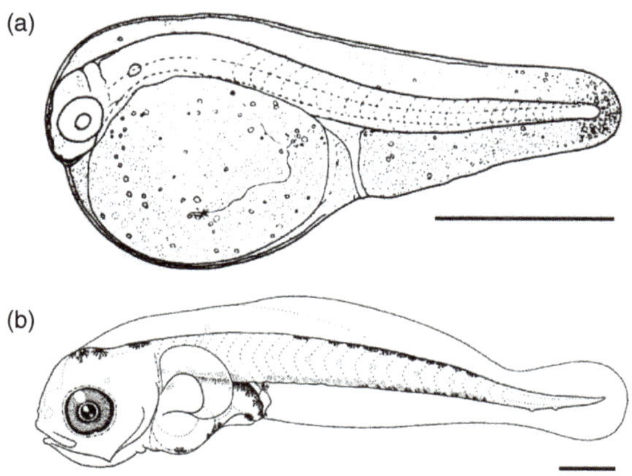

Figure 2.11 Developmental stages in scorpaenoid fishes (Scorpaenoidei). (a) *Scorpaena miostoma* Günther, stage of hatching, 1.55 mm SL, age 35 h from fertilisation (22–24°C); (b) *Sebastes auriculatus* Girard, 1 day after hatching, 5.2 mm TL. Scale bars = 0.5 mm. Reproduced from Pavlov & Emel'yanova (2013) with kind permission of 'Interperiodica'.

and egg volume in 42 oviparous fish species from 26 families showed that the first parameter ranged from 35 eggs to more than six million eggs, and the second parameter ranged from 0.05 to 68.09 mm^3 (Elgar 1990). However, the variability among other fishes is even larger. For example, the absolute fecundity of sunfish *Mola mola* (L.) can reach 300 million eggs,

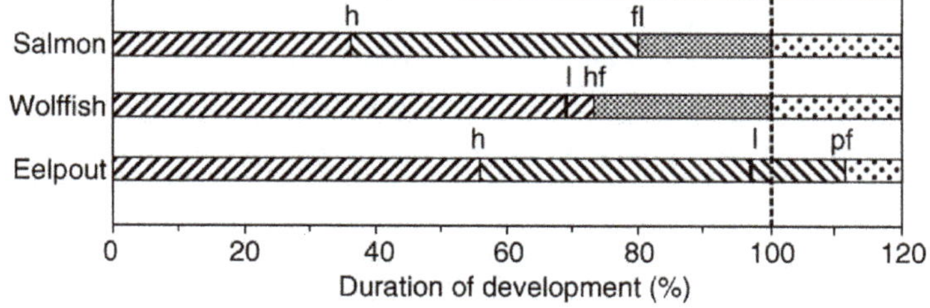

Figure 2.12 Relative duration of main intervals of early ontogeny in several marine and anadromous fishes. Duration of development from egg activation to the beginning of juvenile state = 100% (in Atlantic herring, Atlantic cod, turbot *Psetta maxima* (L.), Atlantic halibut *Hippoglossus hippoglossus* L., Atlantic salmon, common wolffish and eelpout, 74, 59, 90, 108, 112, 177, and 107 days, respectively). H, hatching; L, beginning of the larval state (appearance of ability to feed orally); F, transition to the exogenous feeding; J, beginning of the juvenile state (appearance of adult characteristics); P, parturition. Reproduced (modified) from Pavlov (1999).

each approximately 1 mm in diameter (Winemiller & Rose 1993, Jonathan *et al.* 2002). In a freshwater fish from the family Eleotridae, *Hypseleotris compressa* (Krefft), the mean diameter of ovulated eggs is 0.26 mm (0.02 mm^3 in volume) (Auty 1978; cited by Makeyeva 2002). In mainly freshwater fishes from the family Ariidae (Siluriformes), the egg diameter is between 10 and 25 mm (Fuiman 1984; cited by Kamler 1992). In the majority of viviparous sharks, fecundity ranges from one to four individuals, while only one offspring (from an egg 7–9 cm in diameter) that is one-quarter the length of the adult fish is produced by the female coelacanth, which is also viviparous. Between species, egg size and fecundity tend to be inversely related (Blaxter 1988).

There is a considerable difference between the egg sizes of marine and freshwater teleosts. Marine fishes tend to produce pelagic eggs, while most freshwater teleosts produce demersal eggs, and very few produce pelagic eggs. Among freshwater teleost families, 60% display some form of parental care in guarding their young, while only 16% of marine teleost families show parental care. Fish species (with few exceptions) do not produce large pelagic eggs. Most pelagic sharks are viviparous, with young over 30 cm in length. Thus, there are mainly two life-history options for pelagic reproduction: (i) viviparity and low fecundity (with the exception of viviparous scorpaenoid fishes with a high fecundity); or (ii) oviparity and high fecundity accompanied by small eggs (Jonathan *et al.* 2002).

According to Kryzhanovskii (1949) and Balon (1985, 1990), egg size and density have a major influence on the type of early ontogeny determining the relative duration of the larval period. Other authors (Soin 1981, Meshkov & Lebedeva 1983) hold the opinion that the features of early ontogeny depend mainly on the cytoplasm-to-yolk ratio. In the smelts, pikes, and salmonids (28 species, including 38 forms), the embryos hatch and the larvae transit to exogenous feeding at more advanced stages with increased yolk size. Positive and significant correlations are found between the yolk size and the total length of the organism at: (i) hatching; (ii) the transition to exogenous feeding; and (iii) the beginning of juvenile state (Pavlov & Osinov 2004).

The second of these correlations is the strongest. In these species, the differences in yolk size (and relative features of early ontogeny) are genetically controlled (at least, at the levels of genera and species): the mean yolk diameter is positively and significantly correlated with the number of functional copies of vitellogenin (the main predecessor of yolk) genes. In the genera *Osmerus* + *Esox* + *Coregonus*, the water content in ovulated oocytes is higher than that in the genera *Salmo* + *Parasalmo* + *Oncorhynchus* (60.8–71.6% versus 48.1–60.3%) (Kamler 1992). In the fishes of the former group, the larvae transit to exogenous feeding at a lower degree of morphological development; moreover, this is apparently connected with substantial differences in the yolk size and, to a lower degree, with a small difference in yolk densities. In addition, the effectivity of energy utilisation to somatic growth (K_2) from egg activation to transition to exogenous feeding in fishes from different ecological groups is approximately the same (Novikov 2000). Thus, both the size and morphological condition of the organism at the transition to exogenous feeding depend mainly on the yolk size.

Egg size, fecundity, age at first sexual maturation and life span, together with some other characteristics, are the main parameters of a reproductive strategy of the population or species with a pronounced adaptive significance (e.g., Schaffer 1974, Sargent *et al.* 1987, Elgar 1990). An optimal egg size is the size at which the maximum number of offspring attains the reproductive age (Wootton 1990). Substantial differences in yolk size (and egg size) have been registered at the subspecies level in many fishes. A geographical variation in egg size (with larger egg sizes in colder waters) has been observed in related forms of marine fish species (Rass 1977). The variation in egg size and the size of the young at identical ontogenetic stages can be clearly shown in species with a wide geographical range. For example, variation in the egg size is registered within the range of walleye pollock *Theragra chalcogramma* (Pallas) (Hinckley 1990). Similar differences are observed in seasonal races: the eggs of summer-spawning White Sea herring are smaller than those in the spring-spawning stock (Khrustaleva 2003).

The basic idea proposed by Nikolsky *et al.* (1973) and Nikolsky (1974) is that fecundity is a population-regulating mechanism, which can be regarded as a response to a certain level of egg and offspring mortality. However, substantial changes in population fecundity can be compensated by almost immeasurable changes in mortality or survival rate (Jones 1973), and therefore fecundity seems, based on this argument, relatively unimportant in the regulation of fish populations. In this case, the danger of recruitment overfishing would be very difficult to explain. A resolution of this paradox can be obtained by considering a conceptual model where increased fecundity serves to increase survival rate, and the superabundance of eggs increases the chances that first-feeding larvae will encounter relatively rare spatiotemporal patches of adequate food (Winemiller & Rose 1993, Rothschild 2000).

Evolutionary changes of fecundity and egg size can be partly explained in the terms of r and K selection (Pianka 1970). In a variable and unpredictable environment with mortality not depending on the population density and at a low intra- and interspecies competition, selection is directed to the production of a larger number of tiny offspring. Under more stable environmental conditions and strong concurrent relationships, selection is directed to the production of a lower number of large and morphologically advanced organisms. In addition to fecundity (i.e., clutch size) and survivorship as the sole components of fitness, Winemiller (1992) included a third axis – the timing of reproduction – as an essential part of the life-history strategy, and suggested a three-dimensional model of life-history evolution.

- Following Winemiller (1992) and Winemiller & Rose (1992), an '*opportunistic strategy*' is associated with early maturation, small eggs, small clutches and continuous spawning. Small fishes often maintain dense populations in frequently altered or disturbed habitats and/or experience a high predation mortality in the adult state. This could be considered the ideal colonising or r-strategy.
- A '*periodic strategy*' describes individuals that delay maturation, thereby allowing growth to a size sufficient to enhance fecundity and adult survival during periods of suboptimal environmental conditions. High fecundity is generally associated with small egg size and low parental investment per offspring. Highly fecund fishes can exploit predictable patterns in time or space by releasing massive numbers of small progeny during the periods in which environmental conditions are most favourable for growth and the survival of larvae. Strong cohorts are recruited periodically when early life stages encounter suitable environmental conditions. This strategy is usual for the majority of commercially exploited species.
- An '*equilibrium strategy*' identifies small- or medium-sized fishes with large eggs, small clutches, well-developed parental care, delayed maturity, and protracted spawning seasons. These traits are largely associated with the K-strategy of adaptation to life in resource-limited or density-dependent environments.

The life cycles of several fish species cannot be confidently referred to any one type of these strategies. For example, pink salmon possesses a low fecundity and a large egg size (equilibrium strategy), but matures early and is characterised by a short life cycle (opportunistic strategy).

2.4.3 Critical intervals in the early ontogeny and formation of a year class

The concept of critical intervals is closely related to the description of fish ontogeny as a saltatory process first applied by Vasnetsov (1953) and Kryzhanovskii *et al.* (1953) and later redefined by Balon (1985, 1990, 2002). Saltatory development depicts ontogeny as a process that consists of a hierarchy of stabilised periods and steps separated by less-stable thresholds that progress in a series of leaps rather than as a continuous sequence of events.

The change in the yolk size leads to the appearance of a new phenotype of the organism at the transition to exogenous feeding, and the main mortality of fish offspring is observed at this developmental interval (e.g., Hjort 1914). The size of larvae at this and subsequent stages is directly connected with their survival. Larger organisms use larger volumes of water for feeding (Blaxter 1986, Webb & Weihs 1986), are more resistant to starvation, feed on prey with a wider size range (Hunter 1981), and avoid predators more effectively (Miller *et al.* 1988). Besides, a more advanced morphological condition of larvae at the transition to mixed feeding leads to shortening of the larval period, and the organism reaches the juvenile state with fewer substantial morphological transformations that lead to a decreased mortality rate.

The egg size is apparently a general, but not single, parameter determining the phenotype of the organism at the transition to exogenous feeding. The phenotype can depend on the relationship between the somatic growth rate and morphological development of the organism. For example, a high level of interspecific variability in egg size is mainly accompanied by a change in the size of the larvae (at similar morphological states) at the beginning of mixed feeding in the representatives of the family Nototheniidae (North 2000). A mode leading to the appearance of another phenotype is a change of the 'embryonisation' level. If this level is high, a larger number of developmental stages occur inside the egg envelope and, therefore, the survival of the organism increases. At the same time, the embryonic development becomes more complex. The highest embryonisation level leads to transition of the organism to exogenous feeding at the juvenile state. The problem of elimination of the larval period most sensitive to environmental impacts is effectively solved in viviparous species.

The stage of transition to exogenous feeding is apparently not critical in fishes with direct and transitory ontogenies, as can be supported by the results on cultivated species. For example, the average survival of the juvenile common wolffish (transitory ontogeny) can reach 100% at the transition to exogenous feeding (Pavlov 2004). Nevertheless, the survival of juvenile cod (indirect ontogeny) is usually less than 5% (Bromage 1995).

Unfortunately, the majority of commercial fish species are characterised by indirect ontogeny, and the hypothesis about the mismatch between the fish larva and its prey as a main critical period for survival (Hjort 1914) remains credible. Based on observations of Atlantic mackerel *Scomber scombrus* L., a hypothesis on the coupling between high larval food abundance, low juvenile growth and strong cohorts was proposed (Ringuette *et al.* 2002). In pikeperch *Stizostedion lucioperca* (L.), prey density at the onset of feeding was a limiting factor, which might contribute to the high variation in year-class strength (Ljunggren 2002). The hypothesis of Hjort has been discussed intensively up to now (see reviews of Nikolsky 1974, Solemdal 1997, Rothschild 2000 and the chapters of this book). A similar, but not identical concept is represented by the 'theory of critical periods' (Svetlov 1960, Vladimirov 1975). According to this concept, hidden defects in the development transferred to the progeny by the parents are realised during the critical developmental intervals. Thus, the anomalies induced spontaneously (and not connected with environmental effects) represent a source of natural mortality.

According to Vladimirov (1975), several critical intervals can be identified in the early ontogeny of fishes. One of these intervals is metamorphosis (Chambers & Leggett 1992, Thorisson 1994). The latter author observed feeding difficulties in Atlantic cod within the size interval 10–14 mm. Based on his hypothesis, in years with unfavourable feeding conditions during metamorphosis it would probably not matter much how successful the first feeding period was. Conversely, in years with good feeding conditions during metamorphosis, the fate of the year class would to a greater extent be determined by the conditions in the first feeding period. This hypothesis is supported by later experiments on the response of Atlantic cod larvae to starvation with age (Jordaan & Brown 2003). In Japanese sea bass *Lateolabrax japonicus* (Cuvier), a critical period also may occur during larva–juvenile transition, which is associated with an ontogenetic dietary shift from smaller to larger prey (Islam & Tanaka 2006). The metamorphosis of the representatives of the family Engraulidae and other fishes – in particular, the development of the caudal fin and notochord flexion – is associated with the start of diel vertical migrations, a change in prey selectivity, and an increased ability to evade plankton nets (Somarakis & Nikolioudakis 2010).

A high mortality of a progeny observed during early ontogeny is accompanied by passive migrations (drift of the eggs, larvae and juveniles). The drift is important for spatial distribution of the young and for the integrity of the migration cycle. Both year-class strength and interpopulation differences are formed during the drift. If the passive migrations occur along the main oceanic currents, the spawning grounds in many populations of commercial fish species can overlap, or they are located close to each other. An area of joint spawning grounds in different commercial fish species can be designated as the *common reproductive zone* (Serebryakov 1989). For example, the synchronous spawning of Atlantic cod, haddock, deepwater redfish *Sebastes mentella* Travin and ocean perch *Sebastes marinus* (L.) can be observed off the Lofoten Islands. The tremendous abundance of commercial fish species and appearance of strong year classes are possible mainly due to the occurrence of spawning grounds located within the common reproductive zones.

2.5 Egg quality

Egg and larval quality are directly connected with the recruitment potential and the strength of a year class. Egg quality in its broad sense is determined by the intrinsic properties of the egg itself and the environment in which the egg develops from fertilisation (Bromage 1995). Egg quality in its strict sense can be defined as the egg's ability to produce viable offspring (Kjørsvik *et al.* 1990). According to the latter definition, egg properties depend on the genotype, as well as on the morphological, chemical and physiological processes occurring in the egg. Some of these characteristics are directly transferred from the mother.

2.5.1 Determinants of egg quality

In hatcheries, egg quality can be preliminarily assessed before artificial insemination. The oocyte batches, including cells substantially ranging in size, having unusual inclusions in the cytoplasm, exhibiting exocytosis of at least a part of cortical alveoli, or oocytes with opaque content, apparently, are not able to be fertilised (Pavlov & Emel'yanova 2008, Emel'yanova & Pavlov, 2012). The quality of pelagic eggs is usually assessed by their ability to float or sink in seawater (despite the uncertainty of this method), and the quality of fertilised eggs is determined based on such criteria as fertilisation rate, percentages of eggs with normal cleavage, subsequent survival up to hatching and the fate of the larvae afterwards.

The potential ability of eggs for subsequent normal development can be assessed just after their activation. In the majority of teleost fishes spawned in fresh water, ovulated oocytes are activated under the influence of water, irrespective of the presence of spermatozoa. In non-inseminated or unfertilised activated eggs, the cytoplasm is concentrated at the animal pole, and the cytoplasmic disc does not undergo divisions, or the cleavage is parthenogenetic. Unfertilised eggs can remain alive until hatching of embryos in fertilised eggs of the same batch. In marine fishes, normal activation of ovulated oocytes, more often, is induced after the penetration of a spermatozoon into the micropyle. In non-inseminated oocytes of marine fishes from various ecological groups, incomplete cortical reaction is observed (accompanied by the formation of the cytoplasmic disc, which does not undergo cleavage or shows parthenogenetic cleavage) or cortical reaction is partial, and cytoplasmic disc is absent (Figure 2.13a–2.13c_1). Following insemination, a completed cortical reaction is registered in normal oocytes (Figure 2.13d), and this reaction is incomplete or partial in the eggs of poor quality unable to be fertilised (Figure 2.13e, 2.13f). Therefore, oocyte quality can be preliminary assessed in several minutes after insemination.

The cell structure at the early stages of cleavage is a good indicator of egg quality. The main types of abnormal cleavage leading to subsequent total mortality of eggs are: (i) mitotic cleavage of nuclei accompanied by incomplete cytotomy; (ii) mitotic cleavage of nuclei without cytotomy; (iii) a weak adhesion between cells; and (iv) desynchronisation of cell divisions. For several commercial marine fish species, significant positive correlations are observed between the proportion of normally cleaved eggs, the hatching rate and the viability of larvae (Shields *et al.* 1997, Pavlov *et al.* 2004a, Rideout *et al.* 2004a). When larvae of turbot *Scophthalmus maximus* L. were reared to the end of metamorphosis, there was also a significant correlation between the ratio of normally cleaved eggs, survival to the juvenile state and juvenile quality (Kjørsvik *et al.* 2003).

Assessment of egg quality in the wild is based mainly on the morphological properties of fertilised eggs. Field studies in the Lofoten area of northern Norway, the most important spawning ground for the Arcto-Norwegian cod, showed that approximately 10% of cod eggs underwent abnormal cleavage (Kjørsvik 1994). Later, the method for the assessment of Atlantic cod egg quality in the same region was modified (Solemdal *et al.* 1998, Makhotin *et al.* 2001), and the experiments showed that cumulative mortality at hatching could exceed 30%.

Other criteria that can be used for the evaluation of egg quality are the egg size (especially the coefficient of variation of egg diameter), chemical content (including pigments, vitamins, some organic and inorganic components), chromosomal anomalies, as well as molecular and cellular characteristics. Poor egg quality can be caused by the desynchronisation in the processes of egg

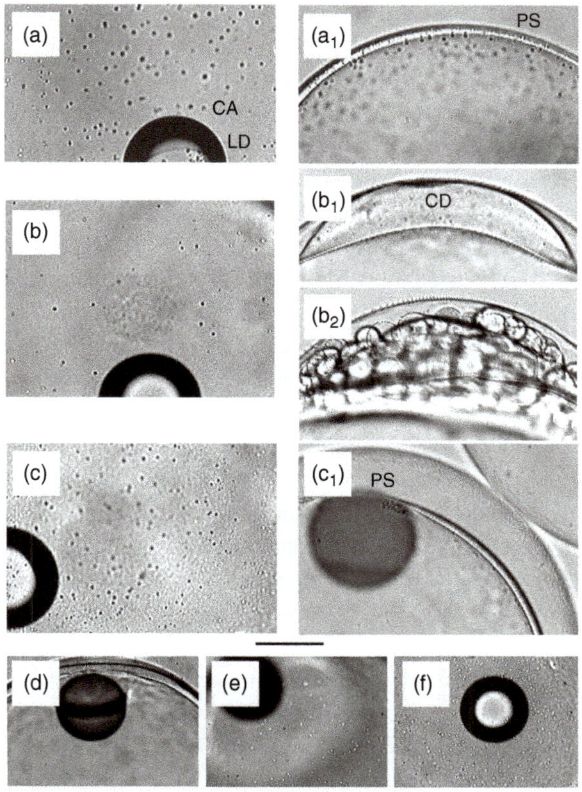

Figure 2.13 Distribution of cortical alveoli and formation of the perivitelline space in ovulated oocytes of *Zebrasoma scopas* (Cuvier). (a, a_1) Oocyte before its placing into water (0 h), presumptive perivitelline space; (b, b_1, b_2) 4 h from placing oocyte into water without insemination, incomplete cortical reaction, single cortical alveoli remain, perivitelline space, uncleaved cytoplasmic disc or parthenogenetic cleavage; (c, c_1) 4 h from placing of oocyte into water without insemination, partial cortical reaction, numerous cortical alveoli remain, perivitelline space, without cytoplasmic disc; (d) 5 min after insemination, completed cortical reaction in the oocyte of good quality; (f) 5 min after insemination, incomplete cortical reaction in the oocyte of poor quality; (d) 5 min after insemination, cortical reaction is not registered in the oocyte of poor quality, partial cortical reaction will occur later. CA, cortical alveoli, CD, cytoplasmic disc, LD, lipid droplet, PS, perivitelline space. Scale bars = 1 mm. Reproduced from Pavlov & Emel'yanova (2008) with kind permission of 'Interperiodica'.

maturation and ovulation (under-ripening or over-ripening) and disturbances in egg maturation (see reviews of Kjørsvik *et al.* 1990, Pavlov *et al.* 2004a, Bobe & Labbé 2010).

2.5.2 *Maternal and paternal effects*

Egg and offspring quality are determined by the genotype and phenotype of the parents. The genetic influences on egg quality can be deduced from studies conducted on cultivated species. For example, the females of rainbow trout producing better-quality eggs in their first spawning season also do so in the subsequent season (Brooks *et al.* 1997). The definition of maternal effect in its broadest sense is the non-genetic contribution of the female to offspring condition. Different interpretations of this term are discussed in a review by Green (2008).

Maternal effects were first investigated on fishes by Nikolsky (1962): the differences in egg size, spawning period and fecundity were associated with the size and condition of the mother. An increase in egg quality with egg size was registered in several species of temperate marine fishes, including Atlantic cod, Atlantic herring, capelin *Mallotus villosus* (Müller) and turbot (Chambers & Leggett 1996). However, a large egg size (and larger size of larvae) is not always associated with increased survival of the young, as has been shown for salmonids (Bromage & Cumaranatunga 1988, Pakkasmaa & Jones 2002), common wolffish (Pavlov 2004) and Atlantic cod (Zhao *et al.* 2001).

Atlantic cod has become a main object for the investigation of maternal effects. This species is a multiple batch spawner, and the reproductive period can be more than 20 years in unexploited stocks. An effect of high fishing pressure is a reduction in average age and body size of the representatives of the stock. Due to possible lower egg and offspring quality from recruit spawners (Solemdal *et al.* 1995), the age and size composition of the spawning population can be an important factor regulating year-class strength. A decreased egg mortality in the second spawners was shown in Norwegian coastal cod from one single broodstock. The egg size increased significantly from the first to the second spawning season, with a smaller increase in third-time spawners kept at the same level of condition (Kjesbu *et al.* 1996). Laboratory studies showed that the hatching success of eggs from second-time spawners was higher (62%) than that of first-time spawners (13%) (Trippel 1998). In Arcto-Norwegian cod collected in the wild and kept in captivity, a reduction in egg mortality by 18% from first to second spawners was registered (Solemdal *et al.* 1998). Incubation of the eggs from this stock over two subsequent spawning seasons with a higher proportion of first-spawning fish in 2001 than in 2000 showed a trend to decreasing egg quality in 2001 (Makhotin *et al.* 2001). In Icelandic cod collected in the wild, eggs of the largest size and with the highest quality tended to be produced by the largest and, in general, the oldest females (Marteinsdottir & Steinarsson 1998).

In the majority of coral reef fishes experiencing high mortality rates in the larval and early juvenile stages, non-genetic parental effect is especially important. Parents of spiny chromis *Acanthochromis polyacanthus* (Bleeker) in good condition commenced breeding earlier, had higher reproductive output, and their eggs exhibited increased survival during embryogenesis when compared to parents in poorer condition (Donelson *et al.* 2008). A paternal effect on egg quality might be expected, but is connected with genetic components. The paternal effects apparently have a lower importance for the egg and early larval development, though they have not been investigated as comprehensively as the maternal effects (see the reviews of Kamler 2002 and Trippel 2003). Studies by Saillant *et al.* (2001) and Rideout *et al.* (2004b) have indicated that the importance of males in the early life history success of marine fishes should be reconsidered.

2.5.3 Change of egg quality over the spawning season

Egg quality is usually highest in the beginning or in the middle of spawning (see the review of Pavlov *et al.* 2004a). For example, the profile of polyunsaturated fatty acids (PUFAs) in common snook *Centropomus undecimalis* (Bloch) changed over the spawning season (37.5–29.4% of the total fatty acid composition of eggs), and egg quality was better in May, June and July than in August and September (Yanes-Roca *et al.* 2009). However, for Atlantic cod the size of eggs from the earliest batches was smaller and the quality poorer than from the following batches (Kjesbu *et al.* 1990, Marteinsdottir & Steinarsson 1998). The mean egg size of cod usually declines throughout the spawning season, as shown both for fish from the broodstock (Kjørsvik 1994, Mangor-Jensen *et al.* 1994) and from the wild (Marteinsdottir & Steinarsson 1998). Other laboratory experiments showed that large cod larvae originating from large eggs had low survival rates, but in delayed feeding groups the larvae from large eggs grew faster than those from small eggs (Zhao *et al.* 2001). A low egg mortality of cod was registered at the end of the spawning season, both for fish collected in the wild and kept in the laboratory and for wild stocks from Lofoten (Solemdal *et al.* 1998, Makhotin *et al.* 2001).

2.6 Influence of environmental factors on reproduction and recruitment

Shifts in biological productivity lead to large variations in stock abundance, even without fishing mortality. The fishing mortality may have relatively little effect on stocks when productivity is increasing, but a substantial effect on stocks when productivity is decreasing (Rothschild 1995, 2000). This is important for fishery management, and brings us to the problem of distinction between the effects of natural and anthropogenic factors, including fishing. However, the mixed nature of such effects is obvious; for example, climate changes represent a consequence of both natural perturbations and human activity.

2.6.1 Natural factors

Hjort (1914) was one of the first scientists to describe periodic variations in the stock abundance of Atlantic cod and Atlantic herring, and to register a connection between solar activity, temperature and the weight of the gonads and liver in cod. Later, periodical fluctuations in stock abundance were described for a large number of fish species from the inland seas of the former Soviet Union (see a review of Nikolsky 1974) and for exploited stocks from the Atlantic and Pacific (see reviews of Rothschild 1995, Klyashtorin & Lyubushin 2007, Klyashtorin *et al.* 2009). In the Norwegian and Barents Seas, strong year classes frequently occur simultaneously for several species, such as herring, cod and haddock. These incidences are presumably related to increased Atlantic inflow and, thereby, to higher temperatures stimulating growth and recruitment of the 0-group (Sætre *et al.* 2002).

Multi-decadal fluctuations in the abundance of fish stocks are correlated with changes in biological production. For example, the decline in zooplankton abundance in the Peru–Chile region during the early 1970s coincided with an abrupt reduction in the biomass of the Peruvian anchovy *Engraulis ringens* (Jenyns). However, the effect on fish production will depend on how closely the productivity of the fish stocks is coupled to primary and secondary production. Small fish larvae with an indirect ontogeny can depend more on primary production than larvae and juveniles with a transitory and direct ontogeny with a diet including comparatively large zooplankton organisms, benthos, or fish larvae. In addition, the abundance of adult planktivorous fishes will be linked linearly with plankton abundance. Numerous physical events may interact with global fluctuations in biological productivity. In particular, El Niño Southern Oscillation events lead to an increase of the surface temperature in the eastern tropical Pacific and suppressing of the inshore upwelling representing the main source for the high biological productivity of the region.

As early life-history stages are extremely sensitive to altered environmental conditions, climate change is expected to have a major effect on the distribution and abundance of fishes through its influence on recruitment. Apart from the direct mortality of early life stages due to intolerable conditions, differences in rates of survival will be attributable to match–mismatch dynamics between the timing of reproduction relative to the production of larval food and predators or the connectivity (retention or transport) between spawning sites and required nursery areas of early life stages (see Chapter 3). Maximum recruitment is often registered at an intermediate temperature during the spawning season in both demersal and pelagic fishes. Fish species are generally plastic in the age and size at which they become sexually mature. An increase in juvenile growth rate caused by increasing temperature may result in a decrease in the length and age at first maturation, affecting the growth of adults as surplus energy can lead to reproduction at an earlier age and smaller size (Pörtner & Peck, 2010). Global warming is accompanied by the appearance of additional habitats suitable for feeding and reproduction that leads to spatial changes in the distribution of marine fish species. For example, surmullet *Mullus surmuletus* L. with normally subtropical range has been observed in the coastal zones of the islands to southwest of Bergen (Norway) since 1992 that is connected with an increased abundance of the species in the English Channel and North Sea caused by a long warming period (Uiblein 2007).

Age and size at metamorphosis is considered a critical determinant of recruitment success (Chambers & Leggett 1992). The larvae of the Hawaiian amphidromous goby *Lentipes concolor* (Gill) settled after shorter planktonic lives and at smaller sizes during months with warmer ocean temperatures (Radtke *et al.* 2001). In flounder *Pleuronectes flesus* (L.), the length at metamorphosis was sensitive to reduced salinity once the critical length was reached. When the larvae moved into estuaries, they sank and were induced to metamorphosis by the reduced salinity. Thus, the metamorphosis was delayed until the animals were near the nursery grounds within the estuary (Hutchinson & Hawkins 2004). In the bluehead wrasse *Thalassoma bifasciatum* (Bloch) of the Caribbean, fish that encountered low-salinity waters early in their larval life had the slowest growth and remained in the plankton for a long period, but exhibited a fast growth after metamorphosis. Fish that did not encounter low salinity exhibited significantly shorter pelagic larval durations, settled at small sizes, and remained smaller during the first week of juvenile life. The mortality increased in the first group due to a prolonged period of pelagic life, and in the second group due to a small size of juveniles. Thus, individuals with contrasting life-history traits may remain in the population due to a balance in the differential selective pressures in the pelagic and reef environments (Sponaugle & Pinkard 2004).

A positive relationship between high wind speeds in April and recruitment of the 0-group is shown for the Norwegian spring-spawning herring. This observation can be explained by increased wind-induced turbulence in the upper layer that improves the feeding conditions for the larvae. The year-class strength of the herring can be forecast with a 78% probability in July (3 months after hatching) by the temperature in the Kola section during winter, together with the frequency of high wind speeds in April and the fledging success of puffins, an important top predator of 0-group herring in northern Norway (Sætre *et al.* 2002).

Owing to the high sensitivity of the germ cells in response to various perturbations of intrinsic and extrinsic origin, gonads can be used as a good indicator of fish health. The response of the fish reproductive system to habitat change can be associated with a decreasing rate of oocyte development and, consequently, with later sexual maturation and a lower number of oocytes (due to their resorption) entering the period of vitellogenesis, as well as a delay of spawning or its rapid termination. For example, a large amount of oocytes with destructive changes is registered in the gonads of a landlocked form of sockeye salmon (kokanee) *Oncorhynchus nerka* (Walbaum) introduced into the Tolmachevskoe Water Reservoir (Kamchatka) (Pavlov E.D. *et al.* 2010). A few females that had reached maturity stage IV had ovaries of different size with partially resorbed oocytes, and sterile gonads were observed in 8% of the individuals. These anomalies can be connected with a low amount of available food and a coldwater habitat that is not usual for the natural range of the species. Numerous anomalies in gonadal condition of females and males are registered in Pacific herring *Clupea pallasi* Valenciennes collected in the region of the Gizhiga Bay and Kamchatka Peninsula, an area which is neither industrialised nor densely populated (Mikodina *et al.* 2005). These anomalies either represent a result of regulatory processes in the population due to its large abundance, or they are connected with global climate change.

2.6.2 Anthropogenic factors

Global strategies for the management of fish stocks can be determined by main life-history patterns (see Section 2.4.2.2). The *opportunistic*-type fishes (Winemiller 1992) are not often exploited commercially, but are very important food resources for piscivores. Given their capacity to compensate heavy losses from all life-history stages (which are typically rather short), a key to their management should be protection from large-scale or chronic perturbations. The maintenance of a minimum density of adult stocks and the protection of spawning and nursery habitats during the short reproductive period are crucial in the management of *periodic* strategists. In *equilibrium* strategists, characterised by relatively high early survivorship and parental care, survival of the progeny is dependent on the condition of adults and the integrity of the adult habitat. Consequently, management of exploited equilibrium fishes should stress habitat quality/stability and maintenance of healthy adult stocks. Owing to a protracted period of egg incubation (or their development inside the female's body in viviparous fishes), equilibrium strategists are especially vulnerable to commercial exploitation and the destruction of spawning sites. Ultimately, the advantage of viviparity can be a disadvantage under the conditions of fishing pressure. For example, a low reproduction rate of sharks with a prolonged gestation period (reaching two or three years) combined with a relatively slow growth rate (many species are not mature before 10–12 years) makes them especially vulnerable (Jennings *et al.* 1998, Frisk *et al.* 2002, 2005, Mabragaña & Cousseau 2004, McClenachan *et al.* 2012).

The commercial exploitation of fish stocks has a major effect on their abundance, age composition and reproductive potential. In addition to the maternal effect (see Section 2.5.2), the diversity of age groups can have a substantial influence on recruitment. Larger repeat spawners of Atlantic cod spawn for a longer period of time compared with both younger fish and first-time spawners (Kjesbu *et al.* 1996, Trippel 1998). The bathymetric and horizontal distributions of eggs from repeat spawners are greater than those from first-time spawners, due to variations in egg buoyancy (Kjesbu *et al.* 1992). Thus, if the spawning stock comprises a number of age groups, the eggs would be spatially widely dispersed. A more dispersed spatial distribution of eggs will lead to a greater opportunity to encounter favourable conditions for survival (Brien *et al.* 2003). The collection of coral reef fishes from the wild for the aquarium trade also can lead to substantial decreases in the age group number and reproductive potential. The life span of representatives of the families Pomacentridae, Acanthuridae and Lutjanidae reaches tens of years; for example, some pomacentrids with a total length of 60 mm may have maximum ages of approximately 16–18 years (Choat & Robertson 2006).

In intensively exploited stocks, a decreased population density and better food availability can lead to compensatory effects, accompanied by changes of sex ratio (an increase in the proportion of females) and fecundity (Nikolsky 1974). In orange roughy *Hoplostethus atlanticus* Collet exploited from 1987, the female spawning stock biomass was 19% of the virgin level in 2006, but the total reproductive potential was estimated as 32% of the virgin level. Average length-standardised fecundity (adjusted for a mean length of 36.86 cm SL across all years) increased from 41 145 eggs in 1992 to 59 236 eggs in 2010, and was positively related to body condition, liver condition, and ovarian lipid levels (Pitman *et al.* 2013).

Intensive fishing pressure can be a reason for a substantial change of age, size and sexual structure of protogynous hermaphrodites. In gag *Mycteroperca microlepis* (Goode & Bean) from the Gulf of Mexico, a reduction in average female and male size and in the proportion of males (from 17% to 2%) was registered between the 1980s and 1991–1993. A fishing selection of males in spawning aggregations can lead to a failed spawning opportunity for females. The size of females at first sexual maturity, and the sex ratio, may indicate the status of exploitation of the stock (Brulé *et al.* 2003a,b). Similar results were obtained for the venus tusk fish *Choerodon venustus* (De Vis) on the Great Barrier Reef. The reef fish, which experienced the highest mortality, underwent sex reversal at a smaller size and younger age than at the other sites (Platten *et al.* 2002).

The relationships between stock size and recruitment have been discussed over a prolonged time period (see the review of Rothschild 2000). If a population falls below some critical size threshold, individuals can experience problems that adversely affect mating success and/or offspring production. At low population density, the individuals may experience delays in seasonal reproduction as more time might be required to find a mate. For batch-spawning fishes, a delay in the release of eggs after ovulation of just a few hours can result in the over-ripening of gametes and a dramatic reduction in viability. A scarcity of potential mates also lessens the potential for mate choice, and individuals may resort to mating with partners of non-preferred phenotypes or genotypes, negatively affecting reproductive success. Newfoundland's Northern cod declined by 99% between 1962 and 2002. Commercial fishing of Northern cod was closed in 1992, but the stock has since shown only small signs of recovery (see Chapter 7). Slow or negligible rates of recovery among cod populations have been attributed to a reduction in per capita population growth rate concomitant with a reduced population size – an association termed 'depensation' in the fisheries literature and 'Allee effect' in the ecological literature. Experiments conducted on captive Atlantic cod were consistent with the hypotheses that: (i) the fertilisation rate declines with abundance and with a reduced number of males per female (Figure 2.14); and (ii) variances in fertilisation rate increase as the population size declines. The former hypothesis identifies one potential mechanism underlying depensation in Atlantic cod, namely that there may be negative genetic consequences for effective population size (Rowe *et al.* 2004).

In water bodies subjected to anthropogenic pressure, the conditions for reproduction are the most important for the life cycles of fishes, and species with certain demands to environmental conditions (including water quality) respond rapidly to changes in the ecological situation (Moiseenko & Lukin 1999). A review of the effects of pollution on the reproduction of fish (Jones & Reynolds 1997) shows that 11 of 19 studies identified a change in behaviour from the norm. Effects on courtship included a change in the frequency of displays, an increased

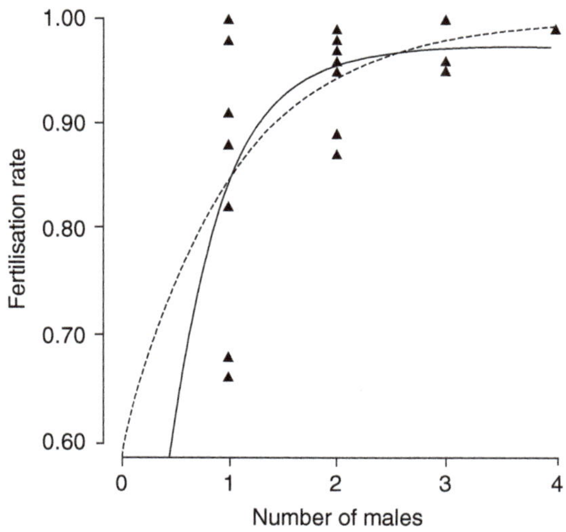

Figure 2.14 Atlantic cod egg fertilisation rate (FR) in individual spawning events involving a single female in relation to the number of males (n) that contributed sperm to the event. Closed triangles represent data provided by Bekkevold *et al.* (2002), the continuous line indicates the exponential asymptotic function that best described the data ($FR = 0.97 \times (1 - e^{-2.02n})$), and dashed line indicates the exponential asymptotic function not forced through the origin ($FR = 1.00 - 0.42 \times e^{1.00n}$). Reproduced from Rowe *et al.* (2004), with kind permission of Oxford University Press.

courtship duration, or the performance of male-like behaviour by masculinised females. Studies on parental care showed decreased nest-building activity, decreased offspring defence, or a change in the division of parental care between sexes. The lability of sex-determination systems in fish makes some species sensitive to environmental pollutants capable of mimicking or disrupting sex hormone actions (see the review of Devlin & Nagahama 2002).

An assessment of the toxic effect of disposed oil and other organic compounds on the Black Sea fishes showed that the most sensitive are pelagic fish species and their eggs, while the demersal species are more resistant (Oven 2004). However, the overall survival of the pelagic and bathypelagic species is higher due to a wider range, a rapid embryonic development, multiple spawning, and high levels of total fecundity. The overall elimination is higher in demersal coastal species with demersal eggs, prolonged embryonic development, and low fecundity.

In addition to general toxic effects on living organisms, the anthropogenic factors – including pollutants – can affect gametogenesis, leading to a disturbance of normal reproduction and the appearance of nonviable progeny. Some of the most usual anomalies in the reproductive cycle include: asymmetrical development and morphological deformation of the gonads; early sexual maturity (or a decrease in growth rate of fish at intensive feeding, accumulation of fat, and retardation of sexual maturity); a change in the duration of the periods of gonadal development; the resorption of sex cells in females and males at various stages; and an increase in the number of fishes omitting spawning and decreased fecundity (Shatunovskii *et al.* 1996, Sharova *et al.* 2003, Akimova & Ruban 2009, Emel'yanova *et al.* 2014).

More serious anomalies (hermaphroditism, sterility, formation of unusual morphological structures, total destruction of sex cells and appearance of dwarf individuals) are registered in fishes from water bodies polluted by heavy metals, mineral oil and radionuclides (Savvaitova *et al.* 1995, Chebotareva *et al.* 1998, Belova & Emel'yanova, 2011). Different reactions of the reproductive system to the influence of negative factors can be explained by changes in the regeneration properties of sex cells at various stages of their development. A predominance of cells at mitotic or meiotic divisions (which are the most sensitive) can lead to a total sterilisation of the gonad (Chmilevskii 2005). A decrease in sperm production has been reported in the males of silver carp *Hypophthalmichthys molitrix* (Valenciennes) and bighead *Aristichthys nobilis* (Richardson) subjected to radiation as a result of the Chernobyl nuclear power plant accident (Verigin *et al.* 1996). An increasing deterioration of the reproductive system has been registered in the progenies of different fish species from third to sixth sequential generations as a result of the mutations accumulation under the influence of radiation (Belova & Emel'yanova 2011). Environmental pollution with substances that disturb reproductive function can lead to irreversible changes. The population abundance may remain fairly high, and hence the effect of enhanced reproduction as a response to population decline does not take place, which eventually leads to a gradual destruction of the population. Pathologies of reproduction can be used as a criterion to assess the state of animals in zones of technogenic disasters (Shilova & Shatunovskii, 2005).

Thus, owing to regeneration properties of the reproductive system in fishes, reproductive strategy changes under the influence of anthropogenic pressure, and the population abundance is maintained at a certain level, thereby preventing the total extinction of populations.

References

Abe, T. & Munehara, H. (2009) Adaptation and evolution of reproductive mode in copulating cottoid species. In: B.G.M. Jamieson (Ed) *Reproductive Biology and Phylogeny of Fishes (Agnathans and Bony Fishes)*. pp. 221–46. Science Publishers, Enfield, NH, USA.

Abou-Seedo, F.S., Dadzie, S. & Al-Kanaan, K.A. (2003) Sexuality, sex change and maturation patterns in the yellowfin seabream, *Acantthopagrus latus* (Teleostei: Sparidae, Houttuyn, 1782). *Journal of Applied Ichthyology*, **19**, 65–75.

Aizenshtadt, T.B. (1984) *Cytology of Oogenesis*. Nauka, Moscow (in Russian).

Akhundov, M.M. (2000) The ecological plasticity of sturgeons sex differentiation. In: B. Norberg *et al.* (Eds) *Proceedings of the 6th International Symposium on the Reproductive Physiology of Fish*. p. 267. Bergen.

Akimova, N.V. & Ruban, G.I. (2009) Anomalies in the development and functioning of the reproductive system in Siberian sturgeon, *Acipenser baerii* Brandt (Acipenseridae), from the Yenisei River. *Biology Bulletin*, **36** (5), 532–6.

Amundsen, T. (2003) Fishes as models in studies of sexual selection and parental care. *Journal of Fish Biology*, **63**, 17–52.

Avise, J.C. & Mank, J.E. (2009) Evolutionary perspectives on hermaphroditism in fishes. *Sexual Development*, **3**, 152–63.

Avise, J.C. & Tatarenkov, A. (2015) Population genetics and evolution of the mangrove rivulus *Kryptolebias marmoratus*, the world's only self-fertilizing hermaphroditic vertebrate. *Journal of Fish Biology*, **87**, DOI: 10.1111/jfb.12741.

Balon, E.K. (Ed.) (1985) *Early Life Histories of Fishes: New Developmental, Ecological and Evolutionary Perspectives. Developments in Environmental Biology of Fishes*. W. Junk, Dordrecht.

Balon, E.K. (1990) Epigenesis of an epigeneticist: the development of some alternative concepts on the early ontogeny and evolution of fishes. *Guelph Ichthyology Reviews*, **1**, 1–42.

Balon, E.K. (2002) Epigenetic processes, when *natura non facit saltum* becomes a myth, and alternative ontogenies a mechanism of evolution. *Environmental Biology of Fishes*, **65**, 1–35.

Belova, N.V. & Emel'yanova, N.G. (2011) State of the reproductive system of Cyprinidae from the Teterev River and Kiev Reservoir after the Chernobyl accident. *Journal of Ichthyology*, **51** (2), 187–97.

Blackburn, D.G. (2005) Evolutionary origins of viviparity in fishes. In: M.C. Uribe & H.J. Grier (Eds) *Viviparous Fishes*. pp. 287–301. New Life Publishing, Homestead, USA.

Blaxter, J.H.S. (1986) Development of sense organs and behavior of teleost larvae with special reference to feeding and predator avoidance. *Transactions of the American Fisheries Society*, **115**, 98–114.

Blaxter, J.H.S. (1988) Pattern and variety in development. In: W.S. Hoar & D.J. Randall (Eds) *Fish Physiology, Vol. XIA*. pp. 377–435. Academic Press, San Diego.

Bobe, J. & Labbé, C. (2010) Egg and sperm quality in fish. *General and Comparative Endocrinology*, **165** (3), 535–48.

Bondarenko, M.V., Krovin, A.S. & Serebryakov, V.P. (2003) *Ranging Year-Class Strength and Survival Rates During Early Life History of the Barents Sea Food Fishes to Establish Biological Reference Points and Evaluate Environmental Effects*. 187 pp. VNIRO Publishing, Moscow.

Brien, L.O., Rago, P.J. & Lough, R.G. (2003) Incorporating early-life history parameters in the estimation of the stock-recruit relationship of Georges Bank Atlantic cod (*Gadus morhua*). *Journal of Northwest Atlantic Fisheries Science*, **33**, 191–205.

Bromage, N. (1995) Broodstock management and seed quality – general considerations. In: N.R. Bromage & R.J. Roberts (Eds) *Broodstock Management and Egg and Larvae Quality*. pp. 1–23. Blackwell Science, Oxford.

Bromage, N. & Cumaranatunga, R. (1988) Egg production in the rainbow trout. In: J.F. Muir & R.J. Roberts (Eds) *Recent Advances in Aquaculture*, Volume 3. pp. 64–138. Croom Helm, London; Timber Press, Portland, OR.

Brooks, S., Tyler, C R. & Sumpter, J P. (1997) Egg quality in fish: what makes a good egg? *Reviews in Fish Biology and Fisheries*, **7** (4), 387–416.

Brulé, T., Déniel, C., Colás-Marrufo, T. & Renán, X. (2003a) Reproductive biology of gag in the southern Gulf of Mexico. *Journal of Fish Biology*, **63**, 1505–20.

Brulé, T., Renán, X., Colás-Marrufo, T., Hauyon, Y., Tuz-Sulub, A. & Déniel, C. (2003b) Reproduction in the protoginius black grouper (*Mycteroperca bonaci* (Poey)) from the southern gulf of Mexico. *Fisheries Bulletin*, **101**, 463–75

Bubley, W.J. & Pashuk, O. (2010) Life history of a simultaneously hermaphroditic fish, *Diplectrum formosum*. *Journal of Fish Biology*, **77**, 679–91.

Chambers, R.C. & Leggett, W.C. (1992) Possible causes and consequences of variation in age and size at metamorphosis in flatfishes (Pleuronectiformes): an analysis at the individual, population, and species levels. *Netherlands Journal of Sea Research*, **29**, 7–24.

Chambers, R.C. & Leggett, W.C. (1996) Maternal influences on variation in egg sizes in temperate marine fishes. *American Zoologist*, **36**, 180–96.

Chebotareva, Yu.V., Savoskul, S.P. & Savvaitova, K.A. (1998) Hermaphroditism in fish of the Norilo-Pyasinsk water system. *Journal of Ichthyology*, **38** (6), 483–87.

Cherfas, N.B. (1987) Gynogenesis in fishes. In: V.S. Kirpichnikov (Ed.) *Genetics and Selection of Fishes*. pp. 309–35. Nauka, Leningrad (in Russian).

Cherfas, N.B. & Emel'yanova, O.V. (1986) Role of distant hybridisation in the appearance of unisexual female complexes in fishes (results of the investigations of natural populations and experiments on the hybridisation). In: V.A. Strunnikov (Ed.) *Developmental Biology and Management by Inheritance*. pp. 82–105. Nauka, Moscow (in Russian).

Chiasson, M., Benfey, T.J. & Martin-Robichaud, D.J. (2008) Gonadal differentiation in Atlantic cod, *Gadus morua* L. and haddock, *Melanogrammus aeglefinus* (L.). *Acta Iihthyologica et Piscatoria*, **38**, (2) 127–33.

Chilke, A.M. (2012) Development of primordial germ cells (PGCs) in an Indian freshwater teleost, *Labeo rohita* (Hamilton). *European Journal of Experimental Biology*, **2** (3), 631–40.

Chmilevskii, D.A. (2005) Reproductive indices of *Oreochromis mossambicus* (Cichlidae) under extreme impacts and their utilization for predicting the reproductive capacities of fish populations. *Journal of Ichthyology*, **45** (3), 258–68.

Choat, J.H. & Robertson, D.R. (2006) Age-based studies on coral reef fishes. In: P.F. Sale (Ed) *Coral Reef Fishes. Dynamics and Diversity in a Complex Ecosystem.* Chapter 3. pp. 57–80. Academic Press, USA.

Devlin, R.H. & Nagahama, Y. (2002) Sex determination and sex differentiation in fish: an overview of genetic, physiological, and environmental influences. *Aquaculture*, **208**, 191–364.

Domínguez-Petit, R., Alonso-Fernández, A. & Saborido-Rey, F. (2011) Incidence and significance of cystic structures in the ovaries of gadoid fish. *Scientia Marina*, **75** (2), 215–21.

Donelson, J.M., McCormick, M.I. & Munday, P.L. (2008) Parental condition affects early life-history of a coral reef fish. *Journal of Experimental Marine Biology and Ecology*, **360**, 109–16.

Durkina, V.B. (2006) Patterns of resorption of vitellogenous oocytes in the striped flounder *Pleuronectes pinnifasciatus*. *Russian Journal of Marine Biology*, **32**, 64–7.

Elgar, M.A. (1990) Evolutionary compromise between a few large and many small eggs: comparative evidence in teleost fish. *Oikos*, **59**, 283–7.

Emel'yanova, N.G., Pavlov, D.A. (2012) *From Oocyte to Larva: Hormonal Induction of Oocyte Maturation and Initial Development of Coral Reef Fishes.* KMK Scientific Press, Moscow.

Emel'yanova, N.G., Pavlov, D.A., Luong Thi Bich Thuan & Vo Thi Ha (2013) Some data on the structure of the gonads of manybar goatfish *Parupeneus multifasciatus* (Mullidae) from the Nha Trang Bay, South China Sea. *Journal of Ichthyology*, **53** (8), 600–9.

Emel'yanova, N.G., Pavlov, D.A., Pavlov, E.D., Luong Thi Bich Thuan & Vo Thi Ha (2014) Anomalies in ovarian condition of manybar goatfish *Parupeneus multifasciatus* (Mullidae) from the coastal zone of south Central Vietnam. *Journal of Ichthyology*, **54** (1), 76–84.

Extavour, C.G. (2007) Evolution of the bilaterian germ line: lineage origin and modulation of specification mechanisms. *Integrative and Comparative Biology*, **47** (5), 770–85.

Fitzhugh, G.R., Shertzer, K.W., Kellison G.T. & Wyanski, D.M. (2012) Review of size- and age-dependence in batch spawning: implications for stock assessment of fish species exhibiting indeterminate fecundity. *Fisheries Bulletin*, **110**, 413–25.

Frisk, M.G., Miller, T.J. & Dulvy, N.K. (2005) Life histories and vulnerability to exploitation of elasmobranchs: inferences from elasticity, perturbation and phylogenetic analyses. *Journal of Northwest Atlantic Fisheries Science*, **35**, 27–45.

Frisk, M.G., Miller, T.J. & Fogarty, M.J. (2002) The population dynamics of little skate *Leucoraja erinacea*, winter skate *Leucoraja ocellata*, and barndoor skate *Dipturus laevis*: predicting exploitation limits using matrix analyses. *ICES Journal of Marine Science*, **59**, 576–86.

Fujimoto, T., Nishimura, T., Goto-Kazeto, R., Kawakami, Y., Etsuro, Y. & Arai, K. (2010) Sexual dimorphism of gonadal structure and gene expression in germ cell-deficient loach, a teleost fish. *Proceedings of the National Academy of Science of United States of America*, **107** (40), 1716–22.

Fujioka, Y. (2001) Thermolabile sex determination in honmoroko. *Journal of Fish Biology*, **59**, 851–61.

Ginzburg, A.S. (1968) *Fertilization in Fish and the Problem of Polyspermy*. Nauka, Moscow. (in Russian, 1972: English translation by Z. Blake, Israel Program for Scientific Translations, Jerusalem.)

Götting, K.S. (1961) Beiträge zur Kenntnis der Grundlagen der Fortpflanzung und zur Fruchtbarkeitsbestimmung bei marinen Teleosteern. *Helgoänder Wissenschaftliche Meeresuntersuchungen*, **8** (1), 1–41.

Green, B.S. (2008) Maternal effects in fish populations. *Advances in Marine Biology*, **54**, 1–105.

Grier, H.J., Uribe, M.C. & Patiño, R. (2008) The ovary, folliculogenesis, and oogenesis in teleosts. In: G.M.B. Jamieson (Ed.) *Reproductive Biology and Phylogeny of Fishes (Agnathans and Bony Fishes).* pp. 25–84. Science Publishers, Enfield, NH, USA.

Grunina, A.S. & Recoubratsky, A.V. (2005) Induced androgenesis in fish: obtaining viable nucleocytoplasmic hybrids. *Russian Journal of Developmental Biology*, **36**, 208–17.

Haug, T., Kjørsvik, E. & Solemdal, P. (1984) Vertical distribution of Atlantic halibut (*Hippoglossus hippoglossus*) eggs. *Canadian Journal of Fisheries and Aquatic Sciences*, **41**, 798–804.

Herpin, A., Rohr, S., Riedel, D., Kluever, N., Raz, E. & Schartl, M. (2007) Specification of primordial germ cells in medaka (*Oryzias latipes*). *BMC Developmental Biology*, **7** (3) 1–10.

Hinckley, S. (1990) Variation of egg size of walleye pollock *Theragra chalcogramma* with a preliminary examination of the effect of egg size on larval size. *Fisheries Bulletin (US)*, **88**, 471–83.

Hjort, J. (1914) Fluctuations in the great fisheries of northern Europe viewed in the light of biological research. *Rapports et Procés-verbaux des Réunions, Conseil International pour L'Exploration de la Mer*, **20**, 1–228.

Hunter, J.R. (1981) Feeding ecology and predation of marine fish larvae. In: R. Lasker (Ed.) *Marine Fish Larvae*. pp. 33–77. Washington Sea Grant Program, University of Washington Press, Seattle.

Hutchinson, S. & Hawkins, L.E. (2004) The relationship between temperature and the size and age of larvae and perimetamorphic stages of *Pleuronectes flesus*. *Journal of Fish Biology*, **65**, 448–59.

Ioganzen, B.G. (1955) To the study of fecundity in fishes. *Trudy Tomskogo Gosudarstvennogo Universiteta*, **31**, 57–68 (in Russian).

Islam M.S. & Tanaka M. 2006. Ontogenetic dietary shift of Japanese sea bass during larva-juvenile transition in Ariake Bay. *Marine Ecology Progress Series*, **323**, 305–10.

Ivankov, V.N. (2001) *Reproductive Biology of Fishes*. Dal'nevostochnyi Gosudarstvennyi Universitet, Vladivostok (in Russian).

Jennings, S., Reynolds, J.D. & Mills, S.C. (1998) Life history correlates of responses to fisheries exploitation. *Proceedings of the Royal Society of London*, **265**, 333–9.

Johannessen, T., Gjøsæter, J. & Moksness, E. (1993) Reproduction, spawning behaviour and captive breeding of the common wolffish *Anarhichas lupus* L. *Aquaculture*, **115**, 41–51.

Johnson, A.D., Richardson, E., Bachvarova, R.F. & Crother, B.I. (2011) Evolution of the germ line-soma relationship in vertebrate embryos. *Reproduction*, **141** (3), 291–300.

Jonathan, A., Freedman, D. & Noakes, L.G. (2002) Why are there no really big bony fishes? A point-of-view on maximum body size in teleosts and elasmobranchs. *Reviews in Fish Biology and Fisheries*, **12**, 403–16.

Jones, J.C. & Reynolds, J.D. (1997) Effects of pollution on reproductive behaviour of fishes. *Reviews in Fish Biology and Fisheries*, **7**, 463–91.

Jones, R. (1973) Density-dependent regulation of the numbers of cod and haddock. *Rapports et Procés-verbaux des Réunions, Conseil International pour L'Exploration de la Mer*, **164**, 156–73.

Jordaan, A. & Brown, J.A. (2003) The risk of running on empty: the influence of age on starvation and gut fullness in larval Atlantic cod (*Gadus morhua*). *Canadian Journal of Fisheries and Aquatic Sciences*, **60**, 1289–98.

Kamler, E. (1992) *Early Life History of Fish. An Energetics Approach*. Chapman & Hall, London.

Kamler, E. (2002) Ontogeny of yolk-feeding fish: an ecological perspective. *Reviews in Fish Biology and Fisheries*, **12** (1), 79–103.

Katasonov, V.Ya. & Gomel'skii, B.I. (1991) *Selection of Fishes and the Bases of Genetics*. Agropromizdat, Moscow (in Russian).

Kendall, A.W. Jr, Ahlstrom, E.H. & Moser, H.G. (1984) Early life stages of fishes and their characters. In: H.G. Moser, W.J. Richards, D.M. Cohen, M.P. Fahay, A.W. Kendall Jr. & S.L. Richardson (Eds) *Ontogeny and Systematics of Fishes. Based on an International Symposium Dedicated to the Memory of Elbert Halvor Ahlstrom, August 15–18, 1983*. pp. 11–22. American Society of Ichthyologists and Herpetologists, Special Publication Number 1, La Jolla, CA.

Khrustaleva, A.M. (2003) Variation in the myomere number in larvae of summer-spawning White Sea herring *Clupea pallasi marisalbi* under the effects of temperature and salinity. *Journal of Ichthyology*, **43**, 390–7.

Kirankumar, S. & Pandian, T.J. (2004) Use of heterologous sperm for the dispermic induction of androgenesis in barbs. *Journal of Fish Biology*, **64**, 1485–97.

Kjesbu, O.S., Kryvi, H., Sundby, S. & Solemdal, P. (1992) Buoyancy variations in eggs of Atlantic cod (*Gadus morhua* L.) in relation to chorion thickness and egg size: theory and observations. *Journal of Fish Biology*, **41**, 581–99.

Kjesbu, O.S., Solemdal, P., Bratland, P. & Fonn, M. (1996) Variation in annual egg production in individual captive Atlantic cod (*Gadus morhua*). *Canadian Journal of Fisheries and Aquatic Sciences*, **53**, 610–20.

Kjesbu, O.S., Witthames, P.R., Solemdal, P. & Greer Walker, M. (1990) Ovulatory rhythm and a method to determine the stage of spawning in Atlantic cod (*Gadus morhua*). *Canadian Journal of Fisheries and Aquatic Sciences*, **47**, 1185–93.

Kjørsvik, E. (1994) Egg quality in wild and broodstock cod *Gadus morhua* L. *Journal of the World Aquaculture Society*, **25** (1), 22–9.

Kjørsvik, E., Hoehne, K. & Reitan, K.I. (2003) Egg and larval quality criteria as predictive measures for juvenile production in turbot (*Scophthalmus maximus* L.). *Aquaculture*, **227** (1–4), 9–20.

Kjørsvik, E., Mangor-Jensen, A. & Holmefjord, I. (1990) Egg quality in fishes. *Advances in Marine Biology*, **26**, 71–113.

Klyashtorin, L.B. & Lyubushin, A. (2007) *Cyclic Climate Changes and Fish Productivity*. VNIRO Publishing, Moscow.

Klyashtorin, L.B., Borisov, V. & Lyubushin, A. (2009) Cyclic changes of climate and major commercial stocks of the Barents Sea. *Marine Biology Research*, **5** (1), 4–17.

Kobayashi, T., Kajiura-Kobayashi, H. & Nagahama, Y. (2003) Germ cells during sex differentiation in the teleost. *Fish Physiology and Biochemistry*, **28**, 157.

Koshelev, B.V. (1984) *Reproductive Ecology of Fishes*. Nauka, Moscow (in Russian).

Koya, Y. & Muñoz, M. (2007) Comparative study on ovarian structures in scorpaenids: possible evolutional process of reproductive mode. *Ichthyological Research*, **54**, 221–30.

Kroon, F.J. & Liley, N.R. (2000) The role of steroid hormones in protogynous sex change in the blackeye goby, *Coryphopterus nicholsii* (Teleostei: Gobiidae). *General and Comparative Endocrinology*, **118**, 273–83.

Kroon, F.J., Munday, P.L. & Pankhurst, N.W. (2003) Steroid hormone levels and bi-directional sex change in *Gobiodon histrio*. *Journal of Fish Biology*, **62**, 153–67.

Kryzhanovskii, S.G. (1949) Ecomorphological principles of development of carps, loaches and catfishes. *Trudy Instituta Morfologii Zhivotnykh Imeni A.N. Severtsova*, **1**, 5–332 (in Russian).

Kryzhanovskii, S.G., Disler, N.N. & Smirnova, E.N. (1953) Ecomorphological principles of development in percids (Percoidei). *Trudy Instituta Morfologii Zhivotnykh AN SSSR*, **10**, 3–138 (in Russian).

Lebrun, C., Billard, R. & Jalabert, B. (1982) Changes in the number of germ cells in the gonads of the rainbow trout (*Salmo gairdneri*) during the 1st 10 post-hatching weeks. *Reproduction Nutrition Development*, **22**, 405–12.

Linares-Casenave, J., Van Eenennaam, J.P. & Doroshov, S.I. (2002) Ultrastructural and histological observations on temperature-induced follicular ovarian atresia in the white sturgeon. *Journal of Applied Ichthyology*, **18**, 382–90.

Lisovenko, L.A. (2000) *Reproduction of Fishes with Discontinuous Oogenesis and Multiple Spawning Based on the Study of Walleye Pollack from Western Kamchatka*. VNIRO, Moscow (in Russian).

Liu, J.-X. & Avise, J.C. (2011) High degree of multiple paternity in the viviparous shiner perch, *Cymatogaster aggregata*, a fish with long-term female sperm storage. *Marine Biology*, **158**, 893–901.

Ljunggren, L. (2002) Growth response of pikeperch larvae in relation to body size and zooplankton abundance. *Journal of Fish Biology*, **60**, 405–14.

Maack, G. & Seoner, H. (2003) Morphological development of the gonads in zebrafish. *Journal of Fish Biology*, **62**, 895–906.

Mabragaña, E. & Cousseau, M.B. (2004) Reproductive biology of two sympatric skates in the southwest Atlantic: *Psammobatis rudis* and *Psammobatis normani*. *Journal of Fish Biology*, **65**, 559–73.

Makeyeva, A.P. (1992) *Embryology of Fishes*. Moscow University, Moscow (in Russian).

Makeyeva, A.P. (2002) Embryonic development of *Micropercops cinctus* (Eleotridae). *Journal of Ichthyology*, **42**, 627–38.

Makeyeva, A.P., Veligorskaya, E. & Emel'yanova, N.G. (1988) Ultrastructure of the primordial germ cells and initial development of the gonads in carp. *Biologicheskie Nauki*, **10**, 37–43 (in Russian).

Makhotin, V., Solemdal, P., Korsbrekke, K. & Salthaug, A. (2001) Types and frequency of malformations and mortality in eggs of Arcto-Norwegian cod: a field study. ICES CM 2001, No. 12, pp. 1–17.

Mangor-Jensen, A., Holm, J.C., Rosenlund, G., Lie, O. & Sandnes, K. (1994) Effects of dietary vitamin C on maturation and egg quality of cod *Gadus morhua* L. *Journal of the World Aquaculture Society*, **25**, 30–40.

Marteinsdottir, G. & Steinarsson, A. (1998) Maternal influence on the size and viability of Iceland cod *Gadus morhua* eggs and larvae. *Journal of Fish Biology*, **52**, 1241–58.

Mattei, X., Siau, Y., Thiaw, T. & Thiam, D. (1993) Peculiarities in the organization of testis of *Ophidion* sp. (Pisces, Teleostei). Evidence for two types of spermatogenesis in teleost fish. *Journal of Fish Biology*, **43**, 931–7.

McClenachan, L., Cooper, A.B., Carpenter, K.E. & Dulvy, N.K. (2012) Extinction risk and bottlenecks in the conservation of charismatic marine species. *Conservation Letters*, **5** (1), 73–80.

Meshkov, M.M. & Lebedeva, O.A. (1983) Egg structure in Teleostei. In: *Ecologo-morphological Investigations of Early Ontogeny in Vertebrates*. pp. 61–70. Nauka, Moscow (in Russian).

Methven, D.A. & Brown, J.A. (1991) Time of hatching affects development, size, yolk volume, and mortality of newly hatched *Macrozoarces americanus* (Pisces: Zoarcidae). *Canadian Journal of Zoology*, **69**, 2161–7.

Micale, V., Maricchiolo, G. & Genovese, L. (2002) The reproductive biology of blackspot sea bream *Pagellus bogaraveo* in captivity. 1. Gonadal development, maturation and hermaphroditism. *Journal of Applied Ichthyology*, **18**, 172–6.

Mikodina, E.V., Sedova, M.A. & Smirnov, A.A. (2005) On abnormal gonads of the Gizhiga–Kamchatka population of the herring *Clupea pallasi* (Clupeidae). *Journal of Ichthyology*, **45** (2), 191–9.

Miller, T.J., Crowder, L.B., Rice, J.A. & Marschall, E.A. (1988) Larval size and recruitment mechanisms in fishes: toward a conceptual framework. *Canadian Journal of Fisheries and Aquatic Sciences*, **45**, 1657–70.

Moiseenko, T.I. & Lukin, A.A. (1999) Fish pathology in polluted water bodies of the Subarctic and its diagnosis. *Journal of Ichthyology*, **39**, 532–43.

Moiseeva, E.B. (1983) Development of the sex glands in round goby *Neogobius melanostomus* (Pallas) (Gobiidae) in the embryonic period. *Voprosy Ikhtiologii*, **23**, 786–96 (in Russian).

Moiseeva, E.B., Mogil'naya, N.A. & Pisarevskaya, I.I. (1988) Development of sex glands in Black Sea golden mullet *Liza aurata* (Risso) at early steps of life cycle. *Voprosy Ikhtiologii*, **28**, 764–72 (in Russian).

Monastyrskii, G.N. (1953) On the types of spawning populations in fishes. In: E.N. Pavlovskii (Ed.) *Ocherki po Obshchim Voprosam Ikhtiologii*. pp. 295–305. AN SSSR, Moscow (in Russian).

Muñoz, M. (2010) Reproduction in Scorpaeniformes. In: K.S. Cole (Ed.) *Reproduction and Sexuality in Marine Fishes: Patterns and Processes*. pp. 65–87. University of California Press, Berkeley, Los Angeles, California.

Muñoz, M., Dimitriadis, C., Casadevall, M., Vila, S., Delgado, E., Lloret, J. & Saborido-Rey, F. (2010) Female reproductive biology of the bluemouth *Helicolenus dactylopterus dactylopterus*: spawning and fecundity. *Journal of Fish Biology*, **77**, 2423–42.

Murua, H. Kraus, G, Saborido-Rey, F, Witthames, P.R., Thorsen, A. & Junquera, S. (2003) Procedures to estimate fecundity of marine fish species in relation to their reproductive strategy. *Journal of Northwest Atlantic Fisheries Science*, **33**, 33–54.

Murua, H. & Saborido-Rey, F. (2003) Female reproductive strategies of marine fish species of the North Atlantic. *Journal of Northwest Atlantic Fisheries Science*, **33**, 23–31.

Nikolsky, G.V. (1962) Concerning the causes of fluctuations in the abundance of fishes. *Voprosy Ikhtiologii*, **1** (4), 659–65. Translation series No. 389, Fisheries Research Board of Canada.

Nikolsky, G.V. (1974) *Theory of the Fish Stock Dynamics as the Biological Base for the Rational Exploitation and Management of Fishery Resources*. Second Edition. Pishchevaya Promyshlennost', Moscow (in Russian).

Nikolsky, G., Bogdanov, A. & Lapin, Y. (1973) On fecundity as a regulatory mechanism in fish population dynamics. *Rapports et Procés-verbaux des Réunions, Conseil International pour L'Exploration de la Mer*, **164**, 174–7.

North, A.W. (2000) Early life history strategies of notothenioids at South Georgia. *Journal of Fish Biology*, **58**, 496–505.

Novikov, G.G. (2000) *Growth and Energetics of Teleost Fishes in Early Ontogeny*. Editorial URSS, Moscow (in Russian).

Oven, L.S. (2004) *Specificity of Development of Sex Cells in Marine Fishes During Spawning as a Parameter of the Spawning Type and Reaction to Environment*. VNIRO, Moscow (in Russian).

Pakkasmaa, S. & Jones, M. (2002) Individual-level analysis of early life history traits in hatchery-reared lake trout. *Journal of Fish Biology*, **60**, 218–25.

Parenti, L.R. & Grier, H.J. (2004) Evolution and phylogeny of gonad morphology in bony fishes. *Integrative and Comparative Biology*, **44**, 333–48.

Pavlov, D.A. (1994) Fertilization in the wolffish, *Anarhichas lupus*: external or internal? *Journal of Ichthyology*, **34** (1), 140–51.

Pavlov, D.A. (1999) Features of transition from larva to juvenile in fishes with different types of early ontogeny. *Environmental Biology of Fishes*, **56**, 41–52.

Pavlov, D.A. (2004) Wolf-fish (*Anarhichas* spp.) In: E. Moksness, E. Kjørsvik, & Y. Olsen (Eds) *Culture of Cold-water Marine Fish*. pp. 454–61. Blackwell Publishing, Oxford.

Pavlov, D.A. (2007) *Morphological Variation in the Early Ontogeny of Teleost Fishes*. GEOS, Moscow (in Russian).

Pavlov, D.A., Burykin, U.B. & Kublik, E.A. (1992) Embryonic and larval development of the mailed sculpin, *Triglops murrayi*. *Journal of Ichthyology*, **32** (6), 62–76.

Pavlov, D.A. & Emel'yanova, N.G. (2008) Morphological criteria of egg quality in marine fishes: activation and cleavage of eggs of *Zebrasoma scopas* (Acanthuridae). *Journal of Ichthyology*, **48**, 533–48.

Pavlov D.A. & Emel'yanova N.G. (2010) Reproductive biology of species from the family Scorpaenidae and transition from oviparity to viviparity in the southern and northern Percomorpha. In: M.C. Uribe & H.J. Grier (Eds) *Viviparous Fishes*. pp. 89–105. New Life Publishing, Homestead, USA.

Pavlov, D.A. & Emel'yanova, N.G. (2013) Transition to viviparity in the order Scorpaeniformes: brief review. *Journal of Ichthyology*, **53** (1), 52–69.

Pavlov, D., Kjørsvik, E., Refsti, T. & Andersen, Ø. (2004a) Brood stock and egg production. In: E. Moksness, E. Kjørsvik, & Y. Olsen (Eds) *Culture of Cold-water Marine Fish*. pp. 129–203. Blackwell Publishing, Oxford.

Pavlov, D.A. & Moksness, E. (1994) Reproductive biology, early ontogeny, and effect of temperature on development in wolffish: comparison with salmon. *Aquaculture International*, **2**, 133–53.

Pavlov, D.A. & Osinov, A.G. (2004) Main features of early ontogeny in salmonids (salmoniformes) and other representatives of the Protacanthopterygii in relation to the phylogeny. *Journal of Ichthyology*, **44** (4), 267–83.

Pavlov, D.A. & Osinov, A.G. (2008) Reproduction and development in the dwarf form of Arctic charr *Salvelinus alpinus* from Lake Davatchan, Transbaikalia. *Journal of Ichthyology*, **48** (1), 96–113.

Pavlov D.A., Vasil'eva, E.D. & Vasil'ev, V.P. (2004b) Embryonic and larval development of spined loaches of the genus *Cobitis* (Cobitidae, Cypriniformes): diploid bisexual species *C. melanoleuca*, triploid clonal form, and hybrid *C. melanoleuca* × *C. taenia*. *Journal of Ichthyology*, **44** (4), 284–99.

Pavlov, E.D., Mikodina, E.V., Sedova, M.A., Emel'yanova, N.G. & Markevich, G.N. (2010) The state of gonads of resident sockeye salmon *Oncorhynchus nerka* from the Tolmachev Reservoir. *Journal of Ichthyology*, **50** (4), 321–29.

Peňáz, M. (2001) A general framework of fish ontogeny: a review of the ongoing debate. *Folia Zoologica*, **50** (4), 241–56.

Persov, G.M. (1975) *Differentiation of Sex in Fishes*. Leningrad University, Leningrad (in Russian).

Pianka, E.R. (1970) On *r* and *K* selection. *American Naturalist*, **104**, 592–7.

Pitman, L.R., Haddy, J.A. & Kloser, R.J. (2013) Fishing and fecundity: The impact of exploitation on the reproductive potential of a deep-water fish, orange roughy (*Hoplostethus atlanticus*). *Fisheries Research*, **147**, 312–19.

Platten, J.R., Tibbetts, I.R. & Sheaves, M.J. (2002) The influence of increased line-fishing mortality on the sex ratio and age of sex reversal of the venus tusk fish. *Journal of Fish Biology*, **60**, 301–18.

Pörtner, H.O. & Peck, M.A. (2010) Climate change effects on fishes and fisheries: towards a cause-and-effect understanding. *Journal of Fish Biology*, **77**, 1745–79.

Radtke, R.L., Kinzie, R.A. & Shafer, D.J. (2001) Temporal and spatial variation in length of larval life and size at settlement of the Hawaiian amphidromous goby *Lentipes concolor*. *Journal of Fish Biology*, **59**, 928–38.

Rass, T.S. (1948) On the periods of life and features of development and growth in fishes. *Izvestiya Akademii Nauk SSSR, Seriya. Biologiya*, **3**, 292–305 (in Russian).

Rass, T.S. (1977) Geographical features of reproduction and development of fishes in different climatic zones. *Trudy Instituta Okeanologii*, **109**, 7–41 (in Russian).

Reinboth, R. (1980) Can sex inversion be environmentally induced? *Biology of Reproduction*, **22**, 49–59.

Rideout, R.M., Trippel, E.A. & Litvak, M.K. (2004a) Predicting haddock embryo viability based on early cleavage patterns. *Aquaculture*, **230** (1/4), 215–28.

Rideout, R.M., Trippel, E.A. & Litvak, M.K. (2004b) Paternal effects on haddock early life history traits. *Journal of Fish Biology*, **64**, 695–701.

Ringuette, M., Castonguay, M., Runge, J.A. & Grogoire, F. (2002) Atlantic mackerel (*Scomber scombrus*) recruitment fluctuations in relation to copepod production and juvenile growth. *Canadian Journal of Fisheries and Aquatic Sciences*, **59**, 646–56.

Rothschild, B.J. (1995) Fish stock fluctuations as indicators of multidecadal fluctuations in the biological productivity of the ocean. In: R.J. Beamish (Ed.) *Climate Change and Northern Fish Populations*. pp. 201–9. Special Publication of the Canadian Journal of Fisheries and Aquatic Sciences 121.

Rothschild, B.J. (2000) 'Fish stocks and recruitment': the past thirty years. *ICES Journal of Marine Science*, **57**, 191–201.

Rowe, S., Hutchings, J.A., Bekkevold, D. & Rakitin, A. (2004) Depensation, probability of fertilization and the mating system of Atlantic cod (*Gadus morhua* L.). *ICES Journal of Marine Science*, **61**, 1144–50.

Ruzen-Range, E. (1980) *Spermatogenesis in Animals*. Mir, Moscow (in Russian).

Sætre, R., Toresen, R. & Anker-Nilssen, T. (2002) Factors affecting the recruitment variability of the Norwegian spring-spawning herring (*Clupea harengus* L.). *ICES Journal of Marine Science*, **59**, 725–36.

Saillant, E., Chatain, B., Fostier, A., Przybyla, C. & Fauvel, C. (2001) Parental influence on early development in the European sea bass. *Journal of Fish Biology*, **58**, 1585–600.

Sargent, R.C., Taylor, P.D. & Gross, M.R. (1987) Parental care and the evolution of egg size in fishes. *American Naturalist*, **129** (1), 32–46.

Sars, G.O. (1869) Report of practical and scientific investigations of the cod fisheries near Lofoten Islands, made during the years 1864–1869. Translated from 'Indberetninger til Departementet for det Indre fra Cand. G.O. Sars om de af ham i aarene 1864-69 anstillede praktisk-videnskabelige undersøgelser angaaende torskefiskeriet i Lofoten', Christiania 1869. Translated by H. Jacobsen in *United States Fish Commission Annual Reports* (1877), **Pt IV**, pp. 565–705.

Savvaitova, K.A., Chebotareva, Yu.V., Pichugin, M.Yu. & Maksimov, V.A. (1995) Anomalies in the structure of fish as the indicators of environment. *Voprosy Ikhtiologii*, **35**, 182–8 (in Russian).

Schaffer, W.M. (1974) Selection for optimal life histories: the effects of age structure. *Ecology*, **55**, 291–303.

Schultz, R.J. (1977) Evolution and ecology of unisexual fishes. *Evolutionary Biology*, **10**, 277–331.

Schulz, R.W., França, L.R., Lareyre, J.J., LeGac, F., Chiarini-Garcia, H., Nobrega, R.H. & Miura, T. (2010) Spermatogenesis in fish. *General and Comparative Endocrinology*, **165**, 390–411.

Selyukov, A.G. (2012) Morphofunctional transformations in fishes of the middle and lower Ob' basin under increasing anthropogenic influence. *Journal of Ichthyology*, **52** (8), 547–65.

Serebryakov, V.P. (1989) Common reproductive zones and drift pattern types of abundant food-fish populations in the North Atlantic. *ICES CM*, **6**, 1–12.

Serebryakov, V.P. (1990) Population fecundity and reproductive capacity of some food fish in relation to year-class-strength fluctuations. *Journal de Conseil International pour L'Exploration de la Mer*, **47**, 267–72.

Sharova, Yu.N., Kaufman, E.S. & Lukin, A.A. (2003) *Oogenesis in Fishes of the European North of Russia at Technological Pollution*. Karel'skii Nauchnyi Tsentr RAN, Petrozavodsk (in Russian).

Shatunovskii, M.I., Akimova, N.V. & Ruban, G.I. (1996) Response of reproductive systems of fishes to anthropogenic impacts. *Journal of Ichthyology*, **36**, 247–56.

Shields, R J., Brown, N P. & Bromage, N R. (1997) Blastomere morphology as a predictive measure of fish egg viability. *Aquaculture*, **155**, 1–12.

Shilova, S.A. & Shatunovskii, M.I. (2005) Ecophysiological indicators of the state of animal populations exposed to damaging factors. *Russian Journal of Ecology*, **36** (1), 27–32.

Soin, S.G. (1968) *Adaptive Features of Development in Fishes*. Moscow University, Moscow. (in Russian)

Soin, S.G. (1981) New classification of the structure of mature fish eggs based on the relationship between yolk and ooplasm. *Ontogenez*, **12** (1), 21–6 (in Russian).

Solemdal, P. (1997) Maternal effects – a link between the past and the future. *Journal of Sea Research*, **37**, 213–27.

Solemdal, P., Kjesbu, O.S. & Fonn, M. (1995) Egg mortality in recruit- and repeat-spawning cod: an experimental study. *ICES CM*, **35**, 1–10.

Solemdal, P., Makhotin, V. & Fonn, M. (1998) Longterm studies on spawning in Arcto-Norwegian cod – mortality pattern of eggs and early larvae. *ICES CM*, **8**, 1–24.

Somarakis, S. & Nikolioudakis, N. (2010) What makes a late anchovy larva? The development of the caudal fin seen as a milestone in fish ontogeny. *Journal of Plankton Research*, **32** (3), 317–26.

Sponaugle, S. & Pinkard, D.R. (2004) Impact of variable pelagic environments on natural larval growth and recruitment of the reef fish *Thalassoma bifasciatum*. *Journal of Fish Biology*, **64**, 34–54.

St Mary, C.M. (1997) Sequential patterns of sex allocation in simultaneous hermaphrodites: do we need models that specifically incorporate this complexity? *American Naturalist*, **150**, 73–97.

Svetlov, P.G. (1960) Theory of critical developmental periods and its importance for understanding the principles of the effect of environment on the ontogeny. In: *Voprosy Tsitologii i Obshchei Fiziologii*. Acad. Nauk SSSR, Moscow, Leningrad (in Russian).

Szabo, T., Szabo, R., Urbanyi, B. & Horvath, L. (2000) Review of the results of common carp (*Cyprinus carpio*) breeding at the large-scale hatchery. *Reproduction in Domestic Animals*, **35**, 89–94.

Thorisson, K. (1994) Is metamorphosis a critical interval in the early life of marine fishes? *Environmental Biology of Fishes*, **40**, 23–36.

Trippel, E.A. (1998) Egg size and viability and seasonal offspring production of young Atlantic cod. *Transactions of the American Fisheries Society*, **127**, 339–59.

Trippel, E.A. (2003) Estimation of male reproductive success of marine fishes. *Journal of Northwest Atlantic Fisheries Science*, **33**, 81–113.

Uiblein, F. (2007) Goatfishes (Mullidae) as indicators in tropical and temperate coastal habitat monitoring and management. *Marine Biology Research*, **3**, 275–88.

Vasil'ev, V.P. (1985) *Evolutionary Karyology of Fishes*. Nauka, Moscow (in Russian).

Vasnetsov, V.V. (1953) Steps in development of bony fishes. In: E.N. Pavloviskii (Ed.) *Ocherki po Obshchim Voprosam Ikhtiologii*. pp. 207–17. Akademiya Nauk SSSR, Moscow (in Russian).

Verigin, B.V., Belova, N.V., Emel'yanova, N.G., Makeyeva, A.P., Vybornov, A.A. & Ryabov, I.N. (1996) Radiobiologic analysis of silver carp (*Hypophthalmichthys molitrix*) from the cooling pond of Chernobyl Nuclear Power Station in the post-disaster period. 3. Results of artificial reproduction of irradiated fish. *Journal of Ichthyology*, **36**, 257–68.

Vladimirov, V.I. (1975) Critical periods in the development of fishes. *Voprosy Ikhtiologii*, **15**, 955–75 (in Russian).

Voskoboinikova, O.S. & Pavlov, D.A. (2006) Larval development of the Amur sleeper *Perccottus glenii* (Perciformes, Gobioidei, Odontobutidae) and the origin of fish of the suborder Gobioidei. *Journal of Ichthyology*, **46** (9), 788–802.

Webb. P.W. & Weihs, D. (1986) Functional locomotor morphology of early life history stages of fishes. *Transactions of the American Fisheries Society*, **115**, 115–27.
Winemiller, K.O. (1992) Life-history strategies and the effectiveness of sexual selection. *Oikos*, **63**, 318–27.
Winemiller, K.O. & Rose, K.A. (1992) Patterns of life-history diversification in North American fishes: implications for population regulation. *Canadian Journal of Fisheries and Aquatic Sciences*, **49**, 2196–218.
Winemiller, K.O. & Rose, K.A. (1993) Why do most fish produce so many tiny offspring? *American Naturalist*, **142**, 584–603.
Witthames, P.R., Thorsen, A., Murua, H., Saborido-Rey, F., Greenwood, L.N., Dominguez, R., Korta, M. & Kjesbu, O.S. (2009) Advances in methods for determining fecundity: application of the new methods to some marine fishes. *Fishery Bulletin*, **107** (2), 148–64.
Wootton, R.J. (1990) *Ecology of Teleost Fishes*. Chapman & Hall, New York.
Wourms, J.P. (1991) Reproduction and development of *Sebastes* in the context of the evolution of piscine viviparity. *Environmental Biology of Fishes*, **30**, 111–26.
Yanes-Roca, C., Rhody, N., Nystrom, M. & Main, K.L. (2009) Effects of fatty acid composition and spawning season patterns on egg quality and larval survival in common snook (*Centropomus undecimalis*). *Aquaculture*, **287**, 335–40.
Yao, Z. & Crim, L.W. (1995) Spawning of ocean pout (*Macrozoarces americanus* L.): evidence in favour of internal fertilization of eggs. *Aquaculture*, **130**, 361–72.
Yoon, C., Kawakami, K. & Hopkins, N. (1997) Zebrafish vasa homologue RNA is localized to the cleavage planes of 2- and 4-cell-stage embryos and is expressed in the primordial germ cells. *Development*, **124**, 3157–65.
Zelenkov, V.M. (1990) Early gametogenesis and sex differentiation in the White Sea herring, *Clupea pallasi marisalbi*. *Voprosy Ikhtiologii*, **30**, 957–62 (in Russian).
Zhao, Y., Chen, Y. & Brown, J.A. (2001) Impacts of egg and larval size on survival and growth of Atlantic cod under different feeding conditions. *Journal of Fish Biology*, **59**, 569–81.

Chapter 3
Recruitment Variability

Edward D. Houde

3.1 Introduction

3.1.1 Recruitment variability

Abundances of fish populations can vary over time by orders of magnitude, and five-to-10-fold variability is usual. The variability is recorded on time scales ranging from millennial to interannual. Interannual scales of variability in abundance for short-lived fishes often are nearly as large as variability seen over decades and centuries, pointing to variability in reproductive success and recruitment variability as primary agents contributing to fluctuations in abundance. In monitored fish stocks, 10-fold interannual variability in recruitment is common, though larger fluctuations occur (Cushing 1973, Pope & Macer 1996). Interannual fluctuations often appear random or even chaotic, and may be superimposed on decadal trends and periods of high or low recruitment. Recruitment levels often are poorly correlated with adult stock abundance, are vaguely related to fishing effort on adults and, in most cases, are unpredictable (Figure 3.1).

During the last three decades of the 20th century, fishery scientists commonly referred to a deficient knowledge of causes of recruitment variability as the "recruitment problem," and lamented the lack of a theoretical basis to explain it or practical means to address it (e.g., Rothschild & Rooth 1982, Houde 2008). Many believed that solving this "problem" would lift fisheries science and management to a new level in an ability to forecast variability and serve management needs. Indeed, prominent scientists during the 1970s viewed the failure to understand or predict recruitment variability as the last great problem in fisheries science (Cushing 1975). It became apparent that a myriad of causes – both physical and biological – acting on early-life stages contribute to recruitment variability (Pepin 1990, Fogarty 1993). Some scientists argued that explaining and predicting recruitment variability based on the analysis of environmental factors is futile (e.g., Walters & Collie 1988). However, compelling arguments also were advanced to promote such research and to identify, evaluate, and quantify causes (Tyler 1992). Recent advances in understanding how ocean variability, environmental factors, adult stock size, and pending climate change affect fish reproductive success have encouraged

Fish Reproductive Biology: Implications for Assessment and Management, Second Edition.
Edited by Tore Jakobsen, Michael J. Fogarty, Bernard A. Megrey and Erlend Moksness.

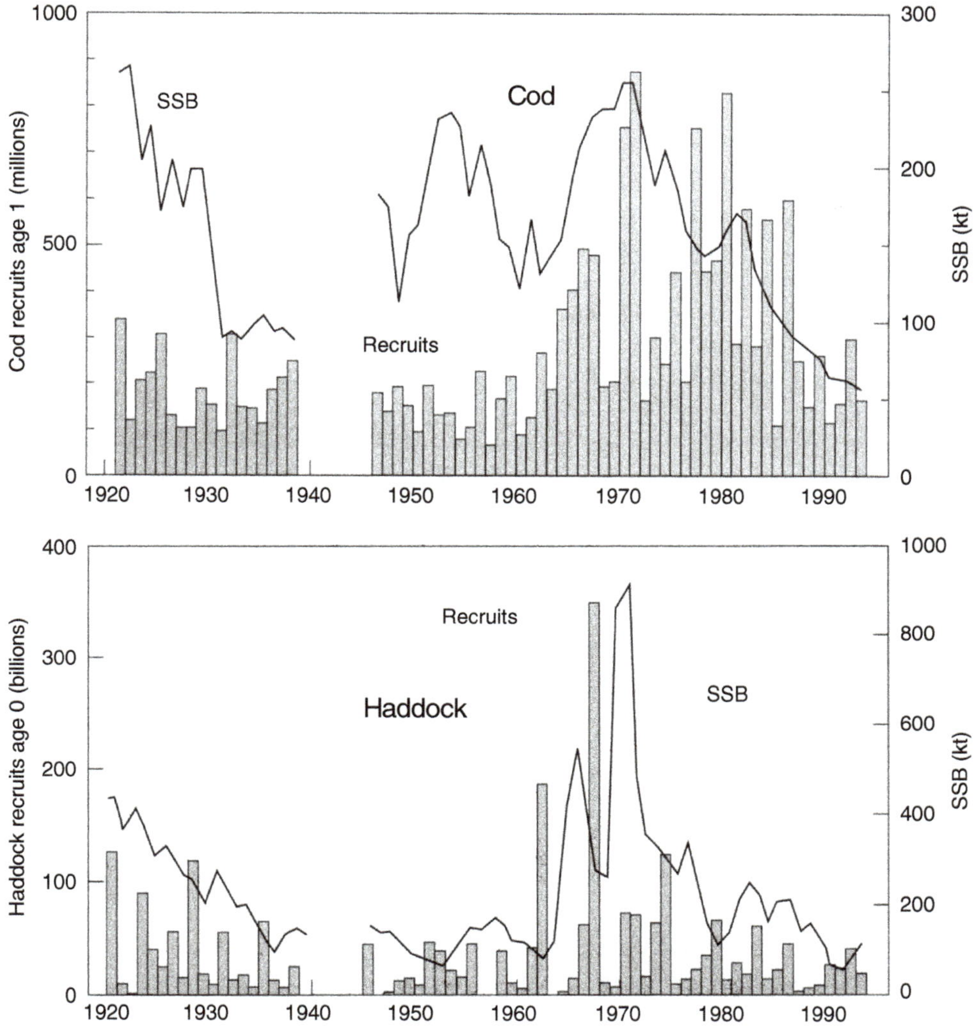

Figure 3.1 Recruitment variability. Interannual variability in recruitment levels (bars) and spawning stock biomasses of Atlantic cod *Gadus morhua* and haddock *Melanogrammus aeglefinus* in the North Sea. Reproduced from Pope & Macer (1996), with permission of Oxford University Press.

efforts to tease apart the complexities of ocean physics and biology and their roles in controlling recruitment (Houde 2008, Peck *et al.* 2012, Ottersen *et al.* 2013, Stige *et al.* 2013).

3.1.2 New technologies, analytical methods and models

Substantial advances in understanding variability have been made in the past two decades, in part because of large, interdisciplinary science programs that broadly addressed the issues of fluctuating abundances of organisms in the sea, especially recruitment variability. This "new" approach, which pooled intellectual, technical and financial resources and focused on a complex problem, enjoyed successes by addressing the "recruitment problem" across relevant time and space scales with new sampling instruments, computing power and modeling capability. Despite advances, it remains difficult to obtain accurate and precise estimates of abundance of early-life stages, their predators, and prey at the spatial scales over which they interact (Heath 1992, Pepin 2004).

Better samplers (nets, acoustics, video, optics) have evolved (Wiebe and Benfield 2003) to allow depth-discrete sampling of ichthyoplankton. These sampling systems, combined with environmental-sensing instrumentation, are now commonly utilized. Analytical tools, especially the emergence of microstructural and chemical analyses of otoliths as methods to reconstruct details of the past lives of fishes, have made it possible not only to estimate age, growth and hatch dates from daily increments, but also to determine cohort-specific survivorship in young fish (Campana 2005). Chemical records in otoliths provide insight into past environmental and growth conditions, ontogenetic migrations or dispersal, as well as providing information contributing to stock identification (Campana *et al.* 1994, 1997, Secor *et al.* 1995, Thorrold *et al.* 1997, 2001, Campana 1999, Swearer *et al.* 1999, Thorrold & Hare 2002). The ability to chemically mark otoliths of larval fishes and to conduct mark–recapture experiments on them is a powerful tool to estimate mortality and growth and to evaluate how environmental factors affect early-life dynamics (e.g., Secor *et al.* 1995, Jones *et al.* 1999). Nutritional biochemistry, especially nucleic acid (RNA/DNA) methodologies have demonstrated value in judging larval condition and growth potential (Ferron & Leggett 1994, Clemmesen 1996, Clemmesen & Doan 1996, Buckley *et al.* 1999, 2008, Calderone *et al.* 2003). Molecular and genetics tools are emerging and will improve identifications, evaluations of stock dynamics, including predation mortality (Fox *et al.* 2012) and an ability to conduct experiments at sea with fish of known stock origins.

3.1.3 Scope of the chapter

In this chapter, knowledge on factors generating variability in fish recruitments is reviewed and synthesized. The focus is on external factors, primarily trophic interactions, hydrographic factors and environmental variables, that act on early-life stages over a range of temporal and spatial scales. It is impossible to appreciate how external factors control recruitment without considering life-history traits, or characteristics of adult spawning stocks (e.g., abundance, biomass and age structure) that operate to regulate recruitment and dampen variability. These internal factors are discussed in detail in Chapters 1, 2, 8, and 9, and noted briefly here. The nature of recruitment variability, relevant hypotheses on its causes, and historical perspectives have been provided in other published literature (e.g., Leggett & Deblois 1994, Browman 1995, Ellertsen *et al.* 1995, Rothschild 2000, Cowan & Shaw 2002, Peck *et al.* 2012). Here, a comprehensive review, discussion, and some personal perspectives are provided on the present state of knowledge.

It is beyond the scope of this chapter to more than briefly note the difficulty of conducting surveys to effectively sample early-life stages in the sea (Heath 1992), but Chapter 5 addresses these issues. The insufficient sampling of early-life stages and a deficit of knowledge constrains our ability to understand or forecast variability in recruitment. Many underlying causes of variability in survival of early-life stages were identified decades ago (e.g., Sette 1943), but rarely quantified. Only in the most recent decades have sufficient data and time series of recruitments become available to undertake meta-analyses and to analyze recruitment time series with confidence (Myers 2001, Myers & Barrowman 1996, Myers *et al.* 1995a,b, 1997).

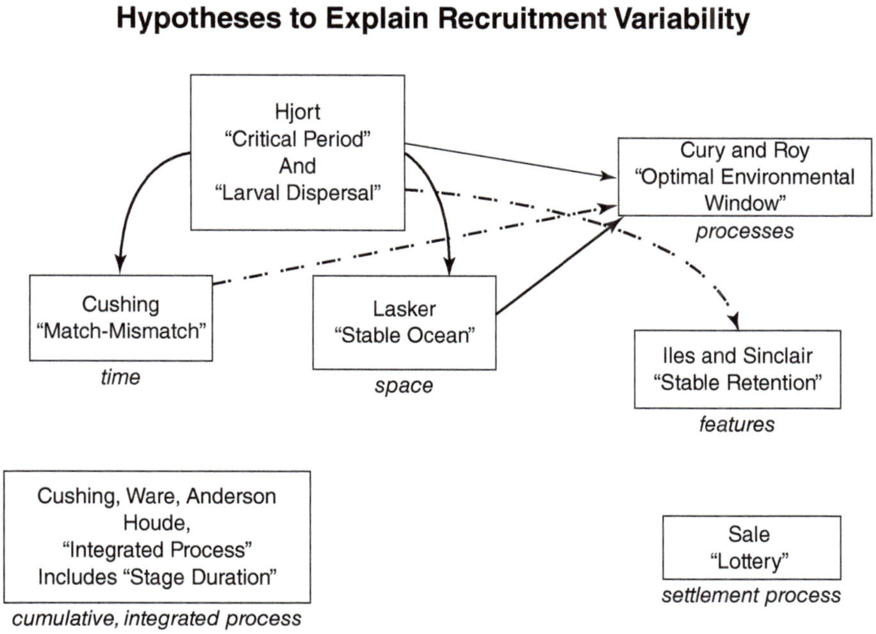

Figure 3.2 Hypotheses to explain recruitment variability in fishes. Most derive from the "*Critical Period*" and "*Aberrant Drift*" hypotheses proposed by Johan Hjort. Solid arrows indicate direct, and broken arrows indirect, derivations from Hjort's hypotheses. The arrow thicknesses indicate the strengths of the relationship. Each hypothesis is described and discussed in the text. Modified from Figure 2 of Houde (2008).

3.2 Theories and hypotheses

Several hypotheses have been proposed to explain the causes of recruitment variability (Figure 3.2). Most derive from insights and hypothesized relationships proposed long ago by Johan Hjort (Hjort 1914, 1926). The proposed mechanisms and processes focus on the early life stages when high and variable mortality rates occur.

3.2.1 Dancing in Hjort's shadow

Despite evidence that recruitment levels in fishes may be controlled, adjusted or regulated throughout the life stages leading from egg to juvenile (Sissenwine 1984, Rothschild 2000), the high mortality observed in the earliest stages (eggs and larvae) suggests that variable mortality in these stages could, for many taxa, be decisive in generating recruitment variability. Hjort (1914, 1926) proposed the enduring *Critical Period* hypothesis in which the fate of year classes hinges on the ability of first-feeding larvae (the critical stage) to find suitable prey in sufficient quantity to support foraging and growth (Figure 3.3). Failure could lead to massive mortality and order-of-magnitude variability in subsequent year-class strength. Confirming or rejecting this hypothesis was a dominant goal of recruitment research during the last half of the 20th century (Kendall & Duker 1998, Anderson 1988, Smith 1994, Cowan & Shaw 2002, Govoni 2005). Levels of available prey often were judged to limit the production of larval-stage fishes but, with notable exceptions, results often failed to support the Critical Period hypothesis,

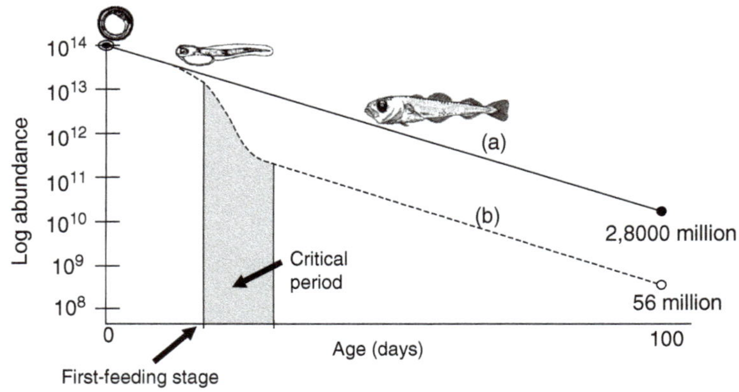

Figure 3.3 Critical period hypothesis, conceptual model. (a) No critical period; constant mortality rate from age 0 to 100 days; (b) Critical period; >90% mortality occurs from starvation during the days at which first-feeding must be established (gray shading). Modified from Figure 3.4 of Houde (2002).

or were equivocal at best (May 1974, Anderson 1988, Leggett & Deblois 1994, Cowan & Shaw 2002), leading many to believe that this hypothesis was by itself insufficient to explain recruitment variability (Houde 2008).

Although less prominent than his Critical Period hypothesis, Hjort (1914, 1926) offered a second hypothesis, proposing that year-class success was controlled by favorable or unfavorable transport of eggs and larvae. While Hjort's second hypothesis received less attention than the Critical Period hypothesis, it gained strong support in recent years as a mechanism that generates interannual variability in recruitment success (Cowan & Shaw 2002). Favorable currents combined with appropriate behavior by larval-stage fishes can transport or retain early-life stages in nursery areas where survival and recruitment are enhanced. While the physics of transport or retention is important to insure connectivity, it is the coupling of biological and

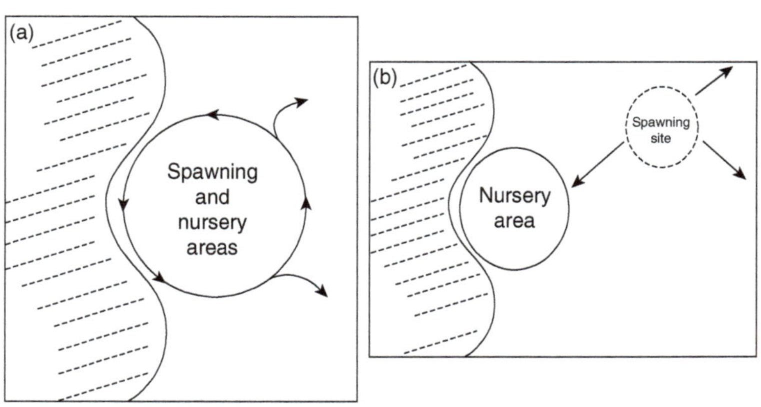

Figure 3.4 Transport, dispersal, and retention processes of early-life stages; two conceptual models. (a) Spawning and nursery areas contiguous. (b) Spawning area isolated from nursery area. Advective and retention pathways to nurseries and dispersal/advective losses are indicated.

physical mechanisms that plays a key role in determining recruitment success (Paris & Cowen 2004, Cowen & Sponaugle 2009, North *et al.* 2009).

3.2.2 *Match and mismatch*

When Cushing (1974, 1975, 1990) extended Hjort's Critical Period hypothesis, his *Match–Mismatch* hypothesis proposed that the timing of larval fish production is critically dependent on schedules and levels of zooplankton production – that is, prey for first-feeding larvae. Cushing noted that spawning by fishes in temperate seas was confined to a narrow time window that often was only a few days or weeks in duration. Spawning is controlled by photoperiod, with proximal cues dependent on temperature, but timing and levels of spring plankton blooms are more temporally variable. Substantial overlap between fish egg and larval production and zooplankton blooms – a *match* – is hypothesized to lead to successful recruitment, while the converse – a *mismatch* – results in year-class failure. There is considerable evidence to support this hypothesis (Leggett & Deblois 1994, Cowan & Shaw 2002), especially in high-latitude seas and especially when both temporal and spatial overlaps occur. Peck *et al.* (2012) have broadly evaluated and synthesized match–mismatch dynamics in the early lives of marine fishes, providing many examples and summarized data. In one example, Atlantic cod *Gadus morhua* recruitments in the Norwegian Arctic were shown to be heavily dependent on the timing and intensity of zooplankton blooms, which vary interannually by several weeks in response to variability in sea temperatures (Ellertsen *et al.* 1989). In the North Sea, variable cod recruitments appear to be tuned to plankton blooms and match–mismatch dynamics (Beaugrand *et al.* 2003).

3.2.3 *Stable ocean*

A substantial and important refinement of Hjort's Critical Period hypothesis was proposed by Lasker (1975, 1978) for upwelling ecosystems. The *Stable Ocean* hypothesis proposes that calm periods between upwelling events, especially in eastern boundary currents, lead to a stratification of the water column and the aggregation of plankton organisms at discontinuities (e.g., pycnocline) where first-feeding fish larvae are able to forage efficiently on high concentrations of planktonic prey. In support of the hypothesis, feeding rates of first-feeding larvae of northern anchovy *Engraulis mordax* were facilitated under stratified conditions, and survival rates were higher when the frequency of calm, low-wind periods was increased in the California Current. However, annual recruitment levels were not related to numbers of calm periods (Peterman & Bradford 1987), indicating that factors acting later in life, on late-larval or juvenile stages, can be decisive.

A significant extension of the Stable Ocean hypothesis was proposed by Cury and Roy (Cury & Roy 1989, Roy *et al.* 1992), termed the *Optimum Environmental Window* hypothesis. This hypothesis, which is supported for anchovies and sardines in upwelling ecosystems, proposes that highest larval-stage survival and, by implication, subsequent recruitment, occurs when winds of intermediate intensity predominate that generate intermediate levels of microturbulence, thereby enhancing the encounter rates between fish larvae and prey and elevating feeding success while limiting offshore dispersal losses of larvae.

3.2.4 *Larval drift and retention*

Hypotheses specifically addressing larval transport and retention mechanisms have gained prominence during recent decades. Their historical roots derive from Hjort's second hypothesis (see above) and also from the Harden-Jones (1968) conceptual model, the "Triangle of Migration", in which successful denatant drift of larvae from spawning grounds to nurseries is hypothesized to lead to high survival and recruitment (see Section 3.3.7.1). The denatant drift is dependent on the behavior of adults in selecting spawning sites and on prevailing circulation patterns. The spawning pattern, larval drift and location of juvenile nursery areas of plaice *Pleuronectes platessa* in the Southern Bight of the North Sea present a particularly supportive example of the "Triangle" hypothesis (Cushing 1975). Whilst there are numerous variations and modifications of the "Triangle" hypothesis, all of them presume that adults spawn in well-defined areas of hydrographic containment that, on average, either retain eggs and larvae, thus minimizing dispersal, or are situated to ensure a favorable transport of eggs and larvae from spawning grounds to nurseries. Year-class failures are hypothesized to occur when there are significant deviations from favorable circulation patterns (Figure 3.4).

The *Member-Vagrant* or *Larval Retention Area* hypothesis of Iles and Sinclair (Iles & Sinclair 1982, Sinclair & Iles 1985, Sinclair 1988) proposes that recruitment success, and broader implications for stock integrity, rest primarily on the reliability of retention and the physical attributes of spawning areas that promote the containment of eggs and larvae. Egg–larval retention areas that define stocks can vary in scale from a few square kilometers (e.g., some Atlantic herring *Clupea harengus* stocks) to large portions of ocean basins in stock-poor species (e.g., eels *Anguilla anguilla* and *A. rostrata* in the Atlantic Ocean). The most forceful arguments supporting this hypothesis are derived from numerous Atlantic herring stocks, in which spawning at well-defined sites occurs by some stocks in each month of the year. Spawning by the herring is site-specific and compliant with physical retention processes, but often not synchronous with plankton bloom dynamics or periodicity – that is, mechanisms supporting the Match–Mismatch hypothesis.

The Larval Retention and Match-Mismatch hypotheses often have been cast as contrasting and, arguably, incompatible alternative mechanisms governing recruitment. Yet, in many instances processes supportive of both hypotheses appear to act simultaneously. For example, the spawning of Atlantic cod and haddock *Melanogrammus aeglefinus* on Georges Bank has strong seasonality that is tuned to spring bloom dynamics; in addition, their pelagic eggs are spawned in waters entrained in a clockwise gyre on the Bank. The gyre serves to retain eggs and larvae in the cross-frontal and gyral circulation over well-mixed, shoal areas on the Bank where juveniles can settle and recruit (Lough *et al.* 1996, 2006, Werner *et al.* 1996, Page *et al.* 1999, Lough & Manning 2001). Haddock reproductive success and recruitment on the Scotian Shelf apparently also benefit from both retentive, although leaky, circulation features and spawning tuned to winter–spring bloom dynamics (Campana *et al.* 1989, Hannah *et al.* 2001, Platt *et al.* 2003, Head *et al.* 2005).

3.2.4.1 *Self-recruitment*

It is now clear that for many fishes (and invertebrates) recruitment depends heavily on local retention processes that minimize dispersal. Self-recruitment refers to settlement and recruitment within or near the spawning site (see Section 3.3.7.2). Most support for the self-recruitment hypothesis is derived from research on marine fishes recruiting to island ecosystems where a substantial fraction of recruits (in some cases 10 to >50%) are derived from local spawning (Cowen *et al.* 2006).

3.2.5 Stage duration and stage-specific dynamics

Fast growth, supported by good nutrition and feeding conditions, in theory acts to counter high predation mortality by reducing the duration of early-life stages (Figure 3.5). This integrated process, operating across early-life stages, was recognized for its potential to control recruitment and its variability (Houde 1987, Anderson 1988, Beyer 1989). Houde (1997b) considered the process and demonstrated by example how variability in growth rates and size- or stage-specific survival generate variability in recruitments.

3.2.5.1 An integrated process

Mortality rates of marine fish larvae are high, often exceeding 10% d^{-1}, and variable (Houde 1989a, 1994), and the cumulative mortality in pre-recruit stages is, accordingly, both high and very variable (Fogarty 1993). Growth rates also show substantial variability (Houde 1989a, 1996, 1997b). Ware (1975) recognized the potential for the control of recruitments via growth-rate variability, and Houde (1987) and Anderson (1988) each elaborated on the mechanism, hypothesizing that good feeding conditions (types and quantity of prey) and favorable environmental conditions (e.g., temperature) can significantly reduce larval stage durations when mortality rates are highest. When considering the causes of variability in recruitment, Fogarty

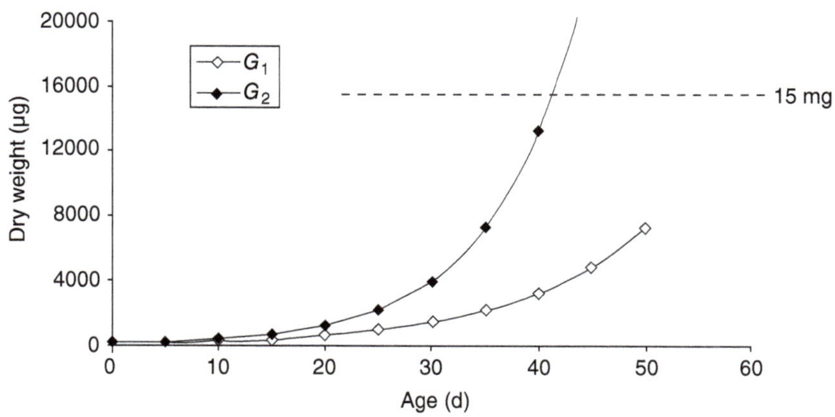

$G_1 = 0.08\ d^{-1}$ $G_2 = 0.12\ d^{-1}$ $W_{met} = 15$ mg $M = 0.10\ d^{-1}$

$W_{met(1)} = 59$ d $W_{met(2)} = 41$ d $S_{met(2)} / S_{met(1)} = 6.2$

Figure 3.5 Effects of growth rate on stage duration and stage-specific survival. Potential differences in recruitment attributed to growth-rate variability during early life. Two weight-specific growth rates are compared at a constant mortality rate. In this hypothetical example, stage duration differs by 18 days as a consequence of the growth-rate difference and there is about a sixfold difference in survival at 15 mg.

(1993) concluded that variability increases under high levels of egg production, a long pre-recruit stage duration and highly variable mortality rates, but that recruitment variability tends to be lower in stocks whose early life stages experience high mortality rates. Houde (1987) noted the wide variability in growth rates of marine fish larvae in laboratory experiments, where weight-specific growth rates of half to twofold the mean are commonly observed, and provided examples of potential recruitment variability attributable to growth-rate variability and its influence on larval stage duration.

Variability in growth rates and larval stage durations may be an indicator of whether primary control over recruitment level is exercised at the larval or juvenile stage. The significance to recruitment of variable growth and mortality during the pre-recruit period can shift or alternate among life stages (eggs, larvae, juveniles) (Rothschild 1986, Anderson 1988). Taxa with short larval stage durations, such as many freshwater fishes and some tropical reef fishes, have a relatively high probability of having the recruitment level set in the juvenile stage, whereas taxa with long larval durations have a relatively high probability of having recruitment set during the larval stage (Houde 1994).

Cushing (1975) believed that the combined dynamics of growth and mortality in early life portrayed an integrated "single process", representing a stock-level response of young fish to competition and predation in a variable environment (see Section 3.4.3.2). To address recruitment variability from this perspective is appealing because it integrates the effects of two strongly size-specific processes (i.e., mortality and growth) that control recruitment. Houde (1989b, 1996, 1997a,b) analyzed these relationships and demonstrated the major consequences to recruitment level of rather small variability in either mortality or growth. Subtle and difficult-to-detect (at least to infer statistical significance) variability in weight-specific growth or daily mortality easily generates order-of-magnitude variability in recruitment, without invoking episodic or dramatic mortality events (Houde 1989b). Beyer (1989) conducted an in-depth analysis of size-specific dynamics of early-life stages, providing mathematical tools and models to evaluate growth and mortality, and their implications for recruitment variability.

3.2.6 *Pelagic versus post-settlement processes*

Most hypotheses addressing recruitment variability emphasize dynamics and energetics processes in the primarily pelagic egg and larval stages of marine fishes. However, the lengthy late larval–juvenile stage may, in many cases, be the stage at which recruitment levels are fixed (see Section 3.5, "Control versus Regulation"). Sissenwine (1984) demonstrated how recruitment variability can be generated in the long juvenile stage, especially for slow-growing fishes with extended juvenile stage durations. The long juvenile stage (from >100 days to years), in which cumulative mortality is both high and variable, provides a broad scope for control and regulation of recruitment.

In the pelagic, pre-settlement stage, larval mortality rates generally are high and can coarsely control recruitment levels. However, it is variability in stage-specific mortality rates, and not the level of mortality, that causes fluctuations in recruitment (Fogarty *et al.* 1991, Houde 1997b). A long, post-settlement, juvenile stage with low but relatively variable mortality rates will generate high and variable cumulative mortality, and variable recruitments (Zijlstra *et al.* 1982, Jones 1991).

	Post-recruitment (PR) competition: Intense	Post-recruitment (PR) competition: Weak
Recruitment modified by PR processes	1. Competition model	2. Predation disturbance models
Recruitment not modified by PR processes	3. Lottery model	4. Recruitment limitation model

Figure 3.6 Four models (hypotheses) that make predictions about factors limiting abundance of coral reef fishes. Here, post-recruitment (PR) refers to juvenile-stage processes after settlement onto the reef. Resource limitation, competition, and predation are major factors operating on the "post-recruitment" stages (models 1 and 2). Larval-stage processes are mostly responsible for recruitment variability in model 4; processes and circumstances at the time of settlement are important in model 3. Modified from Figure 1 of Jones (1991).

Recruitment[1] in demersal fishes from coral-reef ecosystems could be controlled in either the pelagic larval stage (pre-settlement or at time of settlement) or during the days and weeks following settlement of late-stage larvae/early juveniles onto reefs (Jones 1991, Sale 1991). Based on extensive observations and field experiments, divergent hypotheses have evolved (Figure 3.6), and debates conducted during the 1980s and 1990s often centered on whether processes during the pelagic (earliest life stages, referred to as "recruitment limited"), settlement, or post-settlement (juvenile stage; referred to as "post-recruitment" in the coral reef literature) stages were most important (Jones 1991).

Sale (1977, 1978, 1991) proposed that major control over recruitment variability occurred at the time of settlement, and that a *Lottery* hypothesis could explain the mechanism. It was hypothesized that the survival of larvae settling from the plankton is a stochastic process determined at the time of settlement by variable numbers of arriving settlers, interspecific competition among potential settlers, amounts of settlement habitat available, and whether the settling habitat is already occupied. This hypothesis and related alternatives proposing strong regulation via predation and density-dependent, post-settlement processes found support. However, in many cases processes operating on larvae during the pelagic, pre-settlement stage were demonstrated to be dominant (Doherty & Williams 1988, Jones 1991, Doherty & Fowler 1994, Doherty 2002). Recruitment levels and abundances of fishes on reefs have been viewed by many as being governed by equilibrium dynamics, with carrying capacity and post-settlement processes dictating recruitment levels. However, this view has largely been replaced as accumulated evidence has indicated that much of the variability is a consequence of processes

[1] "Recruitment" in much of the literature on coral reef fishes refers to the numbers of settlers establishing themselves on the reef, rather than to the number of individuals that survive to enter fishery at catchable size.

operating during the pelagic, pre-settlement, larval stage (Doherty & Fowler 1994, Jones & McCormick 2002).

With hindsight, it now seems improbable that pelagic, settlement, or post-settlement stages can be categorically declared as the stage at which recruitment level and variability are fixed. Jones (1991) essentially reached this conclusion years ago. Circumstances at the time of settlement are key for demersal fishes, and the settlement process can result in high and variable recruit abundances, depending on the availability of space and pressures from competition, and predation (Hixon & Carr 1997). As evidence has accumulated, it is apparent and not surprising that processes governing recruitment variability on tropical reefs are similar to those in temperate marine ecosystems.

3.3 Physics and hydrography

Circulation patterns and physical features define conditions and set boundaries for processes that affect the survival and recruitment of young fish. A classification of such processes includes: (1) hydrographic properties, especially temperature, salinity and dissolved oxygen; (2) circulation, currents and related processes associated with transport and retention; and (3) physical features associated with the properties and processes.

The critical spatial scales of processes and features controlling early-life dynamics and recruitment differ among taxa with diverse life histories. These spatial scales can range from thousands of kilometers in oceanic gyres (e.g., albacore *Thunnus alalunga* and anguillid eels, e.g., *Anguilla* spp.) to kilometers in turbidity-maximum zones of estuaries (e.g., anadromous fishes, *Alosa* spp., Osmeridae, Moronidae). Stocks of fishes have evolved so that early-life stages develop and recruit within defined physical domains bounded by hydrographic and topographic features that ensure: (i) directed transport or local retention (Heath 1992, Cowen 2002, Sponaugle *et al.* 2002, Cowen & Sponaugle 2009); (ii) cross-shelf delivery (e.g., Shanks 1988, 1995); (iii) selective tidal-stream transport in estuaries (Forward & Tankersley 2001); and (iv) maintenance of stock integrity (Sinclair 1988). Small-scale features and frontal processes operating at the kilometer scale may be important to retain eggs and larvae, such as Estuarine Turbidity Maximum zones (Laprise & Dodson 1989a,b, 1990, Sirois & Dodson 2000, North & Houde 2001, 2003, 2006). Finer-scale processes that link physics with larval behavior, such as vertical migratory behavior (Heath & Maclachlan 1987, Heath *et al.* 1987, Lough & Potter 1993, Lough *et al.* 1996, Bennett *et al.* 2002, Huebert *et al.* 2011), or slicks and internal waves (Shanks 1983, 1988, 1995) emphasize the importance of small-scale processes critical to larval survival and recruitment.

3.3.1 Hydrography and hydrodynamics

Aquatic ecosystems are heterogeneous, three-dimensional seascapes in which eggs of fishes are spawned and larvae forage for food and are preyed upon while entrained in dispersive habitats that may transport or retain them. The youngest stages of most fishes are constrained in their ability to forage or select favored environmental conditions because they are weak swimmers and thus are largely (but not completely!) passively responsive to physical processes in their pelagic surroundings. Temperature, salinity, dissolved oxygen, turbulence and circulation patterns, operating on micro- to ocean-basin scales, control potentials for growth, exposure to predation, and propensity for retention in, or dispersal from, nursery grounds. Hydrographic conditions may impose physiological constraints on the production of larval fish. For example, temperature may affect stage durations of eggs and larvae and growth rates of larvae, thus contributing to variability in spatial distributions of life stages and dispersal patterns (Peck *et al.* 2012). Favorable environments are defined by linked processes operating across spatial scales that support the nutrition of early-life stages. In this regard, Bakun (1996, 2006) described a triad of processes supporting larval production, referred to as: (i) enrichment; (ii) concentration; and (iii) retention processes.

Larval stages of fishes swim sufficiently well to at least partly select habitat in their immediate surroundings, especially through tidally-selected or diel rhythms in vertical migratory behavior, and thus may direct their transport or retention in stratified systems (Norcross & Shaw 1984, Boehlert & Mundy 1988, Miller 1988, Forward & Tankersly 2001, Leis 2007). In estuaries and coastal systems, larvae often utilize tidal dynamics to achieve retention or transport (selective tidal stream transport) (Weinstein *et al.* 1980, Rowe & Epifanio 1994). Selective behavior is largely controlled by vertical migrations tuned to the tides (Forward *et al.* 1999, Hare *et al.* 2005). In unstratified systems, for example coastal habitats adjacent to reefs, other behaviors may be important, including directed swimming guided by olfactory or auditory cues (Leis, 2006, 2007).

Hydrographic variables and associated factors, such as wind regimes, often are correlated with recruitment levels, but predicting recruitment or variability from hydrographic variables has historically been notoriously difficult and unreliable. Simple, single-variable correlates of recruitment with hydrographic or physical oceanographic variables may remain reliable over decades only under unusual conditions (Drinkwater & Myers 1987). For example, Nelson *et al.* (1977) correlated the recruitment of Atlantic menhaden *Brevoortia tyrannus* to coastal estuaries with Ekman transport, but this apparently successful approach failed in the decade following its publication. The failure of such correlations does not mean there were no significant relationships between recruitment and an environmental factor; rather, they demonstrate that shifts in ecosystem properties occur, for example in response to changing climate and weather.

3.3.2 Linking physics and hydrography to biology

Pelagic eggs are passive particles, and the small larvae of most marine fishes have little swimming ability; consequently, their ability to actively select hydrographic conditions or circulation/dispersal features favorable for survival is limited. Nonetheless, even the smallest larvae are capable of vertical migration and selection of depths where favorable hydrographic conditions, prey availability, and circulation favor their survival.

Physical processes and features acting across a broad range of spatial scales can control larval dispersal and trophodynamic interactions (Figure 3.7). Circulation patterns and their variability direct the dispersal of early-life stages. Features – especially vertical stratification and fronts – provide gradients and boundary conditions that serve as terms of reference for early-life stages in the pelagic environment. In moving from the open sea towards shelf and estuarine ecosystems, the scope and variability of hydrographic factors (e.g., salinity, temperature, dissolved oxygen) and habitats (e.g., bottom depth, topography, fronts, plumes, demersal features) become increasingly diverse and are, at least potentially, key factors controlling recruitment. Werner *et al.* (1997) reviewed the role of hydrodynamics in generating recruitment variability,

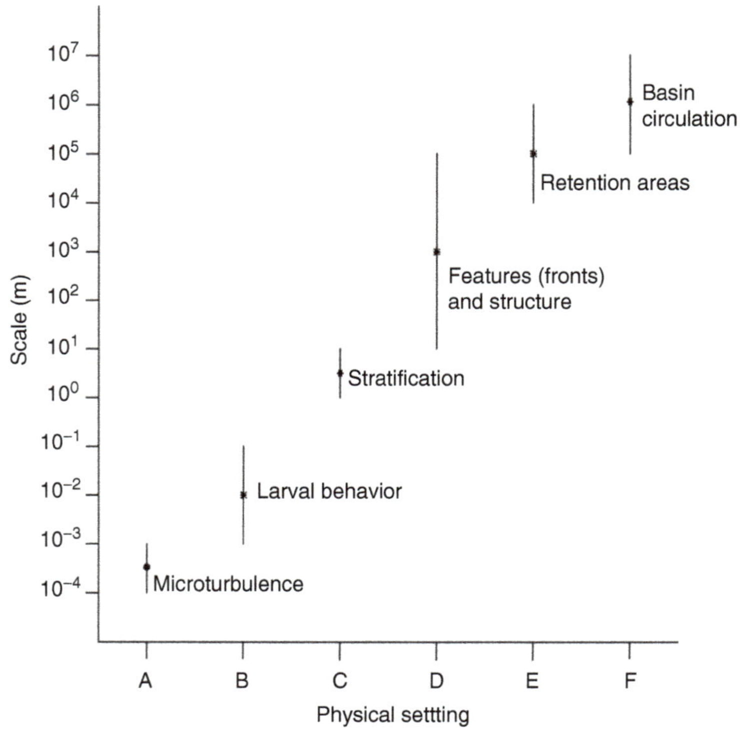

Figure 3.7 Spatial scales, features and mechanisms linking physics and biology to processes contributing to recruitment variability. The mechanisms and features range from millimeters to thousands of kilometers. All are potentially important contributors to recruitment variability.

emphasizing that processes operating across many spatial scales (e.g., eastern/western boundary currents, shelf-estuary interactions, fine-scale encounter-enhancing mechanisms such as microturbulence) are important, although quantifying specific processes is challenging.

3.3.3 *Temperature*

Temperature can directly (physiological) and indirectly (ecological) control virtually all processes in the early lives of poikilotherms, including fishes. Temperature controls various processes that range from an individual's metabolic rate to population-level stock productivity. Interannual and seasonal variabilities of temperatures in spawning and nursery areas often are accompanied by variability and shifts in regional climate and weather patterns (Stige *et al.* 2006). Decadal and longer trends in regional and global climate, and especially recent global increases in temperatures, will affect fisheries productivity, including recruitment processes in many taxa (Ottersen *et al.* 2013).

For some taxa, average recruitment levels and, ultimately, stock abundances will decline or increase in response to directional shifts in an ecosystem's properties precipitated by temperature change. Spawning sites and nursery habitats may shift as climate changes, particularly at the boundaries of geographical ranges (Cushing 1982, Blaxter 1992, Rombough 1996, Shuter *et al.* 2002, Drinkwater 2006). Together, temperature and body size are major factors controlling dynamic and energetics processes in the early lives of fishes. Growth rates, stage durations, metabolism, levels of activity, and susceptibility to predation of early-life stages are temperature-dependent. On a global scale, climate change defined by increases in ocean temperature may already be contributing to shifts in spawning areas and times, and in nursery areas of Atlantic cod and other fishes (Brander 1996, Perry *et al.* 2005, Hare & Able 2007).

Temperature can control prey available to fish larvae through its effect on timing of the production of key planktonic prey (i.e., Match–Mismatch hypothesis) (Peck *et al.* 2012). For example, there is evidence that variable cod recruitments in the North Sea and other regions are a consequence of periodic (e.g., North Atlantic Oscillation) (Stige *et al.* 2006) and long-term temperature variability that affects the timing, abundance and sizes of larval cod prey (Beaugrand *et al.* 2003), and that warming trends over the past two decades, combined with low adult stock levels, have resulted in low recruitments in the North Sea (O'Brien *et al.* 2000). Temperature directly controls the consumption rates of early-life stages. Taken together, prey availability and consumption potential may be the two most important variables controlling the growth and production of young fish. Furthermore, temperature controls not only the feeding rates and growth rates of larval and juvenile pre-recruit fish, but also the consumption and growth rates of their predators, prey and competitors.

In larval fishes, rates of development, metabolism, consumption and growth all increase quite dramatically as temperature increases ($Q_{10} \approx 2.5$–3.0) (Blaxter 1992, Rombough 1996), and fall precipitously when high-temperature tolerance thresholds are exceeded. Small, planktonic poikilotherms such as fish larvae have little ability to regulate, compensate for, or avoid temperature variability, except where vertical migration may allow larvae to seek more favorable temperatures.

Growth rates are temperature-dependent (Figure 3.8) and, as a consequence, the survival of larval fishes is heavily influenced by temperature variability through its effects on growth rates, which control stage durations and sizes-at-age (Houde 1989a, Pepin 1991, Houde & Zastrow 1993) (Figure 3.9, Plate 2). Early-life stages of many fishes, especially taxa from mid and high latitudes, have surprisingly broad temperature tolerances (ranges often of 10–20 °C; Rombough 1996) and, while able to develop and grow over broad temperature ranges, the growth and survival potentials are optimized at favorable temperatures. Growth and production often peak (and stage duration is shortest) at temperatures close to the thermal tolerance maximum (Rombough 1996, Folkvord 2005).

Concern about climate change and its effects on ecosystems and fish populations have prompted analyses of their implications for recruitment variability. In a meta-analysis, Myers (1998) found consistently positive correlations between temperatures and levels of recruitment for marine and freshwater fish stocks living near the high-latitude boundaries of their ranges. In contrast, fish stocks near the low-latitude boundaries often have early-life survival rates and recruitment levels that are inversely related to temperature. Plaice in the Southern Bight of the North Sea provides a good example of such relationships (Harding *et al.* 1978, Van der Veer & Witte 1999) (Figure 3.10) in which an approximately fourfold recruitment variability is attributable to temperatures experienced during the egg–larval stages.

For well-studied Atlantic cod, there is concern that changing climate and rising temperatures are affecting levels of recruitment and may lead to shifts in spawning sites in coming decades (Planque & Fox 1998, Planque & Fredou 1999, O'Brien *et al.* 2000, Drinkwater 2002, Beaugrand *et al.* 2003). Rising temperatures may physiologically limit the survival and growth of young cod or act indirectly through temporal–spatial shifts in plankton productivity

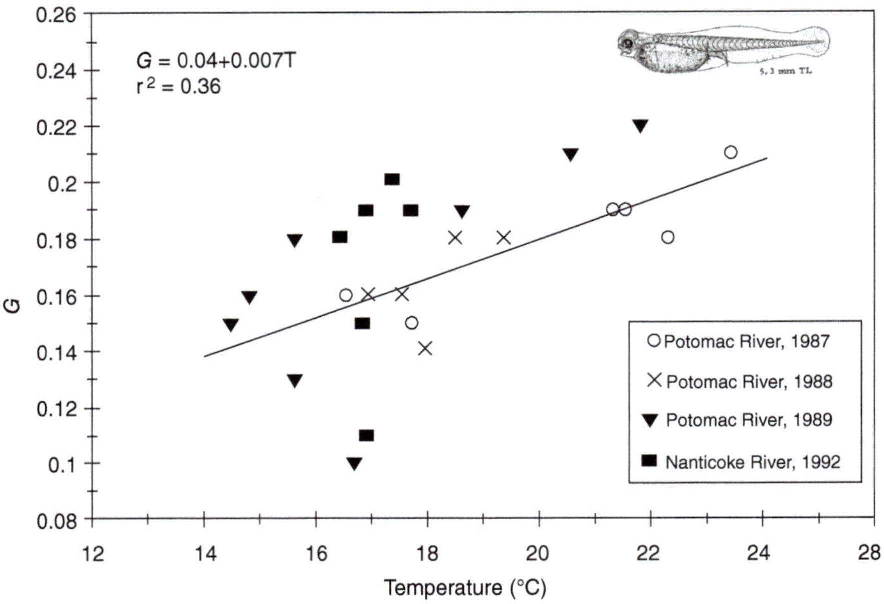

Figure 3.8 Relationship between weight-specific growth rates and temperature for striped bass *Morone saxatilis* larval-stage cohorts in the Potomac and Nanticoke Rivers, tidal tributaries of Chesapeake Bay. Larvae can survive and grow at temperatures from 12 to 24 °C. Stage durations of larval cohorts growing at 12 °C are nearly twice as long as those at 24 °C.

or predator abundances. Temperature is associated with either negative or positive trends in cod recruitment success over its broad geographical range (Planque & Fredou 1999, Sundby 2000, Drinkwater 2002, Stige *et al.* 2006). Recruitment success is positively related to temperature in the northern parts of cod's range, while the reverse is true in the southern (and warmer) parts of its range (Planque & Fox 1998, Sundby 2000) (Figure 3.11 and 3.12). Planque and

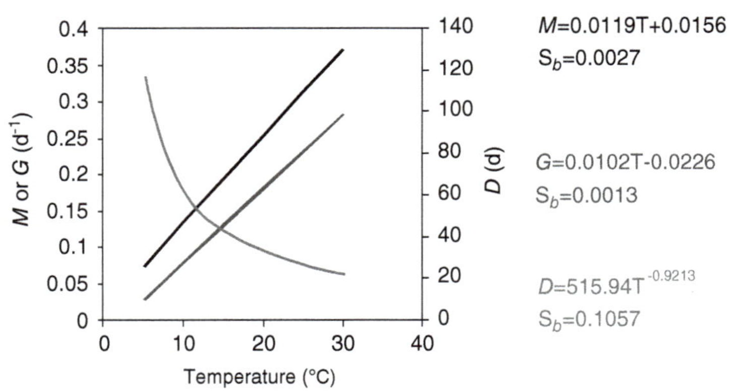

Figure 3.9 Summary figure of *M*, *G*, and *D* for marine fish larvae in relation to temperature, summarized from metadata analysis by Houde (1989). Lines are the linear regression fits for weight-specific growth (*G*) and daily mortality (*M*), and a power model regression fit for Larval Stage Duration (days, *D*) with respect to temperature. S_b = standard error of the regression coefficient. For a color version of this figure, please see Plate 2 in the color plate section.

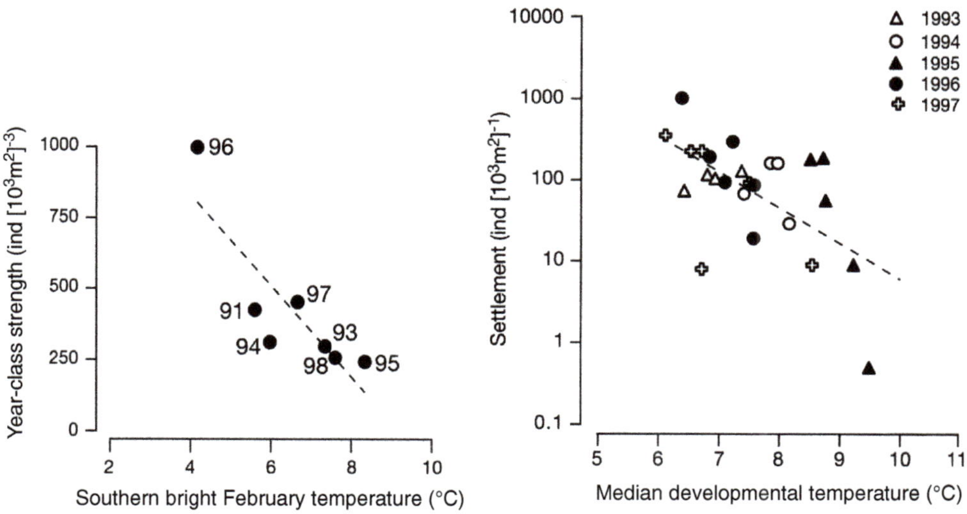

Figure 3.10 Interannual variability in year-class strength of young-of-the-year plaice *Pleuronectes platessa* and intra-annual variability in levels of abundance of cohorts of settling juveniles with respect to temperature in the Southern Bight of the North Sea. Note the strong inverse relationships. Reproduced from Van de Veer & Witte (1999), with permission of Inter-Research.

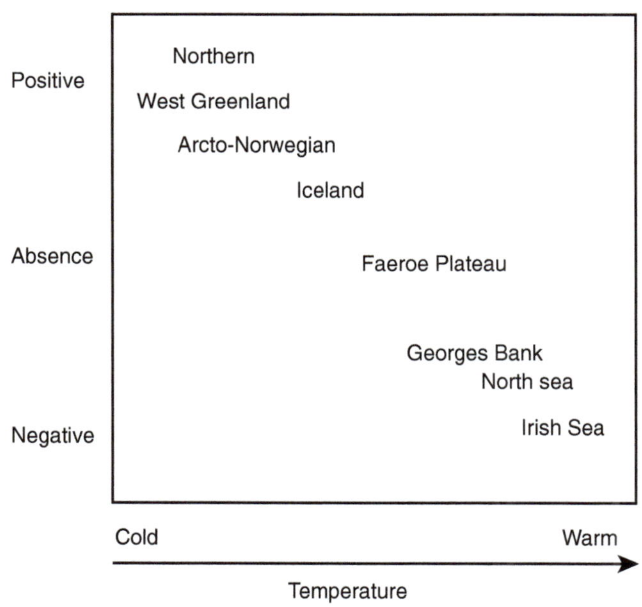

Figure 3.11 Recruitment anomalies of Atlantic cod *Gadus morhua* stocks with respect to temperature. Stocks from high (colder) latitudes have positive recruitment anomalies when temperatures are above average during the egg–larval stages, while the reverse is true for stocks at more southerly (warmer) latitudes. Reproduced from Planque & Fox (1998), with permission of Inter-Research.

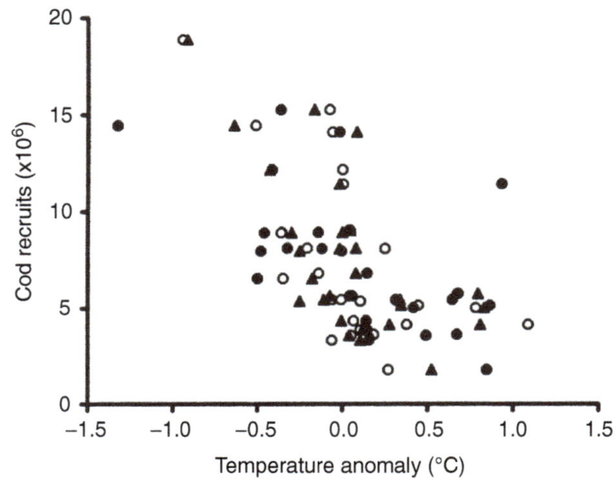

Figure 3.12 Inverse relationship between recruitment levels of Atlantic cod *Gadus morhua* in the Irish Sea and sea surface temperatures during the egg and larval stages. Recruitments vary approximately fourfold over the range of temperature anomalies. Reproduced from Planque & Fox 1998, with permission of Inter-Research.

Fredou (1999) concluded that two factors – adult cod abundance and temperature – explained a large fraction of the variance in recruitments of Atlantic cod.

Survival variability during early-life stages is the probable driving mechanism that translates into recruitment variability associated with temperature effects. The relationship between recruitment in some Atlantic cod stocks and temperature (Planque & Fox 1998, Planque & Fredou 1999, Sundby 2000, Beaugrand *et al.* 2003) supports this hypothesis, although specific mechanisms are poorly known. Since the1980s, recruitment levels of cod in the North Sea have declined approximately fivefold. Temperatures have increased in this period (Beaugrand *et al.* 2003), and the timing and level of copepod production have shifted to cause a "mismatch" in the production of cod larvae and copepod prey. Variability in zooplankton production and correlated trends in cod recruitments support the Match–Mismatch hypothesis. Sundby (2000) argued that recruitment success depended on temperature and its control over the production of a favored prey of cod larvae, the copepod *Calanus finmarchicus*, in spawning and nursery areas. In contrast, a recent synthesis of recruitment variability for many North Atlantic fish stocks, while documenting a strong temperature effect on decadal recruitment dynamics for many species (e.g., Atlantic herring and plaice) found weaker or non-significant effects for cod (Ottersen *et al.* 2013).

In another example, recruitment fluctuations in yellowtail flounder *Limanda ferruginea* in the Middle Atlantic Bight off North America were demonstrated to be temperature-dependent. Recruitments were strongly and inversely correlated with the North Atlantic Oscillation (NAO) winter index. Notably, high recruitments are associated with cooler than average water temperatures during the post-settlement juvenile stage (Sullivan *et al.* 2005). The NAO and other ocean-climate oscillations (e.g., ENSO and the Pacific Decadal Oscillation) that can signal regime shifts and major changes in ocean productivity (Ware & Thomson 2005) often are associated with temperature changes and correlated trends and variability in fish recruitments

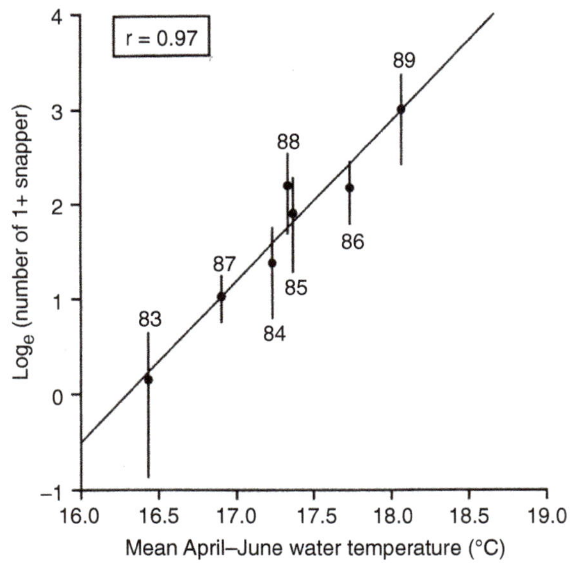

Figure 3.13 Relationship between recruitment level (age 1+) of New Zealand snapper *Pagrus auratus* and April–June sea surface temperature in the year that recruits were hatched. Recruitments varied 17-fold from 1983 to 1989. Reproduced from Francis (1993), with permission of Blackwell Publishing.

(Ottersen *et al.* 2001, Van der Veer *et al.* 2000b, Sullivan *et al.* 2005, Stige *et al.* 2006, Ottersen *et al.* 2013).

Effects of temperature on recruitment levels and variability can be remarkably strong. For example, in New Zealand snapper *Pagrus (= Chrysophrys) auratus*, year-class strength was heavily dependent on the post-spawning sea-surface temperature during fall months (Francis 1993). Observed recruitments varied 17-fold, and fall temperatures explained 94% of this variability (Figure 3.13). In the snapper, variable cumulative mortality during the larval, and possibly juvenile, stages was strongly correlated with temperature. Francis (1993) offered three temperature-related hypotheses to explain the observed variability. The first of these, the larval *Stage–Duration* hypothesis, was supported by otolith-increment analysis on larval cohorts (Francis 1994), whereby larval stage durations decreased rapidly as temperatures increased, and sea surface temperature explained 63% of the variance in stage durations (Figure 3.14). The second and third hypotheses (Francis 1993) were untested but have merit. The second hypothesis proposes that slow-growing larvae and juveniles in years with low fall temperatures were forced to overwinter at small size and suffered high mortality. The third hypothesis proposed that a combination of factors expressed in hypotheses 1 and 2 acts to control year-class strength. Recent research on the snapper in Australia indicated that the growth rates and survival rates of larvae were higher in years of high larval abundance. Such years were characterized by favorable temperatures for growth and high levels of prey availability (Murphy *et al.* 2013).

For anadromous striped bass in Chesapeake Bay, young-of-the-year abundances at approximately 150 days post-hatch vary more than 30-fold. Larval-stage cohorts experiencing favorable temperatures have a higher potential for recruitment (Figure 3.15). Seasonal timing of

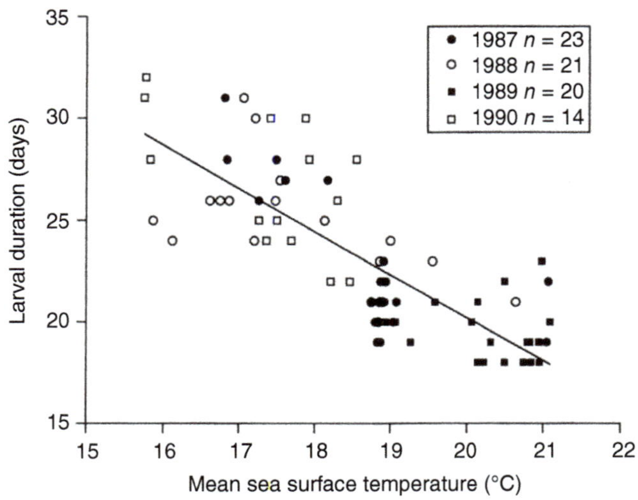

Figure 3.14 Relationship between larval stage duration and sea surface temperature during the larval stage for New Zealand snapper *Pagrus auratus*. Temperature explained 63% of the variability in larval stage duration. Reproduced from Francis (1994), with kind permission of Springer Science and Business Media.

temperatures favorable for spawning, larval growth and survival, and variability in temperatures, combined with favorable freshwater flows and zooplankton prey production, are critical to the recruitment success of striped bass (Rutherford & Houde 1995, Secor & Houde 1995, Martino & Houde 2010).

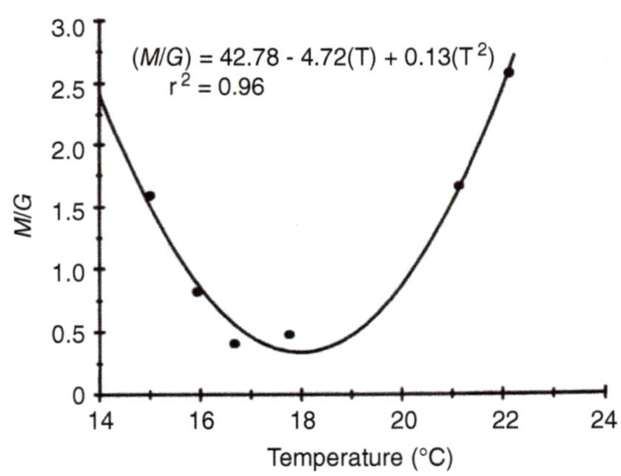

Figure 3.15 Larval stage *M*/*G* for cohorts of striped bass *Morone saxatilis* in the Patuxent River sub-estuary of Chesapeake Bay and its relationship to temperature during the first 25 days post-hatch. *M*/*G* is inversely related to recruitment potential in striped bass. Survival of larvae was highest at 17–19 °C. *M*, daily instantaneous mortality; *G*, weight-specific growth rate; T, temperature. Reproduced from Houde (1997a), with kind permission of Springer Science and Business Media.

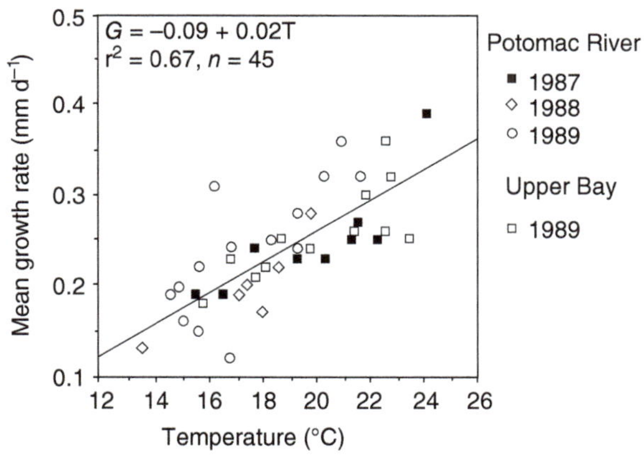

Figure 3.16 Back-calculated, cohort-specific, mean growth rates of striped bass *Morone saxatilis* larvae in Chesapeake Bay sub-estuaries in relation to temperature during the first 20 days post-hatch. Reproduced from Rutherford & Houde (1995).

3.3.3.1 Growth and production

In laboratory experiments, weight-specific growth rates (*G*) of many marine fish larvae vary more than threefold in response to temperature (Blaxter 1992, Rombough 1996). In a synthesis analysis of growth rates of marine fish larvae, Houde and Zastrow (1993) found that larvae of taxa from warm seas grew much faster than larvae from cooler regions. Mean values of *G* in their across-taxa synthesis increased by approximately 0.01 (i.e., ~1%) per °C increase in temperature (Figure 3.9, Plate 2).

Growth rates of individual taxa respond strongly to temperature. For example, reported values of *G* for larval Atlantic herring *Clupea harengus* varied approximately 10-fold over a 6–17 °C range (Houde 1989a). In addition, a more than fivefold variability in *G* was recorded for larvae of Atlantic cod (4–14 °C range) and walleye pollock *Theragra chalcogramma* (5.0–9.5 °C range) (Houde 1989a). Growth rates of larval striped bass in an estuary increased from 0.19 to 0.39 mm d^{-1} (equivalent to 13.9 to 29.7% d^{-1} weight-specific growth) in a 14–24 °C range (Rutherford & Houde 1995) (Figure 3.16).

In the tropics, where there is relatively little temperature variability, temperature still can exercise important control over growth. For example, a larval goatfish *Upeneus tragula* grew 30% faster at 30 °C than at 25 °C (McCormick & Molony 1995). Also, temperature explained more of the variability in growth (30%) than either zooplankton food supply (3.5%) or chlorophyll-*a* (4.1%) in larvae of the tropical damselfish *Pomacentrus coelestis* (Meekan *et al.* 2003). Recruitment was directly related to water temperature, and was the most important variable in explaining recruitment variability in bluehead wrasse *Thalassoma bifasciatum* in the Florida Keys, although other factors added to variability in the complex process of recruitment in this reef species (Sponaugle *et al.* 2006).

Jordaan and Kling (2003) demonstrated that growth rates of laboratory-reared Atlantic cod larvae varied widely and peaked at an intermediate experimental temperature (7.9 °C). However, maximum acceleration in growth rate (mm d^{-1} °C^{-1}) (Figure 3.17), defined by the authors as the most efficient temperature for growth, occurred at a lower temperature (4.2 °C). Maximizing growth rate, whether efficient or not, minimizes stage duration and may reduce

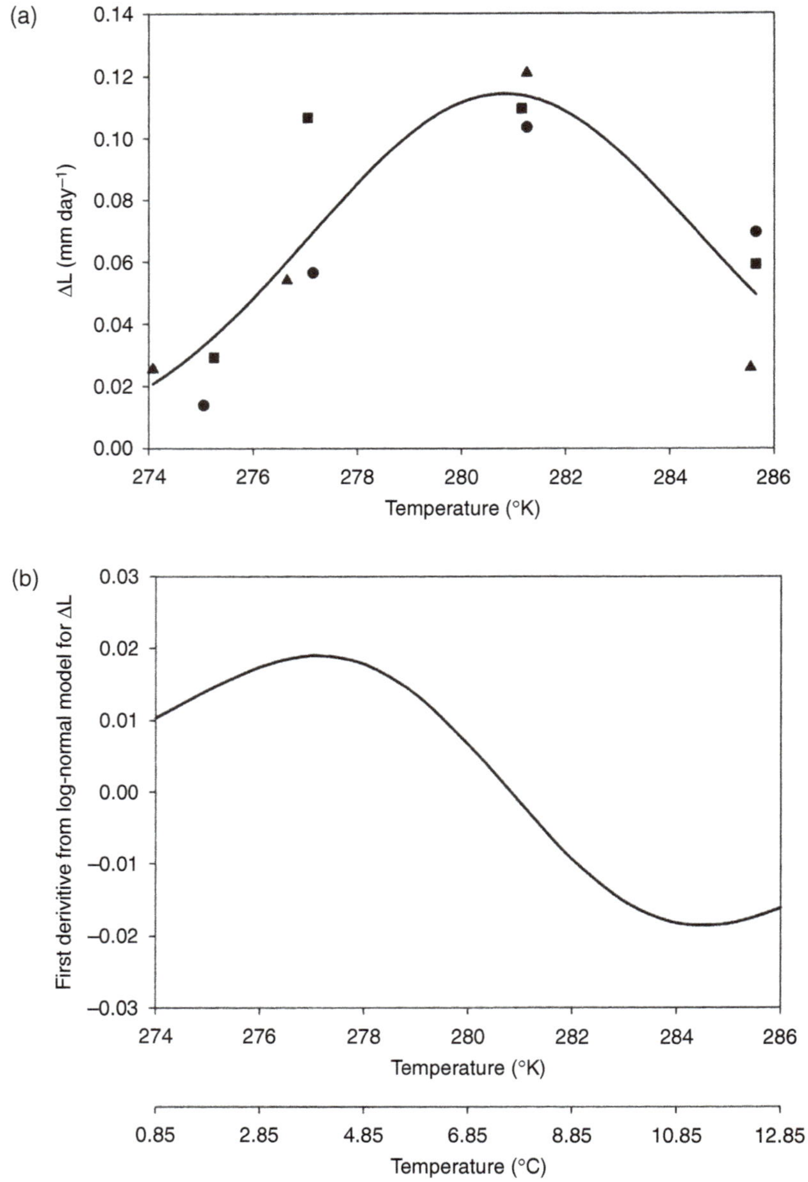

Figure 3.17 Growth of Atlantic cod *Gadus morhua* larvae in relation to temperature, based on laboratory experiments. (a) Growth rates; (b) First derivative of the growth rates (i.e., accelerations) with respect to temperature. Growth rate was maximum at 7.9 °C and accelerating fastest at 4.2 °C. Reproduced from Jordaan & Kling (2003), with permission of Stony Brook University.

the time exposed to predation mortality. For cod, days in the larval stage could vary several-fold for the range of growth rates and temperatures reported by Jordaan & Kling (2003) or by Houde (1989a). Temperatures that maximize the efficiency of growth potentially could reduce the instantaneous rate of predation mortality if larvae at those temperatures spent less time foraging, thereby reducing risk of encounters with predators.

Folkvord (2005) reviewed and analyzed the growth of Atlantic cod larvae based on experiments and field surveys, and concluded that survivors (i.e., potential recruits) in the sea had grown at rates that were near maximum for given temperatures. Growth of cod larvae is sensitive to temperature and body size (Otterlei *et al.* 1999, Folkvord 2005). Folkvord's growth model, when applied to cod larvae from Georges Bank, indicated that +1 or –1 °C shifts in sea temperature over a 70-day period could generate a 50% increase or 40% decrease, respectively, in the mean weight of survivors. These results support the contention that shifting temperature regimes associated with climate change and variability could have strong effects on early-life growth and recruitment variability (Ottersen *et al.* 2001).

While temperature effects on growth and production can be estimated with confidence in laboratory experiments (e.g., Atlantic cod; Otterlei *et al.* 1999, Steinarsson & Bjornsson 1999, Jordaan & Kling 2003) and numerous other species (Margulies 1989, Leach & Houde 1999, Dou *et al.* 2005), confirmation of effects in the sea is less common than might be expected because temperature interacts with hydrographic and biotic variables. Takahashi & Watanabe (2005) found that temperature was more important than prey density in controlling the growth rates of Japanese anchovy *Engraulis japonicus*. In another example, Suthers & Sundby (1993) reported that pelagic juveniles of Norwegian cod at an identical age differed markedly in size as a consequence of early-life residence in water masses of differing temperatures. Sea temperature was the main driver of growth variability in pre-settlement stages of three regional stocks of plaice in the eastern Atlantic that experience differing hatch dates, larval-stage duration and settlement dates (Comerford *et al.* 2013). In this case, models incorporating temperature and region accounted for 73% of the variability in the growth of plaice larvae. In Baltic Sea sprat *Sprattus sprattus* there is a strong, positive correlation between recruitment level and late summer sea temperatures (Baumann *et al.* 2006), but this relationship depends on favorable transport and circulation patterns during the larval and juvenile stages. Variability in the growth of larval cod and haddock on Georges Bank is related to both increasing temperature and photoperiod, and in this case the photoperiod appears to be the dominant controlling variable (Buckley *et al.* 2006). Jenkins and King (2006) reported increased growth and survival of larvae of King George whiting (*Sillaginodes punctata*) under favorable temperature conditions, noting that these conditions may have resulted from varying currents and circulation features, and increased productivity of plankton prey.

In community-level, across-taxa analyses, temperature largely explains the span of growth rates and stage durations observed in larval fishes (Houde 1989a, Morse 1989, Pepin 1991). Such analyses demonstrate how magnitudes, averages and ranges of temperature affect dynamic rates at the community, broad taxonomic and ecosystem levels (Houde & Zastrow 1993), providing insight into potential effects on the productivity and recruitment of young fish. Within a ~3 to 30 °C range, the average weight-specific growth of typical marine fish larvae increases directly with temperature, and ranges from 0.01 to 0.28 (Figure 3.9, Plate 2). Whilst important for defining limits and broad relationships, cross-taxa approaches such as those of Houde & Zastrow (1993) do not explain effects of temperature on species-specific variability in growth.

3.3.3.2 Mortality

Mortality rates of early-life stages are more variable than growth rates (Houde 1997b), and less clearly linked to temperature. At the ecosystem and community (across-taxa) levels there is a direct correlation between mortality rate and temperature (Houde 1989a, Morse 1989, Pepin 1991, Houde & Zastrow 1993). The reported daily instantaneous mortality (*M*) of larval fishes ranges from approximately 0.01 to >1.0 and, similar to weight-specific growth, increases by approximately 0.01 d^{-1} for each 1 °C increase in temperature (Figure 3.9, Plate 2) (Houde 1989a). As a generality, larval fishes from low latitudes and warm seas experience higher daily mortality rates than larvae from high latitudes and colder waters. The highly correlated *M* and *G* of larval fishes increase at similar rates with respect to temperature. Consequently, the net effect of temperature on survival and recruitment potentials is in large part neutralized because cumulative mortality at the end of the larval stage is similar for fish larvae from tropical or temperate ecosystems (Houde 1989a, Pepin 1991).

At temperatures above or below the range of physiological tolerance, mortality in early-life stages can be directly attributable to temperature. Such mortalities may be common in small freshwater and estuarine ecosystems that are poorly insulated against temperature variability associated with local and regional weather events. Exposure to lethal temperatures is less likely in shelf seas and the ocean.

Under most circumstances, temperature-related mortality owes primarily to stresses that act indirectly by controlling physiological rates, prey consumption, swimming activity, encounters with predators, or possibly diseases that are temperature-dependent. These indirect effects influence mortality through controls on growth rate and stage duration, or by altering the behaviors of early-life stages (e.g., via effects on swimming speeds and behaviors that affect encounter rates between predators and prey).

In plaice, which tends to recruit successfully in years of low winter temperature, mortality rates of eggs in the Southern Bight of the North Sea increased directly with temperature (Van der Veer *et al.* 2000a). Keller & Klein-MacPhee (2000) conducted temperature–effect experiments on winter flounder *Pseudopleuronectes americanus* in mesocosms, and found that larval survival was best at cool temperatures despite higher growth in warm mesocosms. Cumulative effects of higher egg mortality and higher losses of larvae to predation in the warm mesocosms resulted in a higher production of juveniles in mesocosms held at cool temperatures. Such direct and indirect effects of temperature may be particularly important in Narragansett Bay, where recruitments of winter flounder have declined for two decades. Climate change and winter warming may be contributing to these declines (Keller & Klein-MacPhee 2000).

Larval cohorts of anadromous striped bass in Chesapeake Bay tributaries suffer daily mortality ranging from $M = 0.02$ to 0.92. Mortality rates and cumulative mortalities are related to temperature. Cohorts with the lowest daily mortalities ($M \leq 0.15$) develop at intermediate temperatures of 17–19 °C (Secor & Houde 1995, Rutherford *et al.* 1997) (see Figure 3.15). In striped bass larvae, high mortalities are associated with slow growth at low temperatures, while increased predation may indirectly lead to high mortality at higher temperatures. In Northeast Arctic cod, the lowest mortality rates of eggs and larvae – and subsequently the highest recruitments – occur in relatively warm years when high copepod production coincides with the hatching of larval cod (support for the Match–Mismatch hypothesis) (Ellertsen *et al.* 1989, Sundby *et al.* 1989, Sundby 2000). In the striped bass and cod examples, temperature-related larval mortalities are mostly indirect, and are attributable to unfavorable environmental stresses, poor feeding conditions and probable higher predation, rather than to direct physiological death.

3.3.3.3 Stage durations

Temperature controls development and growth rates of early-life stages and, consequently, also exercises control over stage durations. Stage durations decline exponentially with increasing temperature (see Figure 3.5 and 3.9) (Houde 1989a, Blaxter 1992, Francis 1994, Jordaan & Kling 2003). This relationship is taken as evidence supportive of the "stage duration hypothesis" to explain recruitment variability. Temperature clearly can control ages at metamorphosis and settlement, and through this mechanism temperature can potentially control recruitments via its effects on cumulative mortality during variable stage durations (Cushing 1975, Houde 1987, Anderson 1988). In marine fishes, egg and larval stage durations can range from a few days to at least one year (Houde 1989a). Stage durations for some species (e.g., Atlantic herring) may vary more than 10-fold, offering ample opportunity to generate variability in survival at recruitment. Pepin and Myers (1991) found that variability in recruitment in many marine fishes was directly related to the larval stage duration. Variability in early-life stage durations is highest in slow-growing fishes from cold seas (Houde 1989a). Consequently, it is probable that the temperature effect on stage durations is a more significant contributor to recruitment variability at high latitudes than in the tropics.

In striped bass, Rutherford & Houde (1995) noted higher growth rates, shorter stage durations and lower cumulative mortalities, leading to a several-fold higher production of late-stage larvae and juveniles in years when temperatures were most favorable in Chesapeake Bay. In another example, the duration of the egg stage of Atlantic cod and haddock in the western North Atlantic may vary from 8 to 30 days over an approximate 15 °C temperature range (Page & Frank 1989). In laboratory experiments, Jordaan & Kling (2003) showed that the stage duration of embryonic and yolk-sac stages of Atlantic cod varied approximately fourfold in the 2–12 °C temperature range (33 versus 8 days, respectively). For the sparid *Pagrus auratus* off New Zealand, the mean larval stage duration varied from 32 to 18 days at temperatures in the range 16–21 °C (see Figure 3.14) (Francis 1994).

3.3.4 Salinity

In estuaries and coastal seas subject to variable freshwater inflow, salinity may contribute directly to variability in the survival and growth of early-life stages of fish, and certainly can define the volume of habitat suitable for survival and growth. In the ocean, salinity is a primary indicator of hydrographic and pelagic habitat conditions through its effects on water column stability, mixing, and stratification, all of which may act indirectly to control recruitment. Unless it is close to the limit of physiological tolerance, salinity may have few direct consequences to the survival of fish early-life stages. In fact, salinity tolerances often are surprisingly broad and tolerance ranges may either expand (e.g., spotted seatrout *Cynoscion nebulosus*; Banks *et al.* 1991) or contract (e.g., rabbitfish *Siganus guttatus*; Young & Duenas 1993) during ontogeny and growth. In chronic exposure, salinity can control osmoregulatory ability, activity levels and metabolic rates (e.g., Atlantic herring and plaice; Almatar 1984) and may thus influence development rates, bioenergetics relationships, and growth in early life (Holliday 1969, Alderdice 1988).

Salinity interacts with temperature to affect rates of development and the survival of embryos and larvae (Alderdice & Forrester 1968, Alderdice & Velsen 1971). Salinity is, in many ways, a "subtle controller"; that is, through its effects on density and water-column stability it can modulate the buoyancy of eggs, thus affecting depth distributions and potentially allowing settlement into unfavorable environments, such as hypoxic bottom waters in estuaries and coastal seas (Nissling & Westin 1991, Nissling *et al.* 1994). The behavior of larvae – for example, vertical migratory behavior – can be modulated by salinity (DeVries *et al.* 1995). Salinity patterns may also serve as water-mass indicators, sometimes defining the predator and prey communities associated with fish early-life stages. Other indirect effects of salinity on recruitment processes and variability include its role in defining spawning areas for adults, especially in anadromous species in estuaries and oceanic species that spawn in specific water masses.

3.3.5 *Stratification*

Much of the ocean is stratified, with denser, colder (often of higher salinity) water underlying a homogeneous surface mixed layer where most plankton productivity and early-life-stage production of fish take place. Thermoclines and pycnoclines, and sometimes also oxyclines, act as boundaries at the base of the mixed layer and may serve as partial barriers to mixing. Shear dynamics along these interfaces also control the distributions and dispersal of early-life stages, and their interactions with predators and prey. The depth of the mixed layer and strength of stratification are largely controlled by winds and wind-induced mixing and turbulence. Although vertical distributions of early-life stages of many fishes have been described and modeled, effects on survival or recruitment attributable to shifts or variability in the structure and stability of the water column are, for the most part, poorly documented. The dispersal, retention and degree of overlap with prey and predators depend on how early-life stages are distributed in the water column and if they conduct vertical migrations.

Fish larvae can undertake substantial (tens of meters) diel vertical migrations, and their vertical distributions may change in response to environmental and hydrographic conditions (Huebert *et al.* 2011). Most marine fish larvae reside in the upper mixed layer (generally in the upper 100 m of the water column), and while local hydrographic conditions may affect vertical distributions, species-specific patterns often are well defined (Neilson & Perry 1990, Palomera 1991, Ropke 1993, Coombs *et al.* 2003). In many cases, vertical distributions of early-life stages are correlated with, and are responsive to, zooplankton prey distributions or to light levels that permit feeding by fish larvae.

The Stable Ocean hypothesis (Lasker 1975, 1978, 1981) evolved from the observed dependence of larval survival in northern anchovy *Engraulis mordax* on water-column stratification and relative stability in the California Current during intermittent periods of calm between wind events and associated upwelling. During the calm periods, planktonic prey of larval anchovy (often dinoflagellates and copepod nauplii) aggregate on the stratified boundary and the feeding conditions for larvae improve. Peterman & Bradford (1987) and Peterman *et al.* (1988) reported that the survival of larval northern anchovy was indeed improved when the number of calm periods increased, but they detected no relationship with recruitment variability, indicating that the recruitment level may be set after the early larval stage, possibly in the late-larval or juvenile-stage. Nevertheless, a reasonably high survival in the larval stage of marine fishes may be a prerequisite for successful recruitment. Modeled dynamics of wind effects on the California Current and its plankton productivity support the proposed relationship between variability in wind regimes, degree of stratification, prey aggregations, and relationships to larval survival (Wroblewski & Richman 1987, Wroblewski *et al.* 1989). In a contrasting case, Coombs *et al.* (2003) found that the concentrations of prey eaten by larval European anchovy *Engraulis encrasicolus* in the Adriatic Sea were enhanced in the surface mixed layer under stratified conditions, but that the mortality rates of the larvae bore no clear relationship to salinity or degree of stratification.

In the Mediterranean Sea, Olivar & Sabates (1997) reported that larvae of European anchovy and most other fish larvae reside above the pycnocline, independent of particular hydrographic conditions. Ropke *et al.* (1993) found vertical distributions of fish larvae with respect to the pycnocline to be taxon- and location-specific in the Arabian Sea. In a stratified estuary, North & Houde (2004) demonstrated a strong propensity for aggregation by eggs and larvae of bay anchovy *Anchoa mitchilli* at the pycnocline, in common with its planktonic prey and jellyfish predators (Figure 3.18).

Stratification and hydrographic conditions at the salt–fresh interface in upper estuaries (e.g., estuarine turbidity maximum zone, ETM) can provide conditions that retain and limit the dispersal or loss of anadromous fish larvae (e.g., striped bass, white perch, alosines) and their zooplankton prey. Years of observed high recruitment of these species in Chesapeake Bay are correlated with an enhanced stratification in the two-layered estuarine circulation that strongly defines the ETM (North & Houde 2001, 2003, Shoji *et al.* 2005b).

3.3.6 Dissolved oxygen

Levels of dissolved oxygen (DO) in the sea generally are well above thresholds and lethal levels that can directly kill early-life stages of fishes. DO levels and variability usually are of minor concern to metabolism, growth and feeding, or predator avoidance behaviors. However, in freshwaters, estuaries, enclosed seas and some shelf seas, persistent or episodic low DO levels contribute to "dead zones" that can cause mortality and limit the production of early-life stages of fishes and their prey, and thus constitute a threat to recruitment (Breitburg 2002). Anoxic and hypoxic ($<2.0\ \mathrm{mg\ l^{-1}}$ DO) conditions in coastal systems often are of anthropogenic origin, the consequence of excess nutrient loading and eutrophication (Holt 2002). Anoxic and hypoxic habitats can limit spawning areas accessible to adults, cause physiological stress or death to embryos and larvae, and affect development rates and activity levels (Werner 2002).

Hypoxia is becoming increasingly common and is a threat to reproductive success and the recruitment of fishes spawning in coastal areas or using hypoxic habitats as nurseries (Breitburg 2002). Atlantic cod in the Baltic Sea is an example that illustrates how low DO can threaten the survival of early-life stages. Adults spawn below the halocline and eggs may sink into hypoxic deep water, suffering high mortality (Wieland *et al.* 1994). The volume of water in the Baltic's Bjornholm Basin, a primary spawning area for cod, with DO levels suitable for cod egg survival, varies nearly 10-fold interannually (MacKenzie *et al.* 1996, 2000), and is an important factor governing recruitment success. Other examples of probable recruitment limitation as a consequence of low DO are reported in the Black Sea (Mee *et al.* 2005) and in Chesapeake Bay (Breitburg *et al.* 1997, Breitburg 2002).

Experimental evidence indicates that early-life stages of marine fishes have quite variable DO tolerance thresholds and lethal levels. Low DO may be directly lethal (usually in the range $1\text{–}3\ \mathrm{mg\ l^{-1}}$), may reduce development and growth rates (often at DO levels only slightly below saturation), and can alter the behaviors of early-life stages (Rombough 1988, Breitburg 2002). At high temperatures, minimum tolerable DO levels may increase as activity levels and food

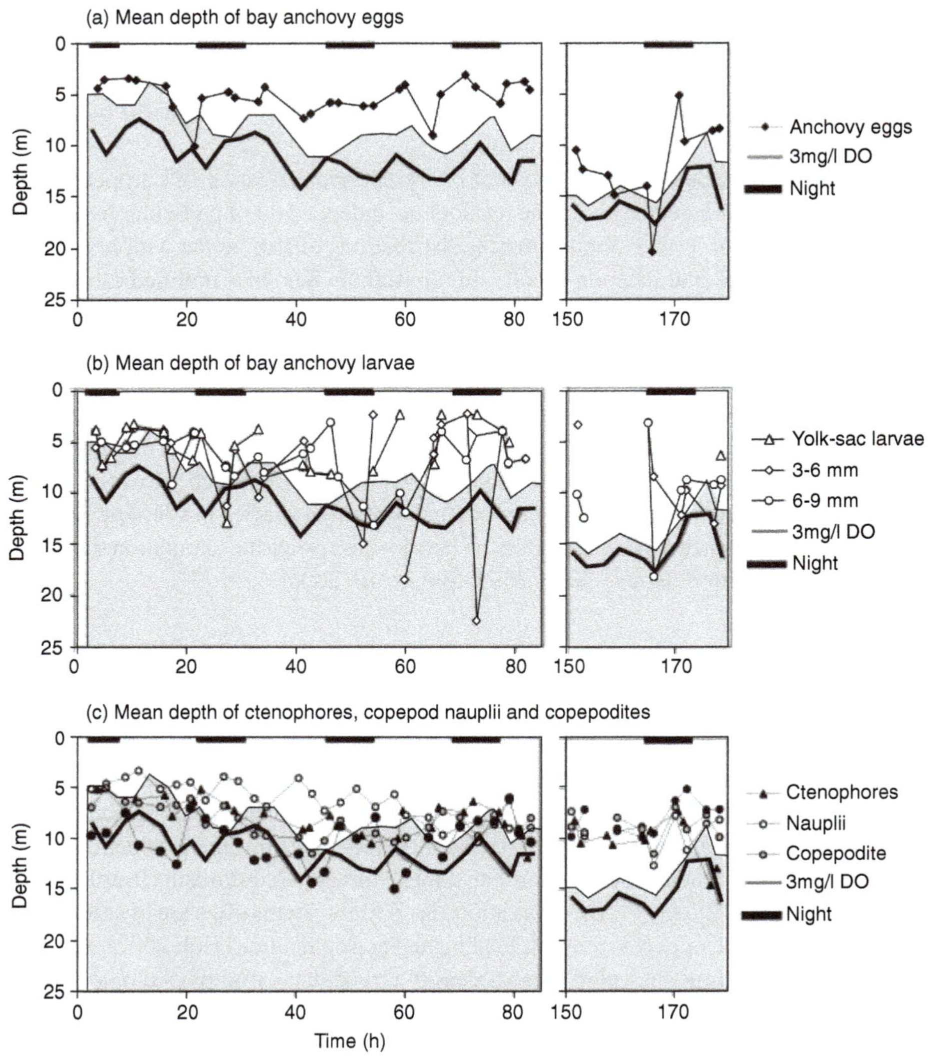

Figure 3.18 Vertical distribution and layering of eggs of bay anchovy *Anchoa mitchilli*, bay anchovy larvae (yolk-sac larvae and four length classes), and the planktonic prey and jellyfish (ctenophore) predators of fish eggs and larvae in Chesapeake Bay. The water column is strongly stratified. Note aggregations of organisms on the pycnocline or oxycline. Reproduced from *Estuarine, Coastal & Shelf Science*, Vol. 60, North & Houde, Distribution & transport of bay anchovy (*Anchoa mitchilli*) eggs & larvae in Chesapeake Bay, pp. 409–29, copyright 2004 with permission of Elsevier.

demand each increase. A low DO also acts indirectly as a stressor by altering trophic interactions (impeding feeding success and behavior, reducing predator avoidance ability, displacing early-life stages from nursery habitats) (Breitburg *et al.* 1997, 1999, Breitburg 2002, Shoji *et al.* 2005a) and increasing physiological stress. Interactions of DO with other environmental or ecological/behavioral factors (e.g., hydrography, predator–prey behaviors, interactions) can have important but difficult to detect effects on early-life survival and recruitment (MacKenzie *et al.* 1996, Breitburg *et al.* 1997, Keister *et al.* 2000).

In affected habitats, hypoxia may be episodic, potentially affecting survival on relatively small spatial-temporal scales (Breitburg 1992, 2002). In other circumstances hypoxia may be chronic, widespread and interannually variable, imposing lethal effects, such as mortality of cod eggs in the Baltic Sea, or sublethal effects on larval growth rates and susceptibility to predation (Breitburg *et al.* 1994, 1999, MacKenzie *et al.* 1996).

3.3.7 Ocean Acidification

Accompanying increasing temperatures, another factor associated with climate change is rising CO_2 levels in the world's oceans and declines in pH. While the effects of low pH were demonstrated to adversely affect development and survival of eggs and larvae of some estuarine and anadromous fishes (Hendrey 1987, Hall *et al.* 1993), most marine fishes that have been studied apparently are little affected by the relatively high pCO_2 and low pH levels expected to be common by the year 2100. In European seabass *Dicentrarchus labrax*, egg and larval ontogeny and survival were impacted by high pCO_2, although the effects were subtle (Pope *et al.* 2013). In two examples from the Baltic sea, the ontogeny and survival of Atlantic herring and Atlantic cod eggs and larvae were apparently unaffected by high pCO_2 and low pH, although the lack of effect in these cases may indicate adaptation or acclimation to the high-CO_2 and low-pH environment of the Baltic (Franke & Clemmesen 2011, Frommel *et al.* 2013). Hurst *et al.* (2012, 2013) found no effect of elevated CO_2 levels on survival of walleye pollock embryos and juveniles. Acidified conditions do have adverse effects on egg and larval survival in some marine fishes, such as summer flounder *Paralichthys dentatus* (Chambers *et al.* 2014).

For some species, the threat of future ocean acidification to the recruitment of marine fishes may be critical. Chambers *et al.* (2014) have listed the effects of ocean acidification in marine fishes, which are expected to: (i) vary among species, life stages, and individuals; (ii) be more pronounced in the youngest life stages; (iii) be subtle yet potentially chronic; and (iv) possibly interact with other stressors. The combined effects of increasing temperatures, declining pH and other subtle effects of climate change on recruitment levels and variability are poorly understood at this time.

3.3.8 Circulation

The dispersal and retention of early-life stages, which depend on circulation patterns and transport pathways, contribute importantly to recruitment variability. The efficacy of these processes also depends on the selection of spawning locations by adults and behavior of early-life stages. The topic is briefly addressed in this chapter.

Major breakthroughs in technologies to measure flow and in modeling circulation have advanced the understanding of processes and development of biophysical coupled models describing transport and retention of early-life stages (Cowen & Sponaugle 2009, North *et al.* 2009). In well-studied ecosystems (e.g., Georges Bank, Shelikof Strait, North Sea), circulation patterns are documented and advective processes that coarsely control the fates of early-life stages have been measured and modeled. Tying the biological dynamics of young fish to circulation pathways and physical features is no simple task, and is challenged by both theoretical and practical problems (Helbig & Pepin 1998a, 1998b). Identifying and describing variability in physical transport mechanisms, while important, may have limited utility in explaining recruitment variability unless circulation features are monitored or modeled at appropriate time and spatial scales. It is increasingly apparent that circulation features alone are unlikely to accurately describe or simulate recruitment dynamics in marine fish stocks (Boucher *et al.* 2013) without considering larval behavior and trophodynamics processes during early life.

3.3.8.1 Pathways

The probable role of circulation in the dispersal of early-life stages was recognized a century ago, and was fundamental to Hjort's second hypothesis on causes of recruitment variability (Hjort 1914, 1926). The hypothesis and related issues undoubtedly were debated early in the 20th century by "Committee A" of the International Council for the Exploration of the Sea (ICES Committee A, the "Migrations Committee") as it addressed causes of variability in the abundance of fish stocks (Smith 1994, Sinclair 1997). Committee A focused its attention on interannual variability in oceanography and hydrography in the hope of identifying causes of an observed more than 10-fold variability in the abundance of fish stocks, which might result from: (1) effects of oceanographic processes on early-life dynamics; or (2) variability in adult fish migrations and distributions (Sinclair 1997). Presumably, stocks evolved to spawn and nurse their young where dispersal or retention of early-life stages would assure recruitment and maintain stock integrity. Circulation features on scales of tens to thousands of kilometers act as guideposts defining spawning migrations and behavior.

On the ocean-basin scale, the anti-cyclonic subtropical gyres define reliable circulation pathways utilized by highly migratory pelagic fishes (e.g., tunas and billfishes), guiding adult migrations, acting to broadly define spawning areas, and delineating the drift of early-life stages. For example, albacore *Thunnus alalunga* spawns over broad areas of the central west Pacific, where its larvae and juveniles are entrained within the broad confines of the North Pacific gyre (Bakun 1996). Catadromous anguillid eels (*Anguilla anguilla* and *A. rostrata*) spawn near the Sargasso Sea in the North Atlantic, and larvae are delivered to North America and Europe by the drift pathway defined by the anti-cyclonic gyre (Sinclair 1988). It is probable that interannual variability in oceanic circulation and dispersal losses account for much of the variability in recruitments in these species.

On regional spatial scales, the conceptual "triangle of migration" model of Harden-Jones (1968), as modified by Cushing (1975), describes the life-history patterns of many fishes from mid- and high-latitude continental shelves (see Section 3.2.4). These fishes (pleuronectiform, gadoid, clupeid) spawn on shelf areas and banks defined by weak gyral circulation. A migration circuit includes adult feeding grounds, spawning areas, an egg–larval drift pathway to a defined nursery, and a pathway for juveniles to reach the adult feeding ground. The "triangle" model portrays a mechanism for the maintenance of stock integrity (for a discussion of history of the concept and its relationship to stock maintenance and closure, i.e., philopatry, see Secor 2002, 2005). The base of the triangle, depicting "denatant" drift of eggs and larvae from spawning to nursery grounds, is a critical element of the model. Aberrant denatant drift can lead to failed recruitment (Cushing 1975).

Plaice in the North Sea is a good example that fits the "triangle of migration" model. Adults migrate to well-defined spawning grounds in the Southern Bight of the North Sea. The denatant drift of larvae is towards shallow, nearshore nursery grounds on the Dutch coast, and juveniles subsequently migrate offshore and recruit to the stock (Cushing 1975). This conceptual model provides a simple description of how a species utilizes circulation features to ensure recruitment and stock maintenance, but it lacks important details of biological and physical processes. Other examples in which recruitment depends on adult migrations and then is anchored by the larval drift include Pacific hake *Merluccius productus* off the west coast of North America (Bailey 1981, Hollowed & Bailey 1989) and walleye pollock *Theragra chalcomgramma* in the Gulf of Alaska (Schumacher & Kendall 1995, Hermann *et al.* 1996, Hinckley *et al.* 2001). The numerous stocks of Atlantic herring also provide good examples (Heath *et al.* 1987, Heath & Rankine 1988), as do stocks of Atlantic cod and haddock that spawn in gyres promoting larval retention, such as the Georges Bank (Lough *et al.* 1994, Werner *et al.* 1996, Lough & Manning 2001, Mountain *et al.* 2003, Boucher *et al.* 2013). In a variation on this theme, adult Northeast Arctic cod migrate from the Barents Sea to spawning sites on the western Norwegian coast. Coastal currents then guide a northerly (denatant) larval drift to nursery grounds in coastal areas of northern Norway (Ellertsen *et al.* 1981, 1990).

Dispersal and transport of early-life stages are expressions of evolved behaviors and life histories that rely on circulation features to support connectivity on appropriate scales, thus reducing recruitment variability and ensuring long-term reproductive success (Cowen *et al.* 2006, 2007). Spawning sites are selected because they provide a reliable link to transport pathways for early-life stages, for example South African anchovy *Engraulis encrasicolus* (Shannon *et al.* 1996, Mullon *et al.* 2002, Huggett *et al.* 2003). The major spawning sites of this anchovy are on the western portion of Agulhas Bank near the southern tip of South Africa. The subsequent transport of eggs and larvae is northward to primary nursery areas off the west coast of South Africa, nearly 500 km from the spawning sites. A return southerly migration of pre-recruits and recruitment to the spawning sites completes the circuit.

3.3.8.2 The self-recruitment process

A combination of physical and biological mechanisms, including appropriate behavior by early-life stages, supports self-recruitment in many taxa (see Section 3.2.4.1) (Cowen 2002, Sponaugle *et al.* 2002, Leis & McCormick 2002, Largier 2004; Cowen & Sponaugle 2009). Retentive gyres, eddies and other circulation features operate over a wide expanse of spatial scales. Adults need not undertake risky and energetically costly migrations to distant spawning sites if circulation features ensure that localized spawning will result in a substantial retention of early-life stages. For species that experience self-recruitment (Figure 3.19), a portion of pre-recruit numbers, which often is the largest fraction, is still transported or broadly dispersed from the spawning location to ensure connectivity in support of metapopulation structure over broad geographic scales (Cowen *et al.* 2000, 2009, Cowen 2002, Doherty 2002, Cowen *et al.* 2006).

Variability in recruitment to island ecosystems results from variability in local circulation and retention features, as well as larger-scale circulation that controls the delivery of some recruits from upstream spawning areas (Cowen 2002). In field experiments, significant proportions (15–89%) of settled juveniles of reef fishes were self-recruited: for example, a wrasse *Thalassoma bifasciatum* based on otolith chemistry signatures (Swearer *et al.* 1999); a damselfish *Pomacentrus amboinensis* based on recaptures of settled juveniles whose otoliths were chemically marked during the embryo stage (Jones *et al.* 1999); a damselfish *Stegastes partitus* based on analysis of hydrography, larval cohorts and vertical migratory behavior by larvae, and supported by a particle-tracking model (Paris & Cowen 2004); and a hamlet *Hypoplectrus*

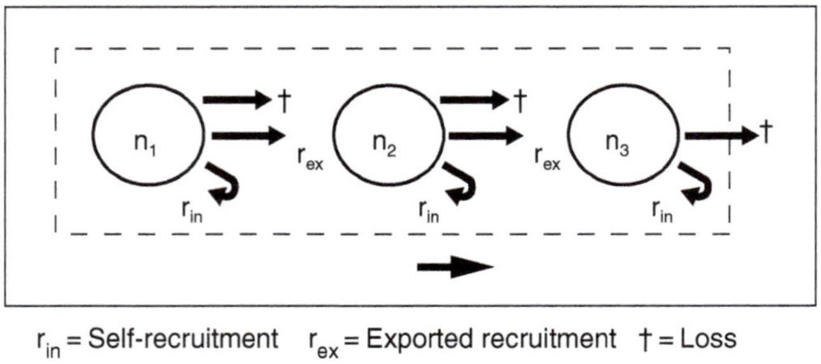

Figure 3.19 The self-recruitment process in fishes. Conceptual model of a metapopulation and its maintenance. Only self-recruitment can populate n_1; n_2 can receive recruits from n_1, self-recruit and export recruits; n_3 can self-recruit and receive recruits from n_2 but cannot export recruits. Reproduced from Hixon & Webster, Density dependence in reef fish populations. In: *Coral Reef Fishes*, Sale, P.F. (Ed.), pp. 303–25, © 2002, with permission of Elsevier.

puella based on genetics analysis (Puebla *et al.* 2009). In a comprehensive study, Basterretxea *et al.* (2013) demonstrated how circulation patterns near Majorca in the Mediterranean Sea resulted in larval retention and the self-recruitment of many coastal fishes. Larval behavior, for example vertical migrations, must play a role in the retention and promotion of self recruitment (Leis 2007). Late-stage larvae of many taxa are now known to have a swimming capability that exceeds the mean velocities of currents in their environment, and their swimming behavior can contribute to self-recruitment (Fisher 2005). In model simulations, Cowen *et al.* (2006) estimated that typical transport distances of larvae of Caribbean fishes were only 10–100 km, thus emphasizing the surprising importance of self-recruitment to island reef ecosystems.

3.3.9 Fronts, features, and processes

Fronts represent boundaries and discontinuities in the sea, and occur at many spatial and temporal scales. Frontal features act with circulation to aggregate and segregate populations, and to promote encounters or interactions of early-life stages with their prey and predators. Fishes often spawn in areas defined by mesoscale fronts and retentive circulation features, an adaptive behavior that has evolved to support recruitment success as well as to maintain integrity of the stock (Sinclair 1988). There are trade-offs associated with aggregation at convergent fronts; for example, feeding opportunities on aggregated plankton prey may be enhanced, but predators on early-life stages also may be in high abundance (Bakun 2006). As such, fronts serve to increase encounter rates between fish larvae and planktonic prey, but also between fish larvae and their predators.

Fronts occur over many spatial scales, and may be either permanent (e.g., shelf-slope fronts) or predictable (e.g., tidal), but they also can be variable (e.g., plume fronts) (Largier 1993, Olson *et al.* 1994, Nakata 1996). Adult fishes may spawn at fronts to promote the retention of eggs, as reported for Japanese sea bass *Lateolabrax japonicus* in the thermohaline coastal front in Tokyo Bay (Nakata 1996). Abundant early-life stages of fish and associated high abundances of plankton organisms that serve as prey, within or at the boundaries of frontal features, suggest that they serve an important role supporting trophodynamics (Munk 1993: Munk *et al.* 1995, 1999), as well as the retention of larvae in a habitat favorable for growth and survival. Passive behaviors by fish larvae in frontal features may be insufficient to assure retention and aggregation. Retention may require appropriate behaviors, such as vertical swimming behavior (Olson *et al.* 1994, Largier 2004).

3.3.9.1 Spatial scales

Features and processes operating at spatial scales ranging from millimeters to thousands of kilometers are important in directing the recruitment process and contributing to variability in recruitments (see Figure 3.7).

Large-scale fronts and systems

Frontal systems associated with oceanic boundaries and the subtropical gyres operate at the largest spatial scales. Intense Western Boundary Currents, the best known being the Kuroshio and Gulf Stream systems, define boundaries and act as major transporters of pelagic organisms, including early-life stages of fishes. Recruitment variability in fishes that occupy coastal areas shoreward of these boundary currents is well studied for sardine and anchovy stocks in the Western Pacific. During the peak years of Japanese sardine *Sardinops melanostictus* recruitment and production in the 1980s, intensive spawning occurred in the Kuroshio front. Large seaward meanders of the Kuroshio were found to be detrimental to recruitments (Nakata *et al.* 1994, 1995), apparently because of the failed transport of larval sardines to nursery areas east of Japan and north of the Kuroshio boundary.

Eastern Boundary Currents define major upwelling systems along the west coasts of continents. Offshore fronts define boundaries of these upwelling systems and nursery areas of many pelagic fishes. Although details differ for particular taxa, reproduction and recruitment of anchovies *Engraulis* spp., sardines *Sardinops* and *Sardina* spp., jack mackerels *Trachurus* spp., and hakes *Merluccius* spp. are keyed to dynamics and features of the coastal upwelling regions (Schwartzlose *et al.* 1999). Breakdowns or failures in upwelling regimes and effects on recruitments, such as the effects of El Niño on Peru anchoveta *Engraulis ringens*, are well documented. El Niños result in lost productivity and recruitment failure of some species, such as anchoveta *E. ringens*.

Meso-scale frontal features

Features at the scale of tens to hundreds of kilometers commonly define the spawning and nursery areas of fishes. Georges Bank is a well-described feature, where an anticyclonic, gyre-like circulation of shallow, well-mixed waters on the Bank supports the reproduction and retention of early-life stages, notably cod and haddock. Spawning is concentrated on the northeast quadrant of Georges Bank, and a large fraction of egg and larval populations is retained on the Bank. In years of favorable winds and high larval retention, recruitment is enhanced (Werner *et al.* 1996, Lough & Manning 2001). Research into tidal mixing, haline and shelf-break fronts in the North Sea (Kiorboe *et al.* 1988, Munk 1993, Munk *et al.* 1995, 1999, 2002), Irish Sea (Lee *et al.* 2005), western Mediterranean (Sabates 1990, Sabates & Olivar 1996), Gulf of St Lawrence (Fortier *et al.* 1992) and Kuroshio Extension (Okazaki *et al.* 2002) provide excellent examples of how such fronts define the distributions of early-life stages in conditions favorable for larval growth (Figure 3.20).

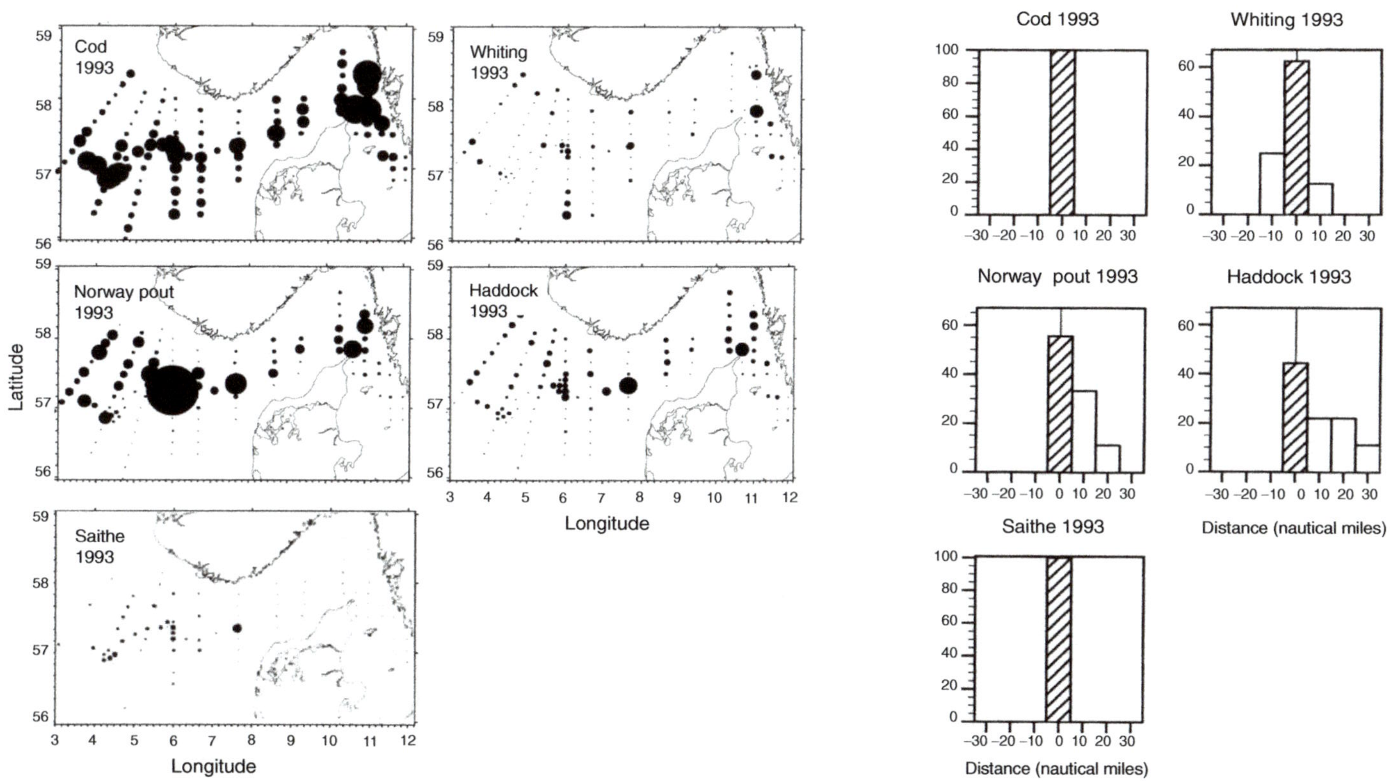

Figure 3.20 Distributions of five species of gadoid fish larvae in the eastern North Sea, 1993. Larvae are concentrated along a well-defined shelf-break front (running west to east). Modified from Munk *et al.* (1999), with permission of Inter-Research.

Growth rates of sprat *S. sprattus*, although not always highest in the front itself, were strongly controlled by the hydrography of a tidal front in the North Sea (Munk 1993). There is strong evidence of aggregation by larvae of some taxa in the large plume front of the Mississippi River. Here, Gulf menhaden *Brevoortia patronus* larvae are five- to tenfold more abundant than in waters landward or seaward of the plume (Govoni *et al.* 1989). Munk *et al.* (1995) listed three probable mechanisms that can account for aggregation and observed assemblage structure of early-life stages at fronts: (i) spawning directly in the frontal zone; (ii) directed horizontal swimming by larvae; and (iii) concentration processes due to convergence of water masses.

Small-scale frontal features

Small-scale frontal features ranging from 10 m to a few kilometers in extent are ubiquitous in aquatic ecosystems, and play important roles in advection, aggregation, retention, orientation, and the facilitation of predator–prey interactions that promote growth and production in early-life stages (Figure 3.21). Examples are tidal fronts, river-plume fronts, Langmuir circulation features, internal waves, thermohaline fronts, estuarine turbidity maxima and vertical stratification features (Iwatsuki *et al.* 1989, Grimes & Kingsford 1996, Nakata 1996, Cowen *et al.* 2000, North & Houde 2001, 2003).

3.3.10 Microturbulence

Physics at micro-scales has the potential to contribute significantly to recruitment variability. Microturbulence (millimeters or less) can increase encounter rates between fish larvae and planktonic prey, increasing the probability of successful feeding (Sundby 1996). Rothschild & Osborn (1988) recognized this possibility and, building in the Gerritsen & Strickler (1977) encounter model, demonstrated how encounter rates between fish larvae and prey could increase in the presence of microturbulence. Most experimental, field and modeling research supports the theory of increased encounters, enhanced prey ingestion and faster growth of larvae under conditions of moderate microturbulence (Sundby & Fossum 1990, MacKenzie *et al.* 1994, Sundby *et al.* 1994, Kiorboe & MacKenzie 1995, Sundby 1997, Dower *et al.* 1997).

Are recruitment levels and variability linked to turbulence and microturbulence? Although theoretical and modeling results are persuasive, there is little direct evidence linking turbulence-enhanced feeding by larvae to explain variability in recruitment. Indirect evidence is supportive, especially in upwelling systems where moderate winds (5–10 m s^{-1}) and associated levels of microturbulence tend to be correlated with high recruitments of clupeoid fishes (Cury & Roy 1989, Roy *et al.* 1992, Bakun 1996, Cowan & Shaw 2002). In these cases, winds of intermediate strength generate intermediate levels of microturbulence that improve feeding conditions for small, first-feeding larvae (i.e., they increase encounter rates without disrupting feeding behavior). For example, Ware & Thomson (1991) found that highest recruitments of Pacific sardine *Sardinops sagax* and Pacific hake in the California Current occurred during decadal periods of moderate (7–8 m s^{-1}) upwelling-favorable winds. Winds in that range generate moderate microturbulence and presumably enhance larval feeding success.

Microturbulence provides a mechanism to increase encounters with prey and thus enhance feeding by fish larvae in ecosystems often thought deficient in levels of prey abundance that

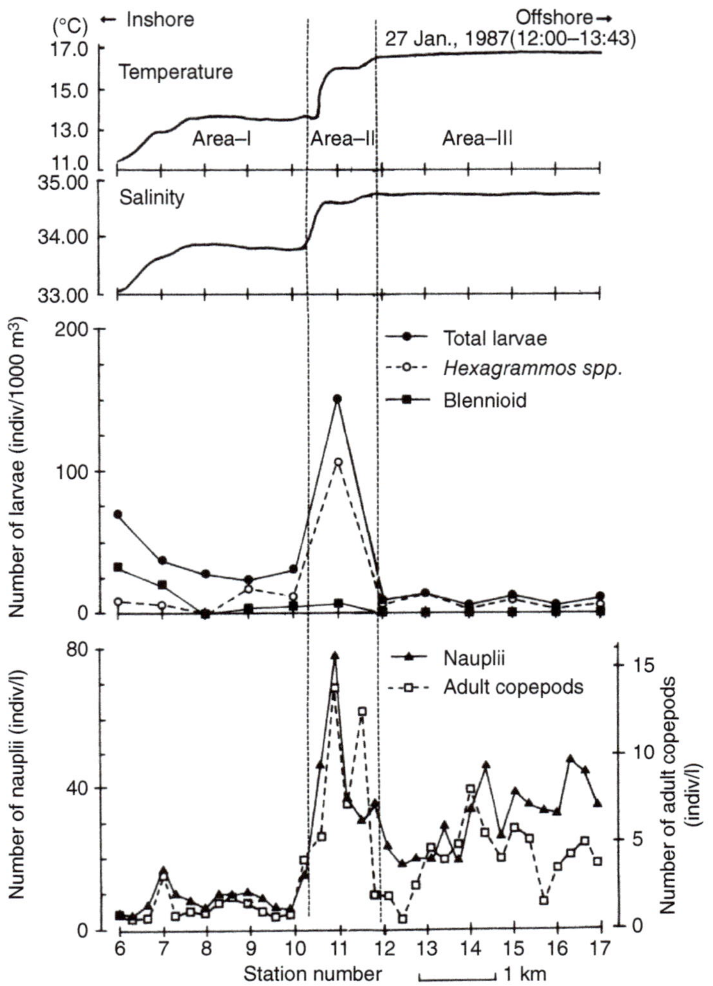

Figure 3.21 Hydrographic conditions and distributions of fish larvae and zooplankton across a thermohaline front in Tokyo Bay. Highest concentrations of ichthyoplankton and zooplankton are in the frontal zone. Reproduced from Iwatsuki *et al.* (1989), with permission of ICES.

could support larval nutrition and growth. Although, in theory, this could lead to increased feeding and growth rates of larvae, microturbulence by itself may be less critical than actual prey levels, the degree of patchiness in prey distributions, and larval foraging behavior. Concentrations of prey organisms in the sea occur in patches at spatial scales of only 0.1 to 1.0 m, and can vary several-fold in concentration (Cassie 1963, Owen 1989).

3.3.11 Climate and weather

Global, regional and local climates control weather patterns and variability that result in variable hydrographic conditions and biological productivity, and ultimately recruitment variability. Weather patterns at regional and ocean-basin scales produce the wind, precipitation and temperature regimes that define physical structure and dynamics. Variability in temperatures, circulation patterns, structure of the water column, and nutrient availability are expressions of weather effects that control levels of primary production and zooplankton production, defining a system's potential to support young fish. Ongoing global climate change is having notable effects on fish distributions and recruitment patterns (Brunel & Boucher 2007) resulting from effects on key environmental and habitat factors that may operate across life stages but often are related to early-life stage dynamics (Petitgas *et al.* 2013).

3.3.11.1 Scales of climate and weather variability

Climate and its variability act on daily to multi-decadal and longer time scales over a broad spectrum of spatial scales to control the productivity of ecosystems and recruitment of fish stocks (Cushing & Dickson 1976, Cushing 1982, Baumgartner *et al.* 1992, Beamish & Noakes 2002, Finney *et al.* 2002, Perry *et al.* 2005, Brander 2010). Decadal climate shifts and associated changes in ecosystems are well known and often referred to as "regime shifts" (e.g., Pacific Decadal Oscillation, North Atlantic Oscillation). Regime shifts operate at ocean-basin scale and can profoundly affect ecosystem productivity and fish recruitment (Francis & Hare 1994, Stenseth *et al.* 2002, Drinkwater 2006). Decadal shifts in recruitment levels can generate 10-fold or greater fluctuations in overall abundance of stocks that perhaps are best documented for clupeoid fishes (Kawasaki 1992, Schwartzlose *et al.* 1999). For sardine species, decadal shifts in abundance resulting from changing levels of recruitment tend to be globally concordant and clearly linked to shifts in oceanographic conditions (McFarlane *et al.* 2002). For sardines and anchovies, regime shifts may in part control alternating, multi-decadal dominance patterns of these taxa (Baumgartner *et al.* 1992, Chavez *et al.* 2003). Duffy-Anderson *et al.* (2005) proposed the term "phase transitions" to describe the shifts in abundance and recruitment success that are precipitated by climate variability, while pointing out the importance of other factors, including life-history characteristics, food-web complexity, and effects of fishing.

Climate variability on sub-decadal timescales, for example the three- to seven-year events associated with warm-water, low-productivity El Niños in boundary currents of the Southeastern Pacific, is associated with alternating high and low recruitments of dominant pelagic fishes. During El Niño events, ocean dynamics, sea temperatures, nutrient availability and productivity shift dramatically in the upwelling ecosystem off Peru and Chile (Barber & Chavez 1983). The effects of El Niño on productivity are strongest in the Peru Current (Arntz 1986), but are felt in ecosystems far removed from the Southeast Pacific (Bailey & Incze 1985, Bailey *et al.* 1995). A strong and well-known expression of El Niño effects on recruitment, and on fishery catches, is seen in the Peru anchoveta, the abundance of which declines more than 10-fold during El Niño years but recovers during years of opposing, cool-water, La Niña conditions (Pauly 1987, Schwartzlose *et al.* 1999) (Figure 3.22).

In many cases, annually variable weather patterns underlie observed 10-fold and greater interannual fluctuations in marine fish recruitments. Atlantic cod and plaice recruitments that vary with respect to temperature (cited above in Section 3.3.3) are good examples. Temperature and freshwater flow each controlled the recruitment of estuarine fishes in the Tagus estuary (Vinagre *et al.* 2009). In another example, the annual recruitment success of anadromous striped bass in Chesapeake Bay and other estuarine nurseries on the east coast of North America is strongly correlated with freshwater flow into the estuary (North & Houde 2001, 2003,

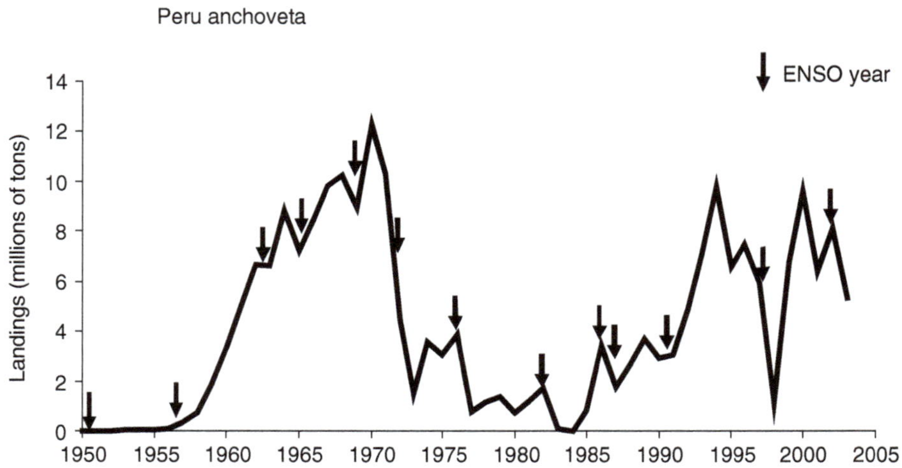

Figure 3.22 Landings of Peru anchoveta *Engraulis ringens* are very variable. Landings are indicative of abundance as affected by recruitment variability in this short-lived species. Low catches and low abundance are associated with El Niño events (indicated by arrows). Developed from FAO fisheries landings data.

Martino & Houde 2004) (Figure 3.23). Recruitments of the shelf-spawning Atlantic menhaden *Brevoortia tyrannus* in Chesapeake Bay follow a contrasting pattern, responding positively to warm, dry, low-precipitation conditions in winter–spring when its late-stage larvae enter estuaries on the east coast of North America (Figure 3.24). Regional weather patterns that

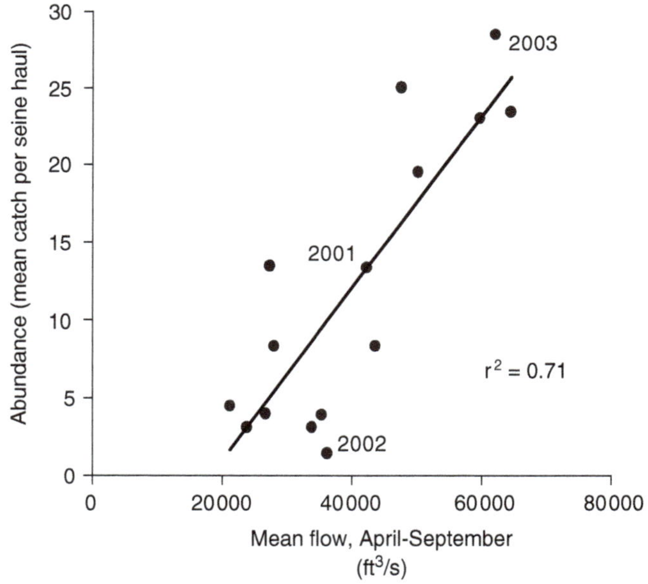

Figure 3.23 Relationship between freshwater flow in spring and summer months and young-of-the-year recruitment levels for anadromous striped bass *Morone saxatilis* in Chesapeake Bay. Years 2001 (average flow), 2002 (low flow) and 2003 (high flow) are highlighted. Modified from Figure 5 of Martino & Houde (2004).

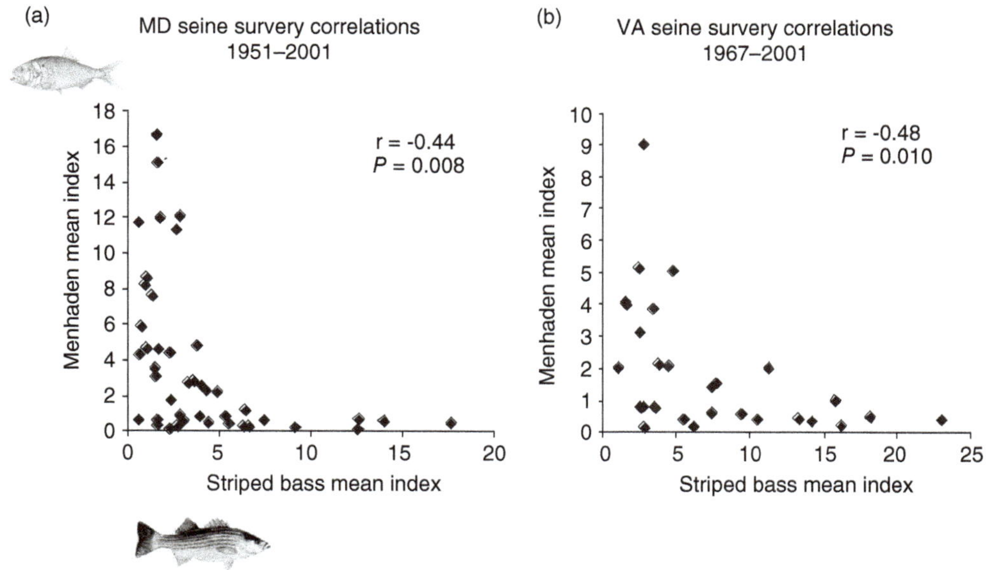

Figure 3.24 Inverse relationship between indices of young-of-the-year abundance of Atlantic menhaden *Brevoortia tyrannus* and striped bass *Morone saxatilis* in Chesapeake Bay. Contrasting climate conditions support high recruitments. Wet and cold weather in late winter–spring is associated with high young-of-the-year recruitment levels for striped bass while warm and dry conditions are best for menhaden. Data from independent (a) Maryland (upper Bay, Maryland Department of Natural Resources, 1951–2001) and (b) Virginia (lower Bay, Virginia Institute of Marine Science, 1967–2001) seine surveys.

coarsely control such contrasting recruitment patterns indicate that variability in frequencies of dominant weather patterns in the months preceding spawning or during the spawning–larval production period control probabilities of high or low recruitment (Kimmel *et al.* 2009, Wood & Austin 2009). For many marine fishes, the relative success of annual recruitment tends to be concordant at approximately 500 km regional scales, but concordance is reduced to only 50 km in freshwater fishes in ecosystems with less connectivity (Myers *et al.* 1995a,c, 1997) and where local weather has bigger impacts (e.g., floods, droughts, weather events).

3.4 Biological (trophodynamic) factors

Hydrography, circulation patterns and water quality provide the environmental mosaic in which early-life stages of fishes live or die. In most circumstances, the success or failure of recruitment is not directly dependent on ambient physical or chemical conditions, but is the result of predator–prey (trophodynamic) interactions mediated by the environment. Interactions between fish larvae, their prey and predators on fish larvae occur at fine scales in the sea, yet few surveys document distributions at those scales (Pepin 2004). Most research on foods and feeding of fish larvae has been aimed at cataloguing prey types, estimating prey concentrations, determining if larvae are feeding selectively on types or sizes of prey, and quantifying consumption. Less attention has been directed to the nutritional quality of prey. Recent reviews of foods and feeding by marine fish larvae are valuable contributions to the state of knowledge (Peck *et al.* 2012, Llopiz 2013). While the major, direct cause of mortality in early life is believed to be from predation (Bailey & Houde 1989), nutritional condition and growth rates of fish larvae are contributing factors.

3.4.1 Prey resources

Central to many recruitment hypotheses (e.g., Critical Period, Match–Mismatch, Stable Ocean) is the apparent low abundance of prey resources available to larval fish. Larval fishes feed primarily, but not exclusively, as carnivores, even in taxa that become herbivores as juveniles and adults. Accumulated evidence now indicates that a wide diversity of prey types may be ingested by larvae (see Peck *et al.* 2012, 2013, Llopiz 2013). Most fish larvae feed predominantly on small, living plankton organisms. Zooplankton, especially stages of copepods, dominate larval diets, but other prey – or the flexibility to utilize alternative prey – can be important. For some fishes, the availability of copepods to larvae is hypothesized to exercise major control over recruitment success, for example Atlantic cod, the recruitments of which in the northeast Atlantic depend on temporal-spatial concordance (i.e., "Match–Mismatch") between larvae and the copepod *Calanus finmarchicus* (reviewed by Sundby 2000) or similar concordance with copepod taxa in other North Atlantic regions (Heath & Lough 2007).

In some taxa, larvae initiate feeding on phytoplankton, especially diatoms and dinoflagellates, and on protozoa (Lebour 1916, Hunter & Thomas 1974, Lasker 1975, Last 1978a,b, Stoecker & Govoni 1984, Van der Meeren 1991, Hunt von Herbing & Gallagher 2000). The diversity of prey in larval diets is higher in fish larvae from low latitudes than in larvae from temperate seas (Peck *et al.* 2012). In Lasker's Stable Ocean hypothesis, motile dinoflagellates were eaten and proposed to be a critical prey resource for the survival of northern anchovy larvae. Fukami *et al.* (1999) and de Figueiredo *et al.* (2007) reported that many larval fishes from Japanese coastal waters and the Irish Sea, respectively, consumed protozoa as a substantial part of their diets. The larvae of some fishes are feeding specialists and may select particular prey types; examples are the appendicularian *Oikopleura* spp. by some pleuronectiform (White 1968, Last 1978a) and other larval fishes (Llopiz 2013). Fish larvae may be the initial prey of first-feeding larvae of Spanish mackerel *Scomberomorus* spp. (Shoji & Tanaka 2001), and are consumed by other scombrid larvae and the larvae of other taxa (Llopiz 2013).

In general, there is a direct relationship between the size of fish larvae and the size of the ingested prey (Shirota 1970, Hunter 1981, Houde 1997a). Selection for prey size is a function of mouth gape, and larvae are categorized as gape-limited predators. The sizes of selected prey differ among fish larvae and may shift during ontogeny (Munk 1997, Pepin & Penney 1997, Sabates & Saiz 2000, Catalan *et al.* 2011). Preferred prey usually is in the range of about 2–10% of the larval body lengths. Munk (1997) indicated that the preferred prey size for Atlantic cod larvae centered on 5.1% of body length, but that preferred prey size for Atlantic herring was only 2.7% of body length. Niche breadth (relative variability of prey sizes in the diet) appears to be narrow and constant during ontogeny for larvae of some, but not all, species (Pearre 1986, Munk 1992, 1997, Pepin & Penney 1997, Sabates & Saiz 2000, Auth 2003, Campfield 2004, Reiss *et al.* 2005, Catalan *et al.* 2011). If niche breadth expands, in theory this signals a wider range and a greater abundance of available prey that potentially could benefit the survival and growth of fish larvae in a prey-limited environment. However, evidence indicates that, while the sizes of consumed prey increase as larvae grow, niche breadth (i.e., the ratio of prey size to predator size, and its variability) may increase, remain relatively constant, or even decline in some cases (Llopiz 2013).

3.4.1.1 *Prey concentrations*

Is prey sufficiently abundant on average to support the feeding of marine fish larvae, or is it limiting for growth, survival, and recruitment? This question is a foundation of Hjort's Critical Period and related hypotheses. In the sea, concentrations of plankton organisms eaten by fish larvae range over at least five orders of magnitude (from $<10^3$ to $>10^7$ per m^3). Early research in the laboratory and in aquaculture trials has suggested that average levels of favored and suitable prey in the sea (e.g., copepod stages), which often range from 1 to 100 per liter, were below concentrations thought capable of supporting larval feeding and growth (Lasker 1975, May 1974, Hunter 1981). To account for this apparent deficit, it was proposed that prey patchiness at various spatial scales, or processes that elevate encounter rates between larval fishes and prey, would explain larval survival under low prey conditions (Lasker *et al.* 1970, O'Connell & Raymond 1970, May 1974, Lasker 1975, Hunter 1972, 1981).

Mechanisms that increase the availability of prey generally depend on a combination of physical processes (aggregating, concentrating) and larval behaviors that raise the probability of encounters with prey. Aggregating mechanisms and larval behavior clearly improve feeding conditions and support higher growth. However, failed or poor feeding reported in experiments conducted during the 1960s and 1970s at prey concentrations commonly found in the sea were most likely in large part artifacts resulting from enclosure in small experimental systems (Houde 1978, Oiestad 1985). Subsequently, improved experimental and culture methods resulted in lower estimates of prey levels required for survival. For example, Munk (1995) found that prey levels of $<10\ l^{-1}$ were adequate to support feeding by Atlantic cod larvae; this level was about an order of magnitude lower than earlier estimates for cod and haddock larvae (Laurence 1974, Solberg & Tilseth 1984). In a recent synthesis, prey concentrations in the sea, ranging from 5 to $10\ l^{-1}$, were reported to support feeding by many fish larvae (Peck *et al.* 2013). As knowledge on diets and feeding has accumulated, it is now apparent that a broader diversity and, therefore, higher concentrations of prey types, are available to fish larvae in the sea (Van der Meeren 1991, Hunt von Herbing & Gallagher 2000, Peck *et al.* 2012).

Prey levels in the sea, while generally capable of sustaining some larval production, are potentially limiting to larval growth. Paradoxically, larval survivors collected at sea typically have grown at near-maximum rates (Folkvord 2005). Such results suggest that survivors experienced above-average feeding conditions attributable to: (i) a high abundance of prey; (ii) prey aggregation by physical processes (e.g., at fronts and from vertical stratification); (iii) larval behaviors that increase encounters with prey; (iv) physical processes that increase encounters with prey, such as microturbulence (see Sections 3.3.10 and 3.4.1.4); or (v) a combination of mechanisms. It also is possible that the fast growth of survivors in part represents the outcome of size- or growth rate-selective predation mortality, or starvation of slower-growing larvae in a cohort at ambient prey levels (see Section 3.4.2).

3.4.1.2 *Ration size and requirements*

Larval fishes are big consumers and demand a large daily ration. On average, marine fish larvae must consume more than 50% of their body weight daily to grow at mean rates observed in surviving larvae (Houde & Zastrow 1993). Weight-specific rations (d^{-1}) exceeding 100% are common (Houde & Schekter 1981, Houde 1994, Fiksen & Folkvord 1999, Wuenschel & Werner 2004). Not surprisingly, rations are temperature-dependent (Figure 3.25). A

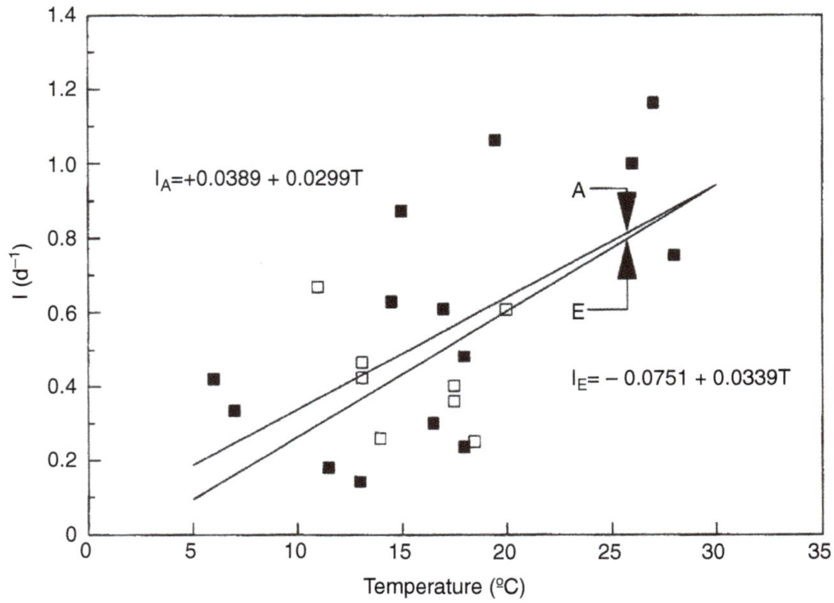

Figure 3.25 Estimated weight-specific, daily rations (I) of fish larvae in relation to temperature, based on a meta-analysis conducted by Houde & Zastrow (1993; see their Figure 15). Rations of fish larvae increase by approximately 3% for each degree increase in temperature. The temperature-adjusted mean daily ration of marine fish larvae is 57% of larval body weight. The two similar regression equations were derived based on different approaches to estimate growth and growth efficiencies, from which I was estimated. Filled symbols represent marine larvae, open symbols represent freshwater larvae.

weight-specific ration of a typical marine fish larva is 35% at 10 °C and this increases to 69% at 20 °C (Houde & Zastrow 1993). The high food requirements to attain average growth lend tacit support to Hjort's Critical Period and other hypotheses that focus on food limitation. In order to grow at moderate to fast rates characteristic of survivors, larvae are committed to foraging more or less constantly when light conditions allow visual feeding.

3.4.1.3 Ontogeny and feeding success

Upon yolk absorption, larval fish have a brief period to begin feeding exogenously on planktonic prey before they reach "the point-of-no-return" (PNR), beyond which the larvae are deprived nutritionally and unable to feed, even if prey becomes available. At the PNR, starvation is irreversible (Blaxter & Hempel 1963, Blaxter 1988). The PNR may be reached in a single day for small larvae in tropical ecosystems, but can be as long as 10–20 days in colder seas. The PNR and time to starvation mortality are reached sooner in species with small larvae than in species with large larvae (Miller *et al.* 1988).

When considered in the context of Hjort's Critical Period hypothesis, the transitional period between yolk nutrition and exogenous feeding may indeed be a critical time for individual larvae, and perhaps for recruitment success. If prey of appropriate size and abundance is scarce, the starvation of weak-swimming, gape-limited larvae will ensue. First-feeding larvae of some taxa, especially clupeoids, have low feeding incidence (Figure 3.26), although the feeding incidence may be higher in other taxa and apparently is higher for first-feeding larvae from tropical

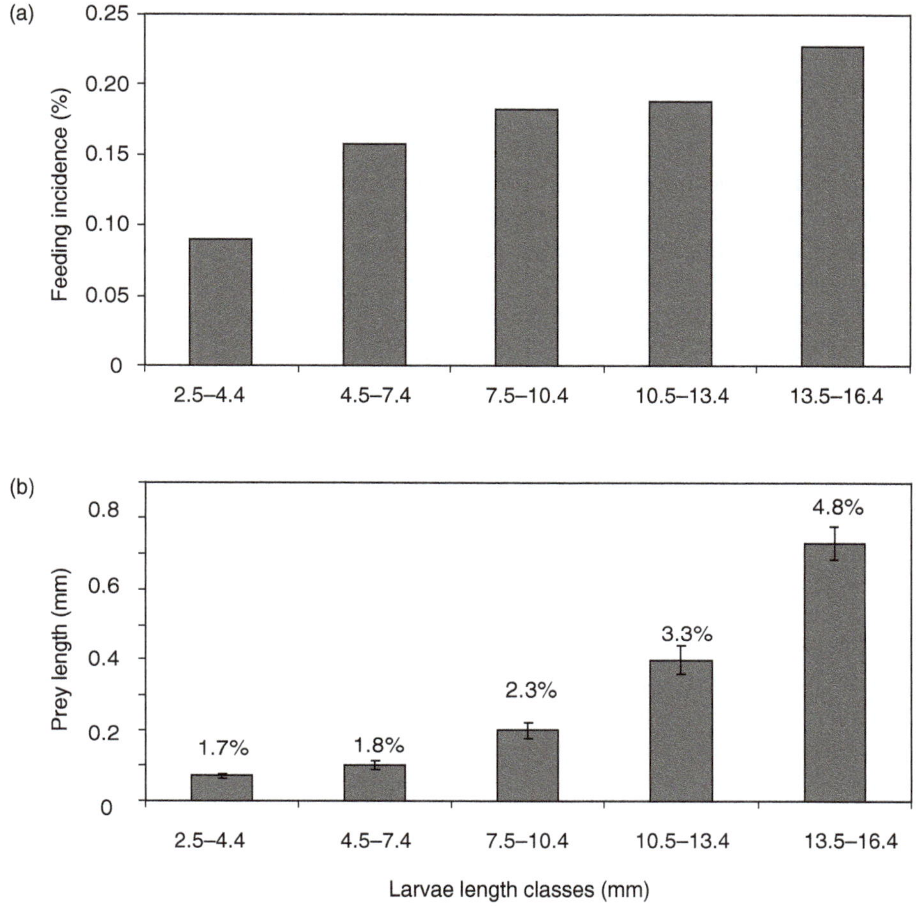

Figure 3.26 (a) Feeding incidence in relation to length of bay anchovy *Anchoa mitchilli* larvae, Chesapeake Bay. Feeding incidence is the proportion of larvae with food in gut. As in many clupeoid larvae, feeding incidence is low, possibly from voiding of gut contents during collection; (b) Prey lengths in relation to bay anchovy larval lengths, Chesapeake Bay. Mean prey length and relative prey size (%) both increase for bay anchovy larvae. Reproduced from data in Auth (2003), with permission of the author.

seas than for larvae from higher latitudes (Llopiz 2013). The percentage of successful prey-capture attempts by first-feeding larvae ranges widely from <10% (Hunter 1981) to >80% (Chesney 2008), and increases as the larvae develop and grow.

Fish larvae are big consumers; for example, based on the synthesized results of Houde & Zastrow (1993) and Houde (1994), a first-feeding cod larva of 50 μg dry weight must consume approximately 60 prey of 0.25 μg dry weight during a 14-h daylight feeding period (i.e., 4.3 prey per hour) to grow at the average rate reported for cod larvae. If the prey-capture success were 25%, then a cod larva must encounter and attempt to eat a minimum of 240 prey per day (17 prey attacks per hour). In laboratory experiments, Munk (1995) determined that cod larvae could consume 0.26 copepod nauplii per minute, with a capture success of 22% at a relatively low prey level of 14 per liter. Expanding Munk's estimate to a 14-h period gives 218 nauplii consumed and 993 attacks (71 per hour) under these experimental conditions. This level of feeding is 4.1-fold higher than the minimum required to support growth at an average level. Munk (1995) emphasized the flexibility in larval cod swimming and feeding behavior in relation to prey density that allows larvae to feed successfully even at prey levels as low as 2 per liter.

In another example, larvae of bay anchovy, a species that lives at high temperatures (25–30 °C) and weigh 25 μg dry weight, must ingest prey weighing 0.25 μg dry weight at 6.9 per hour during a 14-h feeding period (weight-specific ration = 97%) to support average growth. At 25% prey capture efficiency, an anchovy larva must attack 27.6 prey per hour. If prey capture success is higher – perhaps as high as 83% as Chesney (2008) indicates – then a bay anchovy larva would need to attack only 8.3 prey per hour to meet minimum ingestion requirements. Houde & Schekter (1981) estimated that the larval weight-specific ration of bay anchovy was >200% under ideal feeding conditions, representing 57.1 and 17.2 prey attacks per hour at 25% or 83% attack success rates, respectively.

The feeding success initially is low, but increases as the larvae grow and develop (see Figure 3.26). With an increased mouth size, the spectrum of prey types and sizes available to be eaten also expands (Hunter 1981, Pearre 1986, Munk 1992, 1997). In the case of bay anchovy, the relative prey size increases from 1.7% to 4.8% of larval length as the larvae grow from 3.5 to 15.5 mm (see Figure 3.26). In some cases the larvae may show a strong selection for preferred prey sizes, especially at high prey concentrations. In Atlantic herring, the preferred prey size is 2.5–3.0% of the larval herring body length (Munk 1992), while cod larvae prefer prey that is approximately 5.0% of their body length (Munk 1995, Fiksen & MacKenzie 2002). Relative variability in prey size (i.e., trophic niche breadth) may be constant, may increase, or may decrease during ontogeny and growth (Pepin & Penney 1997, Llopiz 2013). For white perch, the prey sizes increased with respect to larval length and ontogenetic state (Campfield 2004, Campfield & Houde 2011), but the niche breadth did not differ significantly over the size range of larvae analyzed (Figure 3.27). In Australian snapper *Chrysophrys auratus* larvae, the trophic niche breadth increased in years of low larval abundance and low prey abundance (Murphy *et al.* 2012)

The threat of starvation may decline during ontogeny because larger and faster-swimming larvae encounter more potential prey, but the availability of large prey with sufficient energy content to support growth and survival may be limiting. The threat of imminent starvation declines during ontogeny, but low growth rates under poor feeding conditions may threaten survival if larval-stage durations are extended while vulnerability to predation remains high (Cushing 1975, Houde 1987, Anderson 1988).

Food-limiting critical periods may at times constrain first-feeding larvae (*sensu* Hjort). However, the metamorphosis period in some fishes also may represent a stage of critical feeding limitation because of ontogenetic changes occurring at that time. For example, feeding incidence declined significantly during metamorphosis in Atlantic cod (Thorisson 1994) and Japanese seabass (Islam & Tanaka 2006). These authors speculated that recruitment levels may be sensitive to reduced feeding success during metamorphosis.

3.4.1.4 Physics and feeding: mechanisms and features

Small planktonic organisms that serve as prey for fish larvae are patchily distributed in the sea. Factors that elevate the probability of encounter with prey and patches of prey favor a

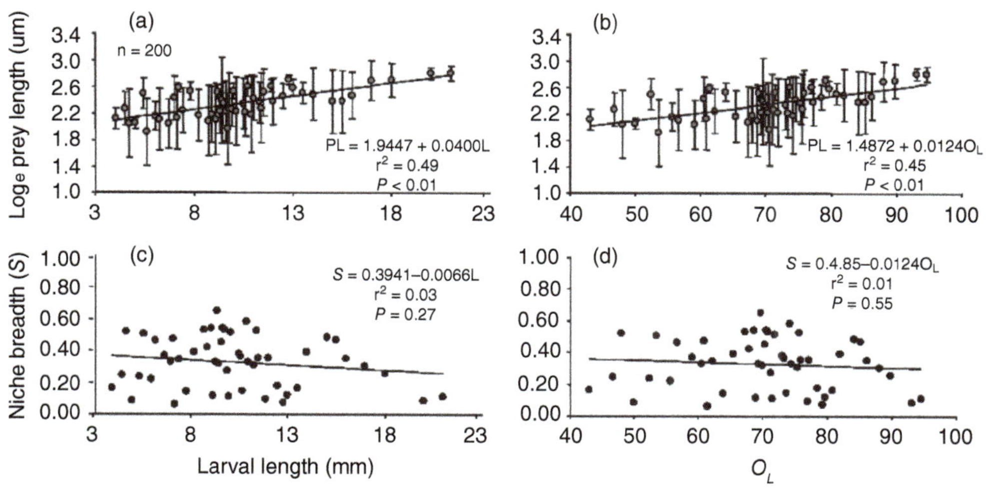

Figure 3.27 (a, b) Size of prey (PL) increases significantly in relation to larval length and ontogenetic state (O_L) for white perch *Morone americana*, in the Patuxent River tidal sub-estuary, Chesapeake Bay; (c, d) Niche breadth (*S*), i.e. relative variability in prey size. *S* does not differ significantly as white perch larvae grow and develop. O_L, the ontogenetic index, is a metric describing per cent completion of larval development (see Fuiman 2002). Reproduced from Campfield (2004), with permission of the author.

fast growth of larvae, while factors that diminish such encounter probability act negatively on growth and condition. Patchiness in prey distributions occurs over broad spatial and temporal scales. Zooplankton suitable as larval prey can exhibit patchy concentrations on spatial scales at least as small as 10 cm (Cassie 1963, Owen 1989), a scale that is virtually never sampled or reported in ichthyoplankton surveys (Pepin 2004). Even in the relatively homogeneous surface mixed layer, turbulence – combined with motility and specific behaviors of plankton organisms – ensures that distributions are not uniform or random, but are patchy in both vertical and horizontal dimensions. Vertical stratification may effectively promote a layering of plankton organisms at relatively high concentrations. For larvae of northern anchovy the pycnocline serves as a layer that aggregates their dinoflagellate prey, and is regarded as evidence in support of the Stable Ocean hypothesis (Lasker 1975, 1978). Numerous frontal features at small (e.g., Langmuir cells) to mesoscale (e.g., shelf-break fronts) can serve a similar purpose by concentrating or aggregating larvae and prey, as well as reducing the probability of dispersal (e.g., Figure 3.18, 3.20, 3.21) that may lower feeding success (Kingsford 1990).

3.4.1.5 Turbulence and microturbulence

Small-scale turbulence (see Section 3.3.10) increases the potential of first-feeding fish larvae to meet their nutritional needs at low prey concentrations. Modeled ingestion rates and field observations indicate that microturbulence can raise the ingestion rates of larvae by three- to fivefold under moderate wind conditions (Rothschild & Osborn 1988, MacKenzie *et al.* 1994, Sundby 1997). Although there are pitfalls in interpreting how microturbulence affects larval behavior, feeding and growth (Browman 1996, Browman & Skiftesvik 1996), it is quite clear that increases in turbulence in response to increasing wind speeds will raise encounter rates between fish larvae and planktonic prey. However, at high wind velocities and levels of

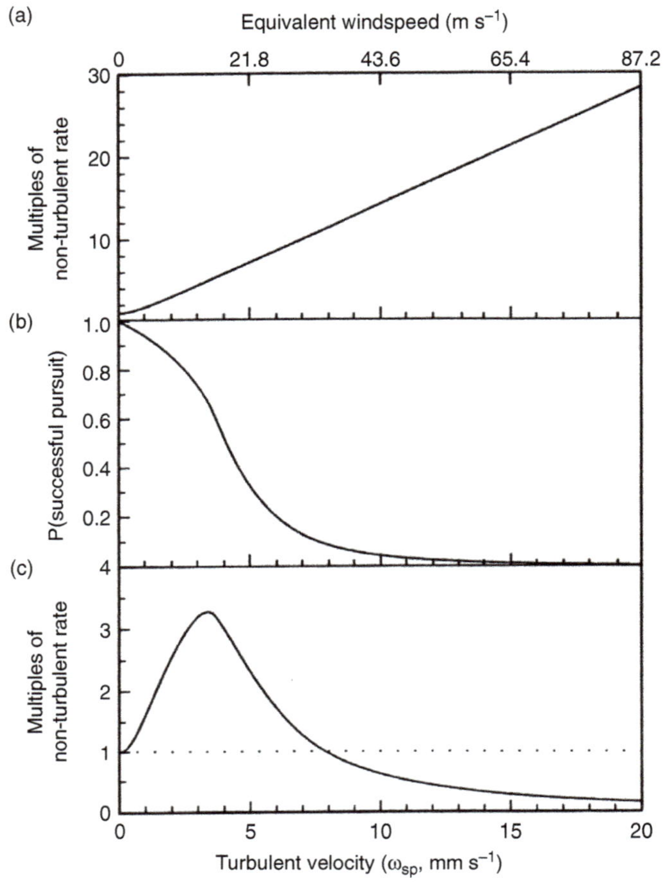

Figure 3.28 Micro-turbulence. Effects of wind-induced microturbulence on (a) encounter rates, (b) successful pursuit of prey, and (c) feeding rates of fish larvae. Intermediate levels of microturbulence provide greatest enhancements to feeding. Reproduced from MacKenzie, B.R., Miller, T.J., Cyr, S. & Leggett, W.C. (1994) Evidence for a dome-shaped relationship between turbulence and larval fish ingestion rates. *shape Limnology and Oceanography*, **38**, 1790–99, © 1994, with permission of the American Society of Limnology and Oceanography, Inc.

microturbulence, the feeding behavior of larvae will be disrupted, resulting in a dome-shaped feeding response (MacKenzie *et al.* 1994), with maximum feeding success at intermediate winds and levels of microturbulence (Figure 3.28). For small larvae, peak feeding success is predicted for wind speeds between 5 and 20 m s^{-1}.

In a field study, Sundby & Fossum (1990) concluded that the contact rate with zooplankton prey and prey capture by Norwegian cod larvae were enhanced more than twofold under moderate winds and microturbulence. In an individual-based model for walleye pollock, modeled consumption peaked at intermediate levels of microturbulence and smallest larvae were the most responsive (Megrey & Hinckley 2001). In the sea, wind speeds and associated microturbulence that maximize feeding by fish larvae (5–10 m s^{-1}) have tended to be lower than in model simulations (10–20 m s^{-1}) or laboratory experiments (15–20 m s^{-1}). These discrepancies may indicate that the models were poorly parameterized, built on poor assumptions regarding prey patchiness and larval behavior (Dower *et al.* 1997), or they may indicate artifacts in small-tank experiments. Galbraith *et al.* (2004) argued that the behavior of fish larvae and search-space geometry in foraging models determines whether modeled microturbulence is considered effective in enhancing feeding and survival. Although most studies have concluded that microturbulence enhances feeding, Reiss *et al.* (2002) detected no effect on consumption by larval silver hake *Merluccius bilinearis* on the Scotian Shelf. In a foraging model for European anchovy, Urtizberea and Fiksen (2013) did not find a positive effect of microturbulence on feeding success in the smallest larvae because of their low capture ability, small size of prey, and small visual volume.

3.4.2 *Predation: a size-selective process*

Predation inflicts mortality on young fish and, as such, can be an important determinant of recruitment variability (Bailey & Houde 1989). With few exceptions, predation in aquatic ecosystems is size-structured; large predators eat smaller prey. Fish larvae are small and vulnerable to a diverse suite of vertebrate and invertebrate predators, most of which feed size-selectively or are gape-limited (Paradis *et al.* 1996, 1999). Size-selective predation often favors the survival of bigger and/or faster-growing fish larvae. The probable relationships and implications were recognized years ago (Ware 1975, Miller *et al.* 1988); namely that if predation is size-selective, the mortality rates of early-life stages are inversely related to size (Peterson & Wroblewski 1984, McGurk 1986, Houde 1997b).

The susceptibility of early-life stages to predation depends on the probability of encounter and ability to avoid capture after an encounter (Bailey & Houde 1989). There is accumulating evidence that smaller and/or slower-growing larvae are more susceptible to predation than larger or faster-growing larvae (Meekan & Fortier 1996, Cowan & Shaw 2002, Takasuka *et al.* 2003, 2004b) and this selective force extends into the juvenile stage (Sogard 1997). In plankton assemblages that include ichthyoplankton, there generally is a broad range of organism sizes, including fish larvae, and a diverse suite of predators (taxa and sizes). In model simulations, predator size and variability in growth rates of larval fishes were among the most important factors controlling survival of the larvae (Cowan *et al.* 1996, 1997, Letcher *et al.* 1996). While predation generally is most intense on the smallest larvae and youngest stages, under some circumstances an abundant predator of a particular size could act as "gatekeeper," creating a bottleneck to recruitment by selectively consuming large larvae or newly settled juveniles in the predator's preferred size range (Leggett & Deblois 1994, Sogard 1997).

3.4.2.1 *Is bigger better? Or, is faster better? Does it matter?*

There is strong evidence supporting the hypothesis that larger size in early life is associated with faster growth, better survival, and metamorphosis at younger age (Houde 1987, 1989b, Anderson 1988, Meekan & Fortier 1996, Takahashi & Watanabe 2004). Predation is hypothesized to selectively remove the smallest and/or slowest-growing larvae. For larval fishes, it is apparent that faster growth in general, and a larger size in many cases, confer advantages for survival and potential to recruit.

Convincing evidence that faster-growing and bigger larvae have a greater potential to recruit comes from otolith microstructure analysis. In this approach, otoliths of juvenile survivors (recruits) are analyzed and the back-calculated growth rates during the larval stage are compared with growth rates and sizes-at-age of the larval population that had been sampled at earlier dates (Hovenkamp 1992, Meekan & Fortier 1996). Results in many cases have demonstrated that the fastest growing and bigger larvae are the survivors. Indirect, theoretical support for the "bigger is better" hypothesis is inferred from size-spectrum theory and size-structured predation in aquatic ecosystems (Kerr & Dickie 2002). Additional support comes from modeling experiments on simulated larval populations exposed to predation. Individual-based models usually indicate that larval populations with high mean growth rates and high variance in growth rates have a greater potential to survive and recruit (Rice *et al.* 1993, Cowan *et al.* 1996, 1997, Letcher *et al.* 1996). However, Letcher & Rice (1997) noted that patchiness of prey could play a dominant role in controlling growth and survival of modeled larval populations, and that under some circumstances cohort survival was not necessarily associated with fastest growth.

Negative evidence for "bigger is better" or "faster is better" is based primarily on laboratory experiments in which larval fish (often in a small size range) are exposed to specific predators (often of fixed size). Under such conditions, either no size preference or even a preference for bigger larvae by the predator sometimes is observed (Litvak & Leggett 1992, Pepin *et al.* 1992). In the sea, fish larvae, their predators and their zooplankton prey coexist as components of diverse, size-structured communities in which there are numerous alternative predator–prey pathways and linkages. Levels of predation mortality usually scale to size (Peterson & Wroblewski 1984); small organisms, including the smallest early-life stages of fishes, are expected to suffer higher mortality rates. Bottlenecks are possible, of course, in which an abundant predator of particular size selectively and intensively predates larger, or faster-growing, larval fish prey. In some reports, little or no evidence of size- or growth-rate selection was detected, based on back-calculated growth rates of a larval cohort, especially under circumstances where environmental conditions were generally favorable for growth, and the predation pressure was low (Robert *et al.* 2007). However, even in these cases there often is evidence of selection for survival of larger or faster- growing individuals during brief periods, or at particular stages, for example size-at-hatch (Raventos & Macpherson 2005) or growth rate at metamorphosis (Searcy & Sponaugle 2001).

Evidence from otolith microstructure analysis on plaice (Hovenkamp 1992), Atlantic cod (Meekan & Fortier 1996, Nielsen & Munk 2004), damselfish *Neopomacentrus filamentosus* (Vigliola & Meekan 2002), a surgeonfish *Acanthurus chirurgus* (Bergenius *et al.* 2002), European anchovy (Allain *et al.* 2003), Japanese anchovy (Takahashi & Watanabe 2004, Takasuka *et al.* 2004b), and Pacific bluefin tuna *Thunnus orientalis* (Tanaka *et al.* 2006, Satoh *et al.* 2013) larvae strongly indicates that faster-growing larvae have a higher potential to recruit. Strong support for "faster is better" and often "bigger is better" emerges from research on tropical reef fishes (Searcy & Sponaugle 2001, Bergenius *et al.* 2002, Shima & Findlay 2002, Vigliola & Meekan 2002, Wilson & Meekan 2002). For example, in the damselfish *N. filamentosus*, a large size at hatch conferred a survival advantage to post-settlement recruits (Vigliola & Meekan 2002), while fast growth of the larval surgeonfish *A. chirurgus* was selectively advantageous for settlement (Bergenius *et al.* 2002) (Figure 3.29). Both size- and growth-selective predation processes operate throughout the egg–larval stage and continue into post-settlement and juvenile stages. For example, the application of otolith microstructure analysis indicated that both size- and growth rate-selective processes operated on larval-stage bluefish *Pomatomus saltatrix* (Hare & Cowen 1997); larger and faster-growing bluefish larvae were

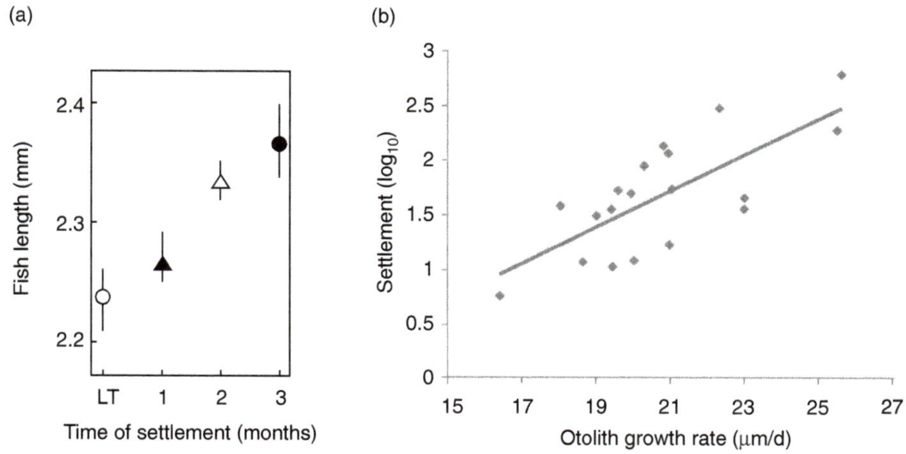

Figure 3.29 Size- and growth-rate selective survival. (a) Back-calculated sizes-at-hatch based on otolith microstructure analysis of settlers of a damselfish *Neopomacentrus filamentosus* collected at time of settlement (LT) and at one, two, and three months after settlement. Larvae that were large at hatch were selected for recruitment. From Vigliola and Meekan (2002; see their Figure 3); (b) Survivor abundance at settlement in a surgeonfish *Acanthurus chirurgus* is directly correlated with otolith growth rate. Reproduced from Bergenius *et al.* (2002), with kind permission of Springer Science and Business Media.

represented disproportionately in the population at successive ages. Searcy & Sponaugle (2001) reported that size- or growth-selective mortality molded the size structure of recruiting populations of juvenile wrasses *Thalassoma bifasciatum* and *Halichoeres bivittatus* (Labridae). In this case, the strongest selection occurred during the brief metamorphosis and early post-settlement periods. Raventos & Macpherson (2005) reviewed research that identified size- and growth-selective mortality during the recruitment process. In their study on two reef fishes *Symphodus roissali* and *S. ocellatus* (Labridae), it was found that only large size-at-hatch consistently acted as a selective force increasing probability of survival. These authors could not explicitly conclude that selection was attributable to predation, although it was the probable agent of mortality. In Atlantic cod, size- and growth-rate selective mortality occur that favor big and fast-growing larvae (Meekan & Fortier 1996, Nielsen & Munk 2004); in this case, mortality of the cod larvae could be attributable to either predation or nutrition-related factors.

The slow-growing fraction of a larval Japanese anchovy population is most vulnerable to cannibalism and small predators, even for larvae that are of the same size (Takasuka *et al.* 2003, 2004a). Takasuka *et al.* (2004b) reported that size- or growth-rate selectivity depended on the taxon and size of predator, but that surviving recruits in the anchovy population had, on average, grown rapidly during the larval stage (Takasuka *et al.* 2004b). Based on laboratory experiments and also observations in the Kuroshio–Oyashio system, Takahashi & Watanabe (2004) concluded that fast-growing larvae of Japanese anchovy had a higher probability to recruit.

3.4.3 *Linking growth, survival, and cohort production*

Recruitment variability derives from responses to combined energetic and numerical processes operating on cohorts during early life (Jones & McCormick 2002). Variability in either mortality or growth rates can generate recruitment variability (Cushing 1975, Houde 1987, 2008). The Stage Duration and related Growth-rate Selection hypotheses recognize the linkage. Favorable environmental conditions promote fast growth and shorten larval stage durations, thus reducing the probability of predation or starvation mortality. Variability in survival and recruitment is generated via variable cumulative mortalities of cohorts (or year classes) during the pre-recruit period. Variability in cumulative mortality ($\Sigma M = M \times t$) and resulting variability in recruitment level are governed by both daily mortality rates (M) and stage durations ($t \approx 1/G$), where G is the weight-specific growth rate (Sissenwine 1984, Houde 2002).

Cumulative mortality (ΣM) of cohorts is linked to variability in both M and G. Levels and variability in M/G during early life express potential variability in stage-specific survival. Processes governing variability in M and G often are size-specific and stage-based. Houde (1997b) outlined an approach to evaluate the size-specific or stage-specific mortality and growth rates of marine fish larvae. Stage-specific mortality rates are an indicator of numerical dynamics (see Beyer 1989), and are linked to bioenergetics processes. Stage-specific mortality is:

$$M_s = (M/G) \cdot \log_e[W_s/W_{s-1}] \quad (1)$$

where W_s and Ws_{-1} are fish weights at two stages. If M/G remains constant, then stage-specific mortality for stages s_{-1} and s, is stable, even if M and G vary. Stage-specific survival (S_s) is a function of M/G:

$$S_s = N_s/N_{s-1} = exp(-M_s) = [W_s/W_{s-1}]^{-M/G} \quad (2)$$

where N_s and N_{s-1} are cohort abundances at two stages. A further link to bioenergetics processes and cohort biomass proliferation is:

$$B_s = B_{s-1}[W_s/W_{s-1}]^{[1-(M/G)]} \quad (3)$$

where B_s and B_{s-1} are cohort biomasses at successive life stages. Werner & Gilliam (1984) discussed M/G in the context of life history strategies as an indicator of fitness, and proposed that increases in this ratio during ontogeny trigger shifts in habitat utilization, increasing survival probability.

Applying these formulations, Houde (1997b) examined the relationships between M, G, and M/G with respect to body mass for the larvae of five species of fish. In general, both M and M/G declined consistently as power functions of larval mass:

$$(M = a \cdot W^b \text{and} \quad M/G = c \cdot W^d) \quad (4)$$

The species-specific exponents in these power models ranged from b = –0.3 to –0.6 (mean = –0.42) for M, and from d = –0.1 to –0.6 (mean = –0.38) for M/G. The relationship of G to body mass, however, was less clear. Weight-specific growth generally decreases as body mass increases during the larval stage (Houde 1997b), but with notable exceptions. For example, G increases in a domed relationship (Figure 3.30) with respect to body mass in Atlantic cod (Otterlei *et al.* 1999, Buckley *et al.* 2006) and striped bass larvae (Houde 1997b).

Most cohorts of larval-stage fishes initially suffer high mortality, and large numerical and biomass losses during a period when $M/G > 1$. Even a small variability in M/G generates a

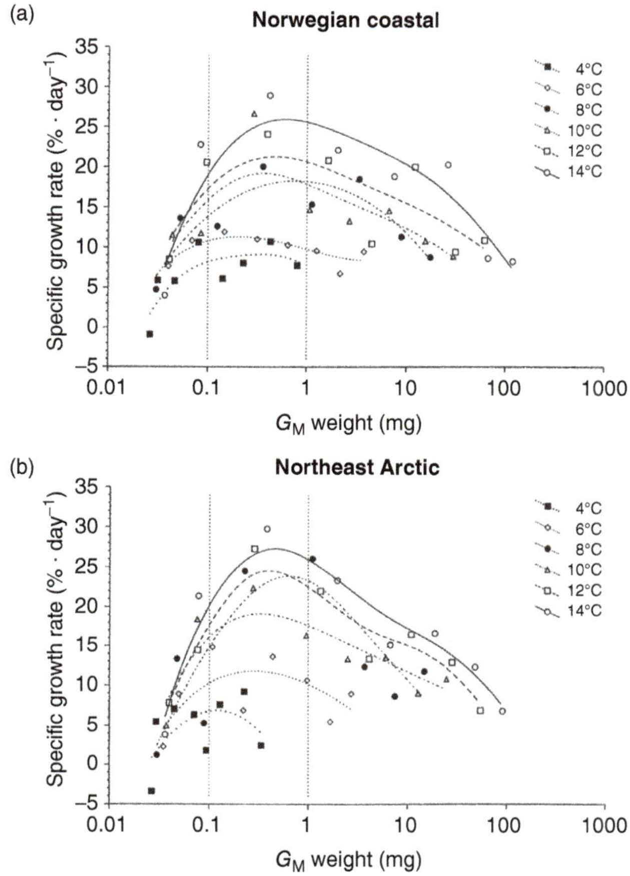

Figure 3.30 Trend in weight-specific growth rate with respect to body mass for laboratory-reared larvae of two stocks of Atlantic cod *Gadus morhua* larvae. (a) Norwegian cod; (b) Northeast Arctic cod. Note the domed relationship, with highest *G* at intermediate larval weights and ages. The domed relationship between *G* and larval weight may not be universal among fish larvae; in many taxa, *G* may decline throughout the larval stage. Also note that *G* is strongly temperature-dependent. Reproduced from Otterlei *et al.* (1999), with permission of Research Press, National Research Council, Canada.

large variability in stage-specific survival in the larval and early juvenile stages (Houde 1989b, 1997b). As cohorts age and grow, the ratio M/G generally declines, a consequence of relatively fast declines in M and slower declines in G (Houde 1996, 1997b). The age or size of a cohort at which M/G shifts from >1 to <1, termed the transition stage (or transition age, transition size) by Houde (1997a), is an indicator of recruitment potential (Figure 3.31). Cohorts making the transition at young ages and/or small sizes generally will have relatively high recruitments.

In first-feeding larval American shad *Alosa sapidissima*, M/G initially is >1.0 and varied by a factor of 1.7 for six year classes, based on an analysis of data by Houde (1997b) for American shad from the Connecticut River (Crecco & Savoy 1985, Crecco *et al.* 1983). There is a strong inverse relationship between the age-0 recruitment index for shad and the transition size or age (i.e., the point where $M/G = 1$). Recruitment success is inversely correlated with the mean level of larval-stage M/G and with the "transition size or age" which occurs relatively early

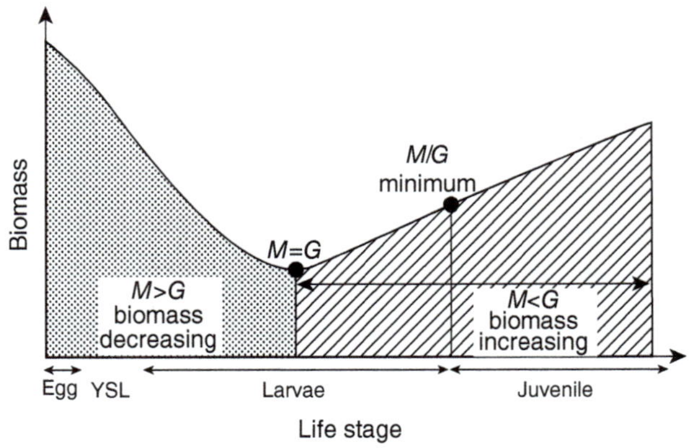

Figure 3.31 Conceptual relationship between cohort biomass and early-life stages of marine fish. Cohort biomass usually declines during the earliest stages, before increasing. Biomass declines when $M > G$ and $M/G > 1.0$, where M is daily instantaneous mortality and G is weight-specific growth rate. The stage (and size or age) at which $M = G$ and $M/G = 1.0$ is a "transition stage." At subsequent stages (sizes, ages), when $M/G < 1.0$, the cohort biomass increases. The "transition stage" can be an indicator of recruitment success; "transition" at earlier (younger, smaller) stages often is inversely correlated with recruitment level. Reproduced from Houde, E.D., Patterns and consequences of selective processes in teleost early life histories. In: Chambers, R.C. & Trippel, E.A. (Eds) *Early Life History & Recruitment in Fish Populations*, 173–95, © 1997, with kind permission of Springer Science and Business Media.

in the larval stage of American shad (Figure 3.32). Recruitments of walleye pollock (Bailey *et al.* 1996a,b) and striped bass (Rutherford *et al.* 1997, Houde 1996) also are inversely related to mean levels of *M/G* in the larval stage and, especially, to the estimated "transition age" at which *M/G* becomes <1.0 (Houde 1997b).

3.4.3.1 *The single process*

In describing the linkage between growth and mortality, and its relationship to recruitment, Cushing (1975) and Cushing & Horwood (1994) referred to the "single process" (see above). In the single process, the growth and mortality of larvae are hypothesized to be governed by the food supply available to larvae, while predation is the presumed proximal mechanism driving the process. Weak, slow-growing, and slow-swimming larvae are presumed to be more vulnerable to predation than better-nourished larvae.

The single process has been modeled under the presumption that either growth, mortality, or both, are density-dependent during the earliest larval stages (Shepherd & Cushing 1980, Cushing & Horwood 1994). Modeled survival and recruitment suggest that competition for limited prey resources during the earliest larval stages could be an important regulator of recruitment.

In broad, across-taxa comparisons, growth and mortality rates of early-life stages are highly correlated (Houde 1989a, 1996, Pepin 1991). Taxa that typically experience high mortality rates also grow fast, and vice-versa. Importantly, it is not the levels of mortality or growth rates that generate variability in recruitment, but the variability in *M* and *G*, and in resulting cumulative mortality during pre-recruit stages (Sissenwine 1984, Fogarty *et al.* 1991, Fogarty 1993). Much of the high and variable cumulative mortality accrues during the larval stage and

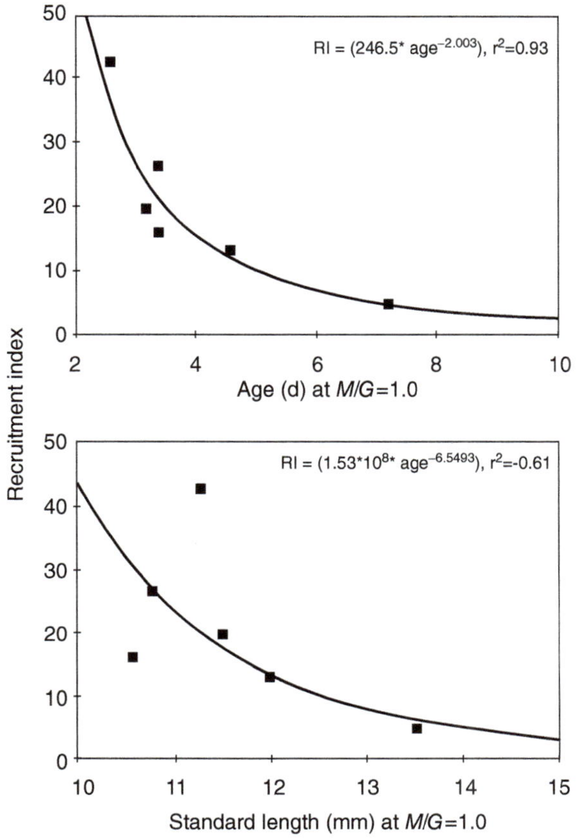

Figure 3.32 American shad *Alosa sapidissima*, Connecticut River. Young-of-the-year recruitment levels are inversely related to the "transition size or age"; that is, size or age at which M/G transitions from >1 to <1.

is mostly density-independent. Surprisingly small differences in cohort (or year class) mortality or growth rates during early life can lead to order-of-magnitude variability in recruitment (Houde 1987, 1989b).

Although, either M or G can generate variability in recruitment, M tends to be more variable than G, leading to the conclusion that for most taxa variability in mortality rate contributes more to variability in recruitment than does variability in growth rate. The coefficients of variation (CV) for M in cohorts of marine fish larvae of four species were two- to fourfold higher than the CV for G (Table 3.1). Larval-stage durations (inversely proportional to G) are much longer and more variable for taxa living in cold environments (Figure 3.9, Plate 2) (Houde 1989a). It is through this mechanism (the Stage Duration hypothesis) that a relatively small variability in G can have major consequences for recruitment variability in high-latitude seas.

3.4.3.2 Identifying and partitioning sources of mortality

Identifying sources of mortality or partitioning mortality in early-life stages into its component causes (e.g., predation, starvation, disease, hydrography and physics) is no simple task.

Table 3.1 Coefficients of variation (CV) in M, G, and M_s for four marine/estuarine fishes during early life. The range in the CVs for year classes where data were available ($n = 5$ to 7) and the average CV are tabulated. Variability is about 1.5- to 4-fold greater for M and M_s than for G. M is daily mortality (d^{-1}); M_s is stage-specific mortality (cumulative M); G is weight-specific growth rate (d^{-1}).

	Coefficients of variation		
	M	G	M_s
American shad			
Range	0.16–0.33	0.08–0.24	0.21–0.45
Mean	0.27	0.14	0.30
Bay anchovy			
Range	0.11–1.45	0.07–0.34	0.42–1.08
Mean	0.67	0.17	0.61
Walleye pollock			
Range	0.24–0.48	0.17–0.41	0.32–0.86
Mean	0.40	0.25	0.52
Striped bass			
Range	0.29–0.76	0.16–0.21	0.37–0.79
Mean	0.51	0.18	0.53

The fraction dying from nutritional deficiency and starvation sometimes can be estimated or indexed from analyses of nutritional condition (e.g., RNA/DNA ratios) (Ferron & Leggett 1994, Clemmesen & Doan 1996, Pepin *et al.* 1999, Buckley *et al.* 2006). Losses from the dispersal of eggs and larvae can confound estimates of mortality unless an entire nursery system is sampled more or less synoptically. Methods have been developed to separate dispersal losses from mortality (McGurk 1989, Hill 1991). Helbig and Pepin (1998a,b) provide cogent arguments and methods to address this issue. In cases where eggs or larvae are dispersed beyond boundaries of nursery areas, such losses effectively lead to mortality (i.e., Hjort's second hypothesis).

3.5 Control and regulation: destabilizing and stabilizing processes

The observed 10-fold and greater variability in fish recruitments may give the impression that fish stocks have little capability to regulate or stabilize abundance. An accumulation of evidence over the past two decades refutes this argument. In fact, given the high fecundities and variable survival rates in early life, it is surprising that interannual variability in recruitments is not greater than observed (Rothschild 1986, Beyer 1989). Regulatory mechanisms must operate during pre-recruit stages, acting to partly stabilize recruitment levels.

Controllers of recruitment level are primarily density-independent and attributable to environmental factors acting on early-life stages. As such, growth and mortality rates are mostly independent of spawning stock abundance or abundance of early-life stages. Density-independent factors coarsely control levels of abundance and can be destabilizing, leading to large, dominant year classes or, conversely, to recruitment failures. Density dependence implies that compensatory mechanisms are significant and may act to regulate recruitment levels. In such cases, the growth or mortality of young fish is regulated by the abundance of adult spawning stock, or constrained by competition among individuals in a cohort of pre-recruits for limited prey and by levels of predation governed by abundance of the pre-recruits.

3.5.1 Environmental controls: density-independent factors

Environmental factors often act most effectively during the earliest life stages (eggs, larvae), leading to substantial variability in survival of early-life stages that translates into variable recruitments (Fogarty *et al.* 1991, Fogarty 1993). Environmental factors often act most strongly during the first 20–50 days post-hatch, controlling survival rates and coarsely defining the potential level of recruitment. Such factors may act directly; for example, death from physiological challenges – temperature, salinity, dissolved oxygen, anthropogenic contaminants; or from physical mechanisms such as unfavorable dispersal. Alternatively, density-independent controls can act indirectly through environmental effects that control abundances of predators and prey of early-life stages. Large, dominant year classes and major recruitment failures are attributable primarily to density-independent factors acting on early-life stages.

3.5.2 Stabilizing mechanisms: density-dependent factors

For decades, fishery scientists and managers sought to model and relate recruitment in fishes as a function of adult stock abundance (see Chapter 1). Historically, these highly variable relationships often led to the dismissal of stock–recruitment relationships as a tool to describe or predict recruitments, or to understand density dependence. Recent synthetic analyses have changed this perspective, confirming the importance of stock–recruitment relationships, particularly at low adult stock levels, as regulators of recruitment (Myers & Barrowman 1996, Myers 2001, 2002). Density dependence and compensation, if present, often act to regulate the dynamics of late-larval and juvenile stages (Cowan *et al.* 2000, Rose *et al.* 2001). Mechanisms include "stock-dependent" processes, such as cannibalism or competition for prey between adults and pre-recruits. Alternatively, "density-dependent" processes are largely regulated by abundances of early-life stages, such as competition for limited prey or higher predation rates at high abundance. Heath & Gallego (2000) listed seven processes that could lead to density dependence and compensatory effects on survival of early-life stages: (i) competition for refuge or territory; (ii) necessity to form schools for protection from predators; (iii) competition for food; (iv) some types of parasitism of larvae and juveniles; (v) cannibalism; (vi) attraction of predators to local abundances of the target species; and (vii) some types of disease. In research directed at recruitment variability in four stocks of fishes in the Northeast Atlantic, Dingsør *et al.* (2007) found that the degree of density dependence in the recruitment process was related to strong species interactions and differing effects of predation, cannibalism, and environmental variables.

If mortality or growth rates are density-dependent during early life, rather small effects on either growth or mortality can have surprisingly strong consequences for recruitments. Although density dependence could regulate recruitment if it were significant during the larval stage – for example, by altering feeding success or vulnerability to predators (Jones 1973, Cushing 1975, Shepherd & Cushing 1980, Rothschild 1986) – there are relatively few cases documenting density-dependent mortality or growth acting on eggs or the youngest larval stages (Cowan *et al.* 2000). Savoy and Crecco (1988) did, however, detect substantial density dependence in American shad *Alosa sapidissima*, in which 24–41% of the cumulative mortality in egg to late-larval stages was judged to be dependent on numbers of spawned eggs (i.e., adult stock).

Compensation and density-dependent regulation are more likely to act in late-larval and newly metamorphosed juveniles than in eggs or smallest larvae. Several field experiments on tropical reef fishes have demonstrated the potential for stabilization of recruitment levels through density-dependent regulation during and after settlement (Doherty 2002, Hixon & Webster 2002). While density dependence may most often be generated via predation, Hixon & Jones (2005) documented cases where competition for limited habitat and prey resources ultimately regulated the recruitment of a reef damselfish *Pomacentrus amboienensis*. Jones & McCormick (2002) did not deny the probable role of density dependence in early lives of reef fishes, but emphasized the complexity of interpreting the interacting numerical and energetics processes that control and regulate recruitment. In this regard, density dependence in mortality of late-larval and juvenile stages of walleye pollock was demonstrated to be mediated by environmental factors, including climate variability (Bailey *et al.* 2005, Ciannelli *et al.* 2004, 2005).

A comprehensive review of field and experimental studies supported the belief that prey availability limits growth, but provided little support for a density-dependent mechanism in the early-larval stage, although density dependence in late-larval and juvenile stages can be substantial (Cowan *et al.* 2000). Rutherford *et al.* (2003) compared the outputs of a Shepherd & Cushing (1980) model and two other models against field data on early-larval cohorts of striped bass from the Potomac River sub-estuary of Chesapeake Bay, and found little evidence that larval growth was density-dependent. In contrast, larvae of southern bluefin tuna *Thunnus maccoyii* in the Indian Ocean grew slowly when in high-abundance patches, exhibiting density-dependent growth and competition for prey in this oligotrophic environment (Jenkins *et al.* 1991). Walleye pollock larvae in years of exceptional abundance may deplete their primary copepod nauplii prey, resulting in density-dependent regulation of growth and an increase in mortality in the late-larval to early juvenile stages (Duffy-Anderson *et al.* 2002).

In a meta-analysis, Myers & Cadigan (1993a,b) demonstrated that survival became significantly density-dependent in young-of-the-year (YOY) juvenile gadoids and some pleuronectiform fishes after the young fish became demersal, but not during the pelagic larval stage. In North Sea haddock, recruitments are linked to density-dependent processes operating on the newly settled juvenile stage (Heath *et al.* 1999, Heath & Gallego 2000). Recruitments of North Sea plaice are coarsely controlled by temperature and other density-independent processes during the pelagic egg and larval stages. Subsequently, compensatory regulation that serves to dampen variability in recruitments occurs at and after settlement via density-dependent predation (by *Crangon* shrimp) on newly settled post-larvae and juveniles (Zijlstra *et al.* 1982, Zijlstra & Witte 1985, Van der Veer 1986, Van der Veer & Bergman 1987, Van der Veer *et al.* 1990).

In striped bass, recruitment is regulated by density-dependent growth and survival of age-0 juveniles in the San Francisco Bay (Kimmerer *et al.* 2000, 2001) and the Chesapeake Bay (Martino & Houde 2012). In Chesapeake Bay and tributaries, growth of age-0 juveniles is inversely related to the availability of prey resources in years experiencing high larval-stage survival and abundance. In these years, juveniles grow slowly and suffer high overwinter mortality (Figure 3.33) that regulates year-class size (Martino & Houde 2012). The observed 50-fold variability in interannual age-0 abundance of striped bass is reduced to three- to fivefold variability in abundance at three years, the age at recruitment.

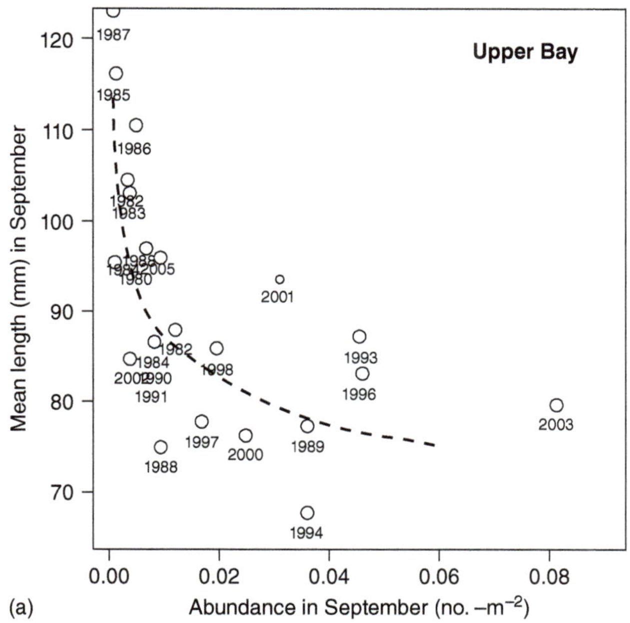

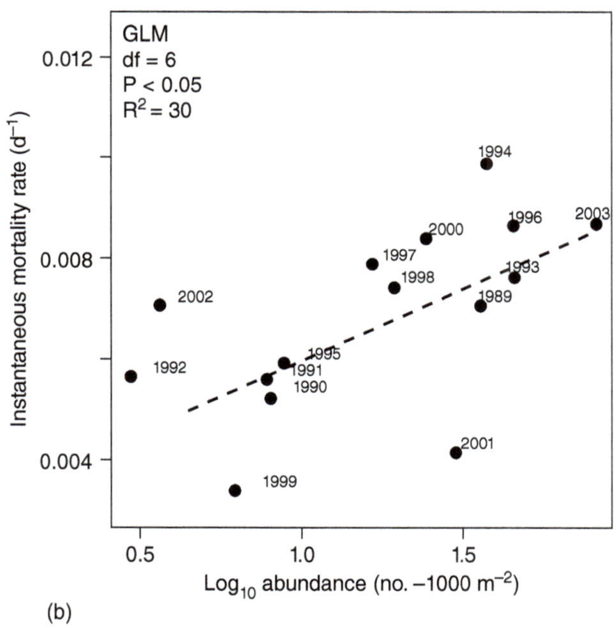

Figure 3.33 Relationships of (a) length (in mm) and (b) instantaneous mortality rate (d^{-1}) in relation to abundance of age-0 juveniles of striped bass in the upper Chesapeake Bay. Figure derived and modified from Martino and Houde (2012; see Figures 2 and 9).

3.5.3 *At what life stage is recruitment fixed?*

Are the observed levels and variability in recruitment explained by variability in survival of a particular early-life stage? Or, does the life stage at which recruitment is set vary among cohorts or year classes? Does the life stage at which recruitment level is determined differ among species?

In many species, such as striped bass, variability in recruitment is generated primarily by effects of environmental factors during the early-larval stage (Rutherford & Houde 1995, Secor & Houde 1995, Rutherford *et al.* 1997, Limburg *et al.* 1999), although there is considerable scope and evidence for compensation during the juvenile stage (Kimmerer *et al.* 2000, Martino & Houde 2012). In Atlantic cod, a large portion of variability in its recruitment in the Baltic Sea is generated by environmental factors, including prey availability, operating on the egg and larval stages (Köster *et al.* 2003). In contrast, Köster *et al.* (2003) determined that recruitment in Baltic sprat was largely determined by environmental factors, including temperature and wind stress, that acted on the late-larval and juvenile stages. In other species, such as northern anchovy (Peterman & Bradford 1987, Peterman *et al.* 1988), Japanese sardine *Sardinops melanostictus* (Watanabe *et al.* 1996, Watanabe 2002), or sprat (Baumann *et al.* 2006), recruitment levels depend most on variability in survival of the late-larval stage or young-of-the-year juveniles. Furthermore, the stage at which recruitment level is determined may differ among years or periodically, possibly related to climate shifts, as observed in walleye pollock from the Gulf of Alaska (Bailey 2000, Bailey *et al.* 2005). In walleye pollock, recent changes in the ecosystem favored predators of juvenile pollock, shifting control of recruitment from the larval stage in the 1980s to juveniles in the 1990s.

Predicting and forecasting the level of recruitment from imprecise estimates of egg or early-stage larval abundances, even when recruitment is in fact controlled by processes operating in those stages, is an uncertain venture. Forecasting based on ichthyoplankton surveys also is problematic because of the difficulty in correctly defining unit stocks and in sampling at appropriate scales (Frank 1997). Inadequate sampling and inaccurate or imprecise estimates of abundance make it difficult to estimate survival of eggs and the earliest larval stages. Projecting future abundances can be risky and uncertain (Pepin 1993).

Although forecasting recruitment based on egg and early larval abundances, and their mortality rates, is highly uncertain, accuracy in forecasting increases at the late-larval or juvenile stage, after most cumulative, pre-recruit mortality has occurred (Bradford 1992, Mertz & Myers 1995). In a review, Bradford (1992) reported that only 5% and 20%, respectively, of recruitment variability could be explained (based on r^2) when the abundance of eggs or first-feeding larval stages was used in forecasting. If abundance of feeding-stage larvae was used to forecast, the explained variability increased substantially, sometimes reaching 50%, although it usually was lower. Most of Bradford's (1992) predictions of recruitment based on early juvenile abundances were reasonably reliable ($r^2 > 0.50$). Bradford & Cabana (1997) concluded that the life stage most important in determining recruitment level was the stage in which most cumulative mortality accrued, and that this often was the juvenile stage. It is notable that predictions of recruitment based on mortality rates of early-life stages were less reliable and precise than predictions based on stage-specific abundances (Bradford 1992). The highest and most variable daily mortality rates occur during the larval stage (typically, 20–90 days in duration), but subsequent cumulative mortality and its variability during the long juvenile, pre-recruit stage (generally >300 days) in some cases is the major determinant of recruitment success (Sissenwine 1984).

Some generalities regarding recruitment and stage-specific survival emerge:

- Poor survival in the egg or early-larval stages usually will lead to weak recruitment.
- Above-average survival in the egg or early-larval stages may lead to strong recruitment, but this outcome is not certain.
- Good survival in the protracted late-larval and juvenile stages can generate strong recruitment, even if survival were below average in the egg and early-larval stages. The long stage duration of late-larval and juvenile stages can generate high and variable cumulative mortality.
- Below-average survival in late-larval or juvenile stages can still result in strong recruitment if survival were exceptionally good in the egg and early-larval stages – that is, coarse controls in earliest stages can prevail.

3.6 A nod to life histories: life styles and recruitment variability

Environmental factors are the "externalities" that exercise major control over recruitment. However, "internalities" owing to life-history patterns and strategies also affect egg production and quality, and reproductive success. These factors, including maternal influences, are emphasized in Chapters 8 and 11. Life histories differ markedly among fish taxa and express both genotypic and phenotypic variability. "Internalities" can play important roles in stabilizing recruitments.

3.6.1 Maternal and paternal effects

Maternal effects on the quality and size of eggs and larvae can make substantial contributions to recruitment variability. Selective fishing, or other human activities that impose selective stresses on a population, can alter the age and size structure of a spawning stock and affect the quantity, quality and sizes of eggs. In general, large (and old) females produce relatively large eggs and larvae, which may confer advantages to larvae in feeding and predator avoidance (Chambers *et al.* 1989, Chambers & Leggett 1996, Solemdal 1997, Trippel *et al.* 1997, Kamler 2005). In experiments on a Pacific rockfish *Sebastes melanops*, Berkeley *et al.* (2004a,b) found that female age was a strong predictor of larval survival and growth, implying that age truncation of the spawning stock through selective fishing could adversely impact recruitment. In striped bass, large females produce larger eggs and larvae (Figure 3.34, Plate 3) that maintain their initial size advantage over larvae hatched from smaller eggs spawned by small females (Zastrow *et al.* 1989, Monteleone & Houde 1990). Many metrics suggest that the larger striped bass larvae produced by big, old females are more fit and more likely to survive than progeny from small females (Figure 3.35, Plate 4).

Paternal effects, while less commonly researched than maternal effects, can be significant (Rideout *et al.* 2004), generally acting early in ontogeny, particularly on early-stage embryos (Kamler 2005). The quality and density of spermatozoa can affect fertilization and hatching success. In a nest-building clownfish, variability in growth and performance of larvae of

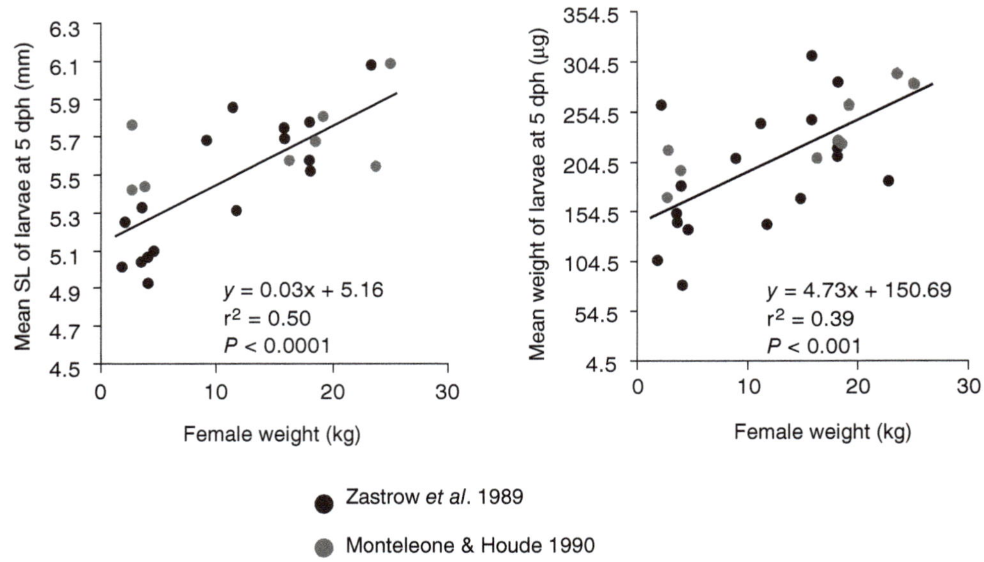

Figure 3.34 Maternal effects. Lengths and dry weights at 5 days post-hatch (day of first feeding) of striped bass *Morone saxatilis* larvae in relation to adult female weight. Larvae from smallest females weigh, on average, only 63% as much as larvae from the largest females. For a color version of this figure, please see Plate 3 in the color plate section. Data reproduced from Zastrow *et al.* (1989), with permission of ICES, and Monteleone & Houde (1990).

Amphiprion melanopus, in which adult males tend the nests, depended more on paternity than maternity (Green & McCormick 2005).

3.6.2 *Is recruitment variability linked to life history strategy?*

While high variability in recruitment is common in all teleosts, levels of variability are in part dependent on life-history characteristics (e.g., age–size structure of the stock, age–size-at-maturity, longevity, fecundity, spawning patterns) (Winemiller & Rose 1992). Cushing (1973) reviewed the scales of recruitment variability in marine fishes, finding typically a 10-fold interannual variability, but with higher levels observed in many taxa. Some fishes, such as haddock, exhibit extraordinary variability in which 1000-fold interannual differences occur (e.g., Figure 3.1).

Short-lived, shoaling pelagic fishes, such as anchovies and sardines, are especially abundant in dynamic and variable upwelling ecosystems in which hydrography and productivity show strong responses to regional weather patterns, El Niño events, ecosystem regime shifts, and climate change. These fishes generally have relatively low fecundities and a low capacity to regulate their recruitments (i.e., a low index of density dependence) (Cushing 1971). Their populations are easily destabilized by stresses of natural or human origin, including fishing. On decadal to century-long time scales, the overall abundances of shoaling pelagic fishes in upwelling ecosystems may vary by three orders of magnitude. These fishes often exhibit low-frequency variability in recruitment levels that translates into major, decadal-scale variability in abundances (Lluch-Belda *et al.* 1989, Kawasaki 1992, Schwartzlose *et al.* 1999, Chavez *et al.* 2003, Checkley *et al.* 2009). Such variability is attributable to interannual, decadal, and

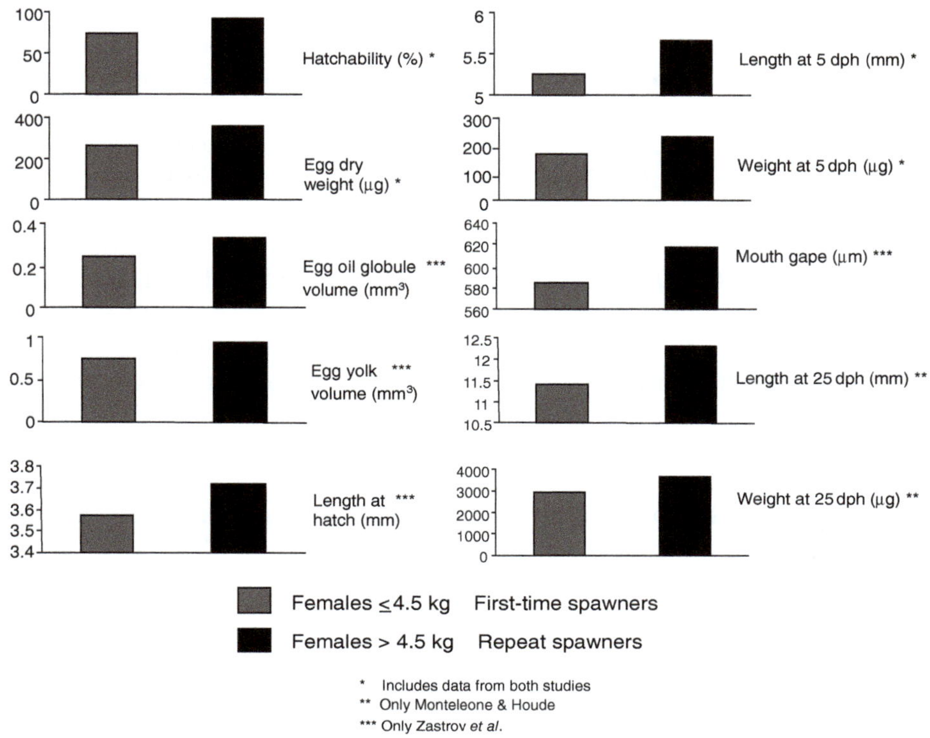

Figure 3.35 Maternal effects. A suite of metrics comparing striped bass *Morone saxatilis* eggs and larvae from small (≤4.5 kg) or large (>4.5 kg) females. In each case, progeny from large females appear to have a size or viability advantage. For a color version of this figure, please see Plate 4 in the color plate section. Data reproduced from Zastrow *et al.* (1989), with permission of ICES, and Monteleone & Houde (1990).

longer fluctuations in recruitment levels, apparently in response to shifts in productivity and carrying capacity of ecosystems (Alheit *et al.* 2009). Examples of such variability and shifts are documented over the past 2000 years for California sardine and northern anchovy from records of fish scales deposited in anoxic sediments (Baumgartner *et al.* 1992, Field *et al.* 2009).

Long-lived, highly fecund species such as flatfishes and gadoids are better able to regulate recruitment levels and compensate for shifts in adult abundance and age structure (Cushing 1971, Rose *et al.* 2001). These stocks, on average, exhibit lower decadal-scale variability in abundance and recruitments (Rijnsdorp *et al.* 1991), although interannual variability still is substantial. In these fishes, protecting age structure to ensure that the spawning stock includes older females is a precautionary step to avoid recruitment failures (Longhurst 2002).

3.6.3 Freshwater and marine contrasts

In contrast to marine fishes, freshwater fishes typically live in habitats and ecosystems that are spatially constrained. Their reproductive success is likely to be more vulnerable to environmental perturbations and weather events that affect survival of young fish (Houde 1994). Freshwater fishes often have specialized reproductive behaviors and other life-history attributes to partly insulate the eggs and larvae from environmental variability and extremes; examples include nesting behaviors, parental care, and substrate spawning (Wootton 1990). In freshwater fishes, recruitment levels of spatially discrete stocks tend to be concordant at approximately 50 km spatial scale, reflecting local-scale weather events and environmental conditions. In contrast, concordance in recruitment levels of marine stocks occurs at spatial scales of 500 km (Myers *et al.* 1995a,b, 1997), in response to regional and ocean-basin variability in weather or climate.

Marine fishes, on average, are more fecund than typical freshwater fishes (Duarte & Alcaraz 1989, Winemiller & Rose 1992) and their eggs and larvae are smaller. In fact, newly hatched larvae of marine fishes weigh 10 times less, on average, than newly hatched freshwater fish larvae (38 versus 360 μg dry weight), but marine and freshwater fishes metamorphose to juveniles at similar weights (Houde 1994). After adjusting for temperature effects, marine fish larvae were found to have higher mean mortality rates than freshwater larvae (0.24 d^{-1} versus 0.16 d^{-1}), but grow at similar weight-specific rates (0.20 d^{-1} versus 0.18 d^{-1}) (Houde 1994). Temperature-adjusted larval stage durations of marine fishes are, on average, relatively long (36 versus 21 days). The high and variable mortality and longer larval stage durations indicate that, on average, the potential for control of recruitment in marine fishes rests more on the larval stage than it does in freshwater fishes.

Freshwater fishes often minimize the dispersal of their eggs and exposure to unpredictable environmental conditions by spawning demersal eggs onto substrates, or in nests on spatially discrete spawning grounds. Some marine fishes also employ this strategy (e.g., herring *Clupea harengus* stocks). Larval drift, which is obligatory for recruitment in many marine fishes (Able 2005), is minimized or scaled back in most freshwater species, controlled partly by the spatial limits of freshwater ecosystems and also by the spawning behaviors of adults. Some marine species have adopted similar behaviors, especially tropical reef fishes. Such species select discrete spawning sites on the reef and may adopt nesting behaviors (e.g., Pomacentridae). Their larvae may utilize directed or retention-promoting, pre-settlement swimming behaviors, some of which include occupying appropriate depths or undertaking directed swimming in response to visual, auditory, or tidal cues. These behaviors can assure substantial "self-recruitment" by reducing risks of long and uncertain larval drift (Cowen *et al.* 2000, 2006, Sponaugle *et al.* 2002, Leis 2006).

Anadromous fishes[2] exhibit specific, often complex spawning migrations and behaviors. Variability in timing of migrations by adults and selection of spawning sites, and vagaries of weather and freshwater flow in rivers where adults spawn, can result in episodic mortalities, in addition to the usual high and variable mortality rates of eggs and larvae (Crecco *et al.* 1986, Crecco & Savoy 1987, Rutherford & Houde 1995, Secor & Houde 1995, Limburg *et al.* 1997, 1999). Adults of anadromous fishes respond to environmental cues, such as high freshwater flow rates and appropriate temperatures that control pre-spawning migrations into estuaries and tidal tributaries. Once committed, the adults, being unable to anticipate future weather and flow events, can only coarsely adjust their spawning times to assure favorable conditions for survival of eggs and larvae. It is this circumstance that sets the stage for generation of recruitment variability in anadromous striped bass (Secor 2000).

[2] Salmonids, and their very special life histories, reproductive and recruitment behaviors and mechanisms are not included in this discussion.

Many anadromous fishes, for example moronids and alosines, spawn in estuarine transition zones. Here, the salt front and features such as estuarine turbidity maxima promote retention of eggs and larvae, minimizing dispersive down-estuary losses while supporting high zooplankton abundances that increase feeding potential of larvae (Laprise & Dodson 1989a,b, 1990, Sirois & Dodson 2000, North & Houde 2003, 2006, Winkler *et al.* 2003, Campfield and Houde 2011). Despite the substantial risks from weather events and other short-term environmental variability on early-life dynamics of anadromous fishes, Rothschild & DiNardo (1987) found that variability in recruitment of anadromous fishes was similar to variability in recruitment of fishes that spawn in the sea.

3.7 Stock and recruitment

A critical evaluation of the relationship between adult stock and recruitment is largely outside the scope of this chapter (these points are addressed in Chapters 1 and 4), but the topic is briefly considered here. For sustainable fisheries and population maintenance, there obviously must be sufficient numbers of adults and egg production, on average, to allow replacement recruitment to occur.

3.7.1 Role of fishing

Fishing reduces the abundance of adults and therefore egg production. Fishing also may selectively remove older, larger, and more fecund females from the stock, thus reducing the overall fitness of the stock. Excessive fishing on spawning stock, if coincident with environmental conditions that are unfavorable for early-life stages, can lead to recruitment failure.

The relationship between numbers of recruits and adult spawning stock usually is poorly defined. Selecting an appropriate stock–recruitment (S–R) model is important to describe the nature of recruitment dependency and variability with respect to spawning stock biomass (SSB), and to partition variability that is responsive to effects of fishing from variability attributable to the environment. While most recruitment variability is generated by environmental factors and their effects on early-life stages (Fogarty 1993), the potential to stabilize recruitment resides in SSB and fecundity, modified by other "internal" mechanisms such as the stock's age structure and nutritional condition that can affect quality of eggs and larvae.

3.7.2 Environmental factors

Incorporating environmental variables into S–R models can improve the models and explain additional variability in recruitment not accounted for in simple S–R models (Iles 1994, Ottersen *et al.* 2013). Typically, weather-related (e.g., precipitation, freshwater inflow, air temperature, wind) or hydrographic-circulation (e.g., temperature, salinity, stratification parameters, transport) variables improve the explanatory power of S–R models. In half of 38 exploited fish stocks analyzed in the North Atlantic, the S–R relationships were improved by the addition of a variable to account for temperature effects (Ottersen *et al.* 2013). As a caution, it is well known that many S–R models with environmental drivers ultimately may fail (Myers 1998) if regime shifts or other major changes in the environment occur that drastically change the productive capacity of the stock or associated biological communities.

Including sea surface temperature, Fraser River discharge and hours of sunlight in a modified Ricker S–R model for Pacific herring (*Clupea harengus pallasi*) improved the fit of recruitment data, increasing the variability explained from 45% to 58% (Stocker *et al.* 1985). In another example, Fargo (1994) explained an additional 10% of recruitment variability in an S–R model for English sole *Pleuronectes vetulus* by including an Ekman transport variable (negative effect). North & Houde (2003) developed modified Ricker S–R models for estuarine striped bass and white perch by incorporating freshwater flow, explaining an additional 41% and 31%, respectively, of observed variability in recruitments. In a more complex application, Sinclair & Crawford (2005) included water transport and herring (prey or predator) variables in a Ricker S–R model for Pacific cod *Gadus macrocephalus*. Together, the transport variable and adult cod abundance explained much of the variability in cod recruitment. Crecco *et al.* (1986) demonstrated that the inclusion of a freshwater-flow variable dramatically improved a Ricker S–R model for anadromous American shad.

3.8 Modeling complex processes

During the last two decades of the 20th century, modeling became a primary tool to investigate complex processes driving recruitment variability. In many circumstances, it is both costly and logistically prohibitive to repeatedly conduct comprehensive at-sea surveys that measure all relevant factors at time and space scales governing the early-life dynamics of marine fishes. The development of sophisticated hydrodynamic models and computing power was instrumental in driving the present emphasis on multidisciplinary modeling experiments and the development of coupled biophysical models as a means to probe and evaluate biological and physical processes that affect recruitment (North *et al.* 2009). Particle-tracking models, some with vertical migratory behavior assigned to the simulated eggs or larvae, can describe or predict larval transport. In a recent example, Churchill *et al.* (2011) applied a three-dimensional hydrodynamic model to Atlantic cod eggs and larvae in the Gulf of Maine, demonstrating that favorable transport and retention occurred under downwelling conditions. More complex models typically address early-life survival and growth in response to hydrodynamic and trophodynamic processes, and less frequently test recruitment hypotheses explicitly. Models are now essential tools for recruitment forecasting in support of stock assessments and management programs.

Individual-based models (IBMs) have proliferated during the past two decades. The IBMs are initialized with large numbers of eggs or larvae and run to simulate and evaluate early-life dynamics, including nutrition and predation processes, the effects of contaminants, and density-dependent mechanisms (Beyer & Laurence 1980, Cowan *et al.* 1993, 1996, 1999, Rice *et al.* 1993, Letcher *et al.* 1996, Letcher and Rice 1997, Rose *et al.* 1993, 1999, 2003, Kristiansen *et al.* 2009, Petrik *et al.* 2009, Daewel *et al.* 2011, Peck & Hufnagl 2012). Large populations with defined biological attributes (e.g., metabolic demands, swimming and feeding behaviors, vertical migratory behavior or selection for depth of occurrence) are subjected to environmental variability and followed through time. The survival and growth of individuals and the population are accounted for at daily (usually) time steps. IBMs are appealing because many of the mechanisms controlling survival and growth operate at the level of the individual, rather than the population. By applying IBMs, survivors can be evaluated to determine attributes that distinguish them from fish that died (Crowder *et al.* 1992).

Many coupled biophysical models now are built on a three-dimensional hydrodynamic model framework into which individual-based biological dynamics are imbedded to simulate variability in transport, foraging success and predation mortality of individuals in early life (Werner *et al.* 1996, 2001, Hinkley *et al.* 1996, Hermann *et al.* 1996, 2001, Heath & Gallego 1998, Brickman *et al.* 2001, Mullon *et al.* 2002, Bartsch & Coombs 2004, Bartsch 2005, North *et al.* 2009). IBMs are particularly effective in linking trophodynamics and larval behavior to hydrodynamic models that simulate circulation patterns and dispersal potential of early-life stages. These individual-based, coupled biophysical models (ICBPMs) have had noted success in explaining the dispersal and retention, growth and survival of eggs and larvae. Miller (2007) reviewed the history and evolution of ICBPMs, noting that published models generally can be categorized as either explanatory (56%) or inferential (31%). Relatively few of the models were developed as hypothesis-testing tools (11%), although the biggest contribution to understanding recruitment variability ultimately may come from that approach. Runge *et al.* (2010) have cogently argued that a research program built upon combined observational data on system productivity and ICBPMs can explain much of the variability in recruitment of marine fish stocks.

3.9 Solving the "recruitment problem"

Historically, fishery scientists have asked, "Is there a solution to the recruitment problem?" However, a better question may be, "Is there a problem?" Research over the past three decades has made substantial strides towards understanding the principal causes of variability in reproductive success and recruitment of fishes and invertebrates (Houde 2008). During the 1970s, many believed that a lack of theory on recruitment and processes contributing to its variability deterred solving the problem. In retrospect, there were hypotheses and theories, but most addressed the "problem" from too narrow a perspective. The new level of understanding gained in recent decades has depended more on advances in technology, interdisciplinary research, and modeling successes than on new theory. It is now clear that no single factor, mechanism, life stage, process, or time-space scale can control recruitment and its variability. Recruitment variability, and also regulation, are the result of a suite of processes acting alone, in concert, and through interactions. Consequently, the "recruitment problem" is not well-defined and has no unique solution. The observed variability in recruitments represents the expression of variable climatic, oceanographic, ecological and anthropogenic (e.g., fishing) factors. There have been enormous gains in understanding how these factors operate and notable success in moving towards a forecasting capability.

With the exception of a few valuable exploited stocks, we may never acquire sufficient knowledge to confidently forecast recruitment levels and variability in most fishes. On the other hand, the relative importance of factors that generate variability has been elucidated in multidisciplinary programs of the past two decades. We now have a stronger appreciation of trends in recruitment, patterns of variability, the role of physical–biological interactions, the ecology of early-life stages, the risks to recruitment from overfishing, and the potential for major shifts in production (dependent on recruitment) associated with regional and global climate change. Much of this new knowledge can be applied by managers to reduce risk of recruitment failures and to predict probable long-term trends in recruitment levels.

For abundant species of high economic value (e.g., Atlantic cod, Atlantic herring, plaice, walleye pollock, striped bass), there have been major advances in describing, modeling and understanding recruitment processes (Ciannelli *et al.* 2004, Bailey *et al.* 2005, 2012, Martino & Houde 2010, Stige *et al.* 2013). Knowledge acquired during the past two decades has brought us close to being able to forecast their recruitment. For economically less-valuable stocks, forecasting may not be possible soon. However, documenting variability and trends in recruitment, and relationships to environmental factors, has real value for resource management, at least in a precautionary approach, and contributes to a fundamental understanding of recruitment variability and probable future trends in stock abundance.

3.10 Conclusions

Hypotheses to explain recruitment variability fall into two general categories: those based on trophodynamics principles; and those based on physical properties of marine systems. The trophodynamics hypotheses generate variability through energetics (nutritional) responses that subsequently translate into numerical responses – that is, nutrition-dependent survival – that exercise control over recruitment levels. Trophodynamics (predator–prey relationships) control energetics, growth, and production. Hypotheses related to physics, hydrography, and circulation generally imply that control is via direct numerical responses, such as losses through larval dispersal or retention associated with relatively high survival. Interactions are important; for example, physical processes often interact with temperature-modulated trophodynamics. Both numerical and energetics processes are critical during the earliest life stages, when larvae are most likely to starve and be dispersed. Energetics processes that lead to increases in body size and shorter stage duration favor recruitment and are important throughout pre-recruit life. Because physics and trophodynamics processes interact, hypotheses based solely on one or the other may be insufficient to explain recruitment variability.

Two fundamental types of recruitment variability are common: (i) interannual variability in recruitment success of 10-fold or more that may appear chaotic or random, in response to environmental variability; and (ii) decadal and longer trends in levels of recruitment reflecting regime shifts, long-term climate change, and shifts in levels of spawning stock biomass.

Often, most variability in recruitment is generated in the first few weeks of life, that is, during the "early" larval stage when coarse controls operate. Substantial adjustments and the tuning of pre-recruit abundances take place in the late larval and juvenile stages. The relatively long durations of these stages may experience high cumulative mortalities and, for some species, these are the decisive stages at which recruitment levels are set. No single life stage can be designated as the stage determining recruitment success. Year-class size. therefore, can be fixed in any of the early-life stages, depending on the trajectory and variability in growth and cumulative mortality in early life.

The types of process that generate variable mortality, and thus control and regulate recruitment, are broadly similar among taxa and in different ecosystems. Dominant factors and life stages that control or regulate abundance differ and are dependent on taxa-specific, life history characteristics. It is the variability in both growth and mortality rates in early life, not their levels, that generates variability in recruitment. Mortality and growth rates, and variability in those rates, decline as larvae grow, with mortality rates generally declining faster than growth rates. Although most of the variability in recruitments is generated as a consequence of variability in mortality rates during early life, stage-specific survival is strongly dependent on the relationship between rates of growth and mortality, and their joint effect on stage durations.

Temperature is the single factor most implicated in contributing to variability in early-life stage dynamics, and also is a key factor controlling spawning times and areas. Overall, the level of temperature and its seasonal/annual patterns and variability play dominant roles in generating variability in the growth and survival of larval and juvenile fish, even in ecosystems where temperature varies relatively little (e.g., tropical ecosystems). Temperature acts directly on young fish through its physiological control over growth, metabolism and swimming, and indirectly via effects on predator–prey interactions. Temperature also plays an important role in defining suitable habitat, through its contribution to hydrography, stratification, and frontal structure. Ongoing global climate change and rising temperatures already are associated with temporal and spatial shifts in spawning by marine fishes and in areas that serve as larval and juvenile nurseries.

Body size, growth rates, and stage durations are important. Mortality rates decline during growth and ontogeny of early-life stages. Size- and growth-rate selective predation shape the recruitment process, with smaller and slower-growing individuals at a selective disadvantage. The dynamics of fishes in early life are expected to conform to size-spectrum theory and, consequently, as a generality "bigger is better" – that is, being large confers survival advantages through a reduced probability of predation. Growth-rate-selective mortality also is important. "Faster is better" because high growth rates reduce stage durations and predation threats to small and vulnerable stages, increasing survival and recruitment potentials.

A proliferation of models, particularly individual-based models coupled to hydrodynamic models, is contributing importantly to describing complex early-life dynamics. Hydrodynamic modeling has advanced rapidly in recent decades to help explain circulation features and patterns, and how they relate to connectivity, transport or retention of early-life stages during the recruitment process. Coupled biophysical models, linking hydrography and transport to trophodynamics in early-life stages, are powerful inferential tools that generate realistic numerical responses to modeled predation mortality and advection losses. Bioenergetics models, often embedded within individual-based models and hydrodynamic models, are effective in simulating growth, stage durations and variability. Lacking in models to date is sufficient knowledge to realistically include the behavior of early-life stages at the fine spatial scales that is required, although notable advances on behavior of early-life stages are now being made.

Recruitments of marine fishes frequently vary 10- to 100-fold interannually. In addition, average recruitment levels and variability are subject to decadal and longer shifts in response to regional and global variability in climate and oceanography. Such shifts have been documented more frequently during recent decades, resulting from heavy fishing that reduced adult stock abundances, thus rendering the stocks more susceptible to environmental variability and ongoing climate change. There also has been a better documentation of fish stock dynamics and their recruitment success as stock assessments have improved. Trends in levels of recruitment often result from regime or phase shifts in ecosystem productivity. When decadal and longer-term trends in recruitment occur, informed and risk-averse fishery management strategies must adapt and recognize that near-term management actions, which typically are keyed to interannual variability in recruitment, may be insufficient to insure sustainability when the reproductive potential of a stock has changed.

No single life stage or process can be cited as the stage or causative agent of recruitment variability. Recruitment is an integrated, cumulative process, subject to a suite of coarse, density-independent controls and finer-scale, density-dependent regulatory processes. Coarse controls via environmental factors (acting on eggs and larvae) may dominate in some taxa and years, but regulatory processes and fine-tuning (often acting on late larvae and juveniles) can be decisive in determining recruitment level. Forecasting recruitment is most confidently accomplished by predicting it based on abundances of late-stage larvae and juveniles. However, success in forecasting based on those life stages does not necessarily mean that recruitment levels were set in the late-larval or juvenile stages, because dynamics in egg and early-larval stages could have driven the outcome.

Although infrequently achieved today, a goal of scientists and fishery managers is to forecast recruitment, based on knowledge of abundances and variability in earliest life stages. Successful forecasting, based on surveys, analysis and modeling, will test present understanding of processes that control and regulate recruitment. Recent successes in forecasting for some well-studied fish stocks, based on combined analysis and modeling of environmental factors and adult stock sizes, are encouraging. Beyond importance for management, success in forecasting also would help to refine recruitment hypotheses and to plan future research on recruitment variability.

References

Able, K.W. (2005) A re-examination of fish estuarine dependence: evidence for connectivity between estuarine and ocean habitats. *Estuarine, Coastal and Shelf Science*, **64**, 5–17.

Alderdice, D.F. (1988) Osmotic and ionic regulation in teleost eggs and larvae. In: Hoar, W.S. & Randall D.J. (Eds) *Fish Physiology, Vol. XI, The Physiology of Developing Fish, Part A, Eggs and Larvae*. pp. 163–251. Academic Press, San Diego.

Alderdice, D.F. & Forrester, C.R. (1968) Some effects of salinity and temperature on early development and survival of the English sole (*Parophrys vetulus*). *Journal of the Fisheries Research Board of Canada*, **25**, 495–521.

Alderdice, D.F. & Velsen, F.P.J. (1971) Some effects of salinity and temperature on early development of Pacific herring (*Clupea pallasi*). *Journal of the Fisheries Research Board of Canada*, **28**, 1545–62.

Alheit, J., Roy, C. & Kifani, S. (2009) Decadal-scale variability in populations. In: D. Checkley, J. Alheit, Y. Oozeki & C. Roy (Eds) *Climate Change and Small Pelagic Fish*. pp. 64–87. Cambridge University Press, Cambridge, UK.

Allain, G.P., Petitgas, P., Grellier, P., Lazure, P. (2003) The selection process from larval to juvenile stages of anchovy (*Engraulis encrasicolus*) in the Bay of Biscay investigated by Lagrangian simulations and comparative otolith growth. *Fisheries Oceanography*, **12**, 407–18.

Almatar, S.M. (1984) Effects of acute changes in temperature and salinity on the oxygen uptake of larvae of herring (*Clupea harengus*) and plaice (*Pleuronectes platessa*). *Marine Biology*, **80**, 117–24.

Anderson, J.T. (1988) A review of size-dependent survival during pre-recruit stages of fishes in relation to recruitment. *Journal of Northwest Atlantic Fisheries Science*, **8**, 55–66.

Arntz, W.E. (1986) The two faces of El Niño 1982–83. *Meeresforschung*, **31**, 1–46.

Auth, T.D. (2003) *Interannual and Regional Patterns of Abundance, Growth and Feeding Ecology of Larval Bay Anchovy (Anchoa mitchilli) in Chesapeake Bay.* Master's Thesis, University of Maryland, College Park, Maryland.

Bailey, K.M. (1981) Larval transport and recruitment of Pacific hake *Merluccius productus*. *Marine Ecology Progress*, Series **6**, 1–9.

Bailey, K.M. (2000) Shifting control of recruitment of walleye pollock *Theragra chalcogramma* after a major climatic and ecosystem change. *Marine Ecology Progress Series*, **198**, 215–24.

Bailey, K.M., Brodeur, R.D. & Hollowed, A.B. (1996a) Cohort survival patterns of walleye pollock, *Theragra chalcogramma*, in Shelikof Strait, Alaska: a critical factor analysis. *Fisheries Oceanography*, **5** (Supplement 1), 179–88.

Bailey, K.M., Brown, A.L., Yoklavich, M.M. & Mier, K.L. (1996b) Interannual variability in growth of larval and juvenile walleye pollock *Theragra chalcogramma* in the western Gulf of Alaska, 1983–91. *Fisheries Oceanography*, **5** (Supplement 1), 137–47.

Bailey, K.M., Ciannelli, L., Bond, N.A., Belgrano, A. & Stenseth, N.C. (2005) Recruitment of walleye pollock in a physically and biologically complex ecosystem: a new perspective. *Progress in Oceanography*, **67**, 24–42.

Bailey, K.M. & Houde, E.D. (1989) Predation on eggs and larvae of marine fishes and the recruitment problem. *Advances in Marine Biology*, **25**, 1–83.

Bailey, K.M. & Incze, L.S. (1985) El Niño and the early life history and recruitment of fishes in temperate marine waters. In: W.S. Wooster & D.L. Fluharty (Eds) *El Niño North: Niño Effects in the Eastern Subarctic Pacific Ocean*. pp. 143–65. Washington Sea Grant WSG–WO 85–3.

Bailey, K.M., Macklin, S.A., Reed, R.K., Brodeur, R.D., Ingraham, W.J., Platt, J.F., Shima, M., Francis, R.C., Anderson, P.J., Royer, T.C., Hollowed, A.B., Somerton, D.A. & Wooster, W.S. (1995) ENSO events in the northern Gulf of Alaska, and effects on selected marine fisheries. *California Cooperative Oceanic Fisheries Investigations Report*, **36**, 78–96.

Bailey, K.M., Zhang, T., Chan, K.S., Porter, S. & Daugherty, A. (2012) Near real-time forecasting of recruitment from larval surveys: application to Alaska pollock. *Marine Ecology Progress Series*, **452**, 205–17.

Bakun, A. (1996) *Patterns in the Ocean. Ocean Processes and Marine Population Dynamics*. California Sea Grant, San Diego, California.

Bakun, A. (2006) Fronts and eddies as key structures in the habitat of marine fish larvae, opportunity: adaptive response and competitive advantage. *Scientia Marina*, **70** (Supplement 2), 105–22.

Banks, M.A., Holt, G.J & Wakeman, J.M. (1991) Age-linked changes in salinity to tolerance of larval spotted seatrout (*Cynoscion nebulosus*, Cuvier). *Journal of Fish Biology*, **39**, 505–14.

Barber, R.T. & Chavez, F.P. (1983) Biological consequences of El Niño. *Science*, **222**, 1203–10.

Bartsch, J. (2005) The influence of spatio-temporal egg production variability on the modeled survival of the early life history stages of mackerel (*Scomber scombrus*) in the eastern North Atlantic. *ICES Journal of Marine Science*, **62**, 1049–60.

Bartsch, J. & Coombs, S.H. (2004) An individual-based model of the early life history of mackerel (*Scomber scombrus*) in the eastern North Atlantic, simulating transport, growth and mortality. *Fisheries Oceanography*, **13**, 365–79.

Basterretxea, G., Catalan, I.A., Jordi, A., Alvarez, I., Palmer, M. & Sabates, A. (2013) Dynamic regulation of larval fish self-recruitment in a marine protected area. *Fisheries Oceanography* **22**, 477–95.

Baumann, H., Hinrichsen, H.-H., Mollmann, C., Koster, F.W., Malzahn, A.M. & Temming, A. (2006) Recruitment variability in the Baltic Sea sprat (*Sprattus sprattus*) is tightly coupled to temperature and transport patterns affecting the larval and early juvenile stages. *Canadian Journal of Fisheries and Aquatic Sciences*, **63**, 2191–201.

Baumgartner, T.R., Soutar, A. & Ferreira-Bartrina, V. (1992) Reconstruction of the history of Pacific sardine and northern anchovy populations over the past two millenia from sediments of the Santa Barbara Basin, California. *California Cooperative Fisheries Investigations Report*, **33**, 24–40.

Beamish, R.J. & Noakes, D.J. (2002) The role of climate in the past, present, and future of Pacific salmon fisheries off the west coast of Canada. In: N.A. McGinn (Ed.) *Fisheries in a Changing Climate*. pp. 231–44. American Fisheries Society, Symposium 32, Bethesda, Maryland.

Beaugrand, G., Brander, K.M., Lindley, J.A., Souissi, S. & Reid, P.C. (2003) Plankton effect on cod recruitment in the North Sea. *Nature*, **426**, 661–4.

Bennett, W.A., Kimmerer, W.J. & Burau, J.R. (2002) Plasticity in vertical migration by native and exotic estuarine fishes in a dynamic low-salinity zone. *Limnology and Oceanography*, **47**, 1496–507.

Bergenius, M.A.J., Meekan, M., Robertson, D.R. & McCormick, M.I. (2002) Larval growth predicts the recruitment success of a coral reef fish. *Oecologia*, **131**, 521–5.

Berkeley, S.A., Hixon, M.A., Larson, R.J. & Love, M.S. (2004a) Fisheries sustainability via protection of age structure and spatial distribution of fish populations. *Fisheries* **29** (8), 23–32.

Berkeley, S.A., Chapman, C. & Sogard, S.M. (2004b) Maternal age as a determinant of larval growth and survival in a marine fish, *Sebastes melanops*. *Ecology*, **85**, 1258–64.

Beyer, J.E. (1989) Recruitment stability and survival – simple size-specific theory with examples from the early life dynamics of marine fish. *Dana*, **7**, 45–147.

Beyer, J.E. & Laurence, G.C. (1980) A stochastic model of larval fish growth. *Ecological Modelling*, **8**, 109–32.

Blaxter, J.H.S. (1988) Pattern and variety in development. In: W.S. Hoar & D.J. Randall (Eds) *Fish Physiology. Volume XI, The Physiology of Developing Fish, Part A*. pp. 1–58. Academic Press, San Diego.

Blaxter, J.H.S. (1992) The effect of temperature on larval fishes. *Netherlands Journal of Zoology*, **42**, 336–57.

Blaxter, J.H.S. & Hempel, G. (1963) The influence of egg size on herring larvae (*Clupea harengus* L.). *Journal du Conseil International pour l'Exploration de la Mer*, **28**, 211–40.

Boehlert. G.W. & Mundy. B.C. (1988) Roles of behavioral and physical factors in larval and juvenile fish recruitment to estuarine nursery areas. *American Fisheries Society*, Symposium, **3**, 51–67.

Boucher, J.M., Chen, C., Sun, Y. & Beardsley, R.C. (2013) Effects of interannual environmental variability on the transport-retention dynamics in haddock *Melanogrammus aeglifinus* larvae on Georges Bank. *Marine Ecology Progress Series*, **487**, 201–15.

Bradford, M.J. (1992) Precision of recruitment predictions from early life stages of marine fishes. *Fishery Bulletin (US)*, **90**, 439–53.

Bradford, M.J. & Cabana, G. (1997) Interannual variability in stage-specific survival rates and the causes of recruitment variation. In: R.C. Chambers & E.A. Trippel (Eds) *Shape, Early Life History and Recruitment in Fish Populations*. pp. 469–93. Chapman & Hall, London.

Brander, K. (1996) Effects of climate change on cod (*Gadus morhua*) stocks. In: C.M. Wood & D.G. McDonald (Eds) *Global Warming: Implications for Freshwater and Marine Fish*. pp. 255–78. Cambridge University Press, Cambridge.

Brander, K.M. (2010) Cod *Gadus morhua* and climate change: processes, productivity and prediction. *Journal of Fish Biology*, **77**, 1899–911.

Breitburg, D.L. (1992) Episodic hypoxia in Chesapeake Bay: interacting effects of recruitment, behavior, and physical disturbance. *Ecological Monographs*, **62**, 525–46.

Breitburg, D. (2002) Effects of hypoxia, and the balance between hypoxia and enrichment, on coastal fishes and fisheries. *Estuaries*, **25**, 767–81.

Breitburg, D.L., Loher, T., Pacey, C.A. & Gerstein, A. (1997) Varying effects of low dissolved oxygen on trophic interactions in an estuarine food web. *Ecological Monographs*, **67**, 489–507.

Breitburg, D.L., Rose, K.A. & Cowan, J.H. Jr (1999) Linking water quality to larval survival: predation mortality of fish larvae in an oxygen-stratified water column. *Marine Ecology Progress Series*, **178**, 39–54.

Breitburg, D.L., Steinberg, N., Dubeau, S., Cooksey, C. & Houde, E.D. (1994) Effects of low dissolved oxygen on predation on estuarine fish larvae. *Marine Ecology Progress Series*, **104**, 39–54.

Brickman, D., Shacke, N.L. & Frank, K.T. (2001) Modelling the retention and survival of Browns Bank haddock larvae using an early life stage model. *Fisheries Oceanography*, **10**, 284–96.

Browman, H.I. (1995) Introduction: commentaries on current research trends in recruitment studies. *Marine Ecology Progress Series*, **128**, 305.

Browman, H.I. (1996) Predator–prey interactions in the sea: commentaries on the role of turbulence. *Marine Ecology Progress Series*, **139**, 301–2.

Browman, H.I. & Skiftesvik, A.B. (1996) Effects of turbulence on the predation cycle of fish larvae: comments on some of the issues. *Marine Ecology Progress Series*, **13**, 309–12.

Brunel, T. & Boucher, J. (2007) Long-term trends in fish recruitment in the north-east Atlantic related to climate change. *Fisheries Oceanography*, **16**, 336–49.

Buckley, L., Caldarone, E. & Ong, T.L. (1999) RNA/DNA ratio and other nucleic acid-based indicators for growth and condition of marine fishes. *Hydrobiologia*, **401**, 265–77.

Buckley, L.J., Caldarone, E.M., Lough, R.G. & St Onge-Burns, J.M. (2006) Ontogenetic and seasonal trends in recent growth rates of Atlantic cod and haddock larvae on Georges Bank: effect of photoperiod and temperature. *Marine Ecology Progress Series*, **325**, 205–26.

Buckley, L.J., Caldarone, E.M. & Clemmesen, C. (2008). Multi-species larval fish growth model based on temperature and fluorometrically derived RNA/DNA ratios: results from a meta-analysis. *Marine Ecology Progress Series* **371**, 221–232.

Calderone, E.M., St Onge-Burns, J.M. & Buckley, L.J. (2003) Relationship of RNA/DNA ratio and temperature to growth in larvae of Atlantic cod *Gadus morhua*. *Marine Ecology Progress Series*, **262**, 229–40.

Campana, S.E. (1999) Chemistry and composition of fish otoliths: pathways, mechanisms and applications. *Marine Ecology Progress Series*, **188**, 263–97.

Campana, S.E. (2005) Otolith science entering the 21st century. *Marine and Freshwater Research*, **56**, 485–95.

Campana, S.E., Fowler, A.J. & Jones, C.M. (1994) Otolith elemental fingerprinting for stock identification of Atlantic cod (*Gadus morhua*) using laser ablation ICPMS. *Canadian Journal of Fisheries and Aquatic Sciences*, **51**, 1942–50.

Campana, S.E., Smith, S.J. & Hurley, P.C.F. (1989) A drift-retention dichotomy for larval haddock (*Melanogrammus aeglefinus*) spawned on Browns Bank. *Canadian Journal of Fisheries and Aquatic Sciences*, **46** (Supplement 1), 93–102.

Campana, S.E., Thorrold, S.R., Jones, C.M., Gunther, D., Tubrett, M., Longerich, H., Jackson, S., Halden, N.M., Kalish, J.M., Piccoli, P., de Pontual, H., Troadec, H., Panfili, J., Secor, D.H., Severin, K.P., Sie, S.H., Thresher, R., Teesdale, W.J. & Campbell, J.L. (1997) Comparisons of accuracy, precision, and sensitivity in elemental assays of fish otoliths using the electron microprobe, proton-induced X-ray emission, and laser ablation inductively coupled plasma mass spectrometry. *Canadian Journal of Fisheries and Aquatic Sciences*, **54**, 2068–79.

Campfield, P.A. (2004) *Ichthyoplankton Community Structure and Feeding Ecology in the Patuxent River Estuarine Transition Zone.* Master's thesis, University of Maryland, College Park, Maryland.

Campfield, P.A. & Houde, E.D. (2011) Ichthyoplankton community structure and comparative trophodynamics in an estuarine transition zone. *Fishery Bulletin (US)*, **109**, 1–19.

Cassie, R.M. (1963) Microdistribution of plankton. *Oceanography and Marine Biology Annual Review*, **1**, 223–52.

Catalan, I.A., Tejedor, A., Alemany, F. & Reglero, P. (2011) Trophic ecology of Atlantic bluefin tuna *Thunnus thynnus* larvae. *Journal of Fish Biology*, **78**, 1545–60.

Chambers, R.C., CandelmoA.C., Habeck, E.A., Poach, M.E., Wieczorek, D., Cooper, K.R., Greenfield, C.E. & Phelan, B.A. (2014) Effects of elevated CO_2 in the early life stages of summer flounder, *Paralichthys dentatus*, and potential consequences of ocean acidification. *Biogeosciences*, **11**, 1613–26.

Chambers, R.C. & Leggett, W.C. (1996) Maternal influences on variation in egg sizes in temperate marine fishes. *American Zoologist*, **36**, 180–96.

Chambers, R.C., Leggett, W.C. & Brown, J.A. (1989) Egg size, female effects, and the correlations between early life history traits of capelin *Mallotus villosus*: an appraisal at the individual level. *Fishery Bulletin (US)*, **87**, 515–23.

Chavez, F.P., Ryan, J., Lluch-Cota, S.E. & Niquen, M. (2003) From anchovies to sardines and back: multidecadal change in the Pacific Ocean. *Science*, **299**, 217–21.

Checkley, D., Alheit, J., Oozeki, Y. & Roy, C. (2009) *Climate change and small pelagic fish.* Cambridge University Press, Cambridge, U.K.

Chesney, E.J. (2008) Foraging behavior of bay anchovy, *Anchoa mitchilli*. *Journal of Experimental Marine Biology and Ecology*, **362**, 117–24.

Churchill, J.H., Runge, J. & Chen, C. (2011) Processes controlling retention of spring-spawned Atlantic cod (*Gadus morhua*) in the western Gulf of Maine and their relationship to an index of recruitment success. *Fisheries Oceanography*, **20**, 32–46.

Ciannelli, L., Chan, K.S., Bailey, K.M. & Stenseth, N.C. (2004) Nonadditive effects of the environment on the survival of a large marine fish population. *Ecology*, **85**, 3418–27.

Cianneli, L., Bailey, K.M., Belgrano, A., & Stenseth, N.C. (2005) Climate change causing phase transition of walleye pollock (*Theragra chalcogramma*) recruitment dynamics. *Proceedings of the Royal Society London B*, **272**, 1735–43.

Clemmesen, C. (1996) Importance and limits of RNA/DNA ratios as a measure of nutritional condition in fish larvae. In: Y. Watanabe, Y. Yamashita & Y. Oozeki Y. (Eds) *Survival Strategies in Early Life Stages of Marine Resources*. pp. 67–82. Balkema, Rotterdam.

Clemmesen, C. & Doan, T. (1996) Does otolith structure reflect the nutritional condition of a fish larva? Comparison of otolith structure and biochemical index (RNA/DNA ratio) determined on cod larvae. *Marine Ecology Progress Series*, **138**, 33–9.

Comerford, S., Brophy, D., Fox, C.J., Taylor, N., van der Veer, H.W., Nash, R.D.M. & Geffen, A.J. (2013) Temperature effect of growth and larval duration of plaice *Pleuronectes platessa* in three regions of the Northeast Atlantic. *Marine Ecology Progress Series*, **476**, 215–26.

Coombs, S.H., Giovanardi, O., Halliday, N.C., Franceschini, G., Conway, D.V.P., Manzueto, L., Barrett, C.D. & McFadzen, I.R.B. (2003) Wind mixing, food availability and mortality of anchovy larvae *Engraulis encrasicolus* in the northern Adriatic Sea. *Marine Ecology Progress Series*, **248**, 221–35.

Cowan, J.H., Jr, Houde, E.D. & Rose, K.A. (1996) Size-dependent vulnerability of marine fish larvae to predation: an individual-based numerical experiment. *ICES Journal of Marine Science*, **53**, 23–37.

Cowan, J.H., Jr, Rose, K.A. & Devries, D.R. (2000) Is density-dependent growth in young-of-the-year fishes a question of critical weight? *Reviews in Fish Biology and Fisheries*, **10**, 61–89.

Cowan, J.H., Jr, Rose, K.A. & Houde, E.D. (1997) Size-based foraging success and vulnerability to predation: selection of survivors in individual-based models of larval fish populations. In: R.C. Chambers & E.A. Trippel (Eds) *Early Life History and Recruitment in Fish Populations*. pp. 357–86. Chapman & Hall, London.

Cowan, J.H., Jr, Rose, K.A., HoudeE.D., Wang, S.-B. & YoungJ. (1999) Modeling effects of increased larval mortality on bay anchovy population dynamics in the mesohaline Chesapeake Bay: evidence for compensatory reserve. *Marine Ecology Progress Series*, **185**, 133–46.

Cowan, J.H., Jr, Rose, K.A., Rutherford, E.S. & Houde, E.D. (1993) Individual-based model of young-of-the-year striped bass population dynamics. II. Factors affecting recruitment in the Potomac River, Maryland. *Transactions of the American Fisheries Society*, **122**, 439–58.

Cowan, J.H. Jr & Shaw, R.F. (2002) *Recruitment*. In: L.A. Fuiman & R.G. Werner (Eds) *Fishery Science: The Unique Contributions of Early Life Stages*. pp. 88–111. Blackwell Science, Oxford.

Cowen, R.K. (2002) Larval dispersal and retention and consequences for population connectivity. In: P.F. Sale (Ed.) *Coral Reef Fishes*. pp. 149–70. Elsevier, New York.

Cowen, R.K., Gawarkiewicz, G., Pineda, J., ThorroldS.R. & WernerF.E.. (2007) Population connectivity in marine systems. *Oceanography* **20**, 14–21

Cowen, R.K., Lwiza, K.M.M., Sponaugle, S., Paris, C.B. & Olson, D.B. (2000) Connectivity of marine populations: open or closed? *Science*, **287**, 857–9.

Cowen, R.K., Paris, C.B. & Srinivasan, A. (2006) Scaling of connectivity in marine populations. *Science*, **311**, 522–7.

Cowen, R.K. and Sponaugle, S. (2009) Larval dispersal and marine population connectivity. *Annual Reviews in Marine Science* **1**, 443–66.

Crecco, V. & Savoy, T. (1985) Effects of biotic and abiotic factors on growth and relative survival of young American shad, *Alosa sapidissima*. *Canadian Journal of Fisheries and Aquatic Sciences*, **42**, 1640–8.

Crecco, V. & Savoy, T. (1987) Effects of climatic and density-dependent factors on intra-annual mortality of larval American shad. *American Fisheries Society Symposium*, **2**, 69–81.

Crecco, V., Savoy, T. & Gunn, L. (1983) Daily mortality rates of larval and juvenile American shad (*Alosa sapidissima*) in the Connecticut River with changes in year-class strength. *Canadian Journal of Fisheries and Aquatic Sciences*, **40**, 1719–28.

Crecco, V., Savoy, T. & Whitworth, W. (1986) Effects of density-dependent and climatic factors on American shad, *Alosa sapidissima*, recruitment: a predictive approach. *Canadian Journal of Fisheries and Aquatic Sciences*, **43**, 457–63.

Crowder, L.B., Rice, J.A., Miller, T.J. & Marschall, E.A. (1992) Empirical and theoretical approaches to size-based interactions and recruitment variability in fishes. In: D.L. DeAngelis & L.J. Gross (Eds) *Individual-based Models and Approaches in Ecology.* pp. 237–55. Chapman & Hall, New York.

Cury, P. & Roy, C. (1989) Optimal environmental window and pelagic fish recruitment success in upwelling areas. *Canadian Journal of Fisheries & Aquatic Sciences*, **46**, 670–80.

Cushing, D.H. (1971) The dependence of recruitment on parent stock in different groups of fishes. *Journal du Conseil pour l'Exploration de la Mer*, **33**, 340–62.

Cushing, D.H. (1973) *Recruitment and Parent Stock in Fishes.* University of Washington Sea Grant Program, WSG 73–1. Seattle, Washington. 197 pp.

Cushing, D.H. (1974) The natural regulation of fish populations. In: F.R. Harden Jones (Ed.) *Sea Fisheries Research.* pp. 399–412. John Wiley & Sons, New York.

Cushing, D.H. (1975) *Marine Ecology and Fisheries.* Cambridge University Press, Cambridge.

Cushing, D.H. (1982) *Climate and Fisheries.* Academic Press, London.

Cushing, D.H. (1990) Plankton production and year-class strength in fish populations: an update of the match/mismatch hypothesis. *Advances in Marine Biology*, **26**, 250–93.

Cushing, D.H. & Dickson, R.R. (1976) The biological response in the sea to climatic changes. *Advances in Marine Biology*, **14**, 1–22.

Cushing, D.H. & Horwood, J.W. (1994) The growth and death of fish larvae. *Journal of Plankton Research*, **16**, 291–300.

Daewel, U., Peck, M.A. & Schrum, C. (2011) Life history strategy and impacts of environmental variability on early life stages of two marine fishes in the North Sea: an individual-based modeling approach. *Canadian Journal of Fisheries and Aquatic Sciences*, **68**, 426–43.

De Figueiredo, G.M., Nash, R.D.M. & Montagnes, D.J.S. (2007) Do protozoa contribute significantly to the diet of larval fish in the Irish Sea? *Journal of the Marine Biological Association of the United Kingdom*, **87**, 843–50.

DeVries, M.C., Forward, R.B., Jr & Hettler, W.F. (1995) Behavioral response of larval Atlantic menhaden *Brevoortia tyrannus* (Latrobe) and spot *Leiostomus xanthurus* (Lacepede) to rates of salinity change. *Journal of Experimental Marine Biology and Ecology*, **185**, 93–108.

Dingsør, G.E., Ciannelli, L., Chan, K.-S., Ottersen, G. & Stenseth, N.C. (2007) Density dependence and density independence during the early life stages of four marine fish stocks. *Ecology*, **88**, 625–34.

Doherty, P.J. (2002) Variable replenishment and the dynamics of reef fish populations. In: P.F. Sale (Ed.) *Coral Reef Fishes.* pp. 327–55. Elsevier, New York.

Doherty, P.J. & Fowler, T. (1994) An empirical test of recruitment limitation in a coral reef fish. *Science*, **263**, 935–9.

Doherty, P.J. & Williams, D.McB. (1988) The replenishment of coral reef fish populations. *Oceanography and Marine Biology Annual Reviews*, **26**, 487–551.

Dou, S.Z., Masuda, R., Tanaka, M. & Tsukamoto, K. (2005) Effects of temperature and delayed initial feeding on the survival and growth of Japanese flounder larvae. *Journal of Fish Biology*, **66**, 362–77.

Dower, J.F., Miller, T.J. & Leggett, W.C. (1997) The role of microscale turbulence in the feeding ecology of larval fish. *Advances in Marine Biology*, **31**, 169–220.

Drinkwater, K.F. (2002) A review of the role of climate variability in the decline of northern cod. In: N.A. McGinn (Ed.) *Fisheries in a Changing Climate.* pp. 113–30. American Fisheries Society, Symposium 32, Bethesda, Maryland.

Drinkwater, K.F. (2006) The regime shift of the 1920s and 1930s in the North Atlantic. *Progress in Oceanography*, **68**, 134–51.

Drinkwater, K.F. & Myers, R.A. (1987) Testing predictions of marine fish and shellfish landings from environmental variables. *Canadian Journal of Fisheries and Aquatic Sciences*, **44**, 1568–73.

Duarte, C.M. & Alcaraz, M. (1989) To produce many small or few large eggs: a size-independent reproductive tactic of fish. *Oecologia*, **80**, 401–4.

Duffy-Anderson, J.T., Bailey, K.M., & Ciannelli, L. (2002) Consequences of a superabundance of larval walleye pollock, *Theragra chalcogramma*, in the Gulf of Alaska in 1981. *Marine Ecology Progress Series*, **243**, 179–90.

Duffy-Anderson, J.T., Bailey, K., Ciannelli, L., Cury, P., Belgrano, A. & Stenseth, N.C. (2005) Phase transitions in marine fish recruitment processes. *Ecological Complexity*, **2**, 205–18.

Ellertsen, B., Fossum, P., Solemdal, P. & Sundby, S. (1989) Relation between temperature and survival of eggs and first-feeding larvae of northeast Arctic cod (*Gadus morhua* L.). *Rapports et Procès-verbaux des Rèunions, Conseil International pour l'Exploration de la Mer*, **191**, 209–19.

Ellertsen, B., Fossum, P., Solemdal, P. & Sundby, S. (1995) The 'critical period' concept – a century of recruitment research. *Marine Ecology Progress Series*, **128**, 306–8.

Ellertsen, B., Fossum, P., Solemdal, P., Sundby, S. & Tilseth, S. (1990) Environmental influence on recruitment and biomass yields in the Norwegian Sea ecosystem. In: K. Sherman, L.M. Alexander & B.D. Gold (Eds) *Large Marine Ecosystems: Patterns, Processes, and Yields*. pp. 19–35. American Association for the Advancement of Science, Washington, DC.

Ellertsen, B., Solemdal, P., Stromme, T., Sundby, S., Tilseth, S. & Westgard, T. (1981) Spawning period, transport and dispersal of eggs from the spawning area of Arcto-Norwegian cod (*Gadus morhua* L.). *Rapports et Procès-verbaux des Rèunions, Conseil International pour l'Exploration de la Mer*, **178**, 260–7.

Fargo, J. (1994) Examining recruitment relationships for Hecate Strait English sole (*Pleuronectes vetulus*). *Netherlands Journal of Sea Research*, **32**, 385–97.

Ferron, A. & Leggett, W.C. (1994) An appraisal of condition measures for marine fish larvae. *Advances in Marine Biology*, **30**, 217–303.

Field, D.B., Baumgartner, T.R., Ferreira, V., Gutierrez, D., Lozano-Montes, H., Salvatecci, R. & Soutar, A. (2009) Variability from scales in marine sediments and other historical records. In: D. Checkley, J. Alheit, Y. Oozeki & C. Roy (Eds) *Climate Change and Small Pelagic Fish*. pp. 45–63. Cambridge University Press, Cambridge, UK.

Fiksen, O. & Folkvord, A. (1999) Modelling growth and ingestion processes in herring *Clupea harengus* larvae. *Marine Ecology Progress Series*, **184**, 273–89.

Fiksen, O. & MacKenzie, B.R. (2002) Process-based models of feeding and prey selection in larval fish. *Marine Ecology Progress Series*, **243**, 151–64.

Finney, B.P., Gregory-Eaves, I., Douglas, M.S.V. & Smol, J.P. (2002) Fisheries productivity in the northeastern Pacific Ocean over the past 2,200 years. *Nature*, **416**, 729–33.

Fisher, R. (2005) Swimming speeds of larval coral reef fishes: impacts on self-recruitment and dispersal. *Marine Ecology Progress Series*, **285**, 223–32.

Fogarty, M.J. (1993) Recruitment in randomly varying environments. *ICES Journal of Marine Science*, **50**, 247–60 .

Fogarty, M.J., Sissenwine, M.P. & Cohen, E.B. (1991) Recruitment variability and the dynamics of exploited marine populations. *Trends in Ecology and Evolution*, **6**, 241–6.

Folkvord, A. (2005) Comparison of size-at-age of larval Atlantic cod (*Gadus morhua*) from different populations based on size- and temperature-dependent growth models. *Canadian Journal of Fisheries and Aquatic Sciences*, **62**, 1037–52.

Fortier, L., Levasseur, M.E., Drolet, R. & Therriault, J.C. (1992) Export production and the distribution of fish larvae and their prey in a coastal jet region. *Marine Ecology Progress Series*, **85**, 203–18.

Forward, R.B., Jr, Reinsel, K.A., Peters, D.S., Tankersley, R.A., Churchill, J.H., Crowder, L.B., HettlerW.F., Warlen, S.M. & Green, M.D. (1999) Transport of larvae through a tidal inlet. *Fisheries Oceanography*, **8** (Supplement 2), 153–72.

Forward, R.B. Jr & Tankersley, R.A. (2001) Selective tidal-stream transport of marine animals. *Oceanography and Marine Biology Annual Review*, **39**, 305–53.

Fox, C.J., Taylor, M.I., van der Kooij, J., Taylor, N., Milligan, S.P., Albaina, A., Pascoal, S., Lallias, D., Maillard, M. & Hunter, E. (2012) Identification of marine fish egg predators using molecular probes. *Marine Ecology Progress Series*, **462**, 205–18.

Francis, M.P. (1993) Does water temperature affect year class strength in New Zealand snapper (*Pagrus auratus*, Sparidae)? *Fisheries Oceanography*, **2**, 65–72.

Francis, M.P. (1994) Duration of larval and spawning periods in *Pagrus auratus* (Sparidae) determined from otolith daily increments. *Environmental Biology of Fishes*, **39**, 137–52.

Francis, R.C. & Hare, S.R. (1994) Decadal-scale regime shifts in the large marine ecosystems of the North-east Pacific: a case for historical science. *Fisheries Oceanography*, **3**, 279–91.

Frank, K.T. (1997) The utility of early life history studies and the challenges of recruitment prediction. In: R.C. Chambers & E.A. Trippel (Eds) *Early Life History and Recruitment in Fish Populations*. pp. 495–512. Chapman & Hall, London.

Franke, A. & Clemmesen, C. (2011) Effect of ocean acidification on early life stages of Atlantic herring (*Clupea harengus* L.). *Biogeosciences Discussion*, **8**, 7097–126.

Frommel, A.Y., Schubert, A., Piatkowski, U. & Clemmesen, C. (2013) Egg and early larval stages of Baltic cod, *Gadus morhua*, are robust to high levels of ocean acidification. *Marine Biology*, **160**, 1825–34.

Fuiman, L.A. (2002) Special considerations of fish eggs and larvae. In: L.A. Fuiman & R.G. Werner (Eds) *Fishery Science: The Unique Contributions of Early Life Stages*. pp. 1–32. Blackwell Publishing, Oxford.

Fukami, K., Watanabe, A., Fujita, S., Yamaoka, K. & Nishijima, T. (1999) Predation on naked protozoan microzooplankton by fish larvae. *Marine Ecology Progress Series*, **185**, 285–91.

Galbraith, P.S., Browman, H.I., Racca, R.G., Skiftesvik, A.B. & Saint-Pierre, J.-F. (2004) Effect of turbulence on the energetics of foraging in Atlantic cod *Gadus morhua* larvae. *Marine Ecology Progress Series*, **281**, 241–57.

Gerritsen, J. & Strickler, J.R. (1977) Encounter probabilities and community structure in zooplankton: a mathematical model. *Journal of the Fisheries Research Board of Canada*, **34**, 73–82.

Govoni, J.J. (2005) Fisheries oceanography and the ecology of early life histories of fishes: a perspective over fifty years. *Scientia Marina*, **69** (Supplement 1), 125–37.

Govoni, J.J., Hoss, D.E. & Colby, D.R. (1989) The spatial distribution of larval fishes about the Mississippi River plume. *Limnology and Oceanography*, **34**, 178–87.

Green, B.S. & McCormick, M.I. (2005) Maternal and paternal effects determine size, growth and performance in larvae of a tropical reef fish. *Marine Ecology Progress Series*, **289**, 263–72.

Grimes, C.B. & Kingsford, M.J. (1996) How do riverine plumes of different sizes influence fish larvae: do they enhance recruitment? *Marine and Freshwater Research*, **47**, 191–208.

Hall, L.W., Jr, Fischer, S.A., Killen, W.D., Ziegenfuss, M.C., Anderson, R.D. & Klauda, R.J. (1993) Assessing the relationships between critical acidic conditions in Maryland coastal plain streams and predicted mortality of early life stages of migratory fish. *Aquatic toxicology*, **26**, 239–72.

Hannah, C G., Shore, J.A. & Loder, J.W. (2001) Seasonal circulation on the western and central Scotian Shelf. *Journal of Physical Oceanography*, **31**, 591–615.

Harden-Jones, F.R. (1968) *Fish Migration*. Edward Arnold, London.

Harding, D., Nichols, J.H. & Tungate, D.S. (1978) The spawning of plaice (*Pleuronectes platessa* L.) in the southern North Sea and English Channel. *Rapports et Procès-verbaux des Rèunions, Conseil International pour l'Exploration de la Mer*, **172**, 102–13.

Hare, J.A. & Able, K.W. (2007) Mechanistic links between climate and fisheries along the east coast of the United States: explaining population outbursts of Atlantic croaker (*Micropogonias undulatus*). *Fisheries Oceanography*, **16**, 31–45.

Hare, J.A. & Cowen, R.K. (1997) Size, growth, development, and survival of the planktonic larvae of *Pomatomus saltatrix* (Pisces: Pomatomidae). *Ecology*, **78**, 2415–31.

Hare, J.A., Thorrold, S., Walsh, H., Reiss, C., Valle-Levinson, A. & Jones, C. (2005) Biophysical mechanisms of larval fish ingress into Chesapeake Bay. *Marine Ecology Progress Series*, **303**, 295–310.

Head, E.J.H., Brickman, D. & Harris, L.R. (2005) An exceptional haddock year class and unusual environmental conditions on the Scotian Shelf in 1999. *Journal of Plankton Research*, **27**, 597–602.

Heath, M.R. (1992) Field investigations of the early life stages of marine fish. *Advances in Marine Biology*, **28**, 1–174.

Heath, M.R. & Gallego, A. (1998) Bio-physical modelling of the early life stages of haddock, *Melanogrammus aeglefinus*, in the North Sea. *Fisheries Oceanography*, **7**, 110–25.

Heath, M.R. & Gallego, A. (2000) *Modelling the Spatial and Temporal Structure of Survivorship to Settlement in North Sea and West of Scotland Haddock*. International Council for the Exploration of the Sea, ICES CM 2000/N,11.

Heath, M.R. & Lough, R.G. (2007) A synthesis of large-scale patterns in the planktonic prey of larval and juvenile cod (*Gadus morhua*). *Fisheries Oceanography*, **16**, 169–85.

Heath, M.R. & MacLachlan, P. (1987) Dispersion and mortality of yolk-sac herring (*Clupea harengus* L.) larvae from a spawning ground to the west of the Outer Hebrides. *Journal of Plankton Research*, **9**, 613–30.

Heath, M.R., Gallego, A., Hislop, J.R.G., Needle, C., Scott, B. & Wright, P.J. (1999) *The Importance of the Late Pelagic and Demersal Settlement Phases for Recruitment Dynamics in the North Sea Haddock*. International Council for the Exploration of the Sea, ICES CM 1999/Y, 11.

Heath, M.R., MacLachlan, P.M. & Martin, J.H.A. (1987) Inshore circulation and transport of herring larvae off the north coast of Scotland. *Marine Ecology Progress Series*, **40**, 11–23.

Heath, M.R. & Rankine, P. (1988) Growth and advection of larval herring (*Clupea harengus* L.) in the vicinity of the Orkney Isles. *Estuarine, Coastal and Shelf Science*, **27**, 547–65.

Helbig, J.A. & Pepin, P. (1998a) Partitioning the influence of physical processes on the estimation of ichthyoplankton mortality rates. I. Theory. *Canadian Journal of Fisheries and Aquatic Sciences*, **55**, 2189–205.

Helbig, J.A. & Pepin, P. (1998b) Partitioning the influence of physical processes on the estimation of ichthyoplankton mortality rates. II. Application to simulated and field data. *Canadian Journal of Fisheries and Aquatic Sciences*, **55**, 2206–20.

Hendrey, G.R. (1987) Acidification and anadromous fish of Atlantic estuaries. *Water, Air, and Soil Pollution*, **35**, 1–6.

Hermann, A.J., Hinckley, S., Megrey, B.A. & NappJ.M. (2001) Applied and theoretical considerations for constructing spatially explicit individual-based models of marine larval fish that include multiple trophic levels. *ICES Journal of Marine Science*, **58**, 1030–41.

Hermann, A.J., Hinckley, S., Megrey, B.A. & Stabeno, P.J. (1996) Interannual variability of the early life history of walleye pollock near Shelikof Strait as inferred from a spatially explicit, individual-based model. *Fisheries Oceanography*, **5**, 39–57.

Hill, A.E. (1991) Advection-diffusion-mortality solutions for investigating pelagic larval dispersal. *Marine Ecology Progress Series*, **70**, 117–28.

Hinckley, S., Hermann, A.J. & Megrey, B.A. (1996) Development of a spatially explicit, individual-based model of marine fish early life history. *Marine Ecology Progress Series*, **139**, 47–68.

Hinckley, S., Hermann, A.J., Mier, K.L. & Megrey, B.A. (2001) Importance of spawning location and timing to successful transport to nursery areas: a simulation study of Gulf of Alaska walleye pollock. *ICES Journal of Marine Science*, **58**, 1042–52.

Hixon, M.A. & Carr, M.H. (1997) Synergistic predation, density dependence, and population regulation in marine fish. *Science*, **277**, 946–9.

Hixon, M.A. & Jones, G.P. (2005) Competition, predation, and density-dependent mortality in demersal marine fishes. *Ecology*, **86**, 2847–59.

Hixon, M.A. & Webster, M.S. (2002) Density dependence in reef fish populations. In: P.F. Sale (Ed.) *Coral Reef Fishes*. pp. 303–25. Elsevier Science, New York.

Hjort, J. (1914) Fluctuations in the great fisheries of northern Europe viewed in the light of biological research. *Rapports et Procès-verbaux des Rèunions, Conseil International pour l'Exploration del la Mer*, **20**, 1–228.

Hjort, J. (1926) Fluctuations in the year classes of important food fishes. *Journal du Conseil International pour l'Exploration de la Mer*, **1**, 1–38.

Holliday, F.G.T. (1969) The effects of salinity on the eggs and larvae of teleosts. In: W.S. Hoar & D.J. Randall (Eds) *Fish Physiology, Vol. I, Excretion, Ionic Regulation and Metabolism*. pp. 293–309. Academic Press, New York.

Hollowed, A.B. & Bailey, K.M. (1989) New perspectives on the relationship between recruitment of Pacific hake *Merluccius productus* and the ocean environment. In: R.J. Beamish & G.A. McFarlane (Eds) *Effects of Ocean Variability on Recruitment and an Evaluation of Parameters used in Stock Assessment Models*. Canadian Special Publication in Fisheries and Aquatic Sciences, **108**, 207–20.

Holt, G.J. (2002) Human impacts. In: L.A. Fuiman & R.G. Werner (Eds) *Fishery Science: The Unique Contributions of Early Life Stages*. pp. 222–42. Blackwell Science, Oxford.

Houde, E.D. (1978) Critical food concentrations for larvae of three species of subtropical marine fishes. *Bulletin of Marine Science*, **28**, 395–411.

Houde, E.D. (1987) Fish early life dynamics and recruitment variability. *American Fisheries Society Symposium*, **2**, 17–29.

Houde, E.D. (1989a) Comparative growth, mortality, and energetics of marine fish larvae: temperature and implied latitudinal effects. *Fishery Bulletin (US)*, **87**, 471–95.

Houde, E.D. (1989b) Subtleties and episodes in the early life of fishes. *Journal of Fish Biology*, **35** (Supplement A), 29–38.

Houde, E.D. (1994) Difference between marine and freshwater fish larvae: implications for recruitment. *ICES Journal of Marine Science*, **51**, 91–7.

Houde, E.D. (1996) Evaluating stage-specific survival during the early life of fish. In: Y. Watanabe, Y. Yamashita & Y. Oozeki (Eds) *Survival Strategies in Early Life Stages of Marine Resources*. pp. 51–66. Balkema, Rotterdam.

Houde, E.D. (1997a) Patterns and consequences of selective processes in teleost early life histories. In: R.C. Chambers & E.A. Trippel (Eds) *Early Life History and Recruitment in Fish Populations*. pp. 173–95. Springer Science and Business Media, Dordrecht.

Houde, E.D. (1997b) Patterns and trends in larval-stage growth and mortality of teleost fish. *Journal of Fish Biology*, **51** (Supplement A), 52–83.

Houde, E.D. (2002) Mortality. In: L.A. Fuiman & R.G. Werner (Eds) *Fishery Science, The Unique Contributions of Early Life Stages*. pp. 64–87. Blackwell Science, Oxford.

Houde, E.D. (2008) Emerging from Hjort's shadow. *Journal of Northwest Atlantic Fisheries Science* **41**, 53–70.

Houde, E.D. & Schekter, R.C. (1981) Growth rates, rations and cohort consumption of marine fish larvae in relation to prey concentrations. *Rapports et Procès-verbaux des Rèunions, Conseil International pour l'Exploration de la Mer*, **178**, 441–53.

Houde, E.D. & Zastrow, C.E. (1993) Ecosystem- and taxon-specific dynamic and energetics properties of larval fish assemblages. *Bulletin of Marine Science*, **53**, 290–335.

Hovenkamp, F. (1992) Growth-dependent mortality of larval plaice *Pleuronectes platessa* in the North Sea. *Marine Ecology Progress Series*, **82**, 95–101.

Huebert, K.B, Cowen, R.K. & Sponaugle, S. (2011) Vertical migrations of reef fish larvae in the Straits of Florida and effects on larval transport. *Limnology and Oceanography*, **56**, 1653–66.

Huggett, J., Freon, P., Mullon, C. & Penven, P. (2003) Modelling the transport success of anchovy *Engraulis encrasicolus* eggs and larvae in the southern Benguela: the effect of spatio-temporal spawning patterns. *Marine Ecology Progress Series*, **250**, 247–62.

Hunt von Herbing, I. & GallagherS.M. (2000) Foraging behavior in early Atlantic cod larvae (*Gadus morhua*) feeding on a protozoan (*Balanion* sp.) and a copepod nauplius (*Pseudodiaptomus* sp.). *Marine Biology*, **136**, 591–602.

Hunter, J.R. (1972) Swimming and feeding behavior of larval anchovy *Engraulis mordax*. *Fishery Bulletin (US)*, **70**, 821–38.

Hunter, J.R. (1981) Feeding ecology and predation of marine fish larvae. In: R. Lasker (Ed.) *Marine Fish Larvae. Morphology, Ecology and Relation to Fisheries.* pp. 34–77. University of Washington Press, Seattle.

Hunter, J.R. & Thomas, G.L. (1974) Effect of prey distribution and density on the searching and feeding behaviour of larval anchovy *Engraulis mordax* Girard. In: J.H.S. Blaxter (Ed.) *The Early Life History of Fish.* pp. 559–74. Springer-Verlag, Berlin.

Hurst, T.P., Fernandez, E.R., Mathis, J.T., Miller, J.A., Stinson, C.M. & Ahgeak, E.F. (2012) Resiliency of juvenile walleye pollock to projected levels of ocean acidification. *Aquatic Biology*, **17**, 247–59.

Hurst, T.P., Fernandez, E.R. & Mathis, J.T. (2013) Effects of ocean acidification on hatch size and larval growth of walleye pollock (*Theragra chalcogramma*). *ICES Journal of Marine Science*, **70**, 812–22.

Iles, T.C. (1994) A review of stock–recruitment relationships with reference to flatfish populations. *Netherlands Journal of Sea Research*, **32**, 399–420.

Iles, T.D. & Sinclair, M. (1982) Atlantic herring: stock discreteness and abundance. *Science*, **215**, 627–33.

Islam, M.S. & Tanaka, M. (2006) Ontogenetic dietary shift of Japanese sea bass during larva-juvenile transition in Ariake Bay. *Marine Ecology Progress Series*, **323**, 305–10.

Iwatsuki, Y., Nakata, H. & Hirano, R. (1989) The thermohaline front in relation to fish larvae. *Rapports et Procès-verbaux des Rèunions, Conseil International pour l'Exploration de la Mer*, **191**, 119–26.

Jenkins, G.P. & King, D. (2006) Variation in larval growth can predict the recruitment of a temperate, seagrass-associated fish. *Oecologia*, **147**, 641–9.

Jenkins, G.P., Young, J.W. & Davis, T.L.O. (1991) Density dependence of larval growth of a marine fish, the southern bluefin tuna, *Thunnus maccoyii. Canadian Journal of Fisheries and Aquatic Sciences*, **48**, 1358–63.

Jones, G.P. (1991) Postrecruitment processes in the ecology of coral reef fish populations: a multifactorial perspective. In: P.F. Sale (Ed.) *The Ecology of Fishes on Coral Reefs*. pp. 294–328. Academic Press, San Diego.

Jones, G.P., Millicich, M.J., Emslie, M.J. & Lunow, C. (1999) Self-recruitment in a coral reef fish population. *Nature*, **402**, 802–4.

Jones, G.P. & McCormick, M.I. (2002) Numerical and energetic processes in the ecology of coral reef fishes. In: P.F. Sale (Ed.) *Coral Reef Fishes*. pp. 221–38. Elsevier Science, New York.

Jones, R. (1973) Density dependent regulation of the numbers of cod and haddock. *Rapports et Procès-verbaux des Rèunions, Conseil International pour l'Exploration de la Mer*, **164**, 156–73.

Jordaan, A. & Kling, L.J. (2003) Determining the optimal temperature range for Atlantic cod (*Gadus morhua*) during early life. In: H.I. Browman & A.B. Skiftesvik (Eds) *The Big Fish Bang*. pp. 45–62. Proceedings of the 26th Annual Larval Fish Conference, Bergen, Norway, 2002. Institute of Marine Research, Bergen.

Kamler, E. (2005) Parent–egg–progeny relationships in teleost fishes: an energetics perspective. *Reviews in Fish Biology and Fisheries*, **15**, 399–421.

Kawasaki, T. (1992) Mechanisms governing fluctuations in pelagic fish populations. *South African Journal of Marine Science*, **12**, 873–9.

Keister, J.E., Houde, E.D. & Breitburg, D.L. (2000) Effects of bottom-layer hypoxia on abundances and depth distributions of organisms in Patuxent River, Chesapeake Bay. *Marine Ecology Progress Series*, **205**, 43–59.

Keller, A.A. & Klein-MacPhee, G. (2000) Impact of elevated temperature on the growth, survival, and trophic dynamics of winter flounder larvae: a mesocosm study. *Canadian Journal of Fisheries and Aquatic Sciences*, **57**, 2382–92.

Kendall, A.W. & Duker, G.J. (1998) The development of recruitment fisheries oceanography in the United States. *Fisheries Oceanography*, **7**, 69–88.

Kerr, S.R. & Dickie, L.M. (2002) *The Biomass Spectrum. A Predator–Prey Theory of Aquatic Production.* Columbia University Press, New York.

Kimmel, D.G., Miller, W.D., Harding, L.W., Jr, Houde, E.D. & Roman, M.R. (2009) Estuarine ecosystem response captured using a synoptic climatology. *Estuaries and Coasts*, **32**, 403–9.

Kimmerer, W.J., Cowan, J.H. Jr, Miller, L.W. & Rose, K.A. (2000) Analysis of an estuarine striped bass (*Morone saxatilis*) population: influence of density-dependent mortality between metamorphosis and recruitment. *Canadian Journal of Fisheries and Aquatic Sciences*, **57**, 478–86.

Kimmerer, W.J., Cowan, J.H., Miller, L.W. & Rose, K.A. (2001) Analysis of an estuarine striped bass population: effects of environmental conditions during early life. *Estuaries*, **24**, 557–75.

Kingsford, M.J. (1990) Linear oceanographic features: a focus for research on recruitment processes. *Australian Journal of Ecology*, **15**, 391–401.

Kiorboe, T. & MacKenzie, B. (1995) Turbulence-enhanced prey encounter rates in larval fish: effects of spatial scale, larval behaviour and size. *Journal of Plankton Research*, **17**, 2319–31.

Kiorboe, T., Munk, P., Richardson, K., Christensen, V. & Paulsen, H. (1988) Plankton dynamics and larval herring growth, drift and survival in a frontal area. *Marine Ecology Progress Series*, **44**, 205–19.

Köster, F.W., Hinrichsen, H.-H., Schnack, D., St John, M.A., Mackenzie, B.R., Tomkiewicz, J., Möllmann, C., Kraus, G., Plikshs, M., Makarchouk, A. & Aro, E. (2003) Recruitment of Baltic cod and sprat stocks: identification of critical life stages and incorporation of environmental variability into stock-recruitment relationships. *Scientia Marina*, **67** (Supplement 1), 129–54.

Kristiansen, T., Lough, R.G., Werner, F.E., Broughton, E.A. & Buckley, L.J. (2009) Individual-based modeling of feeding ecology and prey selection of larval cod on Georges Bank. *Marine Ecology Progress Series*, **376**, 227–43.

Laprise, R. & Dodson, J.J. (1989a) Ongogenetic changes in the longitudinal distribution of two species of larval fish in a turbid well-mixed estuary. *Journal of Fish Biology*, **35** (Supplement A), 39–47.

Laprise, R. & Dodson, J.J. (1989b) Ontogeny and importance of tidal vertical migrations in the retention of larval smelt *Osmerus mordax* in a well-mixed estuary. *Marine Ecology Progress Series*, **55**, 101–11.

Laprise, R. & Dodson, J.J. (1990) The mechanism of retention of pelagic tomcod, *Microgadus tomcod*, larvae and juveniles in the well-mixed part of the St Lawrence estuary. *Environmental Biology of Fishes*, **29**, 293–302.

Largier, J. L. (1993) Estuarine fronts: how important are they? *Estuaries*, **16**, 1–11.

Largier, J. (2004) The importance of retention zones in the dispersal of larvae. In: J.B. Shipley (Ed.) *Aquatic Protected Areas as Fisheries Management Tools*. pp. 105–22. American Fisheries Society, Symposium 42, Bethesda, Maryland.

Lasker, R. (1975) Field criteria for survival of anchovy larvae: the relation between inshore chlorophyll maximum layers and successful first feeding. *Fishery Bulletin (US)*, **73**, 453–62.

Lasker, R. (1978) The relation between oceanographic conditions, and larval anchovy food in the California Current: identification of factors contributing to recruitment failure. *Rapports et Procès-verbaux des Rèunions, Conseil International pour l'Exploration de la Mer*, **173**, 212–30.

Lasker, R. (1981) Factors contributing to variable recruitment of the northern anchovy (*Engraulis mordax*) in the California Current: contrasting years, 1975 through 1978. *Rapports et Procès-verbaux des Rèunions, Conseil International pour l'Exploration de la Mer*, **178**, 375–88.

Lasker, R., Feder, H.M., Theilacker, G.H. & May, R.C. (1970) Feeding, growth, and survival of *Engraulis mordax* larvae reared in the laboratory. *Marine Biology*, **5**, 345–53.

Last, J.M. (1978a) The food of four species of pleuronectiform larvae in the eastern English Channel and southern North Sea. *Marine Biology*, **45**, 359–68.

Last, J.M. (1978b) The food of three species of gadoid larvae in the eastern English Channel and southern North Sea. *Marine Biology*, **48**, 377–86.

Laurence, G.C. (1974) Growth and survival of haddock (*Melanogrammus aeglefinus*) larvae in relation to planktonic prey concentrations. *Journal of the Fisheries Research Board of Canada*, **31**, 1415–19.

Leach, S.D. & Houde, E.D. (1999) Effects of environmental factors on survival, growth, and production of American shad larvae. *Journal of Fish Biology*, **54**, 767–86.

Lebour, M.V. (1916) The food of post-larval fish. *Journal of the Marine Biological Association of the United Kingdom*, **11**, 433–69.

Lee, O., Nash, R.D.M. & Danilowicz, B.S. (2005) Small-scale spatio-temporal variability in ichthyoplankton and zooplankton distribution in relation to a tidal mixing front in the Irish Sea. *ICES Journal of Marine Science*, **62**, 1021–36.

Leggett, W.C. & Deblois, E. (1994) Recruitment in marine fishes: is it regulated by starvation and predation in the egg and larval stages? *Netherlands Journal of Sea Research*, **32**, 119–34.

Leis, J.M. (2006) Are larvae of demersal fishes plankton or nekton? *Advances in Marine Biology*, **51**, 59–141.

Leis, J.M. (2007) Behaviour as input for modelling dispersal of fish larvae: behaviour, biogeography, hydrodynamics, ontogeny, physiology and phylogeny meet hydrography. *Marine Ecology Progress Series*, **347**, 185–93.

Leis, J.M. & McCormick, M.I. (2002) The biology, behavior, and ecology of the pelagic, larval stage of coral reef fishes. In: P.F. Sale (Ed.) *Coral Reef Fishes*. pp. 171–99. Elsevier Science, New York.

Letcher, B.H. & Rice, J.A. (1997) Prey patchiness and larval fish growth and survival: inferences from an individual-based model. *Ecological Modelling*, **95**, 29–43.

Letcher, B.H., Rice, J.A., Crowder, L.B. & Rose, K.A. (1996) Variability in survival of larval fish: disentangling components with a generalized individual-based model. *Canadian Journal of Fisheries and Aquatic Sciences*, **53**, 787–801.

Limburg, K.E., Pace, M.L. & Arend, K.K. (1999) Growth, mortality, and recruitment of larval *Morone* spp. in relation to food availability and temperature in the Hudson River. *Fishery Bulletin (US)*, **97**, 80–91.

Limburg, K.E., Pace, M.L., Fischer, D. & Arend, K.K. (1997) Consumption, selectivity, and use of zooplankton by larval striped bass and white perch in a seasonally pulsed estuary. *Transactions of the American Fisheries Society*, **126**, 607–21.

Litvak, M.K. & Leggett, W.C. (1992) Age and size-selective predation on larval fishes: the bigger-is-better hypothesis revisited. *Marine Ecology Progress Series*, **81**, 13–24.

Llopiz, J. (2013) Latitudinal and taxonomic patterns in the feeding ecologies of fish larvae: A literature synthesis. *Journal of Marine Systems*, **109-110**, 69–77.

Lluch-Belda, D., Crawford, R.J.M, KawasakiT., MacCall, A.D., Parrish, R.H., Schwartzlose, R.A. & Smith, P.E. (1989) World-wide fluctuations of sardine and anchovy stocks: the regime problem. *South African Journal of Marine Science*, **8**, 195–205.

Longhurst, A. (2002) Murphy's law revisited: longevity as a factor in recruitment to fish populations. *Fisheries Research*, **56**, 125–31.

Lough, R.G., Calderone, E.M., Rotunno, T.K, Broughton, E.A., Burns, B.R. & Buckley, L.J. (1996) Vertical distribution of cod and haddock eggs and larvae, feeding and condition in stratified and mixed waters on southern Georges Bank, May 1992. *Deep-Sea Research II*, **43**, 1875–904.

Lough, R.G., Hannah, C.G., Berrien, P., Brickman, D., Loder, J.W. & Quinlan, J.A. (2006) Spawning pattern variability and its effect on retention, larval growth and recruitment in Georges Bank cod and haddock. *Marine Ecology Progress Series*, **310**, 193–212.

Lough, R.G. & Manning, J.P. (2001) Tidal-front entrainment and retention of fish larvae on the southern flank of Georges Bank. *Deep-Sea Research II*, **48**, 631–44.

Lough, R.G. & Potter, D.C. (1993) Vertical distribution patterns and diel migrations of larval and juvenile haddock *Melanogrammus aeglefinus* and Atlantic cod *Gadus morhua* on Georges Bank. *Fishery Bulletin (US)*, **91**, 281–303.

Lough, R.G., Smith, W.G., Werner, F.E., Loder, J.W., Page, F.H., Hannah, C.G., Naimie, C.E., Perry, R.I., Sinclair, M. & Lynch, D.R. (1994) Influence of wind-driven advection on interannual variability in cod egg and larval distributions on Georges Bank: 1982 vs 1985. *ICES Marine Science Symposium*, **198**, 356–78.

MacKenzie, B.R., Hinrichsen, H.-H., Plikshs, M., Wieland, K. & Zezera, A.S. (2000) Quantifying environmental heterogeneity: habitat size necessary for successful development of cod *Gadus morhua* eggs in the Baltic Sea. *Marine Ecology Progress Series*, **193**, 143–56.

MacKenzie, B., St John, M. & Wieland, K. (1996) Eastern Baltic cod: perspectives from existing data on processes affecting growth and survival of eggs and larvae. *Marine Ecology Progress Series*, **134**, 265–81.

MacKenzie, B.R., Miller, T.J., Cyr, S. & Leggett, W.C. (1994) Evidence for a dome-shaped relationship between turbulence and larval fish ingestion rates. *Limnology and Oceanography*, **38**, 1790–99.

Margulies, D. (1989) Effects of food concentration and temperature on development, growth, and survival of white perch, *Morone americana*, eggs and larvae. *Fishery Bulletin (US)*, **87**, 63–72.

Martino, E.J. & Houde, E.D. (2004) *Environmental Controls and Density Dependent Constraints in the Recruitment Process of Striped Bass* Morone saxatilis *in the Estuarine Transition Zone of Chesapeake Bay.* International Council for the Exploration of the Sea, ICES CM 2004/J, 05.

Martino, E.J. & Houde, E.D. (2010) Recruitment of striped bass in Chesapeake Bay: spatial and temporal environmental variability and availability of zooplankton prey. *Marine Ecology Progress Series*, **409**, 213–28.

Martino, E.J. & Houde, E.D. (2012) Density-dependent regulation of year-class strength in age-0 juvenile striped bass (*Morone saxatilis*). *Canadian Journal of Fisheries and Aquatic Sciences*, **69**, 430–46.

May, R.C. (1974) Larval mortality in marine fishes and the critical period concept. In: J.H.S. Blaxter (Ed.) *The Early Life History of Fish*. pp. 3–19. Springer-Verlag, New York.

McCormick, M.I. & Molony, B.W. (1995) Influence of water temperature during the larval stage on size, age, and body condition of a tropical reef fish at settlement. *Marine Ecology Progress Series*, **118**, 59–68.

McFarlane, G.A., Smith, P.E., Baumgartner, T.R. & Hunter, J.R. (2002) Climate variability and Pacific sardine populations and fisheries. In: N.A. McGinn (Ed.) *Fisheries in a Changing Climate*. pp. 195–214. American Fisheries Society, Symposium 32, Bethesda, Maryland.

McGurk, M.D. (1986) Natural mortality of marine pelagic fish eggs and larvae: role of spatial patchiness. *Marine Ecology Progress Series*, **34**, 227–42.

McGurk, M.D. (1989) Advection, diffusion and mortality of Pacific herring larvae *Clupea harengus pallasi* in Bamfield Inlet, British Columbia. *Marine Ecology Progress Series*, **51**, 1–18.

Mee, L.D., Friedrich, J. & Gomoiu, M.T. (2005) Restoring the Black Sea in times of uncertainty. *Oceanography*, **18**, 100–11.

Meekan, M.G. & Fortier, L. (1996) Selection for fast growth during the larval life of Atlantic cod *Gadus morhua* on the Scotian Shelf. *Marine Ecology Progress Series*, **137**, 25–37.

Meekan, M.G., Carleton, J.H., McKinnon, A.D., Flynn, K. & Furnas, M. (2003) What determines the growth of tropical reef fish larvae in the plankton: food or temperature? *Marine Ecology Progress Series*, **256**, 193–204.

Megrey, B.A. & Hinckley, S. (2001) Effect of turbulence on feeding of larval fishes: a sensitivity analysis using an individual-based model. *ICES Journal of Marine Science*, **58**, 1015–29.

Mertz, G. & Myers, R. (1995) Estimating the predictability of recruitment. *Fishery Bulletin (US)*, **93**, 657–65.

Miller, J.M. (1988) Physical processes and the mechanisms of coastal migrations of immature marine fishes. *American Fisheries Society Symposium*, **3**, 68–76.

Miller, T.J. (2007) Contribution of individual-based coupled physical biological models to understanding recruitment in marine fish populations. *Marine Ecology Progress Series*, **347**, 127–38.

Miller, T.J., Crowder, L.B., Rice, J.A. & Marschall, E.A. (1988) Larval size and recruitment mechanisms in fishes: toward a conceptual framework. *Canadian Journal of Fisheries and Aquatic Sciences*, **45**, 1657–70.

Monteleone, D.M. & Houde, E.D. (1990) Influence of maternal size on survival and growth of striped bass *Morone saxatilis* Walbaum eggs and larvae. *Journal of Experimental Marine Biology and Ecology*, **140**, 1–11.

Morse, W.W. (1989) Catchability, growth, and mortality of larval fishes. *Fishery Bulletin (US)*, **87**, 417–46.

Mountain, D., Berrien, P. & Sibunka, J. (2003) Distribution, abundance and mortality of cod and haddock eggs and larvae on Georges Bank in 1995 and 1996. *Marine Ecology Progress Series*, **263**, 247–60.

Mullon, C., Cury, P. & Penven, P. (2002) Evolutionary individual-based model for the recruitment of anchovy (*Engraulis capensis*) in the southern Benguela. *Canadian Journal of Fisheries and Aquatic Sciences*, **59**, 910–22.

Munk, P. (1992) Foraging behaviour and prey size spectra of larval herring *Clupea harengus*. *Marine Biology*, **80**, 149–58.

Munk, P. (1993) Differential growth of larval sprat *Sprattus sprattus* across a tidal front in the eastern North Sea. *Marine Ecology Progress Series*, **99**, 17–27.

Munk, P. (1995) Foraging behaviour of larval cod (*Gadus morhua*) influenced by prey density and hunger. *Marine Biology*, **122**, 205–12.

Munk, P. (1997) Prey size spectra and prey availability of larval and small juvenile cod. *Journal of Fish Biology*, **51** (Supplement A), 340–51.

Munk, P., Larsson, P.O., Danielsen, D. & Moksness, E. (1995) Larval and small juvenile cod *Gadus morhua* concentrated in the highly productive areas of a shelf break front. *Marine Ecology Progress Series*, **125**, 21–30.

Munk, P., Larsson, P.O., Danielsen, D. & Moksness, E. (1999) Variability in frontal zone formation and distribution of gadoid fish larvae at the shelf break in the northeastern North Sea. *Marine Ecology Progress Series*, **177**, 221–33.

Munk, P., Wright, P.J. & Pihl, N.J. (2002) Distribution of the early larval stages of cod, plaice and lesser sandeel across haline fronts in the North Sea. *Estuarine, Coastal and Shelf Science*, **55**, 139–49.

Murphy, H.M., Jenkins, G.P., Hamer, P.A. & Swearer, S.E. (2012) Interannual variation in larval survival of snapper (*Chrysophrys auratus*, Sparidae) is linked to diet breadth and prey availability. *Canadian Journal of Fisheries and Aquatic Sciences*, **69**, 1340–51.

Murphy, H.M., Jenkins, G.P., Hamer, P.A. & Swearer, S.E. (2013) Interannual variation in larval abundance and growth in snapper *Chrysophrys auratus* (Sparidae) is related to prey availability and temperature. *Marine Ecology Progress Series*, **487**, 151–62.

Myers, R.A. (1998) When do environment–recruitment correlations work? *Reviews in Fish Biology and Fisheries*, **8**, 285–305.

Myers, R.A. (2001) Stock and recruitment: generalizations about maximum reproductive rate, density dependence, and variability using meta-analytic approaches. *ICES Journal of Marine Science*, **58**, 937–51.

Myers, R.A. (2002) Recruitment: understanding density-dependence in fish populations. In: P.J.B. Hart & J.D. Reynolds (Eds) *Handbook of Fish Biology and Fisheries, Volume 1, Fish Biology*. pp. 123–48. Blackwell Science, Oxford.

Myers, R.A. & Barrowman, N.J. (1996) Is fish recruitment related to spawner abundance? *Fishery Bulletin (US)*, **94**, 707–24.

Myers, R.A., Barrowman, N.J. & Thompson, K.R. (1995a) Synchrony of recruitment across the North Atlantic: an update (or 'now you see it, now you don't!'). *ICES Journal of Marine Science*, **52**, 103–10.

Myers, R.A., Bridson, J. & Barrowman, N.J. (1995b) Summary of worldwide stock and recruitment data. *Canadian Technical Report of Fisheries and Aquatic Sciences*, **2024**. 327 pp.

Myers, R.A. & Cadigan, N.G. (1993a) Density-dependent juvenile mortality in marine demersal fish. *Canadian Journal of Fisheries and Aquatic Sciences*, **50**, 1576–90.

Myers, R.A. & Cadigan, N.G. (1993b) Is juvenile natural mortality in marine demersal fishes variable? *Canadian Journal of Fisheries and Aquatic Sciences*, **50**, 1591–8.

Myers, R.A., Mertz, G. & Barrowman, N.J. (1995c) Spatial scales of variability in cod recruitment in the North Atlantic. *Canadian Journal of Fisheries and Aquatic Sciences*, **52**, 1849–62.

Myers, R.A., Mertz, G. & Bridson, N.J. (1997) Spatial scales of inter-annual recruitment variations of marine, anadromous, and freshwater fish. *Canadian Journal of Fisheries and Aquatic Sciences*, **54**, 1400–7.

Nakata, H. (1996) Coastal fronts and eddies: their implications for egg and larval transport and survival processes. In: Y. Watanabe, Y. Yamashita & Y. Oozeke (Eds) *Survival Strategies in Early Life Stages of Marine Resources*. pp. 227–41. Balkema, Rotterdam.

Nakata, K., Hada, A. & Matsukawa, Y. (1994) Variations in food abundance for Japanese sardine larvae related to the Kuroshio meander. *Fisheries Oceanography*, **3**, 39–49.

Nakata, K., Zenitani, H. & Inagake, D. (1995) Differences in food availability for Japanese sardine between the frontal region and the waters on the offshore side of Kuroshio. *Fisheries Oceanography*, **4**, 68–79.

Neilson, J.D. & Perry, R.I. (1990) Diel vertical migrations of juvenile fish: an obligate or facultative process? *Advances in Marine Biology*, **26**, 115–68.

Nielsen, R. & Munk, P. (2004) Growth pattern and growth dependent mortality of larval and pelagic juvenile North Sea cod *Gadus morhua*. *Marine Ecology Progress Series*, **278**, 261–70.

Nelson, W.R., Ingham, M.C. & Schaaf, W.E. (1977) Larval transport and year-class strength of Atlantic menhaden, *Brevoortia tyrannus*. *Fishery Bulletin (US)*, **75**, 23–42.

Nissling, A., Solemdal, P., Svensson, M. & Westin, L. (1994) Survival, activity and feeding ability of Baltic cod (*Gadus morhua*) yolk-sac larvae at different salinities. *Journal of Fish Biology*, **45**, 435–45.

Nissling, A. & Westin, L. (1991) Egg mortality and hatching rate of Baltic cod (*Gadus morhua*) in different salinities. *Marine Biology*, **111**, 29–32.

Norcross, B.L. & Shaw, R.F. (1984) Oceanic and estuarine transport of fish eggs and larvae: a review. *Transactions of the American Fisheries Society*, **113**, 153–65.

North, E.W. & Houde, E.D. (2001) Retention of white perch and striped bass larvae: biological–physical interactions in Chesapeake Bay estuarine turbidity maximum. *Estuaries*, **24**, 756–69.

North, E.W. & Houde, E.D. (2003) Linking ETM physics, zooplankton prey, and fish early-life histories to striped bass *Morone saxatilis* and white perch *M. americana* recruitment. *Marine Ecology Progress Series*, **260**, 219–36.

North, E.W. & Houde, E.D. (2004) Distribution and transport of bay anchovy (*Anchoa mitchilli*) eggs and larvae in Chesapeake Bay. *Estuarine, Coastal and Shelf Science*, **60**, 409–29.

North, E.W. & Houde, E.D. (2006) Retention mechanisms of white perch (*Morone americana*) and striped bass (*Morone saxatilis*) early-life stages in an estuarine turbidity maximum: an integrative fixed-location and mapping approach. *Fisheries Oceanography*, **15**, 429–50.

North, E.W., Gallego, A. & Petitgas, P. (2009) Manual of recommended practices for modeling physical-biological interactions during fish early life. *ICES Cooperative Research Report* **295**, 1–110.

O'Brien, C.M., Fox, C.J., Planque, B. & Casey, J. (2000) Climate variability and North Sea cod. *Nature*, **404**, 142.

O'ConnellC.P. & RaymondL.P. (1970) The effect of food density on survival and growth of early post-yolk-sac larvae of the northern anchovy (*Engraulis mordax* Girard) in the laboratory. *Journal of Experimental Marine Biology and Ecology*, **5**, 187–97.

Oiestad, V. (1985) Predation on fish larvae as a regulatory force, illustrated in mesocosm studies with large groups of larvae. *NAFO Scientific Council Studies*, **8**, 25–32.

Okazaki, Y., Nakata, H. & Kimura, S. (2002) Effects of frontal eddies on the distribution and food availability of anchovy larvae in the Kuroshio Extension. *Marine and Freshwater Research*, **53**, 403–10.

Olivar, M.P. & Sabates, A. (1997) Vertical distribution of fish larvae in the north-west Mediterranean Sea in spring. *Marine Biology*, **129**, 289–300.

Olson, D.B., Hitchcock, G.L., Mariano, A.J., Ashjian, C.J., Peng, G., Nero, R.W. & Podesta, G.P. (1994) Life on the edge: marine life and fronts. *Oceanography*, **7**, 52–60.

Otterlei, E., Nyhammer, G., Folkvord, A. & Stefansson, S.O. (1999) Temperature- and size-dependent growth of larval and early juvenile cod (*Gadus morhua*): a comparative study of Norwegian coastal cod and northeast Arctic cod. *Canadian Journal of Fisheries and Aquatic Sciences*, **56**, 2099–111.

Ottersen, G., Planque, B., Belgrano, A., Post, E., Reid, P.C. & Stenseth, N.C. (2001) Ecological effects of the North Atlantic Oscillation. *Oecologia*, **28**, 1–14.

Ottersen, G., Stige, L.C., Dulrant, J.M., Chan, K.-S., Rouyer, T.A., Drinkwater, K.F. & Stenseth, N.C. (2013) Temporal shifts in recruitment dynamics of North Atlantic fish stocks: effects of spawning stock and temperature. *Marine Ecology Progress Series* **480**, 205–25.

Owen, R.W. (1989) Microscale and finescale variations of small plankton in coastal and pelagic environments. *Journal of Marine Research*, **47**, 197–240.

Page, F.H. & Frank, K.T. (1989) Spawning time and egg stage duration in northwest Atlantic haddock (*Melanogrammus aeglefinus*) stocks with emphasis on Georges and Browns Bank. *Canadian Journal of Fisheries and Aquatic Sciences*, **46** (Supplement 1), 68–81.

Page, F.H., Sinclair, M., Naimie, C.E., Loder, J.W., Losier, R.J., Berrien, P.L. & Lough, R.G. (1999) Cod and haddock spawning on Georges Bank in relation to water residence times. *Fisheries Oceanography*, **8**, 212–26.

Palomera, I. (1991) Vertical distribution of eggs and larvae of *Engraulis encrasicolus* in stratified waters of the western Mediterranean. *Marine Biology*, **111**, 37–44.

Paradis, A.R., Pepin, P. & Brown, J.A. (1996) Vulnerability of fish eggs and larvae to predation: review of the influence of the relative size of prey and predator. *Canadian Journal of Fisheries and Aquatic Sciences*, **53**, 1226–35.

Paradis, A.R., Pepin, M. & Pepin, P. (1999) Disentangling the effects of size-dependent encounter and susceptibility to predation with an individual-based model for fish larvae. *Canadian Journal of Fisheries and Aquatic Sciences*, **56**, 1562–75.

Paris, C.B. & Cowen, R.K. (2004) Direct evidence of a biophysical retention mechanism for coral reef fish larvae. *Limnology and Oceanography*, **49**, 1964–79.

Pauly, D. (1987) Managing the Peruvian upwelling ecosystem: a synthesis. In: D. Pauly & I. Tsukayama (Eds) *The Peruvian Anchoveta and its Upwelling Ecosystem: Three Decades of Change*. pp. 325–42. ICLARM Studies and Reviews 15, Callas, Peru, Instituto del Mar del Peru.

Peck, M.A., Huebert, K.B. & Llopiz, J.K. (2012) Intrinsic and extrinsic factors driving match-mismatch dynamics during the early life history of marine fishes. *Advances in Ecological Research*, **47**, 177–302.

Peck, M.A. & Hufnagl, M. (2012) Can IBMs explain why most larvae die in the sea? Model scenarios and sensitivity analyses reveal research needs. *Journal of Marine Systems*, **93**, 77–93.

Peck, M.A., Reglero, P., Takahashi, M. & Catalan, I.A. (2013) Life cycle ecophysiology of small pelagic fish and climate-driven changes in populations. *Progress in Oceanography*, **116**, 220–45.

Pearre, S. Jr (1986) Ratio-based trophic niche breadths of fish, the Sheldon spectrum, and the size-efficiency hypothesis. *Marine Ecology Progress Series*, **27**, 299–314.

Pepin, P. (1990) Biological correlates of recruitment variability in North Sea fish stocks. *Journal du Conseil International pour l'Exploration de la Mer*, **47**, 89–98.

Pepin, P. (1991) Effect of temperature and size on development, mortality, and survival rates of the pelagic early life history stages of marine fish. *Canadian Journal of Fisheries and Aquatic Sciences*, **48**, 503–18.

Pepin, P. (1993) An appraisal of the size-dependent mortality hypothesis for larval fish: comparison of a multispecies study with an empirical review. *Canadian Journal of Fisheries and Aquatic Sciences*, **50**, 2166–74.

Pepin, P. (2004) Early life history studies of prey–predator interactions: quantifying the stochastic individual responses to environmental variability. *Canadian Journal of Fisheries and Aquatic Sciences*, **61**, 659–71.

Pepin, P. & Myers, R.A. (1991) Significance of egg and larval size to recruitment variability of temperate marine fish. *Canadian Journal of Fisheries and Aquatic Sciences*, **48**, 1820–8.

Pepin, P., Evans, G.T. & Shears, T.H. (1999) Patterns of RNA/DNA ratios in larval fish and their relationship to survival in the field. *ICES Journal of Marine Science*, **56**, 697–706.

Pepin, P. & Penney, R.W. (1997) Patterns of prey size and taxonomic composition in larval fish: are there general size-dependent models? *Journal of Fish Biology*, **51** (Supplement A), 84–100.

Pepin, P., Shears, T.H. & deLafontaine, Y. (1992) Significance of body size to the interaction between a larval fish (*Mallotus villosus*) and a vertebrate predator (*Gasterosteus aculeatus*). *Marine Ecology Progress Series*, **81**, 1–12.

Perry, A.L., Low, P.J., Ellis, J.R. & Reynolds, J.D. (2005) Climate change and distribution shifts in marine fishes. *Science*, **308**, 1912–15.

Peterman, R.M. & Bradford, M.J. (1987) Wind speed and mortality rate of a marine fish, the northern anchovy (*Engraulis mordax*). *Science*, **235**, 354–6.

Peterman, R.M., Bradford, M.J., Lo, N.C.H. & Methot, R.D. (1988) Contribution of early life stages to interannual variability in recruitment of northern anchovy (*Engraulis mordax*). *Canadian Journal of Fisheries and Aquatic Sciences*, **45**, 8–16.

Peterson, I. & Wroblewski, J.S. (1984) Mortality rate of fishes in the pelagic ecosystem. *Canadian Journal of Fisheries and Aquatic Sciences*, **41**, 1117–20.

Petitgas, P., Rijnsdorp, A.D., Dickey-Collas, M., Engelhard, G.H., Peck, M.A., Pinnegar, J.K., Drinkwater, K., Huret, M. & Nash, R.D.M. (2013) Impacts of climate change on the complex life cycles of fish. *Fisheries Oceanography*, **22**, 121–39.

Petrik, C.M., Kristiansen, T., Lough, R.G. & Davis, C.S. (2009) Prey selection by larval haddock and cod on copepods with species-specific behavior: an individual-based model analysis. *Marine Ecology Progress Series*, **396**, 123–43.

Planque, B. & Fox, C.J. (1998) Interannual variability in temperature and the recruitment of Irish Sea cod. *Marine Ecology Progress Series*, **172**, 101–5.

Planque, B. & Fredou, T. (1999) Temperature and the recruitment of Atlantic cod (*Gadus morhua*). *Canadian Journal of Fisheries and Aquatic Sciences*, **56**, 2069–77.

Platt, T., Fuentes-Yaco, C. & Frank, K.T. (2003) Spring algal bloom and larval fish survival. *Nature*, **423**, 398–9.

Pope, E.C., Ellis, R.P., Scolamacchia, M., Scolding, J.W.S., Keay, A., Chingombe, P., Shields, R.J., Wilcox, R., Speirs, D.C., Wilson, R.W., Lewis, C. & Flynn, K.J. (2013) Is the perceived resiliency of fish larvae to ocean acidification masking more subtle effects? *Biogeosciences Discussion*, **10**, 17043–70.

Pope, J.G. & Macer, C.T. (1996) An evaluation of the stock structure of North Sea cod, haddock, and whiting since 1920, together with a consideration of the impacts of fisheries and predation effects on their biomass and recruitment. *ICES Journal of Marine Science*, **53**, 1157–69.

Puebla, O., Bermingham, E. & Guichard, F. (2009) Estimating dispersal from genetic isolation by distance in a coral reef fish (*Hypoplectrus puella*). *Ecology*, **90**, 3087–98.

Raventos, N. & Macpherson, E. (2005) Effect of pelagic larval growth and size-at-hatching on post-settlement survivorship in two temperate labrid fish of the genus *Symphodus*. *Marine Ecology Progress Series*, **285**, 205–11.

Reiss, C.S., Anis, A., Taggart, C.T., Dower, J.F. & Ruddick, B. (2002) Relationships among vertically structured *in situ* measures of turbulence, larval fish abundance and feeding success and copepods on Western Bank, Scotian Shelf. *Fisheries Oceanography*, **11**, 156–74.

Reiss, C., McLaren, I., Avendano, P. & Taggart, C. (2005) Feeding ecology of silver hake larvae on the Western Bank, Scotian Shelf, and comparison with Atlantic cod. *Journal of Fish Biology*, **66**, 703–20.

Rice, J.A., Miller, T.J, Rose, K.A., Crowder, L.B., Marschall, E.A., Trebitz, A.S. & DeAngelis, D.L. (1993) Growth rate variation and larval survival: inferences from an individual-based size-dependent predation model. *Canadian Journal of Fisheries and Aquatic Sciences*, **50**, 133–42.

Rideout, R.M., Trippel, E.A. & Litvak, M.K. (2004) Paternal effects on haddock early life history traits. *Journal of Fish Biology*, **64**, 695–701.

Rijnsdorp, A.D., Daan, N., Van Beek, F.A. & Heessen, H.J.L. (1991) Reproductive variability in North Sea plaice, sole and cod. *Journal du Conseil International pour l'Exploration de la Mer*, **47**, 352–75.

Robert, D., Castonguay, M. & Fortier, L. (2007) Early growth and recruitment in Atlantic mackerel *Scomber scombrus*: discriminating the effects of fast growth and selection for fast growth. *Marine Ecology Progress Series*, **337**, 209–19.

Rombough, P.J. (1988) Respiratory gas exchange, aerobic metabolism, and effects of hypoxia during early life. In: W.S. Hoar & D.J. Randall (Eds) *Fish Physiology, Volume XI. The Physiology of Developing Fish, Part A, Eggs and Larvae*. pp. 59–161. Academic Press, San Diego.

Rombough, P.J. (1996) The effects of temperature on embryonic and larval development. In: C.M. Wood & D.G. McDonald (Eds) *Global Warming: Implications for Freshwater and Marine Fish*. pp. 177–223. Cambridge University Press, Cambridge.

Ropke, A. (1993) Do larvae of mesopelagic fishes in the Arabian Sea adjust their vertical distribution to physical and biological gradients? *Marine Ecology Progress Series*, **101**, 223–35.

Ropke, A., Nellen, W. & Piatkowski, U. (1993) A comparative study on the influence of the pycnocline on the vertical distribution of fish larvae and cephalopod paralarvae in three ecologically different areas of the Arabian Sea. *Deep-Sea Research II*, **40**, 801–19.

Rose, K.A., Cowan, J.H. Jr, Clark, M.E., Houde, E.D. & Wang, S.-B. (1999) An individual-based model of bay anchovy population dynamics in the mesohaline region of Chesapeake Bay. *Marine Ecology Progress Series*, **185**, 113–32.

Rose, K.A., Cowan, J.H. Jr, Houde, E.D. & Coutant, C.C. (1993) Individual-based modeling of environmental quality effects on early life stages of fishes: a case study using striped bass. *American Fisheries Society, Symposium*, **14**, 125–45.

Rose, K.A., Cowan, J.H. Jr, Winemiller, K.O., Myers, R.A. & Hilborn, R. (2001) Compensatory density dependence in fish populations: importance, controversy, understanding and prognosis. *Fish and Fisheries*, **2**, 293–327.

Rose, K.A., Murphy, C.A., Diamond, S.L., Fuiman, L.A. & ThomasP. (2003) Using nested models and laboratory data for predicting population effects of contaminants on fish: a step toward a bottom-up approach for establishing causality in field studies. *Human and Ecological Risk Assessment*, **9**, 231–57.

Rothschild, B.J. (1986) *Dynamics of Marine Fish Populations*. Harvard University Press, Cambridge.

Rothschild, B.J. (2000) 'Fish stocks and recruitment': the past thirty years. *ICES Journal of Marine Science*, **57**, 191–201.

Rothschild, B.J. & DiNardo, G.T. (1987) Comparison of recruitment variability and life history data among marine and anadromous fishes. *American Fisheries Society Symposium*, **1**, 531–46.

Rothschild, B.J. & Osborn, T.R. (1988) Small-scale turbulence and plankton contact rates. *Journal of Plankton Research*, **10**, 465–74.

Rothschild, B.J. & Rooth, C. (Eds) (1982) *Fish Ecology III. A Foundation for REX, a Recruitment Experiment.* University of Miami, Technical Report No. 82008. 389 pp.

Rowe, P.M. & Epifanio, C.E. (1994) Tidal stream transport of weakfish larvae in Delaware Bay, USA. *Marine Ecology Progress Series*, **110**, 105–14.

Roy, C., Cury, P. & Kifani, S. (1992) Pelagic fish recruitment success and reproductive strategy in upwelling areas: environmental compromises. *South African Journal of Marine Science*, **12**, 135–46.

Runge, J.A., Kovach, A.I., Churchill, J.H., Kerr, L.A., Morrison, J.R., Beardsley, R.C., Belinsky, D.L., Chen, C., Cadrin, S.X., Davis, C.S., Ford, K.H., Grabowski, J.H., Howell, W.H., Ji, R., Jones, R. J., Pershing, A.J., Record, N.R., Thomas, A.C., Sherwood, G.D., Tallack, S.M.L. and Townsend, D.W. (2010) Understanding climate impacts on recruitment and spatial dynamics of Atlantic cod in the Gulf of Maine: Integration of observations and modeling. *Progress in Oceanography*, **87**, 251–63.

Rutherford, E.S. & Houde, E.D. (1995) The influence of temperature on cohort-specific growth, survival, and recruitment of striped bass, *Morone saxatilis*, larvae in Chesapeake Bay. *Fishery Bulletin (US)*, **93**, 315–32.

Rutherford, E.S., Houde, E.D. & Nyman, R.M. (1997) Relationship of larval-stage growth and mortality to recruitment of striped bass, *Morone saxatilis*, in Chesapeake Bay. *Estuaries*, **20**, 174–98.

Rutherford, E.S., Rose, K.A. & Cowan, J.H. Jr (2003) Evaluation of the Shepherd & Cushing (1980) model of density-dependent survival: a case study using striped bass (*Morone saxatilis*) larvae in the Potomac River, Maryland, USA. *ICES Journal of Marine Science*, **60**, 1275–87.

Sabates, A. (1990) Changes in the heterogeneity of mesoscale distribution patterns of larval fish associated with a shallow coastal haline front. *Estuarine, Coastal and Shelf Science*, **30**, 131–40.

Sabates, A. & Olivar, M.P. (1996) Variation of larval fish distributions associated with variability in the location of a shelf-slope front. *Marine Ecology Progress Series*, **135**, 11–20.

Sabates, A. & Saiz, E. (2000) Intra- and interspecific variability in prey size and niche breadth of myctophiform fish larvae. *Marine Ecology Progress Series*, **201**, 261–71.

Sale, P.F. (1977) Maintenance of high diversity in coral reef fish communities. *American Naturalist*, **111**, 337–59.

Sale, P.F. (1978) Coexistence of coral reef fishes – a lottery for living space. *Environmental Biology of Fishes*, **3**, 85–102.

Sale, P.F. (1991) Reef fish communities: open nonequilibrial systems. In: P.F. Sale (Ed.) *The Ecology of Fishes on Coral Reefs*. pp. 564–98. Academic Press, San Diego.

Satoh, K., Tanaka, Y., Masujima, M., Okazaki, M., Kato, Y., Shono, H. & Suzuki, K. (2013) Relationship between the growth and survival of larval Pacific bluefin tuna, *Thunnus orientalis*. *Marine Biology*, **160**, 691–702.

Savoy, T.F. & Crecco, V.A. (1988) The timing and significance of density-dependent and density-independent mortality of American shad, *Alosa sapidissima*. *Fishery Bulletin (US)*, **86**, 467–82.

Schumacher, J. & Kendall, A.W. Jr (1995) An example of fisheries oceanography: walleye pollock in Alaskan waters. *Reviews of Geophysics (Supplement)*, **33** (Suppl. 2), 1153–63.

Schwartzlose, R.A., Alheit, J., Bakun, A., Baumgartner, T.R., Cloete, R, Crawford, R.J.M., Fletcher, W.J., Green-Ruiz, Y., Hagen, E., Kawasaki, T., Lluch-Belda, D., Lluch-Cota, S.E., MacCall, A.D., Matsuura, Y., Nevarez-Martinez, M.O., Parrish, R.H., Roy, C., Serra, R., Shust, K.V., Ward, M.N. & Zuzunaga, J.Z. (1999) Worldwide large-scale fluctuations of sardine and anchovy populations. *South African Journal of Marine Science*, **21**, 289–347.

Searcy, S.P. & Sponaugle, S. (2001) Selective mortality during the larval–juvenile transition in two coral reef fishes. *Ecology*, **82**, 2452–70.

Secor, D.H. (2000) Spawning in the nick of time? Effect of adult demographics on spawning behaviour and recruitment in Chesapeake Bay striped bass. *ICES Journal of Marine Science*, **57**, 403–11.

Secor, D.H. (2002) Historical roots of the migration triangle. *ICES Marine Science Symposia*, **215**, 323–9.

Secor, D.H. (2005) Fish migration and the unit stock: three formative debates. In: S. Cadrin, K. Friedland & J. Waldman (Eds) *Stock Identification Methods*. pp. 17–44. Elsevier, Amsterdam.

Secor, D.H. & Houde, E.D. (1995) Temperature effects on the timing of striped bass egg production, larval viability, and recruitment potential in the Patuxent River (Chesapeake Bay). *Estuaries*, **18**, 527–44.

Secor, D.H., Houde, E.D. & Monteleone, D.M. (1995) A mark-release experiment on larval striped bass, *Morone saxatilis*, in a Chesapeake Bay tributary. *ICES Journal of Marine Science*, **52**, 87–101.

Sette, O.E. (1943) *Studies on the Pacific Pilchard or Sardine (Sardinops caerulea). Structure of a Research Program to Determine how Fishing Affects the Resource.* US Department of Interior, Special Scientific Report 19, Washington, DC.

Shanks, A.L. (1983) Surface slicks associated with tidally forced internal waves may transport pelagic larvae of benthic invertebrates and fishes shoreward. *Marine Ecology Progress Series*, **13**, 311–15.

Shanks, A.L. (1988) Further support for the hypothesis that internal waves can cause shoreward transport of larval invertebrates and fish. *Fishery Bulletin (US)*, **86**, 703–14.

Shanks, A.L. (1995) Mechanisms of cross-shelf dispersal of larval invertebrates and fish. In: L. McEdward (Ed.) *Ecology of Marine Invertebrate Larvae*. pp. 323–67. CRC Press, Boca Raton.

Shannon, L.J., Nelson, G., Crawford, R.J.M. & Boyd, A.J. (1996) Possible impacts of environmental change on pelagic fish recruitment: modeling anchovy transport by advective processes in the southern Benguela. *Global Change Biology*, **2**, 407–20.

Shepherd, J.G. & Cushing, D.H. (1980) A mechanism for density-dependent survival of larval fish as the basis of a stock-recruitment relationship. *Journal du Conseil International pour l'Exploration de la Mer*, **39**, 160–7.

Shima, J.S. & Findlay, A.M. (2002) Pelagic larval growth rate impacts benthic settlement and survival of a temperate reef fish. *Marine Ecology Progress Series*, **235**, 303–9.

Shirota, A. (1970) Studies on the mouth size of fish larvae. *Bulletin of the Japanese Society for Scientific Fisheries*, **36**, 353–68.

Shoji, J., Masuda, R., Yamashita, Y. & Tanaka, M. (2005a) Effect of low dissolved oxygen concentrations on behavior and predation rates on red sea bream *Pagrus major* larvae by the jellyfish *Aurelia aurita* and by juvenile Spanish mackerel *Scomberomorus niphonius*. *Marine Biology*, **147**, 863–8.

Shoji, J., North, E.W. & Houde, E.D. (2005b) The feeding ecology of *Morone americana* larvae in the Chesapeake Bay estuarine turbidity maximum: the influence of physical conditions and prey concentrations. *Journal of Fish Biology*, **66**, 1328–41.

Shoji, J. & Tanaka, M. (2001) Strong piscivory of Japanese Spanish mackerel larvae from their first feeding. *Journal of Fish Biology*, **59**, 1682–5.

Shuter, B.J., Carline, R.F. & McGinn, N.A. (2002) An overview of the symposium. In: N.A. McGinn (Ed.) *Fisheries in a Changing Climate*. pp. xiii–xx. American Fisheries Society, Symposium 32, Bethesda, Maryland.

Sinclair, A.F. & Crawford, W.R. (2005) Incorporating an environmental stock-recruitment relationship in the assessment of Pacific cod (*Gadus macrocephalus*). *Fisheries Oceangraphy*, **14**, 138–50.

Sinclair, M. (1988) *Marine Populations. An Essay on Population Regulation and Speciation.* Washington Sea Grant Program, University of Washington Press, Seattle.

Sinclair, M. (1997) Prologue: recruitment in fish populations: the paradigm shift generated by ICES Committee A. In: R.C. Chambers & E.A. Trippel (Eds) *Early Life History and Recruitment in Fish Populations*. pp. 1–27. Chapman & Hall, London.

Sinclair, M. & Iles, T.D. (1985) Atlantic herring (*Clupea harengus*) distributions in the Gulf of Maine-Scotian Shelf area in relation to oceanographic features. *Canadian Journal of Fisheries and Aquatic Sciences*, **42**, 880–7.

Sirois, P. & Dodson, J.J. (2000) Influence of turbidity, food density and parasites on the ingestion and growth of larval rainbow smelt *Osmerus mordax* in an estuarine turbidity maximum. *Marine Ecology Progress Series*, **193**, 167–79.

Sissenwine. M.P. (1984) Why do fish populations vary? In: R.M. May (Ed.) *Exploitation of Marine Communities*. pp. 59–94. Springer-Verlag, Berlin.

Smith, T.D. (1994) *Scaling Fisheries*. Cambridge University Press, Cambridge.

Sogard, S.M. (1997) Size-selective mortality in the juvenile stage of teleost fishes: a review. *Bulletin of Marine Science*, **60**, 1129–57.

Solberg, T. & Tilseth, S. (1984) Growth, energy consumption and prey density requirements in first feeding larvae of cod (*Gadus morhua*). In: E. Dahl, D.S. Danielsen, E. Moksness & P. Solemdal (Eds) *The Propagation of Cod* (*Gadus morhua*). pp. 145–166. Flødevigen Rapportser 1, Arendal, Norway.

Solemdal, P. (1997) Maternal effects – a link between the past and the future. *Journal of Sea Research*, **37**, 213–27.

Sponaugle, S., Cowen, R.K., Shanks, A., Morgan, S.G., Leis, J.M., Pineda, J., Boehlert, G.W., Kingsford, M.J., Lindeman, K.C., Grimes, C. & Munro, J.L. (2002) Predicting self-recruitment in marine populations: biophysical correlates and mechanisms. *Bulletin of Marine Science*, **70**, 341–75.

Sponaugle, S., Grorud-Colvert, K. & Pinkard, D. (2006) Temperature-mediated variation in early life history traits and recruitment success of the coral reef fish *Thalassoma bifasciatum* in the Florida Keys. *Marine Ecology Progress Series*, **308**, 1–15.

Steinarsson, A. & Bjornsson, B. (1999) The effects of temperature and size on growth and mortality of cod larvae. *Journal of Fish Biology*, **55** (Supplement A), 100–9.

Stenseth, N.C., Mysterud, A., Ottersen, G., Hurrell, J.W., Chan, K.-S. & Lima, M. (2002) Ecological effects of climate fluctuations. *Science*, **297**, 1292–6.

Stige, L.C., Ottersen, G., Brander, K., Chan, K.-S. & Stenseth, N.C. (2006) Cod and climate: effect of the North Atlantic Oscillation on recruitment in the North Atlantic. *Marine Ecology Progress Series*, **325**, 227–41.

Stige, L.C., Hunsicker, M.E., Bailey, K.M., Yaragina, N.A. & Hunt, G.L., Jr (2013) Predicting fish recruitment from juvenile abundance and environmental indices. *Marine Ecology Progress Series*, **480**, 245–61.

Stocker, M., Haist, V. & Fournier, D. (1985) Environmental variation and recruitment of Pacific herring (*Clupea harengus pallasi*) in the Strait of Georgia. *Canadian Journal of Fisheries and Aquatic Sciences*, **42** (Supplement 1), 174–80.

Stoecker, D.K & Govoni, J.J. (1984) Food selection by young larval gulf menhaden (*Brevoortia patronus*). *Marine Biology*, **80**, 299–306.

Sullivan, M.C., Cowen, R.K. & Steves, B.P. (2005) Evidence for atmosphere-ocean forcing of yellowtail flounder (*Limanda ferruginea*) recruitment in the Middle Atlantic Bight. *Fisheries Oceanography*, **14**, 386–99.

Sundby, S. (1996) Turbulence-induced contact rates in plankton: the issue of scales. *Marine Ecology Progress Series*, **139**, 305–7.

Sundby, S. (1997) Turbulence and ichthyoplankton: influence on vertical distributions and encounter rates. *Scientia Marina*, **61** (Supplement 1), 159–76.

Sundby, S. (2000) Recruitment of Atlantic cod stocks in relation to temperature and advection of copepod populations. *Sarsia*, **85**, 277–98.

Sundby, S., Bjorke, H., Soldal, A.V. & Olsen, S. (1989) Mortality rates during the early life stages and year-class strength of northeast Arctic cod (*Gadus morhua* L.). *Rapports et Procès-verbaux des Rèunions, Conseil International pour l'Exploration de la Mer*, **191**, 351–8.

Sundby, S., Ellertsen, B. & Fossum, P. (1994) Encounter rates between first-feeding cod larvae and their prey during moderate to strong turbulent mixing. *ICES Marine Science Symposium*, **198**, 393–405.

Sundby, S. & Fossum, P. (1990) Feeding conditions of Arcto-Norwegian cod larvae compared to the Rothschild–Osborn theory of small-scale turbulence and plankton contact rates. *Journal of Plankton Research*, **12**, 1153–62.

Suthers, I.M. & Sundby, S. (1993) Dispersal and growth of pelagic juvenile Arcto-Norwegian cod (*Gadus morhua*), inferred from otolith microstructure and water temperature. *ICES Journal of Marine Science*, **50**, 261–70.

Swearer, S.E., Caselle, J.E., Lea, D.W. & Warner, R.R. (1999) Larval retention and recruitment in an island population of a coral-reef fish. *Nature*, **402**, 799–802.

Takahashi, M. & Watanabe, Y. (2004) Growth rate-dependent recruitment of Japanese anchovy *Engraulis japonicus* in the Kuroshio-Oyashio transitional waters. *Marine Ecology Progress Series*, **266**, 227–38.

Takahashi, M. & Watanabe, Y. (2005) Effects of temperature and food availability on growth rate during late larval stage of Japanese anchovy (*Engraulis japonicus*) in the Kuroshio-Oyashio transition region. *Fisheries Oceanography*, **14**, 223–35.

Takasuka, A., Aoki, I. & Mitani, I. (2003) Evidence of growth-selective predation on larval Japanese anchovy *Engraulis japonicus* in Sagami Bay. *Marine Ecology Progress Series*, **252**, 223–38.

Takasuka, A., Aoki, I. & Mitani, I. (2004a) Three synergistic growth-related mechanisms in the short-term survival of larval Japanese anchovy *Engraulis japonicus* in Sagami Bay. *Marine Ecology Progress Series*, **270**, 217–28.

Takasuka, A., Oozeki, Y., Kimura, R., Kubota, H. & Aoki, I. (2004b) Growth-selective predation hypothesis revisited for larval anchovy in offshore waters: cannibalism by juveniles versus predation by skipjack tuna. *Marine Ecology Progress Series*, **278**, 297–302.

Tanaka, Y., Satoh, K., Iwahashi, M. & Yamada, H. (2006) Growth-dependent recruitment of Pacific bluefin tuna *Thunnus orientalis* in the northwestern Pacific Ocean. *Marine Ecology Progress Series*, **319**, 225–35.

Thorisson, K. (1994) Is metamorphosis a critical interval in the early life of marine fishes? *Environmental Biology of Fishes*, **40**, 23–36.

Thorrold, S.R. & Hare, J.A. (2002) Otolith applications in reef fish ecology. In: P. Sale (Ed.) *Coral Reef Fishes*. pp. 243–64. Elsevier Science, New York.

Thorrold, S.R., Jones, C.M. & Campana, S.E. (1997) Response of otolith microchemistry to environmental variations experienced by larval and juvenile Atlantic croaker (*Micropogonias undulatus*). *Limnology and Oceanography*, **42**, 102–11.

Thorrold, S.R., Latkoczy, C., Swart, P.K. & Jones, C.M. (2001) Natal homing in a marine fish metapopulation. *Science*, **291**, 297–9.

Trippel, E.A., Kjesvu, O.S. & Solemdal, P. (1997) Effects of adult age and size structure on reproductive output in marine fishes. In: R.C. Chambers & E.A. Trippel (Eds) *Early Life History and Recruitment in Fish Populations*. pp. 31–62. Chapman & Hall, London.

Tyler, A.V. (1992) A context for recruitment correlations: why marine fisheries biologists should look for them. *Fisheries Oceanography*, **1**, 97–107.

Urtizberea, A. & Fiksen, ø. (2013) Effects of prey size structure and turbulence on feeding and growth of anchovy larvae. *Environmental Biology of Fishes*, **96**, 1045–63.

Van der Meeren, T. (1991) Algae as first food for cod larvae, *Gadus morhua* L.: filter feeding or ingestion by accident? *Journal of Fish Biology*, **39**, 225–37.

Van der Veer, H.W. (1986) Immigration, settlement, and density-dependent mortality of a larval and early postlarval 0-group plaice (*Pleuronectes platessa*) population in the western Wadden Sea. *Marine Ecology Progress Series*, **29**, 223–36.

Van der Veer, H.W., Berghahn, R., Miller, J.M. & Rijnsdorp, A.D. (2000a) Recruitment in flatfish, with special emphasis on North Atlantic species: progress made by the flatfish symposia. *ICES Journal of Marine Science*, **57**, 202–15.

Van der Veer, H.W. & Bergman, M.J.N. (1987) Predation by crustaceans on a newly settled 0-group plaice *Pleuronectes platessa* population in the western Wadden Sea. *Marine Ecology Progress Series*, **35**, 203–15.

Van der Veer, H., Geffen, A.J. & Witte, J.I.J. (2000b) Exceptionally strong year classes in plaice *Pleuronectes platessa*: are they generated during the pelagic stage only, or also, in the juvenile stage? *Marine Ecology Progress Series*, **199**, 255–62.

Van der Veer, H.W., Pihl, L. & Bergman, M.J.N. (1990) Recruitment mechanisms in North Sea plaice *Pleuronectes platessa*. *Marine Ecology Progress Series*, **64**, 1–12.

Van der Veer, H.W. & Witte, J.I.J. (1999) Year-class strength of plaice *Pleuronectes platessa* in the Southern Bight of the North Sea: a validation and analysis of the inverse relationship with winter seawater temperature. *Marine Ecology Progress Series*, **184**, 245–57.

Vigliola, L. & Meekan, M.G. (2002) Size at hatching and planktonic growth determine post-settlement survivorship of a coral reef fish. *Oecologia*, **131**, 89–93.

Vinagre, C., Santos, F.D., Cabral, H.N. & Costa, M.J. (2009) Impact of climate and hydrology on juvenile fish recruitment towards estuarine nursery grounds in the context of climate change. *Estuarine, Coastal and Shelf Science*, **85**, 479–86.

Walters, C.J. & Collie, J.S. (1988) Is research on environmental factors useful to fisheries management? *Canadian Journal of Fisheries and Aquatic Sciences*, **45**, 1848–54.

Ware, D.M. (1975) Relation between egg size, growth and natural mortality of larval fish. *Journal of the Fisheries Research Board of Canada*, **32**, 2503–12.

Ware, D.M. & Thomson, R.E. (1991) Link between long-term variability in upwelling and fish production in the northeast Pacific Ocean. *Canadian Journal of Fisheries and Aquatic Sciences*, **48**, 2296–306.

Ware, D.M. & Thomson, R.E. (2005) Bottom-up ecosystem trophic dynamics determine fish production in the Northeast Pacific. *Science*, **308**, 1280–4.

Watanabe, Y. (2002) Resurgence and decline of the Japanese sardine population. In: L.A. Fuiman & R.G. Werner (Eds) *Fishery Science: The Unique Contributions of Early Life Stages*. pp. 243–57. Blackwell Publishing, Oxford.

Watanabe, Y., Zenitani, H. & Kimura, R. (1996) Causes of population decline in the Japanese sardine (*Sardinops melanostictus*): overfishing or early mortality? In: Y. Watanabe, Y. Yamashita & Y. Oozeki (Eds) *Survival Strategies in Early Life Stages of Marine Resources*. pp. 83–94. Balkema, Rotterdam.

Weinstein, M.P., Weiss, S.L., Hodson, R.G. & Gerry, L.R. (1980) Retention of three taxa of postlarval fishes in an intensively flushed tidal estuary, Cape Fear River, North Carolina. *Fishery Bulletin (US)*, **78**, 419–36.

Werner, E.E. & Gilliam, J.F. (1984) The ontogenetic niche and species interactions in size-structured populations. *Annual Review of Ecology and Systematics*, **15**, 393–425.

Werner, F.E., Perry, R.I., Lough, R.G. & Naimie, C.E. (1996) Trophodynamic and advective influences on Georges Bank larval cod and haddock. *Deep-Sea Research II*, **43**, 1793–822.

Werner, F.E., Quinlan, J.A., Blanton, B.O. & Luettich, R.A. Jr (1997) The role of hydrodynamics in explaining variability in fish populations. *Journal of Sea Research*, **37**, 195–212.

Werner, F.E., Quinlan, J.A., Lough, R.G. & Lynch, D.R. (2001) Spatially-explicit individual based modeling of marine populations: a review of the advances in the 1990s. *Sarsia*, **86**, 411–21.

Werner, R.G. (2002) Habitat requirements. In: L.A. Fuiman & R.G. Werner (Eds) *Fishery Science: The Unique Contributions of Early Life Stages*. pp.161–82. Blackwell Science, Oxford.

White, R.J. (1968) Importance of appendicularians as food of larval plaice (*Pleuronectes platessa* L.) off Helgoland. *Berichte Deutschen Wissenschaftlichen Kommission fur Meeresforschung*, **19**, 288–91.

Wiebe, P.H. & Benfield, M.C. (2003) From the Hensen net toward four-dimensional biological oceanography. *Progress in Oceanography*, **56**, 7–136.

Wieland, K., Waller, U. & Schnack, D. (1994) Development of Baltic cod eggs at different levels of temperature and oxygen content. *Dana*, **10**, 163–77.

Wilson, D.T. & Meekan, M.G. (2002) Growth-related advantages for survival to the point of replenishment in the coral reef fish *Stegastes partitus* (Pomacentridae). *Marine Ecology Progress Series*, **231**, 247–60.

Winemiller, K.O. & Rose, K.A. (1992) Patterns of life-history diversification in North American fishes: implications for population regulation. *Canadian Journal of Fisheries and Aquatic Sciences*, **49**, 2196–218.

Winkler, G., Dodson, J.J., Bertrand, N., Thivierge, D. & Vincent, W.F. (2003) Trophic coupling across the St Lawrence River estuarine transition zone. *Marine Ecology Progress Series*, **251**, 59–73.

Wood, R.J. & Austin, H.M. (2009) Synchronous multidecadal fish recruitment patterns in Chesapeake Bay, USA. *Canadian Journal of Fisheries and Aquatic Sciences*, **66**, 496–508.

Wootton, R.J. (1990) *Ecology of Teleost Fishes*. Chapman & Hall, London.

Wroblewski, J.S. & Richman, J.G. (1987) The non-linear response of plankton to wind mixing events-implications for survival of larval northern anchovy. *Journal of Plankton Research*, **9**, 103–23.

Wroblewski, J.S., Richman, J.G. & Mellor, G.L. (1989) Optimal wind conditions for the survival of larval northern anchovy, *Engraulis mordax*: a modeling investigation. *Fishery Bulletin (US)*, **87**, 387–95.

Wuenschel, M.J. & Werner, R.G. (2004) Consumption and gut evacuation rate of laboratory-reared spotted seatrout (Sciaenidae) larvae and juveniles. *Journal of Fish Biology*, **65**, 723–43.

Young, P.S. & Duenas, C.E. (1993) Salinity tolerance of fertilized eggs and yolk-sac larvae of the rabbitfish *Siganus guttatus* (Bloch). *Aquaculture*, **112**, 363–77.

Zastrow, C.E., Houde, E.D. & Saunders, E.H. (1989) Quality of striped bass (*Morone saxatilis*) eggs in relation to river source and female weight. *Rapports et Procès-verbaux des Rèunions, Conseil International pour l'Exploration de la Mer*, **191**, 34–42.

Zijlstra, J.J., Dapper, R. & Witte, J.I.J. (1982) Settlement, growth and mortality of post-larval plaice (*Pleuronectes platessa*) in the western Wadden Sea. *Netherlands Journal of Sea Research*, **15**, 250–72.

Zijlstra, J.J. & Witte, J.I.J. (1985) On the recruitment of 0-group plaice in the North Sea. *Netherlands Journal of Zoology*, **35**, 360–76.

Chapter 4

Effects of Fishing on the Population

Marie-Joëlle Rochet and Lise Marty

Le capitaine Nemo me montra de la main cet amoncellement prodigieux de pintadines, et je compris que cette mine était véritablement inépuisable, car la force créatrice de la nature l'emporte sur l'instinct destructif de l'homme.[1]

Jules Verne, Vingt mille lieues sous les mers, 1869

4.1 Introduction

At the outset of the 21st century, Jules Verne's optimism is no longer in vogue. Fisheries literature seems to be an inexhaustible mine of gloomy accounts of depleted stocks, devastated habitats and marine ecosystems shifted towards less-diverse and less-productive states. However, if we think about the power of the exploitative pressures exerted by fisheries for centuries, the creative forces of Nature must have been strong and inventive for us still to contemplate exploiting marine resources. Populations compensate for fishing mortality by increased productivity, targeted animals learn how to escape fishing gears and hide in refuges, and species replace one another. Depletion or major changes in resources have occurred after decades to centuries of increasing pressures, or in conjunction with environmental changes. In the meantime, most populations and ecosystems have resisted fishing, and our hope of continuing exploitation relies on an ability to understand this resistance and its limits, so as to utilise it better for managing fisheries. This chapter is devoted to the devices that Nature deploys to counteract mankind's destructive power. The focus is on populations, a unit for which specific mechanisms have emerged along evolutionary history. Effects of fishing on broader systems such as food webs or communities have recently been extensively studied (Jennings & Kaiser 1998, Hall 1999, ICES 2000), and it is more difficult to underpin them by a compensation theory.

[1] Captain Nemo pointed to that enormous mass of pearl oysters, and I realised that this was, indeed, an inexhaustible mine, for nature's creative powers, it seemed, were greater than the power of man to destroy.

Twenty Thousand Leagues Under The Sea, by Jules Verne. Quote from Penguin Classics English edition (1994), translated by Mendor T. Brunetti.

Fish Reproductive Biology: Implications for Assessment and Management, Second Edition.
Edited by Tore Jakobsen, Michael J. Fogarty, Bernard A. Megrey and Erlend Moksness.

Traditionally, fisheries science aims at predicting how populations will react to various management options (Beverton & Holt 1957, Hilborn & Walters 1992). Apart from short time frames, this relies mainly on estimating stock–recruitment relationships from historical stock data. The high variability of annual recruitment makes predictions uncertain; this is the reason why so much effort has been devoted to understand early life history and the effects of environmental uncertainty on larval and juvenile survival (see Chapters 1, 3, and 5). However, variability in recruitment is also partly due to variability in the population reproductive potential. Exploitation changes individual growth and reproduction, resulting in changes in reproductive potential. This is one important way populations react to fishing – that is, what we need to know to make projections.

This chapter first elaborates on the mechanisms by which fishing should affect population dynamics and characteristics. It then reviews the observed changes in life-history traits, and how the various mechanisms responsible for these changes have been investigated. Finally, the ways the available knowledge can be used for management advice are presented.

4.2 Why should fishing affect populations? Theoretical expectations

The ways in which fishing should affect populations are quite easy to infer. Fishing removes individuals, thereby decreasing population abundance, and hence, its geographic range or its density, or both. Decreased density might increase the resources available to the remainder of the population, improving their condition, growth and reproduction. However, at very low densities the probability of finding mates might be so low that reproduction would be impaired. Moreover, fishing is generally selective, targeting only part of the population. In many instances, fishing mortality is highest for larger-sized fish, and this induces changes in the length and age structure of the population, which might affect its per capita reproductive output if the age- or size-specific reproduction schedule is not uniform. In addition, if larger and older fish bear a high mortality rate, their contribution to the population's reproduction will be small compared with that of younger and smaller individuals. If there is individual variability in growth and reproductive schedule, and if these traits are genetically transmitted or 'heritable' (i.e., the offspring resemble their parents), then the proportion of early-maturing and small-sized individuals in the population will increase across generations. Other forms of fishing selectivity will similarly create selective pressure and promote evolutionary changes in other traits related to reproduction or growth, such as behaviour or migration patterns. Finally, some fishing gears might affect fish habitat, impairing key processes in their life cycle.

Although the reasoning is simple, the resultant picture looks somewhat intricate. Whether, for example, fish size will decrease or increase in an exploited population will depend on the respective magnitude and time scales of the different mechanisms. In addition, this is further complicated by the trade-offs between life-history traits, and the links between life-history and behavioural traits. That is, life-history traits are not independent but linked by genetic, biophysical, energetic and physiological constraints, as well as environmental settings and evolutionary history (Roff 1992, Stearns 1992). Realised combinations of life-history traits are sometimes called life-history strategies (Roff 1992) and constrain the effects of fishing on populations. Energetic status and physiology also constrain animal behaviour, which can thereby evolve owing to selection on correlated life-history traits (Heino & Godø 2002, Uusi-Heikkilä *et al.* 2008). Some theoretical studies that attempted to predict fishing effects on populations in a more formal way than the qualitative reasoning above are summarised in this section.

4.2.1 *Direct effects of fishing*

Fishing removes selected parts of populations, inducing changes in population abundance and size and age structure. This forms the basis of classical fisheries theory (Beverton & Holt 1957), especially structured-population models and stock-assessment methods, which rely on tracking abundance changes in size- or age-classes to estimate fishing mortality. This results in changes in the average length of fish in the catch and the population, which can be used to estimate total mortality (Beverton & Holt 1957, Ault *et al.* 2005). This also induces changes in the age and length distribution of the fish that contribute to reproduction (Figure 4.1). Because reproductive efficiency increases with age and size in many fish species, there is increasing concern that effects of these changes on the reproductive potential of populations have not been accounted for (Trippel *et al.* 1997). First, fecundity often increases with female body size faster than linearly (Raitt 1932), and female age adds a positive effect on this relationship (Horwood *et al.* 1986, Bobko & Berkeley 2004); hence, old and large spawners have a much larger contribution to egg production than their younger and smaller counterparts. Second, older and larger females spawn larger eggs with a higher probability of hatching and

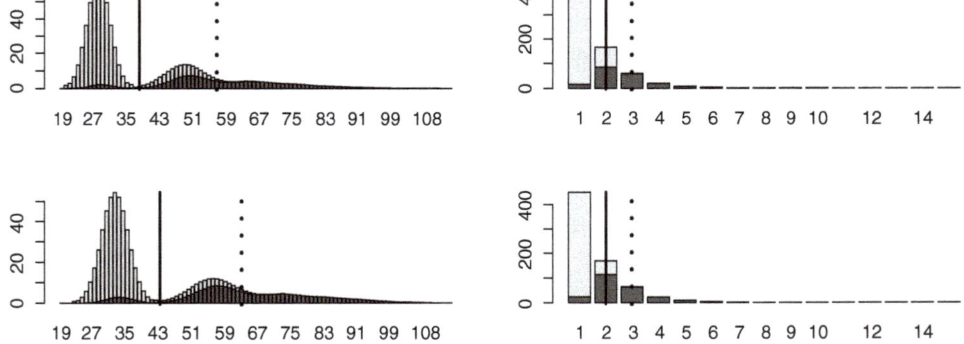

Figure 4.1 Direct and density-dependent effect of fishing on a population. Simulated length (left) and age (right) distributions of a cod-like population: unfished (top), fished with no compensation (middle) and fished with density-dependent growth and a maturation reaction norm (bottom). Open bars, immature fish; dark bars, mature fish. The solid line shows the average length/age of fish in the population; the dotted line shows the average length/age of spawners in the population.

higher subsequent larval survival than do small and young females (see reviews in Trippel *et al.* 1997; see also Kamler 2005, Marteinsdottir & Steinarsson 1998, Trippel 1998, Berkeley *et al.* 2004a). This maternal effect might even result in lifelong inter-individual differences in growth and reproduction (Reznick 1991, Chambers & Leggett 1996). Third, older and larger females span a longer spawning season, enhancing the probability that offspring will meet favourable environments (Hutchings & Myers 1993, Trippel *et al.* 1997, Rideout *et al.* 2005). Overall, the direct effect of fishing on the reproductive potential of populations might be of significant magnitude (Murawski *et al.* 2001).

It is intuitive that fishing should affect the reproductive potential of hermaphroditic populations (where all individuals begin life as one sex and subsequently change sex) more than gonochoric populations, as increased mortality due to fishing might suppress the sex expressed later in life and dramatically affect the sex ratio. This has been confirmed by simulations, with sex change depending, or not depending, on population age structure (Huntsman & Schaaf 1994, Armsworth 2001, Alonzo & Mangel 2005).

Fishing may also decrease effective population size, which is predicted to reduce additive genetic variance because of faster genetic drift. Fishing can change stock geographic distribution, typically reducing its spatial range when abundance declines (Fisher & Frank 2004, Overhotz 2002). An explanation for the positive correlation between abundance and distribution is that, as population abundance increases, more individuals will choose to settle in marginal habitats, where intraspecific competition and habitat suitability are lower. Conversely, as abundance declines, population at the edge of its range, where the population growth rate is negative and immigration positive ('sink' habitat), will become extinct, because remaining individuals no longer disperse from core areas where the population growth rate is positive (Shepherd & Litvak 2004).

4.2.2 Fishing as a change of fish environment

Because natural populations seldom become extinct, nor explode beyond limits, the concept of population regulation is rooted in early ecological theory (Sinclair 1989). Regulation arises both from food limitation and density-dependent processes, i.e. processes whose rate changes with population density or abundance (Sinclair 1989), such as maturation rate or disease mortality. Compensatory processes increase population growth as population density decreases. Depensatory processes, which act in the reverse direction and can thus precipitate population collapse through positive feedback, have been the focus of increasing interest in recent years (Lierman & Hilborn 2001). Compensatory mechanisms are an essential prerequisite for a population to sustain exploitation. If all vital rates were fixed or randomly varying within bounds, the additional mortality caused by fishing would inevitably lead to extinction. With regard to population models used in fisheries science, compensation is accounted for, such as in production models that assume that the population growth rate has a maximum with regard to population abundance (Schaefer 1954), and in dome-shaped or asymptotic stock-recruitment models (Ricker 1954, Beverton & Holt 1957). However, there has been endless debate about the particular processes responsible for compensation and the question of their relative influence on population dynamics, compared with environmental fluctuations (Jones 1989, Fogarty *et al.* 1991). Compensation for fishing mortality could take place in the early life history, with increased larval and juvenile survival leading to increased recruitment when egg production from a depleted population is low; this formed the basis of classical stock–recruitment theory.

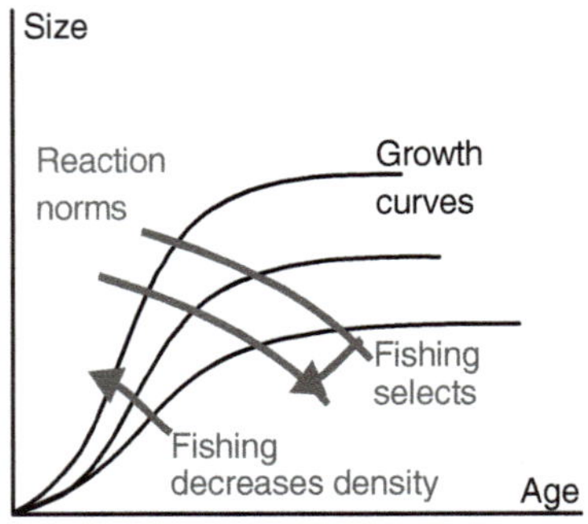

Figure 4.2 The maturation reaction norm is hypothesised to be characteristic of the genetic composition of a population. Environmental variability (resources, temperature...) results in growth variability. When their growth curve intersects with the reaction norm, individuals mature. Fishing decreases population size and hence permits faster growth (due to density dependence), leading to earlier maturation. Selective fishing of larger fish will select for a lower maturation reaction norm, leading to earlier maturation at smaller sizes. For a colour version of this figure, please see Plate 5 in the colour plate section.

On the other hand, fishing could be compensated for in the adult stage by lower mortality from other causes, and/or by improved reproductive capacity enhancing individual egg production at low population densities (Jones 1989). The latter in turn might result from earlier maturation, increased fecundity at maturity, and/or a steeper increase of fecundity with age.

Actually, life-history traits of fishes are assumed to be essentially plastic, that is, they are partly shaped by environmental conditions (Wootton 1984). This has allowed fish populations to survive in fluctuating environments, and has contributed to their resilience to fishing. Theoretical ecology and genetics have developed the concept of 'phenotypic plasticity', which occurs when individuals bearing a common genotype but raised in different environments exhibit different phenotypes (Roff 1992). A typically plastic trait of fishes is their scope for growth, which is widely recognised as being highly dependent both on food availability and temperature (Iles 1974). Whereas plastic growth alone cannot compensate for increased mortality, it can indirectly contribute to compensation through its influence on survival and reproduction (Lorenzen & Enberg 2001, Rose *et al.* 2001). At low densities, fish would grow larger and faster on the basis of increased food availability, thus incurring a lower predation mortality (which generally is assumed to be size-dependent) and producing earlier larger amounts of progeny (Figure 4.1, lower). The latter has been developed into the concept of reaction norm for age and size at maturation. A reaction norm is the genetically determined phenotypic response curve or function to environmental variation through plasticity. The reaction norm for age and size at maturation (or, in short, the maturation reaction norm) is defined as a conceptual line in the age–size plan where individuals develop maturation, when their growth curve crosses this line (Figure 4.2, Plate 5). Optimisation methods have been used to predict how an organism encountering stress that would slow its growth would alter its age at maturation to keep its fitness as high as possible (Stearns & Koella 1986). The shape of the maturation reaction norm that maximises fitness (i.e., the contribution to future generations) depends on the way that mortality, fecundity and growth trajectories are related (Stearns & Crandall 1984, Stearns & Koella 1986, Perrin & Rubin 1990, Marty *et al.* 2011). Predicting fishing-induced plastic changes in age and size at maturation in any particular fish population requires knowledge of these relationships. In many cases, this knowledge will be only partly available, and the relationships will have to be assumed. Nevertheless, the general prediction is that, if fishing affects individual growth rates through density dependence, this should be translated into changes in age and size at maturation, with potential consequences for individual fecundity through maternal effect, which should be investigated.

Fishing can also trigger behavioural responses through learning or phenotypic plasticity, such as gear avoidance (Albert *et al.* 2003) and modified migration route (Jørgensen *et al.* 2008).

4.2.3 Fishing as a cause of evolution

Concern that selective fishing might elicit genetic changes in fish populations was raised at an early point by fisheries scientists. Miller (1957) speculated that the differential removal of faster-growing fish might select for slow growth, and the angling of less-intelligent fish would select for intelligence. Borisov (1979) warned that increased fishing mortality considerably increased the relative contribution to reproduction of fish maturing early, thus selecting for early maturation. A number of theoretical studies have been devoted to the quantitative prediction of these selective effects of fishing, employing three broad categories of approach: (i) phenotypic approaches (optimisation, game theory and adaptive dynamics); (ii) quantitative genetics approaches; and (iii) eco-genetic models.

Optimisation studies assume that selection only keeps the organisms with the greatest contribution to future generations, or fitness. Thus, those that remain after selection should have the trait value that maximises their fitness. For example, the age at maturation that maximises the total number of eggs spawned by an individual undergoing fishing mortality, as compared with actual age at maturation, indicates the selection pressure exerted by fishing. One difficulty lies in the definition of an appropriate measure of fitness, which has been the topic of debate (Kozlowski 1993, Benton & Grant 2000). In addition, because of frequency dependence, genotypes contribute in different ways to the environment of the population (e.g., they remove different shares of resources), a scenario that optimisation models cannot address. In contrast, game theory considers that natural selection tends to favour the most successful genotype: selection should favour the phenotype with the Evolutionarily Stable Strategy (ESS) – that is, the phenotype which, once established, cannot be invaded by a mutant (Brommer 2000). Unifying the two previous approaches, adaptive dynamics provides a theoretical framework to take account of explicit ecological scenarios with density- and frequency-dependence.

These phenotypic approaches neglect genetic mechanisms underlying evolution, and thereby do not allow a prediction of the speed of evolutionary dynamics, nor to assert whether genetic variance will be sufficient for the population to reach the evolutionary endpoint. Quantitative genetics addresses the effects of trait genetic variance and genetic correlations between traits to predict evolutionary dynamics: the change in a character in one generation is determined by the difference between the distribution of the character in the parent population, and in the fraction allowed to reproduce (selection differential), multiplied by the proportion of the character that is transmitted to offspring (heritability) (Munch *et al.* 2005). This can be further complicated by trade-offs – that is, phenotypic correlations between traits which might cause evolution in traits that are not selected for, just because they are correlated with a trait under selection pressure (e.g., size at different ages; Law 1991, Law 2000). Conversely, the selection of one trait could be severely constrained by genetic correlations with non-selected traits (Smith *et al.* 1993). The main predictions of these theoretical studies concern fishing-induced evolutionary changes in growth and age at maturation. Nevertheless, quantitative genetics cannot be used to predict long-term evolution because it assumes constant genetic variance, considering that new genetic variance through mutation and recombination balances the loss of genetic variability due to selection. Any unbalance between these processes contributes to evolution of genetic variance, which alters predictions.

The recent development of eco-genetic models allows accounting for all of individual life history, environmental stochasticity, population dynamic and environmental feedback influencing trait phenotypic expression and evolution, together with the underlying genetic mechanisms (Dunlop *et al.* 2009). Inheritance can be described according to quantitative genetic principles (e.g., Enberg *et al.* 2009) or gametic transmission (e.g., Kuparinen & Hutchings 2012). Eco-genetic models can thus be used to predict the rate, transient trajectory and endpoint of mean trait evolution, and the evolution of genetic variance under density- and frequency-dependent ecological scenarios.

According to life-history theory, uniformly elevated mortality is supposed to favour individuals with early maturation and high reproductive investment. Moreover, fishing gears are for most of the time size-selective; regulations such as minimum landing size or minimum mesh size, as well as economic incentives, often create a preference towards large fish. As a consequence, harvesting pressure could generate an evolutionary response towards slower growth and smaller size; this has been studied mainly by quantitative genetics methods (Favro *et al.* 1979, Law & Rowell 1993). General results about the evolutionary consequences of age-specific predation were obtained by optimisation (Law 1979) and game theory (Michod 1979). To summarise, increased mortality at or after a given age class will favour increased reproductive effort before this age class, whereas increased mortality before an age class will select for decreased reproductive effort after this age class. As a consequence, the evolution of age at first reproduction in an exploited population will depend on the fraction of the population targeted by exploitation. This is illustrated by the well-known example of the Northeast Arctic cod. This population supports both a spawner fishery, targeting only mature fish on the spawning grounds and, more recently, a feeder fishery, directed towards both immature and mature fish. The spawner fishery creates a light selection pressure for later maturation, whereas the feeder fishery selects for earlier age at maturation (Law 1979). Generally, it is expected that size-selective fishing with recruitment before age at first reproduction will select for earlier age at maturation, and here again decreased yield from the population (Law & Grey 1989, Stokes *et al.* 1993). The selection effects might be stronger in seasonal fisheries (Murphy & Rodhouse 1999).

However, the effect of fishing on growth is not straightforward to predict, because growth is influenced by several behavioural, physiological and developmental processes that can be affected by fishing (Enberg *et al.* 2012). Increased mortality can favour fish with a higher investment in both growth and reproduction, because growing faster increases the chance to mature before capture (see for instance Eikeset *et al.* 2013). The overall direction of evolutionary changes in growth would therefore depend on the strength of selection acting in both directions. Taking into account different trade-offs linking natural mortality with life-history traits and behaviour, Jørgensen & Fiksen (2010) predicted a fishing-induced evolution towards faster growth. Other models have predicted the evolution towards either faster or slower growth, depending on the size-selectivity of the fishery (Andersen & Brander 2009, Dunlop *et al.* 2009). The rate of evolution might be quite fast (0.9 to 2.2 haldanes, which measure the rate of change in units of standard deviation per generation, for 23 stocks; Devine *et al.* 2012), although selection differentials, trade-offs and heritability might moderate this rate (Law & Rowell 1993, Law 2000, Andersen & Brander 2009). Simulations suggest that the most vulnerable stocks might be those with a short spawning or fishing season (Williams & Shertzer 2005). The main expected consequence of this selection response is that fishing will select for decreased yields (Stokes *et al.* 1993), although evolutionary changes may not always be detrimental to economic yields, and may even sometimes be beneficial (Eikeset *et al.* 2013). Another concern is that fishing-induced evolution may hamper stock recovery (Enberg *et al.* 2009), in particular because size-selective predation mortality on smaller animals is expected to increase natural mortality. Another mechanism which could prevent the population from bouncing back when fishing mortality is released are trade-offs between life-history traits and survival, such as the cost of reproduction increased by earlier maturation (Jørgensen & Fiksen 2010, Jørgensen & Holt 2013).

Earlier age at maturation and decreased yield are still predicted when phenotypic plasticity is taken into account (Ernande *et al.* 2004). Adaptive dynamics simulations predict that, when both immature and mature individuals are harvested, the reaction norm for age and size at maturation is shifted towards earlier ages and smaller sizes and its slope becomes less steep. Reverse effects are predicted when mature individuals only are harvested. More complex selection patterns emerge when, in addition, consumer-resource dynamics are included in the model (De Roos *et al.* 2006). However, impact of harvesting is still in the same direction. A size-structured (as opposed to age-structured) model also provided consistent predictions, together with a rather fast estimate of size at maturation evolutionary rate in the Baltic cod population (Andersen *et al.* 2007).

Behavioural and physiological traits may also affect an individual's probability of capture, and hence its fitness, especially in fisheries using passive fishing gears (angling, gillnetting, trapping), which may remove bold, active personalities from harvested populations. Because these behavioural traits are heritable and positively correlated with somatic growth rate, this is expected to create a direct selection against fast-growing individuals (Uusi-Heikkilä 2008). Acting on top of size-selectivity, this selective pressure suggests that fishing-induced evolutionary changes could occur faster in populations with different behavioural types, and because of the positive correlation with productivity traits, rises extra concern for population recruitment, and hence viability (Biro and Stamps 2008). Moreover, differential selection acting on behavioural traits involved in vulnerability could negatively impact catchability and reduce fishable biomass (Rijnsdorp *et al.* 2008).

4.3 Estimating fishing effects: the evidence

4.3.1 Population studies

Evidence that life-history traits change in exploited populations has accumulated for decades. The most documented effect is a decrease in age-at-maturity, as reviewed by Trippel (1995) and Rochet (2000a). Evidence has further accumulated since then (Gunderson 1997, Harris & McGovern 1997, Morgan & Colbourne 1999, Gerritsen *et al.* 2003, Armstrong *et al.* 2004, Bobko & Berkeley 2004, Yoneda & Wright 2004). Earlier maturation occurs in either sex or both, in populations of any taxon, geographical area and ecological setting, with diverse exploitation histories. From these and some additional papers, it emerges that length-at-age

(Bolle *et al.* 2004) and maturity-at-length (Cardinale & Modin 1999, Silva *et al.* 2006) were also found to vary, although in a less consistent way. Fecundity has been less often examined, and was generally found to increase under the effect of fishing.

These results are not surprising. Whereas fishing is predicted to decrease age at maturity in most instances by all mechanisms listed above, the expectations for the other traits depend on which process will dominate among direct effects, density dependence and selective pressure. So the results are consistent with expectations. Moreover, many other factors are suspected to influence variability in fish population properties. As a consequence, in most single-population analyses it is difficult to prove that the observed changes are due to fishing selection rather than to any other cause, and to estimate the amount of change due to fishing. A mirror example can illustrate this issue. The population of walleye pollock off northern Japan decreased in abundance from 1985 to 1990, but then increased again to its former level until 1995. During the period 1990–1995, age at maturity was younger, individual fecundity at age was higher, and survival of eggs to recruitment was lower, compared to 1985–1989 (Hamatsu *et al.* 2004). It might be hypothesised that compensatory and/or selective effects of fishing during the late 1980s led to the earlier and increased egg production of the cohorts born when the population was at its lowest. This was probably at the price of smaller and lower-quality eggs, both because mothers were younger and because of the trade-off between egg size and fecundity. The lower survival of these smaller eggs would explain the sharp decrease in the steepness of the spawner–recruit relationship between the 1980s and the 1990s. However, the authors of the study focused their analyses on the changes in sea-surface temperature, and concluded that the decadal change in the ocean environment had caused a change in the reproduction and recruitment of walleye pollock. The lesson is that conclusions about causes of population changes are likely to be biased by the authors' interests, and that as many candidate factors as possible should be investigated together (Sinclair *et al.* 2002a). The risk is that such studies will have a low power, because time series of consistently measured traits are generally short (one to two decades) and the number of candidate factors can be rather high, with not all of them showing contrast over the period examined. This is the reason why long-term studies or alternative methods such as comparative or experimental approaches will probably be more informative.

An outstanding example of a long-term population study is the analysis of long-term effects of fishing in North Sea plaice by Rijnsdorp (1992). The study aim was to investigate population changes over the 20th century, with the contrast provided by a long history of heavy exploitation that was substantially reduced during the Second World War. Changes in somatic growth, reconstructed from otoliths, were mainly ascribed to density dependence, in juvenile as well as adult plaice (Rijnsdorp & van Leeuwen 1992). The maturation of male and female plaice was found to occur at much lower length and age by the end of the century than at the beginning (Rijnsdorp 1989). Maturation was found to depend on growth and temperature on the coastal nursery ground, but these factors did not explain all variations in length at first maturity. The remaining variability could be satisfactorily explained by the selective effect of fishing, as estimated from the patterns of fishing mortality in relation to age (Rijnsdorp 1993a,b). Fecundity at length increased during the century, but the ovary weight–body size relationship did not change, which suggested that reproductive investment was stable and that increased fecundity was mainly explained by smaller egg size (Rijnsdorp 1991). Further analyses of changes in body weight during the spawning period failed to reveal any evidence of fisheries-induced change in reproductive investment since 1960 (Rijnsdorp *et al.* 2005). Hence, this

Table 4.1 Fishing effects on fish life-history traits, estimated by a comparative study across 84 populations. Effect of moderate fishing: average ratio between traits of weakly exploited populations ($F/M < 1$) and those of moderately exploited populations ($1 \leq F/M < 2$). Effect of strong fishing: ratio between weakly exploited populations and strongly exploited populations ($2 \leq F/M$). All traits were log-transformed, so estimated effects are ratios (×), except the slope of the fecundity–length relationship, where the effect is a difference (+).

Trait	Effect of moderate fishing	Effect of strong fishing
Age at maturity	×0.94	×0.77
Length at maturity relative to adult length	×1.06	×1.12
Fecundity at maturity	×4.3	×5.2
Egg diameter	×0.73	×0.63
Slope of the fecundity–length relationship	+0.14	+0.49

comprehensive study concluded that all speculated mechanisms interact to determine variations in population characteristics.

In hermaphroditic populations, fishing or high adult mortality induce changes in growth, size and age at sex change (Krug 1998, DeMartini *et al.* 2005, McBride & Richardson 2007), but this is not always sufficient to prevent changes in sex ratio and a drastic depletion of males in some populations (McGovern *et al.* 1998, Hawkins & Roberts 2003). Spawning behaviour and the mechanism of sex transition might be determining for the sensitivity of these species to fishing (Coleman *et al.* 1996, Armsworth 2001).

4.3.2 Comparative approaches

To escape the curse of confounding effects in population studies, comparative approaches can be used (Harvey & Pagel 1991). Basically, comparisons establish the generality of phenomena, and these approaches are increasingly used as comparative methods are developed for all organisation levels (Cole *et al.* 1991). The accumulation of anecdotes across many populations makes a strong argument that effects cannot all be due to confounding factors. Comparative approaches were used by Trippel (1995) to prove that fishing decreases age at maturity in cod, haddock and other species, and by Brander (2007) to highlight the link between changes in growth and population declines in cod. Moreover, the method can be used to estimate the magnitude of effects across populations. The magnitude of fishing effects on five life-history traits was estimated across 84 fish populations by Rochet *et al.* (2000), using this method. Traits were chosen to be potentially affected by fishing, to be crucial to the population dynamics, and measurable. Data were gathered from the literature, with a thorough check of data quality. Populations were classified into three groups according to fishing pressure, based on the ratio of fishing to natural mortality. A specific method was developed to estimate the difference in traits between these groups, taking account of the phylogenetic relationships between the population analysed, and the influence of length on all traits (Cornillon *et al.* 2000). Significant fishing effects on all five traits were found (Table 4.1): fishing decreases age at maturity and egg size, while increasing fecundity at first reproduction and the slope of the fecundity–length relationship. Length at maturity, scaled by an index of adult size, was slightly increased. This provided an appraisal of fishing effects magnitude across the 84 populations analysed. However, these populations are not a sample of exploited populations over the world, but are highly biased by information availability. These estimations were obtained independently of the mechanisms causing the changes. An average 25% reduction of age at maturity in strongly exploited populations, compared with weakly exploited populations, could result from the direct effect of fishing, phenotypic plasticity or genetic evolution. Other methods have to be used to disentangle these causes.

A more recent comparative study compiled 41 time series of age and/or length at maturity for 26 North Atlantic declining populations (Hutchings & Baum 2005). Age and length at maturity were found to decrease in all cases, on average by 21% and 13% respectively (over various time periods). Two recent meta-analyses found that the rate of phenotypic changes in maturation indices (age and length at 50% maturity and probabilistic maturation reaction norms; see section 4.4.3.3) were correlated with fishing intensity (Sharpe & Hendry 2009, Devine *et al.* 2012).

Compensation for fishing mortality were found to vary among taxonomic groups with different life-history strategies. Clupeiformes do not increase their initial reproductive effort, whereas Gadiformes, Perciformes and Pleuronectiformes mature earlier and/or at a larger size and increase their fecundity at first reproduction (Rochet 2000a, Rochet *et al.* 2000). Consistently, the changes in age and length at maturity estimated by Hutchings & Baum (2005) were lower among pelagic populations (which were all clupeids) than among demersal ones (gadoids and flatfish).

4.4 Understanding the changes: the processes

The most powerful method for understanding the mechanisms underlying changes in life-history traits is the experimental approach (see Section 4.4.1). However, because the results cannot be directly generalised to wild populations, field studies are also needed. Many have aimed at quantifying the magnitude of density-dependent processes in life-history traits variability (Section 4.4.2). More recently, progress has also been made in estimating evolutionary changes (Section 4.4.3).

4.4.1 Experimental results

A unique means of distinguishing genetic from plastic changes in life-history traits is provided by common garden experiments. Fish displaying different phenotypes are reared under similar conditions, such that any remaining differences between phenotypes are caused by differences in genotypes. Maternal effects might also influence life-history traits, especially in the early stages, but these can be eliminated if the fish are able to reproduce in captivity; the phenotypes can then be measured on offspring. Several litters must be obtained to check for genetic diversity in parents. Rearing tractable species with a small size and a short life cycle makes the required replications easy, and several generations can be obtained within a short time. This has allowed several selective fishing experiments to be conducted in which replicate populations are exposed to various harvest regimes.

Silliman (1975) grew two batches of *Tilapia mossambica* under similar conditions. One batch was unselectively fished by removing 10–20% individuals every two months, whilst in the other batch only the largest fish that could not pass through 25-mm vertical slots were removed. After 77 months, males in the selectively fished population grew significantly slower and to a smaller size than those from the control population, whereas no difference was found in females. Although the design was not perfectly clean (e.g., there was no replicate, and wild fish were added during the experiment to increase genetic variability), this was an early experimental demonstration that size-selective fishing can induce genetic changes in growth.

Much more can be learned from model species that have been extensively studied both in the wild and in tanks. Trinidadian guppies (*Poecilia reticulata*), freshwater 15- to 35-mm fish which mature within a few weeks, have been extensively studied by Reznick and colleagues (for reviews, see Reznick 1993 and Reznick & Ghalambor 2005). Common-garden experiments across ranges of food availability provided data about growth-dependent age-size at maturation and age-fecundity reaction norms. These studies were conducted on first-generation offspring to avoid maternal effects. The maturation reaction norms were either boomerang-shaped or linear, with slow-growing fish maturing later at a smaller size. The magnitude of the difference could be as large as twofold age and half size at maturation for the lowest compared to the highest feeding level (Reznick 1993). Moreover, Reznick took advantage of guppies inhabiting streams with, or without, size-selective predators, so as to mimic size-selective effects of fishing. Guppies that lived in high-predation sites were smaller at maturity and produced more and smaller offspring, and also had a smaller size and higher reproductive allocations. All of these findings were consistent with the expected and observed effects of fishing. Moreover, experimental evolution was induced by manipulating the distribution of guppies and their predators in natural streams. High mortality selected within a few years for earlier maturity at a smaller size (Reznick & Ghalambor 2005).

Another extensively studied model species is Atlantic silverside (*Menidia menidia*) (Conover *et al.* 2005). These small marine fish, which inhabit nearshore waters along the east coast of North America, provided the first discovery of environment-dependent sex determination (Conover & Kynard 1981). The fish display a countergradient growth variation, that is, their capacity for growth increases with latitude, and this genetic variation counteracts the decrease in temperature and duration of the growing season that would otherwise cause less growth in the North (Conover and Present 1990). The main selection pressure responsible for this pattern is the strong winter size-selective mortality that dominates in high latitudes, and favours a fast growth rate to ensure a large body size and better survival. At low latitudes, winter mortality is reduced and rapid growth entails more costs (e.g., predation risks) than benefits (Conover & Present 1990, Munch *et al.* 2003). With this knowledge of growth determinants and spatial patterns, a size-selective harvest experiment was conducted (Conover & Munch 2002) in which populations were subjected to large, small, and random size-selective harvests of adults over four generations. The large-harvested populations initially produced the highest yield but quickly evolved a lower yield than controls. This was caused by the selection of genotypes with slower larval and adult growth. The reverse was true for the small-harvested populations (Figure 4.3) (Conover & Munch 2002). Moreover, this experiment also permitted measurements of genetic correlations between adult length at harvest and early life-history traits known to influence recruitment (Munch *et al.* 2005), and adult traits (Walsh *et al.* 2006). Selection on adult size resulted in significant genetic changes in fecundity, consumption rate and growth efficiency, vertebral number, and predator avoidance behaviour (Walsh *et al.* 2006). Changes were also evident in egg diameter, independent of maternal effects, but not in the other juvenile traits examined. The resulting impact on recruitment was estimated to be low and mainly driven by selective changes in fecundity (Munch *et al.* 2005). These results provided strong evidence that life-history traits are tightly linked and that selective fishing can have numerous side effects. Using the strains of Atlantic silverside selected by Conover & Munch (2002), Conover *et al.* (2009) tested the reversibility of fisheries-induced evolution. However, in this case the authors relaxed size selection for five generations

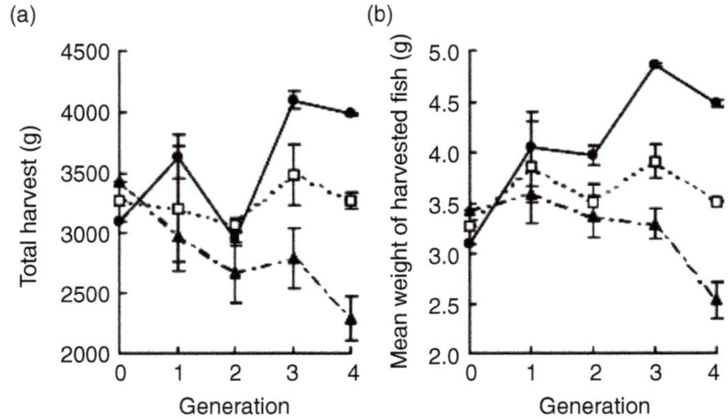

Figure 4.3 Trends in average total weight harvested (a) and mean weight of harvested individuals (b) across multiple generations of size-selective exploitation. Closed circles, small-harvested lines; open squares, random-harvested lines; closed triangles, large-harvested lines. Reproduced from Conover, D.O. & Munch, S.B., Sustaining fisheries yield over evolutionary time scales. *Science*, **297**, 94–6, © 2002 with permission of AAAS.

over all lines and found a significant rebound in size whenever populations had evolved a smaller size and a lower growth rate, though the reversal was 2.5-fold slower than in the first experiment.

A recent selection experiment went one step further and demonstrated a genetic basis for phenotypic response to selection, using lines of Trinidadian guppy artificially selected for small or large size (van Wijk *et al.* 2013). In this case, a phenotypic response was detected together with changes in candidate genetic loci collected for their influence on standard body length, while neutral markers (microsatellites) were sampled to assess genetic changes against the baseline of genetic drift.

Together, these studies provided convincing evidence that selective exploitation can drive genetic evolution in exploited populations, though one problem here is the generality of these findings (Conover *et al.* 2005, Reznick & Ghalambor 2005). First, selective forces might be much more diverse in the wild, and gene flow from neighbour populations might slow or even counteract the effect of selective fishing. However, for the two model species, the importance of size-selective mortality in the wild has been demonstrated. The importance of gene flow will depend on the particular structure of the populations. In addition, model species are typically small-sized and short-lived – characteristics which are opposite to those of many commercial species. Their life-history strategies might display different responses to environmental pressure. However, some characteristics of the model species are widespread among fish species, for example growth-dependent maturation reaction norm or countergradient growth variation. This provides more reasons to consider that the results obtained for silversides and guppies could be generalised to many exploited species. One pending question is how the rates of evolution in the wild might differ from those in tanks, owing to gene flow and complexity of selective forces. The high rates of evolution in the manipulated natural guppy populations, which were very close to natural processes, suggest that evolution can be very fast in response to a changing environment (Reznick *et al.* 1997).

Other studies have examined the effect of fishing on behavioural trait evolution. Biro & Post (2008) used two strains of juvenile trout, one corresponding to fast-growing and bold genotypes, and the other one to slow-growing and shy genotypes, and stocked them in two replicate lakes to simulate an intensive commercial gillnet fishery. It was found that, independent of size, 50% of the fast-growing/bold genotypes were harvested against 30% of the slow-growing/shy genotypes, thus providing evidence of a direct selection against fast-growing genotypes in addition to an indirect selection due to size-selectivity. The results of this study demonstrated that gillnet vulnerability may depend more on fish behaviour than on body size, and questioned the relevance of minimum size limit regulations for the management of this stock. Fishing-induced selection acting on behavioural differences has also been documented in largemouth bass, a popular recreational fish species. In an artificial selection experiment, largemouth bass were selected over three generations for either high or low vulnerability to angling. It has been shown that angling vulnerability was indeed heritable (with a heritability of 0.15; Philipp *et al.* 2009). Using the two selected lines established in this experiment (high-vulnerability and low-vulnerability), high-vulnerability fish were found to have a lower absolute growth rate (Redpath *et al.* 2009), but a higher resting cardiac activity, metabolic requirement, and male parental investment (Cooke *et al.* 2007). Despite the lower absolute growth rate and smaller adult body size, males from the high-vulnerability strain still had the highest reproductive fitness thanks to a higher mating success and a greater intensity of parental care, which improved brood survival (Sutter *et al.* 2012). This could have negative consequences for recruitment in harvested populations in which some behavioural traits (such as parental care) are correlated with reproductive fitness and fishing vulnerability.

4.4.2 Density dependence

Direct measurement of compensation can be difficult, because of the high measurement error of most field data, and the lack of sufficient data at extreme population densities (Rose *et al.* 2001). A single correlative study showing that length at a given age is correlated with population density is not convincing, because of potential confounding factors that were, or were not, reported in the study. For example, if both temperature and population density varied monotonically over the observation period, the effects of temperature and density on mean length at age would be indistinguishable (Walters & Wilderbuer 2000, Wieland 2005). Similarly, in North Sea sole, density dependence in growth is suspected to be partially confounded by the effect of beam trawling that induces both decreased density and increased food availability to sole (de Veen 1976, Millner & Whiting 1996). However, the accumulation of such studies, again, is a strong indication that density dependence occurs in both juvenile and adult stages in fish and marine invertebrates. The tracking of density dependence implies an investigation of supposedly fully reversible changes, so that it can be investigated in both increasing and declining populations. The ideal material to demonstrate density dependence would be a population showing both trends successively, with a full reversal of the studied characteristic, but data-series encompassing such a sequence are rare. Comparisons across zones with contrasted densities can also provide information, especially for less-mobile animals (Weinberg 1998). Growth was found to be density-dependent in nine out of 16 populations from a variety of fish taxa in a comparative study by Lorenzen & Enberg (2001). In addition, these authors showed by simulation that the estimated levels of growth density dependence in the recruited phase were sufficient to account for population regulation. Due to trade-offs between growth and reproduction, density-dependent changes in growth will be reflected in reproductive potential. For example, fast-growing fish have been shown to mature earlier in populations of herring (Hubold 1978, Brophy & Danilowicz 2003), cod (Holdway & Beamish 1985, Chen & Mello 1999), and hake (Helser & Almeida 1997). However, the interest in reaction norms has recently developed, with attempts to estimate their changes through time, and this will be reported below (see Section 4.4.3.3). Moreover, density dependence in reproduction, independent of that in growth, has been observed, for example in flatfish populations (Rijnsdorp 1994, Nash *et al.* 2000). Clearly, density dependence – either mediated by growth or not – cannot be dismissed as a compensation mechanism for fishing, even if further studies should investigate more explicitly the underlying mechanisms (Rose *et al.* 2001).

4.4.3 Evolutionary changes

Recent progress has been made in: (i) demonstrating and estimating the selective pressure exerted by fishing on exploited populations; and (ii) proving the evolutionary nature of phenotypic changes observed in populations undergoing such selective pressure. A particular effort has been devoted to the estimation of changes in maturation reaction norms, which thus deserve a particular subsection.

4.4.3.1 Selection pressure

Several methods have been used to measure the amount or size-selective pressure exerted by fishing. The basic idea is to be able to compare the size distributions of animals of the parent population with that of the fraction allowed to reproduce. For example, Miller & Kapuscinski (1994) compared mean size at age in a lake walleye population before fishing seasons and in subsequent spawning runs. However, in species without massive spawning aggregations or migrations, the spawner fraction cannot be easily sampled. In this case, the size distribution of the animals caught by the fishery before reproducing can help to reconstruct the selection differential (Law 2000). Hanson & Chouinard (1992) identified the fast-growing fish as those of the upper quartile of the length distribution at each age, and the slow-growers as the lower quartile. They estimated the proportions of fish taken by the fishery from each of these quartiles in the population, as determined from research surveys, to compare the pattern of size-selective fishing between two periods with contrasted growth rates in the southern Gulf of St Lawrence cod. Sinclair *et al.* (2002b) reconstructed growth histories of individual fish using back-calculations based on otoliths. They estimated size-selective mortality by comparing the back-calculated length distributions of fish of a given cohort caught in separate years. The mean back-calculated length, decreasing as age at capture increased, suggested that smaller individuals had a lower mortality than their larger counterparts, so that there was a selection for small size. This effect involves the well-known 'Lee's phenomenon', or 'phenomenon of apparent change in growth rate' (Lee 1912). By Lee's phenomenon, lengths back-calculated from hard pieces (e.g., scales or otoliths) are smaller than mean lengths observed in the population. This was ascribed as soon as the early 20th century to size-selective mortality, and has been found in innumerable fish populations (though not in all populations investigated) during the past 100 years (Jones 1956). This would suggest a ubiquitous selection pressure in exploited populations. However, Campana (1990) showed that there is a bias in the method normally used to back-calculate lengths from hard pieces, that relies on an incorrect assumption that the otolith–fish length relationship is independent of fish growth rate. Although Sinclair *et al.* (2002b) corrected for this bias, they still found differences in the back-calculated lengths of fish caught in separate years. These findings mean that the bias in back-calculation does not always account for Lee's entire phenomenon. The suspicion of a pervasive long-term size selection still holds, and a retrospective meta-analysis of Lee's phenomenon corrected for back-calculation bias would help to appraise its occurrence and magnitude. Another method to estimate size-selective mortality is to compare the mean growth rate of recaptured cohorts with the individual growth rate of tagged fish. This was achieved by Kristiansen & Svåsand (1998) with released cod, whereby individual fish were seen to have a much faster growth rate than the apparent mean growth rate of the recaptured cohort, and suggesting that mortality at sea had eliminated the faster-growing fish. In addition, survival at sea was lower for larger fish. Using both acoustic telemetry and traditional tags, Olsen & Moland (2011) estimated natural and harvest mortality in Atlantic cod from the Norwegian Skagerrak coast. The selection differentials determined showed that fisheries favoured the survival of smaller fish, while natural selection had the reverse effect.

Although only the latter could distinguish between natural and fishing mortality, all studies found evidence for size-selective mortality. Size selection can favour fast growth (St Lawrence cod in the 1970s: Hanson & Chouinard 1992, Sinclair *et al.* 2002b, Swain *et al.* 2007) or slow growth (coastal Norwegian cod: Kristiansen & Svåsand 1998; St Lawrence cod in the 1990s: Swain *et al.* 2007, Alpine whitefish: Nusslé *et al.* 2009). Even more complicated patterns of size selectivity can be found, such as disruptive selection with fast and slow growth favoured, while intermediate growth rates are selected against (St Lawrence cod in the 1980s: Sinclair *et al.* 2002b). Selection patterns are closely related to the selectivity of the fishing gears (Miller & Kapuscinski 1994), and to the match between patterns in the size segregation of fish on the fishing grounds and concentration of fishing effort (Sinclair *et al.* 2002b). All of this is complicated by the fact that small fish are discarded in many fisheries (Alverson *et al.* 1994, Kelleher 2005), and most of these are killed. This results in a less-strong size-selective pressure than that estimated from the difference between population and landings, and generally assumed in theoretical studies. For example, in a small area of the Bay of Biscay, the minimum size of *Nephrops* caught (and discarded) was equal to the minimum size on the ground (data described in Trenkel *et al.* 2007a, Trenkel *et al.*, 2008). Given the spread and magnitude of discarding practices, this might seriously affect selective pressure, to an extent that is difficult to appraise. However, size selectivity exists at least in some fisheries, as evidenced by back-calculations on hard pieces.

In the Atlantic cod from the Norwegian Skagerrak coast population, it has also been shown that individuals with a strong diel vertical migration, alternating between shallow (foraging arena) and deep-water (refuge) habitats, had a higher vulnerability to passive fishing gears (traps, gillnet and handline), underlying the importance of fish behaviour for harvest selection (Olsen *et al.* 2012).

Atlantic salmon entering rivers earlier in the season bear a higher fishing mortality than late-running fish (Thorley & Youngson 2007), and running time has been shown to be linked with sex (Pérez *et al.* 2005) and with size at maturation, correlated with a genetic marker (Consuegra *et al.* 2005), providing strong evidence of differential mortality of genetic types. This selective fishing has been shown to result in shifts in run timing and declines in body size, sea age, and life span, and a change in mitochondrial DNA frequencies in Irish and Iberian salmon populations (Consuegra *et al.* 2005, Quinn *et al.* 2006). Estimating selection pressure on age at maturity is not straightforward, as it requires a population model to compare the relative contribution of spawners of different ages to egg production (Arlinghaus *et al.* 2009). As maternal effects might be determining for offspring quality and survival to an extent that is poorly quantified, the final contributions might be very difficult to estimate. Neglecting the latter component, Borisov (1979), Rijnsdorp (1993b) and Stokes & Blythe (1993) found significant selection pressure for earlier maturity in Arcto-Norwegian cod and in five North Sea stocks. However, maternal effects might mitigate this. Owing to their anadromous and semelparous life history, Pacific salmon provide adequate materials for estimating selection pressure on maturation. All fish that migrate into freshwater are indeed maturing individuals, which allows direct estimation of selection differentials in age and size at maturation. Kendall *et al.* (2009) analyzed and compared 57 years of sockeye salmon data collected when fish migrate ('escapement' data) with fishery catch data and found that larger and older fish were more vulnerable in most years.

4.4.3.2 Evidence for evolutionary changes in exploited fish populations

Evolutionary changes are difficult to demonstrate. To prove that genetic changes occurred, genetic materials have to be analysed, and the relevant tools have been only recently developed (Guinand *et al.* 2004). In addition, proving the genetic origin of a change in a given trait would require knowing which genes determine this trait, which is generally not the case. However, significant progress is being made in this field.

Before the development of population genetic tools, a comprehensive study concluded that size-selective fishing was the most probable factor for changes in average size and age of Pacific salmon (Ricker 1981). All Pacific salmon species are semelparous: individuals mature only once and then die. The study took advantage of consistent trends across numerous populations within species, whereas the particular characteristics of the life cycles and exploitation of each species explained the differences among species. Pink salmon all mature at age 2 years and are caught by trolls and gill nets, which remove fish of larger than average size. The catch from all populations but one showed a significant decrease in average weight between 1951 and 1975. By contrast, chum salmon are mostly taken as a bycatch with other species, and not all gears select the largest fish in this species. As a result, the changes in size were much less consistent than in pink salmon, and average age in the catch was increased between 1957 and 1972. Ricker (1981) checked for the potential influence of environmental factors on the observed changes and rejected them. Estimates of mean size heritability based on selection differentials and observed changes were of plausible magnitude. Although this study does not prove the genetic origin of the changes, it contains strong arguments for massive fishing-induced evolutionary changes in these species.

Other exceptional material is provided by the study of Norwegian grayling, a small freshwater salmonid. A system of five populations that were successively founded by dispersal or human introduction have been recently analysed for life-history variation (Haugen & Vøllestad 2001). Adult and juvenile traits diverged over nine to 22 generations, and this was mainly ascribed to adaptive evolution. The genetic nature of the divergence in juvenile traits was checked by a common-garden experiment and found to be related to habitat; divergence in adult traits was most probably caused by differences in fishing intensity. Genetic analyses of microsatellite DNA, which is evolutionarily neutral, showed that the divergence among populations was caused both by strong founder effects (the small number of dispersing individuals determining the initial gene pool) and by natural selection (Koskinen *et al.* 2002). In the initial population, eight generations of size-selective gillnet fishing induced a steady decrease in length-at-age one to five years, as well as in age and length at maturity (–0.33 year and –0.18 mm per 10 years). However, these trends were reversed when fishing was relaxed (Haugen & Vøllestad 2001).

More generally, recent findings show that human- or environment-induced contemporary evolution occurs at much faster rates than previously assumed in diverse taxa of insects, birds, mammals and fishes (Thompson 1998, Stockwell *et al.* 2003). For example, Pacific salmon introduced in New Zealand showed a rapid genetic divergence in life-history traits within 30 generations (Quinn *et al.* 2001). In marine-exploited populations, archived hard pieces such as scales and otoliths can be used to reconstruct time series of genetic information. Polymorphic enzyme diversity was significantly decreased in six years in the newly exploited orange roughy population in New Zealand (Smith *et al.* 1991). Microsatellite diversity was significantly decreased in a closed New Zealand snapper population over the 50 years of its exploitation, but not in a neighbouring open population (Hauser *et al.* 2002). The strongly exploited Flamborough Head cod population showed fluctuations in microsatellite diversity across 44 years that can be related to the history of exploitation and migrations in this small open population (Hutchinson *et al.* 2003). Whilst these are not changes in phenotypic traits, it is relevant to fisheries management because it provides evidence of fishing-induced genetic modification. In addition, both latter studies estimated that effective population size – the number of adults that effectively contribute to the next generation – was five orders of magnitude less than the population size. This very low ratio of effective size to census size has been reported in several other commercially important fish populations (Turner *et al.* 2002, Hoarau *et al.* 2005, Poulsen *et al.* 2006). Thus, in depleted populations, effective size might be very low and a loss of genetic diversity would be expected. However, various processes probably determine effective size, including life history and various mortality sources, and further research is required to determine whether these low ratios are a serious concern (Hoarau *et al.* 2005, Poulsen *et al.* 2006, Turner *et al.* 2006).

4.4.3.3 Age and size at maturation reaction norms

Recently, several studies have attempted to demonstrate the selective effect of fishing on reproductive schedule, taking account of phenotypic plasticity, based on field data. The basic assumption is that the maturation reaction norm (see Figure 4.2, Plate 5) would be genetically determined. In order to distinguish phenotypic plasticity from evolution (or genetic changes) in a changing phenotypic trait, the reaction norm could be estimated at various points in time. If the reaction norm changed, this would be an indication of a genetic modification. Starting from the initial deterministic concept of the maturation reaction norm in which the onset of sexual maturity is fully determined by the age and size of an individual, Heino *et al.* (2002a) drew attention to the large individual variability of maturation within a population. They outlined that maturation is a complex process influenced by factors other than age and size, such as current resource availability or body reserves. This was their motivation to introduce the probabilistic maturation reaction norm (PMRN) as the age- and size-dependent probability of maturing. Methods based on logistic regression have been developed to estimate these probabilities from survey data with age and length as covariates, when age at first reproduction is archived as a special mark on the otolith of individual fish (Heino *et al.* 2002a), or not (Barot *et al.* 2004a). Large sample sizes are required in the latter case (100 individuals per age class for each year class).

Downward shifts in the reaction norm – that is, a tendency to mature earlier at a smaller size – have been found in Georges Bank and Gulf of Maine cod between 1970 and 2000 (Barot *et al.* 2004b), in Northeast Arctic cod between 1932 and 1998 (Heino *et al.* 2002b), in collapsing cod populations off Labrador and Newfoundland between 1975 and 1990 (Olsen *et al.* 2004), where signs of reversal appeared after the closure of the fisheries between 1993 and 2000 (Olsen *et al.* 2005), and in Southern Gulf of St Lawrence cod between 1950 and 2000 (Swain 2011). The latter trend has been shown to be concomitant with a temporal increase in natural mortality, which prevented both biomass and age at maturation recoveries despite little fishing for the past 15 years (Swain 2010). Similar trends in PMRN were also shown in North Sea gadoids from 1975 to 2009 (Marty *et al.* 2014), in North Sea plaice from 1955 to 1995 (Grift *et al.* 2003) and sole from 1963 to 1996 (Mollet *et al.* 2006), and in the declining American plaice population off Newfoundland between 1970 and 1995 (Barot *et al.* 2005). However, in the latter case the decline in population may not have been due to fishing (Morgan *et al.* 2002). Evidence was less strong in Norwegian spring-spawning herring, where growth-related phenotypic plasticity dominated the changes in maturity (Engelhard & Heino 2004), and was absent in two lake smallmouth bass populations between 1937 and 1990 (Dunlop *et al.* 2005), and in North Sea Norway pout from 1975 to 2009 (Marty *et al.* 2014). Interpreting these results in order to differentiate evolution from plastic response relies on two hypotheses: (i) within a population, growth variability mainly results from plasticity in response to the environment, not from genetic diversity; and (ii) growth encompasses most environmental effects on age and size at maturation. These remain assumptions and the ability of the reaction norm approach to identify the cause of maturation trends is controversial (Marshall & Browman 2007). Many early life-history events are suspected to influence maturation (Thorpe 2007, Wright 2007), and further investigations are required to better disentangle environment from genetic factors (Berner & Blanckenhorn 2007, Morita & Fukuwaka 2007). Whatever the mechanism, all of these studies have provided strong evidence of fishing-induced changes in exploited populations. Experimental investigations (Suquet *et al.* 2005) or the analysis of genetic markers (Guinand *et al.* 2004) could be applied to evaluate the evolutionary component of these changes.

4.4.4 Summary of evidence

To summarise the findings of Sections 4.3 and 4.4, all expected effects of fishing have been observed (Figure 4.4). Fish growth has generally been found to respond plastically to many intricate factors, including environmental influences and density dependence. In addition, fishing has been shown to exert a size-selective pressure, although it might be less strong than generally assumed, owing to discarding practices. Fishing-induced selection has been shown to result in evolutionary changes in size at age in several field studies and size-selective harvest experiments. Most of these changes were, as expected, toward smaller size and/or slower growth, resulting in decreasing yields from the populations. Changes were detectable over a

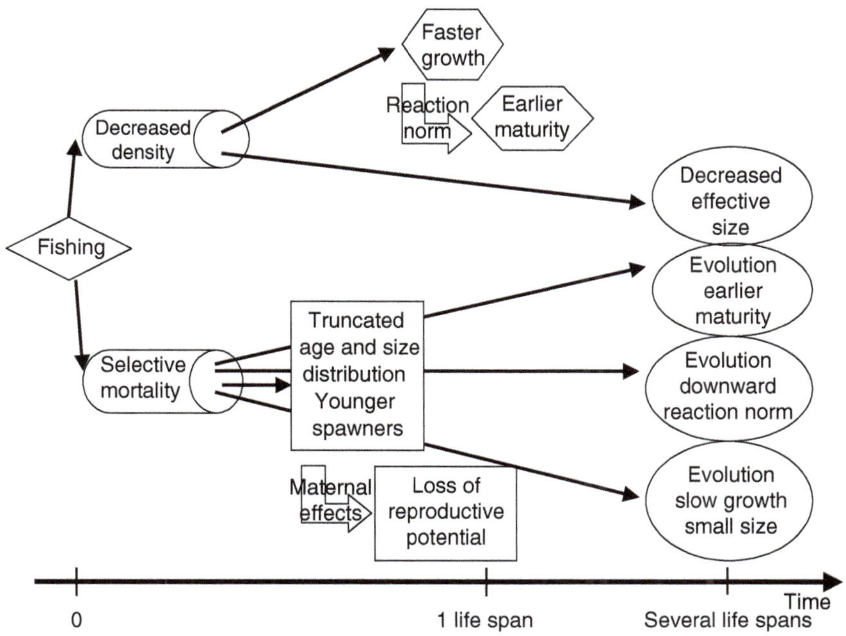

Figure 4.4 A summary of the most common effects of fishing on populations. At present the number of life spans necessary for fishing-induced evolution is not really known. Note that the time axis is somehow elastic as fishing reduces the life span. Squares represent direct effects; diamonds represent plastic changes; bubbles represent genetic changes.

few generations both in experiments and in the field, and evolutionary changes cannot be qualified as a long-term effect of fishing. Rapid evolution creates cumulative changes on time scales that are almost comparable to those of plastic responses, still complicating the interpretation of correlative studies. Multiple regressions to explain growth variations have to be carefully designed, including in the analysis all explanatory variables that are suspected to play a role in that particular population (Sinclair *et al.* 2002a). Studies with a list of potential factors selected based on data availability included in an exploratory approach are deemed to be inconclusive.

Age at maturity decreases in exploited populations, and this is the best proven effect of fishing. These changes arise from the direct effect of fishing (age–structure truncation), and from growth-related and other phenotypic plasticity. In addition, a part of these changes can probably be ascribed to evolution. The magnitude and rate of changes found in experimental populations make a strong case for an evolutionary decrease of generation time in exploited populations. Changes in length at maturity are less unanimous, as this trait is related to both growth rate and age at maturity. When both decrease, length at maturity decreases as well, and this was the finding in a majority of studies, although in some cases length at maturity could also increase in exploited populations. Changes in other reproductive traits such as fecundity or reproductive investment are less well documented, but there is evidence for both plastic and evolved changes in exploited and experimental populations. Investigations of selection on behavioural trait are still scarce in commercial fisheries, although this could intensify selective pressure on correlated life-history traits and explain why theoretical estimates of evolution speed are systematically higher than empirical estimates (Audzijonyte *et al.* 2013).

4.5 Fishing effects and management advice

Fishing effects on populations introduce bias and uncertainty in stock size estimates. In this section, the extent of these errors is first examined, and how they could be corrected for in stock assessments and projections. How life-history traits themselves could be used as indicators for fisheries management is then discussed. Finally, the management tools available to mitigate fishing effects on populations are examined.

4.5.1 Assessment, diagnostics and projections

Concern has been raised that variations in life-history traits which are not taken into account in stock assessments and projections result in erroneous estimates of current, reference and future stock size and reproductive potential. First, estimates of past and current spawning stock biomass are incorrect, and their variability is underestimated when they are based on fixed maturity ogives (Ulltang 1996, Bromley 2003) or fixed growth curves (Walters & Wilderbuer 2000). Neglecting size-selective mortality when growth varies both phenotypically and between individuals overestimates yield per recruit and spawning per recruit under high fishing pressure, thus overestimating optimal fishing levels (Parma & Deriso 1990). The concept of spawning per recruit itself is ambiguous because, if growth and reproduction are density-dependent, the spawning biomass of a cohort is not proportional to the number recruited. Thus, the associated reference points such as F_{low}, F_{med}, F_{high} and $F_{\%\ spr}$ might not be meaningful (Rochet 2000b).

As a solution, it has been suggested that more biological information should be incorporated into the estimates of spawning stock biomass (ICES 2003, Kell & Bromley 2004). Incorporating variability in maturity at age, sex ratio, and potential egg production related to changes in length-at-age, as well as the differential survival of eggs spawned by older compared to younger females, leads to more refined estimates of population reproductive potential, and a better perception of stock productivity (Murawski *et al.* 2001, Morgan & Brattey 2005). As it might be costly to monitor all biological processes on a yearly basis, proxies have been suggested, such as indices of growth, condition or lipid content (Marshall *et al.* 1998, 2000), or age diversity of females in the population (Marteinsdottir & Thorarinsson 1998). Rapid and cost-effective methods are being developed to estimate fecundity (Thorsen & Kjesbu 2001).

All of this is fine as long as attention is focused on stock assessment. With regards to diagnostics and projections, the problem is more complex because establishing reference points and projecting future stock states implies modelling the mechanisms that underlie growth and reproduction changes. From the review above, it was apparent that even a change in as simple a characteristic as average length is the result of a combination of the direct effect of selective fishing, density-dependent and environment-driven growth, and selective pressure for slower growth and smaller size. As these various mechanisms act on different time scales, predicting future changes without knowing which one has been (or will be) the most determining one for the population of interest is more than a challenge. An alternative might be to use multivariate time-series analysis (Needle *et al.* 2001), at least for short-term projections, provided that sufficiently long time series are available. In both cases, assuming an underlying mechanism or using time-series modelling, estimating the associated parameters might be difficult because the available data do not contain the appropriate information, and thus model identifiability might be poor (Parma & Deriso 1990, Rochet 2000b). For example, many attempts have been made to improve stock–recruitment relationships by incorporating more biological information into the stock reproductive potential estimates (Marshall *et al.* 1999, Cardinale & Arrhenius 2000, Marshall *et al.* 2000, Marteinsdottir & Begg 2002, Marshall *et al.* 2006, Óskarsson & Taggart 2006), and this should play an increasing role in medium-term projections as multi-annual advice is increasingly required, for example by the European Union (ICES 2003). However, the overall predictive capacity of the resulting models remains poor, even if they bring significant improvements to standard spawning stock biomass estimates. Currently, stock–recruitment relationships are rarely used in stock projections. The use of proxies such as lipid content for fecundity will be useful only when the relationship between fecundity and the proxy is well established (e.g., $r^2 > 0.5$ in the simulation settings used by De Oliveira *et al.* 2006).

As for predicting evolutionary changes, the problem is similar to those described above, with more poorly known processes and parameters to be incorporated into models: 'Unfortunately, when it comes to predicting evolutionary responses in any particular fishery, the devil is in the stock-specific details of life-history strategy, harvest selection gradients, and genetic variances and covariances.' (Conover *et al.* 2005).

The idea that fishing should differently affect populations with varying life histories has been largely invoked to advocate life-history-based harvest policies (Clark 1991, Beverton 1992, Winemiller & Rose 1992). Indeed, fishing effects on life-history traits have been found to differ between orders with different life-history strategies, with compensation in reproduction lower in Clupeiforms than in higher orders (Rochet 2000a, Rochet *et al.* 2000). Life-history theory also predicts that species, or populations, with a fast growth rate, a late maturation at large size, and a weak reproductive investment are particularly sensitive to harvest-induced evolutionary changes in age and size at maturation. This has been empirically supported for four species of North Sea Gadoids (Marty *et al.* 2014). Higher compensation is not sufficient, and populations with later maturity and larger size incur higher decreases in abundance (Jennings *et al.* 1998) and have a lower growth rate at low abundances, that is, a lower potential for recovery (Denney *et al.* 2002). However, the way to translate this into policy is not straightforward. Conceptual management frameworks for broad categories of life-history strategies have been suggested (King & McFarlane 2003, Young *et al.* 2006, Laugen *et al.* 2012) but, to the present authors' knowledge, are rarely implemented. Attempts to design biological reference points have been more related to individual traits (Clark 1991) or to the shape of stock–recruit relationships (Williams & Shertzer 2003) than to life-history strategies *per se.* This might be because the underlying trade-offs are not yet well understood and quantified, leading to a poor predictive power of population responses (Rose 2005).

4.5.2 Indicators

An alternative to incorporating life-history traits variations in reproductive stocks estimates to improve their accuracy and the perception of the actual state of the stock, would be to use these traits as indicators themselves (Munkittrick & Dixon 1989, Rochet 2000c). This would be less costly, and the information content and uncertainty of the unit pieces may be easier to appraise than their combination. These indicators alone do not allow the projection of future states, but lack of projection may not be much worse than uncertain projections based on poorly estimated reproductive stock. Life-history traits are indicators of the renewal potential of a population;

for example, age at maturity determines generation time. As such, they are complementary to the estimation of stock size. Assume that the managers of the Northwest Atlantic cod stocks during the late 1980s had been provided with estimates of spawning stock size, and, in addition, been informed that median age at maturity was sharply decreasing (Olsen *et al.* 2004), as well as size and condition (Dutil *et al.* 1999). These 'red signs' may have led to a more appropriate management action than stock projections alone.

Useful indicators should be sensitive and responsive to fishing, easy to understand and to estimate, and it should be possible to determine reference values (targets or thresholds not to be exceeded) for them (Rice & Rochet 2005). From the review above, clearly the most sensitive life-history trait is age at maturity, which decreases in exploited populations, whatever the mechanism. Its use as an indicator of fishing pressure was suggested by Trippel (1995). Rochet (2000c) showed, for two flatfish populations, that it is sensitive to fishing, and that variations in age at maturity contribute largely to population response to fishing; moreover, age at maturity is determining for population dynamics as it is closely linked to generation time. It was also found to be the most responsive indicator for evaluating exploitation status of walleye in a comparative study of 10 Minnesota lakes (Gangl & Pereira 2003). Length at maturity also responds to fishing (Gangl & Pereira 2003), but in a less consistent way, although it could provide a surrogate when age data are not available. In that case, length at maturity might be a useful indicator, provided that indicators of changes in size or growth would also be available to help interpret its changes. Age and length at maturity are usually estimated by visual examination of the gonads of individual fish caught in scientific surveys or sampled at the market; the latter approach proved reliable, while reducing the cost of sampling itself (Bromley 2003). The difficulty lies in sampling the appropriate time and space to span the entire age or size range of maturing fish. In addition, macroscopic examination might systematically overestimate the proportion of mature individuals (Vitale *et al.* 2006), thereby underestimating age and length at 50% maturity. In the case of Baltic cod, the bias in the proportion mature is larger for younger age classes, which means that if the proportion of younger spawners in the population increases, so too will the bias in estimates of age and length at maturity and spawning stock biomass. This would be a serious concern if this bias were to be found in many species, but in that case models based on hepatosomatic and gonadosomatic indices might be used as a proxy for proportion mature (Vitale *et al.* 2006). The precision and accuracy of these methods, and their usefulness for populations other than Baltic cod, have still to be investigated. Other indicators of the population's reproductive potential, such as the ratio of multiparous to primiparous spawners or the sex ratio, have been proposed, but their potential as management tools is not yet well established (Fogarty & Gendron 2004, Marshall *et al.* 2006).

Clearly, size and growth also respond to fishing, but in a still less-consistent way than length at maturity, because growth variability is also driven by multiple environmental factors. Therefore, the interpretation of changes in length-based or growth indicators will be difficult in the absence of indicators for these factors, as they cannot be used in their own right as indicators of fishing impacts. Indicators of individual growth such as average growth rate on an age range, parameters of a growth model, or average length at a given age have been proposed and tested, and proved to be sensitive to exploitation (Gangl & Pereira 2003). However, estimating these indicators usually requires costly and time-consuming age reading from hard pieces. By contrast, length-based indicators are easy to estimate and underpinned by well-established theories (Shin *et al.* 2005). These will encapsulate both direct effects of fishing on the size- and age-structure of the population, and fishing-induced changes in growth. Average length in the population has been used in many studies (McClanahan & Kaunda-Arara 1996, Ault *et al.* 1998), and has proved to be well estimated relative to most other population indicators, to be sensitive to fishing (Trenkel & Rochet 2003), and to serve as a useful complement to indicators of changes in population size (Haedrich & Barnes 1997, Rochet *et al.* 2005). Other descriptors of the length distribution of the population, such as an index of maximum length (Jennings *et al.* 1999, Shin *et al.* 2005), the proportion of small or large fish, and length-distribution percentiles and variance could provide useful complements, and deserve further examination.

A difficulty common to life-history-based indicators is the definition of reference points. What should be the median age at maturity in a given population, and how should it be decided under which threshold the median age should not be allowed to decrease? The current value of median age could be used as a parameter in a population model to determine a limit reference point for fishing mortality rate (Rochet 2001). Alternatively, the median age could be used in combined indicators frameworks where a reference point for each indicator is not required (Rochet *et al.* 2005, Trenkel *et al.* 2007b). Another difficulty with these indicators is their responsiveness – that is, the time scale of their response to fishing and to management actions. The various mechanisms underlying life-history trait variations act on different time scales, although these might not be as contrasted as previously assumed and, hence, responsiveness to fishing might be variable. Moreover, all changes are not reversible on similar time scales. Whereas, density-dependent changes should be reversed over a generation time, it might take evolutionary time to reverse the genetic effects of selective fishing (Browman 2000, Law 2000). This is another point to be taken into account in the interpretation of these indicators.

4.5.3 Management tools and strategies

Assuming that managers would be appropriately informed about changes in life-history traits in populations, what could they do? There is no obvious ground to mitigate the direct and plastic changes in traits, which should be taken as pieces of information on stock status. The optimal harvesting of a year class with density-dependent growth was investigated, as 'thinning' the population would improve the growth potential of the remaining fish (Hannesson 1986). However, most likely because of the complexity of growth determinism, this had little practical or theoretical follow-up. By contrast, genetic changes induced by selective fishing are a serious concern, because: (i) they generally reduce the scope for growth, hence potential yield and reproductive potential; (ii) they might be reversible on long time frames only; and (iii) they decrease the genetic diversity of the populations, hence their potential to further evolve if conditions change (Stokes *et al.* 1993, Heino 1998, Browman 2000, Law 2000). Many pleas have been made to manage this problem, but fewer ideas have emerged about appropriate management methods. The ideas pertain to two categories: (i) setting total allowable catches (TACs) or effort limits that take fishing-induced evolution into account; and (ii) reducing the mortality of large fish, either by size or mesh limits, or by designating protected marine areas.

Evolutionary stable optimal harvesting strategies (ESOHS) will maximise the yield from an exploited population, with the life-history parameters that will result from evolutionary changes induced by this strategy (Law & Grey 1989). Fishing that targets spawners only selects for delayed maturation and a larger yield, and an ESOHS aimed at maximising the yield will optimise this gain. The price to pay is short-term losses in yield. By contrast, targeting both spawners and immature fish selects for earlier maturation and a lower yield –an ESOHS cannot reverse this (Heino 1998). By simulating the 'co-evolution' between an evolving stock and an 'evolutionary enlightened manager', long-term optimal harvest rates after successive evolutionary steps in age at maturity can be obtained. Clearly, from an evolutionary perspective, harvesting on the spawning grounds is preferable to harvesting irrespective of maturity, though from other points of view it might not be reasonable, because, for example, spawner fisheries are prone to overexploitation (Heino 1998). In addition, management strategies accounting for life-history trait evolution are faced with multitrait evolution. A spawner fishery is 'good' for age and size at maturation, but 'bad' for growth as it would induce slower growth.

To limit the loss of genetic diversity in exploited populations, management objectives such as maintaining diversity among and within populations could be achieved by monitoring and setting reference points for the number of populations, the relative size of populations, and their effective size (Kenchington *et al.* 2003). Much progress has still to be made along these lines.

To avoid the catching of large fish might change the patterns of selective pressure inducing evolutionary change, and leave among the population more old and large spawners with a high reproductive potential. Whilst this approach has been widely advocated (Heino 1998, Browman 2000, Berkeley *et al.* 2004b, Birkeland & Dayton 2005, Law 2007), it may be of limited feasibility. Berkeley *et al.* (2004b) listed three management methods to protect old fish:

- Exploitation levels could be reduced, but to an extent that might not be economically acceptable to be effective (Berkeley *et al.* 2004b).
- Slot size limits might set both a minimum and a maximum size for capture or retention. Model results have suggested that a maximum size might not be effective to restore an already evolved size at maturation, but it would be an option to preserve size at maturation in a healthy stock (Andersen *et al.* 2007). From a practical standpoint, minimum landing sizes have proven less efficient than expected, even when they are accompanied by mesh size regulations, because optimal gear configuration varies among species (Halliday & Pinhorn 2002). This would most likely also apply to maximum size limits. A possible future approach would be the development of fishing methods, gears and strategies that avoid killing large fish.
- The introduction of marine reserves, where fishing is prohibited. This management tool may benefit fisheries by protecting 'source populations' which provide larval supply or adult spillover through density-dependent habitat selection into harvested areas. Baskett *et al.* (2005) developed a model to assess the potential of marine reserves to prevent size-at-maturation evolution induced by heavy fishing, and marine reserves provided an equivalent protection to harvest limitations or size limits. The preservation of diversity among populations based on marine reserves has been reported, but their effect on the preservation of genetic diversity within populations is still to be investigated. Moreover, the benefits to be expected from marine reserves compared with conventional fishery management tools depend on the life-history of the exploited species and the characteristics of the fisheries targeting them (Hilborn *et al.* 2004). As a consequence, there is no reason to expect that marine reserves of an acceptable size will be effective to preserve a wide diversity of resources.

More generally, three categories of 'evolutionary engineering' approaches have been suggested to slow human-induced evolution: (i) reduce variation in fitness-related traits; (ii) reduce directional selection; and (iii) reduce heritability of a fitness-related trait (Palumbi 2001). Whereas the first and third approaches will be difficult to implement in fisheries, reducing directional selection could be achieved by varying selection over time and space (as is done, for example, with herbicide rotation to prevent resistance evolution). Encouraging a diversity of gears exerting a variety of selection pressures, or closing exploitation for each species in turn, might be other options to be investigated in multispecies fisheries.

4.6 Conclusion: future challenges

Pervasive changes in the life histories of marine resources have occurred. A cod is no longer a cod – if a cod is to be a large, long-lived highly iteroparous fish. Many resources now have a squid-like life style, with the majority of the population consisting of recruits that spawn and die within one or a few years. Phenotypic diversity, both within and among species, is low compared with before, such as in 1869, when Jules Verne was writing *Twenty Thousand Leagues Under The Sea.* What about genetic diversity? What scope is there for re-diversifying and further evolution? Clearly, one of the major challenges for fisheries science in the forthcoming years will be to disentangle the mechanisms responsible for these changes. We have to determine which part of these changes is reversible, and on which time frames. Some people advocate that rebuilding marine ecosystems should be the new objective of fisheries management (Pitcher & Pauly 1998), but to what extent is this possible? Is it even possible to keep exploiting the systems as they are, or will the accumulated loss of diversity result in further collapses and ecosystem shifts, whatever future management practices? This depends on the magnitude of the remaining diversity, and on the mechanisms still available to restore it. We also have much to learn from quantitative genetics and evolutionary biology.

Obviously, changes in populations will result in changes in food webs and communities, and on their dynamics. If the turnover of populations is faster and if, in addition, large predator species are largely depleted (Myers & Worm 2003), the turnover of ecosystems will also speed up. This can have consequences on system variability and the reversibility of changes, which are the focus of increasing research (Scheffer & Carpenter 2003, Harris & Steele 2004).

On the management side, all of this is a plea for more precautionary approaches. The creative power of Nature – that is, the potential for compensation and evolution in exploited populations – is strong and has probably contributed to the false impression that marine resources were inexhaustible. This belief was spread in the 19th century, and some fishing theories developed in the 20th century relied on compensation assumptions. Because the mechanisms for adaptation are various and intricate, they have concealed each other and the progress in knowledge has been slow. However, we now know that the compensation potential is not infinite, and we have reached the limits in several instances. How far the remaining resources are from the limits is uncertain, which calls for precaution.

4.7 Acknowledgements

The authors thank Dan Duplisea for Jules Verne's translation, and Benoît Mesnil for the simulated length frequencies in Figure 4.1. Bruno Ernande and Verena Trenkel gave useful comments on an earlier draft of the manuscript.

References

Albert, O.T., Harbitz, A. & Høines, Å.S. (2003) Greenland halibut observed by video in front of survey trawl: behaviour, escapement, and spatial pattern. *Journal of Sea Research*, **50**, 117–27.

Andersen, K.H. & Brander, K. (2009) Expected rate of fisheries-induced evolution is slow. *Proceedings of the National Academy of Sciences*, **106**, 11657–60.

Alonzo, S.H. & Mangel, M. (2005) Sex-change rules, stock dynamics, and the performance of spawning-per-recruit measures in protogynous stocks. *Fishery Bulletin*, **103**, 229–45.

Alverson, D.L., Freeberg, M.H., Murawski, S.A. & Pope, J.G. (1994) A global assessment of bycatch and discards. *FAO Fisheries Technical Paper*, **339**, 233 pp.

Andersen, K.H., Farnsworth, K.D., Thygesen, U.H. & Beyer, J.E. (2007) The evolutionary pressure from fishing on size at maturation of Baltic cod. *Ecological Modelling*, **204**, 246–52.

Arlinghaus, R., Matsumura, S. & Dieckmann, U. (2009) Quantifying selection differentials caused by recreational fishing: development of modeling framework and application to reproductive investment in pike (*Esox lucius*). *Evolutionary Applications*, **2**, 335–55.

Armstrong, M.J., Gerritsen, H.D., Allen, M., McCurdy, W.J. & Peel, J.A.D. (2004) Variability in maturity and growth in a heavily exploited stock: cod (*Gadus morhua* L.) in the Irish Sea. *ICES Journal of Marine Science*, **61**, 98–112.

Armsworth, P.R. (2001) Effects of fishing on a protogynous hermaphrodite. *Canadian Journal of Fisheries and Aquatic Sciences*, **58**, 568–78.

Audzijonyte, A., Kuparinen, A. & Fulton, E.A. (2013) How fast is fisheries-induced evolution? Quantitative analysis of modelling and empirical studies. *Evolutionary Applications*, **6**, 585–95.

Ault, J.S., Bohnsack, J.A. & Meester, G.A. (1998) A retrospective (1979–1996) multispecies assessment of coral reef fish stocks in the Florida Keys. *Fishery Bulletin*, **96**, 395–414.

Ault, J.S., Smith, S.G. & Bohnsack, J.A. (2005) Evaluation of average length as an estimator of exploitation status for the Florida coral-reef fish community. *ICES Journal of Marine Science*, **62**, 417–23.

Barot, S., Heino, M., Morgan, M.J. & Dieckmann, U. (2005) Maturation of Newfoundland American plaice (*Hippoglossoides platessoides*): long-term trends in maturation reaction norms despite low fishing mortality? *ICES Journal of Marine Science*, **62**, 56–64.

Barot, S., Heino, M., O'Brien, L. & Dieckmann, U. (2004a) Estimating reaction norms for age and size at maturation when age at first reproduction is unknown. *Evolutionary Ecology Research*, **6**, 659–78.

Barot, S., Heino, M., O'Brien, L. & Dieckmann, U. (2004b) Long-term trends in the maturation reaction norm of two cod stocks. *Ecological Applications*, **14**, 1257–71.

Baskett, M.L., Levin, S.A., Gaines, S.D. & Dushoff, J. (2005) Marine reserve design and the evolution of size at maturation in harvested fish. *Ecological Applications*, **15**, 882–901.

Benton, T.G. & Grant, A. (2000) Evolutionary fitness in ecology: comparing measures of fitness in stochastic, density dependent environments. *Evolutionary Ecology Research*, **2**, 769–89.

Berkeley, S.A., Chapman, C. & Sogard, S.M. (2004a) Maternal age as a determinant of larval growth and survival in a marine fish, *Sebastes melanops*. *Ecology (New York)*, **85**, 1258–64.

Berkeley, S.A., Hixon, M.A., Larson, R. & Love, M.S. (2004b) Fisheries sustainability via protection of age structure and spatial distribution of fish populations. *Fisheries*, **29**, 23–32.

Berner, D. & Blanckenhorn, W.U. (2007) An ontogenetic perspective on the relationship between age and size at maturity. *Functional Ecology*, **21**, 505–12.

Beverton, R.J.H. (1992) Patterns of reproductive strategy parameters in some marine teleost fishes. *Journal of Fish Biology*, **41**, 137–60.

Beverton, R.J.H. & Holt, S.J. (1957) *On the Dynamics of Exploited Fish Populations*. Fishery Investigations, Series II, **19**, Her Majesty's Stationery Office, London. 533 pp.

Birkeland, C. & Dayton, P.K. (2005) The importance in fishery management of leaving the big ones. *Tree*, **20**, 356–8.

Biro, P.A. & Post, J.R. (2008) Rapid depletion of genotypes with fast growth and bold personality traits from harvested fish populations. *Proceedings of the National Academy of Sciences of the United States of America*, **105** (8), 2919–22.

Biro, P.A. & Stamps, J.A. (2008) Are animal personality traits linked to life-history productivity? *Trends in Ecology & Evolution*, **23**, 361–8.

Bobko, S.J. & Berkeley, S.A. (2004) Maturity, ovarian cycle, fecundity, and age-specific parturition of black rockfish (*Sebastes melanops*). *Fishery Bulletin*, **102**, 418–29.

Bolle, L.J., Rijnsdorp, A.D., van Neer, W., Millner, R.S., van Leeuwen, P.I., Ervynck, A., Ayers, R. & Ongenae, E. (2004) Growth changes in plaice, cod, haddock and saithe in the North Sea: a comparison of (post-)medieval and present-day growth rates based on otolith measurements. *Journal of Sea Research*, **51**, 313–28.

Borisov, V.M. (1979) The selective effect of fishing on the population structure of species with a long life cycle. *Journal of Ichthyology*, **18**, 896–904.

Brander, K.M. (2007) The role of growth changes in the decline and recovery of North Atlantic cod stocks since 1970. *ICES Journal of Marine Science*, **64**, 211–17.

Bromley, P.J. (2003) The use of market sampling to generate maturity ogives and to investigate growth, sexual dimorphism and reproductive strategy in central and south-western North Sea sole (*Solea solea* L.). *ICES Journal of Marine Science*, **60**, 52–65.

Brommer, J.E. (2000) The evolution of fitness in life-history theory. *Biological Reviews*, **75**, 377–404.

Brophy, D. & Danilowicz, B.S. (2003) The influence of pre-recruitment growth on subsequent growth and age at first spawning in Atlantic herring (*Clupea harengus* L.) *ICES Journal of Marine Science*, **60**, 1103–13.

Browman, H.I. (2000) 'Evolution' of fisheries science. *Marine Ecology Progress Series*, **208**, 299–313.

Campana, S.E. (1990) How reliable are growth back-calculations based on otoliths? *Canadian Journal of Fisheries and Aquatic Sciences*, **47**, 2219–27.

Cardinale, M. & Arrhenius, F. (2000) The influence of stock structure and environmental conditions on the recruitment process of Baltic cod estimated using a generalized additive model. *Canadian Journal of Fisheries and Aquatic Sciences*, **57**, 2402–9.

Cardinale, M. & Modin, J. (1999) Changes in size-at-maturity of Baltic cod (*Gadus morhua*) during a period of large variations in stock size and environmental conditions. *Fisheries Research*, **41**, 285–95.

Chambers, R.C. & Leggett, W.C. (1996) Maternal influences on variation in egg sizes in temperate marine fishes. *American Zoologist*, **36**, 180–96.

Chen, Y. & Mello, L.G.S. (1999) Growth and maturation of cod (*Gadus morhua*) of different year classes in the Northwest Atlantic, NAFO subdivision 3Ps. *Fisheries Research*, **42**, 87–101.

Clark, W.G. (1991) Groundfish exploitation rates based on life history parameters. *Canadian Journal of Fisheries and Aquatic Sciences*, **48**, 734–50.

Cole, J., Lovett, G. & Findlay, S. (Eds) (1991) *Comparative Analyses of Ecosystems*. Springer, New York. 375 pp.

Coleman, F.C., Koenig, C.C. & Collins, L.A. (1996) Reproductive styles of shallow-water groupers (Pisces: Serranidae) in the eastern Gulf of Mexico and the consequences of fishing spawning aggregations. *Environmental Biology of Fishes*, **47**, 129–41.

Conover, D.O. & Present, T.M. (1990) Countergradient variation in growth rate: compensation for length of the growing season among Atlantic silversides from different latitudes. *Oecologia*, **83**, 316–24.

Conover, D.O., Arnott, S.A., Walsh, M.R. & Munch, S.B. (2005) Darwinian fishery science: lessons from the Atlantic silverside (*Menidia menidia*). *Canadian Journal of Fisheries and Aquatic Sciences*, **62**, 730–7.

Conover, D.O. & Kynard, B.E. (1981) Environmental sex determination: interaction of temperature and genotype in a fish. *Science*, **213**, 577–9.

Conover, D.O. & Munch, S.B. (2002) Sustaining fisheries yield over evolutionary time scales. *Science*, **297**, 94–6.

Consuegra, S., García de Leániz, C., Serdio, A. & Verspoor, E. (2005) Selective exploitation of early running fish may induce genetic and phenotypic changes in Atlantic salmon. *Journal of Fish Biology*, **67** (Supplement A), 129–45.

Cooke, S.J., Suski, C.D., Ostrand, K.G., Wahl, D.H. & Philipp, D.P. (2007) Physiological and behavioral consequences of long-term artificial selection for vulnerability to recreational angling in a teleost fish. *Physiological and Biochemical Zoology*, **80**, 480–90.

Cornillon, P.-A., Pontier, D. & Rochet, M.J. (2000) Autoregressive models for estimating phylogenetic and environmental effects: accounting for within-species variations. *Journal of Theoretical Biology*, **202**, 247–56.

De Oliveira, J.A.A., Roel, B.A. & Dickey-Collas, M. (2006) Investigating the use of proxies for fecundity to improve management advice for western horse mackerel, *Trachurus trachurus*. *ICES Journal of Marine Science*, **63**, 25–35.

De Roos, A.M., Boukal, D.S. & Persson, L. (2006) Evolutionary regime shifts in age and size at maturation of exploited fish stocks. *Proceedings of the Royal Society of London B*, **273**, 1873–80.

de Veen, J.F. (1976) On changes in some biological parameters in the North Sea sole. *Journal du Conseil International pour l'Exploration de la Mer*, **37**, 60–90.

DeMartini, E.E., Friedlander, A.M. & Holzwarth, S.R. (2005) Size at sex change in protogynous labroids, prey body size distributions, and apex predator densities at NW Hawaiian atolls. *Marine Ecology Progress Series*, **297**, 259–71.

Denney, N.H., Jennings, S. & Reynolds, J.D. (2002) Life-history correlates of maximum population growth rates in marine fishes. *Proceedings of the Royal Society of London B, Biological Sciences*, **269**, 2229–37.

Devine, J.A., Wright, P.J., Pardoe, H.E. & Heino, M. (2012) Comparing rates of contemporary evolution in life-history traits for exploited fish stocks. *Canadian Journal of Fisheries and Aquatic Sciences*, **69**, 1105–20.

Dunlop, E.S., Heino, M. & Dieckmann, U. (2009) Eco-genetic modeling of contemporary life-history evolution. *Ecological Applications*, **19**, 1815–34.

Dunlop, E.S., Shuter, B.J. & Ridgway, M.S. (2005) Isolating the influence of growth rate on maturation patterns in the smallmouth bass (*Micropterus dolomieu*). *Canadian Journal of Fisheries and Aquatic Sciences*, **62**, 844–53.

Dutil, J.-D., Castonguay, M., Gilbert, D. & Gascon, D. (1999) Growth, condition, and environmental relationships in Atlantic cod (*Gadus morhua*) in the northern Gulf of St Lawrence and implications for management strategies in the Northwest Atlantic. *Canadian Journal of Fisheries and Aquatic Sciences*, **56**, 1818–31.

Eikeset, A.M., Richter, A., Dunlop, E.S., Dieckmann, U. & Stenseth, N.C. (2013) Economic repercussions of fisheries-induced evolution. *Proceedings of the National Academy of Sciences of the United States of America*, **110**, 12259–64.

Enberg, K., Jørgensen, C., Dunlop, E.S., Heino, M. & Dieckmann, U. (2009) Implications of fisheries-induced evolution for stock rebuilding and recovery. *Evolutionary Applications*, **2**, 394–414.

Enberg, K., Jørgensen, C., Dunlop, E.S., Varpe, Ø., Boukal, D.S., Baulier, L., Eliassen, S. & Heino, M. (2012) Fishing-induced evolution of growth: concepts, mechanisms and the empirical evidence. *Marine Ecology*, **33**, 1–25.

Engelhard, G.H. & Heino, M. (2004) Maturity changes in Norwegian spring-spawning herring *Clupea harengus*: compensatory or evolutionary responses? *Marine Ecology Progress Series*, **272**, 245–56.

Ernande, B., Dieckmann, U. & Heino, M. (2004) Adaptive changes in harvested populations: plasticity and evolution of age and size at maturation. *Proceedings of the Royal Society of London B, Biological Sciences*, **271**, 415–23.

Favro, L.D., Kuo, P.K. & McDonald, J.F. (1979) Population-genetic study of the effects of selective fishing on the growth rate of trout. *Journal of the Fisheries Research Board of Canada*, **36**, 552–61.

Fisher, J.A. & Frank, K.T. (2004) Abundance-distribution relationships and conservation of exploited marine fishes. *Marine Ecology Progress Series*, **279**, 201–13.

Fogarty, M.J. & Gendron, L. (2004) Biological reference points for American lobster (*Homarus americanus*) populations: limits to exploitation and the precautionary approach. *Canadian Journal of Fisheries and Aquatic Sciences*, **61**, 1392–403.

Fogarty, M.J., Sissenwine, M.P. & Cohen, E.B. (1991) Recruitment variability and the dynamics of exploited marine populations. *Trends in Ecology and Evolution*, **6**, 241–6.

Gangl, R.S. & Pereira, D.L. (2003) Biological performance indicators for evaluating exploitation of Minnesota's large-lake walleye fisheries. *North American Journal of Fisheries Management*, **23**, 1303–11.

Gerritsen, H.D., Armstrong, M.J., Allen, M., McCurdy, W.J. & Peel, J.A.D. (2003) Variability in maturity and growth in a heavily exploited stock: whiting (*Merlangius merlangus* L.) in the Irish Sea. *Journal of Sea Research*, **49**, 69–82.

Grift, R.E., Rijnsdorp, A.D., Barot, S., Heino, M. & Dieckmann, U. (2003) Fisheries-induced trends in reaction norms for maturation in the North Sea plaice. *Marine Ecology Progress Series*, **257**, 247–57.

Guinand, B., Lemaire, C. & Bonhomme, F. (2004) How to detect polymorphisms undergoing selection in marine fishes? A review of methods and case studies, including flatfishes. *Journal of Sea Research*, **51**, 167–82.

Gunderson, D.R. (1997) Spatial patterns in the dynamics of slope rockfish stocks and their implications for management. *Fishery Bulletin*, **95**, 219–30.

Haedrich, R.L. & Barnes, S.M. (1997) Changes over time of the size structure in an exploited shelf fish community. *Fisheries Research*, **31**, 229–39.

Hall, S.J. (1999) *The Effect of Fishing on Marine Ecosystems and Communities*. Fish Biology and Aquatic Resources Series. Blackwell Science, Oxford. 274 pp.

Halliday, R.G. & Pinhorn, A.T. (2002) A review of the scientific and technical bases for policies on the capture of small fish in North Atlantic groundfish fisheries. *Fisheries Research*, **57**, 211–22.

Hamatsu, T., Yabuki, K. & Watanabe, K. (2004) Decadal changes in reproduction of walleye pollock (*Theragra chalcogramma*) off the Pacific coast of northern Japan. *Fisheries Oceanography*, **13** (Suppl. 1), 74–83.

Hannesson, R. (1986) Optimal thinning of a year-class with density dependent growth. *Canadian Journal of Fisheries and Aquatic Sciences*, **43**, 889–92.

Hanson, J.M. & Chouinard, G.A. (1992) Evidence that size-selective mortality affects growth of Atlantic cod (*Gadus morhua* L.) in the southern Gulf of St Lawrence. *Journal of Fish Biology*, **41**, 31–41.

Harris, P.J. & McGovern, J.C. (1997) Changes in the life history of red porgy, *Pagrus pagrus*, from the southeastern United States, 1972–1994. *Fishery Bulletin*, **95**, 732–47.

Harris, R.P. & Steele, J.H. (Eds) (2004) Regime shifts in the ocean. Reconciling observations and theory. *Progress in Oceanography*, **60**, 2–4.

Harvey, P.H. & Pagel, M.D. (1991) *The Comparative Method in Evolutionary Biology*. Oxford Series in Ecology and Evolution. Oxford University Press, Oxford. 239 pp.

Haugen, T.O. & Vøllestad, L.A. (2001) A century of life-history evolution in grayling. *Genetica*, **112–13**, 475–91.

Hauser, L., Adcock, G.J., Smith, P.J., Bernal Ramírez, J.H. & Carvalho, G.R. (2002) Loss of microsatellite diversity and low effective population size in an overexploited population of New Zealand snapper (*Pagrus auratus*). *Proceedings of the National Academy of Sciences of the United States of America*, **99**, 11742–7.

Hawkins, J.P. & Roberts, C.M. (2003) Effects of fishing on sex-changing Caribbean parrotfishes. *Biological Conservation*, **115**, 213–26.

Heino, M. (1998) Management of evolving fish stocks. *Canadian Journal of Fisheries and Aquatic Sciences*, **55**, 1971–82.

Heino, M., Dieckmann, U. & Godø, O.R. (2002a) Measuring probabilistic reaction norms for age and size at maturation. *Evolution*, **56**, 669–78.

Heino, M., Dieckmann, U. & Godø, O.R. (2002b) *Reaction Norm Analysis of Fisheries-induced Adaptive Change and the Case of the Northeast Arctic Cod.* ICES CM 2002/Y:14, 14.

Heino, M. & Godø, O.R. (2002) Fisheries-induced selection pressures in the context of sustainable fisheries. *Bulletin of Marine Science*, **70**, 639–56.

Helser, T.E. & Almeida, F.P. (1997) Density dependent growth and sexual maturity of silver hake in the north-west Atlantic. *Journal of Fish Biology*, **51**, 607–23.

Hilborn, R., Stokes, K., Maguire, J.-J., Smith, T., Botsford, L.W., Mangel, M., Orensanz, J., Parma, A., Rice, J., Bell, J., Cochrane, K.L., Garcia, S., Hall, S.J., Kirkwood, G.P., Sainsbury, K., Stefansson, G. & Walters, C. (2004) When can marine reserves improve fisheries management? *Ocean and Coastal Management*, **47**, 197–205.

Hilborn, R. & Walters, C.F. (1992) *Quantitative Fisheries Stock Assessment. Choice, Dynamics and Uncertainty.* Chapman & Hall, New York. 570 pp.

Hoarau, G., Boon, E., Jongma, D.N., Ferber, S., Palsson, J., Van der Veer, H.W., Rijnsdorp, A.D., Stam, W.T. & Olsen, J.L. (2005) Low effective population size and evidence for inbreeding in an overexploited flatfish, plaice (*Pleuronectes platessa* L.). *Proceedings of the Royal Society of London B*, **272**, 497–503.

Holdway, D.A. & Beamish, F.W.H. (1985) The effect of growth rate, size, and season on oocyte development and maturity of Atlantic cod (*Gadus morhua* L.). *Journal of Experimental Marine Biology and Ecology*, **85**, 3–19.

Horwood, J.W., Bannister, R.C.A. & Howlett, G.J. (1986) Comparative fecundity of North Sea plaice (*Pleuronectes platessa* L.). *Proceedings of the Royal Society of London B, Biological Sciences*, **228**, 401–31.

Hubold, G. (1978) Variations in growth rate and maturity of herring in the North sea in the years 1955–1973. *Rapports et Procès-verbaux des Réunions du Conseil International pour l'Exploration de la Mer*, **172**, 154–63.

Huntsman, G.R. & Schaaf, W.E. (1994) Simulation of the impact of fishing on reproduction of a protogynous grouper, the graysby. *North American Journal of Fisheries Management*, **12**, 41–52.

Hutchings, J.A. & Baum, J.K. (2005) Measuring marine fish biodiversity: temporal changes in abundance, life history and demography. *Philosophical Transactions of the Royal Society B*, **360**, 315–38.

Hutchings, J.A. & Myers, R.A. (1993) Effect of age on the seasonality of maturation and spawning of Atlantic cod, *Gadus morhua*, in the Northwest Atlantic. *Canadian Journal of Fisheries and Aquatic Sciences*, **50**, 2468–74.

Hutchinson, W.F., van Oosterhout, C., Rogers, S. & Carvalho, G.R. (2003) Temporal analysis of archived samples indicates marked genetic changes in declining North Sea cod (*Gadus morhua*). *Proceedings of the Royal Society of London B*, **270**, 2125–32.

ICES (2000) Ecosystem effects of fishing. Proceedings of an ICES/SCOR Symposium held in Montpellier, France, 16–19 March 1999. *ICES Journal of Marine Science*, **57**, 465–792.

ICES (2003) *Report of the Study Group on Growth, Maturity and Condition in Stock Projections.* ICES CM 2003/D:01 Ref. ACFM, C, G, H, WGMG, 85 pp.

Iles, T.D. (1974) The tactics and strategy of growth in fishes. In: F.R. Harden Jones (Ed.) *Sea Fisheries Research*. pp. 331–45. Elek Science, London.

Jennings, S., Greenstreet, S.P.R. & Reynolds, J.D. (1999) Structural change in an exploited fish community: a consequence of differential fishing effects on species with contrasting life histories. *Journal of Animal Ecology*, **68**, 617–27.

Jennings, S. & Kaiser, M.J. (1998) The effects of fishing on marine ecosystems. *Advances in Marine Biology*, **34**, 201–351.

Jennings, S., Reynolds, J.D. & Mills, S.C. (1998) Life history correlates of responses to fisheries exploitation. *Proceedings of the Royal Society of London, B*, **265**, 333–9.

Jones, R. (1956) *Lee's Phenomenon of 'Apparent Change in Growth-rate' with Particular Reference to Haddock and Plaice.* International Commission for the Northwest Atlantic Fisheries Special Publication, **1**, 229–42.

Jones, R. (1989) Towards a general theory of population regulation in marine teleosts. *Journal du Conseil International pour l'Exploration de la Mer*, **45**, 176–89.

Jørgensen, C., Dunlop, E.S., Opdal, A.F. & Fiksen, Ø. (2008) The evolution of spawning migrations: state dependence and fishing-induced changes. *Ecology*, **89**, 3436–48.

Jørgensen, C. & Fiksen, Ø. (2010) Modelling fishing-induced adaptations and consequences for natural mortality. *Canadian Journal of Fisheries and Aquatic Sciences*, **67**, 1086–97.

Jørgensen, C. & Holt, R.E. (2013) Natural mortality: Its ecology, how it shapes fish life histories, and why it may be increased by fishing. *Journal of Sea Research*, **75**, 8–18.

Kamler, E. (2005) Parent-egg-progeny relationships in teleost fishes: an energetics perspective. *Reviews in Fish Biology and Fisheries*, **15**, 399–421.

Kell, L.T. & Bromley, P.J. (2004) Implications for current management advice for North Sea plaice (*Pleuronectes platessa* L.): Part II. Increased biological realism in recruitment, growth, density dependent sexual maturation and the impact of sexual dimorphism and fishery discards. *Journal of Sea Research*, **51**, 301–12.

Kelleher, K. (2005) *Discards in the World's Marine Fisheries: An Update*. FAO Fisheries Technical Paper, 470. 131 pp.

Kenchington, E., Heino, M. & Nielsen, E.E. (2003) Managing marine genetic diversity: time for action? *ICES Journal of Marine Science*, **60**, 1172–6.

Kendall, N.W., Hard, J.J. & Quinn, T.P. (2009) Quantifying six decades of fishery selection for size and age at maturity in sockeye salmon. *Evolutionary Applications*, **2**, 523–36.

King, J.R. & McFarlane, G.A. (2003) Marine fish life history strategies: applications to fishery management. *Fisheries Management and Ecology*, **10**, 249–64.

Koskinen, M.T., Haugen, T.O. & Primmer, C.R. (2002) Contemporary fisherian life-history evolution in small salmonid populations. *Nature*, **419**, 826–30.

Kozlowski, J. (1993) Measuring fitness in life-history studies. *Trends in Ecology and Evolution*, **8**, 84–5.

Kristiansen, T.S. & Svåsand, T. (1998) Effect of size-selective mortality on growth of coastal cod illustrated by tagging data and an individual-based growth and mortality model. *Journal of Fish Biology*, **52**, 688–705.

Krug, H.M. (1998) Variation in the reproductive cycle of the blackspot seabream, *Pagellus bogaraveo* (Brünnich, 1768) in the Azores. Arquipélago. *Life and Marine Sciences*, **16A**, 37–47.

Kuparinen, A. & Hutchings, J.A. (2012) Consequences of fisheries-induced evolution for population productivity and recovery potential. *Proceedings of the Royal Society B: Biological Sciences*, **279**, 2571–2579.

Laugen, A.T., Engelhard, G.H., Whitlock, R., Arlinghaus, R., Dankel, D.J., Dunlop, E.S., Eikeset A.M., Enberg, K., Jørgensen, C., Matsumara, S., Nusslé, S., Urbach, D., Baulier, L., Boukal, D.S., Ernande, B., Johnston, F.D., Mollet, F., Pardoe, H., Therkildsen, N.O., Uusi-Heikkilä, S., Vainikka, A., Heino, M., Rijnsdorp, A.D. & Dieckmann, U. (2012) Evolutionary impact assessment: accounting for evolutionary consequences of fishing in an ecosystem approach to fisheries management. *Fish and Fisheries*, **15**, 65–96.

Law, R. (1979) Optimal life histories under age-specific predation. *The American Naturalist*, **114**, 399–417.

Law, R. (1991) On the quantitative genetics of correlated characters under directional selection in age-structured populations. *Philosophical Transactions of the Royal Society of London B*, **331**, 213–23.

Law, R. (2000) Fishing, selection, and phenotypic evolution. *ICES Journal of Marine Science*, **57**, 659–68.

Law, R. (2007) Fisheries-induced evolution: present status and future directions. *Marine Ecology Progress Series*, **335**, 271–7.

Law, R. & Grey, D.R. (1989) Evolution of yields from populations with age-specific cropping. *Evolutionary Ecology*, **3**, 343–59.

Law, R. & Rowell, C.A. (1993) Cohort-structured populations, selection responses, and exploitation of the North Sea cod. In: T.K. Stokes, J.M. McGlade& R. Law (Eds) *The Exploitation of Evolving Resources*, pp. 155–73. Lecture Notes in Biomathematics, 99. Springer-Verlag, Berlin.

Lee, R.M. (1912) An investigation into the methods of growth determination in fishes. *Conseil Permanent International pour l'Exploration de la Mer, Publications de Circonstance*, **63**, 1–34.

Lierman, M. & Hilborn, R. (2001) Depensation: evidence, models and implications. *Fish and Fisheries*, **2**, 33–58.

Lorenzen, K. & Enberg, K. (2001) Density dependent growth as a key mechanism in the regulation of fish populations: evidence from among-population comparisons. *Proceedings of the Royal Society of London B, Biological Sciences*, **269**, 49–54.

Marshall, C.T. & Browman, H. (2007) Disentangling the causes of maturation trends in exploited fish populations. *Marine Ecology Progress Series*, **335**, 249–310.

Marshall, C.T., Kjesbu, O., Yaragina, N.A., Solemdal, P. & Ulltang, Ø. (1998) Is spawner biomass a sensitive measure of the reproductive and recruitment potential of Northeast Arctic cod? *Canadian Journal of Fisheries and Aquatic Sciences*, **55**, 1766–83.

Marshall, C.T., Needle, C.L., Thorsen, A., Kjesbu, O.S. & Yaragina, N.A. (2006) Systematic bias in estimates of reproductive potential of an Atlantic cod (*Gadus morhua*) stock: implications for stock–recruit theory and management. *Canadian Journal of Fisheries and Aquatic Sciences*, **63**, 980–94.

Marshall, C.T., Yaragina, N.A., Ådlandsvik, B. & Dolgov, A.V. (2000) Reconstructing the stock–recruit relationship for Northeast Arctic cod using a bioenergetic index of reproductive potential. *Canadian Journal of Fisheries and Aquatic Sciences*, **57**, 2433–42.

Marshall, C.T., Yaragina, N.A., Lambert, Y. & Kjesbu, O.S. (1999) Total lipid energy as a proxy for total egg production by fish stocks. *Nature*, **402**, 288–90.

Marteinsdottir, G. & Begg, G.A. (2002) Essential relationships incorporating the influence of age, size and condition on variables required for estimation of reproductive potential in Atlantic cod *Gadus morhua*. *Marine Ecology Progress Series*, **235**, 235–56.

Marteinsdottir, G. & Steinarsson, A. (1998) Maternal influence on the size and viability of Iceland cod *Gadus morhua* eggs and larvae. *Journal of Fish Biology*, **52**, 1241–58.

Marteinsdottir, G. & Thorarinsson, K. (1998) Improving the stock–recruitment relationship in Icelandic cod (*Gadus morhua* L.) by including age diversity of spawners. *Canadian Journal of Fisheries and Aquatic Sciences*, **55**, 1372–7.

Marty, L., Dieckmann, U., Rochet, M.J. & Ernande, B. (2011) Impact of environmental covariation in growth and mortality on evolving maturation reaction norms. *The American Naturalist*, **177**, E98–E118.

Marty, L., Rochet, M.J. & Ernande, B. (2014) Temporal trends in age and size at maturation of four North Sea gadid species: cod, haddock, whiting and Norway pout. *Marine Ecology Progress Series*, **497**, 179–97.

McBride, R.S. & Richardson, A.K. (2007) Evidence of size-selective fishing mortality from an age and growth study of hogfish (Labridae: *Lachnolaimus maximus*), a hermaphroditic reef fish. *Bulletin of Marine Science*, **80**, 401–17.

McClanahan, T.R. & Kaunda-Arara, B. (1996) Fishery recovery in a coral-reef marine park and its effect on the adjacent fishery. *Conservation Biology*, **10**, 1187–99.

McGovern, J.C., Wyanski, D.M., Pashuk, O., Manooch II, C.S. & Sedberry, G.R. (1998) Changes in the sex ratio and size at maturity of gag, *Mycteroperca microlepsis*, from the Atlantic coast of the southeastern United States during 1976–1995. *Fishery Bulletin*, **96**, 797–807.

Michod, R.E. (1979) Evolution of life histories in response to age-specific mortality factors. *The American Naturalist*, **113**, 531–50.

Miller, L.M. & Kapuscinski, A.R. (1994) Estimation of selection differentials from fish scales: a step towards evaluating genetic alteration of fish size in exploited populations. *Canadian Journal of Fisheries and Aquatic Sciences*, **51**, 774–83.

Miller, R.B. (1957) Have the genetic patterns of fishes been altered by introductions or by selective fishing? *Journal of the Fisheries Research Board of Canada*, **14**, 797–806.

Millner, R.S. & Whiting, C.L. (1996) Long-term changes in growth and population abundance of sole in the North Sea from 1940 to the present. *ICES Journal of Marine Science*, **53**, 1185–95.

Mollet, F.M., Kraak, S.B.M. & Rijnsdorp, A.D. (2006) *Fisheries-induced Evolutionary Changes in Maturation Reaction Norms in North Sea Sole* (*Solea solea*). ICES CM 2006/H,14, 22 pp.

Morgan, M.J. & Brattey, J. (2005) Effect of changes in reproductive potential on perceived productivity of three Northwest Atlantic cod (*Gadus morhua*) stocks. *ICES Journal of Marine Science*, **62**, 65–74.

Morgan, M.J., Brodie, W.B. & Kulka, D.W. (2002) Was over-exploitation the cause of the decline of the American plaice stock off Labrador and northeast Newfoundland? *Fisheries Research*, **57**, 39–49.

Morgan, M.J. & Colbourne, E.B. (1999) Variation in maturity-at-age and size in the three populations of American plaice. *ICES Journal of Marine Science*, **56**, 673–88.

Morita, K. & Fukuwaka, M. (2007) Why age and size at maturity have changed in Pacific salmon. *Marine Ecology Progress Series*, **335**, 289–94.

Munch, S.B., Mangel, M. & Conover, D.O. (2003) Quantifying natural selection on body size from field data: winter mortality in *Menidia menidia*. *Ecology*, **84**, 2168–77.

Munch, S.B., Walsh, M. & Conover, D.O. (2005) Harvest selection, genetic correlations, and evolutionary changes in recruitment: one less thing to worry about? *Canadian Journal of Fisheries and Aquatic Sciences*, **62**, 802–10.

Munkittrick, K.R. & Dixon, D.G. (1989) A holistic approach to ecosystem health assessment using fish population characteristics. *Hydrobiologia*, **188/189**, 123–35.

Murawski, S.A., Rago, P.J. & Trippel, E.A. (2001) Impacts of demographic variation in spawning characteristics on reference points for fishery management. *ICES Journal of Marine Science*, **58**, 1002–14.

Murphy, E.J. & Rodhouse, P.G. (1999) Rapid selection effects in a short-lived semelparous squid species exposed to exploitation: inferences from the optimisation of life-history functions. *Evolutionary Ecology*, **13**, 517–37.

Myers, R.A. & Worm, B. (2003) Rapid worldwide depletion of predatory fish communities. *Nature*, **423**, 280–3.

Nash, R.D.M., Witthames, P.R., Pawson, M. & Alesworth, E. (2000) Regional variability in the dynamics of reproduction and growth of Irish Sea plaice, *Pleuronectes platessa* L. *Journal of Sea Research*, **44**, 55–64.

Needle, C., O'Brien, C.M. & Darby, C.D. (2001) *Ogive Characterisation in Medium-term Stock Projections.* ICES CM 2001/ V:19, 6 pp.

Nusslé, S., Bornand, C.N. & Wedekind, C. (2009) Fishery-induced selection on an Alpine whitefish: quantifying genetic and environmental effects on individual growth rate. *Evolutionary Applications*, **2**, 200–8.

Olsen, E.M., Heino, M., Lilly, G.R., Morgan, M.J., Brattey, J., Ernande, B. & Dieckmann, U. (2004) Maturation trends indicative of rapid evolution preceded the collapse of northern cod. *Nature*, **428**, 932–5.

Olsen, E.M., Lilly, G.R., Heino, M., Morgan, M.J., Brattey, J. & Dieckmann, U. (2005) Assessing changes in age and size at maturation in collapsing populations of Atlantic cod (*Gadus morhua*). *Canadian Journal of Fisheries and Aquatic Sciences*, **62**, 811–23.

Olsen, E.M. & Moland, E. (2011). Fitness landscape of Atlantic cod shaped by harvest selection and natural selection. *Evolutionary Ecology*, **25**, 695–710.

Olsen, E.M., Heupel, M.R., Simpfendorfer, C.A. & Moland, E. (2012). Harvest selection on Atlantic cod behavioral traits: implications for spatial management. *Ecology and Evolution*, **2**, 1549–62.

Óskarsson, G.J. & Taggart, C. (2006) Fecundity variation in Icelandic summer-spawning herring and implications for reproductive potential. *ICES Journal of Marine Science*, **63**, 493–503.

Overhotz, W.J. (2002) The Gulf of Maine-George Bank Atlantic herring (*Clupea harengus*): spatial pattern analysis of the collapse and recovery of a large marine fish complex. *Fisheries Research*, **57**, 237–54.

Palumbi, S.R. (2001) Humans as the world's greatest evolutionary force. *Science*, **293**, 1786–90.

Parma, A.M. & Deriso, R.B. (1990) Dynamics of age and size composition in a population subject to size-selective mortality: effects of phenotypic variability in growth. *Canadian Journal of Fisheries and Aquatic Sciences*, **47**, 274–89.

Pérez, J., Izquierdo, J.I., de la Hoz, J. & Garcia-Vazquez, E. (2005) Female biased angling harvests of Atlantic salmon in Spain. *Fisheries Research*, **74**, 127–33.

Perrin, N. & Rubin, J.F. (1990) On dome-shaped norms of reaction for size-to-age at maturity in fishes. *Functional Ecology*, **4**, 53–7.

Philipp, D.P., Cooke, S.J., Claussen, J.E., Koppelman, J.B., Suski, C.D. & Burkett, D.P. (2009) Selection for vulnerability to angling in largemouth bass. *Transactions of the American Fisheries Society*, **138**, 189–99.

Pitcher, T.J. & Pauly, D. (1998) Rebuilding ecosystems, not sustainability, as the proper goal of fishery management. In: T.J. Pitcher, P.J.B. Hart & D. Pauly (Eds) *Reinventing Fisheries Management*. pp. 311–29. Fish and Fisheries Series, 23. Kluwer Academic Publishers, Dordrecht.

Poulsen, N.A., Nielsen, E.E., Schierup, M.H., Loeschke, V. & Grønkjær, P. (2006) Long-term stability and effective population size in North Sea and Baltic Sea cod (*Gadus morhua*). *Molecular Ecology*, **15**, 321–31.

Quinn, T.P., Kinnison, M.T. & Unwin, M.J. (2001) Evolution of chinook salmon (*Oncorhynchus tshawytscha*) populations in New Zealand: pattern, rate, and process. *Genetica*, **112–113**, 493–513.

Quinn, T.P., McGinnity, P. & Cross, T.F. (2006) Long-term declines in body size and shifts in run timing of Atlantic salmon in Ireland. *Journal of Fish Biology*, **68**, 1713–30.

Raitt, D.S. (1932) *The Fecundity of the Haddock*. Fisheries, Scotland, Scientific Investigations, I.

Redpath, T.D., Cooke, S.J., Arlinghaus, R., Wahl, D.H. & Philipp, D.P. (2009) Life-history traits and energetic status in relation to vulnerability to angling in an experimentally selected teleost fish. *Evolutionary Applications*, **2**, 312–23.

Reznick, D.N. (1991) Maternal effects in fish life histories. In: E.C. Dudley (Ed.) *The Unity of Evolutionary Biology*. 4th International Congress of Systematic and Evolutionary Biology. pp 780–93. Discorides Press, Portland.

Reznick, D.N. (1993) Norms of reaction in fishes. In: T.K. Stokes, J.M. McGlade & R. Law (Eds) *The Exploitation of Evolving Resources*. pp. 72–90. Lecture Notes in Biomathematics, 99. Springer-Verlag, Berlin.

Reznick, D.N. & Ghalambor, C.K. (2005) Can commercial fishing cause evolution? Answers from guppies (*Poecilia reticulata*). *Canadian Journal of Fisheries and Aquatic Sciences*, **62**, 791–801.

Reznick, D.N., Shaw, F.H., Rodd, F.H. & Shaw, R.G. (1997) Evaluation of the rate of evolution in natural populations of guppies (*Poecilia reticulata*). *Science*, **275**, 1934–7.

Rice, J. & Rochet, M.J. (2005) A framework for selecting a suite of indicators for fisheries management. *ICES Journal of Marine Science*, **62**, 516–27.

Ricker, W.E. (1954) Stock and recruitment. *Journal of the Fisheries Research Board of Canada*, **11**, 559–623.

Ricker, W.E. (1981) Changes in the average size and average age of Pacific Salmon. *Canadian Journal of Fisheries and Aquatic Sciences*, **38**, 1636–56.

Rideout, R.M., Trippel, E.A. & Litvak, M.K. (2005) Effects of egg size, food supply and spawning time on early life history success of haddock *Melanogrammus aeglefinus*. *Marine Ecology Progress Series*, **285**, 169–80.

Rijnsdorp, A.D. (1989) Maturation of male and female North Sea plaice (*Pleuronectes platessa* L.). *Journal du Conseil International pour l'Exploration de la Mer*, **46**, 35–51.

Rijnsdorp, A.D. (1991) Changes in fecundity of female North Sea plaice (*Pleuronectes platessa* L.) between three periods since 1900. *ICES Journal of Marine Science*, **48**, 253–80.

Rijnsdorp, A.D. (1992) *Long-term Effects of Fishing in North Sea Plaice*. Universiteid van Amsterdam. 220 pp.

Rijnsdorp, A.D. (1993a) Fisheries as a large-scale experiment on life-history evolution: disentangling phenotypic and genetic effects in changes in maturation and reproduction of North Sea plaice, *Pleuronectes platessa* L. *Oecologia*, **96**, 391–401.

Rijnsdorp, A.D. (1993b) Selection differentials in male and female North Sea plaice and changes in maturation and fecundity. In: T.K. Stokes, J.M. McGlade & R. Law (Eds) *The Exploitation of Evolving Resources*. pp. 19–36. Lecture Notes in Biomathematics, 99, Springer-Verlag, Berlin.

Rijnsdorp, A.D. (1994) Population-regulating processes during the adult phase in flatfish. *Netherlands Journal of Sea Research*, **32**, 207–23.

Rijnsdorp, A.D., Grift, R.D. & Kraak, S.B.M. (2005) Fisheries-induced adaptive change in reproductive investment in North Sea plaice (*Pleuronectes platessa*)? *Canadian Journal of Fisheries and Aquatic Sciences*, **62**, 833–43.

Rijnsdorp, A.D. & van Leeuwen, P.I. (1992) Density-dependent and -independent changes in somatic growth of female North Sea plaice *Pleuronectes platessa* between 1930 and 1985 as revealed by back-calculation of otoliths. *Marine Ecology Progress Series*, **88**, 19–32.

Rijnsdorp, A.D., Poos, J.J., Quirijns, F.J., HilleRisLambers, R., De Wilde, J.W. & Den Heijer, W.M. (2008) The arms race between fishers. *Journal of Sea Research*, **60**, 126–38.

Rochet, M.-J. (2000a) A comparative approach to life history strategies and tactics among four orders of Teleost fishes. *ICES Journal of Marine Science*, **57**, 228–39.

Rochet, M.-J. (2000b) Does the concept of spawning per recruit make sense? *ICES Journal of Marine Science*, **57**, 1160–74.

Rochet, M.-J. (2000c) May life history traits be used as indices of population viability? *Journal of Sea Research*, **44**, 145–57.

Rochet, M.-J. (2001) *Decorrelating the Estimation of Indicators and their Reference Points: The Use of Survey Data.* ICES CM 2001/ P:20, 8 pp.

Rochet, M.-J., Cornillon, P.A., Sabatier, R. & Pontier, D. (2000) Comparative analysis of phylogenetic and fishing effects in life history patterns of Teleost fishes. *Oikos (Copenhagen)*, **91**, 255–70.

Rochet, M.-J., Trenkel, V., Bellail, R., Coppin, F., Le Pape, O., Mahé, J.-C., Morin, J., Poulard, J.C., Schlaich, I., Souplet, A., Vérin, Y. & Bertrand, J.A. (2005) Combining indicator trends to assess ongoing changes in exploited fish communities: diagnostic of communities off the coasts of France. *ICES Journal of Marine Science*, **62**, 1647–64.

Roff, D.A. (1992) *The Evolution of Life Histories: Theory and Analysis.* Chapman & Hall, New York. 550 pp.

Rose, K.A. (2005) Lack of relationship between simulated fish population responses and their life history traits: inadequate models, incorrect analysis, or site-specific factors? *Canadian Journal of Fisheries and Aquatic Sciences*, **62**, 886–902.

Rose, K.A., Cowan, J.A.J., Winemiller, K.O., Myers, R. & Hilborn, R. (2001) Compensatory density dependence in fish populations: importance, controversy, understanding and prognosis. *Fish and Fisheries*, **2**, 293–327.

Schaefer, M.B. (1954) Some aspects of the dynamics of populations important to the management of the commercial marine fisheries. *Inter-American Tropical Tuna Commission Bulletin*, **I**, 27–56.

Sharpe, D.M. & Hendry, A.P. (2009) Life history change in commercially exploited fish stocks: an analysis of trends across studies. *Evolutionary Applications*, **2**, 260–75.

Scheffer, M. & Carpenter, S.R. (2003) Catastrophic regime shifts in ecosystems: linking theory to observation. *Trends in Ecology and Evolution*, **18**, 648–56.

Shepherd, T.D. & Litvak, M.K. (2004) Density-dependent habitat selection and the ideal free distribution in marine fish spatial dynamics: considerations and cautions. *Fish and Fisheries*, **5**, 141–52.

Shin, Y.-J., Rochet, M.-J., Jennings, S., Field, J. & Gislason, H. (2005) Using size-based indicators to evaluate the ecosystem effects of fishing. *ICES Journal of Marine Science*, **62**, 384–96.

Silliman, R.P. (1975) Selective and unselective exploitation of experimental populations of *Tilapia mossambica*. *Fishery Bulletin*, **73**, 495–507.

Silva, A., Santos, M.B., Caneco, B., Pestana, G., Porteiro, C., Carrera, P. & Stratoudakis, Y. (2006) Temporal and geographic variability of sardine maturity at length in the northeastern Atlantic and the western Mediterranean. *ICES Journal of Marine Science*, **63**, 663–76.

Sinclair, A.F., Swain, D.P. & Hanson, J.M. (2002a) Disentangling the effects of size-selective mortality, density, and temperature on length-at-age. *Canadian Journal of Fisheries and Aquatic Sciences*, **59**, 372–82.

Sinclair, A.F., Swain, D.P. & Hanson, J.M. (2002b) Measuring changes in the direction and magnitude of size-selective mortality in a commercial fish population. *Canadian Journal of Fisheries and Aquatic Sciences*, **59**, 361–71.

Sinclair, A.R.E. (1989) Population regulation in animals. In: J.M. Cherrett, A.D. Bradshaw, F.B. Goldsmith, P.J. Grubb & J.R. Krebs (Eds) *Ecological Concepts. The Contribution of Ecology to an Understanding of the Natural World.* pp 197–241. Blackwell Scientific Publications, Oxford.

Smith, P.J., Francis, R.I.C.C. & McVeagh, M. (1991) Loss of genetic diversity due to fishing pressure. *Fisheries Research*, **10**, 309–16.

Smith, R.H., Linton, L.A. & Sibly, R.M. (1993) Trade-offs and genetic correlations among life-history traits: theory and simulation. In: T.K. Stokes, J.M. McGlade & R. Law (Eds) *The Exploitation of Evolving Resources*. pp.128–44. Lecture Notes in Biomathematics, 99. Springer-Verlag, Berlin.

Stearns, S.C. (1992) *The Evolution of Life Histories.* Oxford University Press, Oxford. 250 pp.

Stearns, S.C. & Crandall, R.E. (1984) Plasticity for age and size at sexual maturity: a life-history response to unavoidable stress. In: G.W Potts & R.J. Wootton (Eds) *Fish Reproduction. Strategies and Tactics.* pp. 13–53. Academic Press, London.

Stearns, S.C. & Koella, J.C. (1986) The evolution of phenotypic plasticity in life-history traits: predictions of reaction norms for age and size at maturity. *Evolution*, **40**, 893–913.

Stockwell, C.A., Hendry, A.P. & Kinnison, M.T. (2003) Contemporary evolution meets conservation biology. *Tree*, **18**, 94–101.

Stokes, T.K. & Blythe, S.P. (1993) Size-selective harvesting and age-at-maturity. II: real populations and management options. In: T.K. Stokes, J.M. McGlade & R. Law (Eds) *The Exploitation of Evolving Resources*. pp. 232–47. Lecture Notes in Biomathematics, 99. Springer-Verlag, Berlin.

Stokes, T.K., McGlade, J.M. & Law, R. (Eds) (1993) *The Exploitation of Evolving Resources.* Lecture Notes in Biomathematics 99. Springer-Verlag, Berlin. 264 pp.

Suquet, M., Rochet, M.-J. & Gaignon, J.-L. (2005) Experimental ecology: a key to understanding fish biology in the wild. *Aquatic Living Resources*, **18**, 251–9.

Sutter, D.A., Suski, C.D., Philipp, D.P., Klefoth, T., Wahl, D.H., Kersten, P., Cooke, S.J. & Arlinghaus, R. (2012) Recreational fishing selectively captures individuals with the highest fitness potential. *Proceedings of the National Academy of Sciences of the United States of America*, **109**, 20960–5.

Swain, D.P., Sinclair, A.F. & Hanson, J.M. (2007) Evolutionary response to size-selective mortality in an exploited fish population. *Proceedings of the Royal Society of London B*, **274**, 1015–22.

Swain, D.P. (2011). Life-history evolution and elevated natural mortality in a population of Atlantic cod (*Gadus morhua*). *Evolutionary Applications*, **4**, 18–29.

Thompson, J.N. (1998) Rapid evolution as an ecological process. *Trends in Ecology and Evolution*, **13**, 329–32.

Thorley, J.L. & Youngson, A.F. (2007) Seasonal variation in rod recapture rates indicates differential exploitation of Atlantic salmon, *Salmo salar*, stock components. *Fisheries Management and Ecology*, **14**, 191–8.

Thorpe, J.E. (2007) Maturation responses of salmonids to changing developmental opportunities. *Marine Ecology Progress Series*, **335**, 285–8.

Thorsen, A. & Kjesbu, O.S. (2001) A rapid method for estimation of oocyte size and potential fecundity in Atlantic cod using a computer-aided particle analysis system. *Journal of Sea Research*, **46**, 295–308.

Trenkel, V.M., Le Loc'h, F. & Rochet, M.-J. (2007a) Small-scale spatial and temporal interactions among benthic crustaceans and one fish species in the Bay of Biscay. *Marine Biology*, **151**, 2207–15.

Trenkel, V.M. & Rochet, M.J. (2003) Performance of indicators derived from abundance estimates for detecting the impact of fishing on a fish community. *Canadian Journal of Fisheries and Aquatic Sciences*, **60**, 67–85.

Trenkel, V.M., Rochet, M.-J. & Mahévas, S. (2008) Interactions between fishing strategies of *Nephrops* trawlers in the Bay of Biscay and Norway lobster diel activity patterns. *Fisheries Management and Ecology*, **15**, 11–18.

Trenkel, V.M., Rochet, M.-J. & Mesnil, B. (2007b) From model-based prescriptive advice to indicator-based interactive advice. *ICES Journal of Marine Science*, **64**, 768–74.

Trippel, E.A. (1995) Age at maturity as a stress indicator in fisheries. *Bioscience*, **45**, 759–71.

Trippel, E.A. (1998) Egg size and viability and seasonal offspring production of young Atlantic cod. *Transactions of the American Fisheries Society*, **127**, 339–59.

Trippel, E.A., Kjesbu, O.S. & Solemdal, P. (1997) Effects of adult age and size structure on reproductive output in marine fishes. In: R.C. Chambers & E.A. Trippel (Eds) *Early Life History and Recruitment in Fish Populations.* pp 29–62. Fish and Fisheries Series, 21. Chapman & Hall, London.

Turner, T.F., Osborne, M.J., Moyer, G., Benavides, M.A. & Alò, D. (2006) Life history and environmental variation interact to determine effective population to census size ratio. *Proceedings of the Royal Society of London B*, **273**, 3065–73.

Turner, T.F., Wares, J.P. & Gold, J.R. (2002) Genetic effective size is three orders of magnitude smaller than adult census size in an abundant, estuarine-dependent marine fish (*Sciaenops ocellatus*). *Genetics*, **162**, 1329–39.

Ulltang, Ø. (1996) Stock assessment and biological knowledge: can prediction uncertainty be reduced? *ICES Journal of Marine Science*, **53**, 659–75.

Uusi-Heikkilä, S., Wolter, C., Klefoth, T., & Arlinghaus, R. (2008) A behavioral perspective on fishing-induced evolution. *Trends in Ecology & Evolution*, **23**, 419–21.

Vitale, F., Svedäng, H. & Cardinale, M. (2006) Histological analysis invalidates macroscopically determined maturity ogives of the Kattegat cod (*Gadus morhua*) and suggests new proxies for estimating maturity status of individual fish. *ICES Journal of Marine Science*, **63**, 485–92.

Walsh, M.R., Munch, S.B., Chiba, S. & Conover, D.O. (2006) Maladaptive changes in multiple traits caused by fishing: impediments to population recovery. *Ecology Letters*, **9**, 142–8.

Walters, G.E. & Wilderbuer, T.K. (2000) Decreasing length at age in a rapidly expanding population of northern rock sole in the eastern Bering Sea and its effect on management advice. *Journal of Sea Research*, **44**, 17–26.

Weinberg, J.R. (1998) Density dependent growth in the Atlantic surfclam, *Spisula solidissima*, off the coast of the Delmarva Peninsula, USA. *Marine Biology*, **130**, 621–30.

Wieland, K. (2005) Changes in recruitment, growth, and stock size of northern shrimp (*Pandalus borealis*) at West Greenland: temperature and density dependent effects at released predation pressure. *ICES Journal of Marine Science*, **62**, 1454–62.

van Wijk, S.J., Taylor, M.I., Creer, S., Dreyer, C., Rodrigues, F.M., Ramnarine, I.W. & Carvalho, G.R. (2013) Experimental harvesting of fish populations drives genetically based shifts in body size and maturation. *Frontiers in Ecology and the Environment*, **11**, 181–7.

Williams, E.H. & Shertzer, K.W. (2003) Implications of life-history invariants for biological reference points used in fishery management. *Canadian Journal of Fisheries and Aquatic Sciences*, **60**, 710–20.

Williams, E.H. & Shertzer, K.W. (2005) Effects of fishing on growth traits: a simulation analysis. *Fishery Bulletin*, **103**, 392–403.

Winemiller, K.O. & Rose, A. (1992) Patterns of life-history diversification in North American fishes: implications for population regulation. *Canadian Journal of Fisheries and Aquatic Sciences*, **49**, 2196–218.

Wootton, R.J. (1984) Introduction: strategies and tactics in fish reproduction. In: G.W. Potts & R.J. Wootton (Eds) *Fish Reproduction: Strategies and Tactics.* pp 1–12. Academic Press, London.

Wright, P.J. (2007) Understanding the maturation process for field investigations of fisheries-induced evolution. *Marine Ecology Progress Series*, **335**, 279–83.

Yoneda, M. & Wright, P.J. (2004) Temporal and spatial variation in reproductive investment of Atlantic cod *Gadus morhua* in the northern North Sea and Scottish west coast. *Marine Ecology Progress Series*, **276**, 237–48.

Young, J.L., Bornik, Z., Marcotte, M.L., Charlie, K.N., Wagner, G.N., Hinch, S.G. & Cooke, S.J. (2006) Integrating physiology and life history to improve fisheries management and conservation. *Fish and Fisheries*, **7**, 262–83.

Part II

Information Critical to Successful Assessment and Management

Chapter 5
Egg, Larval, and Juvenile Surveys

Nancy C.H. Lo, Paul E. Smith, and Motomitsu Takahashi

5.1 Introduction

Pre-recruit stages of fish (eggs, larvae, and juveniles) may be surveyed to determine distribution and abundance and, for well-known species, survival and dispersal may also be estimated. Distribution and abundance time series can be used for indices of spawning biomass. It is simple in theory to obtain an absolute, instantaneous estimate of spawning biomass: one needs only estimate the area-specific numbers of eggs or larvae produced and a simultaneous and representative value of eggs (or larvae by live spawners) produced per unit adult biomass. Practical barriers for applying the theory differ by species. Identity of eggs and larvae, rates of growth and survival, and the ovarian histology of egg or larval production must be known. For schooling spawners, heterogeneity of eggs is the principal barrier. For demersal spawners or live bearers, accounting for the growth rate, mortality, and transport of the spawn may be more problematic.

Post-settlement demersal and reef juveniles may be amenable to quantitative projection of recruitment. For pelagic schooling juveniles, population stage estimation is exacerbated by lack of sufficient capturing capacity, and inaccessible and patchy distribution patterns in brood waters. Echo sounders and sonar, combined with analogous commercial fishing techniques, may be required for purposes of recruitment projection.

The quantitative approaches to spawning biomass estimation and recruitment projection mentioned above are ambitious for any one species. There is much to be gained for initiating the ecosystem management approach in instituting these surveys: the most abundant members of the fish community will be represented in the samples as well as the zooplankton predators and producers, which may control or support the fish populations. It is beyond the scope of this chapter to describe all the ancillary benefits to ichthyoplankton and juvenile sampling but there is a massive literature on distribution and abundance surveys, which can be examined and applied.

Fish Reproductive Biology: Implications for Assessment and Management, Second Edition.
Edited by Tore Jakobsen, Michael J. Fogarty, Bernard A. Megrey and Erlend Moksness.

5.2 General considerations

To sample fish eggs and larvae effectively, proper nets and efficient sampling designs are required. The optimal sampling strategies depend on the spatial distribution, or degree of patchiness, and the density of larvae. Both the patchiness and mean density contribute to variance of larval counts. The patchiness of fish larvae was best described by a U-shaped distribution of "patchiness" up to larvae of 15 mm (Hewitt 1981, Matsuura & Hewitt 1995). The patchiness is high at both ends: the embryonic stage patchiness is residual to the spawning and fertilization process; while this pattern continues for the larger larvae, to this is added the patterns of survival and re-aggregation. The density of larvae is much lower than that of the embryonic stage. Smaller nets are preferred for eggs and yolk-sac larvae, because the evasion or avoidance and the costs of sorting and staging are reduced. As larvae grow, due to their dispersion and avoidance of the net tow, nets filtering large volumes of water, like Bongo (Smith & Richardson 1977), are necessary. The minimum point on the U-curve is the point at which the sample volume for the stage is not an issue. Different nets used for different early life stages off California and other parts of the world are vertical tows like CalVET for the eggs (Smith *et al.* 1985), and oblique nets like Bongo for larvae, and a subsurface commercial or small high-speed trawl net for late larvae and juveniles (Methot 1986, Takahashi *et al.* 2001).

The survey design for most ichthyoplankton surveys is fixed line and station, like CalCOFI (California Cooperative Oceanic and Fisheries Investigations) surveys off the California coast where either the station or the line has been used as the sampling unit. A variation of the line and station design is a randomized spacing of lines used for combined acoustic and egg production surveys for South African anchovy *Engraulis capensis* (Gilchrist), and sardine *Sardinops ocellatus* (Pappe) (Armstrong *et al.* 1988, Jolly & Hampton 1990, Hampton 1996, Van der Lingen & Huggett 2003).

In addition, adaptive sampling survey designs (Thompson & Seber 1996, Smith *et al.* 2004) have been used for larvae of Pacific hake *Merluccius productus* (Ayres) (Lo *et al.* 1997) and eggs of Pacific sardine *Sardinops sagax* (Jenyns) (Lo *et al.* 2001, Smith *et al.* 2004, Stratoudakis *et al.* 2006). The former is a restrictive adaptive sampling where the adaptation is stopped by a pre-determined criterion because of the fixed survey time. The latter is an adaptive allocation survey design where the survey area was post-stratified based on data observed during the survey. These kinds of adaptive survey designs are most efficient for the patchy population at low population level (Lo *et al.* 2001).

For the juvenile survey of Japanese sardine *Sardinops melanostictus* (Temminck & Schlegel), in the Pacific coastal waters off northern Japan, the sampling design is fixed line and station with east-west survey transects between the Kuroshio Extension and the Oyashio, between 35—40°N (Yasuda 2003). The actual survey area may vary depending on the locations of the two currents.

Conversion of survey information to management decisions is a multi-faceted problem of perceiving bias, improving precision, and controlling costs of field, laboratory, and analytical procedures. A basic task is fusing counts of organisms per unit area with measurements of factors that control their abundance in samples. Estimation of the area of distribution of a stage in the life cycle, such as spawn or recruit, requires a survey which surrounds the population and where the boundary is characterized by samples with no organisms of the investigated stage present. A most useful procedure has been developed using a weighted negative binomial analysis which preserves the characteristics of counts but gathers factors with continuous distributions into a single factor affecting the probability of the observation. Thus, the counts retain binomial and Poisson features in which the mean and variance are related (equal in Poisson) while the effects of factors can be better modeled by the negative binomial distribution because it is a Poisson distribution with a parameter which follows a gamma distribution (Johnson & Kotz 1969). Most of the organisms in these populations are found in aggregations, often with a few Poisson-distributed organisms dispersed around the basic aggregation owing to interactions of ocean movements and the planktonic habit. Most of the factors involved relate to variations in sampling procedure, variations in growth rate, changing susceptibility to capture and retention by the given sampler with simple smooth relationships like the representative depth, space and time. Both the scale of the sample and the number of samples can be designed so that the number of collected organisms be minimized as this controls the variance function (variance can be a multiple of the mean) but the size of the sample must be sufficient so that the precise presence-absence fraction approaches 50% as an ideal ($p = q$).

Since the ideal procedure we describe will be quite expensive, it will be crucial that the definition of the objectives of management be tuned and accommodated to this expense with the realization that ambiguity will be the result of imprecise evaluations of the fish stocks. We shall also use the ideal measurements as an occasional check on indices of relative abundance, which may be acceptable for most management objectives most of the time.

While this work is geared to monitoring populations, the technology can also be used to describe and monitor habitats using suitable approximations of transport of the populations across the habitat boundaries.

5.2.1 Precision

The conceptual basis for the sampling strategy is to obtain unbiased and precise estimates. This includes provision for "counts" and "measurements" of factors in the final calculation. A characteristic of "counts" is that the variance of counts is a function of the mean: for example, randomly distributed objects may be described by the Poisson distribution in which the variance is equal to the mean. When "counts" are derived from aggregated populations, the variance is normally larger than its mean and can be a function of the mean. Therefore, for patchy distribution of counts, in terms of sampling strategy, it is advantageous to keep the mean count small by using a proper sampler, e.g. a net with a small opening for fish eggs, like CalVET net and a net with a larger opening with the capacity of filtering a larger amount of water for fish larvae, like Bongo.

The population mean can be written as a product of the mean for the positive tows and the proportion of positive tows, e.g. $E(x) = E(x|$ positive tows) * P(positives). The proportion of positive tows can be estimated from the binomial distribution, for example. Sampling net tows designed to keep the mean low entail lowering the per cent positive, with a countervailing loss of precision of the product of mean for the positive tows and per cent positive based on the binomial distribution in terms of coefficient of variation. The variance of the estimates of $E(x)$ depends on the variance of the mean given positive tows and the variance of proportion of positive tows and the covariance of the two.

At the stage of eggs, the variance was driven primarily by the mean given positive tows due to the skewed distribution of egg counts with few extremely high densities of eggs because of its patchiness. After eggs become larvae, the surviving larvae disperse and the proportion of positives increases. The counts for positive tows become less skewed and may even approach the Poisson distribution in the rare last stages of larvae. Some older larvae aggregate and the proportion of positive tows are reduced and yet the counts are low due to mortality. This will lead to the familiar U-shaped distribution of "patchiness" (Hewitt 1981, Matsuura & Hewitt 1995) which is high at the embryonic end of the population curve because the means are high, decreases as larvae grow and disperse, and becomes high at the surviving larvae end because the per cent positive is low. While the variance of egg stage is likely driven by the egg counts, the variance of older larvae is driven by the low proportion of positives. One could conclude that the minimum point on the U-curve is the point at which the sample volume is not an issue for the variance of estimates, because most likely the distribution of counts of larvae is less patchy and could be that of Poisson distribution. Thus the deviations from the Poisson, or even the negative binomial, control the precision of the embryonic stages and the low probability binomial controls the precision of the late larval stages. From a strategic point of view, the use of a single scale of sampler, even if representative and accurate, could be too large for a precise estimate of the embryonic stages and too small for a precise estimate of the larger surviving larvae. We assume this disparity continues into the juvenile stage; thus the scale of the sampler must change: we currently recommend that the eggs and yolk sac larvae be sampled with CalVET, feeding larvae to 14 mm be sampled with an oblique Bongo, and juveniles be sampled with a short tow of a commercial trawl fitted with small mesh in the cod end or a subsurface trawl.

5.2.2 Accuracy

In combination with the statistics of "counts" as controlling precision, one needs to consider factors which may introduce biases in order to obtain an accurate or unbiased estimate. Sampling in early life history of pelagic organisms is subject to the following possible biases: the duration of time spent in the stage (Hunter 1976, Zweifel & Lasker 1976), retention of small organisms near the size of the net mesh (Lo 1983), evasion or avoidance by larger mobile organisms of the entire sampling apparatus as affected by size and time of day, and finally, changes in the length caused by time since preservation, manipulation, and fixation of delicate tissues (Theilacker 1980, Zweifel & Smith 1981).

The development of embryonic stages is controlled by temperature (Ahlstrom 1948). The duration is a non-linear function of temperature owing to protracted durations in the cold. The later stages of the ichthyoplankton are likely to be influenced by the amount, quality, and distribution of esculent particles. There is an empirical summary of duration ranges in Butler *et al.* (1993). The catch appears to be proportional to the duration. This could have the obverse effect of causing an appearance of more larvae where the conditions for growth and development are worst. Also, starvation and slow growth may change the ability to evade the path of the net, but also could increase the risk to predation. It would appear from these biases that a vertically stratified tow is necessary for the interpretation of survival and growth estimates, until shown otherwise.

Retention by the meshes of the net is usually treated as an increase with size of fish from *as low as 0 to as high as 1* as a function of length or width or a multi-parameter power or logistic function. It may also be affected by the average speed of the net and the degree of pulsations of pressure in the net due to swell, adjustments of ship speed, or adjustments of wire retrieval.

Evasion of the net by larvae and juveniles could be controlled by the speed of tow, the construction of the devices preceding the net, and by light level as affected by depth and time of day. As mentioned above, the physiological condition may interact with the ability to evade the net and may range from total evasion in healthy, schooled late larvae and juveniles to total capture by randomly distributed unhealthy or recently dead larvae.

It is assumed that the objects to be sampled are "patchy"—heterogeneously distributed in time and space—at a few dominant scales. For the Pacific sardine, it was hypothesized that the principal pattern was determined by the spawning segment of a fish school which was adapted to high rates of external fertilization in the high seas (Smith 1973, Mangel & Smith 1990). At a smaller scale, there is the pattern of the spawn produced in an hour by an individual female. At larger scales, the movements of individual schools may be concentrated into school groups around mesoscale features of the production of the food and suitable patches of acceptable temperatures at persistent fronts and eddies.

An important statistical advance in the egg production survey is the allocation of quantitative samples according to the density of eggs taken from towed or mounted surface pumps (Continued Underwater Fish Egg Sampler—CUFES) (Checkley *et al.* 1997, Lo *et al.* 2001).

5.3 Egg production surveys

Two major egg production methods have been developed in recent years: the daily egg production and the daily fecundity reduction. The daily egg production method was developed for the northern anchovy *Engraulis mordax* (Girard), which spawns every few days up to a week. The daily fecundity reduction method was developed for the Dover sole *Microstomus pacificus* (Lockington), in the eastern Pacific (Hunter & Lo 1997), which spawns once per year over a protracted period of months.

The daily egg production method was developed in the late 1970s (Parker 1980, Lasker 1985). This method has been used for many pelagic fish populations around the world (Hunter & Lo 1997; Table 5.1). New methods of estimating P0 have been developed, like generalized linear models (GLM) or generalized additive models (GAM) (Borchers *et al.* 1997, Stratoudakis *et al.* 2006), and Empirical Bayesian estimates (Lo *et al.* 2005).

The CUFES is a relatively new device that provides high resolution spatial maps of fish eggs by sieving the eggs from water pumped from a fixed depth of 3 m while the ship is underway. It was first used to sample eggs of menhaden *Brevoortia tyrannus,* and pinfish *Lagodon rhomboids,* off the coast of the eastern US in the mid-1990s. Since the introduction of the CUFES (Checkley *et al.* 1997, 2000, van der Lingen *et al.* 1998, Watson *et al.,* 1999), it has been used as a routine sampler for eggs of sardine and anchovy in recent years (Lo *et al.* 2001, Stratoudakis 2002, 2003, Stratoudakis *et al.* 2006). Since 1996, it has been used for sardine and anchovy eggs off South Africa, the western coast of the US, Mexico, Peru, Chile, France, Spain, and Portugal (Checkley *et al.* 2000).

Data of sardine eggs collected with CUFES are incorporated in various ways, depending on the survey design, to estimate the daily egg production (Hill *et al.* 1999, Lo *et al.* 2001) (Figure 5.1, Plate 6). The onshore and offshore positions are responsive to the sea temperature distribution. Currently, CUFES is most useful in determining the spawning grounds and allocating CalVET net tows in routine daily egg production method (DEPM) egg surveys off California and off Portugal, Spain and France (Stratoudakis *et al.* 2006), using an adaptive allocation survey design (Lo *et al.* 2001, Pepin *et al.* 2006).

Table 5.1 Regions and species where the egg production methods have been applied.

Region	Species	Key reference
Chesapeake Bay, USA	*Anchoa mitchilli*	Rilling *et al.* 1995
Pearl Harbor, Hawaii, USA	*Encrasicholina purpurea*	Somerton *et al.* 1993
Argentina	*Engraulis anchoita*	Sanchez *et al.* 1996
Benguela Current, South Africa	*Engraulis capensis*	Shelton *et al.* 1993
Western Mediterranean Sea, Spain	*Engraulis encrasicolus*	Palomera & Pertierra 1993
Bay of Biscay, Spain	*Engraulis encrasicolus* *Engraulis encrasicolus*	Motos & Santiago 1990; Santiago & Sanz 1992; Somarakis *et al.* 2002 and 2004
Black Sea, Russia	*Engraulis encrasicolus ponticus*	Arkhipov *et al.* 1992
Southwest Korea	*Engraulis japonicus*	Kim & Lo 2001
California Current, Southern California, USA	*Engraulis mordax*	Bindman 1986
Humboldt Current, Peru	*Engraulis ringens*	Santander *et al.* 1984
North Chile	*Engraulis ringens*	Rojas & Oliva 1993
Central-south Chile, Central Aegean, and Ionian seas	*Engraulis ringens*	Cubillos *et al.* 2005
North Spain	*Sardina pilchardus*	Perez *et al.* 1989; Garcia *et al.* 1992
Portugal	*Sardina pilchardus*	Cunha *et al.* 1992
Brazil	*Sardinella brasilliensis*	Alheit 1993
California Current, Alta & Baja California, USA, and Mexico	*Sardinops sagax*	Lo *et al.* 1996
California Current, the Oregon Coast, USA	*Sardinops sagax*	Bentley *et al.* 1996
Magdalina Bay, Baja California, Mexico	*Sardinops sagax*	Torres-Villegas 1986
Humboldt Current, Peru	*Sardinops sagax*	Carrasco, personal communication
Western Australia	*Sardinops sagax*	Fletcher *et al.* 1996
North Eastern Atlantic, ICES Survey	*Scomber scombrus*	Priede & Watson 1993
Northern New Zealand	Snapper *(Pagrus auratus, Sparidae)*	Zeldis & Francis 1998
Western Australia	Snapper *(Pagrus auratus)*	Jackson & Cheng 2001
Central-south Chile	*Strangomera bentincki* (sardine)	Cubillos *et al.* 2005
Pacific	*Encrasicholina sp.*	Somerton *et al.* 1993; Milton *et al.* 1995
Oceania (Australia)	*Sardinops sagax*	Gaughan *et al.* 2004
Baltic sea	*Sprattus sprattus*	Kraus & Koster 2004
Japan	*Scomber japonicus*	Watanabe *et al.* 1999

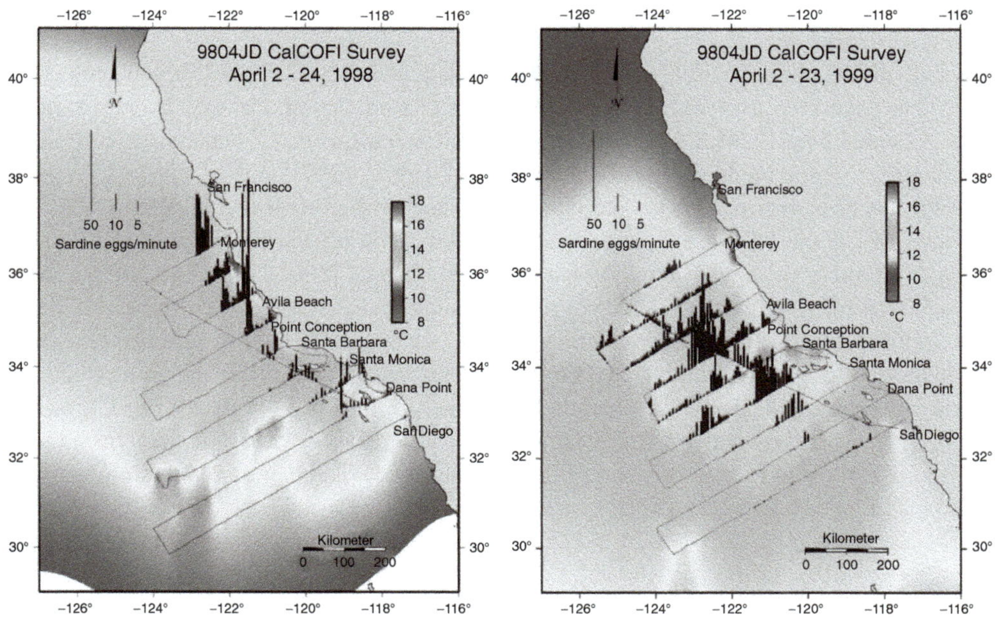

Figure 5.1 Sardine egg pattern from continuous underwater fish egg sampler (CUFES) in April 1998 and April 1999. For a color version of this figure, please see Plate 6 in the color plate section.

For fish with determined fecundity, like Dover sole (Lo *et al.* 1992, 1993, Hunter & Lo 1997) that spawn near the ocean floor at depths of 600–1500 m, a net tow, like the oblique deep Bongo (DBOBL), was necessary to collect eggs in all depths. It is also important to estimate the vertical distribution of eggs sampled because egg development rate is a function of temperature, which varies with depth. A Pareto mortality curve was used to model the Dover sole eggs.

Data collected by CUFES may be used as an index of fish abundance, to study spawning habitats, to estimate the patch size of eggs (Lo *et al.* 2001, Curtis 2004), and if converted to the numbers of eggs in the full-water-column, they may be used to estimate fish biomass (Hunter & Lo 1997, Pepin *et al.* 2006). In the DEPM, spawning biomass is calculated from the number of staged eggs taken in plankton samples and the daily fecundity of the parents from trawl samples. Since the introduction of CUFES in DEPM surveys in the mid-1990s, the survey design has been changed to efficiently use data collected by CUFES to allocate additional CalVET samples to estimate the egg production at age 0 (Hunter & Lo 1997, Lo *et al.* 2001, Stratoudakis 2002, 2003, Stratoudakis *et al.* 2006).

The three primary uses of CUFES in DEPM ichthyoplankton surveys are: (1) To determine the boundary of area of high density eggs and thus efficiently allocate CalVET net tows (Smith *et al.* 1985) in routine DEPM surveys for, in particular, sardine eggs. (2) To determine the scale and pattern of patch size of eggs (Lo *et al.* 2001, Curtis 2004). The patch size can then be used to determine the optimal distance between transect lines for design-based surveys. (3) To guide the trawl samples for estimating adult reproductive parameters. The three primary uses of CUFES in DEPM surveys and some issues associated with using CUFES off California and off Spain and Portugal in recent years are described below.

5.3.1 *Adaptive allocation sampling*

Since its introduction, CUFES (Lo *et al.* 2001, Stratoudakis 2002, 2003, Stratoudakis *et al.* 2006) has been used as a routine sampler for eggs of sardine and anchovy. Data of sardine eggs collected with CUFES have been incorporated in various ways, depending on the survey design, to estimate the daily egg production (Hill *et al.* 1999, Lo *et al.* 2001). Currently, CUFES has been most useful in determining the spawning grounds and allocating CalVET net tows in routine DEPM egg surveys off California and off Portugal, Spain and France, using an adoptive allocation survey design similar to Lo *et al.* (2001). In this design, the starting and stop rules for extra CalVET net tows are slightly different between these two areas. Off California, CalVET tows are added in areas where they were not pre-assigned if sardine egg densities in two consecutive CUFES collections were higher than 1 egg/min while off Iberia, the threshold is one egg/collection from either CUFES or CalVET. CalVET tows are stopped after two consecutive CUFES samples have an egg density less than the threshold off California and equal to zero off Iberia (Stratoudakis 2003). Daily egg production was computed based on data from the CalVET samples.

5.3.2 *The patch size of sardine eggs*

The distribution of most fish eggs is patchy. Understanding the degree of patchiness is useful in survey designs and in learning the behavior of spawning fish schools. However, the measurement of the patchiness of fish eggs requires sampling with the distance of net tows shorter than the diameter of patch size. CUFES is ideal for such a study because samples are taken continuously and the interval between collections is easily controlled. The National Marine Fisheries Service conducted its first ichthyoplankton survey using CUFES off California in 1996 (Lo *et al.* 2001). The patch size of sardine eggs and eggs of each of three age groups collected by CUFES were estimated by the ranges of variograms (Cressie 1991). The patch sizes for 1-day, 2-day, and 3-day old sardine eggs were 14.8 km (8 nmi), 18.5 km (10 nmi), and 22.2 km (12 nmi). The increase of diameter of sardine egg patches with age confirms the gradual dispersion of patches of sardine eggs with time (Smith 1973). The distance between transect lines greater than 22 km, the maximum of the patch size, prevents correlation among transect lines if a design-based survey is used. In addition to the above "larger" scale of patch sizes, studies had been conducted to examine fine scale spatial patterns of sardine and anchovy (Barange & Hampton 1997, Curtis 2004). Curtis (2004) found that off California Pacific sardine eggs showed spatial structure, while northern anchovy eggs did not on the scale of 0.75–2.5 km.

5.3.3 *Allocation of adult samples using CUFES*

CUFES data collected aboard were also used as guidance together with acoustic signals to allocate trawl samples for estimating reproductive parameters off California in 2002 and 2004 due to patchy distribution of Pacific sardine (Lo & Macewicz 2004). For example, during the 2004 DEPM survey for Pacific sardine off California, surface trawling (0–6 fathoms depth) was conducted in potential adult sardine areas as identified by the presence of eggs >1 sardine egg per minute in the CUFES collection and when acoustic signals identified fish schools. Using CUFES as a guide to allocate trawl samples is recommended in particular if the population level is low and/or the fish schools are patchy. To obtain unbiased estimates of adult reproductive parameters, it is recommended that trawl hauls be taken in the area of low egg density to ascertain the reason for low egg density: lack of adults or lack of spawning adults.

In addition to using the egg production to obtain spawning biomass, the spatial distribution of egg production can also be used to identify the spawning center. For example, off Japan, a ring net with 45 cm diameter and 0.335 mm mesh aperture has been used to collect eggs of small pelagic fish on a monthly basis since 1978. The ring net was towed vertically from 150 m depth to the surface. Monthly egg abundance was calculated from the average density of eggs in a $30'' \times 30''$ grid (latitude × longitude) and was summed up as a year egg production. The spatial distribution of egg abundance has been used to identify the center of the spawning grounds; for example, the spawning grounds for Japanese sardine moved from coastal waters shallower than 200 m depth on the continental shelf to offshore waters across the axis of the Kuroshio with the increase in the adult abundance in the 1980s (Watanabe *et al.* 1996).

5.4 Larval survival surveys

The early life stages of several fish species, e.g. northern anchovy, jack mackerel *Trachurus symmetricus* (Ayres), Pacific hake, and short belly rockfish *Sebastes jordani* (Gilbert) (Ralston *et al.* 2003), have been extensively studied as they are a link between the present adult stock and future recruitment to the adult stock. Larval density or larval production have been used as population indices for various species off California or used to directly estimate the spawning biomass because fish eggs were not available, as in live bearing rock fish, when fish eggs are difficult to identify, or when the location of fish eggs are inconvenient for the nets to reach, e.g. Pacific hake. In recent years, Pacific sardine has reappeared off the west coast of the American continent, and the yolk-sac larvae of sardine and sardine eggs at age were used to model the embryonic mortality curve (Lo *et al.* 1996). Among the larval mortality curves, two representative functions are exponential $P_t = P_h \exp(-zt)$ and Pareto functions $P_t = P_h (t+1)^{-\beta}$ where P_h is the daily larval production at hatching /10 m^2 and t is the age of larvae from hatching (days). The former assumes a constant instantaneous mortality rate used for shortbelly rockfish ($z(t) = c$) while the latter assumes the instantaneous mortality rates decrease with age ($z(t) = \beta/(t+1)$) used for Pacific hake larvae. An understanding of the accuracy and precision of estimates of daily larval production and larval mortality rates is necessary, e.g. to obtain an unbiased mortality curve.

The precision and bias of estimates of larval production were evaluated (Lo *et al.* 1989, Hauser & Sissenwine 1991, Gunderson 1993). Both Lo *etal.* (1989) and Hauser & Sissenwine (1991) showed that the bias of the estimate of mortality rate increases with the degree of change of mortality rates among larval stages, and is positively correlated with the bias of growth rates. Lo *et al.* (1989) found that a bias of the instantaneous mortality rate (IMR) and larval production would be introduced if only data from the inshore area were included because large larvae were dispersed to the offshore area. The results justify interannual comparisons of larval anchovy mortality rates when interannual variation in larval growth is less than twofold. The results also indicate that the sample size required for adequate precision of estimates of mortality rates is modest compared with that required for adequate representation of the spawning season and larval habitat.

One of the most advanced usages of larval surveys is for estimating spawning biomass based on larval production per unit area, larval production per weight of female, and estimating spawning area. Ralston *et al.* (2003) noted that in contrast to egg production surveys, where the temperature history of the embryo must be known to back-estimate the time and abundance of production, live born rockfish larvae have a birth check on the otolith providing a direct count of the age from otolith rings without reference to ambient temperature. This approach for rockfish is also made more precise because the fertilization is internal, months before spawning, so the spawning act does not require mass aggregation for spawning and fertilization as is the case for the pelagic schooling clupeoids.

5.5 Juvenile surveys

Ecological studies for small pelagic fishes, such as anchovy and sardine, have focused on eggs and early larvae collected by plankton nets (Scofield 1934, Nakai & Hattori 1962) and late juveniles and adults collected from fishery catch or mid-water trawl surveys (Mais 1974). Standard length (SL) of anchovy and sardine collected by the plankton nets and the mid-water trawl nets ranged from 5 to 20 mm and from 50 to 160 mm, respectively. The late larvae and early juveniles of 15 to 60 mm have been missing links in studies on recruitment systems in anchovy and sardine due to their swimming speed and ability to escape from predator attacks (5–35 mm SL). In addition, fish re-school in the late larval and early juvenile stages (30–70 mm SL) in the captive studies (van Olst & Hunter 1970, Hunter 1972, Folkvord & Hunter 1986). Therefore late larvae and early juveniles are the most difficult stages to capture by regular nets. Frame trawl nets were developed to catch late larvae and early juveniles off California and have been used off South Africa while off Japan, subsurface trawls have been used to collect juveniles successfully (Methot 1986, Aoki *et al.* 2000, van der Lingen & Huggett 2003, Oozeki *et al.* 2005a).

Frame trawl systems are exemplified by the Methot trawl (Methot-Isaacs-Kidd trawl), which is a micronekton net designed to sample pelagic larval and juvenile fish that avoid plankton nets and pass through the mesh of large mid-water trawls. The fish length collectable by this net is in the range of 15–60 mm for northern anchovy and overlaps the larvae collected by the plankton nets and the juveniles collected by the trawl nets. Body length of anchovy and sardine collected by the frame trawls off Japan, however, was only up to *c.* 35 mm SL, corresponding to the larval stage.

Subsurface trawl nets have been used to collect late larvae and early juveniles (20–100 mm SL) of Japanese anchovy *Engraulis japonicus* (Temminck & Schlegel) and Japanese sardine in May and June in the eastern offshore waters off northern Japan—the Kuroshio-Oyashio transition region (Figure 5.2, Plate 7) (Takahashi *et al.* 2001, Nishida *et al.* 2005, Yatsu *et al.* 2005, Takahashi *et al.* 2008). The juvenile survey has been conducted for forecasting recruitment abundance of small pelagic fishes by the Fisheries Research Agency, Japan, since 1996. The trawl net has a 25 × 25 m opening, various-sized mesh in the body, resembling commercial nets, and 1 cm stretch mesh at the cod end. Although collecting samples with large commercial-type trawl nets is inferior in quantitative accuracy compared with plankton nets and frame trawl nets, relative catch-per-unit-effort is available for quantitative examination if the sampling protocol and net specifications are fixed.

Growth and development during late larval and early juvenile stages in the Kuroshio-Oyashio transition region have been considered to be key factors for regulating recruitment abundance to the adult stocks of Japanese anchovy and sardine (Watanabe *et al.* 1995,

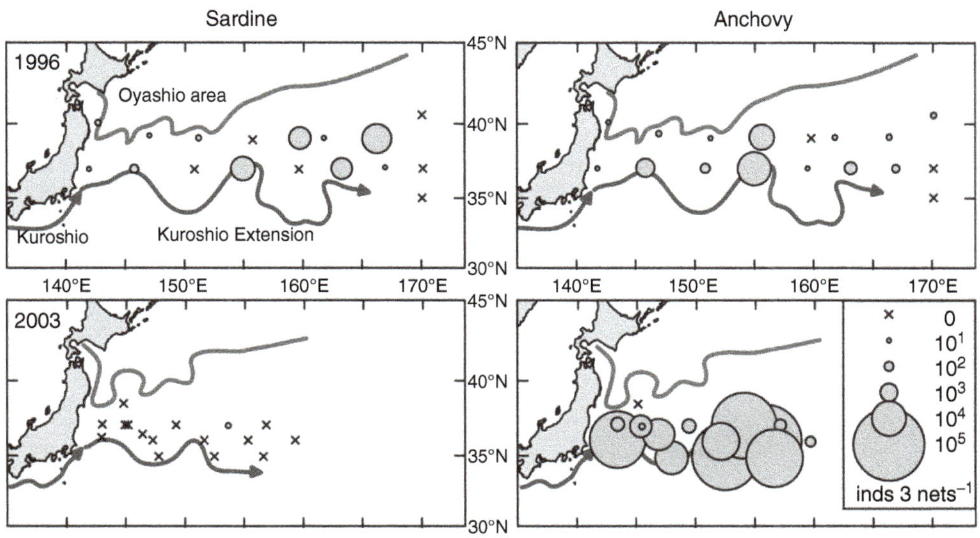

Figure 5.2 Number of late larval and juvenile sardine and anchovy (20–100 mm standard length) in the Kuroshio-Oyashio transition region in May-June 1996 and 2003. One circle is the number offish collected in three trawl hauls/night. For a color version of this figure, please see Plate 7 in the color plate section.

Takahashi & Watanabe 2004, Takahasi *et al.* 2008). For the juvenile survey, east-west survey transects were set between the Kuroshio Extension and the Oyashio, 35–40°N (Yasuda 2003), from late spring to early summer as the peak spawning time is in February and March based on egg samples collected in the 1980s (Watanabe *et al.* 1996). The trawl net was towed for 30 min with a ship speed of 3–4 knots. Three trawl hauls were made per night in the stratum shallower than 25 m and in total 50–60 trawl hauls were made during the survey.

The density of sardine juveniles in the Kuroshio-Oyashio transition region was similar to that of anchovy in 1996, but extremely low in 2003 (Figure 5.2, Plate 7). Sardine biomass in the Pacific waters off Japan attained 20 million metric tonnes in 1987, and then dramatically decreased to 800 thousand tonnes in 1996 and 100 thousand tonnes in 2003 (Nishida *et al.* 2005). Meanwhile, anchovy biomass increased from 350 thousand tonnes in 1996 to 2.6 million tonnes in 2003 (Oozeki *et al.* 2005b). Density of anchovy juveniles in the transition region increased with the stock biomass.

Survey for pre-spawning anchovy off South African waters can establish the lifelong fecundity of the cohort (Figure 5.3). Anchovy larvae smaller than 34 mm in total length appeared offshore of the continental shelf (dotted lines in the figure), while juveniles larger than 35 mm in total length were distributed in the inner continental shelf of the west coast of South Africa in March 1998. The cross-shelf distribution pattern annually was consistent. Anchovy eggs were found from September to February with a peak in December (summer) during the period 1995–2001 (van der Lingen & Huggett 2003).

Date of hatching of juveniles can be back-calculated from the age of juveniles. Distributions of hatching date of juveniles and pre-recruits of northern anchovy (55–120 mm SL) collected from the fishery catch from October to December 1978 and collected by mid-water trawl in November 1979 off California were estimated from otolith daily increments and were compared with early larval production estimated based on samples collected by Bongo net

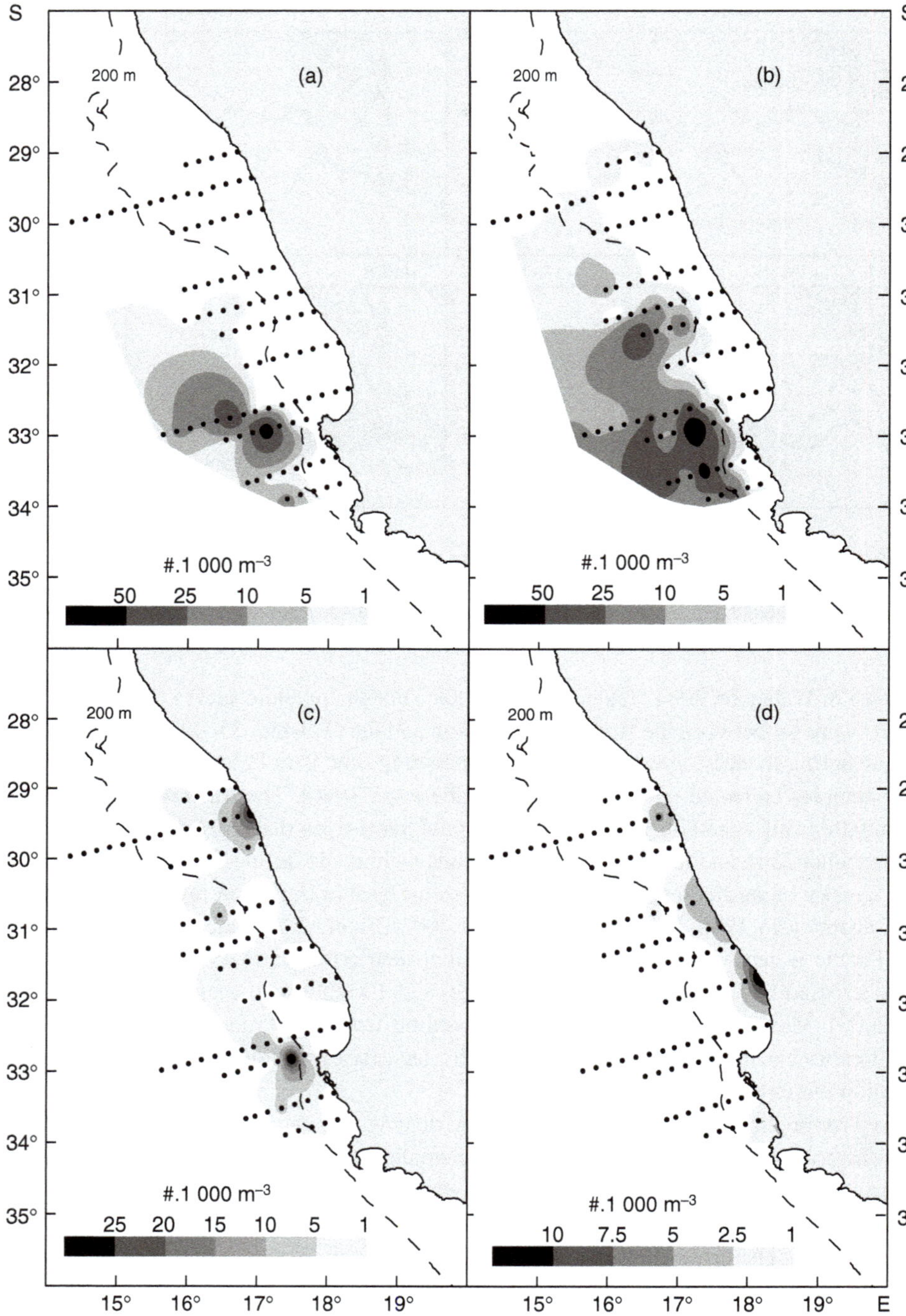

Figure 5.3 Positions of late larval and juvenile anchovy in Autumn off South Africa (March 1998). (a) Larvae less than 20 mm. (b) Larvae 20–34 mm. (c) Juveniles 35–50 mm. (d) Juveniles greater than 50 mm. Lengths are total lengths and the scale of shading used decreases with size. Reproduced from van der Lingen & Huggett (2003), with permission of the Institute of Marine Research.

during the CalCOFI cruise from December 1977 to May 1979 (Methot 1983). Hatching dates of the juveniles collected in autumn were mainly in March and April, while larval production showed that main spawning occurred from January to March. This implies that larvae hatched during the second half of the spawning season had a higher probability of surviving to juvenile stage.

Juvenile surveys can establish a link between late larvae and new recruits. Growth and development rates through the late larval and early juvenile stages have been considered to be positively related to survival rates and recruitment abundance in many marine fishes (Houde 1987, 1989, Anderson 1988). Lo *et al.* (1989) examined, based on a numerical model, effects of stage-specific vital rates on population growth for northern anchovy and Pacific sardine in the California Current region through egg to adult stages and demonstrated that growth rates during larval stages are determinants for population growth in both northern anchovy and Pacific sardine. Butler (1989) examined growth trajectories of northern anchovy based on otolith daily increments between 1980 and 1984, including El Nino events in 1982–1983 when water temperature increased and zooplankton biomass decreased, and found that growth rates during the juvenile stage declined in the El Nino years compared with other years, while no differences were found in the larval growth rates. Butler (1989) suggested that juvenile growth was reduced due to food decline during El Nino events.

In the western North Pacific, growth trajectories of juvenile anchovy and sardine in the Kuroshio-Oyashio transition region can reveal mechanisms of alternate fluctuations between anchovy and sardine populations. Egg abundance of Japanese sardine in the Pacific coastal waters off southern and central Japan increased with the abundance of spawning adults in the early 1980s, when the population level was high (Watanabe *et al.* 1996). However, recruitment abundances dramatically decreased after 1988 (Watanabe *et al.* 1995), even though egg abundance increased in the late 1980s. This indicates that survival rates in the late larval and juvenile stages have become extremely low since 1988 and the recruitment failure caused collapse of the sardine population. Most of the eggs spawned near the Kuroshio axis were transported northeastward to the Kuroshio Extension region within 2 weeks (Heath *et al.* 1998). Watanabe (2002), therefore, hypothesized that growth and development rates regulate survival rates during late larval and early juvenile stages and subsequent recruitment in the Kuroshio-Oyashio transition region. Takahashi *et al.* (2008) examined relationships between growth rate of juveniles for Japanese sardine collected in the transition region and recruitment abundance of age-0 fish calculated by virtual population analysis between 1996 and 2003 and found that growth rate in the early juvenile stage positively correlated with the recruitment abundance.Takahashi & Watanabe (2004) compared otolith increments during early life stages for Japanese anchovy between recruited survivors and the original juveniles and found that early juveniles with faster growth rates had higher probability of successful recruitment to the adult stock in the transition region. Growth rates of sardine juveniles decreased from 1996 to 2002, while those of anchovy juveniles increased with the decrease of sardine growth rate in the transition region (Takahashi *et al.*, unpublished). Northward expansion of the Kuroshio Extension waters into the transition region simultaneously changed the local environment to be favorable for anchovy, but not for sardine between 1996 and 2002. Increased temperature with rich food enhanced growth and survival for anchovy. In contrast, declined food availability diminished growth and survival for sardine. Similar changes in water structures in the transition region between 1996 and 2002 were observed in the late 1980s, when the sardine population began to collapse and the anchovy population started to increase.

5.6 Management

For Pacific sardine off California, various stock assessment models have been used, among which were an age-structured stock assessment model, CANSAR-TAM (Catch-at-age ANalysis for SARdine—Two Area Model) (Deriso *et al.* 1996, Hill *et al.* 1999) in 1994–2003 and an age structure assessment program (ASAP) in 2004–2007. The "Stock Synthesis 2" (SS2) is being considered for sardine assessment (Methot 2005). Among fishery-independent time series from ichthyoplankton surveys off central and southern California are: (1) proportion-positive stations of sardine eggs and larvae from the CalCOFI survey; (2) DEPM spawning biomass (Mt) (Lo *et al.* 1996); and (3) spawning area (nmi^2) from CalCOFI and DEPM survey data (Barnes *et al.* 1997). Items number (2) and (3) are informative only when the population is at a low level. Thus in recent years, they were not used as inputs in the stock assessment.

The proportion of positive collections of sardine egg and larvae from Bongo samples collected in January-April CalCOFI surveys between 1984 and 2003 was estimated by a generalized additive model (GAM) with logistic regression (Lo & Methot 1989, Deriso *et al.* 1996, Barnes *et al.* 1997, Hill *et al.* 1998). The independent variables included year, month, standard Ca1COFI line, and inshore/offshore location (Barnes *et al.* 1997). These indices tend to "saturate" as sardine abundance increases (Hill *et al.* 1998) and the degree of saturation in the index was incorporated in the assessment models, like CANSAR-TAM.

Spawning areas are calculated for 1983–2003 using all available geographic information on sardine egg distribution from Ca1COFI and DEPM cruises collected by Bongo nets, Ca1VET nets for years from 1983, and CUFES from 1996 using Geographic Information System software (ArcViewT, GIS) (Barnes *et al.* 1997, Hill *et al.* 1998). The area of each polygon drawn around major egg concentrations was calculated and the sum of polygon areas gives the total spawning area (nmi^2) per calendar year. Predicted values fit well to the observed data and residuals were not serially correlated.

For Japanese sardine in the Pacific coastal waters off Japan, stock assessment has been conducted by cohort analysis based on commercial catch-at-age data. The outcome has been tuned by fishery-independent data, such as egg production in the Pacific waters off southern and central Japan, density of juveniles on the nursery grounds in the Kuroshio-Oyashio transition region, and density of immature fish on the feeding and over-wintering grounds, except for 2005 (Nishida *et al.* 2005).

The density of early juveniles in spring in the Kuroshio-Oyashio transition region has been used as an appropriate index for forecasting recruitment abundance to the adult stocks since 1996 (Nishida *et al.* 2005). Because of the warm water intrusion of the Kuroshio from the south and a cold water intrusion of the Oyashio from the north (Figure 5.2, Plate 7), the Kuroshio-Oyashio transition region is dominated by mesoscale eddies, streamers, and complex thermohaline fronts. In order to estimate actual distribution of densities of the early juveniles, the sea surface temperature (SST) specific densities of the juveniles are weighted by SST zones in the survey area.

Immature sardines have been collected using a gill net in early summer and fall on the feeding grounds and using a mid-water trawl net in winter in the over-wintering area off central and northern Japan since 2001. Acoustics also has been used for estimating the standing stocks in addition to the trawl net for the over-wintering immature fish.

Pacific hake is a migratory species. Adults spawn off California and Baja California, Mexico in the winter and migrate north as far as British Columbia, Canada in the summer for feeding. In

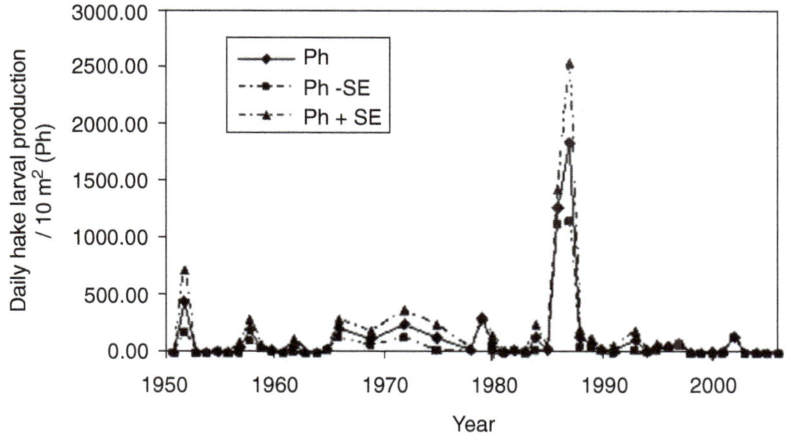

Figure 5.4 Hake larval production/10 m^2 off the area from San Diego to San Francisco, US, in January-April from 1951 to 2006.

the autumn, adult hake move back south (Hollowed 1992). Time series of hake larval density were reported by Hollowed (1992) for the period 1961–1985 based on data collected from the CalCOFI survey. An updated time series from 1951 to 2006 was constructed for the area from San Diego to San Francisco (Figure 5.4) (Lo 2007). Similar to the estimation procedures used for anchovy larval production, all larval abundance data were adjusted to conform to the following standard conditions: no extrusion (Lenarz 1972, Zweifel & Smith 1981, Lo 1983), no day-night difference in avoidance (Hewitt & Methot 1982), and a constant water volume filtered per unit depth (Ahlstrom 1948). The density of larvae divided by the duration in each length category yields estimated larval production/10 m^2 (Hewitt & Methot 1982). This time series, to be updated each year, is a candidate of fishery independent population index to be incorporated into the stock assessment of Pacific hake off the northern American continent.

5.7 Remote sensing

Remote sensing is useful in delimiting the habitat or describing the scale and pattern of fisheries resources. For example, night visual surveys of luminescence may reveal the schools and school behavior of near surface fishes. Aggregations may be detected below the surface by echo sounders and near the surface by sonar or LIDAR (Light Detection and Ranging) surveys (Lo *et al.* 2000; Figure 5.5). Where temperature or chlorophyll limits are known, fish distribution limits may be described by satellite sensing. The satellite image is synoptic over the area and the airborne sensors are continuous along tracks and these allow the scale of the distribution to be described at high speed relative to ship mounted remote sensors.

5.8 Species assemblages and water masses

The growing emphasis on ecosystem analysis in the management of marine fisheries can be substantially aided by ichthyoplankton and juvenile fish surveys. Boundaries and proximities

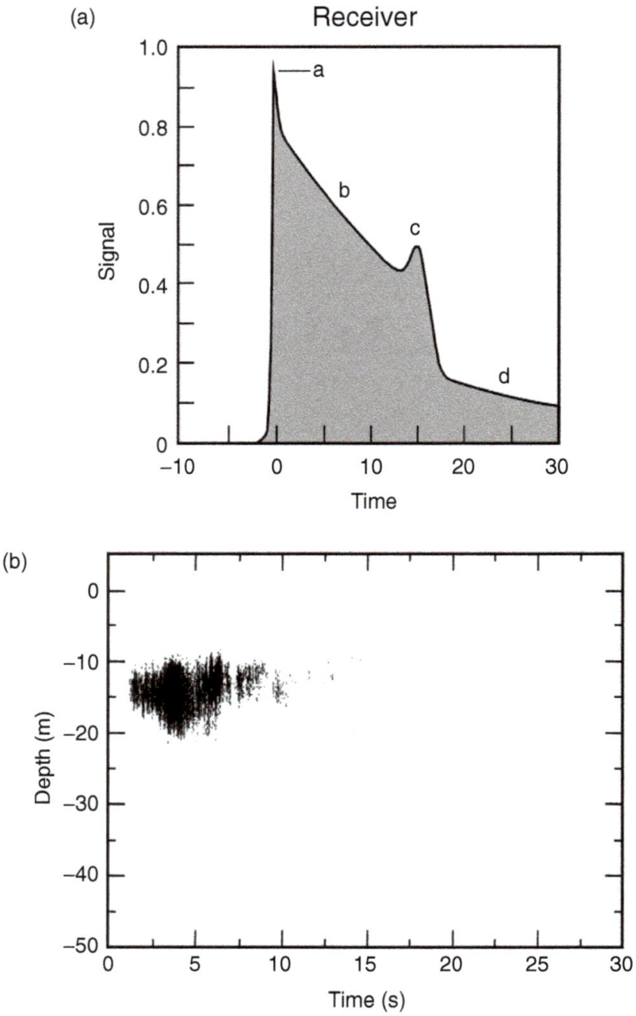

Figure 5.5 (a) The amplitude return of a single pulse, showing the form of a single returning pulse where "a" is the reflection from the water surface, "b" is the decrease in signal as it attenuates with water depth, "c" is the reflection from a fish school, and "d" is the "shadow" cast by the school. (b) Successive LIDAR images of a large Pacific sardine school detected during a cooperative validation study (Hunter & Churnside 1995).

among species can be described (Smith & Moser 2003, Aceves-Medina *et al.* 2004) and in particular, temporal changes in the larval populations of fished and unfished species can be used to detect environmental shifts that may accompany population depletion and natural variations (Hsieh *et al.* 2005).

5.9 Summary

Surveys and identification of fish larvae have led to great advances in detailing the oceanic distribution of the spawning areas of fishes over the past century. Increasing life history knowledge from laboratory rearing has allowed the institution of productivity studies of a few species and this has subsequently been applied to obtaining instantaneous, absolute estimates of spawning biomass by fishery independent means. As such, these techniques can be used to calibrate acoustic and population dynamics estimates of population biomass. Table 5.1 lists the areas of the ocean in which spawning production estimation is increasingly used for fisheries management.

There are strong barriers to precise quantification of spawning biomass which originate in the seasonality, schooling, and batch structure of spawning. Added to this is the expense of small-scale sampling over wide ocean areas. CUFES at the surface has furnished valuable guidance for regional allocation of samples. Yet, personnel costs for laboratory analysis and data management are significant.

To the cost of estimation of egg production per unit area of the sea is added the cost of obtaining adult samples for a simultaneous estimate of egg production per unit of adult biomass. The age composition, length specific weight and fecundity, and histologically determined batch fecundity and ovarian activity are further tasks to be completed routinely. The CUFES survey as well as acoustic or aerial remote surveys of juveniles and adults can be further used for allocating the required direct samples by analogs to commercial trawls.

While the costs of spawning biomass estimation are high, the procedures are rich in physiological, behavior, and population dynamics information, which will need to be applied to the optimal management of each species and the ecosystem management of all the effects and demands of human use.

References

Aceves-Medina, G., Jimez-Rosenberg, S.P.A., Hinojosa-Medina, A., Funes-Rodriguez, R., Saldierna-Martinez, R.J. & Smith, P.E. (2004) Fish larvae assemblages in the Gulf of California. *Journal of Fish Biology*, **65**, 832–47.

Ahlstrom, E. (1948) A record of pilchard eggs and larvae collected during surveys made from 1939 to 1941. *US Fisheries and Wildlife Service Special Scientific Report*, **54**. 82 pp.

Alheit, J. (1993) Use of the daily egg production method for estimating biomass of clupeoid fishes: a review and evaluation. *Bulletin of Marine Science*, **53**, 750–67.

Anderson, J.T. (1988) A review of size-dependent survival during pre-recruit stages of fishes in relation to recruitment. *Journal of Northwest Atlantic Fisheries Science*, **8**, 55–66.

Aoki, I., Miura, T., Imai, N. & Komatsu, T. (2000) Sampling large larvae and juveniles of pelagic fish with a frame-type midwater trawl. *Nippon Suisan Gakkaishi*, **66**, 10–17.

Arkhipov, A.G., Koval'chuk, L.A., Chashchin, A.K. & Yankauskas, V. (1992) Statisticheskij analiz mnogoletnikh nablyudenij raspredeleniya anchousa *Engraulis encrasicolus ponticus* v Chernom more. *Voprosy Ikhtiologii*, **32**, 176–82.

Armstrong, M., Shelton, P., Hampton, I., Jolly, G. & Melo, Y. (1988) Egg production estimates of anchovy biomass in the southern Benguela. *California Cooperative Oceanic Fisheries Investigation Report*, **29**,137–57.

Barange, M. & Hampton, I. (1997) Spatial structure of co-occurring anchovy and sardine populations from acoustic data: implications for survey design. *Fisheries Oceanography*, **6**, 94–108.

Barnes, J.T., Yaremko, M., Jacobson, L., Lo, N.C.H. & Stehly, J. (1997) Status of the Pacific sardine *(Sardinops sagax)* resource in 1996. NOAA-NMFS-SWFSC-237.

Bentley, P.J., Emmett, R.L., Lo, N.C.H. & Moser, H.G. (1996) Egg production of Pacific sardine *(Sardinops sagax)* off Oregon in 1994. *California Cooperative Oceanic Fisheries Investigation Report*, **37**, 193–200.

Bindman, A.G. (1986) The 1985 spawning biomass of the northern anchovy. *California Cooperative Oceanic Fisheries Investigation Report*, **27**, 16–24.

Borchers, D.L., Buckland, S.T., Priede, I.G. & Almandi, S. (1997) Improving the precision of the daily egg production method using generalized additive models. *Canadian Journal of Fisheries and Aquatic Sciences*, **54**, 2727–42.

Butler, J.L. (1989) Growth during the larval and juvenile stages of the northern anchovy, *Engraulis mordax*, in the California Current during 1980–84. *Fisheries Bulletin (US)*, **87**, 645–52.

Butler, J.L., Smith, P.E. & Lo, N.C.H. (1993) The effect for natural variability of life-history parameters on anchovy and sardine population growth. *California Cooperative Oceanic Fisheries Investigation Report*, **34**, 104–11.

Checkley, D.M. Jr, Dotson, R.C. & Griffith, D.A. (2000) Continuous, underway sampling of eggs of Northern anchovy *(Engraulis mordax)* and Pacific Sardine *(Sardinops sagax)* in Spring 1996 and 1997 off Southern and Central California. Deep Sea Research, the second volume of a special series on the California Current. *Deep-Sea Research II*, **47**, 1139–55.

Checkley, D.M. Jr, Ortner, P.B., Settle, L.R. & Cummmings, S.R. (1997) A continuous, underway fish egg sampler. *Fisheries Oceanography*, **6**, 58–73.

Cressie, N. (1991) *Statistics for Spatial Data.* John Wiley & Sons, New York. 900 pp.

Cubillos, L.A., Ruiz, R.P., Ruiz, P., Nunez, S.P., Claramunt, G., Oliva, J., Oyarzun, C., Gacitua, C.S., Sepulveda, A. & Castro, L. (2005) Spawning daily egg production and spawning biomass of common sardine, *Strangomera bentincki,* and anchoveta, *Engraulis ringens,* off central South Chile in 2002. Globec Report No 22. Report of the SPACC Meeting on small pelagic fish spawning habitat dynamics and the daily egg production method (DEPM).

Cunha, M.E., Figueiredo, I. & Farinha, A. (1992) Estimation of sardine spawning biomass off Portugal by the Daily Egg Production Method. *Boletin del Instituto Espafiol de Oceanografia*, **8**, 139–53.

Curtis, K.A. (2004) Fine scale spatial pattern of Pacific sardine *(Sardinops sagax)* and northern anchovy *(Engraulis mordax)* eggs. *Fisheries Oceanography*, **13**, 239–54.

Deriso, R.B., Barnes, J.T., Jacobson, L.D. & Arenas, P.R. (1996) Catch-at-age analysis for Pacific sardine *(Sardinops sagax),* 1983–1995. *California Cooperative Oceanic Fisheries Investigations Report*, **37**, 175–87.

Fletcher, W.J., Lo, N.C.H., Hayes, E.A., Tregonning, R.J. & Blight, S.J. (1996) Use of the daily egg production method to estimate the stock size of Western Australian sardines *(Sardinops sagax). Marine and Freshwater Research*, **47**, 819–25.

Folkvord, A. & Hunter, J.R. (1986) Size-specific vulnerability of northern anchovy, *Engraulis mordax,* larvae to predation by fishes. *Fisheries Bulletin*, **84**, 859–69.

Garcia, A., Perez, N., Lo, N.C.H., Lago de Lanzos, A. & Sola, A. (1992) The Egg Production Method applied to the spawning biomass estimation of sardine, *Sardina pilchardus* (Walb.) on the North Atlantic Spanish coast. *Boletin del Instituto Espanol de Oceanografia*, **8**, 123–38.

Gaughan, D.J., Leary, T.I, Mitchell, R.W. & Wright, I.W. (2004) A sudden collapse in distribution of pacific sardine *(Sardinops sagax)* off southwestern Australia enables an objective re-assessment of biomass estimates. *Fisheries Bulletin*, **102**, 617–33.

Gunderson, D.R. (1993) *Surveys Of Fisheries Resources.* John Wiley & Sons, New York. 248 pp.

Hampton, I. (1996) Acoustic and egg-production estimates of South African anchovy biomass over a decade: comparisons, accuracy, and utility. *ICES Journal of Marine Science*, **53**, 493–500.

Hauser, J.W. & Sissenwine, M.P. (1991) The uncertainty in estimates of the production of larval fish derived from samples of larval abundance. *ICES Journal of Marine Science*, **48**, 23–32.

Heath, M., Zenitani, H., Watanabe, Y., Kimura, R. & Ishida, M. (1998) Modelling the dispersal of larval Japanese sardine, *Sardinops melanostictus,* by the Kuroshio Current in 1993 and 1994. *Fisheries Oceanography*, **7**, 335–46.

Hewitt, R. (1981) The early life history of fish: recent studies. *Rapports et Proces-verbaux des Reunions du Conseil International pour l'Exploration de la Mer*, **178**, 229–36.

Hewitt, R.P. & Methot, R.D. Jr (1982) Distribution and mortality of northern anchovy larvae in 1978 and 1979. *California Cooperative Oceanic Fisheries Investigation Report*, **23**, 226–45.

Hill, K.T., Jacobson, L.D., Lo, N.C.H., Yaremko, M. & Dege, M. (1999) Stock assessment of Pacific sardine for 1998 with management recommendations for 1999. Marine Region, Admin. Report 99-4. California Department of Fish and Game.

Hill, K.T., Yaremko, M., Jacobson, L.D., Lo, N.C.H. & Hanan, D.A. (1998) Stock assessment and management recommendations for Pacific sardine in 1997. Marine Region, Admin. Report 98-5. California Department of Fish and Game.

Hollowed, A.B. (1992) Spatial and temporal distributions of Pacific hake, *Merlucciusproductus,* larvae and estimates of survival during early life stages. *California Cooperative Oceanic Fisheries Investigations Report*, **33**, 100–23.

Houde, E.D. (1987) Fish early life dynamics and recruitment variability. *American Fisheries Society Symposium*, **2**, 17–29.

Houde, E.D. (1989) Subtleties and episodes in the early life of fishes. *Journal of Fish Biology*, **35**(Suppl. A), 29–38.

Hsieh, C., Reiss, C., Watson, W., Allen, M.J., Hunter, J.R., Lea, R.N., Rosenblatt, R.H., Smith, P.E. & Sugihara, G. (2005) A comparison of long-term trends and variability in populations of larvae of exploited and unexploited fishes in the southern California region: a community approach. *Progress in Oceanography*, **67**, 160–85.

Hunter, J.R. (1972) Swimming and feeding behavior of larval anchovy *Engraulis mordax. Fisheries Bulletin (US)*, **70**, 821–38.

Hunter, J.R. (1976) Culture and growth of Northern anchovy, *Engraulis mordax,* larvae. *Fisheries Bulletin (US)*, **74**, 81–8.

Hunter, J.R. & Churnside, J.H. (1995) Airborne fishery assessment technology: a NOAA workshop report. National Marine Fisheries Service, Southwest Fisheries Science Center Administration Report LJ-95-02. 33 pp.

Hunter, J.R. & Lo, N.C.H. (1997) The daily egg production methods of biomass estimation: some problems and potential improvements. Ozeanografika, Boletin de la Sociedad de Oceanografica de Gipuzkoa, Medalla de Oro de la Ciudad de San Sebastian, No 2:9-40.

Jackson, G. & Cheng, Y.W. (2001) Parameter estimation with egg production surveys to estimate snapper, *Pagrus auratus,* biomass in Shark Bay, Western Australia. *Journal of Agricultural Biological and Environmental Statistics*, **6**, 243–57.

Johnson, N.L. & Kotz, S. (1969) *Discrete Distributions.* John Wiley & Sons, New York. 328 pp.

Jolly, G.M. & Hampton, I. (1990) A stratified random transect design for acoustic surveys of fish stocks. *Canadian Journal of Fisheries and Aquatic Sciences*, **47**, 1282–91.

Kim, J. & Lo, N.C.H. (2001) Temporal variation of seasonality of egg production and the spawning biomass of pacific anchovy, *Engraulis japonicus,* in the southern waters of Korea in 1983–1994. *Fisheries Oceanography*, **10**, 297–310.

Kraus, G. & Koster, F.W. (2004) Estimating Baltic sprat *(Sprattus sprattus balticus S.)* population sizes from egg production. *Fisheries Research*, **69**, 313–29.

Lasker, R (Ed.) (1985) An egg production method for estimating spawning biomass of pelagic fish: application to the northern anchovy *(Engraulis mordax).* US Department of Commerce, National Oceanic and Atmospheric Administration Technical Report NMFS 36. 99 pp.

Lenarz, W.H. (1972) Mesh retention of *Sardinops caerulea* and *Engraulis mordax* by plankton nets. *Fisheries Bulletin (US)*, **70**, 839–48.

Lo, N.C.H. (1983) Re-examination of three parameters associated with anchovy egg and larval abundance: temperature dependent incubation time, yolk-sac growth rate and egg and larval retention in mesh nets. NOAA-TM-NMFS-SWFC-31. NOAA Technical Memorandum NMFS, US Department of Commerce, National Oceanic and Atmospheric Administration, National Marine Fisheries Service, Southwest Fisheries Center. 33 pp.

Lo, N.C.H. (2007) Daily larval production of Pacific hake *(Merluccius productus)* off California in 1951–2006. *California Cooperative Oceanic Fisheries Investigations Report*, **48**, 147–64.

Lo, N.C.H., Green Ruiz, Y.A., Cervantes, M.J, Moser, H.G. & Lynn, R.J. (1996) Egg production and spawning biomass of Pacific sardine *(Sardinops sagax)* in 1994, determined by the daily egg production method. *California Cooperative Oceanic Fisheries Investigations Report*, **37**, 160–74.

Lo, N.C.H., Griffith, D. & Hunter, J.R. (1997) Using a restricted adaptive cluster sampling to estimate Pacific hake larval abundance. *California Cooperative Oceanic Fisheries Investigations Report*, **38**, 103–13.

Lo, N.C.H., Hunter, J.R. & Charter, R. (2001) Use of a continuous egg sampler for ichthyoplankton surveys: application to the estimation of daily egg production of Pacific sardine *(Sardinops sagax)off* California. *Fisheries Bulletin (US)*, **99**, 554–71.

Lo, N.C.H., Hunter, J.R. & Churnside, J.H. (2000) Modeling statistical performance of an airborne lidar survey system for anchovy. *Fisheries Bulletin (US)*, **98**, 264–82.

Lo, N.C.H., Hunter, J.R. & Hewitt, R.P. (1989) Precision and bias of estimates of larval mortality. *Fisheries Bulletin (US)*, **87**, 399–416.

Lo, N.C.H., Hunter, J.R., Moser, H.G. & Smith, P.E. (1993) A daily fecundity reduction method of biomass estimation with application to Dover sole *Microstomus pacificus*. *Bulletin of Marine Science*, **53**, 842–63.

Lo, N.C.H., Hunter, J.R., Moser, H.G., Smith, P.E. & Methot, R.D. (1992) The daily fecundity reduction method: a new procedure for estimating adult fish biomass. *ICES Journal of Marine Science*, **49**, 209–15.

Lo, N.C.H. & Macewicz, B. (2004) Spawning biomass of Pacific sardine *(Sardinops sagax)* off California in 2004 and 1995. SWFSC Administration Report, La Jolla, LJ-04-08. 30 pp.

Lo, N.C.H., Macewicz, B.J. & Griffith, D.A. (2005) Spawning biomass of Pacific sardine *(Sardinops sagax),* from 1994–2004 off California. *California Cooperative Oceanic Fisheries Investigation Report*, **46**, 93–112.

Lo, N.C.H. & Methot, R.D. (1989) Spawning biomass of the northern anchovy in 1988. *California Cooperative Oceanic Fisheries Investigations Report*, **30**, 18–31.

Mais, K. (1974) Pelagic fish survey in the California Current. *California Department of Fish and Game Fisheries Bulletin*, **162**, 1–79.

Mangel, M. & Smith, P.E. (1990) Presence-absence sampling for fisheries management. *Canadian Journal of Fisheries and Aquatic Science*, **47**, 1875–87.

Matsuura, Y. & Hewitt, R. (1995) Changes in the spatial patchiness of Pacific mackerel, *Scomber japonicus*, larvae with increasing age and size. *Fisheries Bulletin (US)*, **93**, 172–8.

Methot, R.D. (1983) Seasonal variation in survival of larval northern anchovy, *Engraulis mordax,* estimated from the age distribution of juveniles. *Fisheries Bulletin (US)*, **81**, 741–50.

Methot, R.D. (1986) Frame trawl for sampling pelagic juvenile fish. *California Cooperative Oceanic Fisheries Investigation Report*, **27**, 267–78.

Methot, R. (2005) User manual for the assessmentprogram stocksynthesis 2 (SS2) Model Version 1.19. April. 47 pp.

Milton, D., Blaber, S. & Rawlinson, N. (1995) Fecundity and egg production of four species of short-lived clupeoid from Solomon Islands, tropical South Pacific. *ICES Journal of Marine Science* **52**, 111–25.

Motos, L. & Santiago, J. (1990) An egg production estimate of biomass of the Bay of Biscay anchovy *(Engraulis encrasicolus* L.) in 1989. ICES CM 1990/H,26.

Nakai, Z. & Hattori, S. (1962) Quantitative distribution of eggs and larvae of the Japanese sardine by year, 1949 through 1951. *Bulletin of the Tokai Regional Fisheries Research Laboratory*, **9**, 23–60.

Nishida, H., Yatsu, A., Ishida, M., Noto, M. & Suda, M. (2005) Stock assessment and evaluation for the Japanese sardine (fiscal year 2003). In: Fisheries Agency and Fisheries Research Agency of Japan (Eds) *Marine Fisheries Stock Assessment and Evaluation for Japanese Waters (Fiscal Year 2003).* pp. 11–45.

Oozeki, Y., Fu, F., Kubota, H., Sugisaki, H. & Kimura, R. (2005a) Newly designed quantitative frame trawl for sampling larval and juvenile pelagic fish. *Fisheries Science*, **70**, 223–32.

Oozeki, Y., Kubota, H., Takasuka, A., Akamine, T. & Shimizu, A. (2005b) Stock assessment and evaluation for the Japanese anchovy (fiscal year 2004). In: Fisheries Agency and Fisheries Research Agency of Japan (Eds) *Marine Fisheries Stock Assessment and Evaluation for Japanese Waters (Fiscal Year 2004)*. pp. 604–28.

Palomera, I. & Pertierra, J.P. (1993) Anchovy spawning biomass estimate by the daily egg production method in 1990 in the Western Mediterranean Sea. *Scientia Marina*, **57**, 243–51.

Parker, K. (1980) A direct method for estimating northern anchovy, *Engraulis mordax,* spawning biomass. *Fisheries Bulletin (US)*, **84**, 541–4.

Perez, N., Garcia, A., Lo, N.C.H. & Franco, C. (1989) The egg production method applied to the spawning biomass estimation of sardine *(Sardinapilchardus,* Walb.) on the North-Atlantic Spanish coasts. ICES CM 1989/H:23.

Pepin, P., Curtis, K.A., Snelgrove, P.V., de Young, B. & Helbig, J.A. (2006) Optimal estimation of catch by the continuous underway fish egg sampler based on a model of the vertical distribution of American plaice *(Hippoglossoides plaessoides)* eggs. *ICES Journal of Marine Science*, **64**, 18–30.

Priede, I.G. & Watson, J.J. (1993) An evaluation of the daily egg production method for estimating biomass of Atlantic mackerel *(Scomber scombrus)*. *Bulletin of Marine Science*, **53**, 891–911.

Ralston, S., Bence, J.R., Eldridge, M.B. & Lenarz, W.H. (2003) An approach to estimating rockfish biomass based on larval production, with application to *Sebastes jordani*. *Fishery Bulletin*, **101**, 129–46.

Rilling, G.C., Houde, E.D. & Trice, T.M. (1995) Temporal and spatial variability in the distribution and dynamics of bay anchovy *(Anchoa mitchilli)* early life stages in the Chesapeake Bay. ICES CM 1995/L:24 Ref. H. Poster. 10 pp.

Rojas, O. & Oliva, J. (1993) Evaluation del Stock Desovante de Anchoveta de la Zona Norte por el Metodo de Production de Huevos. Programa de evaluation directa de recursos pelagicos de la zona Norte. Convenio Empresas Pesqueras del Norte e Instituto de Fomento Pesquero. 50 pp.

Sanchez, R., Pajaro, M. & Macchi, G. (1996) The application of the Daily Egg Production Method to the assessment of the SW Atlantic anchovy *(Engraulis anchoita)*, spawning biomass off Argentina. ICES CM 1996/H:29.

Santander, H., Alheit, J. & Smith, P.E. (1984) Estimation de la biomasa de la poblacion desovante de anchoveta peruana *Engraulis ringens* en 1981 por aplicacion del 'Metodo de production de huevos'. *Boletin Instituto del Mar del Peru*, **8**, 213–50.

Santiago, J. & Sanz, A. (1992) Egg production estimates of the Bay of Biscay anchovy, *Engraulis encrasicolus* (L.), spawning stock in 1987 and 1988. *Boletin del Instituto Espanol de Oceanografia*, **8**, 225–30.

Scofield, E.C. (1934) Early life history of the California sardine *(Sardina caerulea)*, with special reference to distribution of eggs and larvae. *Californian Department of Fish and Game Fisheries Bulletin*, **41**, 1–48.

Shelton, P.A., Armstrong, M.J. & Roel, B.A. (1993) An overview of the application of the daily egg production method in the assessment and management of anchovy in the southeast Atlantic. *Bulletin of Marine Science*, **53**, 778–94.

Smith, D.R., Brown, J. & Lo, N. (2004) Application of adaptive sampling to biological population. In: B. Thompson (Ed.) *Sampling Rare or Elusive Species*. Island Press, Washington DC.

Smith, P.E. (1973) The mortality and dispersal of sardine eggs and larvae. *Rapports du Congres de la Comission Internationale pour l'Exploration Scientifique de la Mer Mediterranee*, **164**, 282–92.

Smith, P.E., Flerx, W. & Hewitt, R.P. (1985) The CalCOFI Vertical egg tow (CalVET). In: R. Lasker (Ed.) *An Egg Production Method for Estimating Spawning Biomass of Pelagic Fish: Application to the Northern Anchovy (Engraulis Mordax)*. pp. 27–32. US Department of Commerce, NOAA Technical Report NMFS 36.

Smith, P.E. & Moser, H.G. (2003) Long-term trends and variability in the larvae of Pacific sardine and associated fish species of the California Current region. *Deep-Sea Research II*, **50**, 2519–36.

Smith, P.E. & Richardson, S.L. (1977) *Standard Techniques for Pelagic Fish Egg and Larva Surveys*. Fisheries Technical Paper No. 175. FAO, Rome.

Somarakis, S., Koutsikopoulos, C., Machias, A. & Tsimenidis, N. (2002) Applying the daily egg production method (DEPM) to small stocks in highly heterogeneous seas. *Fisheries Research*, **55**, 193–204.

Somarakis, S., Palomera, I., Garcia, A., Quintanilla, L., KoutsikopoulosC., Uriarte, A. & Motos, L. (2004) Daily egg production of anchovy in European waters. *ICES Journal of Marine Science*, **61**, 944–58.

Somerton, D.A., Kobayashi, D.R. & Landgraf, K.C. (1993) Stock assessment of Nehu, *Encrasicholina purpurea,* using the egg production method. *Bulletin of Marine Science*, **53**, 768–77.

Stratoudakis, Y. (2002) Report of the study group on the estimation of spawning stock biomass of sardine and anchovy. Lisbon, Portugal, 22-25 October 2001. ICES CM 2002/G:01. 57 pp.

Stratoudakis, Y. (2003) Report of the study group on the estimation of spawning stock biomass of sardine and anchovy. 23-27 June 2003. ICES CM 2003/G:17. 107 pp.

Stratoudakis, Y., Bernal, M., Ganlas, K. & Uriarte, A. (2006) The daily egg production method: recent advances, current applications, and future challenges. *Fish and Fisheries*, **7**, 35–57.

Takahashi, M., Nishida, H., Yatsu, Y. & Watanabe, Y. (2008) Year-class strength and growth rates after metamorphosis of Japanese sardine *Sardinops melanostictus* in the western North Pacific Ocean during 1996–2003. *Canadian Journal of Fisheries and Aquatic Science*, **65**, 1425–34.

Takahashi, M. & Watanabe, Y. (2004) Growth rate dependent recruitment of Japanese anchovy, *Engraulis japonicus,* in the Kuroshio-Oyashio transitional waters. *Marine Ecology Progress Series*, **266**, 227–38.

Takahashi, M., Watanabe, Y., Kinoshita, T. & Watanabe, C. (2001) Growth of larval and early juvenile Japanese anchovy, *Engraulis japonicus,* in the Kuroshio-Oyashio transition region. *Fisheries Oceanography*, **10**, 235–47.

Theilacker, G.H. (1980) Changes in body measurements of larval northern anchovy, *Engraulis mordax,* and other fishes due to handling and preservation. *Fisheries Bulletin (US)*, **78**, 685–92.

Thompson, S.K. & Seber, G.A.F. (1996) *Adaptive Sampling.* John Wiley & Sons, New York. 265 pp.

Torres-Villegas, J.R. (1986) Evaluation de *Sardinops Sagax* por el Metodo de Production de Huevos, en Bahfa Magadalena, BCS, Mexico. Master of Science Thesis CICIMAR-IPN, La Paz, BCS, Mexico. 116 pp.

van der Lingen, C.D., Checkley, D. Jr, Barange, M., Hutchings, L. & Osgood, K. (1998) Assessing the abundance and distribution of eggs of sardine, *Sardinops sagas,* and round herring, *Etrumenus whiteheadi,* on the western Agulhas Bank, South Africa, using a continuous, underway fish egg sampler. *Fisheries Oceanography*, **7**, 35–47.

van der Lingen, C.D. & Huggett, J.A. (2003) The role of ichthyoplankton surveys in recruitment research and management of South African anchovy and sardine. In: H.I. Browman & A.B. Skiftesvik (Eds) *The Big Fish Bang.* Proceedings of the 26th Annual Larval Fish Conference. Institute of Marine Research, PO Box 1870 Nordnes, N-5817 Bergen, Norway.

van Olst, J.C. & Hunter, J.R. (1970) Some aspects of the organization of fish schools. *Journal of the Fisheries Research Board of Canada*, **27**, 1225–38.

Watanabe, Y. (2002) Resurgence and decline of the Japanese sardine population. In: L.A. Fuiman & R.G. Werner (Eds) *Fishery Science.* pp. 243–57. Blackwell Publishing, Oxford.

Watanabe, C., Hanai, T., Meguro, K., Ogina, R. & Kimura, R. (1999) Spawning biomass estimates of chub mackerel *Scomber japonicus* of Pacific subpopulation off central Japan by a daily egg production method. *Nippon Suisan Gakkaishi*, **65**, 695–702.

Watanabe, Y., Zenitani, H. & Kimura, R. (1995) Population decline of the Japanese sardine *Sardinops melanostictus* owing to recruitment failures. *Canadian Journal of Fisheries and Aquatic Science*, **52**, 1609–16.

Watanabe, Y., Zenitani, H. & Kimura, R. (1996) Offshore expansion of spawning of the Japanese sardine, *Sardinops melanostictus,* and its implication for egg and larval survival. *Canadian Journal of Fisheries and Aquatic Science*, **53**, 55–61.

Watson, W., Charter. R.L., Moser, H.G., Vetter, R.D., Ambrose, D.A., Charter, S.R., Robertson, L.I., Sandknop, E.M., Lynn, E.A. & Stannard, J. (1999) Fine-scale distributions of planktonic fish eggs in the vicinities of Big Sycamore Canyon and Vandenberg Ecological Reserves, and Anacapa and San Miguel Island. *California Cooperative Oceanic Fisheries Investigation Report*, **40**, 128–53.

Yasuda, I. (2003) Hydrographic structure and variability in the Kuroshio-Oyashio transition area. *Journal of Oceanography*, **59**, 389–402.

Yatsu, A., Sassa, C., Moku, M. & Kinoshita, T. (2005) Night-time vertical distribution and abundance of small epipelagic and mesopelagic fishes in the upper 100 m layer of the Kuroshio-Oyashio transition zone in spring. *Fisheries Science*, **71**, 1280–6.

Zeldis, J.R. & Francis, R.I.C.C. (1998) A daily egg production method estimate of snapper, orange roughy, and hoki in New Zealand. *Bulletin of Marine Science*, **53**, 864–90.

Zweifel, J.R. & Lasker, R. (1976) Prehatch and posthatch growth of fishes—a general model. *Fisheries Bulletin*, **74**, 609–21.

Zweifel, J.R. & Smith, P.E. (1981) Estimates of abundance and mortality of larval anchovies (1951–75). *Rapports et Proces-verbaux des Reunions du Conseil International pour l'Exploration de la Mer*, **178**, 248–59.

Chapter 6
Stock Identification

Gavin A. Begg and Steven X. Cadrin

6.1 Introduction

A fundamental issue in the assessment and conservation of living marine resources is understanding population structure and identifying stock units. The identification of stocks is necessary for fisheries management, including allocations of catch between competing fisheries or sectors, recognition and protection of nursery and spawning areas, and for the development of optimal harvest and monitoring strategies (Kutkuhn 1981, Grimes *et al.* 1987, Smith *et al.* 1990, Begg *et al.* 1999a, Sloan *et al.* 2014). However, despite the long-term recognition and significance of stock identification for successful assessment and management of exploited fisheries (Hjort 1914, Beverton & Holt 1957, Sinclair 1988), it was often ignored because of pragmatic and historical difficulties in defining stock boundaries. Although most assessment methods model the dynamics of closed populations and assume homogeneous life-history characteristics, stock boundaries are often established without any reference to the composition and integrity of biological stock units, with respect to reproductive isolation, individual spawning components, sub-stocks or metapopulations. With the frequent failure of fisheries management to prevent overfishing, mandates for a precautionary approach, and the focus on stock rebuilding and implementation of effective harvest strategies, there has been a concerted effort to redress these shortcomings, leading to increased interest in stock identification and the need to preserve population diversity and spatial complexity (Stephenson *et al.* 2001).

Reproductive biology has a two-way relationship with stock identification research:

- Reproductive isolation and spatial variation in reproductive processes are critical aspects of defining stocks (e.g., Begg 2005, McBride 2014a).
- Population structure is an essential consideration for researching and monitoring reproductive biology.

The study of recruitment dynamics illustrates the relationship between stock identification and reproductive biology. An abundance of recruiting year classes is typically modeled and predicted as a function of the mature biomass or egg production at the time the year class was spawned. Such stock–recruit relationships implicitly assume that all recruits were produced by the same spawning stock (i.e., no recruits to the stock were spawned by another stock). The independence of recruitment processes among stocks is also used to delineate stocks, whereby

Fish Reproductive Biology: Implications for Assessment and Management, Second Edition.
Edited by Tore Jakobsen, Michael J. Fogarty, Bernard A. Megrey and Erlend Moksness.

groups of fish from locations with similar recruitment patterns are likely to comprise a single stock, and those from locations with different patterns probably represent separate stocks (Clark *et al.* 1982, Koslow 1984, Koslow *et al.* 1987, Waldman *et al.* 1988, Thompson & Page 1989, Dawson 1994). The degree of asynchrony in recruitment and other reproductive life-history traits provides useful information on the disparity and separation of adjacent stocks that is beneficial to fisheries management (Casselman *et al.* 1981). Statistically significant and temporally stable differences in population statistics between fish stocks can be accepted as evidence that different environments – and hence different locations – are occupied throughout the life history of the fish (Ihssen *et al.* 1981), while spatial and temporal divergence in spawning behaviors may manifest reproductive isolation and hence, genotypic separation of stocks.

Similar to other reproductive traits, intraspecific geographic differences in recruitment provide an indirect basis for stock separation (Begg & Waldman 1999). Recruitment is a critical process regulating the productivity and long-term sustainability of a stock (Haddon 2001); hence, the importance of understanding the factors governing recruitment variability and year-class strength. At the core of this understanding are the basic foundations or population structures from which progeny originate. A diverse population structure comprised of multiple spawning components provides insurance against recruitment failure and the overall impacts of localized stock depletions or extirpations (DeYoung & Rose 1993, Hutchings *et al.* 1993, Begg & Marteinsdottir 2000). The reproductive potential and productivity of individual stocks and their relative contribution to a fishery also shift with time due to stock-specific differences in recruitment and exploitation rates (Waldman & Fabrizio 1994). The recruitment of early life-history stages in stocks can provide information on year-class strength and stock resilience, as well as stock relatedness (Begg 2005). Information on the origins of early life-history stages and fundamental reproductive traits such as spatial and temporal patterns of spawning, therefore, is necessary for understanding the mechanisms responsible for population structuring and recruitment variability (Marteinsdottir *et al.* 2000a, Begg 2005, Hare 2005).

In this chapter, the significance of stock identification for the successful assessment and management of exploited fisheries is illustrated, and its implicit links with reproductive biology and recruitment dynamics are demonstrated. A brief synopsis of the current understanding of the stock concept is provided and various methods used in stock identification research are described. Key stock structure considerations for understanding reproductive biological processes are also discussed, particularly with respect to the diversity of population spatial scales and their relevance to management. In conclusion, insights into the implications of stock structure for conserving reproductive potential are provided, demonstrating the importance of spatial complexity to reproductive success and, in turn, the necessity to match appropriate scales of management with those of biology.

6.2 Stock definitions and stock identification methods

General definitions of the term "stock" are offered by Booke (1981), as a "... group of fish that maintains and sustains itself over time in a definable area", and by Waples (1998) as "... a group of organisms whose demographic/genetic trajectory is largely independent of other such groups". More precise definitions relate to the type of resource management that is needed. For the purpose of conserving genetic diversity and minimizing the risk of extinction, the appropriate management unit is a "genetic stock" or "a reproductively isolated unit, which is genetically different from other stocks" (Carvalho & Hauser 1994). Genetic differences are less important for modeling populations on ecological time scales (e.g., decades), in contrast to demographics or ontogenetic rates (e.g., growth, maturity, reproduction) which are more important. For the purpose of managing sustainable yields from fisheries, the appropriate management unit is a "phenotypic stock," a group that maintains "… characteristics which are expressed in one or more ways depending on the type of environment" (Booke 1981) or a "harvest stock," a "… locally accessible fish resource in which fishing pressure on one resource has no effect on the abundance of fish in another contiguous resource" (Gauldie 1988). Given the dependence of stock definition on management objectives, the relative merits of the various methodological approaches to stock identification also depend on the task at hand.

Approaches to identifying stocks vary widely, with each method offering unique sensitivities and perspectives on population structure. Technical methods can be categorized into genetic analyses, examination of phenotypic traits, evaluation of movement rates, environmental signals, and distributional studies; all of which have experienced rapid advancement during the past decade (Cadrin *et al.* 2005, 2014). Although a brief synopsis is provided on these methods, more detailed reviews have been provided by Ihssen *et al.* (1981), Pawson & Jennings (1996), Begg & Waldman (1999), and Cadrin *et al.* (2014).

6.2.1 *Life-history traits*

Geographic patterns in life-history traits (e.g., distribution, growth rates, and reproductive rates) are the most commonly available information for stock identification, because they are routinely sampled to support stock assessment. In a review of stock identification of Atlantic United States fishery resources, McBride (2014b) reported that life history characteristics form the basis of most stocks.

6.2.1.1 Distributional analyses

Geographic Information Systems and spatial analyses greatly enhance the power of distributional analyses, offering the most basic information on "harvest stocks" or self-sustaining fishery resources. The geographic distribution of natural resources may either be continuous throughout the species range, or discrete with no apparent connection among components. Furthermore, the temporal pattern of abundance among spatial components reflects the degree of connectivity among putative stocks, which is important for fishery and resource management. A harvest stock is more influenced by recruitment and mortality within the stock area than by migration to and from adjacent areas. The concept of harvest stocks has an implicit time element, in that the rate of exchange between areas is not adequate to rebuild the resource in a depleted area within a period that is acceptable to resource managers. For example, fishery managers may wish to rebuild depleted stocks within 10 years (DOC 1996). If the movement between adjacent areas is not sufficient to rebuild an overfished area in a decade, the spatial components should be managed as separate stocks, regardless of genetic or phenotypic homogeneity.

As an example of the value of distributional analyses, Begg *et al.* (1999) examined distribution plots of haddock (*Melanogrammus aeglefunus*) eggs, larvae, juveniles, and adults to determine areas of concentration and potentially indicate stock structure (Figure 6.1). Distributions of juvenile and adult haddock in United States waters suggested discrete components in the Gulf of Maine and on the eastern and western Georges Bank. Distributions of haddock egg and larval stages were not as discrete as those of adults and juveniles, but distributions of eggs suggested separate spawning groups in the Gulf of Maine and on Georges Bank. Haddock larvae formed a single Georges Bank group, with larvae extending down into the Middle Atlantic Bight. Larvae were rare in the Gulf of Maine, and egg and larval distributions did not indicate a division on Georges Bank. Distributional analyses supported separate management of haddock in the Gulf of Maine and on Georges Bank, which is consistent with the current United States management system.

Trends in abundance over time are also important for determining stock structure. For example, an exploratory analysis of abundance data from trawl surveys indicated two geographically distinct groups of yellowtail flounder (*Limanda ferruginea*) off the northeastern United States, with different patterns of abundance and biomass over time between 1967 and 2000 (Cadrin 2010). The major pattern of abundance by geographic stratum and year indicated by cluster analysis was a difference between northern and southern survey strata, with the southern abundance peaking in the early and late 1980s, whereas the northern abundance generally increased during the 1990s (Figure 6.2). The boundary between the two major clusters was on southwestern Georges Bank, where survey catches reflected both southern and northern peaks in abundance (Figure 6.3, Plate 8). These results confirmed earlier observations on fishery landings and supported the hypothesis that movements between the northern and southern areas may not be adequate to replenish the depleted southern resource within a desirable time frame for management (McBride & Brown 1980, Brown *et al.* 1987). Therefore, patterns of abundance and biomass over time suggested separate "harvest stocks" in northern and southern waters with a transition zone on southwestern Georges Bank.

Spatiotemporal patterns may also vary by life-history stage or demographics. Field & Ralston (2005) investigated spatial patterns in recruitment for three species of rockfish in the California Current System (chilipepper rockfish *Sebastes goodie*, widow rockfish *S. entomelas*, and yellowtail rockfish *S. flavidus*; Figure 6.4, Plate 9). The results showed that some strong recruitment events occurred in all areas (e.g., 1984 recruitment of chilipepper rockfish, 1970 recruitment of widow rockfish), and others were more local.

6.2.1.2 Ontogenetic rates

Phenotypic traits, such as rates of growth and maturity, are influenced by genetic variation and environmental conditions. The phenotypic stock definition is less conservative than the genetic stock definition, because it allows for some mixing among stocks. However, partial isolation is enough that geographic differences in phenotype persist. Despite environmental influences, ontogenetic rates determine many population attributes (e.g., reproduction, fecundity, longevity, size structure) that are related to population dynamics (e.g., intrinsic rate of increase, carrying capacity, productivity, resilience), and determine how each stock responds to exploitation. Therefore, for the purpose of fishery management and resource conservation, groups with persistent differences in growth or reproductive dynamics should be modeled and managed separately, regardless of genetic homogeneity.

An example of phenotypic stock identification that is based on geographic patterns of growth and maturity was described by Begg *et al.* (1999), who found temporally stable differences in

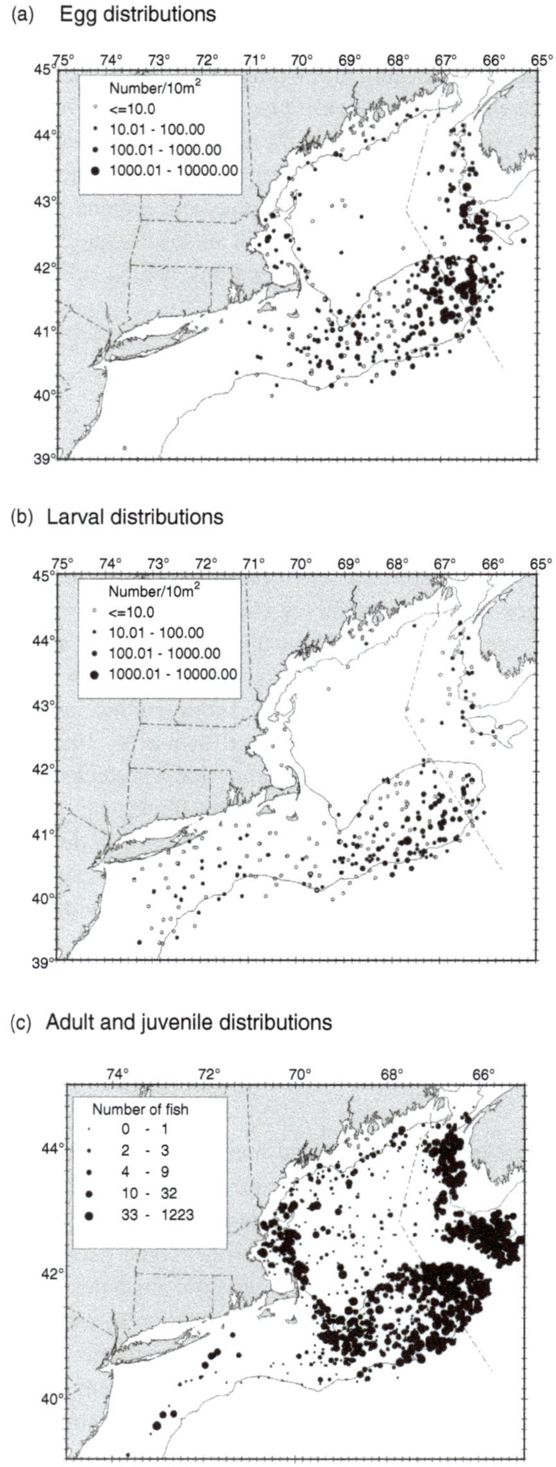

Figure 6.1 Distribution of haddock (a) eggs, (b) larvae, and (c) adults and juveniles from icthyoplankton and bottom trawl spring surveys, 1977–1988. Reproduced from Begg *et al.* 1999b.

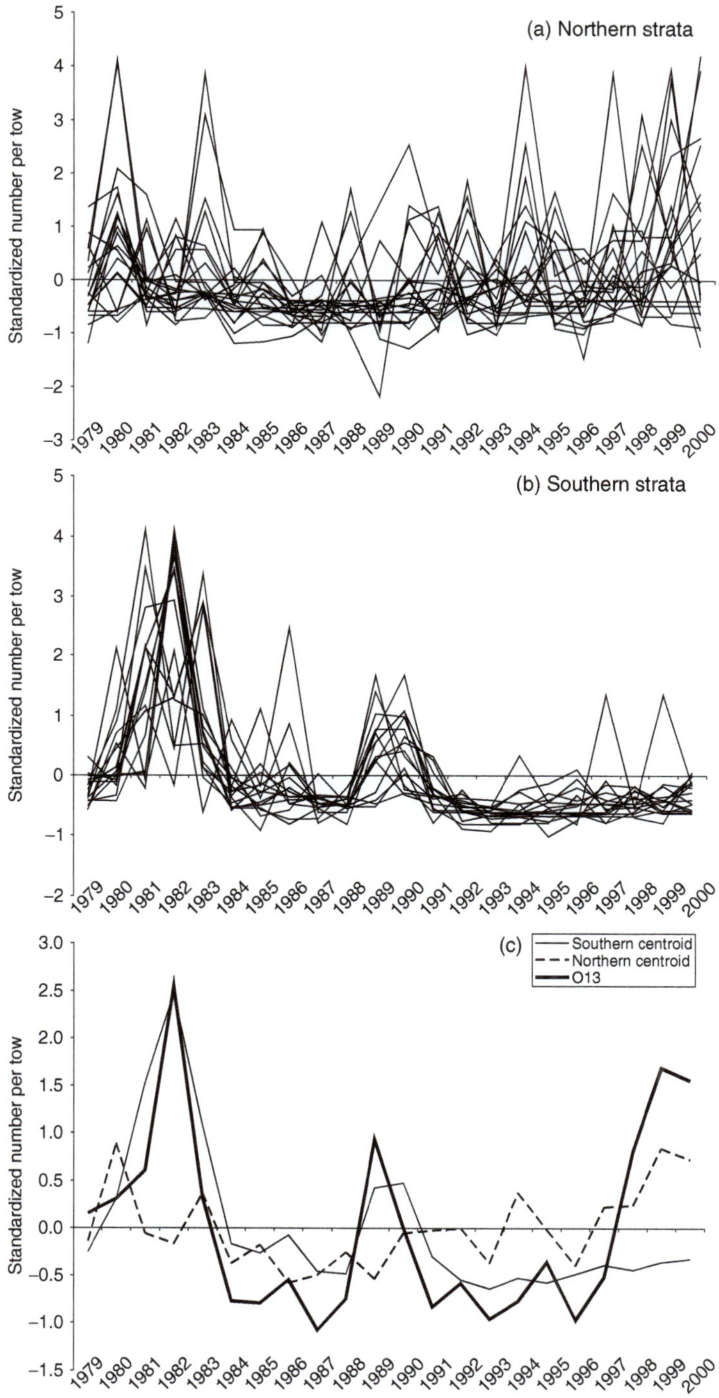

Figure 6.2 Standardized number per tow of yellowtail flounder by geographic survey stratum. (a) Northern strata; (b) southern strata; (c) "transitional" stratum O13 showing consistent temporal patterns of abundance in the southern area (abundance peaks in the early and late 1980s), less consistency in northern areas (general increase in abundance in the 1990s), and a transitional stratum that reflects both patterns. Reproduced from Cadrin 2010, with permission of the author.

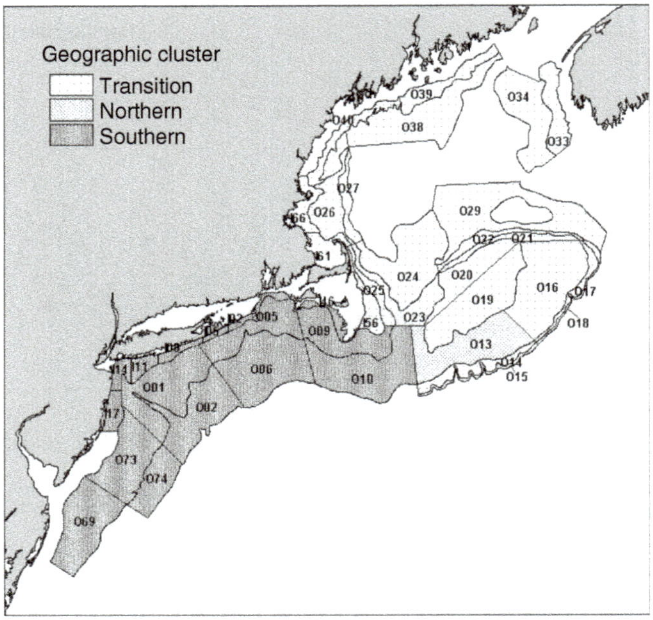

Figure 6.3 Northeast Fisheries Science Center survey strata grouped by clusters with different temporal patterns of yellowtail flounder abundance and biomass over time, 1979–2000. Reproduced from Cadrin 2010, with permission of the author. For a color version of this figure, please see Plate 8 in the color plate section.

growth rates of yellowtail flounder between the Cape Cod grounds in the Gulf of Maine and Georges Bank (Figure 6.5). A geostatistical analysis of the same data indicated two phenotypic stocks of yellowtail flounder off the northeastern United States (Cadrin 2010). A composite score of ontogenetic development rate was derived by averaging the standardized score of each of four variables (female age-2 maturity, male age-2 maturity, female age-2 length, male age-2 length) for each cell in the geographic grid. A spatial analysis of age-2 growth and maturity showed a pattern of relatively low proportion mature and small size at age in the Gulf of Maine, in contrast to relatively high proportion mature and large size at age on Georges Bank (Figure 6.6, Plate 10). Geographic patterns in the composite developmental score showed a boundary located from the Great South Channel to Nantucket Shoals that delineates fast-developing yellowtail flounder from their slow-developing counterparts.

6.2.1.3 Reproductive traits

Many reproductive life-history traits have been used to describe stock dynamics and provide the basis for stock differentiation (Begg 2005, McBride 2014a), including: timing, duration, and location of spawning (Hutchings *et al.* 1993, Begg 1998); median or mean age, length, and weight at maturity (Beacham 1983, O'Brien 1990, Griffiths 1997); egg weight, size, viability, and fecundity relationships (Bradford & Stephenson 1992, Marteinsdottir *et al.* 2000a); proportion of recruit and repeat spawners (Rochet 2000); and maternal effects and reproductive potential (Trippel 1999, Marteinsdottir & Begg 2002). Individual stocks can develop phenotypic and genotypic differences in these traits over time due to reproductive isolation (Waldman

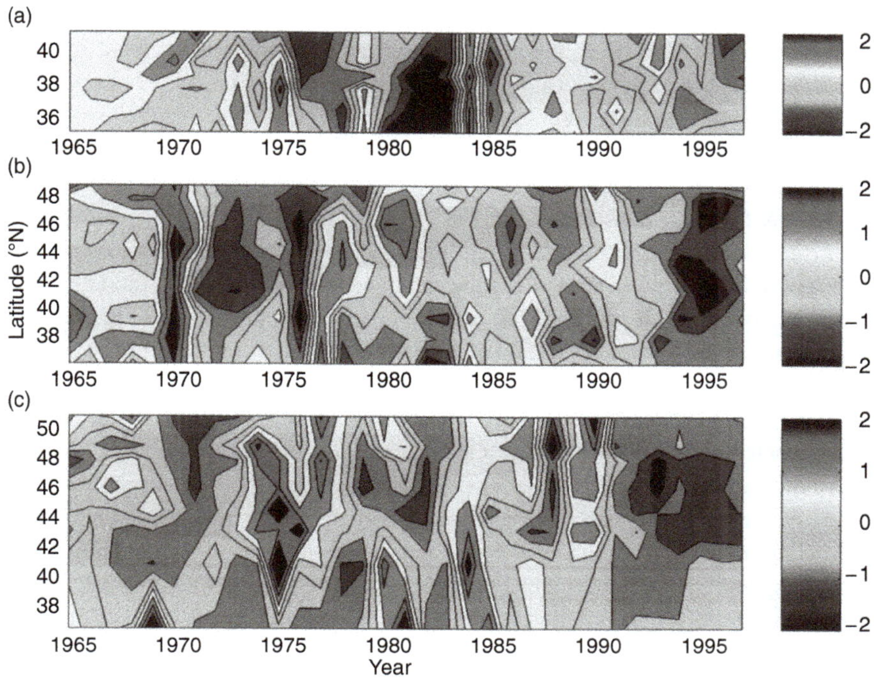

Figure 6.4 Log recruitment deviations plotted over space and time for (a) chilipepper rockfish (*Sebastes goodei*), (b) widow rockfish (*S. entomelas*), and (c) yellowtail rockfish (*S. flavidus*) in the California Current System. Reproduced from Field & Ralston 2005, with permission of NOAA Fisheries. For a color version of this figure, please see Plate 9 in the color plate section.

et al. 1988), which arise from diverse environmental conditions, differential selection pressure, and evolutionary divergence through drift and local adaptation (Dizon *et al.* 1992, Waldman 1999).

Differences in the timing and location of spawning provide a valuable criterion for stock identification because they can result in reproductive isolation among stocks by restricting gene flow to a level that effectively isolates stock units (Iles & Sinclair 1982, Dizon *et al.* 1992, Bailey *et al.* 1999, Kovach *et al.* 2010). Reproductive isolation among stocks is necessary for the formation and maintenance of stock integrity which can be derived from concurrent spawning of stocks in geographically dispersed locations (Horrall 1981) or sympatric stocks with different spawning seasons (Kovach *et al.* 2010, Zemeckis *et al.* 2014). For example, groups of beaked redfish (*Sebastes mentella*) with common nursery grounds, but different adult habitats and spawning grounds, apparently diverged relatively rapidly to form distinct genetic stocks (Cadrin *et al.* 2010). Other studies have also implied stock discreteness for a range of species based on differential spawning times and locations of adult spawning fish (Sinclair & Tremblay 1984, Page & Frank 1989, Hutchings *et al.* 1993).

Differences in spawning and hatch-day distributions of eggs and larvae have also been used to demonstrate differential stock or spawning components (Begg & Marteinsdottir 2000, Marteinsdottir *et al.* 2000b, Bruce *et al.* 2001, Gaughan *et al.* 2001; Figure 6.7). Discrete larval distributions linked to particular geographical regions or hydrological features provide a mechanism for stock structure, imprinting, and spawning site fidelity (O'Boyle *et al.* 1984,

Stephenson 1991). Egg and larval surveys frequently provide information which assists with stock identification because stock integrity depends upon spawning fish from different stocks being separated in space or time, even if they mix at other stages of their life history (Pawson & Jennings 1996). For example, several genetically distinct stocks of herring (*Clupea harengus*) in the northwest Atlantic Ocean have been determined by the number of geographically stable spawning and larval retention areas, where the stocks spawn in relatively discrete geographic locations (Iles & Sinclair 1982, Stephenson 1991). Eggs identified from such surveys provide a direct or immediate response to spawning, while larvae provide an indication of movement to nursery grounds. Prediction of larval movements using oceanographic models can further indicate the extent to which progeny from different spawning stocks are dispersed and separated, and in turn, assist the identification of stock-specific spawning locations which provide recruitment to specific nursery grounds (Pawson & Jennings 1996, Hare 2005).

Overall, the investigation of life-history parameters provides a useful and relatively inexpensive approach for stock identification (Chang *et al.* 2013), where often data for estimating these parameters are routinely collected as part of an ongoing fisheries monitoring program. More recently, there has been an interest at the individual organism level, and how this translates to population or stock differences. For example, the influence of animal physiology and behavior on population and reproductive dynamics, and the link between reproductive output (Young *et al.* 2006), is an information gap that may further our understanding of population structure.

6.2.2 *Morphological variation*

Morphological correlates to life-history variation can also be used to identify phenotypic stocks. Morphological methods include morphometry (e.g., size and shape of body or skeletal parts), meristics (e.g., number of vertebrae, fin rays, gill rakers), and pattern analysis of calcified structures (e.g., growth rings). Morphological approaches have advanced with the continuing improvements in multivariate analyses and relatively new tools such as image analysis (Cadrin & Friedland 1999, 2005, Cadrin 2005).

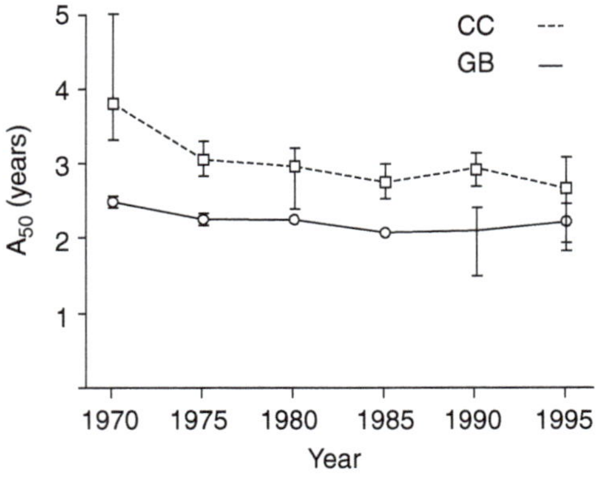

Figure 6.5 Age at 50% maturity of female yellowtail flounder (±95% confidence intervals) for the Cape Cod (CC) and Georges Bank (GB) stocks from 1970 to 1997. Reproduced from Begg *et al.* 1999b.

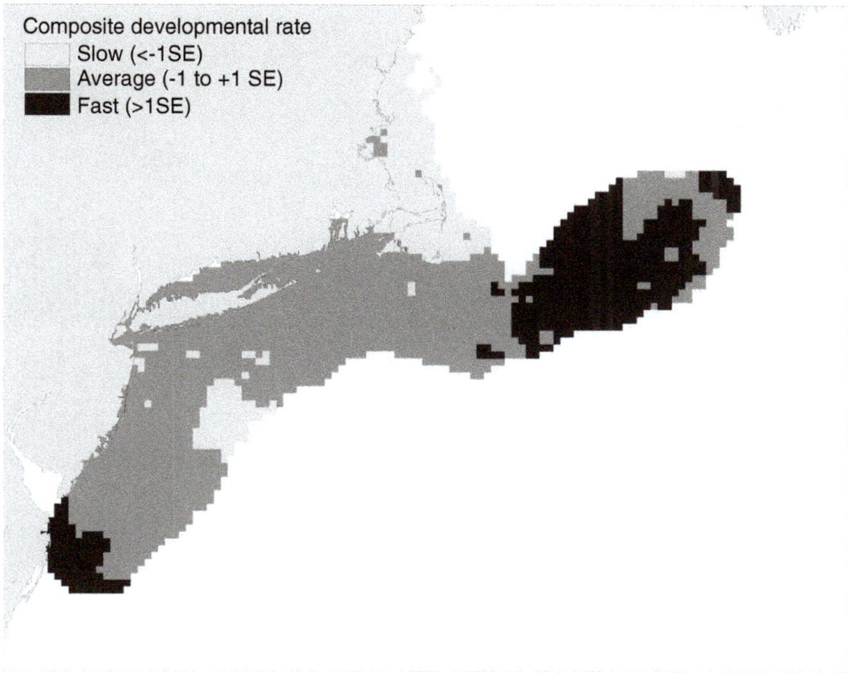

Figure 6.6 Geographic patterns of composite growth and maturity of age-2 yellowtail, expressed as slow (< −1 standard error), average and fast (>1 standard error). Reproduced from Cadrin 2010, with permission of the author. For a color version of this figure, please see Plate 10 in the color plate section.

An example of morphometric variation that is used for stock discrimination and stock composition analysis is Atlantic herring off New England. Samples of post-spawning herring were classified into their respective spawning groups using discriminant analysis of morphometric characters with about 90% accuracy, and the discriminant function was used to determine the stock composition of a mixed-stock winter fishery (Armstrong & Cadrin 2001, Bolles 2006).

Another example of morphological stock identification is the detection of the age at which Atlantic salmon (*Salmo salar*) emigrate from rivers to marine environments, by employing scale pattern analysis. The growth and spacing of the scale circuli increase as smolts move from freshwater to seawater (Figure 6.8), such that the relative timing of an important life-history event is recorded in the morphology of the scales.

6.2.3 Environmental signals

The rapid advancement of chemical analysis of microstructures allows insights into the environmental history of marine species. Hard parts (e.g., otoliths) provide a record of the specimen's chemical environment since birth, and consequently research workers can infer where a specimen was spawned, developed, and spent its adult life. Such life-history information is critical for determining group membership and fidelity to spawning areas.

A benchmark study on evaluating natal homing was provided by Thorrold *et al.* (2001), who analyzed geochemical patterns in otoliths to determine the juvenile and adult habitats of weakfish (*Cynoscion regalis*). Weakfish migrate to estuaries in spring to spawn, but estuarine

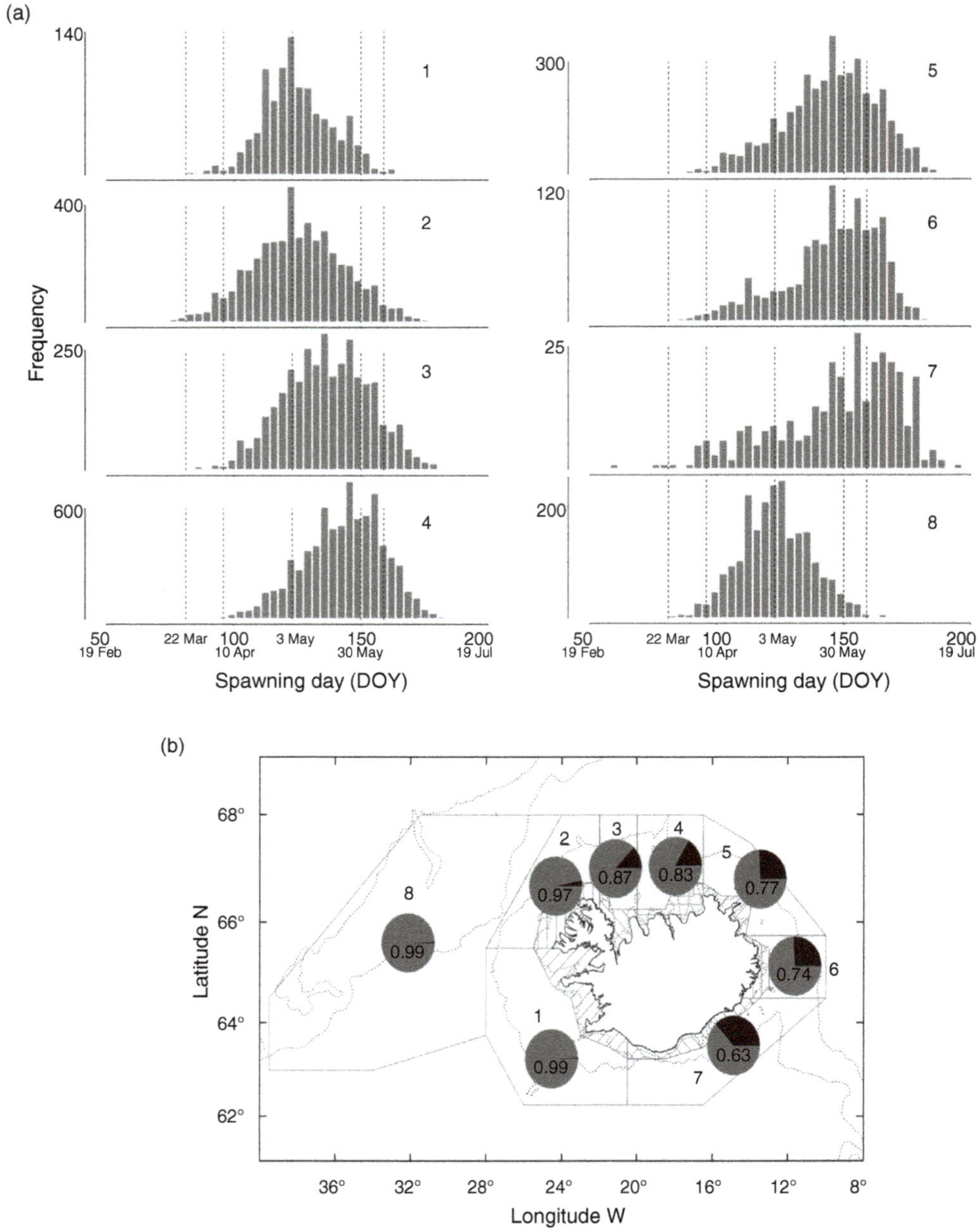

Figure 6.7 Spawning day distributions (DOY) of juvenile (0-group) Atlantic cod (*Gadus morhua*) used to show separate stock components in Icelandic waters. (a) Spawning day distributions of 0-group cod sampled in eight offshore regions (1970–1998). Dashed vertical lines: mean spawning day and 99th and 95th percentiles for main spawning grounds (Region 1); (b) Proportion of 0-group cod in each region predicted to have originated from main spawning grounds. Proportions are based on number of cod spawned at a later DOY than that estimated for the 99th percentile. Results show that large numbers of the juvenile population may originate from other spawning components besides those of the main spawning grounds in the south. Reproduced from Begg & Marteinsdottir 2000.

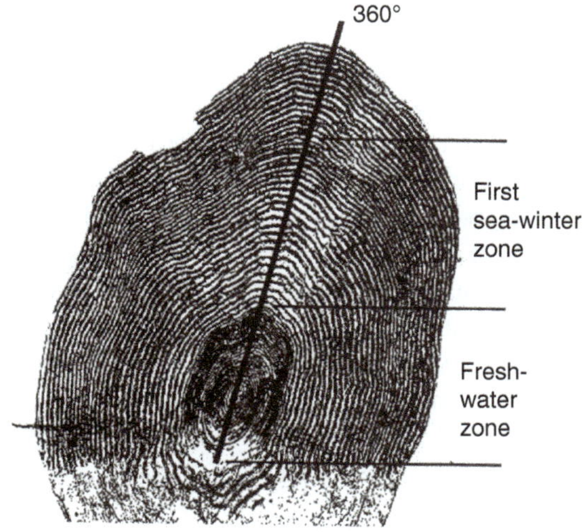

Figure 6.8 Salmon scale with 360° transect marked. Reproduced from Friedland *et al.* 1994, with permission.

spawning groups mix in winter feeding grounds. Thorrold *et al.* (2001) estimated 60–81% fidelity to natal estuaries based on geochemical signatures (Figure 6.9).

A more traditional approach to environmental analyses is to examine rates of parasitic infection, with specimens from different areas typically hosting different parasite species (MacKenzie & Abaunza 1998). Although environmental signals do not necessarily indicate genetic stocks, they are essentially natural tags and are important for understanding population dynamics, such as stock–recruitment patterns and self-sustaining stocks, which are more relevant for typically short- to mid-term, ecological time-scales of assessment and management.

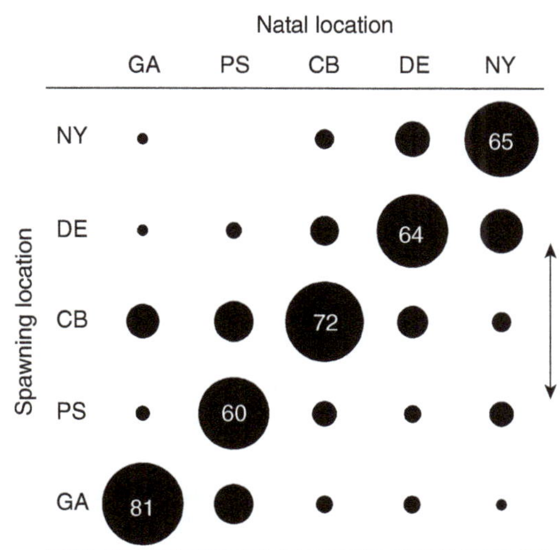

Figure 6.9 Rates of natal homing in spawning weakfish. Reproduced from Thorrold *et al.* 2001.

6.2.4 *Genetic variation*

The most direct evidence for reproductive isolation within a population is genetic divergence, as indicated by differences in frequencies of genetic characteristics. Physical or behavioral barriers can result in reproductive isolation, allowing for the development of genetic differences through genetic drift (Flannery *et al.* 2013). The fields of molecular analyses and breeding experiments continue to progress, providing a wide array of heritable characteristics for stock identification (Mariani & Bekkevold 2014). The different genetic characteristics offer insights into the degree of isolation, as well as the duration of stock separation, with recent studies showing greater genetic structuring than previously thought, even across small spatial scales (Ruzzante 1998, Hauser & Carvalho 2008). For example, variation in chromosome morphology among stocks indicates major genetic divergence formed over thousands of years of reproductive isolation. By contrast, differences in mitochondrial DNA characters, which have faster mutation rates, indicate more recent isolation (Carvalho & Hauser 1994). Allozymes, which are products of protein synthesis, reveal differences in genetic expressions among groups that may result from reproductive isolation or local selection of associated phenotypic traits. Many types of nuclear DNA characteristics (e.g., single-copy and repetitive sequences, random amplified polymorphic DNA and amplified length polymorphic DNA) also offer a wide variety of characteristics with different mutation rates and exposure to selection. Mitochondrial DNA is maternally inherited, and in contrast to nuclear DNA patterns, can reveal female-based isolating mechanisms (e.g., natal homing by females, but not males). Microsatellite characters (repetitive nuclear DNA sequences) are a favored approach to identifying stocks because they are not selected characteristics and are highly variable. Although there is a range of heritable characteristics that can be used for stock identification, mixed-stock analysis using genotypic frequencies has been identified as a reliable method for estimating stock composition (Cadrin *et al.* 2005, Flannery *et al.* 2013); albeit, like most stock assessments, current population genetic inferences are based on relatively simplistic population models (Palsbøll *et al.* 2007).

The most comprehensively studied fishes, with respect to genetic analyses, are the Pacific salmonids. Waples *et al.* (1990) suggested that the amount of genetic information on Pacific salmonids was exceeded only by the information on genetics of *Drosophila* and human. Genetic stock identification is used to estimate the composition of genetic groups in mixed-stock salmonid fisheries. The process involves collection of "baseline" samples from source populations, sampling fisheries and statistical analysis of stock composition (Waples *et al.* 1990; Figure 6.10).

6.2.5 *Tagging*

One mechanism of reproductive isolation in marine species is fidelity to spawning areas. Conversely, movement among areas allows reproductive mixing and homogenizes vital rates among groups. The most frequent approach to understanding movement is through tagging studies. The traditional method, which is still frequently applied, involves marking specimens on spawning grounds and monitoring time at large and the location of recapture to observe individual movements. The approach has expanded technologically, with electronic tags now being used to monitor environmental variables such as temperature, depth and light to determine movement trajectories or telemetric tags which transmit information on actual positions.

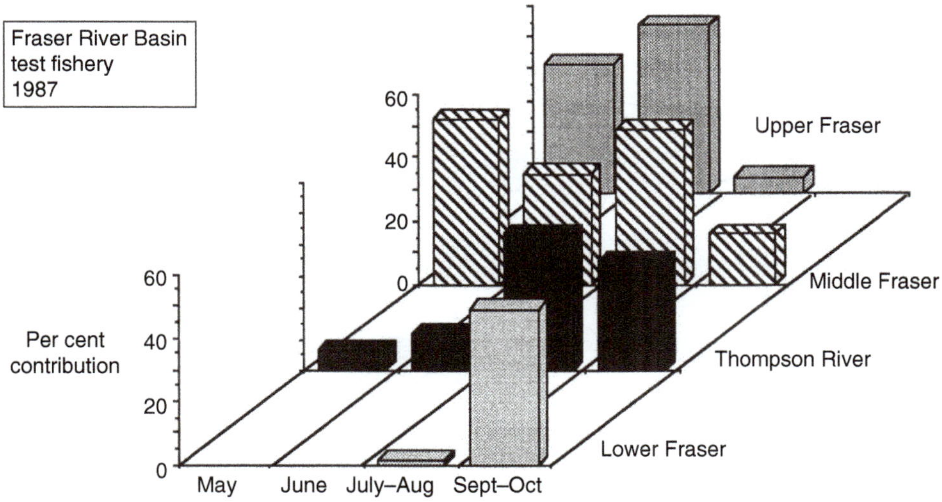

Figure 6.10 Contributions of four Chinook salmon stocks to mixed-stock fisheries, by season, as estimated by genetic stock identification and stock composition analysis. Reproduced from Waples *et al.* 1990, with permission.

The population modeling of Atlantic bluefin tuna (*Thunnus thynnus*) illustrates the role of tagging information in stock identification. A comprehensive review of information on genetics, spawning grounds, spawning seasons, age at maturity, larval dispersion and movements led to the conclusion that the resource has two primary spawning stocks – the Mediterranean Sea, and the Gulf of Mexico (National Research Council 1994, Rooker *et al.* 2007). However, stock assessments are sensitive to the degree of mixing assumed between the two spawning stocks (Figure 6.11; National Research Council 1994, Taylor *et al.* 2011), and therefore precise estimates of movement are required for accurate population modeling.

Another approach to studying the movement of marine fish and invertebrates is through early life-history studies (Hare 2005, Hare & Richardson 2014). Many marine species have pelagic eggs and larvae that can be dispersed by ocean currents. To determine movement rates, by early life-history stages or juveniles and adults, is important for monitoring connectivity among the population components and for identifying stocks. Jones *et al.* (1999) marked the otoliths of developing embryos of damselfish (*Pomacentrus amboinensis*) to demonstrate, for the first time, that juveniles from a coral reef fish population can return to their natal reef, while Secor (2005) has provided an excellent review of how larval dispersion affects population structure of Atlantic eel species (*Anguilla* spp.).

An emerging approach to interdisciplinary stock identification is that of simulation and management strategy evaluation (Kerr & Goethel 2014), whereby all available information on spatial population structure and fishing patterns can be used to develop a spatially explicit operating model. Operating models can simulate alternative stock structure scenarios and evaluate implications for stock assessment and fishery management. For example, Secor *et al.* (2009) examined the implications of entrainment (i.e., learned spawning migration) versus spawning site fidelity for population dynamics of Atlantic herring, while Kerr *et al.* (2014) examined the implications of a mismatch between spatial management units of United States cod resources versus recently revised perceptions of biological stock structure. Operating models can also

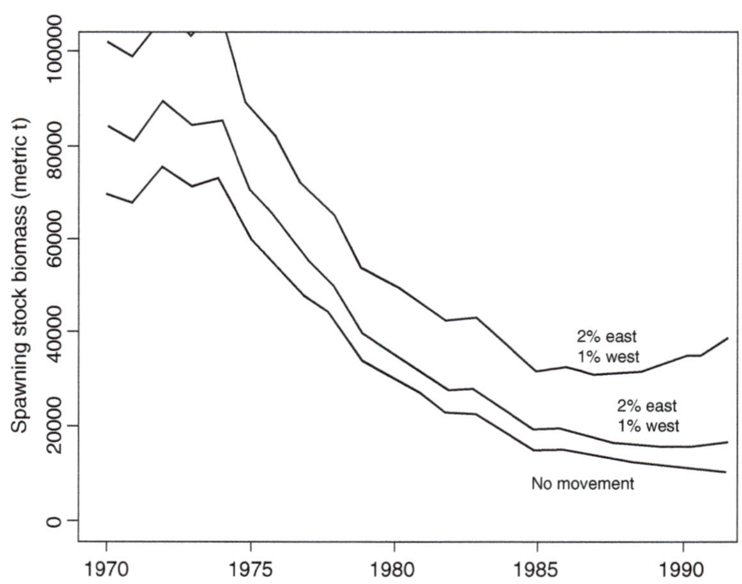

Figure 6.11 Sensitivity in estimates of spawning stock biomass of Atlantic bluefin tuna to assumptions about annual emigration rates from the eastern and western stock components. Modified from NRC 1994.

produce simulations to evaluate the performance of alternative management procedures in the context of spatial complexity (Kritzer & Liu 2014).

6.2.6 Interdisciplinary analysis

During the history of stock identification, new methods have been developed and promoted as better approaches to determining population structure, often leading to equivocal information from competing methodological camps. However, when the results from each approach are viewed in the context of what precise aspect of stock structure they reveal, a more holistic view with multiple perspectives is possible, providing more reliable information for resource management. As new methods continue to emerge, the results obtained will be considered along with those from traditional approaches to improve the ability to study stock structure. Despite the increasing power and number of stock identification approaches, the evaluation of stocks and their boundaries remains a balance of considering reproductive isolation, the homogeneity of vital rates within stocks, and movement among stocks. Recent reviews have agreed that the most comprehensive and effective strategy for stock identification is to integrate results from disparate methods and disciplines to form conclusions about population structure that are consistent with the various approaches employed (Hohn 1997, Coyle 1998, Begg & Waldman 1999, Abaunza *et al.* 2008b, Papetti *et al.* 2013).

6.3 Stock structure considerations for reproductive biology

Information on reproductive biology provides direct insights to understanding the mechanisms responsible for the formation and maintenance of a species' underlying stock structure (Begg

1998). The underpinning of the biological or "genetic" definition of a stock is that they are self-sustaining or reproductively isolated units, with members of each putative stock exhibiting homogeneous traits (Ihssen *et al.* 1981, Hilborn & Walters 1992). This definition depends on our knowledge of spawning behavior and other reproductive traits that are necessary for the formation and maintenance of stock structure. Conversely, information on stock structure is an essential consideration for understanding patterns in reproductive biology. Knowledge of a species' stock structure and spatial complexity across its geographic distribution identify the times and locations of individual spawning components, or the source populations from which progeny originate. This information assists in understanding patterns in reproductive biology and recruitment, and directs when and where samples need to be sourced to characterize these patterns. Importantly, it also identifies the relevant scale of population diversity that needs to be assessed according to the relevant scale of management. An emerging concept in reproductive biology is the physiology–life history nexus paradigm (Ricklefs & Wikelski 2002), where physiological structure and function at the individual level influence life-history characteristics and dynamics at the population level, which in turn, may lead to a better understanding of reproductive strategies and spatial structure (Young *et al.* 2006, Lowerre-Barbieri *et al.* 2011).

6.3.1 Diversity of spatial scales

Anthropogenic, biophysical, and ecological processes affecting reproductive life-history traits function at a range of spatial and temporal scales that govern the formation and maintenance of stock structure (Begg 2005). These traits have been used successfully for stock identification at a diversity of spatial scales, although which trait to use will ultimately depend on the management objective and relevant scale for assessment. The terms "stock" and "population" are often used interchangeably throughout the literature, although population tends to refer to the genetic construct of the stock concept. During more recent years, there has been a growing recognition of the diversity and spatial complexity in marine populations, where management units are typically comprised of multiple spawning components that are interconnected via dispersal and migration as metapopulations (Smedbol & Wroblewski 2000, Kritzer & Liu 2014). Consequently, metapopulations are composites of local populations (i.e., spawning components) between which individuals move (Hanski 1991, Tyler & Rose 1994, Hanski & Simberloff 1997). For example, coral reef fish typically form metapopulations, because most are relatively sedentary and site-attached to individual reefs after settlement, with distinct subpopulations interconnected by larval dispersal (Kritzer 2001). The degree of separation between subpopulations or spawning components may range from slight to almost complete isolation, depending on proximity, geographic barriers, and dispersive capabilities (Harrison & Taylor 1997, Smedbol & Wroblewski 2000). Stock components that spawn in temporal or spatial proximity most likely exhibit a greater connectivity, and thus are more closely related than those from more isolated components (Smedbol & Stephenson 2001).

The maintenance of spatial and temporal integrity among spawning components is important for population structuring, particularly where spawning is adapted to the physical dispersive properties of a geographic location (Heath 1992). Hydrological conditions can minimize the mixing of eggs and larvae between neighboring stocks, effectively maintaining genetic discreteness and reproductive isolation (Iles & Sinclair 1982, Palumbi 1994). Alternatively, mixing between stocks may occur during egg, larval or juvenile stages with subsequent re-segregation later in life (Swain *et al.* 1980), leading to potential differences among stocks in a range of reproductive and morphological traits (Begg 2005).

Reproduction plays a fundamental role in the dispersal or retention of progeny through their initial spatial and temporal placement (Hare & Cowen 1993), and assists in defining the stock structure of a species, because genetic isolation must involve the spawning life-history stage to restrict gene flow to the level that effectively isolates stock units (Iles & Sinclair 1982). Spatial and temporal discontinuities in spawning and larval distribution, therefore, are critical in the definition of stock structure (Smedbol & Stephenson 2001, Hare 2005). The spatial distributions and spawning times of a stock may represent evolutionary adaptations to circulation patterns (Parrish *et al.* 1981, Sinclair 1988), which assist in maintaining reproductive isolation and stock integrity. For example, spawning times and locations of gadoid stocks are usually associated with well-defined circulation or hydrological features that enhance the retention of eggs and larvae (Hutchings *et al.* 1993, Page *et al.* 1999), as are the spawning locations of Atlantic herring, which are hypothesized to be dependent on the geographic extent of oceanographic larval retention areas (Iles & Sinclair 1982, O'Boyle *et al.* 1984, Sinclair 1988, Stephenson 1991). Furthermore, the integration of genetic data with oceanographic models provides insights into the patterns of larval dispersal and fine-scale population structure (Hauser & Carvalho 2008, Selkoe *et al.* 2008). Oceanographic data streams now being collected as part of the integrated marine observing systems around the world will provide a key source of information to assist in discerning the mechanisms driving population structure and the link with reproductive biology (Hill *et al.* 2010, Anonymous 2013).

Individual spawning components are difficult to discern from traditional stock identification techniques, and have generally been of limited use at this scale (Stephenson 1999). However, this should not be seen as a reason for management inaction. Indeed, Stephenson (1999) leads the growing support for the default management scenario in accordance with the precautionary approach to assume stock diversity, and manage the situation accordingly. Furthermore, with increased sophistication in the array of available stock identification techniques, mounting evidence for finer spatial population structure supports the population diversity paradigm moving from one of theory to actuality (Begg *et al.* 1999a, Smedbol & Stephenson 2001, Bergenius 2007).

6.3.2 *Sampling and inference*

Effective stock assessment and fishery management require the mapping of population structure to define the appropriate unit of analysis. In essence, depending on the question at hand, the appropriate scale of management needs to be matched with that of biology and population structure. The collection and analysis of life-history data on spawning adult fish and their progeny provides a direct relationship to reproductive isolation and stock discreteness (Begg *et al.* 1999b, Begg 2005). Individual fish sampled from putative stocks during their respective spawning season maximize stock discreteness, which may otherwise be obscured by spatial overlap and stock mixing during other times of the year (Casselman *et al.* 1981). Spawning (i.e., ripe and running) fish collected from assumed spawning locations would overcome the potential problem of stock mixing (Stephenson 1991), and refine spawning periodicity. Likewise, given the imminent release of eggs once hydration occurs, the collection location of spawning females with hydrated eggs should closely approximate spawning locations (Hutchings *et al.* 1993). Data collected on individual fish during the spawning season from putative stocks could also be used to estimate age and length at maturity and other reproductive traits that provide insights to recruitment and productivity (Begg 2005).

Reproductive life-history traits are useful in discerning the underlying biological characteristics that shape stock structure. These traits provide insight into the isolating mechanisms that are responsible for the maintenance of stock integrity, and are critical inputs for defining the productivity and discreteness of a stock. Inexplicably linked with reproductive life-history traits are those related to recruitment and early life-history stages which are postulated to be the principal determinants of year-class strength (Cushing 1969, Campana *et al.* 1989, Mertz & Myers 1994). Prior to analysis, samples should be stratified according to sampling year, age group, sex, and so forth, or standardized with respect to length or age, depending on the particular variable under scrutiny, to minimize biases and potential errors (Begg 2005). It is important that all efforts are made to avoid sampling biases or problems that may render it difficult to synthesize stock identification studies (Abaunza *et al.* 2008a). Appropriate statistical analyses can then be conducted to determine if there is a need to account for any sampling biases, before conclusions regarding stock status are made. Reproductive traits should also be examined among stocks over consistent and extended time frames, because erroneous results concerning the stock structure of a species could be derived because of the plasticity of these traits to the effects of changing environmental conditions and fishing pressures (Begg *et al.* 1999b). Failure to account for temporal variation or other sampling biases may result in falsely attributing differences between stocks to a stock effect, whereas these differences may in fact be reflective of differences in sampling attributes (Begg & Waldman 1999). Further discussions on sampling strategies and considerations of stock structure analyses are available in the reviews by Fabrizio (2005) and Abaunza *et al.* (2014), while implications for an holistic stock identification study of horse mackerel (*Trachurus trachurus*) are reported by Abaunza *et al.* (2008a).

6.4 Implications of stock structure for conserving reproductive potential

The identification of stocks and their composite spawning components, as well as the recognition and acceptance of the importance of managing spatial diversity, are the first steps to conserving the reproductive potential and long-term productivity of a species. Reproductive potential represents the ability of a stock to produce viable progeny and subsequent recruits (Trippel 1999), incorporating individual and stock effects of its spawning components (Marteinsdottir & Begg 2002, Hilborn *et al.* 2003). Stocks comprised of geographically dispersed, multiple spawning components are more likely to optimize conditions conducive to progeny survival and recruitment success because they have a greater chance of some components encountering favorable conditions in an unpredictable environment (Sinclair 1988, DeYoung & Rose 1993, Marteinsdottir *et al.* 2000b). Spawning diversity is important for successful recruitment and year-class strength as it disperses the mortality risk of early life-history stages over the spatial and temporal ranges they originate, thereby enhancing survival by decreasing the potential for mismatch, while increasing the potential for retention or dispersal to favorable nursery grounds (Sinclair 1988, Cushing 1990, Begg & Marteinsdottir 2000). Accurate stock identification and determination of the relative contribution of individual spawning components to the overall productivity of a stock complex, therefore, facilitate our understanding of reproductive potential and recruitment variability (Hutchings *et al.* 1993, Marteinsdottir *et al.* 2000a). Conversely, differentiating environmental and stock effects on early life-history stages, and the relationship between spawning stock biomass and recruitment, are important steps in understanding the factors that govern recruitment dynamics and stock structure (Ricker 1954, Beverton & Holt 1957, Myers *et al.* 1995).

An understanding of stock structure is also vital to designing appropriate management strategies for fisheries where multiple stocks are differentially exploited (Ricker 1981). A disregard of stock structure and ineffective fisheries management can lead to changes in the biological attributes and productivity rates of a species (Altukhov 1981, Ricker 1981, Smith *et al.* 1991), and lead to overfishing and the depletion of less-productive stocks (Ying *et al.* 2011, Kerr *et al.* 2014). However, most stock assessment methods upon which management advice is based model the dynamics of closed populations and assume homogeneous life-history characteristics. Misleading results can be produced if several closed populations or a portion of a population are the components actually modeled, when a closed population is assumed (Cadrin & Friedland 1999). As a result, an inherent level of uncertainty concerning the actual stock structure being modeled generally prevails in stock assessment models (National Research Council 1994), which invariably cascades through to management. Although information on stock structure and finer-scale spatial dynamics can create more uncertainty and greater data demands for management – particularly when it contradicts historically established management boundaries – to ignore such information can contribute to erroneous and ineffective management (Begg *et al.* 1999a). An understanding of reproductive biology and stock structure dynamics, therefore, is necessary for assessing the full impacts of exploitation and in devising appropriate management strategies (Hammer & Zimmermann 2005).

One aspect of fisheries management that can be improved by more explicit consideration of stock structure is the development of recovery plans to rebuild the reproductive potential of depleted resources. Harvest strategies that promote recovery and sustainable harvests should have separate reference points applicable to the different stocks across a fishery, or alternatively be based on the most vulnerable stock or those where the greatest reproductive output occurs (Sloan *et al.* 2014). Stock structure is an important consideration for rebuilding offshore resources, as illustrated by the recovery of Georges Bank herring (Overholtz & Friedland 2002) and Gulf of Maine cod (Zemeckis *et al.* 2014). Conventional rebuilding plans focus on biomass and age structure, but spatial structure is also an important demographic aspect to monitor and conserve.

Stock enhancement and artificial breeding programs are common approaches used to augment declining fish stocks (Neff *et al.* 2011). Stock enhancement programs of anadromous resources are most successful when streams are stocked with spawners from the same genetic stock, or at least one spawner that has similar adaptations needed to survive and reproduce in the stratum to which it is stocked (Cadrin 2005).

Marine protected areas (MPAs) provide an alternate management strategy for conserving the reproductive potential and population structure of an exploited stock. Traditional fisheries management has been based on single-species and stock-specific strategies that have typically failed to encompass the diversity of population structures becoming increasingly evident (Begg *et al.* 2005). However, with the difficulties involved in matching scales of biology to those of management, MPAs provide an holistic systems approach to enable the preservation of population spatial complexity. MPAs provide an efficient ecosystem-based management strategy that embodies the directives of the precautionary approach, fulfilling the dual functionality of biodiversity conservation and sustainable utilization of selected resources (Agardy 1997, Bohnsack 2000). MPAs provide an array of potential benefits, particularly those related to the preservation of population diversity and insurance against recruitment and management failure (Roberts & Polunin 1991, Bohnsack 1998, Agardy 2000), provided that their size and location encompass the diversity of spawning components within a stock complex. It must be remembered, however, that while MPAs provide potential conservation benefits within their boundaries, they should be used in concert with more traditional management tools to ensure conservation of reproductive potential and population diversity outside their boundaries (Beverton & Holt 1957, Polacheck 1990).

6.5 Conclusions

Stock identification is essential for the successful assessment and management of marine resources, but is rarely accounted for because of logistical issues or historical practice and organizational inertia against change. Stock boundaries, therefore, are often considered without reference to actual population structure and biological integrity of individual spawning components. Hence, although at face value the inclusion of stock structure information in the assessment and management process is fundamental and, in principle is relatively straightforward, it is in fact an area of research where it has proven difficult to bridge the gap between basic research findings and practical management applications. However, during recent years there has been a growing recognition of the importance of population diversity and stock complexity to the overall productivity and long-term sustainability of exploited stocks. In turn, this has led to an increased interest in understanding the ecological functions of populations at different spatial scales, and the need to match these with the appropriate scales of management.

Stock structure is a central tenet of resiliency of a population to exploitation. A diverse population structure provides an insurance against recruitment failure and the overall impacts of localized stock depletions or extirpations (DeYoung & Rose 1993, Hutchings *et al.* 1993, Begg & Marteinsdottir 2000). Accurate stock identification, therefore, is a prerequisite for deciphering the complexities responsible for recruitment variation. Information on the origins of progeny, together with spatial and temporal patterns in spawning, is critical for understanding the mechanisms responsible for population structuring and recruitment variability (Marteinsdottir *et al.* 2000a, Hare 2005). This relationship demonstrates the implicit connection between reproductive biology and stock identification, where reproductive isolation and spatial variation in reproductive processes are critical aspects of defining stocks and, in turn, population structure is an essential consideration for researching and monitoring reproductive biology.

Despite its obvious importance, stock identification remains a challenge to both scientists and managers alike, with a wide array of techniques and conflicting terminologies and interpretations (Cadrin *et al.* 2005, 2014). The spatial and temporal scales of management interest define the stock identification technique to be used, although the strongest inferences are drawn from a suite of complementary techniques (Begg & Waldman 1999). Irrespective of these challenges, incorporating stock structure and spatial diversity into the assessment and management process should be viewed as a necessity rather than simply a point of interest.

References

Abaunza, P., Murta, A.G., Campbell, N., Cimmaruta, R., Comesaña, A.S., Dahle, G., Gallo, E., García Santamaría, M.T., Gordo, L.S., Iversen, S.A., MacKenzie, K., Magoulas, A., Mattiucci, S., Molloy, J., Nascetti, G., Pinto, A.L., Quinta, R., Ramos, P., Ruggi, A., Sanjuan, A., Santos, A.T., Stransky, C. & Zimmermann, C. (2008a) Considerations on sampling strategies for an holistic approach to stock identification: the example of the HOMSIR project. *Fisheries Research*, **89**, 104–13.

Abaunza, P., Murta, A.G., Campbell, N., Cimmaruta, R., Comesaña, A.S., Dahle, G., García Santamaría, M.T., Gordo, L.S., Iversen, S.A., MacKenzie, K., Magoulas, A., Mattiucci, S., Molloy, J., Nascetti, G., Pinto, A.L., Quinta, R., Ramos, P., Sanjuan, A., Santos, A.T., Stransky, C. & Zimmermann, C. (2008b) Stock identity of horse mackerel (*Trachurus trachurus*) in the northeast Atlantic and Mediterranean Sea: integrating the results from different stock identification approaches. *Fisheries Research*, **89**, 196–209.

Abaunza, P., Murta, A.G. & Stransky, C. (2014) Sampling for interdisciplinary analysis. In: S.X. Cadrin, L.A. Kerr & S. Mariani (Eds) *Stock Identification Methods. Applications in Fishery Science*. pp. 475–498. Elsevier Academic Press, Amsterdam.

Agardy, T.S. (1997) *Marine Protected Areas and Ocean Conservation*. Academic Press, San Diego.

Agardy, T.S. (2000) Information needs for marine protected areas: scientific and societal. *Bulletin of Marine Science*, **66**, 875–88.

Altukhov, Y.P. (1981) The stock concept from the viewpoint of population genetics. *Canadian Journal of Fisheries and Aquatic Sciences*, **38**, 1523–38.

Anonymous (2013) U.S. Integrated Ocean Observing System (U.S. IOOS) 2013 Report to Congress.

Armstrong, M.P. & Cadrin, S.X. (2001) Morphometric variation among spawning groups of the Gulf of Maine-Georges Bank herring complex. In: F.J. Funk, J. Blackburn, D. Hay, A.J. Paul, R. Stephenson, R. Toresen & D. Witherell (Eds) *Herring: Expectations for the New Millennium*. pp. 575–90. University of Alaska Sea Grant, AK-SG-01-04, Fairbanks.

Bailey, K.M., Quinn, T.J., II, Bentzen, P. & Grant, W.S. (1999) Population structure and dynamics of walleye pollock, *Theragra chalcogramma*. *Advances in Marine Biology*, **37**, 179–255.

Beacham, T.D. (1983) Variability in median size and age at sexual maturity of Atlantic cod, *Gadus morhua*, on the Scotian Shelf in the northwest Atlantic Ocean. *Fishery Bulletin*, **81**, 303–21.

Begg, G.A. (1998) Reproductive biology of school mackerel (*Scomberomorus queenslandicus*) and spotted mackerel (*S. munroi*) in Queensland east-coast waters. *Marine and Freshwater Research*, **49**, 261–70.

Begg, G.A. (2005) Life history parameters. In: S.X. Cadrin, K.D. Friedland & J.R. Waldman (Eds) *Stock Identification Methods. Applications in Fishery Science*. pp. 119–50. Elsevier Academic Press, Burlington.

Begg, G.A., Friedland, K.D. & Pearce, J.B. (1999a) Stock identification and its role in stock assessment and fisheries management: an overview. *Fisheries Research*, **43**, 1–8.

Begg, G.A., Hare, J.A. & Sheehan, D.D. (1999b) The role of life history parameters as indicators of stock structure. *Fisheries Research*, **43**, 141–63.

Begg, G.A., Mapstone, B.D., Williams, A.J., Adams, S., Davies, C.R. & Lou, D.C. (2005) Multivariate life-history indices of exploited coral reef fish populations used to measure the performance of no-take zones in a marine protected area. *Canadian Journal of Fisheries and Aquatic Sciences*, **62**, 679–92.

Begg, G.A. & Marteinsdottir, G. (2000) Spawning origins of pelagic juvenile cod *Gadus morhua* inferred from spatially explicit age distributions: potential influences on year-class strength and recruitment. *Marine Ecology Progress Series*, **202**, 193–217.

Begg, G.A. & Waldman, J.R. (1999) An holistic approach to fish stock identification. *Fisheries Research*, **43**, 35–44.

Bergenius, M.A.J. (2007) *Stock Structure of a Coral Reef Fish,* Plectropomus Leopardus: *Identification and Implications for Harvest Strategy Evaluation.* PhD Thesis, James Cook University, Queensland.

Beverton, R.J.H. & Holt, S.J. (1957) *On the Dynamics of Exploited Fish Populations.* Fisheries Investigation Series II, **19**, MAFF, London. 533 pp.

Bohnsack, J.A. (1998) Application of marine reserves to reef fisheries management. *Australian Journal of Ecology*, **23**, 298–304.

Bohnsack, J.A. (2000) A comparison of the short-term impacts of no-take marine reserves and minimum size limits. *Bulletin of Marine Science*, **66**, 635–50.

Bolles, K.L. (2006) *Morphometric Discrimination of Atlantic Herring (Clupea harengus) in the Northwestern Atlantic Ocean.* MS Thesis, University of Massachusetts.

Booke, H.E. (1981) The conundrum of the stock concept – are nature and nurture definable in fishery science? *Canadian Journal of Fisheries and Aquatic Sciences*, **38**, 1479–80.

Bradford, R.G. & Stephenson, R.L. (1992) Egg weight, fecundity, and gonad weight variability among northwest Atlantic herring (*Clupea harengus*) populations. *Canadian Journal of Fisheries and Aquatic Sciences*, **49**, 2045–54.

Brown, B.E., Darcy, G.H. & Overholtz, W. (1987) Stock assessment/stock identification: an interactive process. In: H.E. Kumpf, R.N. Vaught, C.B. Grimes, A.G. Johnston & E.L. Nakamura (Eds) *Proceedings of the Stock Identification Workshop 5–7 November 1985, Panama, City Beach, Florida. NOAA Technical Memorandum NMFS-SEFC-199.* pp. 1–23. United States Government Printing Office, Panama City.

Bruce, B.D., Evans, K., Sutton, C.A., Young, J.W. & Furlani, D.M. (2001) Influence of mesoscale oceanographic processes on larval distribution and stock structure in jackass morwong (*Nemadactylus macropterus*: Cheilodactylidae). *ICES Journal of Marine Science*, **58**, 1072–80.

Cadrin, S.X. (2005) Morphometric landmarks. In: S.X. Cadrin, K.D. Friedland & J.R. Waldman (Eds) *Stock Identification Methods. Applications in Fishery Science.* pp. 153–72. Elsevier Academic Press, Burlington.

Cadrin, S.X. (2010) Interdisciplinary analysis of yellowtail flounder stock structure off New England. *Reviews in Fisheries Science*, **18**, 281–99.

Cadrin, S.X., Bernreuther, M. Daníelsdóttir, A.K., Hjörleifsson, E., Johansen, T., Kerr, L., Kristinsson, K., Mariani, S., Nedreaas, K., Pampoulie, C., Planque, B., Reinert, J., Saborido-Rey, F., Sigurdsson, T. & Stransky, C. (2010) Population structure of beaked redfish, *Sebastes mentella*: evidence of divergence associated with different habitats. *ICES Journal of Marine Science*, **67**, 1617–30.

Cadrin, S.X. & Friedland, K.D. (1999) The utility of image processing techniques for morphometric analysis and stock identification. *Fisheries Research*, **43**, 129–39.

Cadrin, S.X. & Friedland, K.D. (2005) Morphometric outlines. In: S.X. Cadrin, K.D. Friedland & J.R. Waldman (Eds) *Stock Identification Methods. Applications in Fishery Science.* pp. 173–83. Elsevier Academic Press, Burlington.

Cadrin, S.X., Friedland, K.D. & Waldman, J.R. (Eds) (2005) *Stock Identification Methods. Applications in Fishery Science.* Elsevier Academic Press, Burlington.

Cadrin, S.X., Kerr, L.A. & Mariani, S. (Eds) (2014) *Stock Identification Methods. Applications in Fishery Science. Second edition.* Elsevier Academic Press, Amsterdam.

Campana, S.E., Frank, K.T., Hurley, P.C.F., Koeller, P.A., Page, F.H. & Smith, P.C. (1989). Survival and abundance of young Atlantic cod (*Gadus morhua*) and haddock (*Melanogrammus aeglefinus*) as indicators of year class strength. *Canadian Journal of Fisheries and Aquatic Sciences*, **46** (Suppl. 1), 171–82.

Carvalho, G.R. & Hauser, L. (1994) Molecular genetics and the stock concept in fisheries. *Reviews in Fish Biology and Fisheries*, **4**, 326–50.

Casselman, J.M., Collins, J.J., Crossman, E.J., Ihssen, P.E. & Spangler, G.R. (1981) Lake whitefish (*Coregonus clupeaformis*) in the Great Lakes region. *Canadian Journal of Fisheries and Aquatic Sciences*, **38**, 1772–89.

Chang, S.-K., DiNardo, G., Farley, J., Brodziak, J. & Yuan, Z.-L. (2013) Possible stock structure of dolphinfish (*Coryphaena hippurus*) in Taiwan coastal waters and globally based on reviews of growth parameters. *Fisheries Research*, **147**, 127–36.

Clark, S.H., Overholtz, W.H. & Hennemuth, R.C. (1982) Review and assessment of the Georges Bank and Gulf of Maine haddock fishery. *Journal of the Northwest Atlantic Fishery Science*, **3**, 1–27.

Coyle, T. (1998) Stock identification and fisheries management: the importance of using several methods in a stock identification study. In: D.A. Hancock (Ed.) *Taking Stock: Defining and Managing Shared Resources*. pp. 173–82. Australian Society for Fishery Biology, Sydney.

Cushing, D.H. (1969) The regularity of the spawning season of some fishes. *Journal du Conseil International pour l'Exploration de la Mer*, **33**, 81–97.

Cushing, D.H. (1990) Plankton production and year-class strength in fish populations: an update of the match/mismatch hypothesis. *Advances in Marine Biology*, **26**, 249–93.

Dawson, P. (1994) *The Stock Structure of Bering Sea Walleye Pollock (Theragra chalcogramma)*. M.Sc. Thesis, University of Washington.

DeYoung, B. & Rose, G.A. (1993) On recruitment and distribution of Atlantic cod (*Gadus morhua*) off Newfoundland. *Canadian Journal of Fisheries and Aquatic Sciences*, **50**, 2729–41.

Dizon, A.E., Lockyer, C., Perrin, W.F., Demaster, D.P. & Sisson, J. (1992) Rethinking the stock concept: a phylogenetic approach. *Conservation Biology*, **6**, 24–36.

DOC (Department of Commerce) (1996) Magnuson-Stevens Fishery Conservation and Management Act as amended through October 11, 1996. NOAA Tech. Mem. NMFS-F/SPO-23.

Fabrizio, M.C. (2005) Experimental design and sampling strategies for mixed-stock analysis. In: S.X. Cadrin, K.D. Friedland & J.R. Waldman (Eds) *Stock Identification Methods. Applications in Fishery Science*. pp. 467–98. Elsevier Academic Press, Burlington.

Field, J.C. & Ralston, S. (2005) Spatial variability in rockfish (*Sebastes* spp.) recruitment events in the California Current System. *Canadian Journal of Fisheries and Aquatic Sciences*, **62**, 2199–210.

Flannery, B.G., Spangler, R.E., Norcross, B.L., Lewis, C.J. & Wenburg, J.K. (2013) Microsatellite analysis of population structure in Alaska eulachon with application to mixed-stock analysis. *Transactions of the American Fisheries Society*, **142**, 1036–48.

Friedland, K.D., Esteves, C., Hansen, L.P. & Lund, R.A. (1994) Discrimination of Norwegian farmed, ranched and wild-origin Atlantic salmon, *Salmo salar* L., by image processing. *Fisheries Management and Ecology*, **1**, 117–28.

Gaughan, D.J., White, K.V. & Fletcher, W.J. (2001) The links between functionally distinct adult assemblages of *Sardinops sagax*: larval advection across management boundaries. *ICES Journal of Marine Science*, **58**, 597–606.

Gauldie, R.W. (1988) Tagging and genetically isolated stocks of fish: a test of one stock hypothesis and the development of another. *Journal of Applied Ichthyology*, **4**, 168–73.

Griffiths, M.H. (1997) The life history and stock separation of silver kob, *Argyrosomus inodorus* in South African waters. *Fishery Bulletin*, **95**, 47–67.

Grimes, C.B., Johnson, A.G. & Fable, W.A., Jr (1987) Delineation of king mackerel (*Scomberomorus cavalla*) stocks along the US east-coast and in the Gulf of Mexico. In: H.E. Kumpf, R.N. Vaught, C.B. Grimes, A.G. Johnston & E.L. Nakamura (Eds) *Proceedings of the Stock Identification Workshop 5–7 November 1985, Panama, City Beach, Florida. NOAA Technical Memorandum NMFS-SEFC-199*. pp. 186–7. United States Government Printing Office, Panama City.

Haddon, M. (2001) *Modelling and Quantitative Methods in Fisheries*. Chapman & Hall, Boca Raton.

Hammer, C. & Zimmerman, C. (2005) The role of stock identification in formulating fishery management advice. In: S.X. Cadrin, K.D. Friedland & J.R. Waldman (Eds) *Stock Identification Methods. Applications in Fishery Science*. pp. 631–58. Elsevier Academic Press, Burlington.

Hanski, I. (1991) Single-species metapopulation dynamics: concepts, models and observations. *Biological Journal of the Linnean Society*, **42**, 17–38.

Hanski, I. & Simberloff, D. (1997) The metapopulation approach, its history, conceptual domain, and application to conservation. In: I. Hanski & M.E. Gilpin (Eds) *Metapopulation Biology: Ecology, Genetics and Evolution*. pp. 5–26. Academic Press, London.

Hare, J.A. (2005) The use of early life stages in stock identification studies. In: S.X. Cadrin, K.D. Friedland & J.R. Waldman (Eds) *Stock Identification Methods. Applications in Fishery Science*. pp. 89–117. Elsevier Academic Press, Burlington.

Hare, J.A. & Cowen, R.K. (1993) Ecological and evolutionary implications of the larval transport and reproductive strategy of bluefish *Pomatomus saltatrix*. *Marine Ecology Progress Series*, **98**, 1–16.

Hare, J.A. & Richardson, D.E. (2014) The Use of Early Life Stages in Stock Identification Studies. In: *Stock Identification Methods*, 2nd edn, pp. 329–64. Elsevier, Academic Press.

Hauser, L. & Carvalho, G.R. (2008) Paradigm shifts in marine fisheries genetics: ugly hypotheses slain by beautiful facts. *Fish and Fisheries*, **9**, 333–62.

Harrison, S. & Taylor, A.D. (1997) Empirical evidence for metapopulation dynamics. In: I. Hanski & M.E. Gilpin (Eds) *Metapopulation Biology: Ecology, Genetics and Evolution*. pp. 27–39. Academic Press, London.

Heath, M.R. (1992) Field investigations of the early life stages of marine fish. *Advances in Marine Biology*, **28**, 1–174.

Hilborn, R. & Walters, C.J. (1992) *Quantitative Fisheries Stock Assessment. Choice, Dynamics and Uncertainty*. Chapman & Hall, London.

Hilborn, R., Quinn, T.P., Schindler, D.E. & Rogers, D.E. (2003) Biocomplexity and fisheries sustainability. *Proceedings of the National Academy of Science of the United States of America*, **100**, 6564–8.

Hill, K., Moltmann, T., Proctor, T. & Simon, A. (2010) The Australian Integrated Marine Observing System: delivering data streams to address national and international research priorities. *Marine Technology Society Journal*, **44**, 65–72.

Hjort, J. (1914) Fluctuations in the great fisheries of northern Europe viewed in the light of biological research. *Rapports et Procès-verbaux des Réunions, Conseil International pour l'Exploration de la Mer*, **20**, 1–228.

Hohn, A.A. (1997) *Design for a Multiple-Method Approach to Determine Stock Structure of Bottlenose Dolphins in the Mid-Atlantic*. NOAA Technical Memorandum NMFS-SEFSC 401.

Horrall, R.M. (1981) Behavioural stock isolating mechanisms in Great Lake fishes with special reference to homing and imprinting. *Canadian Journal of Fisheries and Aquatic Sciences*, **38**, 1481–96.

Hutchings, J.A., Myers, R.A. & Lilly, G.R. (1993) Geographic variation in the spawning of Atlantic cod, *Gadus morhua*, in the northwest Atlantic. *Canadian Journal of Fisheries and Aquatic Sciences*, **50**, 2457–67.

Ihssen, P.E., Booke, H.E., Casselman, J.M., McGlade, J.M., Payne, N.R. & Utter, F.M. (1981) Stock identification: materials and methods. *Canadian Journal of Fisheries and Aquatic Sciences*, **38**, 1838–55.

Iles, T.D. & Sinclair, M. (1982) Atlantic herring: stock discreteness and abundance. *Science*, **215**, 627–33.

Jones, G.P., Milicich, M.J., Emslie, M.J. & Lunow, C. (1999) Self-recruitment in a coral reef fish population. *Nature*, **402**, 802–4.

Kerr, L.A. & Goethel, D.R. (2014) Simulation modeling as a tool for synthesis of stock identification information. In: S.X. Cadrin, L.A. Kerr & S. Mariani (Eds) *Stock Identification Methods. Applications in Fishery Science*. pp. 499–532. Elsevier Academic Press, Amsterdam.

Kerr, L.A., Cadrin, S.X., & Kovach, A.I. (2014) Consequences of a mismatch between biological and management units on our perception of Atlantic cod off New England. *ICES J. Marine Science*, **71** (6), 1366–81.

Koslow, J.A. (1984) Recruitment patterns in northwest Atlantic fish stocks. *Canadian Journal of Fisheries and Aquatic Sciences*, **41**, 1722–9.

Koslow, J.A., Thompson, K.R. & Silvert, W. (1987) Recruitment of northwest Atlantic cod (*Gadus morhua*) and haddock (*Melanogrammus aeglefinus*) stocks: influence of stock size and climate. *Canadian Journal of Fisheries and Aquatic Sciences*, **44**, 26–39.

Kovach, A.I., Breton, T.S., Berlinsky, D.L., Maceda, L. & Wirgin, I. (2010) Fine-scale spatial and temporal genetic structure of Atlantic cod off the Atlantic coast of the USA. *Marine Ecology Progress Series*, **410**, 177–95.

Kritzer, J.P. (2001) *Patterns of Spatial Variation in the Population Biology of a Coral Reef Fish and Implications for Metapopulation Dynamics*. PhD Thesis, James Cook University, Queensland.

Kritzer, J.P. & Liu, O.R. (2014) Fishery management strategies for addressing complex spatial structure in marine fish stocks. In: S.X. Cadrin, L.A. Kerr & S. Mariani (Eds) *Stock Identification Methods. Applications in Fishery Science*. pp. 29–58. Elsevier Academic Press, Amsterdam.

Kutkuhn, J.H. (1981) Stock definition as a necessary basis for cooperative management of Great Lakes fish resources. *Canadian Journal of Fisheries and Aquatic Sciences*, **38**, 1476–8.

Lowerre-Barbieri, S.K., Brown-Peterson, N.J., Murua, H., Tomkiewicz, D.M., Wyanski, D.M. & Saborido-Rey, F. (2011) Emerging issues and methodological advances in fisheries reproductive biology. *Marine and Coastal Fisheries: Dynamics, Management, and Ecosystem Science*, **3**, 32–51.

MacKenzie, K. & Abaunza, P. (1998) Parasites as biological tags for stock discrimination of marine fish: a guide to procedures and methods. *Fisheries Research*, **38**, 45–56.

Mariani, S. & Bekkevold, D. (2014) The nuclear genome: neutral and adaptive markers in fisheries science. In: S.X. Cadrin, L.A. Kerr & S. Mariani (Eds) *Stock Identification Methods. Applications in Fishery Science*. pp. 297–328. Elsevier Academic Press, Amsterdam.

Marteinsdottir, G. & Begg, G.A. (2002) Essential relationships incorporating the influence of age, size and condition on variables required for estimation of reproductive potential in Atlantic cod *Gadus morhua* stocks. *Marine Ecology Progress Series*, **235**, 235–56.

Marteinsdottir, G., Gudmundsdottir, A., Thorsteinsson, V. & Stefánsson, G. (2000a) Spatial variation in abundance, size composition and viable egg production of spawning cod (*Gadus morhua* L.) in Icelandic waters. *ICES Journal of Marine Science*, **57**, 824–30.

Marteinsdottir, G., Gunnarsson, B. & Suthers, I.M. (2000b) Spatial variation in hatch date distributions and origin of pelagic juvenile cod in Icelandic waters. *ICES Journal of Marine Science*, **57**, 1182–95.

McBride, R.S. (2014a) The continuing role of life history parameters to identify stock structure. In: S.X. Cadrin, L.A. Kerr & S. Mariani (Eds) *Stock Identification Methods. Applications in Fishery Science*. pp. 77–108. Elsevier Academic Press, Amsterdam.

McBride, R.S. (2014b) Managing a marine stock portfolio: stock identification, structure, and management of 25 fishery species along the Atlantic Coast of the United States. *North American Journal of Fisheries Management*, **34**, 710–34.

McBride, M.M. & Brown, B.E. (1980) *The Status of the Marine Fishery Resources of the Northeastern United States*. NOAA Technical Memorandum. NMFS-F/NEC-5.

Mertz, G. & Myers, R.A. (1994) Match/mismatch predictions of spawning duration versus recruitment variability. *Fisheries Oceanography*, **3**, 236–45.

Myers, R.A., Barrowman, N.J., Hutchings, J.A. & Rosenberg, A.A. (1995) Population dynamics of exploited fish stocks at low population levels. *Science*, **269**, 1106–8.

National Research Council (NRC) (1994) *An Assessment of Atlantic Bluefin Tuna*. National Academy Press, Washington DC.

Neff, B.D., Garner, S.R. & Pitcher, T.E. (2011) Conservation and enhancement of wild fish populations: preserving genetic quality versus genetic diversity. *Canadian Journal of Fisheries and Aquatic Sciences*, **68**, 1139–54.

O'Boyle, R.N., Sinclair, M., Conover, R.J., Mann, K.H. & Kohler, A.C. (1984) Temporal and spatial distribution of ichthyoplankton communities of the Scotian Shelf in relation to biological, hydrological, and physiographic features. *Rapports et Procès-verbaux des Réunions, Conseil International pour l'Exploration de la Mer*, **183**, 27–40.

O'Brien, L. (1990) *Effects of Fluctuations in Stock Abundance upon Life History Parameters of Atlantic Cod, Gadus Morhua, for the 1970–1987 Year Classes from Georges Bank and the Gulf Of Maine*. M.Sc. Thesis, University of Washington.

Overholtz, W.J. & Friedland, K.D. (2002) Recovery of the Gulf of Maine-Georges Bank Atlantic herring (*Clupea harengus*) complex: perspectives based on bottom trawl survey data. *Fishery Bulletin*, **100**, 593–608.

Page, F.H. & Frank, K.T. (1989) Spawning time and egg stage duration in northwest Atlantic haddock (*Melanogrammus aeglefinus*) stocks with emphasis on Georges and Browns Bank. *Canadian Journal of Fisheries and Aquatic Sciences*, **46** (Suppl. 1), 68–81.

Page, F.H., Sinclair, M., Naimie, C.E., Loder, J.W., Losier, R.J., Berrien, P.L. & Lough, R.G. (1999) Cod and haddock spawning on Georges Bank in relation to water residence times. *Fisheries Oceanography*, **8**, 212–26.

Palsbøll, P.J., Berube, M. & Allendorf, F.W. (2007) Identification of management units using population genetic data. *Trends in Ecology and Evolution*, **22**, 11–16.

Palumbi, S.R. (1994) Genetic divergence, reproductive isolation, and marine speciation. *Annual Review of Ecology and Systematics*, **25**, 547–72.

Papetti, C., Di Franco, A., Zane, L., Guidetti, P., De Simone, V., Spizzotin, M., Zorica, B., Kec, V.C. & Mazzoldi, C. (2013) Single population and common natal origin for Adriatic *Scomber scombrus* stocks: evidence from an integrated approach. *ICES Journal of Marine Science*, **70**, 387–98.

Parrish, R.H., Nelson, C.S. & Bakiun, A. (1981) Transport mechanisms and reproductive success of fishes in the California Current. *Biological Oceanography*, **1**, 175–203.

Pawson, M.G. & Jennings, S. (1996) A critique of methods for stock identification in marine capture fisheries. *Fisheries Research*, **25**, 203–17.

Polacheck, T. (1990) Year around closed areas as a management tool. *Natural Resource Modeling*, **4**, 327–54.

Ricker, W.E. (1954) Stock and recruitment. *Journal of the Fisheries Research Board of Canada*, **11**, 559–623.

Ricker, W.E. (1981) Changes in the average size and average age of Pacific salmon. *Canadian Journal of Fisheries and Aquatic Sciences*, **38**, 1636–56.

Ricklefs, R.E. & Wikelski, M. (2002) The physiology/life history nexus. *Trends in Ecology and Evolution*, **17**, 462–8.

Roberts, C.M. & Polunin, N.V.C. (1991) Are marine reserves effective in management of reef fisheries? *Reviews in Fish Biology and Fisheries*, **1**, 65–91.

Rochet, M.-J. (2000) May life history traits be used as indices of population viability? *Journal of Sea Research*, **44**, 145–57.

Rooker, J.R., Alvarado Bremer, J.R., Block, B.A., Dewar, H., De Metrio, G., Corriero, A., Kraus, R.T., Prince, E.D., Rodríguez-Marín, E. & Secor, D.H. (2007) Life history and stock structure of Atlantic bluefin tuna (*Thunnus thynnus*). *Reviews in Fisheries Science*, **15**, 265–310.

Ruzzante, D.E. (1998) A comparison of several measures of genetic distance and population structure with microsatellite data: bias and sampling variance. *Canadian Journal of Fisheries and Aquatic Sciences*, **55**, 1–14.

Secor, D.H. (2005) Fish migration and the unit stock: three formative debates. In: S.X. Cadrin, K.D. Friedland & J.R. Waldman (Eds) *Stock Identification Methods. Applications in Fishery Science*. pp. 17–44. Elsevier Academic Press, Burlington.

Secor, D.H., Kerr, L.A. & Cadrin, S.X. (2009) Connectivity effects on productivity, stability, and persistence in a herring metapopulation model. *ICES Journal of Marine Science*, **66**, 1726–32.

Selkoe, K.A., Henzler, C.M. & Gaines, S.D. (2008) Seascape genetics and the spatial ecology of marine populations. *Fish and Fisheries*, **9**, 363–77.

Sinclair, M. (1988) *Marine Populations: An Essay on Population Regulation and Speciation*. University of Washington Press, Seattle.

Sinclair, M. & Tremblay, M.J. (1984) Timing of spawning of Atlantic herring (*Clupea harengus*) populations and the match-mismatch theory. *Canadian Journal of Fisheries and Aquatic Sciences*, **41**, 1055–65.

Sloan, S.R., Smith, A.D.M., Gardner, C., Crosthwaite, K., Triantafillos, L., Jeffries, B. & Kimber, N. (2014). National guidelines to develop fishery harvest strategies. FRDC Report – Project 2010/061. Primary Industries and Regions, South Australia.

Smedbol, R.K. & Stephenson, R.L. (2001) The importance of managing within-species diversity in cod and herring fisheries of the north-western Atlantic. *Journal of Fish Biology*, **59** (Suppl. A), 109–28.

Smedbol, R.K. & Wroblewski, J.S. (2000) *Metapopulation Theory and Northern Cod Population Structure: Interdependency of Subpopulations in Recovery of a Groundfish Population*. Canadian Stock Assessment Secretariat Research Document 2000/087.

Smith, P.J., Francis, R.I.C.C. & McVeagh, M. (1991) Loss of genetic diversity due to fishing pressure. *Fisheries Research*, **10**, 309–16.

Smith, P.J., Jamieson, A. & Birley, A.J. (1990) Electrophoretic studies and the stock concept in marine teleosts. *Journal du Conseil International pour l'Exploration de la Mer*, **47**, 231–45.

Stephenson, R.L. (1991) Stock discreteness in Atlantic herring: a review of arguments for and against. In: V. Wespestad, J. Collie & E. Collie (Eds) *Proceedings of the International Herring Symposium, Anchorage, Alaska, October 23–25, 1990 (9th Lowell Wakefield Fisheries Symposium)*. pp. 659–66. University of Alaska, Fairbanks.

Stephenson, R.L. (1999) Stock complexity in fisheries management: a perspective of emerging issues related to population sub-units. *Fisheries Research*, **43**, 247–9.

Stephenson, R.L., Clark, K.J., Power, M.J., Fife, F.J. & Melvin, G.D. (2001) Herring stock structure, stock discreteness, and biodiversity. In: F. Funk, J. Blackburn, D. Hay, A.J. Paul, R. Stephenson, R. Toresen & D. Witherell (Eds) *Herring: Expectations for a New Millenium*. pp. 559–71. University of Alaska Sea Grant, Ak-SG-01-04, Fairbanks.

Swain, A., Parrish, B.B. & Horsted, S.A. (1980) Tagging of salmon smolts in European rivers with special references to recaptures off West Greenland in 1972 and earlier years. *Rapports et Procès-verbaux des Réunions, Conseil International pour l'Exploration de la Mer*, **176**, 93–113.

Taylor, N., McAllister, M., Lawson, G., Carruthers, T. & Block, B. (2011) Atlantic Bluefin Tuna: A Novel Multistock Spatial Model for Assessing Population Biomass. *PLoS ONE*, **6** (12), e27693. doi:10.1371/journal.pone.0027693.

Thompson, K.R. & Page, F.H. (1989) Detecting synchrony of recruitment using short, autocorrelated time series. *Canadian Journal of Fisheries and Aquatic Sciences*, **46**, 1831–8.

Thorrold, S.R., Latkoczy, C., Swart, P.K. & Jones, C.M. (2001) Natal homing in a marine fish metapopulation. *Science*, **291**, 297–9.

Trippel, E.A. (1999) Estimation of stock reproductive potential: history and challenges for Canadian Atlantic gadoid stock assessments. *Journal of the Northwest Atlantic Fishery Science*, **25**, 61–81.

Tyler, J.A. & Rose, K.A. (1994) Individual variability and spatial heterogeneity in fish population models. *Reviews in Fish Biology and Fisheries*, **4**, 91–123.

Waldman, J.R. (1999) The importance of comparative studies in stock analysis. *Fisheries Research*, **43**, 237–46.

Waldman, J.R. & Fabrizio, M.C. (1994) Problems of stock definition in estimating relative contributions of Atlantic striped bass to the coastal fishery. *Transactions of the American Fisheries Society*, **123**, 766–78.

Waldman, J.R., Grossfield, J. & Wirgin, I. (1988) Review of stock discrimination techniques for striped bass. *North American Journal of Fisheries Management*, **8**, 410–25.

Waples, R.S. (1998) Separating the wheat from the chaff: patterns of genetic differentiation in high gene flow species. *Journal of Heredity*, **89**, 438–50.

Waples, R.S., Winans, G.A., Utter, F.M. & Mahnken, C. (1990) Genetic approaches to the management of Pacific salmon. *Fisheries*, **15**, 19–25.

Ying, Y., Chen, Y., Lin, L. & Gao, T. (2011) Risks of ignoring fish population spatial structure in fisheries management. *Canadian Journal of Fisheries and Aquatic Sciences*, **68**, 2101–20.

Young, J.L., Bornik, Z.B., Marcotte, M.L., Charlie, K.N., Wagner, G.N., Hinch, S.G., & Cooke, S.J. (2006) Integrating physiology and life history to improve fisheries management and conservation. *Fish and Fisheries*, **7**, 262–83.

Zemeckis, D.R., Martins, D., Kerr, L.A. & Cadrin, S.X. (2014) Stock identification of Atlantic cod (*Gadus morhua*) in US waters: an interdisciplinary approach. *ICES Journal of Marine Science*, doi: 10.1093/icesjms/fsu032.

Chapter 7
Stock Assessment Models and Predictions of Catch and Biomass

John G. Pope

7.1 Introduction

Stock assessment models and predictions of catch and biomass are handmaidens to fisheries management. Their primary purpose is to help manage fisheries rationally. However, they also help us understand the biology of fish stocks by quantifying the key factors, such as spawning potential and recruitment that lay at the heart of reproductive biology.

To manage rationally, fisheries managers need to know the likely consequence of the management measures they may choose to adopt. The consequences of the choice will be biological in so much as they affect the fish. However, the consequences will also be economic or social in the way they affect fishers and associated industries, and perhaps legal or political in so far as they allocate resources between different groups or different countries. In this book, we are only concerned with the biological outcomes, but the economic, social and legal-political environment within which management operates always has an influence on the form and nature of the advice required. This will obviously affect the biological outcomes. Also, the nature and quality of the data on which advice is based depends upon the priorities that Government or other agencies give to managing fisheries.

To provide advice to managers requires that fisheries scientists are able in some sense to predict what future outcomes may result from specific management actions. Because predictions are always uncertain, this implies indicating the range of possible outcomes and the probability of their occurrence. Moreover, since fisheries are renewable resources, such predictions need to indicate consequences both on short and on longer time scales. To do this typically requires an understanding of the current and past states of the fish stock(s) on which fisheries are conducted. This encompasses the major biological processes (growth, mortality, maturation and recruitment) that determine the dynamics of the stock and provides a base from which predictions can be made.

This chapter thus describes the approaches that fisheries scientists use to understand stock dynamics, to estimate past–present stock states, and to predict future states and catches. It focuses particularly on the prediction of stock abundance and stock biomass, and lays

Fish Reproductive Biology: Implications for Assessment and Management, Second Edition.
Edited by Tore Jakobsen, Michael J. Fogarty, Bernard A. Megrey and Erlend Moksness.

particular emphasis on the need to estimate and predict the recruitment and spawning potential of stocks. The methods by which stock dynamics are understood and by which estimates are made are usually described in mathematical terms. However, as mathematics are a 'barrier' to understanding for many people, in this chapter we will – as far as possible – use words and pictures rather than mathematics.

7.2 Fish stocks, management measures and types of advice

7.2.1 How fish stocks work

Most commercially important fish live for a number of years. As they grow older they grow bigger, become sexually mature, and produce eggs or live young. Throughout life, varying proportions die, with death occurring both from natural causes, such as disease or predation, and as a result of fishing. If, on average, the stock is to remain at the same size, the losses of stock weight to death must be balanced by gains from the growth of the surviving stock, and by young fish being born and growing up to recruit to the stock. Such steady-state situations are what those in fisheries management typically want to attain, because only then is the fishery living from the interest generated by the stock rather than from continuously 'mining down' the capital of stock biomass to give quicker but unsustainable returns.

'Steady state' is an abstract concept. In practice, the number of young fish being born each year can vary widely as the survival success is affected not only by the physical and biotic environment in which they grow, but also by the number and quality of eggs produced by the mature part of the stock. Processes of growth and natural mortality can also be influenced by the physical and biotic environment. Finally, of course, the size and age structure of the stock can be affected by how much fishing harvests different ages. In reality, fish stocks are sculpted by all of these forces, but only the effects of fishing are under immediate human supervision.

My own 'platonic ideal' of a fish stock is the Northeast Arctic cod. Figure 7.1 (Plate 11) shows the biomass-at-age of the Northeast Arctic cod, or rather, because this has not been directly observed, it shows the biomass that results from a particular reconstruction of the stock made by Pope *et al.* (2001). This is chosen rather than the ICES working group assessment (ICES 2005, ICES 2012a) because it penetrates further back into the earlier years of the fishery. How such models are constructed is discussed later in this chapter, but for the time being we will assume it gives us a fair picture of how the stock has changed through time.

One key measurement of any fish stock is the size of the spawning population. Northeast Arctic cod currently become mature at about age 7, so the upper shaded part of the biomass approximately represents the biomass of spawning fish (spawning stock biomass, SSB) in recent years. However, in the earlier years of the series, when the biomass was higher, fish grew more slowly and only first spawned at about age 11 (Rollefsen 1954, Jørgensen 1990, 1992). It is noticeable that these older ages form very little of the recent SSB. Moreover, in practice bigger fish are more effective spawners (see Marshall *et al.* 1998, Marteinsdottir & Steinarsson 1998; and Chapter 8), so a true appreciation of spawning potential requires the biomass to be weighted more heavily towards the older ages. It seems likely that the spawning output of the stock has become reduced in the more recent years. How has this happened, and whether it matters, are critical questions to ask when managing fisheries.

The concern with the size of SSB is because, if it were too small, the stock might not be able to adequately reproduce itself. Figure 7.2 (Plate 12) shows the same picture as Figure 7.1

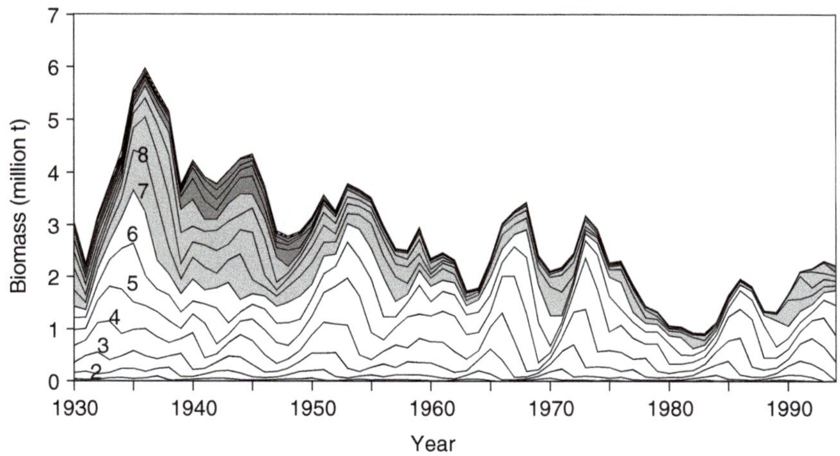

Figure 7.1 Biomass by age of the Northeast Arctic cod (after Pope *et al.* 2001). Ages are shown from 0 nearest the axis to 15 furthest from axis. Unshaded ages are those which are typically immature. Dark shaded ages are those (11–15) that were mature in all years. Lighter shaded ages (7–10) are those that have only been mature in more recent years. For a color version of this figure, please see Plate 11 in the color plate section.

(Plate 11), but the information is organised in terms of the broods of fish born in different years. It is apparent that the biomass of this stock is dominated by the strongest year classes (year broods). These are highlighted in the figure, for example the 1950 year class. These large year classes occur intermittently and are often followed by poor year classes, even though the SSB remained much the same size. Clearly, SSB is not the only determinant of year-class strength. It is also clear that the age structure of the stock varies substantially over the years, sometimes being dominated by small fish and sometimes by larger fish. It seems that the concept of a steady state, and thus of sustainability, can only be understood in some long-term average sense.

Death rates and growth rates also affect the biomass of fish. Like most commercial fish stocks, the Northeast Arctic cod over the years illustrated was not in a steady state at all. The amount of fishing has varied considerably through time, from a period of low fishing intensity

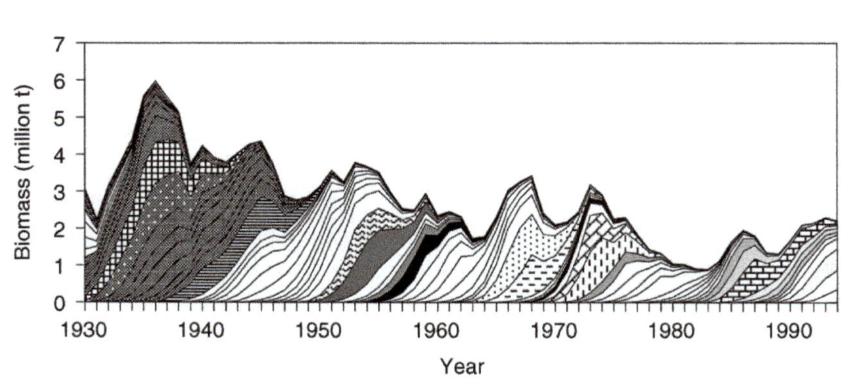

Figure 7.2 Biomass of Northeast Arctic cod by year class (after Pope *et al.* 2001). The large contributions of the 1929, 1930, 1937, 1949, 1950, 1963, 1964, 1969, 1970 and 1983 year classes are highlighted with background patterns. For a color version of this figure, please see Plate 12 in the color plate section.

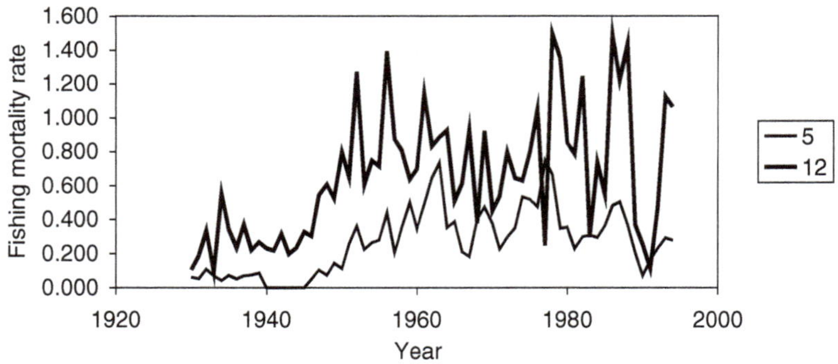

Figure 7.3 Fishing mortality rate for Northeast Arctic cod at age 5 (light line) and age 12 (heavy line) (after Pope *et al.* 2001).

during the 1940s at the time of World War II to a period of high intensity in the late 1980s. Figure 7.3 shows how the fishing mortality rate has changed through time on juvenile fish and on adult fish. Although both fishing rates have increased since 1940, those on the younger fish have risen more. As a consequence of increased fishing mortality, survival has varied through time, owing both to increased fishing intensity and to compensatory changes in natural mortality rate on younger fish. The latter are due to cannibalism (which is accounted for in this model) being reduced as the stock became smaller. But other changes also occurred. As the stock became smaller, then growth increased somewhat and the age at which fish first became mature tended to decrease. The overall decrease in biomass between 1940 and 1980 clearly results from the increased exploitation level. This has been partially, but not fully, compensated by changes in growth rate and by changes in age at maturity.

Let us consider the fishing mortality rate in more detail. Fishing and natural mortality rate are the two typical components of total mortality rate. These mortality rates are concepts of fisheries population dynamics that are rather difficult for a non-specialist to grasp. In mathematical terms, total mortality rate is defined as the rate of decrease of population number with time. An alternative definition as the natural logarithm of the survival rate from all causes is scarcely more intelligible! Perhaps the most understandable definition of fishing mortality rate is as the proportion of the average population that is caught. Similarly, total mortality rate is the total proportion of the average population that die, and natural mortality rate is the proportion that die from non-fisheries causes. However, these are sneaky definitions because the average population is itself affected by the level of the total mortality rate and becomes smaller if the rate is higher. This is why mortality rates can be larger than 1.0, and it can be seen from Figure 7.3 that the fishing mortality rate was higher than 1.0 in some years for the Northeast Arctic cod.

Perhaps this can be better explained as follows. I am writing this sitting in my kitchen in the summer. Two flies are drifting around the room (they represent the average population). Last week my wife hung a sticky flypaper to the ceiling and now 20 dead flies are stuck to it. These 20 are the catch, so the flypapering mortality rate measured over a week is 20/2, that is 10 per week. Notice that the average population is affected by the mortality rate. If the sticky flypaper was not there, then there would be far more flies buzzing around my head. If I took out a fly-spray and instantly killed the other two, the mortality rate (just for flies I hope) would have become infinite. You can see from this illustration that basically survival goes

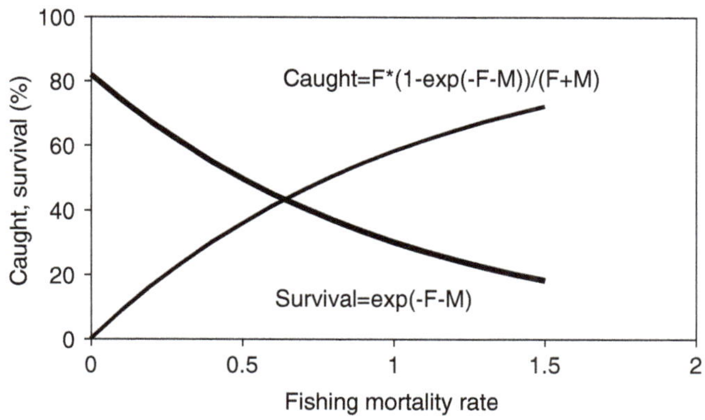

Figure 7.4 The relationship of the proportion surviving a year (heavy line) and the proportion caught in a year (light line) to fishing mortality rate (F) when natural mortality rate (M) = 0.2.

down towards 0 asymptotically (progressively approaching while never quite getting there) as fishing mortality rate goes up, while catch proportion increases asymptotically towards 1. Just how this happens also depends on what deaths there are from natural causes, the natural mortality rate. Figure 7.4 shows how the percentage surviving and the percentage being caught changes with fishing mortality rate when natural mortality rate is 0.2. Perhaps this scientific formulation of fishing mortality rate is confusing but its advantages are considerable since we can add mortality rates. Hence, for example, if a fleet size doubled or my wife put up a second flypaper, I might expect the mortality rates from those causes to double. In the latter case I might expect the population of live flies to roughly half but I would not expect the catch of 20 to go up by much.

7.2.2 How do fisheries affect fish stocks?

From the example of the Northeast Arctic cod it seems fairly likely that one effect of fishing is to reduce biomass and to change the age structure. But how does this happen, and what influence does this have on how the stock should be managed? In practice, these questions are often first answered in terms of the consequences of some simple hypotheses about the main processes that influence biomass. A simple initial set consists of those adopted by Beverton & Holt (1957). These are:

(1) Fish grow in length according to the von Bertalanffy growth equation.
(2) Growth in weight is isometric so that weight changes as length cubed.
(3) On the ages of fish subject to fishing, natural mortality is constant on all ages for all years.
(4) Fishing mortality rate has a fixed value for a given year above some age of first capture.
(5) Recruitment is constant, regardless of spawning stock size (or perhaps has a statistical distribution with a constant mean).
(6) Fish mature at a fixed age; thereafter spawning potential is proportional to biomass of these ages.

With assumptions (1) to (5), Beverton & Holt were able to derive a mathematical equation for yield per recruit as a function of fishing mortality rate and age of first capture. Including

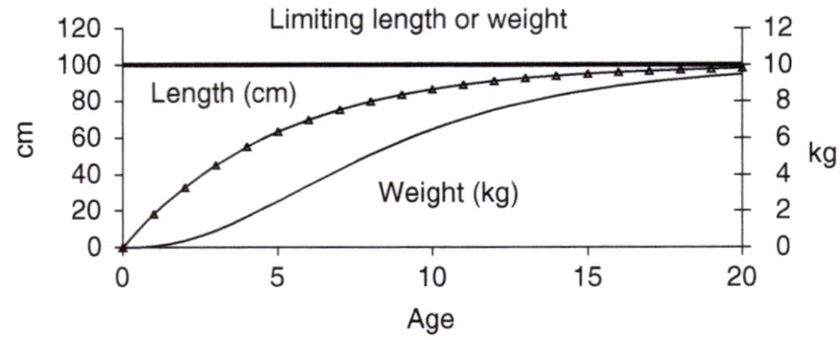

Figure 7.5 An example of von Bertalanffy growth in length with the corresponding growth in weight.

assumption (6) enables SSB per recruit to be calculated. The mathematics can be found in many standard textbooks, and therefore we omit them here but outline the approach graphically.

The growth characteristics are shown in Figure 7.5. Assumption (1) implies that length initially increases rapidly but increases become smaller as the fish approaches some limiting size (100 cm in the example in Figure 7.5). Assumption (2) leads to weight increases, which are initially small, and speed up and finally slow down again to approach a limiting weight (10 kg in the example shown in Figure 7.5).

Let us now think about survival. At a particular age an individual fish may either still be a survivor or already be dead. We may think about the probability of it surviving to various ages. Figure 7.6 illustrates the actions of assumption (3) and assumption (4) on its chances of surviving to each age. These assumptions imply that after a fish recruits (i.e., it might in principle be taken by the fishery, taken here as age 1), but before it reaches the age of first capture when it actually starts being caught (taken here as age 3) it will only risk death from natural mortality (taken here at about 20% a year). Hence, it has about an 80% chance of surviving each year during that period, and its chance of survival declines slowly between ages 1 and 3. However, after the age of first capture it is at risk of death from both natural and fishing mortality. Hence, its chance of surviving a year is less (only about 50% in the example of Figure 7.6). The chance of its survival then rapidly declines as its age increases.

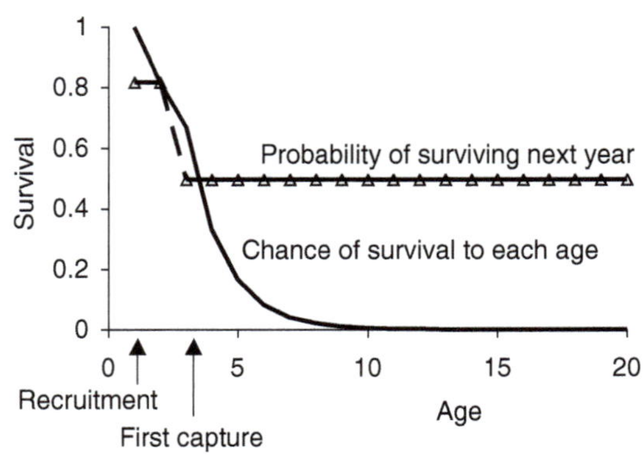

Figure 7.6 Probability by year (triangles) of surviving the next year and chance of survival from the age of recruitment (age 1) to each age for an age of first capture (3) and a fishing mortality rate (0.5) and a natural mortality rate (0.2).

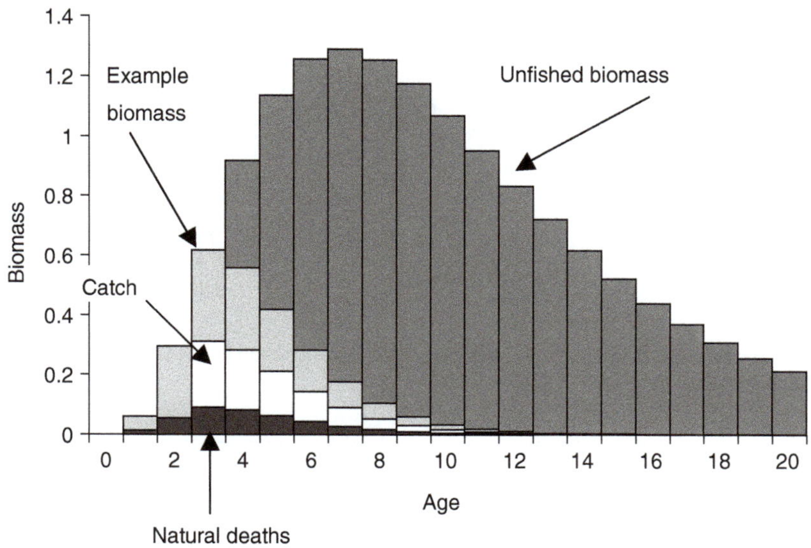

Figure 7.7 The evolution of biomass (light grey), catch (white) and natural deaths-at-age (black) in the example, together with the equivalent biomass (dark grey) there would have been at each age had there been no fishing. For a color version of this figure, please see Plate 13 in the color plate section.

In this illustration there is less than a 20% chance of surviving to age 5, and only a very small chance (about 0.5%) of surviving beyond age 10. Obviously, if the fishing mortality rate is diminished or the age of first capture increased then a fish's chances of surviving to a particular age increases.

If we multiply these survival probabilities by the number of recruits, then we arrive at the average numbers surviving. If in the example there were 1000 recruits, then about 165 would survive to age 5 and only about five to age 10. Similarly, if we had only one recruit we may conveniently think of the survival probabilities as fractional numbers surviving, even though biologically a surviving fraction of a fish (half a fish?) makes no sense. If we multiply these fractional survivors by the weight-at-age, then we get the biomass (number × weight) of the surviving (fractional) fish. These biomass-at-age results are shown in Figure 7.7 (Plate 13) (the total height of the blue bars). Figure 7.7 (Plate 13) also shows what the biomass would (according to the assumptions) have been in an unfished state (the total height of the pink bars). With no fishing there will be substantially more biomass in the older age groups. Clearly, fishing at rates typically seen in stocks such as the Northeast Arctic cod could both reduce biomass substantially and change the age structure. Therefore, it must influence the potential yield that can be achieved. Figure 7.7 (Plate 13) also shows the part of the biomass that would die naturally each year (red) and the part (white) that would be caught (about one-third).

If we add up the catches at each age in Figure 7.7 (Plate 13), we get the total yield (in this case about 0.82 kg) that on average one fish would have provided; this is called the *yield per recruit*. We can remake these calculations for various ages of first capture and various levels of fishing mortality rate and build up a contour plot of how the yield per recruit changes as we change these control variables. Such a surface, called a *yield isopleth diagram*, is shown in Figure 7.8. This indicates that, with the particular values of assumptions we have chosen, yield per recruit would be higher if the age of first capture were age 6 rather than age 1. It also

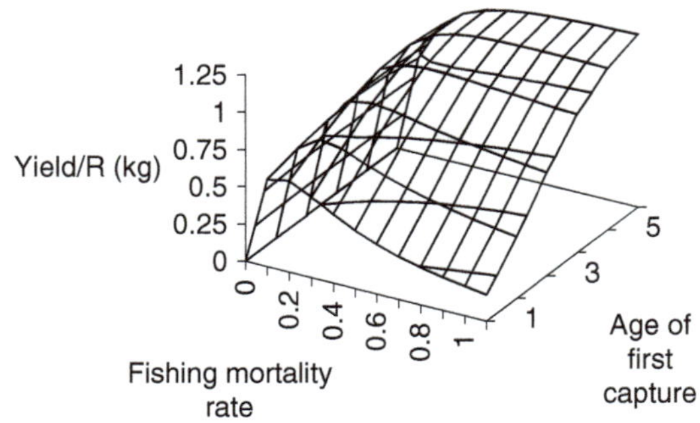

Figure 7.8 Yield isopleth diagram for the example. Yield as a function of the fishing mortality rate and of the age of first capture.

indicates that the curves peak at rather higher levels of fishing mortality rate and become flatter-topped as the age of first capture increases. The fishing mortality that gives the maximum yield per recruit for a given age of first capture is called F_{max}.

If assumption (5) is true, then these plots also indicate how we can obtain the best yield from the system. Certainly, this would be convenient because the inputs needed to quantify assumptions (1) to (4) are relatively undemanding. However, it is at this point that we will start questioning our assumptions, starting with assumption (5). Is it really possible that the stock will be able to produce the same number of recruits however hard we fish the stock? Common sense, bitter experience, and the burden of other chapters of this book all suggest it is not.

Rather than be constant, we might at first imagine that the number of recruits would be proportional to the size of the spawning stock. However, with a little thought we soon realise that if bigger stocks give more recruits, more recruits in turn lead to proportionality bigger stocks and so on, *ad infinitum*. Clearly, in a pre-fishery world this would have led to infinite-sized stocks, but as this did not happen recruitment per stock must reduce as stocks become bigger. Just how this happens – and how the relationship deviates from strict proportionality and gives the shape of the recruits per spawning stock curve (R-SSB curve) – is of course a major question for fisheries science, and for this book.

Figure 7.9 shows the shapes of two possible forms of the relationship between recruitment and spawning stock size. The first form – the Beverton & Holt R-SSB curve – has what

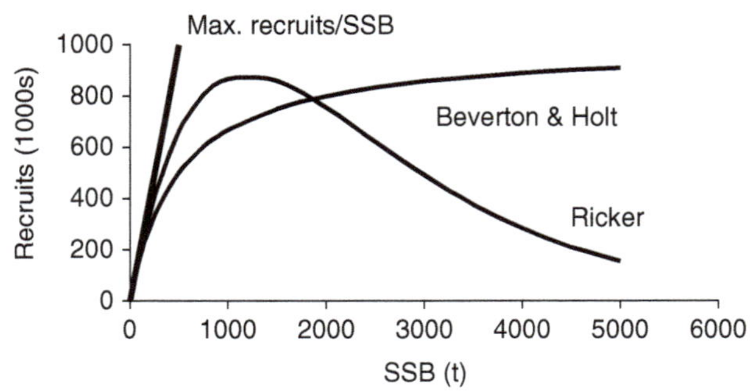

Figure 7.9 Examples of the Beverton & Holt and of the Ricker recruitment–stock relationships. SSB, spawning stock biomass.

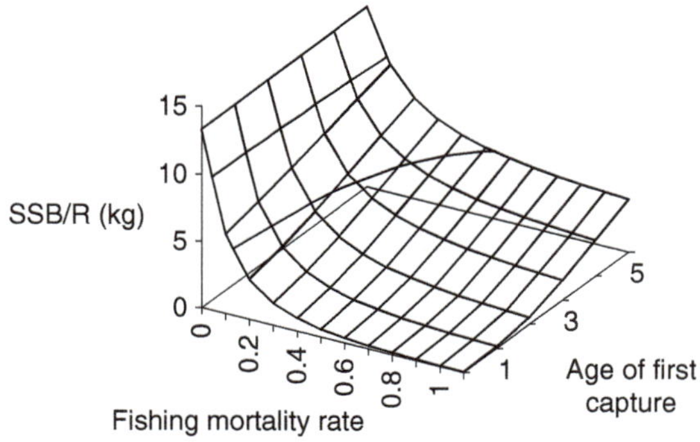

Figure 7.10 Spawning stock biomass per recruit (SSB/R) for the example as a function of fishing mortality rate and of age of first capture.

mathematicians call an asymptotic form. With this form, at low stock sizes recruitment increases almost linearly with increases in spawning stock size, but as the latter increases to high levels the resulting recruitment increases more and more slowly as it approaches the maximum level it would attain at infinite stock size. Thus, if spawning stock is always high enough to provide recruitment close to the maximum, then recruitment is almost constant and assumption (5) effectively holds. The second form of curve – the Ricker curve – peaks at some intermediate level of SSB and declines asymptotically towards zero at higher levels of SSB. This might result if an overcrowding of pre-recruits were to reduce survival, or if cannibalism occurred with adults feeding on pre-recruits. Clearly with this curve, assumption (5) does not hold at all.

A common feature of both curves is that they have a finite slope at the origin. This slope, as indicated in Figure 7.9 by the line 'Max. recruits/SSB', is their most important feature. It indicates that a low stock size can produce only a certain maximum number of recruits per unit weight (here, two recruits per kilogram). In turn, these recruits will produce a certain SSB as they grow up and reproduce themselves. If the amount of SSB they produced is less than the SSB that produced them (i.e., if SSB per recruit <1/(Max. recruits/SSB)), then the stock will produce fewer recruits in the next generation, and so on, until the stock declines to zero.

Now, let us consider the spawning biomass per recruit for the example. Suppose in Figure 7.7 (Plate 13) that fish at and above age 4 were mature. If we add up the biomass of age 4 or older, we obtain the aggregate biomass that would be produced by one fish under the particular combination of age of first maturity and fishing mortality examined. This is termed the spawning stock biomass per recruit (SSB/R). As with yield/recruit, we can build up a picture of how this looks for various combinations of age of first capture and fishing mortality rate.

Such a picture of SSB/R is shown in Figure 7.10. For higher ages of first capture, there is a minimum level of SSB/R, regardless of how high the fishing mortality rate becomes. This is because fish have a chance to spawn before they become available to capture. If the age of first capture is lower than the age of maturity, SSB/R approaches zero as fishing mortality increases. In this example, when the age of first capture is 1 or 2, the SSB/R goes below 0.5 kg per recruit for the higher levels of fishing mortality shown. Hence, if the recruit/stock curves shown in Figure 7.9 were true, we would expect both the stock and the recruitment to decline to zero in these cases.

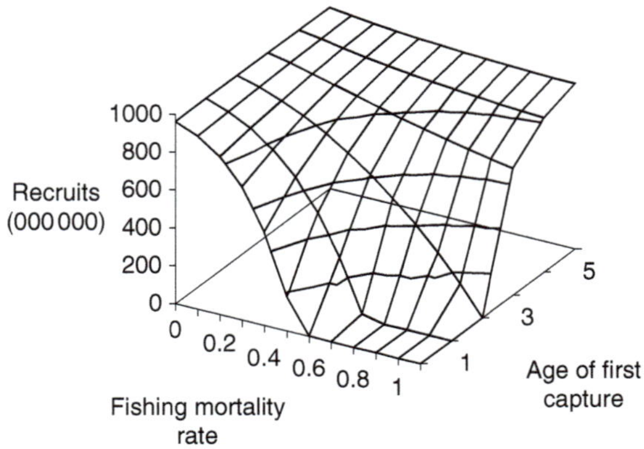

Figure 7.11 Steady-state recruitment as a function of fishing mortality and of age of first capture.

If we know the SSB/R for a particular fishing regime and the form of the R-SSB curve, then it is possible to solve either graphically (see Sissenwine & Shepherd 1987) or mathematically (see Pope 2003) for the steady-state recruitment that would result. If we take the Beverton & Holt R-SSB curve shown in Figure 7.9 as being the correct one for our theoretical stock, then the resulting long-term recruitment at each level of stock is shown in Figure 7.11. As predicted, recruitment is zero for the higher levels of fishing mortality shown when the age of first capture is 1 or 2. However, it only declines slowly with fishing mortality when the age of first capture is 6. Obviously, the situation would have been more complex if we had adopted the Ricker curve, but it would have shown the same hole in recruitment. It is worth noting that recruitment can (in the long term) decline dramatically if fishing mortality rate is too high.

If we multiply the number of recruits at each fishing mortality rate/age of first capture combination (shown in Figure 7.11) by the equivalent yield per recruit (shown in Figure 7.8), then we obtain the overall yield. For our example this is shown in Figure 7.12. On comparing Figures 7.8 and 7.12, we see clearly that taking account of the changes in recruitment alters our perception of how yield would change with exploitation. We note that, for a given age of

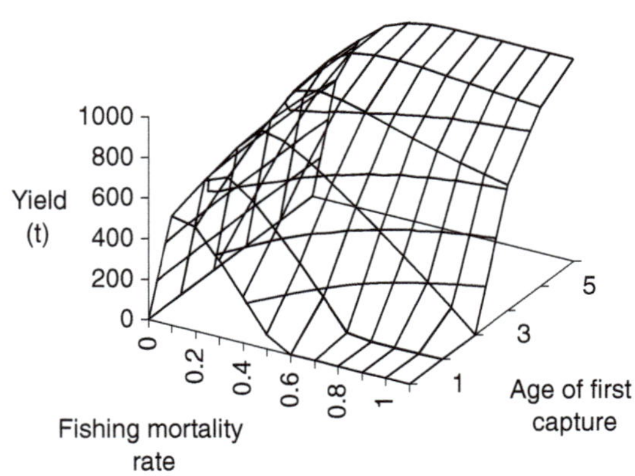

Figure 7.12 Overall steady-state yield as a function of fishing mortality rate and of age of first capture.

first capture, fishing mortality F_{msy} that gives maximum sustainable yield (MSY) may well be different from the fishing mortality F_{max} that gives a maximum sustainable yield per recruit, MSY/R. This is particularly the case when the age of first capture is low. We also see the need to challenge the simplifying assumptions that lead to Figure 7.8.

7.2.3 *Why fisheries need to be managed, and how can it be achieved?*

Figure 7.12 suggests that too much fishing reduces catch. We might therefore ask why would fishers work harder to catch less? In most industries, free markets give signals to both producers and consumers that lead to an optimisation of production. For producers, profitable operations signal them to increase production capacity, while unprofitable operations signal them to reduce production. For such industries, the markets give the right signals, but because fisheries are based on wild fish which are not owned by the fishers, they are subject to the 'Tragedy of the commons' (Hardin 1968). As a result, free markets left to themselves lead to more production capacity, for example, more fishing boats, more fishermen, more fish plants and so on, than are needed for the economically optimal exploitation of a stock.

We may simply understand this as follows. The job of fishers is to catch fish and bring them to market. A given intensity of fishing catches a proportion of the stock. If the intensity is increased, for example by increasing the number or the size or the efficiency of fishing boats, then a bigger proportion of the stock is caught. Initially, this will give bigger catches, but in the long run increasing the proportion of the stock being caught and killed will decrease the size of the fish stock. Often, this will reduce its productivity and thus the long-term yield of the stock. However, the annual profits earned from fishing reflect the short-term catch. Thus, fishers continue to receive a short-term profit signal to increase fishing capacity beyond levels that will be profitable or even sustainable in the longer term. They may also receive signals to catch fish at suboptimal sizes. Even worse, if any individual fisher realises these long-term problems and as an individual shows restraint, the common-property nature of unregulated fisheries means that he/she just leaves more room for others to increase their capacity until such time as no profit exists.

Because of the failure of markets to warn fishers when they are fishing too much, fisheries need to be managed so as to be sustainable both biologically and economically, as well as socially. Successful management thus has to take into account biological, economic, and social aspects. Biological management requires that the stock is not exploited too heavily or at too small a size. Economic management stresses the provision of market incentives for good practice and disincentives to discourage too much fishing, while social management stresses the need for successful management to be inclusive and consensual rather than imposed by 'them from above'. If it is to be successful, fisheries management needs to satisfy all of these requirements. This chapter, however, focuses on the biological aspects while acknowledging the equal importance of the other aspects.

In practice, much management is biological in the sense that, while the objectives of management are economic and/or social, these are underpinned by the health of the fish stock or stocks on which the fishers depend. Moreover, many of the tools of management are framed in terms of what fish may be caught, by what gear and how much fish may be caught, when and where fish may be caught, and by whom fish may be caught. These tools may act to reduce the general level of fishing and/or to avoid wasteful fishing practices such as catching fish at too small a size. Many of these are framed in biological terms and consequently require biological understanding (assessment) of the fish stock to be managed, and also biological advice as to what precisely the measure should be. The type of advice, its cost, and the frequency with which it is needed depends upon the measures chosen. To be effective they need to control both the size at which fish are caught (the age of first capture in the example) and the rate at which they are caught.

Technical conservation measures are management tools that regulate how, where, and when a fishery is to be conducted, but they do not specify how much fishing is allowed or who can do it. Examples of technical conservation measures are:

- The prohibition of harmful capture methods (e.g., poison or explosives).
- Regulations as to how certain gears must be rigged and used (e.g., mesh regulations and attachment restrictions for trawls and gill nets, hook size, escape hatches in pots, etc.).
- Restrictions on the size, or breeding state of fish (fish is understood to include shellfish) that may be landed (e.g., minimum landing sizes, restrictions on landing berried female lobsters).
- Restrictions on the mix of species that can be taken by certain classes of gear (e.g., by-catch limits).
- Restrictions on where or when the fishery can be conducted (e.g., closed seasons or closed areas including Marine Protected Areas; MPAs).
- Restrictions on what is to be done with unwanted catch (e.g., discard bans or requirements).

For the most part such measures are designed to avoid wasteful practices and to ensure that fish are not caught at too small a size. Some are based on common sense coupled with a basic understanding of the biology and the behaviour of fish. For example, knowledge of the size at first spawning might give an indication of a sensible minimum landing size or mesh size for a single-species fishery. Similarly, knowledge of where the nursery grounds of a stock are might indicate a suitable closed area, or simple knowledge of the spawning season might indicate the time for a closure. In many cases, however, the questions posed by technical measures can be quite complex. This is particularly the case if measures have to be chosen that affect fisheries based on a mix of species. Such questions may need to be addressed if we are seeking to explain the detailed effects of proposed measures on each stock in the fishery. Management measures may also need to be considered in terms of how they could change the biological interactions among species. Such questions may in some cases be more complex than current analysis or data can readily answer, although there is a long tradition in fisheries science of trying to provide partial answers to such questions (e.g., Beverton & Holt 1957, ICES 1984, 1986, 1989, Pope 1991).

With technical measures it is necessary to evaluate the change in selection of fish by size. Advice will include a quantification of any short-term losses in landings and any changes in efficiency that will occur, and will also indicate any long-term gains in landings, catch rate, and spawning stock. For economical, social and political reasons, it may also be necessary to indicate what the changes in revenue will be, when they will occur, and which groups of fishers will benefit and which will suffer the largest losses. In practice, these changes are often expressed relative to what would have occurred if the status quo had been maintained. Such relative advice is less dependent on a precise evaluation of the current status of the stock and thus often less demanding of precision in data or estimates.

In terms of Figure 7.12, technical conservation measures act on one of the two dimensions of fishing mortality, notably the one concerned with controlling the age of first capture. Direct conservation measures act on the other dimension by trying to limit the fishing mortality rate. Hence, direct measures limit the amount of fishing effort deployed (input controls) or limit catches (output controls).

Input controls act to limit the fishing mortality rate by controlling the level of fishing activity in some fashion. Examples of input controls are:

- Limited entry schemes.
- Effort quotas.
- Closed seasons or areas designed to limit fishing effort or to reduce its efficiency.

Such approaches control the scale of fishing mortality rate but do not directly influence how it bears on different age groups. Clearly, an essential requirement for an effort control is to be able to evaluate the change in fishing mortality rate it will cause. As with technical measures, advice is needed on how catches and catch rate will change in both the short and long term, but these can be presented relative to the status quo. Hence, again the precision of short-term catch forecasts is less critical and such advice is not needed on an annual basis. However, a typical effect of any effort restriction is to encourage greater efficiency in the boats that remain fishing, and thus for fishing mortality produced per unit of effort to increase through time. Because of this creep in efficiency, there is a need to monitor how fishing mortality rate changes through time.

Output controls, such as total allowable catches (TACs), are designed to limit fishing mortality rates to some desired level, and for this purpose the precision of short-term catch forecast is critical. Moreover, since biomass changes from year to year, it is clearly necessary that the catch quota must also be adjusted. Hence, these measures tend to require the most precise data and often need to be prepared on an annual cycle. It is from stocks managed in this fashion that we are likely to find the best data for studies of reproductive biology.

7.2.4 Time scales of predictions

Advice on how fish stocks will react to management measures requires that predictions of their future state be made. Typically, such advice is needed at short-, medium- and long-term time scales. Short-term predictions are those that can be made based mostly on the known current status of the stock. The time span will differ depending on how rapidly results are dominated by the strength of unobserved recruitment. Clearly, this will be only a short period if age of first capture is low and mortality rates are high. It may be longer where the age of first capture is higher, where recruiting year classes are measured by surveys several years before they enter the fisheries, and where the mortality rate is low. Long-term predictions are those where sufficient time has elapsed for the stock to have adjusted to the new management regime so that it is no longer influenced by the past management. Medium-term predictions are those made for the intervening years.

What is needed for short-term predictions? By definition, short-term predictions are those based on the current state of the stock. Thus, they need to be based on knowledge of that state, and of how this will evolve under any proposed fishing mortality regime given the expected natural mortality rates and expected growth rates.

A fully age-based approach would proceed by first estimating the current numbers at each age, and then calculating what proportion of each of these would survive into the next year, how many new recruits would enter the fishable part of the stock, what the average weight would be at each age, what proportion would be mature, and what proportion would be caught. Summary results could then be prepared which showed how the catch, the total biomass, the fishable biomass and the spawning biomass would evolve under different options of fishing mortality.

Having estimated the state in the next year, these calculations can then be continued for the year after, and so on. Such predictions based on measured stock components start to break down when the incoming recruitment is no longer estimated from data but has to be guessed using either average values of recruitment or estimates based on a stock–recruitment relationship. Once substantial parts of the biomass are from ages whose biomass is based on estimated values of recruitment, predictions become increasingly subject to the error in these estimates. Moreover, as forward predictions are made relative to the supposed state of the stock, small errors in estimating that state become magnified. For all these reasons, after a few years the predictions become intrinsically variable and are best made in a probabilistic fashion showing the range of plausible answers rather than a single point estimate. This is the main feature of medium-term predictions. Iterating the predictions further forward still results in the distribution of results to be expected in the long term. Because the parameters – recruitment, growth and natural mortality – that make up the prediction are often affected by environmental factors (e.g., climate) or biotic factors (e.g., predation or food availability), such predictions may have to be for a particular regime, or need to take the effects of such variation into account.

So, all management approaches have some need for prediction of future stock states! However, we do not need to use the same methods or models to solve for all management questions!

7.2.5 *Take-home message*

Fishing is one of the major forces that affect the size of fish stocks and the subsequent harvest. It has the potential to reduce stocks to very small sizes, if they are fished in a way that does not give due consideration to the need to maintain adequate spawning potential through time. Management of fishing can seek both to adapt the size at which fish are caught, or the intensity with which they are harvested. Clearly, the management tools adopted need both to reflect the objectives of managers and to reflect such data and understanding of the stock as are available. Both commodities are expensive, and thus the use of expensive tools is most applicable to large valuable fisheries such as that for the Northeast Arctic cod. For smaller fisheries simpler, more pragmatic approaches may be appropriate.

7.3 The assessment problem and approaches to its solution

7.3.1 *Understanding the past to describe relevant processes*

The application of a model of a fish stock such as the yield model in Section 7.2.2 requires that we estimate the specific parameters associated with the various assumptions. To make predictions of the near future of a fish stock also requires us to understand its current status. Ideally such knowledge will include:

(1) The stock's abundance by size and age structure.
(2) The weight and condition of fish at various sizes or ages.
(3) What proportion of the stock is mature.
(4) The current schedule of fishing mortality and natural mortality rates by age.

We also need to know the main processes that may affect future developments. Processes likely to be of importance are:

(5) How fish grow, and how this is affected by changes in the biotic or abiotic environment?
(6) What affects the rates at which they die from natural causes?
(7) What is the pattern of fishing mortality-at-age, how is this affected by the fishing of different fleets and gears? How might this be affected by changes to fishing gear such as mesh changes?
(8) How does the proportion of mature fish change with age?
(9) How much does recruitment vary? Can some of its variation be attributed to changes in stock size or marine climate or environment?

The commonest route to answering these questions is by understanding the past history of the stock. Because all fish stocks are to some extent unique, this is certainly a desirable source of insight into how a stock behaves under exploitation. Although it will not provide full information into situations that the stock has not previously encountered during the period for which data are available, such understanding might be gained by analogy with similar stocks.

Some of the above questions can be answered from the results of a consistent time-series of well-conducted surveys. For example, questions (2), (3), (5) and (8), and possibly also question (9), might be addressed in this way. However, questions such as (1), (4), (6), (7) and (9) typically require a reconstruction of the past population dynamics of the stock. Such a reconstruction is made using models in order to interpret how the stock works. These models are needed because the key components, population numbers and mortality rates are, in most cases, not directly observable in an unbiased fashion by a survey. In what follows, therefore, we focus on the problems of estimating these more elusive components.

7.3.2 *Estimating population-at-age*

7.3.2.1 Simple age-based techniques: virtual population analysis and cohort analysis

As seen in Figure 7.2 (Plate 12), reconstructing the past structure of the stock is helpful both in understanding how the stock has behaved in the past and how the fishery affects it, and in providing a route for estimating the current status and the parameters needed to predict its future. The problem then is to estimate the past size and age structure of the stock. Age-based approaches to doing this have been developed over time. The perception that fish stocks had an age structure dates back to the studies of Petersen (1892) and of Hjort (1914). The systematic ageing of fish using ring structures on hard parts has a long history in our science (e.g., Jackson 2007), and hence for some stocks such as the Northeast Arctic cod it is possible to have quite long time-series of catch-at-age.

Early attempts to estimate past populations assumed that the backward sum of catches in number from a cohort (the fish born in one year) provided a minimum estimate of the population that was present at each age. Fry (1949) described the approach and indicated potential

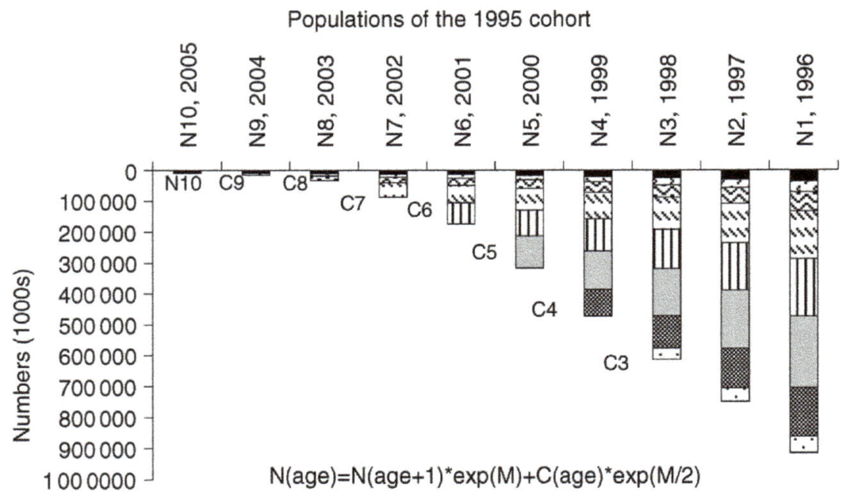

Figure 7.13 Cohort analysis on the 1995 cohort of Northeast Arctic cod. Different shadings indicate the contribution from the catch at each age as they are projected backward through time using the cohort analysis formula shown. For a color version of this figure, please see Plate 14 in the color plate section.

biases, and this idea was refined by Jones (1961) and particularly by Gulland (1965) and Murphy (1965). Their approaches solve the question: 'if I know the population numbers surviving at the end of a particular year and the natural mortality rate and the catch numbers of the cohort during the year, then what was the population number at the beginning of the year?', by solving the fishing mortality rate in the year. Essentially, the problem is to find a value of fishing mortality rate (F) in the two equations shown in Figure 7.4 such that they give the ratio that is observed between the catch numbers and the survivors. Having solved this, the survival equation can then be reversed to estimate the initial population from the population of survivors. The population numbers at the beginning of the year are of course the survivors from the previous year, so the process can be continued back down the cohort to the youngest age for which age data are available. Gulland's method, termed Virtual Population Analysis (VPA), while distinctly more convoluted than Murphy's Cohort Analysis, was the approach adopted in the ICES area, where most catch-at-age data were available at that time. It was first used on the Northeast Arctic cod, and the empirical finding was that the reconstruction of cohort[1] numbers in this fashion converged towards a similar answer on the younger ages, even though considerable uncertainty existed about the population number used to initiate the process. I provided a simple approximation to the process (Pope 1972). This had an explicit equation for the preceding year's population numbers that allows for the action of natural mortality. This approach, that was also (confusingly?) called Cohort Analysis, clarified what the VPA model was doing and why it converged backwards in time.

Figure 7.13 (Plate 14) shows how this works, and also give the formula. The figure uses the catch-at-age data from the 1995 year class of Northeast Arctic cod. These fish were age

[1] A Cohort originally meant a grouping of six hundred (480 soldiers and 120 non-combatants) in a Roman Legion, but is often adopted to indicate an age group, or in fisheries terms a year class. All the fish born in a particular Cohort may then be traced through time.

1 in 1996, age 2 in 1997 and so on, up to age 9 in 2004, the last year for which data were available at the time of writing (ICES 2005). Each entire column represents the population estimates of this cohort in each year (and hence each age). The columns are partitioned into the contribution made by the estimate of the fish that survive until age 10 at the beginning of 2005 (the black-filled cells) and the contribution of the catch numbers at each age in successive years. Notice particularly that if we have all of the catch from the stock accounted for, then the population estimates are the estimates of the total numbers of fish in the sea. This feature of estimating absolute population size makes the VPA/Cohort Analysis and other catch-age-based approaches particularly valuable, both for providing management advice and for making studies of recruitment processes. Such methods also provide estimates of fishing mortality rate by age and by year.

The analysis starts on the left of Figure 7.13 (and Plate 14) with a guess as to the population surviving to be age 10 in 2005 (N10, 2005; coloured solid black). It then proceeds backward in time, adding in the catch for each age of the cohort and making adjustments for natural mortality rate to obtain the population at each age. For example, the catch of age 5 (coloured blue in Plate 14) first enters in the N5 column in the year 2000. In each of the earlier years (those to the right) its contribution grows because it is adjusted to account for the erosion of population by natural mortality rate. Note that, for the same reason, the N10 contribution (coloured black) also becomes bigger in absolute size as the analysis moves to younger ages. However, the proportion it contributes to the total number grows less as younger ages are approached because this number becomes dominated by the contribution from the catches in various years. This leads to two insights. First, even quite large errors in the guess of N10 become relatively less important as the analysis moves backward through time. Second, the more of the stock has been removed by fishing between a population-at-age estimate and the guess of N10, the less the impact of any error in N10 on the estimate. Thus, the method converges proportionally towards the correct answer the further back in time you go, and it converges faster when fishing mortality is high. These insights can be generalised by theory.

So far we have assumed that the catches at each age are known exactly. If they are biased, for example because catch is under-reported, then the population estimates will be biased. If they are variable owing to sampling error but not biased, then another insight that can be gleaned from the figure is that population estimates are made up of functions of the catches at older ages. Thus, the coefficient of variation of the population estimates will be smaller than the coefficient of variation of the catches. Roughly speaking, the coefficient of variation of the population estimates for the younger ages might be half of the coefficient of variation of the catch data. Since the coefficient of variation of catch-at-age data for the better sampled stocks is seldom less than 20%, estimates of populations are likely to have coefficients of variation of about 10% or more. However, because year classes vary by factors of 10 in many stocks, and by more than 100-fold in some, such precision is certainly enough to tell good, bad and average year classes apart.

If our only concern was to study the history of a stock, then VPA techniques alone might be enough. We could simply discard the results for the most recent years and oldest ages, and the answers might be good enough for our purposes. However, if we have to predict forward then we need to know the current status. Hence, we need to know N10 and the populations from the other ages in the most recent year (2005 in the example) with more precision. Clearly, since N(10, 2005) was just a guess, VPA does not give an answer for this. Moreover, you might note that we have used up the available nine catch data to estimate the nine populations-at-age, and therefore we have no data left to estimate N10. Hence, we either need to find more data or use restrictive assumptions to estimate the populations in the most recent year. We also need to make assumptions in order to estimate populations for the oldest age group available in each year. These estimates are required to initiate the analysis on cohorts that have passed right through our data set. These latter estimates will particularly influence the estimates of SSB we make. This is because the SSB each year is based upon the estimates of the populations of all the older ages, which are quite closely coupled to our assumptions about numbers in the oldest age group.

How then, might we estimate the populations in the most recent year? Let us imagine for the moment that the catch-at-age data were exact. Let us also imagine that we also had a time series of exact catch-rate-by-age data from an annual fishing survey. It is reasonable to suppose that a consistently conducted survey catches a fixed proportion of the population at each age. This proportion of the total population that the survey catches is called the *catchability coefficient*. For example, to estimate the numbers at age 3 in 2005, the estimate of the population at age 3 in 1998 (N3, 1998) is used to calibrate the age 3 catchability coefficient from the survey result for that age. This is then used to calibrate the survey result at age 3 in 2005 to estimate the population at age 3 in 2005 (N3, 2005).

A cheaper alternative to survey catch-at-age data as a relative abundance measure are the time series of catch rates at age per unit fishing effort of commercial fishing vessels. This may often be derived from commercial landings data. Ideally, this measure – which usually is termed catch per unit effort (CPUE) – also reflects the relative abundance of fish at age and can be used instead of, or as well as, survey data. Because it is usually based on many more hauls than survey data it is potentially less variable, but because it is usually taken throughout the year it normally first needs to be standardised to remove the effect of major factors that can affect catchability, such as area fished, season and species targeted. There is also the strong possibility that vessel efficiency will increase with time and with the adoption of new technologies. For some species – particularly those caught in pelagic purse seine fisheries – it is also possible that catchability is correlated with species abundance. Where this occurs, catch per unit effort data should be avoided as they do not measure species abundance.

The approach, to calibrating catchability suggested above, supposes that both survey catch-at-age (or CPUE) and total fisheries catch-at-age are measured exactly. In reality, both are subject to considerable sampling error, and the process proposed above would serve to amplify these errors. A better approach would be to estimate the catchability at age 3 for all the years for which the survey (or CPUE) data are available and to estimate the mean and variance of the catchability estimates. We could then use the mean value of the catchability to calibrate the survey to obtain N3, 2005. Moreover, the variance would give us some idea of how well the survey data fitted the VPA results. Were there two surveys (or both survey and CPUE data), we could then use the variance obtained this way from each to form a weighted average of the separate values of N3, 2005 that we obtained from the two surveys. Done sensibly, the combined result should be less variable than the result we would have obtained from each survey taken separately. This sort of thinking was at the basis of many of the early methods of what has been called VPA 'tuning'.[2]

[2] And it came to pass in these days that we held a meeting at the Lowestoft Fisheries Laboratory entitled 'the well tempered Virtual Population Analysis' (cf., Johann Sebastian Bach's 'The Well-Tempered Clavier') and the tuning name stuck.

There are some complications that need to be mentioned. First, taking the simple variance assumes that the catchability coefficient has normal variance. This does not seem very likely since both survey and commercial catch-at-age data tend to have errors that are roughly proportionate to their size. Consequently, taking the mean and variance of the logarithm of the catchability coefficient is likely to be more sensible. Moreover, if we change the values of N10 then the values that go to make up the VPA estimates of the population at age 9, 8, and so on will be altered, and hence the calibration of the catchability coefficient for these ages. Hence, the calculation will need repeating a number of times (called *iteration*) until the population numbers no longer change (have converged). Another problem is the older ages. Normally, otoliths are read up to some maximum age (a_{max}), for example age 10 in this case. Fish older than that are simply placed in what is called a 'plus group' (e.g., 11+). Obviously, the populations we estimate for the younger ages of cohorts that have reached a_{max} at some year in the time series are dependent on the guesses we make of the population at a_{max} (N a_{max}). Survey (or CPUE) catchability estimates cannot help us here because we do not know any of them, and so we have to use assumptions to estimate these populations. The usual approach is to assume that the fishing mortality on these ages is similar to that on younger ages in the same year. Since fish are often of a similar size and maturity stage at these ages, this seems a reasonable starting assumption, but it is not necessarily correct. We can of course find fishing mortality at each age and year by using the survival equation and considering the ratio between its populations at successive ages in the cohort. These then form the basis of our assumption about fishing mortality rate on the oldest age and hence, via the capture equation, population on the oldest ages.

A characteristic of such *ad hoc* tuning methods was that the VPA results were treated as though they were exact. Thus, any variation found was treated as being due to the other data sets. Of course, this assumption was not true but at the time these methods were developed it seemed that total catch-at-age data would be much less variable than the survey abundance at age data (or at that time far more often commercial catch-at age-per-unit-of-effort data) used to provide relative population time series. Moreover, the assumption that the catch-at-age data were correct simplified the method and meant that computer programs could calculate the results in a reasonable time. This was a serious consideration at the time when computers were much slower than they are today and had limited memory This technological advance is a process negatively correlated with this aged author's declining mental powers!

7.3.2.2 A simple separable model of fishing mortality rate

As computers became more powerful, it became appropriate to develop models that treat catch-at-age data in the same way as other data sets. To do this required more restrictive hypotheses about the various parameters, particularly fishing mortality rate. A simple example of a more restrictive hypothesis is the assumption that fishing mortality-at-age can be split into an age component, which is the same for all years, and a year component, which is the same for all ages (Pope & Shepherd 1982). The product of these two factors gives the fishing mortality rate by age. You might think of the year factor as being proportional to fishing intensity (called *fishing effort*), and that the age factor is a selectivity factor appropriate for the type of gear being used. Hypothesis (3) in Section 7.2.2 was an example of an utterly simple, knife-edged selectivity function (0.0 below and 1.0 above some age of first capture).

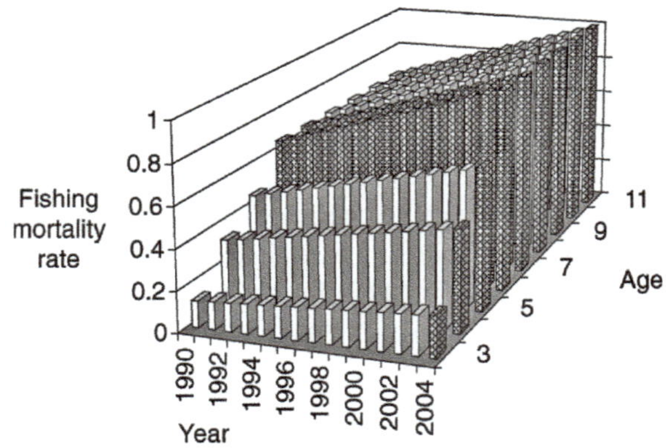

Figure 7.14 Fishing mortality rates based on an assumed separable pattern of year and age effects. The highlighted columns in year 2004 (for which the year factor was 1.0) indicate the values of the age factor assumed for each age, while the highlighted columns for age 6 (for which the age factor was 1.0) indicate the values of the year factor assumed for each year.

Figure 7.14 shows an example where the fishing mortality rate is the product of a year effect that increases by a constant amount each year from 1990 (where it is 0.65) and 2004 (where it is 1.0) and an age effect that is 0.2 for age 3 and increases to 1.0 by age 6 and then stays at this level on older ages. Let us further assume the recruitment of each year class at age 3 after 1990 is 1000, and also that in 1990 the sizes of the population of each age was also 1000. We will further assume that there is a constant natural mortality rate of 0.2 for all ages and years. With these assumptions the population at each age can be predicted. With this and the fishing and natural mortality rates at each age and year, we can predict the catch numbers for each age and year. Essentially, this would just be a slightly more detailed analysis than that we have already seen in Figures 7.6 and 7.7 for the Beverton & Holt model. Figure 7.15 shows the catches-at-age resulting from our assumptions.

We can compare these modelled catch-at-age estimates with the observed catch-at-age data from a real fish stock. Figure 7.16 shows such data for the Northeast Arctic cod for these

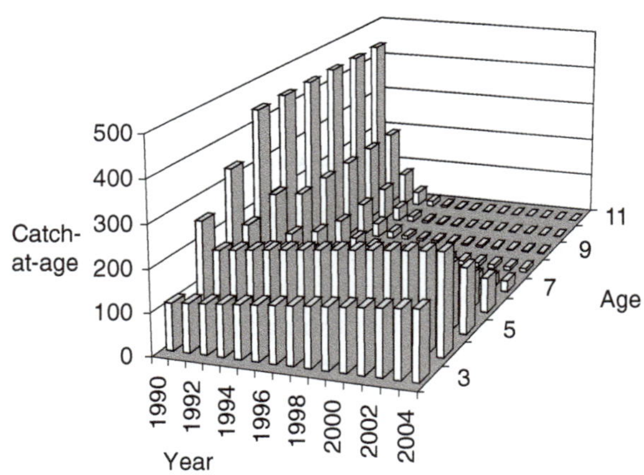

Figure 7.15 Illustration of what the catch-at-age would be if fishing mortality rate was as shown in Figure 7.14, and the initial population of each cohort at the age and year it entered the figure was 1000.

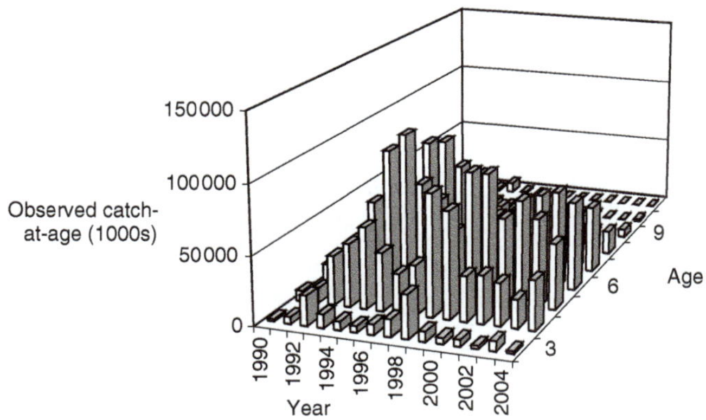

Figure 7.16 Observed catch-at-age data for the Northeast Arctic cod (years 1990–2004, ages 3–11).

ages and years. Clearly, our predictions in Figure 7.15 do not look anything like the real data. However, we can use a minimisation routine to adjust our assumptions about the year and age effects that combine to estimate fishing mortality rate and also the year-class size until our predicted catch-at-age fits the real data set as well as possible. First, though, we need to ask what we mean by 'fit as well as possible'. What is our criterion for the best fit? Minimising the sum of the squares of the difference between the predicted and the observed catch-at-age might be one idea. However, in proportional terms this would try to fit the large values well and do a poor job at fitting the small values. A better idea might be to minimise the sum of squares of difference between the logarithms of the observed and the expected values. This would try to fit errors that were in the same proportion to the observed values, whether these were large or small. In statistical terms, this is equivalent to assuming that all the data have the same coefficient of variation.

If we minimise in this way, we obtain modelled values of the catch that look quite like the data. These catch-at-age estimates (Figure 7.17) are similar to but not quite the same as the real data (Figure 7.16). The differences between the logarithms of the observed and the modelled values are called the *residuals*. Their average value (root mean square) in this example is about

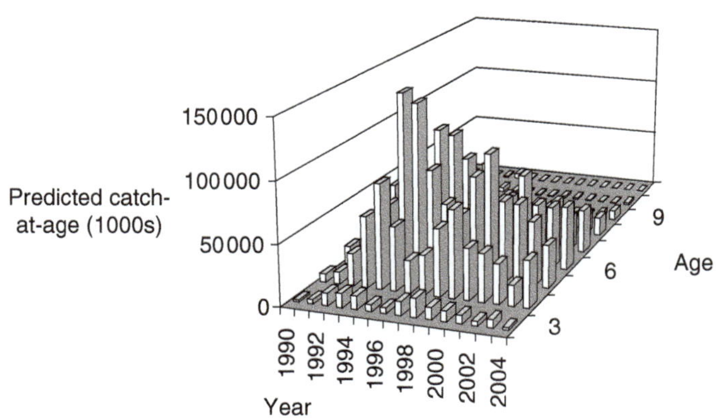

Figure 7.17 Predicted catch-at-age data for the Northeast Arctic cod (years 1990–2004, ages 3–11) obtained by using the separable model.

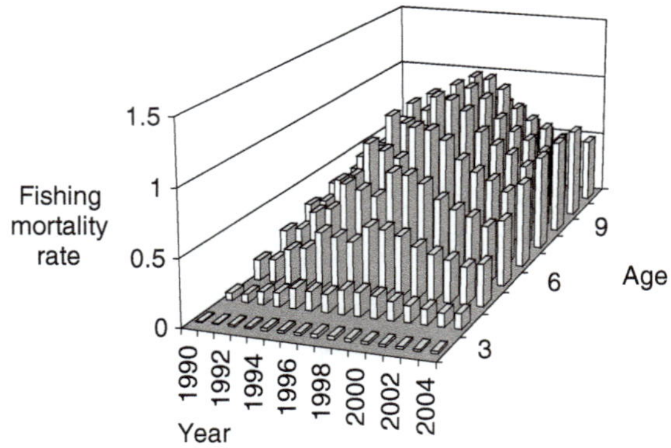

Figure 7.18 Predicted fishing mortality rate-at-age data for the Northeast Arctic cod (years 1990–2004, ages 3–11) obtained by using the separable model with the year effect in 2004 held at 0.5 and the age effects on ages 6 and 11 held at 1.

0.22, which is equivalent to a ratio of the modelled values being 24% higher or 20% lower than the observed value. Figure 7.18 shows the fishing mortality rates that correspond to this fit that are clearly different from the initial assumptions shown in Figure 7.14. Figure 7.19 shows the corresponding fitted population numbers at each age and year. It is clear that the initial size of each year-class as it enters the table is very different from our initial assumption of them all being 1000. Extending such an analysis for more years and all the ages would provide the basis for estimating recruitment for each cohort and the population building blocks that are needed for estimating biomass. These population estimates coupled with weight-at-age and maturity-at-age data from surveys would allow us to estimate biomass and spawning biomass each year as we saw in Figure 7.2.

Several interesting points emerge from this example. The first point is historical. When I first proposed a method similar to this in the 1970s, it took me a month or more to write the computer code and one run took a matter of some hours. The present example took about half an hour to write as a spreadsheet model, and minimisation runs using the solver function in EXCEL took just a second or two. This is a tribute to the improved speed of computers and the ease of using modern computer packages. It is not surprising therefore that the number of

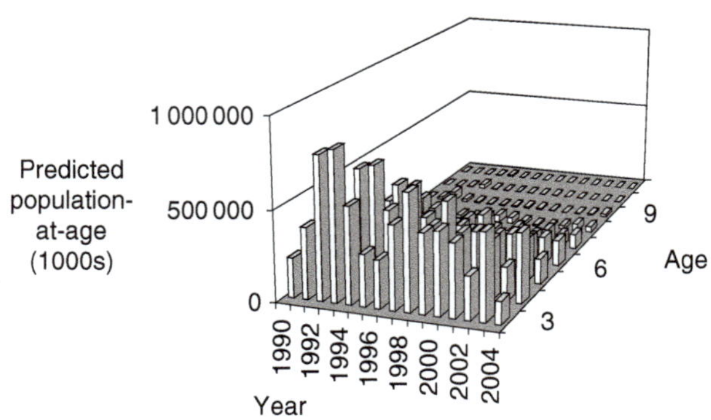

Figure 7.19 Predicted population number-at-age data for the Northeast Arctic cod (years 1990–2004, ages 3–11) obtained by using the separable model with the same constraints as Figure 7.18.

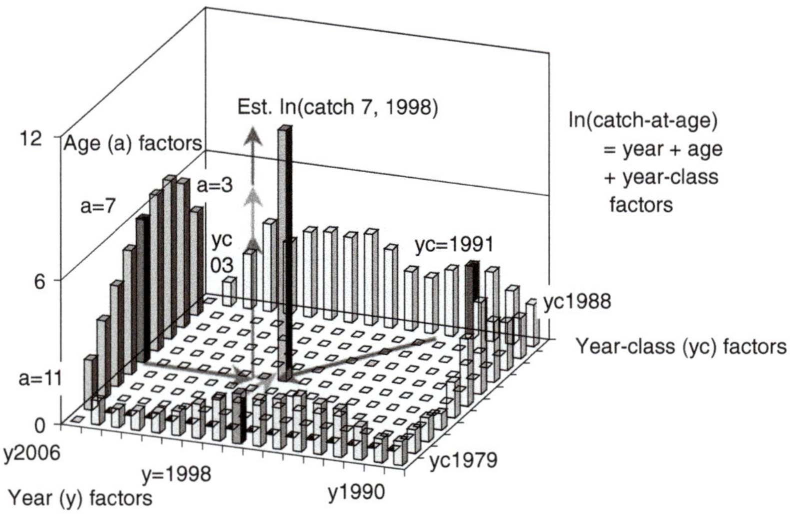

Figure 7.20 An illustration of how the additive effects of the Shepherd–Nicholson model estimates the natural logarithm of the catch-at-age 7 in year 1998 by the summing an age 7 effect, a year 1998 effect and a year-class 1991 effect. All other ln(catches-at-age)(these are not shown) are similarly explained by the sums of the appropriate age factors (shown along the left edge), year factors (shown along the bottom edge) and year-class factors (shown along both the top and the right-hand edges). For a color version of this figure, please see Plate 15 in the color plate section.

methods used has ballooned through time; scientists in various areas are able to model their own stocks in the way they feel best mirrors what is going on and the information they have. Therefore, this section cannot be exhaustive; it just indicates how a simple approach works. If you need more detail, it is time to read a textbook on the subject (e.g., Hilborn & Walters 1992, Quinn & Deriso 1999).

The second point is that, as I found when I first developed this approach in the 1970s, the method does not give a very clear signal as to what is the right value of the year effect in the last year or the age effect on the last age. For example, the sum of squares is at a minimum if the year effect for 2005 is set to 1.5. This value in fact sounds rather high. However, the sum of squares increases by less than 5% of the minimum if we set the year effect anywhere between 0.7 and 3.0. Clearly, the minimum obtained is not sharply focused on any specific level of year effect in 2005. This means that we only vaguely know the mortality rates or the populations in the last year. The same is also true for the age effect on the last age. It seems that the trend in fishing intensity with year or age is not well defined by catch-at-age data alone. This finding is closely analogous to what we found from cohort analysis.

The reason for this being the case is that the logarithms of catch-at-age data can be well approximated by the sum of year factors, age factors, and year-class factors. This formulation is described by Shepherd & Nicholson (1991), and is a very useful tool for examining fisheries data. Figure 7.20 (Plate 15) shows how the year factor for 1998, the age factor for age 7 and the year-class factor for the 1991 year-class sum up to give the ln(catch) estimate of 7-year-old fish in 1998. Fitting this simple additive Analysis of Variance model to our Northeast Arctic cod data explains all but a residual sum of squares of 7.4 which is only a little bigger than the minimum of 6.2 achieved by the separable model. However, the

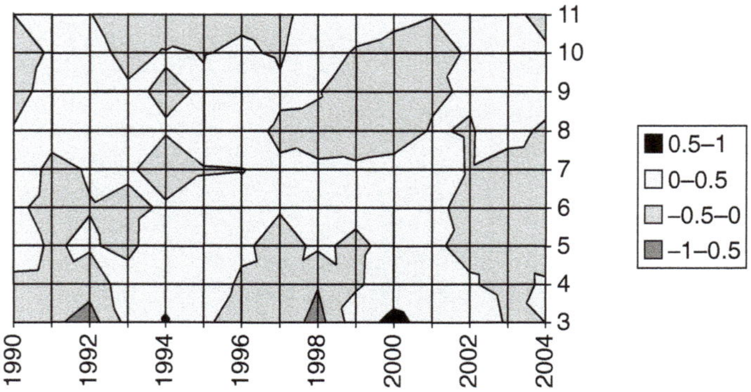

Figure 7.21 Residuals of the fit to catch-at-age data for Northeast Arctic cod (years 1990–2004, ages 3–11) found using the separable model. For a color version of this figure, please see Plate 16 in the color plate section.

year class that the catch of a cohort of fish belongs to is simply the year they were caught minus their age. Hence, if a term (h*year) is added to each year factor, a term –h*age is added to each age factor and –h*(year-age) to each year-class factor, then these all cancel out and we have the same fit as before. Clearly, there is what statisticians call aliasing between these factors; we could get the same answers with an upward trend in the year factor and a balancing downward trend by age and by year-class factors. Since this additive model has much the same structure as the separable model with factors for year class, age and year, a similar aliasing is likely to occur with the separable model. We can have a positive trend with the year factors coupled with a negative trend with age and year-class factors and get much the same fit. Hence, to tie these age and year factors down either we need to have auxiliary data or to make assumptions about how fishing mortality changes with age and with year.

The third point is that by plotting out the residuals for the separable model by age and year (Figure 7.21; Plate 16), we see some tendency for high values and low values to clump together. This might suggest that the age effect is changing between years and that perhaps a more detailed model is needed than our simple – one selection pattern fits all years – separable model. This would certainly be the case if specific fleets in successive years fished more or less on particular ages of fish. In the case of Northeast Arctic cod, this is rather likely because the TAC of the stock is mostly shared between Russia, which catches more young fish in their sector, and Norway, which catches somewhat older fish in their sector. The TAC is based on all ages of fish but the national shares are fixed. Hence, in years when there are less young fish, the Russian fleet will have to fish harder for their share of the TAC and, similarly, the Norwegian fleet will have to fish harder if there was a period with fewer old fish. This might well disrupt the pattern of the age factor between periods of years. Thus, more complex models of fishing mortality rate may be appropriate.

The fourth point is that Figure 7.21 (Plate 16) also shows that the largest residuals all occur on age 3 fish, which suggests that our assumption of equal log variance is incorrect. Thus, more elaborate statistical models of variability might be more appropriate.

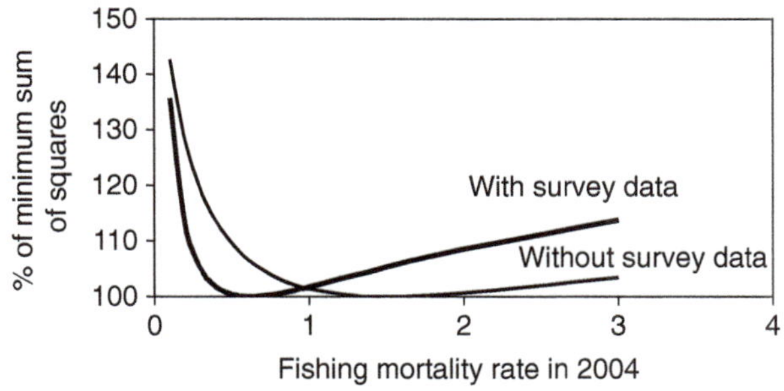

Figure 7.22 Percentage change in the sum of squares function for the Northeast Arctic cod fitted with and without survey data.

7.3.2.3 Integrated models

Let us now consider including auxiliary data (e.g., Deriso *et al.* 1985, Gavaris 1988, Megrey 1989, Methot 1990). For use in our example, one possible set is the Norwegian acoustic survey of the Barents Sea and Lofoten area. This survey provides estimates of relative population size from ages 3 to 11. However, on the older ages some estimates are zeros that are difficult to use when we are considering log differences. Hence, we will just use the results from ages 3 to 9 from years 1990 to 2004. We can fit these data to the model we have already constructed simply by multiplying the population estimates at each age (as those in Figure 7.19) by an age-specific catchability term. We can then derive the sum of squares of the differences between the logarithms of the observed and the modelled survey results, and add these to the sum of squares that we already have from the catch-at-age data. If we felt the survey data were more reliable or less reliable than the catch-at-age data, we might weight their contribution to the sum up or down accordingly. For the moment, we will just take the sum of the two sums of squares and use the minimisation routine to adjust our old inputs and our new catchability-at-age assumptions to give an overall best fit.

If we allow the minimisation routine to change all our input parameters, including the fishing mortality rate in the last year, then this last fishing mortality rate is estimated at about 0.61. If we hold this parameter fixed at a range of values, we find that the minimum is now rather better defined than before: the sum of squares is within 5% of the minimum when this value is between 0.5 and 1.4. Thus, we have narrowed the range of values we might accept for fishing mortality rate in 2004, although the range is still quite wide. Figure 7.22 compares how the sum of squares changes with our estimate of fishing mortality rate in 2004, and also shows this for the separable model we looked at above.

7.3.2.4 Using auxiliary data

In the case of the Northeast Arctic cod, there is in fact more than one survey series available. Fitting all of these simultaneously would perhaps result in a better estimate, and this is what the relevant ICES working group does (ICES 2005). Calculated on the same basis as our estimate, their equivalent combined 2004 year estimate would be about 0.52. A problem with using more than one survey series is deciding how much weight to give to the various ages of the different surveys. Various computer packages have been developed over the years to provide solutions to this problem. When using these, it is important to consider their diagnostics carefully. It is also worth fitting each data set separately and making plots of the sorts we saw in Figure 7.22. For instance, two survey series may give quite different results! In this case, one might believe either by itself but not believe both at the same time. This should make us think hard about the data and how they are collected. Might we have some reason to suppose that a particular survey was biased? For example, has the gear used been changed during the time-series? Has the sampling protocol always been the same? If the data are catch-per-unit-effort data from commercial fleets, has the fleets' fishing behaviour changed? It is certainly much more important to focus on the quality of the data than on what the model says! The old saying 'garbage in, garbage out' remains true, however clever the method!

7.3.2.5 Time-series models

We have just seen that, even with auxiliary data, quite a wide range of plausible answers are possible. Are there other approaches that might be used to narrow them down? If we fit the last model with the fishing mortality rate in 2004 at 1.4 (recall this was about the upper believable value), then it is noticeable that this is much higher than the rates estimated for the previous years. The highest of these was 0.87 in 1999. Higher levels of fishing mortality rate require either that fishing effort has increased (more boats or more working time per boat), or that the fish have become more catchable. Both explanations are possible, but for a stable fishery like that for the Northeast Arctic cod we might expect things to stay much the same from year to year. We could introduce this idea, by forming a sum of squares of the difference between the logarithms of fishing mortality rate in successive years. If we do this without fitting the auxiliary data, we could minimise the sum of squares obtained from the catch-at-age data together with this measure of annual variability in fishing mortality rate. However, we should ask ourselves how much weight it should be given. A first guess might be times nine, because the year effects are applied over nine ages. Using this weighting, we obtain an estimate of year effect in 2004 of 0.66. If we try higher weightings, the fishing mortality rates in all years gradually converge to a common mean value. Using assumptions about how fishing mortality rate might change provides an alternative approach to auxiliary data. However, this approach requires a good understanding of how the fishery works. There might be plausible reasons why the fishing mortality rate might jump between two years.

We noted above that our separable model of fishing mortality rate may well be too simple. This might be improved by fitting models to catch-at-age data from different fleets. Each of these fleets could then have a separate year effect and age-based selection pattern. It may also be that our biological model is too simple. In particular, the assumption that natural mortality rate is constant may be challenged. In the case of Northeast Arctic cod, the working group responsible for its assessment calculates the amount of cod eaten. Much of the predation occurs below age 3 but in some years predation on age 3 fish can be considerable. Over the time period considered, heavy cannibalism occurred between 1995 and 1997. Considerable effort has been put into quantifying predation in the Northeast Arctic ecosystem, as well as in the North Sea and Baltic Sea ecosystems (see Pope 1991, Sparholt 1991).

In an analysis that combines both the constraint on fishing mortality and auxiliary data, we could also include other types of data that provide information on population (e.g., egg survey results), on fishing intensity (e.g., tagging experiments), or selection pattern (selectivity experiments). We might also include other assumptions. An example would be fitting the parameters of the recruitment–stock relationship directly in our model rather than treating this as a secondary investigation, to be conducted once a time series of annual SSB and recruitment values had been established. Similarly, the age effect might be constrained to having a specific functional form rather than allowing it to be freely estimated at each age.

As we use more data and constraints in our model, it becomes increasingly difficult to decide how to weight the different types of information and constraints. Moreover, the model that fits the various components becomes more of a black box, the results of which are difficult to understand. Too much fitting to one sort of data may cause the model to balloon out in some unconnected dimension. These points would seem to argue for a careful fitting of data separately as well as in combined models to better understand, which data sets are mutually compatible and which are contradictory.

7.3.2.6 Maximum likelihood

The problem of how to weight different data sets remains. One increasingly used approach is to replace least-squares estimates with maximum-likelihood estimates. The idea is to take the likelihood of seeing all the events that have occurred given a set of parameters and then to find the parameter set that maximises this likelihood. Often, this is framed as maximising log likelihood because likelihood estimates are often products of tiny numbers. That is to say, they represent the least unlikely interpretation. Taking the logarithm is therefore convenient and because the logarithm of a function has its maximum at the same parameter values as the function, it does not change the solution.

At times, maximum-likelihood solutions may in fact be the same thing as a least-squares solution. In our first separable fishing mortality example, we minimised the sum of squares of the fit of our model to the logarithms of the catch-at-age data. The unstated assumption here is that the data have a log normal distribution for all ages with the same variance. If that were the case, then the probability of seeing particular values of the logarithm of each data point would lay on a normal curve with a mean that is given by the modelled estimate of the logarithm of catch-at-age. The likelihood is the joint probability of all the data. This is calculated assuming that the probability of each is described by these bell-shaped normal distributions. Taking the logarithm of this function has two effects. First, it changes the product into a sum. Second, it changes the shape of the bell-shaped normal curves into parabolas, and the probability of each observed data value becomes proportional to minus the square of the difference between the logarithms of the observed value of catch-at-age and its expected value. If variances were different these terms would be weighted by the inverse of the variance, but if our assumption that all variances are equal holds then we are back to the equivalent of the least-squared solution. Of course, this convergence of the two approaches is not true for all statistical distributions; the use of the maximum likelihood approach scores best when more complex distributions than the normal distribution are postulated. It also scores well if the precise distribution to expect from the nature of the data is known. An example is estimating a selection curve by length from experiments with a covered cod-end. Suppose we had data on how many fish were caught and how many passed through the mesh into the cod-end cover. At each length there is a probability of being caught and a complementary probability of passing the mesh, and we know how many fish faced this choice. The numbers retained should then follow what is called a *binomial distribution* – that is, the distribution of how many heads you would get from tossing that number of coins when each coin had the same probability of coming up heads as the retention probability. Another example of a known distribution is when fish are simply sampled at random for age. In this case, their age distribution should follow a multinomial distribution. However, most of the assessment data we see, such as catch-at-age data, are typically built up from the results of multiple samples collected under different national sampling schemes. In these circumstances, it is virtually impossible to derive a theoretical sampling distribution that would describe combined data. We are then left with observing the resulting data and assuming sensible statistical distributions.

Whichever approach is used, there is virtue in giving the opportunity to individual auxiliary data sets to fit the model as well as possible. When this is done, the residual variation represents just the random errors (noise) about the fit and how noisy the data are should suggest the weight to apply.

7.3.2.7 Bayesian methods

Another approach is based upon Bayesian statistics. This is a slightly different approach to traditional frequentist statistics, as Bayesian statistics endeavour to calculate how our belief changes in the light of new evidence. They thus calculate the probability that we give to various hypotheses (or estimates) in the light of new evidence (data). We thus move from some statement of our prior belief in a hypothesis (or an estimate) given as a prior probability distribution, via the data, to some revised statement of belief given as a probability distribution.

In practice, in fisheries assessment we often have only rather vague prior beliefs as to the value of particular parameters. These parameters may be of two types: (i) those that the data tells us about rather clearly (such as the size of a particular recruitment some years ago); and (ii) points such as natural mortality rate, about which typical assessment data such as catch-at-age data tell us rather little. In the case of the former, it is likely that the evidence in the data will heavily outweigh any prior belief we may have, while in the latter case the posterior probability distribution may be little changed from the prior probability. As an example, we might think that for Northeast Arctic cod (at least older than age 5) the natural mortality is about 0.2, but we could also believe it to be 0.1 or 0.3, more extreme values probably being unlikely. On the whole, it is unlikely we would change our opinion in the light of typical data, unless perhaps we had some rather restrictive beliefs in other parameters such as the change in catchability with size. What makes the Bayesian approach particularly useful is that it gives us revised probability distributions of the parameters that we can estimate such as year-class strength, which are tempered by our lack of certainty about key, but not very estimable, parameters such as natural mortality rate. Moreover, from these probability distributions we can build up probability distributions of constructs of our parameters such as next year's catch.

As a very simple example of this approach, we may use the Shepherd–Nicholson model reanalysis of the 1990–2004 ages 3 to 11 data from Northeast Arctic cod. We saw this earlier in Section 7.3, and particularly in Figure 7.20. For the year, age and year-class parameters of this model, we will probably have rather little prior information. We recall that these factors are all logarithmic terms and add together, and are thus subject to scaling factors that need to be hard-wired. This is usually done by setting the factor for the first year class (the 1979 year class) and for the first year effects (1990) to zero. Prior distributions for the other year class and year factors might therefore reasonably be set at zero, which would imply a belief that they might well be the same as the first year. However, we might expect year class to vary over several orders of magnitude while the year effects (which are basically a measure of

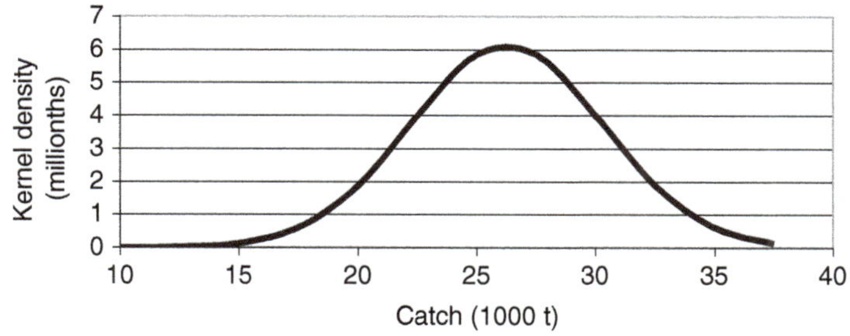

Figure 7.23 Example of a posterior distribution plot (smoothed kernel density from a sample of 40 000) of catch weight in 2005 of Northeast Arctic cod aged 3–11 resulting from a Bayesian fitting of the Shepherd–Nicholson model to catch-at-age data between 1990 and 2004, ages 3–11.

fishing intensity) would normally be expected to be less variable. The age effects are scaled to the general level of the logarithms of catch-at-age data, but their relative sizes reflect both selection effects and the effect of cumulative mortality with age. Thus, differences between ages may reflect the total mortality. We might therefore attempt to set up priors that reflect how high or low a total mortality we might expect. However, it is easier just to assume we do not know and give the age effects uninformative priors. Similarly, we might well give the year effects uninformative priors. Finally, the variance of how well the model fits could be given a prior. Here, we do have some feeling that this would probably have a coefficient of variation more than 10% but probably less than 50%. Fitting a Bayesian model on this basis results in the posterior probability distribution for the 2005 catch shown in Figure 7.23.

7.3.2.8 Delay-difference approaches

A useful feature of the von Bertalanffy growth equation (see Section 7.2) is that it can be rearranged so that the length of a fish next year can be expressed as a simple linear regression of its length this year. The constants of this relationship are just expressions of the growth rate and the asymptotic size, so they are unaffected by the age or size of the fish. Traditionally, this relationship was used in the Ford–Walford plot approach to estimating the growth parameters. It is an example of using a recurrence relationship; that is to say defining something in terms of its past values.

Cohort analysis (explained above in Section 7.3) is also an example of a recurrence relationship. In its usual formulation it works backwards (see Figure 7.13; Plate 14), but of course this can be reversed to define population numbers next year in terms of population and catch this year. The coefficients of this relationship are given in terms of the exponential of natural mortality rate. With assumption (3) we considered natural mortality rate to be the same for all ages.

If we use these two recurrence relationships together, and recall that biomass is weight × number, we can write a recurrence relationship for the biomass of a year class of fish next year in terms of its biomass and number and catch this year. Moreover, with our assumptions of von Bertalanffy growth and constant natural mortality-at-age, the relationship has the same multipliers for all ages, so we can add the formulae for each separate year class together. This provides a recurrence relationship for next year's total biomass in terms of the biomass, catch, recruitment and population numbers of this year. There are various ways of doing this but the general approach of writing biomass and/or numbers in terms of the previous year's results are called delay-difference methods.

In fisheries science, delay-difference approaches go back to the Russell equation. This states that next year's biomass is equal to this year's biomass, less the weight loss to catch and to natural deaths and plus the weight gain due to growth and to recruitment. However, it was Deriso (1980) who showed how to convert from an age-based to a biomass-based model along the lines indicated above. The attraction of this class of models is that we have retained our biological understanding of what goes into the relationship without needing to describe the precise age structure of the stock.

To make the conversion, it is easier to assume – as did Deriso – that the Ford–Walford plot applies to growth in weight rather than length. However, while this often works well for year-to-year predictions, it is not strictly correct and may therefore break down when extrapolated to give long-term relationships. Fournier & Doonan (1987) and Pope (2003) show ways it may be written using the Ford–Walford plot to give the relationship between successive years' lengths. These formulations indicate that, instead of describing the structure of the population in terms of numbers-at-age, we can define it in terms of total numbers and the first four statistical moment descriptors of its length distribution (mean, standard deviation, skew and kurtosis). My earlier report (Pope 2003) tracked these measures as the sum of products of numbers by lengths raised to powers between 0 and 4, which for convenience I call *protomoments*. As well as the statistical moments, these provide stock numbers and stock biomass, and thus pretty well everything we really need to track the stock behaviour both in the short and the long term.

An advantage of having the five protomoments to describe the size distribution is that we may use them to form polynomials, which can closely approximate quantities, such as the SSB, that are weighted differentially by fish size. While protomoments do not estimate recruitment directly, it does come into their formulation and they can thus be used to calibrate recruitment results from surveys. Therefore, we could get much the same biological understanding from a delay-difference model as we do from an age-based model, but with a considerable saving in the routine ageing of fish. These savings should make these models particularly useful as an approach for understanding and managing smaller fish stocks whose catch value does not justify expensive science programmes.

The method can be used to interpret past catch and survey data in a fashion similar to VPA or Cohort Analysis, and to project future catches in the light of this interpretation. Figure 7.24 (Plate) 17 shows estimates of the five protomoments for the Southern Newfoundland (3Ps) cod. These show both an interpretation of the past and a forward projection under the assumption of unchanged catch until 2010. With an understanding of the recruitment–stock relationship, it is also possible to estimate long-term yield using protomoments. A yield isopleth surface for the 3Ps cod calculated by the protomoment model is shown in Figure 7.25. The previous isopleth surfaces we saw in Figures 7.8 and 7.12 were plotted using the fishing mortality rate and age of first capture as the horizontal axes. The protomoment isopleth surface is instead plotted with horizontal axes of the proportion of the 3+ biomass harvested in one direction, and the ratio of the average weight of fish in the catch to the average weight of fish in the sea in the other direction. However, the former measure is closely related to fishing mortality rate while the latter measure is clearly related to gear selectivity, because a higher ratio clearly means that larger fish are being selected.

It is interesting that the curves seen at each ratio of the average weight of fish in the catch to average weight of fish in the sea have a parabolic form against harvest rate. This clearly links

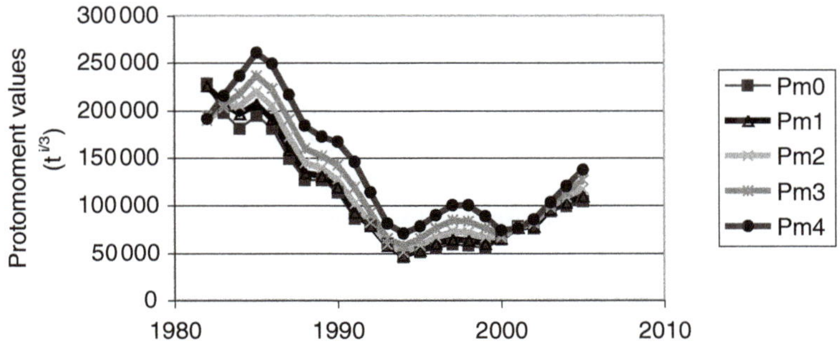

Figure 7.24 Protomoments (i = 0 to 4) of the Southern Newfoundland (3Ps) cod fitted to past data and projected on the assumption of a 15 K TAC from 2003 to 2008. For a color version of this figure, please see Plate 17 in the color plate section.

this approach to general production models such as the Schaefer model that fits a parabolic yield curve to catch and effort data. Such production models have the advantage of simplicity in the way they address the main management question of the right amount of fishing to obtain maximum yield. Moreover, in such models fish growth and recruitment are implicitly density-dependent. However, being based only on catch and effort data they have far less recourse to other data sets, and as Hilborn & Walters (1992) explain, they are indeterminate if faced by a 'one-way trip'. Moreover, while they represent a pragmatic and simple approach to providing fisheries management advice, they do not provide estimates of the underlying biological processes, and thus for the purpose of this book they do not readily help us understand reproductive biology.

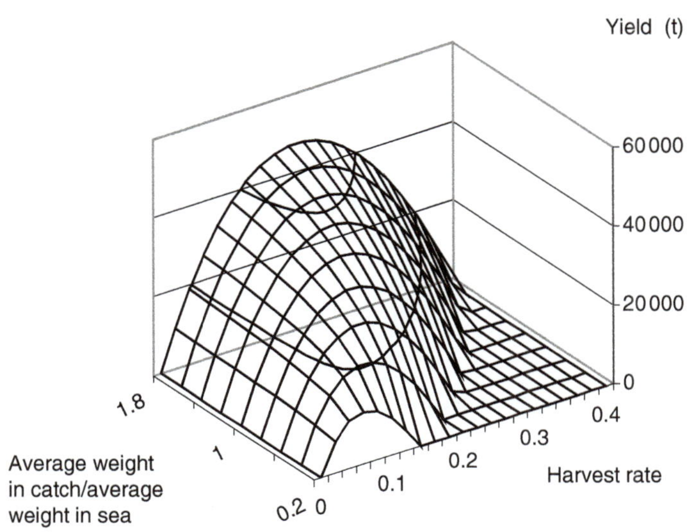

Figure 7.25 Long-term yield surface for Southern Newfoundland cod.

7.3.3 *Management procedures*

The idea behind management procedures is for managers to agree in advance a precise procedure for how the management plan will be applied in given circumstances. The management plan might, for example, be to set an annual TAC according to a procedure of calculating it in a set way using only pre-specified data sets. In advance of a specific procedure being chosen, realistic simulations can be made of a series of various candidate procedures based upon different methods and data set combinations. These simulations allow the managers to see how well each of the candidate procedures would succeed in managing the stocks according to their various objectives for the fishery. The managers can then chose the rule that best matches their criteria of management success.

In practice, the rules adopted in such an approach may be quite simple. One simple example might be: if the trend in catch rate from a groundfish survey has been upward for the past five years, then increase the TAC by some fraction of its upward trend. Indeed, it seems that fairly simple rules often work best. This does not mean that the detailed assessments we have discussed previously are useless, as they will certainly be used in making the realistic simulations needed to test various rules. Moreover, we will still need them to better understand the stock's biology. However, it may well be that the procedure testing shows that a simple rule performs better than a complex assessment technique for the purpose of providing management bodies with advice on the value of the annual TAC.

7.3.4 *Take home message*

Stock assessment methods are designed to estimate those aspects of fish stocks that are difficult to observe directly, such as fishing mortality rate, absolute population size and recruitment. In particular, it has always seemed to me that if we do not understand what proportion of fish are being caught from a stock then we are in a confusing situation. We do not know if a certain catch is a small proportion of a big stock or a very big proportion of a small stock, and thus we do not know whether it would be better to expand or to contract the fishery. It is this confusion that fisheries assessment methods must seek to resolve.

There are many methods available, and some scientists are passionate advocates of particular approaches. One thing is agreed however – that all need good data to function properly and many will be seriously affected by biased inputs such as catch data that has been distorted by the fishing industry mis-reporting or discarding fish. Consequently, it is vital to get to know the fishery and the fishermen so that you can judge how good or bad the data are that you are using.

7.4 A few case histories

We have seen in the previous sections how stocks might react to exploitation under our simple assumptions. We have also seen how factors such as stock size, recruitment, spawning stock size and exploitation rate that cannot be directly observed can be estimated. We will now see how these results are used to show how real stocks have reacted to extended exploitation. Do the simple assumptions of Section 7.2 hold, or do they need to be modified? How in particular does recruitment react to spawning stock size and to other factors? It is easiest to look for these factors on stocks for which we have many years of data and that have experienced a range of exploitation levels and of climatic conditions.

7.4.1 *Northeast Arctic cod*

We will start with the 'platonic ideal fish stock', the Northeast Arctic cod, which we have already seen in the examples in this chapter.

Data from 1946 onward are available on the ICES website in the report of the Northeast Arctic Working Group (ICES 2005), which provides their interpretation in population and mortality terms. Catch-at-age data and other data for this stock can be extended back at least to 1930 (Pope *et al.* 2001). Both data sets give nominal recruitment and spawning stock series. However, the fecundity of these cod is known to improve with size and with feeding conditions (Chapter 8), and so these nominal SSB series are not the best index of spawning potential. To create the SSB time series, it is necessary to consider how the maturity-at-ages series and weights-at-age have changed over time. These changes indicate what may be a density-dependent increase in growth rate due to a reduced stock density and a corresponding lowering of the age of first maturity. A more worrying hypothesis would be that this response was due to fisheries providing a selective advantage to younger spawning elements of the gene pool (e.g., Heino & Gødo 2002, Kenchington *et al.* 2003). Certainly, both the weights-at-age have increased and the age of 50% maturity has decreased from about age 11 in 1948 to about age 7 at the present time. This suggests that hypotheses of constant growth and constant maturity schedule implied by assumptions (2) and (6) of Section 7.2.2 must be rejected and some density-dependent model put in their place, if long-term yield is to be correctly estimated.

It is also well known that climate, as measured by the temperature of the standard Kola hydrographic section, has a dominant effect on cod recruitment. This is not particularly surprising for a stock at the northern end of its range (Loeng 1989a,b). Some authors (e.g., Nilssen *et al.* 1994) have suggested that spawning stock effects on recruitment can only be revealed after the effects of temperature and cannibalism are stripped out of the recruitment time series. The backward extension of the analysis to 1930 also suggests that climate and cannibalism by older ages of cod most strongly modifies the recruitment series. Consequently, it is not surprising that a straightforward plot of recruitment on spawning stock size (Figure 7.26) tends to show much variation in recruitment at any specific size of spawning stock observed.

From Figure 7.26 it is nevertheless noticeable that the lowest values of recruitment are associated with the lowest level of spawning stock. If the recruitment series is corrected for

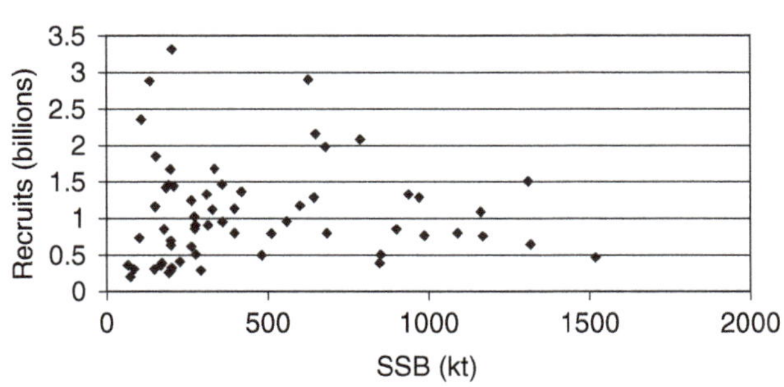

Figure 7.26 Stock–recruitment scatter for Northeast Arctic cod (after the analysis of Pope *et al.* 2001).

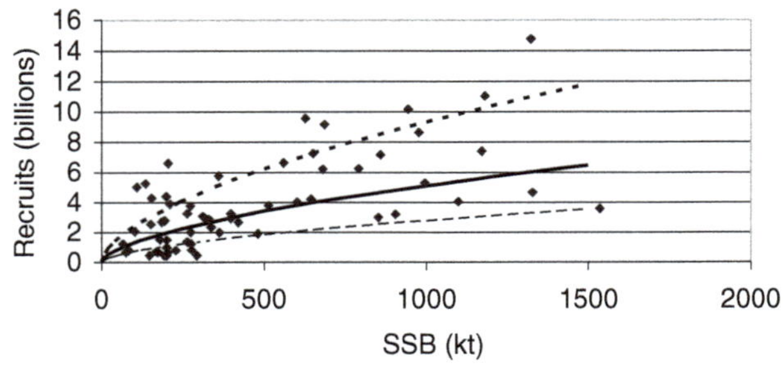

Figure 7.27 Stock–recruitment scatter for Northeast Arctic cod (1930–1990) (after the analysis of Pope *et al.* 2001), including cannibalism effects. The three power curve stock–recruitment relationships shown are drawn at average temperature (solid dark line) and with temperature 1 °C higher (dotted line) and 1 °C lower (dashed line).

temperature and cannibalism, then the relationship between spawning stock and recruitment becomes more apparent (Figure 7.27 shows the corrected figure from Pope *et al.* 2001). We might note that the effects of the reduction of spawning stock biomass are most likely felt when environmental conditions are also most adverse. Because these conditions are not generally predictable far into the future, upper limits on fishing mortality and lower limits on spawning stock biomass had best be set at levels that will allow the stock to survive and reproduce through the most adverse environmental situations that can reasonably be predicted. Time has moved on since the analyses was made on which Figures 7.1–7.3 and 7.26–7.27 were based. ICES (2012a) provides a more up-to-date analysis made along similar lines, while the most recent assessments are available at the ICES website (http://www.ices.dk/publications/Pages/default.aspx). Since 2004 a harvest control rule has been agreed for this stock, and this has served to reduce the mortality rate both on older fish and younger fish to about half of those seen in Figure 7.3. This reduced fishing rate, together with some strong year classes, has resulted in a very strong growth of Spawning Stock Biomass, and in 2011 this was almost 2 million tonnes, which is a level higher than that seen in the 1930–1995 series. Despite the reduction in fishing intensity, catches have now recovered to levels broadly comparable to those of earlier years. Thus, the recent management of this stock is a success story and underlines the basic thesis of fisheries population dynamics that too much fishing is counterproductive.

7.4.2 Northern cod of Newfoundland

The Northern cod of Newfoundland provides a clear example of the consequences of the spawning potential of a stock being heavily reduced by fishing prior to a period of adverse environment.

Much has been written about the collapse of this stock. As of 2005, the stock status summary provided by DFO (2006) gave the following assessments: the stock biomass was reduced from about 3 million tonnes in the 1960s to a small remnant in the 1990s, and remains at a very depleted level. The biomass of this remnant is difficult to estimate but seems likely to be substantially less than 100 000 tonnes in the early years of the 21st century.

In the productive period during the late 1970s, the fishing pressure exerted seemed, when viewed on a yield-per-recruit basis, to be sustainable. However, a similar fishing pressure was

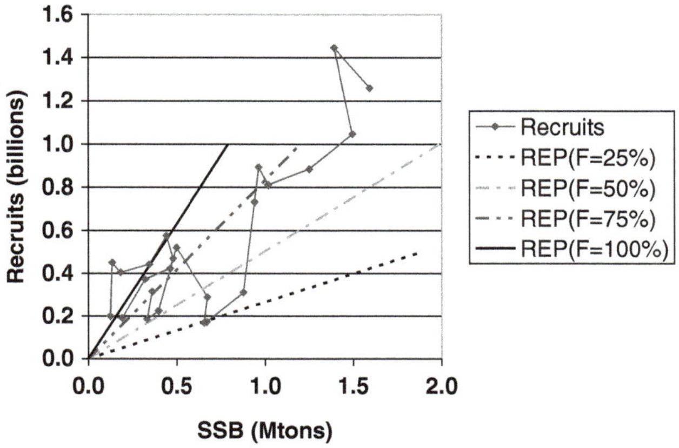

Figure 7.28 Recruitment plotted against spawning stock biomass for the Northern cod of Newfoundland (year classes 1962–1983). The replacement lines (REP%) are drawn at multiples of 100%, 75%, 50% and 25% of the average fishing mortality rate between 1962 and 1989. This fishing mortality rate averaged 0.47 between ages 5 and 10. For a color version of this figure, please see Plate 18 in the color plate section.

exerted certainly during the late 1980s and the 1990s. Water temperatures, measured off St John's Harbour, in some years of this period fell lower than –1.5 °C, and consequently growth rates and fish condition were notably low during this cold, unproductive period. However, an analysis of the recruitment process suggested that the average fishing intensity was not sustainable even in the productive period. Figure 7.28 (Plate 18) illustrates this, showing that the replacement line based on the average level of fishing in the period 1962–1989 was higher than the recruitment SSB points for all but a very few years. This means that the average future SSB that would be produced by a given recruitment would be lower than the SSB that spawned it. In short, the stock was not replacing itself at this average mortality rate. Figure 7.28 also shows the replacement lines associated with various fractions of the average mortality rate. It would seem likely that the rate would have had to be no more than 50% of the average exploitation rate to have been sustainable, and even lower to have been entirely safe.

As of 2005, recruitment had remained poor despite some amelioration of environmental conditions. The situation of a depleted spawning stock seemed still to be exacerbated by a lack of forage fish, predominantly capelin, and by a total mortality that remains quite high. In the offshore area, while some of this may be due to by-catch mortality in other fisheries, some is likely due to elevated levels of natural mortality rate, which includes predation by seals. This predation mortality results in a biomass removal, which would have been only a minor and probably undetectable component of the cod's mortality rate at higher stock sizes, but which might represent a substantial part of the stock biomass, and hence create a high mortality rate on cod when stock biomass is low. Thus, this cod stock had possibly fallen into a predator trap from which it will need exceptional circumstances to escape, followed by continual protection during any regeneration.

Clearly, from a management point of view these are circumstances best avoided, since the road back to stock health is long, difficult, unpopular, and uncertain. Scientifically, the experience of this stock shows that the simple assumptions of Section 7.2 must be rejected in favour of models that predict the response of the stock to changing spawning stock size and environmental conditions. In recent years, the situation of this stock has improved somewhat (DFO

2013), with the geographical distribution having spread offshore and the condition and age structure of the fish having improved. Surveys suggest that there has been some increase in offshore biomass up to at least until 2008, though after that it has remained without clear trend. However, despite these gradual improvements the stock is still far below the limit reference point set for spawning stock biomass, and recovery has a long, long way to go. This emphasises how long stock rebuilding can take, as this stock has been in severe trouble since the early 1990s.

7.4.3 North Sea herring

A more encouraging picture is seen in the case of the North Sea herring. Figure 7.29 shows the trends in both SSB and recruitment. This stock complex, which is formed of a number of geographically separate spawning components, suffered stock–recruitment collapse during the late 1960s and early 1970s due to heavy fishing reducing stock sizes. This was also a time, which the continuous plankton recorder time series suggests was a period of diminished *Calanus* production. A moratorium on fishing in the late 1970s and an increase in the environmental signal in the 1980s led to a partial recovery of this stock's spawning potential. This situation faltered during the 1990s but was corrected by a reduction in fishing intensity, and the spawning stock is now higher than at any time since the early 1960s. Despite this, recruitment has been low in the most recent years and would imply a need for cautious management. A more recent assessment (ICES 2012b) also indicates that, despite lower fishing mortality rates in recent years and a high spawning stock biomass, recruitment to the stock in the past decade has been roughly half of what it was in the mid 1980s to the mid 1990s, and it would seem that other factors are at play than the effect of the fishery. However, it could be that managing the stock with a low fishing mortality and high SSB has averted a collapse such as that seen after 1960 or for the Newfoundland cod.

Perhaps the most important question with the North Sea herring is the effect that it might have on other stocks. The recruitment in gadoid stocks was generally higher during the period of the herring collapse, and has tended to be lower since their resurgence. Figure 7.30 (Plate 19) shows the catches of major components of the fish biomass in the North Sea from 1902 to 1990. It is uncertain whether the switch seen between pelagic fish abundance and demersal fish abundance between 1960 and 1970 was a direct result of reduced herring predation on eggs and/or larvae of other species, or the effect of changes in concomitant variables such as

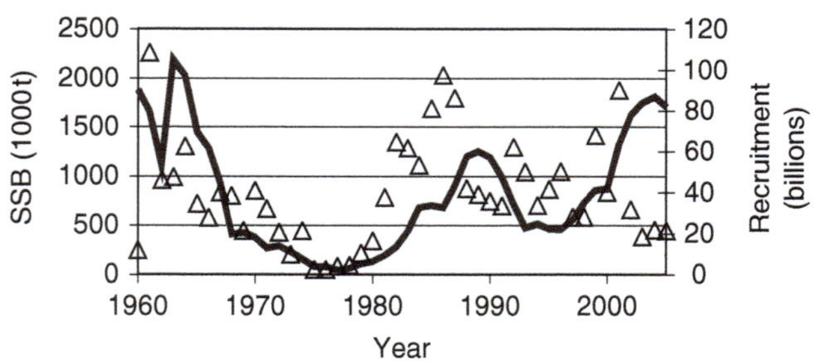

Figure 7.29 Time trends of recruitment (triangles) in billions and SSB (solid line) in 1000 tonnes for North Sea Herring. After ICES 2006.

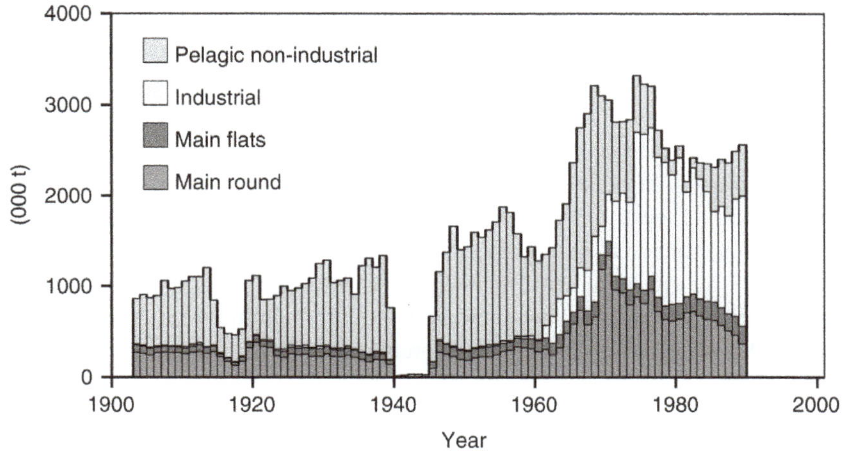

Figure 7.30 Trends of major components of North Sea catch 1902–1989. For a color version of this figure, please see Plate 19 in the color plate section.

zooplankton abundance or timing. If it is a direct predation effect, then the correct management of herring would need to be considered together with that of other species.

7.4.4 *Baltic stocks*

The need to consider environment and species interactions is also a prime need in understanding the main cod, herring, and sprat stocks of the Baltic Sea. The Baltic is a brackish sea that flushes with saltier water from the North Sea at infrequent intervals. Following flushing events, the environment seems to favour cod, which preys heavily on sprat and herring and tends to suppress the former in particular. In the periods between flushing events, the surface waters gradually become fresher and prevent mixing of the waters in the deeper basins, which therefore become anoxic. In these fresher conditions, the cod finds it harder to reproduce itself, and the herring and sprat seem to do better. Since sprat and, to a lesser extent, herring prey on cod eggs and larvae, this in turn tends to further suppress cod. It would seem that this is an area where sustainable fishing arguments seem secondary to environmental and multispecies effects. However, some minimum abundance of each species is still a basic requirement of management. Species with multiple-year spawning strategies such as cod and herring are probably hedging their bets with respect to surviving in a variable environment by potentially living to old ages. The effect that fishing has in terms of reducing the age structure of populations may well work against this strategy and cause fishery management problems. ICES (2013) provides an overview of these problems.

7.4.5 *Take home message*

The example stocks give some idea of how the estimates of stock abundance, fishing mortality rate and recruitment are used in providing the background understanding of real-life fisheries. These now exist for many stocks, and are available on the web for detailed study at sites like that of the ICES.

The chosen examples suggest that the simple models that we started with are inadequate, and that generally the assumption of constant life-history parameters is suspect. More specifically, the assumption that recruitment is fluctuating around a constant mean is not tenable. In various examples, we see that recruitment can be influenced by the marine environment, by the effects of other species in the system and by the spawning potential of the species in question.

The interrelationship of these factors seems complex, and the original aim of describing these in mathematical equations now seems somewhat optimistic. Nevertheless, the simple equations act as parables that may guide us away from undesirable outcomes, even if they cannot be trusted to provide an exact map of how to manage fisheries optimally. Perhaps all we can hope to do is to achieve either something like my 'minimum sustainable whinge' that seeks to avoid the worst pitfalls or Hilborn's 'pretty good yield'.

7.5 Incorporating understanding of the recruitment process into predictions

7.5.1 What should we understand about the stock–recruitment problem?

For a number of fish stocks (e.g., the Northern cod of Newfoundland or the North Sea herring) that have been fished to dangerously low levels, we have a fairly clear idea of the shape of the stock–recruitment relationship. However, for other stocks such as the Northeast Arctic cod we may just see it, when we have cleaned away the noise and modifications created by other processes. In other cases, we may suspect it is lurking under the noise but cannot see it clearly.

It is also important to remember that the stock–recruitment relationship in our examples is not just subject to noise from environmental or multispecies processes, but its shape may also be modified by these processes (see the various lines on Figure 7.27). Thus, a stock size or a level of exploitation that appeared safe based upon past experience may prove unsafe in extreme conditions.

Most importantly, we must realise that, when a stock has lost the ability to adequately reproduce itself, rebuilding it is likely to be a slow and painful process. My old colleague John Shepherd had a dictum '... that if you know the stock–recruitment relationship of your stock, then you have lost!'. In the light of the examples, this seems horribly true! So it is certainly wise to assume the relationship is there, even if it is non-apparent, and hence to be cautious and adopt policies that do not depress the stock's reproductive capacity too far.

But how far is too far? Clearly, a better understanding of stock–recruitment processes is desirable but is not always available from assessment results alone. Thus, process studies designed to clarify these need to be encouraged, as do studies designed to provide wider generalisations across stocks.

7.5.2 How can we incorporate our knowledge into our advice process?

Clearly, whilst we must include stock–recruitment considerations in our assessments, it will only rarely be possible to use a well-established stock–recruitment relationship. In lacking this knowledge, using a variety of approaches seems wise. Some are simple. For example:

- Be very cautious if the stock produces one or more exceptionally low recruitments in succession.
- Be very cautious if its spawning areas contract or change or if subcomponents are lost.
- If possible, arrange the gear selectivity and fishing mortality rates so as to give fish the chance to spawn at least once or a few times before they are fished.
- Alternatively, allow fish to run a gauntlet of fishing during some period but ensure that they are protected against capture after passing a certain age.
- Closing spawning grounds to fishing would be a way of reinforcing such measures.

For choosing fishing mortality rates, the simple SSB per recruit plots are helpful and sensible targets such as $F_{33\%}$ (the fishing mortality rate that reduces spawning stock biomass to one-third of its unfished value) might be adopted. Where extensive time series of fishing mortality rate, recruitment and spawning stock exist, pragmatic choices of fishing mortalities and spawning stock sizes that have proved safe in the past can be used to establish limit reference points. However, the point made above that levels that seemed safe in productive times may prove unsafe in unproductive times needs to be kept clearly in mind. Some measure of recent productivity is thus a useful measure to have to hand.

The F_{med} (also known as F_{rep}) Sissenwine & Shepherd (1987) measure is another way of suggesting levels of fishing mortality that should be safe, provided that other factors affecting recruitment remain constant. It has the virtue (Jakobsen 1992) that it is a measure which is insensitive to the choice of natural mortality rate in the analysis. This measure is based on choosing as the replacement line for a stock–recruitment plot the line through the origin that has equal numbers of observations above and below it. Figure 7.28 (Plate 18) shows four replacement lines for the Northern cod of Newfoundland. The REP (F = 75%) line corresponds approximately to the F_{med} replacement line, since there are 10 observations above it and 10 below, and two more-or-less on the line. However, in the light of subsequent history of this stock, the F_{med} of 0.35 that this predicts would seem excessive for this stock and the REP (F = 50%) which corresponds to a fishing mortality of about 0.235 would seem the upper limit of what could be considered prudent when the stock has recovered sufficiently to be fished. The reason that F_{med} would not have proven prudent for this stock was (as has been explained in Section 7.4.2) is because the collapse occurred during a cold, unproductive period when any fishing mortality rate would have been too much. Indeed, the stock continued to decline during the early years of the moratorium on fishing it.

Stock–recruitment relationships are often incorporated into advice by providing limit reference points for spawning stock size and for fishing mortality rates. The intention is to provide biological limits beyond which it would be imprudent for management to stray. Such limits are an essential part of the precautionary approach to fisheries management (FAO 1995). Obviously, stock–recruitment relationships should also help to provide advice on target levels of stock and fishing mortality rate, but this requires more knowledge than we may have.

7.5.3 *Take home message*

Do the best you can, but never stop worrying! Since I have more or less retired, I will let you do the worrying from now on and go out and prune the roses. May you and your stocks both 'Live long and prosper'!

References

Beverton, R.J.H. & Holt, S.J. (1957) *On the Dynamics of Exploited Fish Populations*. Fisheries Investigation Series II, **19**. MAFF, London.

Deriso, R.B. (1980) Harvesting strategies and parameter estimation for an age-structured model. *Canadian Journal of Fisheries and Aquatic Science*, **37**, 268–82.

Deriso, R.B., Quinn, T.J., II & NealP.R. (1985) Catch-at-age analysis with auxiliary information. *Canadian Journal of Fisheries and Aquatic Science*, **42**, 815–24.

DFO (2006) *Stock Assessment of Northern (2J3KL) Cod in 2006*. Department of Fisheries and Oceans, Canada. Scientific Advisory Secretariat Scientific Advisory Report. 2006/015. http://www.dfo-mpo.gc.ca/csas/Csas/status/2006/SAR-AS2006_015_E.pdf

DFO (2013) Stock Assessment of Northern (2J3KL) Cod in 2013. Department of Fisheries and Oceans, Canada. Scientific Advisory Secretariat Scientific Advisory Report. 2013/014. http://www.dfo-mpo.gc.ca/csas-sccs/Publications/SAR-AS/2013/2013_014-eng.pdf

FAO (1995) Precautionary approach to fisheries. Part 1: Guidelines on the precautionary approach to capture fisheries and species introductions. Elaborated by the Technical Consultation on the Precautionary Approach to Capture Fisheries (Including Species Introductions). Lysekil, Sweden, 6–13 June 1995 (a scientific meeting organized by the Government of Sweden in cooperation with FAO). FAO Fisheries Technical Paper No. 350, Part 1. 52 pp.

Fournier, D.A. & Doonan, I.J. (1987) A length-based stock assessment method utilizing a generalized delay-difference model. *Canadian Journal of Fisheries and Aquatic Science*, **39**, 1195–207.

Fry, F.E.J. (1949) Statistics of a lake trout fishery. *Biometrics*, **5**, 27–67.

Gavaris, S. (1988) An adaptive framework for the examination of population size. Canadian Atlantic Fisheries Scientific Advisory Committee (CAFSAC) Research Document 88/29. 12 pp.

Gulland, J.A. (1965) Estimation of mortality rates. Annex to Arctic Fisheries Working Group Report (meeting in Hamburg, January 1965). ICES, CM 1965, Doc. No. 3 (mimeograph).

Hardin, G. (1968) The Tragedy of the Commons. *Science*, **162**, 1243–8.

Heino, M. & Gødo, O.R. (2002). Fisheries induces selective pressures in the context of sustainable fisheries. *Bulletin of Marine Science*, **70**, 639–56.

Hilborn, R. & Walters, C.J. (1992) *Quantitative Fisheries Stock Assessment: Choice, Dynamic and Uncertainty*. Chapman & Hall, New York.

Hjort, J. (1914) Fluctuations in the great fisheries of northern Europe viewed in the light of biological research. *Rapports et Procès-verbaux des Réunions, Conseil International pour l'Exploration de la Mer*, **20**, 1–228.

ICES (1984) Report of the Ad Hoc Multispecies Assessment Working Group, 18–22 June 1984, Copenhagen. ICES CM 1984/Assess:20.

ICES (1986) Report of the Ad Hoc Multispecies Assessment Working Group, 13–19 November 1985, Copenhagen. ICES CM 1986/Assess:9.

ICES (1989) Report of the Multispecies Assessment Working Group, 7–16 June 1989, Copenhagen. ICES CM 1989/Assess:20.

ICES (2005) Report of the Arctic fisheries Working group (AFWG) 19–28 April 2005, Murmansk, Russia. ICES CM 2005/ACFM:20.

ICES (2006) Report of Herring Assessment WG for the Area South of 62° N. (HAWG) 14–23 March, Copenhagen. ICES CM 2006/ACFM: 20.

ICES (2012a) Report of the Arctic Fisheries Working Group 2012 (AFWG), 20 - 26 April 2012, ICES Headquarters, Copenhagen. ICES CM 2012/ACOM:05. 6ŝŖ pp.

ICES (2012b) Report of the Herring Assessment Working Group for the Area South of 62 N (HAWG), 13 - 22 March 2012, Copenhagen, Denmark. ICES CM 2012/ACOM:06. 835 pp.

ICES (2013) Report of the Workshop on Integrated/Multispecies Advice for Baltic Fisheries (WKMULTBAL), 6–8 March 2012, Charlottenlund, Denmark. ICES CM 2012/ACOM:43. 112 pp.

Jackson, J.R. (2007) Earliest references to age determination of fishes and their early application to the study of fisheries. *Fisheries*, **32**, 321–8.

Jakobsen, T. (1992) Biological reference points for North-East Arctic cod and haddock. *ICES Journal of Marine Science*, **49**, 155–66.

Jones, R. (1961) The assessment of the long term effects of changes in gear selectivity and fishing effort. *Marine Research*, **2**, 19 pp.

Jørgensen, T. (1990) Long-term changes in age at sexual maturity of Northeast Arctic cod (*Gadus morhua*). *Journal du Conseil International pour l'Exploration de la Mer*, **46**, 235–48.

Jørgensen, T. (1992) Long-term changes in growth of Northeast Arctic cod (*Gadus morhua* L.). *Journal du Conseil International pour l'Exploration de la Mer*, **49**, 263–77.

Kenchington, E., Heino, M. & Nielson, E.E. (2003) Managing marine genetic diversity: a time for action? *ICES Journal of Marine Science*, **60**, 1172–6.

Loeng, H. (1989a) The influence of temperature on some fish population parameters in the Barents Sea. *Journal of Northwest Atlantic Fisheries Science*, **9**, 103–13.

Loeng, H. (1989b) Ecological features of the Barents Sea. In: L. Rey & V. Alexander (Eds) *Proceedings of the Sixth Conference of Comité Arctique International: 13–15 May 1985.* pp. 327–65. E.J. Brill, Leiden.

Marshall, C.T., Kjesbu, O.S., Yaragina, N.A., Solemdal, P. & Ulltang, O. (1998) Is spawner biomass a sensitive measure of the reproductive and recruitment potential of Northeast Arctic cod? *Canadian Journal of Fisheries and Aquatic Science*, **55**, 1766–783.

Marteinsdottir, G. & Steinarsson, A. (1998) Maternal influence on the size and viability of Iceland cod *Gadus morhua* eggs and larvae. *Journal of Fish Biology*, **52**, 1241–58.

Megrey, B.A. (1989) Review and comparison of age-structured stock assessment models from theoretical and applied points of view. *American Fisheries Society Symposium*, **6**, 8–48.

Methot, R.D. (1990) Synthesis model: an adaptive framework for analysis of diverse stock assessment data. *International North Pacific Fisheries Commission Bulletin*, **50**, 259–77.

Murphy, G.I. (1965) A solution of the catch equation. *Journal of the Fisheries Research Board of Canada*, **22**, 191–202.

Nilssen, E.M., Pedersen, T., Hopkins, C.C.E., Thyholdt, K. & Pope, J.G. (1994) Recruitment variability and growth of Northeast Arctic cod: influence of physical environment, demography and predator–prey energetics. *ICES Marine Science Symposium*, **198**, 449–70.

Petersen, C.J.G. (1892) Fiskensbiologiske forhold i Holboek Fjord, 1890–1891. Beretning fra de Danske Biologiske Station for 1890 (1891) 1:121–83. (Original not seen, summarised in Ricker, W.E. (1975) Computation and interpretation of biological statistics of fish populations. *Bulletin of the Fisheries Research Board of Canada*, **191**, 1–382.)

Pope, J.G. (1972) An investigation of the accuracy of virtual population analysis using cohort analysis. *Research Bulletin of the International Commission for Northwest Atlantic Fisheries*, **9**, 65–74.

Pope, J.G. (1991) The ICES Multispecies Assessment Working Group: evolution, insights and future problems. In: N. Daan & M.P. Sissenwine (Eds) *Multispecies Models Relevant to Management of Living Resources*. pp. 22–33. ICES Marine Science Symposia, 193.

Pope, J.G. (VI) (2003) Golden ages or magic moments? *Natural Resource Modeling*, **16**, 439–64.

Pope, J.G., Large, P. & Jakobsen, T. (2001) Revisiting the influences of parent stock, temperature and predation on the recruitment of the Northeast Arctic cod stock, 1930–1990. *ICES Journal of Marine Science*, **58**, 967–72.

Pope, J.G. & Shepherd, J.G. (1982) A simple method for the consistent interpretation of catch-at-age data. *Journal du Conseil International pour l'Exploration de la Mer*, **40**, 176–84.

Quinn, T.J. (II) & Deriso, R.B. (1999) *Quantitative Fish Dynamics*. Oxford University Press, New York.

Rollefsen, G. (1954) Observations on the cod and cod fisheries of Lofoten. *Journal du Conseil International pour l'Exploration de la Mer*, **136**, 40–7.

Shepherd, J.G. & Nicholson, M.D. (1991) Multiplicative modelling of catch-at-age data, and its application to catch forecasts. *Journal du Conseil International pour l'Exploration de la Mer*, **47**, 284–94.

Sissenwine, M.P. & Shepherd, J.G. (1987) An alternative perspective in recruitment overfishing and biological reference points. *Canadian Journal of Fisheries and Aquatic Science*, **44**, 913–18.

Sparholt, H. (1991) Multispecies assessment of Baltic fish. In: N. Daan & M.P. Sissenwine (Eds) *Multispecies Models Relevant to Management of Living Resources*. pp. 64–79. ICES Marine Science Symposia, 193.

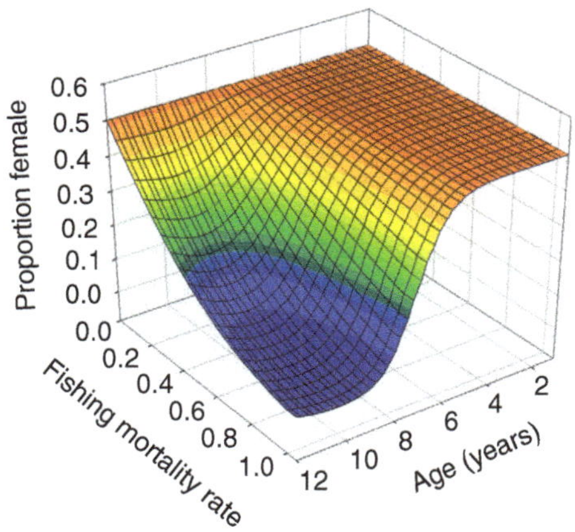

Plate 1 Proportion of females in a population as a function of age and fishing mortality when females exhibit faster growth and males and females experience identical natural mortality rates. The sex ratio at birth is assumed to be 1:1.

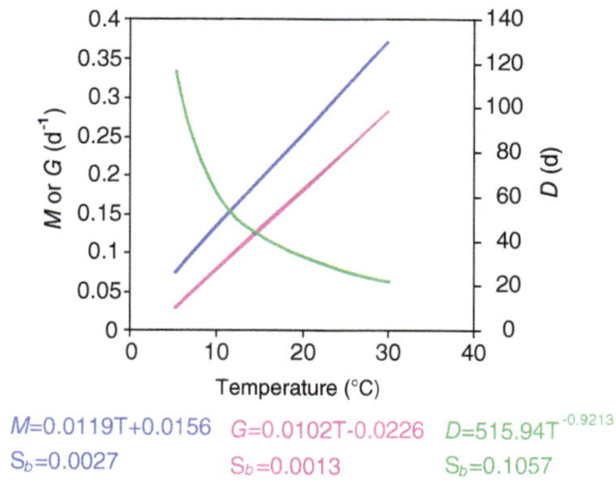

Plate 2 Summary figure of *M*, *G*, and *D* for marine fish larvae in relation to temperature, summarised from metadata analysis by Houde (1989). Lines are the linear regression fits for weight-specific growth (*G*) and daily mortality (*M*), and a power model regression fit for Larval Stage Duration (days, *D*) with respect to temperature. S_b = standard error of the regression coefficient.

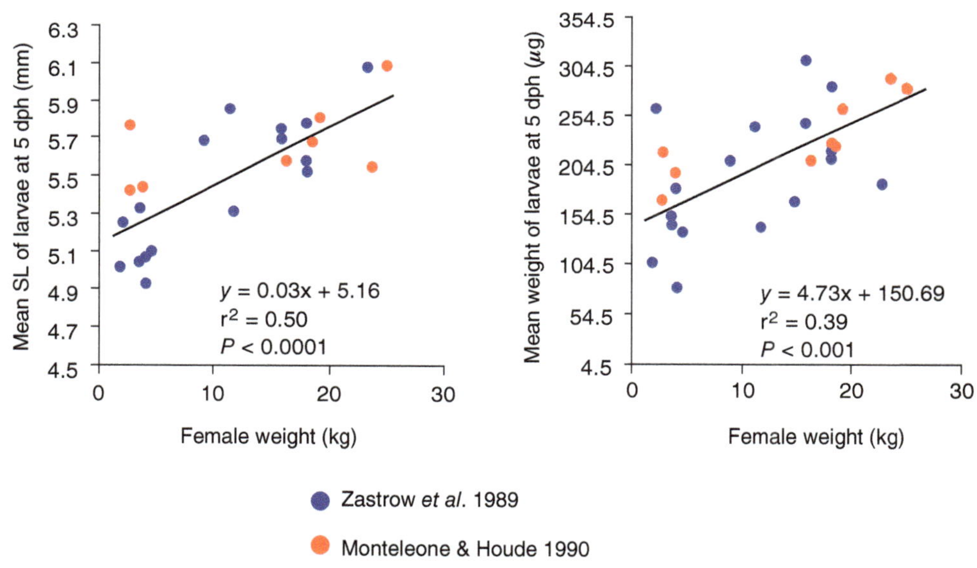

Plate 3 Maternal effects. Lengths and dry weights at 5 days post-hatch (day of first feeding) of striped bass *Morone saxatilis* larvae in relation to adult female weight. Larvae from smallest females weigh, on average, only 63% as much as larvae from the largest females. Data reproduced from Zastrow *et al.* (1989), with permission of ICES, and Monteleone & Houde (1990).

Fish Reproductive Biology: Implications for Assessment and Management, Second Edition.
Edited by Tore Jakobsen, Michael J. Fogarty, Bernard A. Megrey and Erlend Moksness.

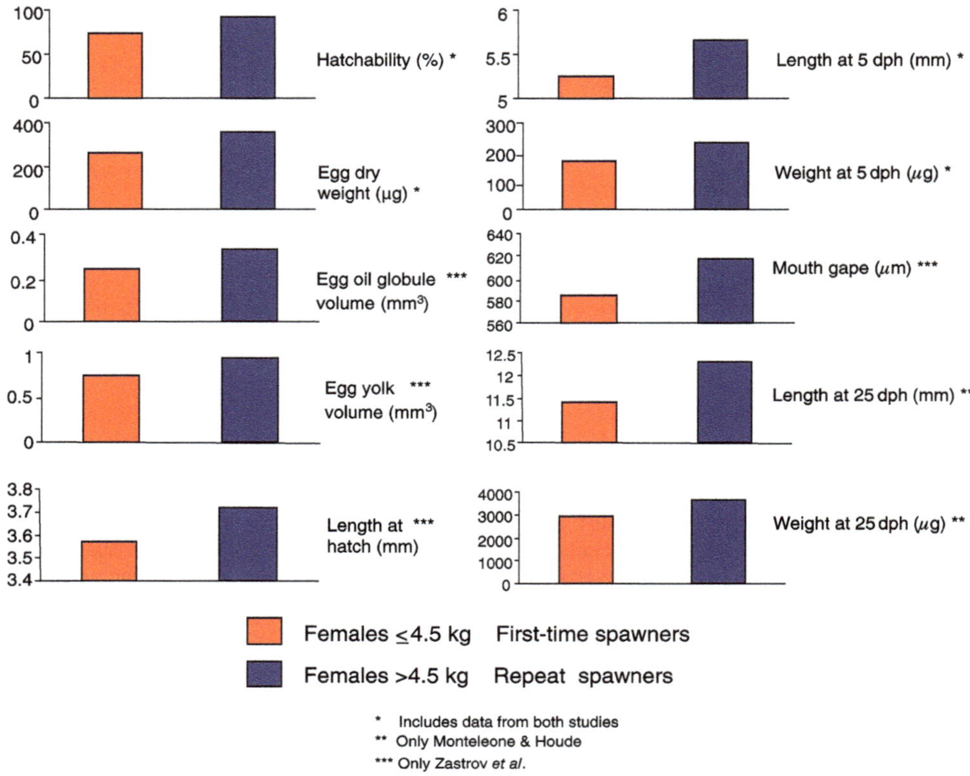

Plate 4 Maternal effects. A suite of metrics comparing striped bass *Morone saxatilis* eggs and larvae from small (≤4.5 kg) or large (>4.5 kg) females. In each case, progeny from large females appear to have a size or viability advantage. Data reproduced from Zastrow *et al.* (1989), with permission of ICES, and Monteleone & Houde (1990).

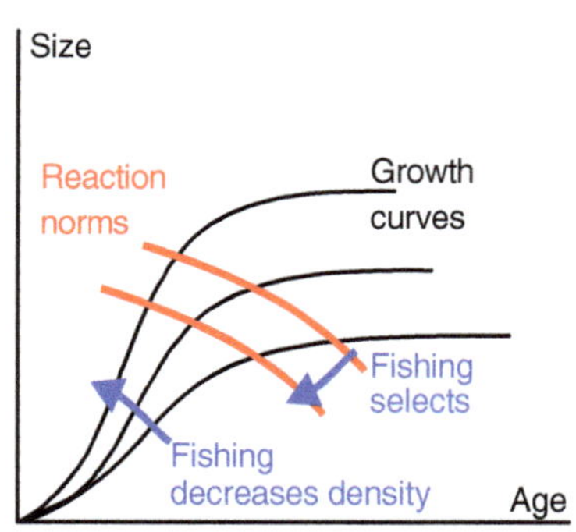

Plate 5 The maturation reaction norm is hypothesised to be characteristic of the genetic composition of a population. Environmental variability (resources, temperature. . .) results in growth variability. When their growth curve intersects with the reaction norm, individuals mature. Fishing decreases population size and hence permits faster growth (due to density dependence), leading to earlier maturation. Selective fishing of larger fish will select for a lower maturation reaction norm, leading to earlier maturation at smaller sizes.

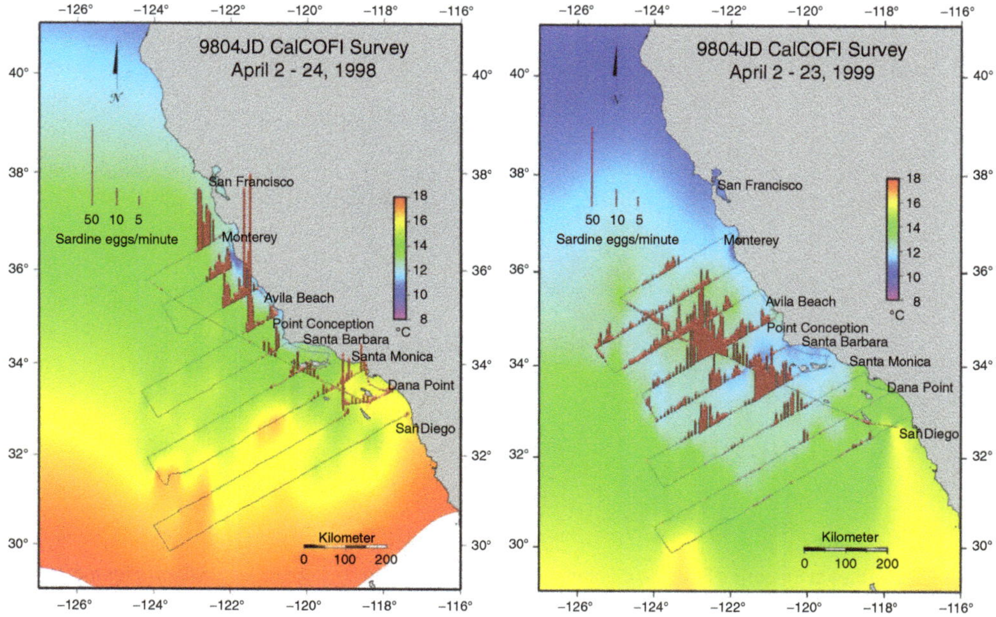

Plate 6 Sardine egg pattern from continuous underwater fish egg sampler (CUFES) in April 1998 and April 1999.

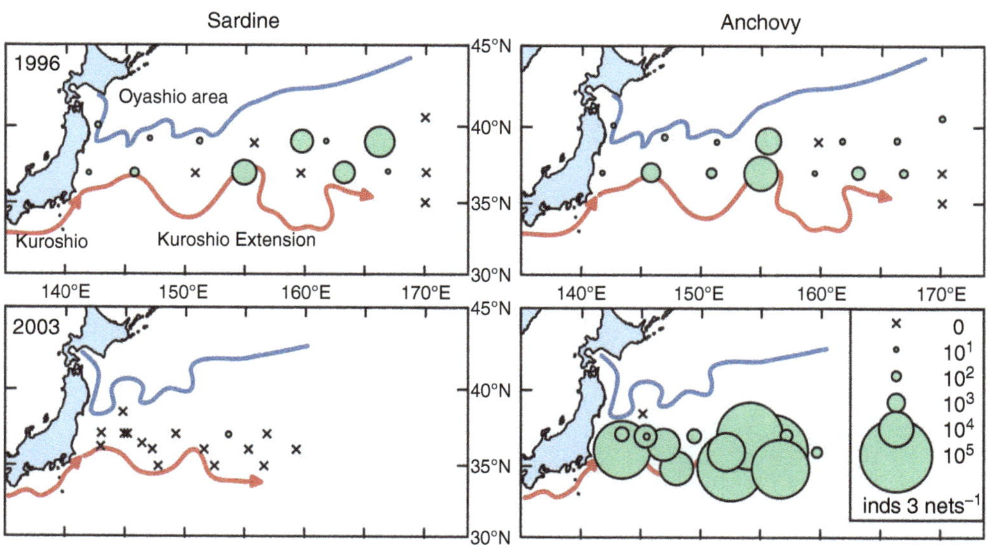

Plate 7 Number of late larval and juvenile sardine and anchovy (20–100 mm standard length) in the Kuroshio-Oyashio transition region in May-June 1996 and 2003. One circle is the number offish collected in three trawl hauls/night.

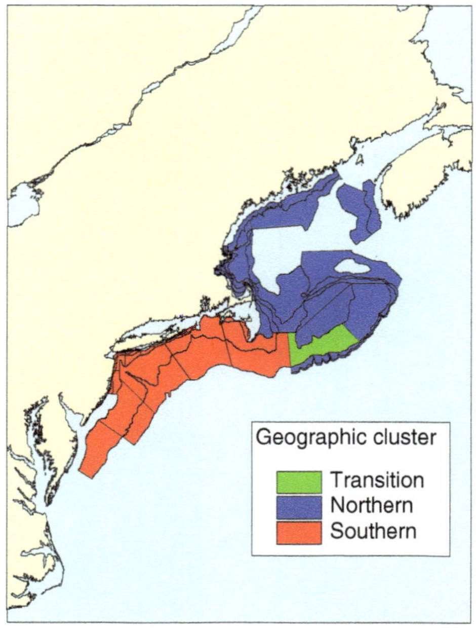

Plate 8 Northeast Fisheries Science Center survey strata grouped by clusters with different temporal patterns of yellowtail flounder abundance and biomass over time, 1979–2000. Reproduced from Cadrin 2010, with permission of the author.

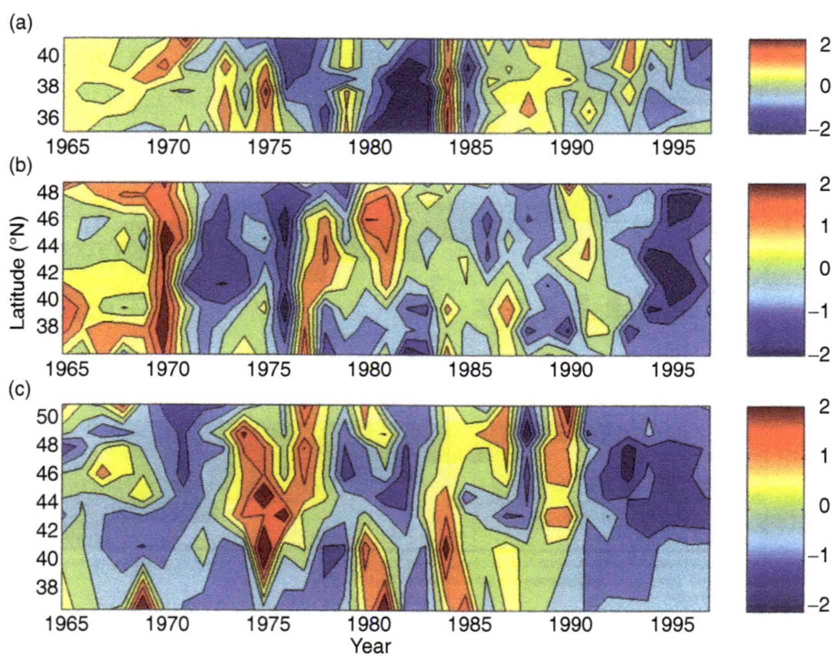

Plate 9 Log recruitment deviations plotted over space and time for (a) chilipepper rockfish (*Sebastes goodei*), (b) widow rockfish (*S. entomelas*), and (c) yellowtail rockfish (*S. flavidus*) in the California Current System. Reproduced from Field & Ralston 2005, with permission of NOAA Fisheries.

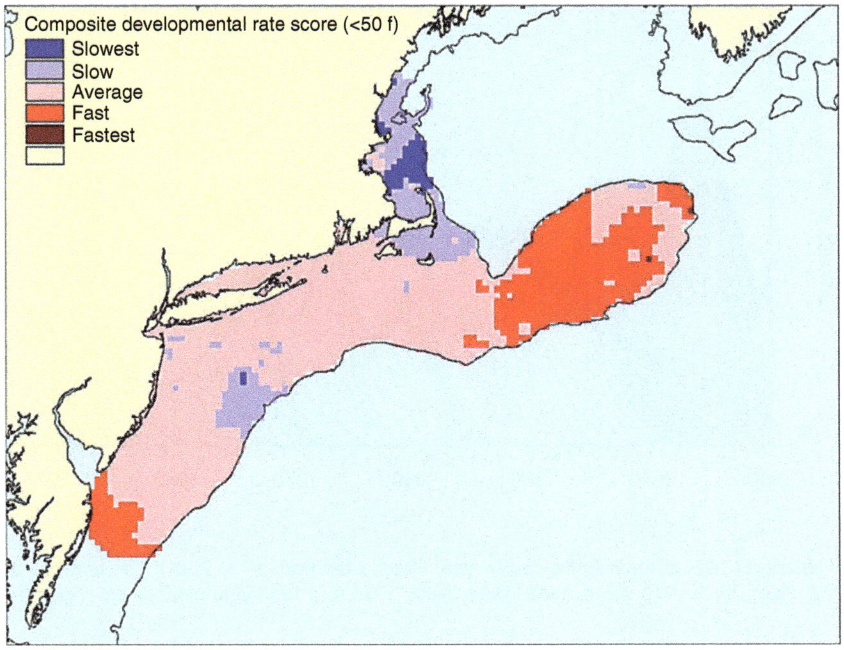

Plate 10 Geographic patterns of composite growth and maturity of age-2 yellowtail, expressed as slow (< −1 standard error), average and fast (>1 standard error). Reproduced from Cadrin 2010, with permission of the author.

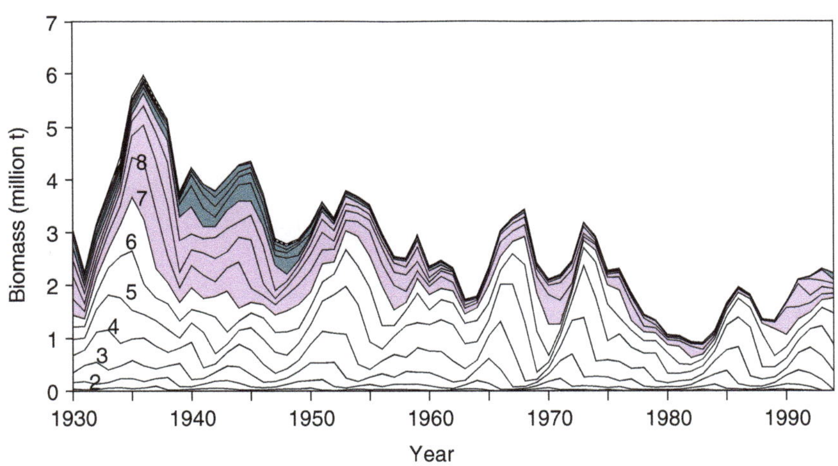

Plate 11 Biomass by age of the Northeast Arctic cod (after Pope *et al.* 2001). Ages are shown from 0 nearest the axis to 15 furthest from axis. Unshaded ages are those which are typically immature. Dark shaded ages are those (11–15) that were mature in all years. Lighter shaded ages (7–10) are those that have only been mature in more recent years.

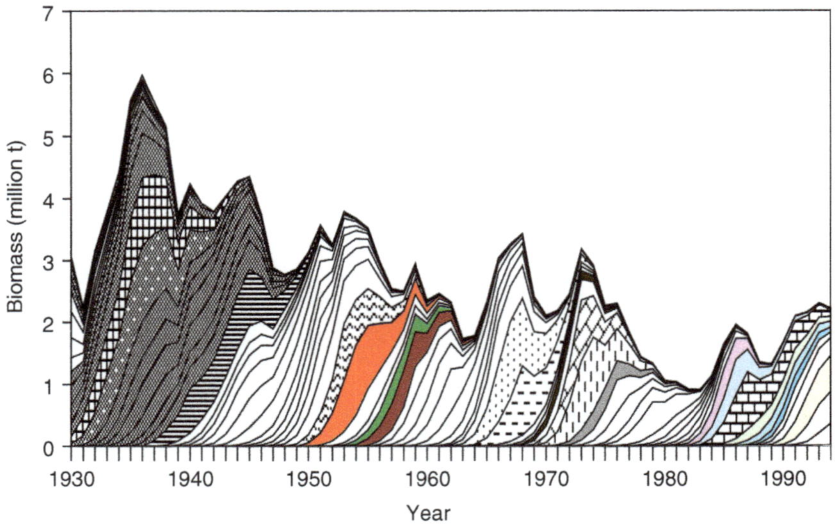

Plate 12 Biomass of Northeast Arctic cod by year class (after Pope *et al.* 2001). The large contributions of the 1929, 1930, 1937, 1949, 1950, 1963, 1964, 1969, 1970 and 1983 year classes are highlighted.

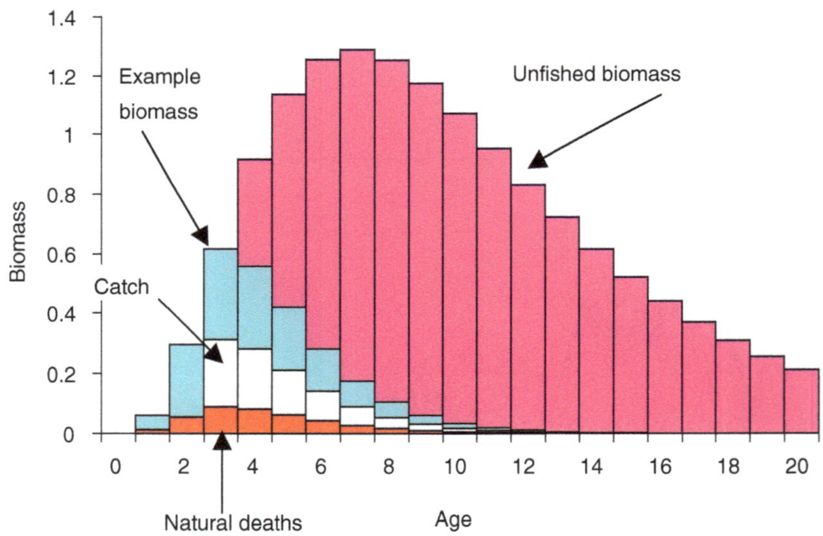

Plate 13 The evolution of biomass (blue), catch (white) and natural deaths-at-age (red) in the example, together with the equivalent biomass (pink) there would have been at each age had there been no fishing.

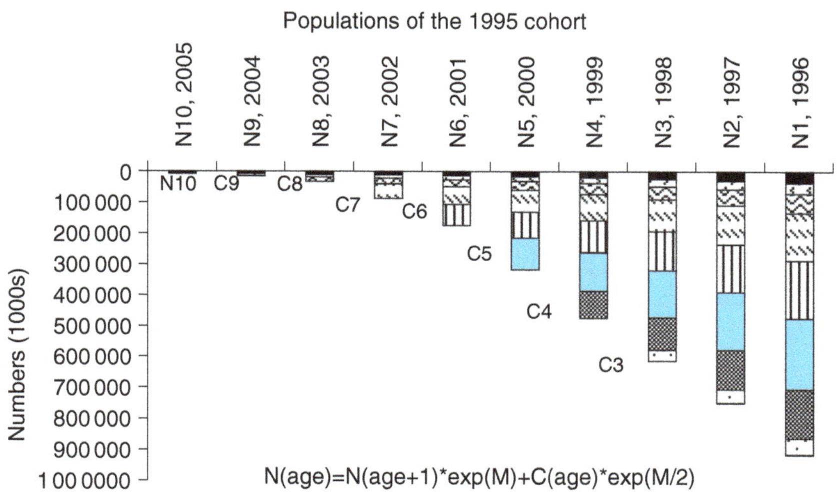

Plate 14 Cohort analysis on the 1995 cohort of Northeast Arctic cod. Different shadings indicate the contribution from the catch at each age as they are projected backward through time using the cohort analysis formula shown.

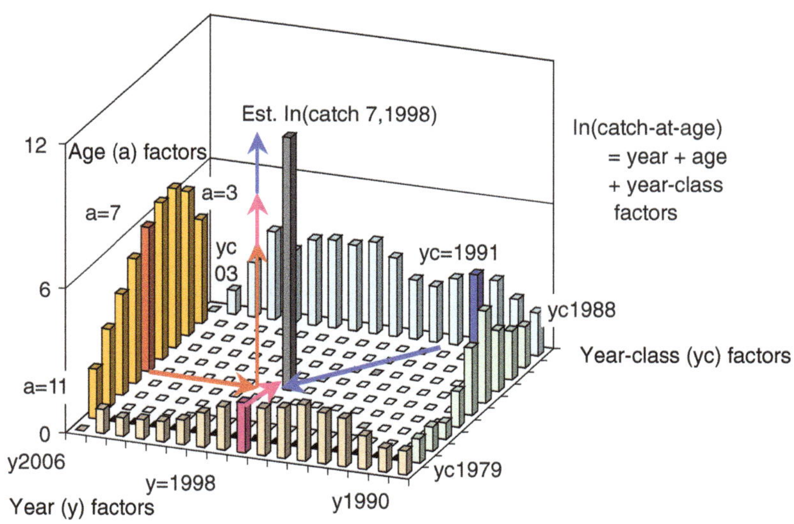

Plate 15 An illustration of how the additive effects of the Shepherd–Nicholson model estimates the natural logarithm of the catch-at-age 7 in year 1998 by the summing an age 7 effect, a year 1998 effect and a year-class 1991 effect. All other ln(catches-at-age)(these are not shown) are similarly explained by the sums of the appropriate age factors (shown along the left edge), year factors (shown along the bottom edge) and year-class factors (shown along both the top and the right-hand edges).

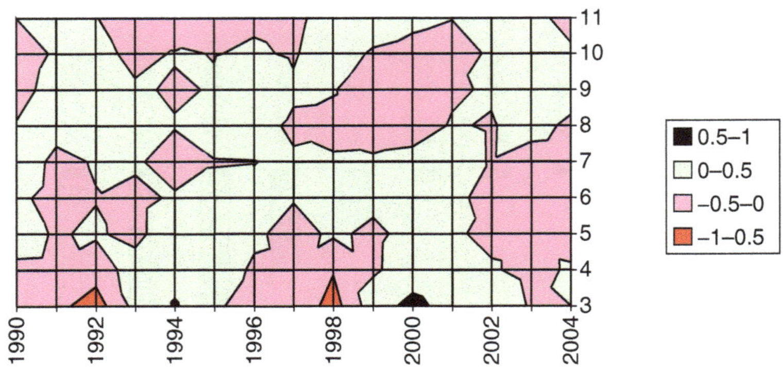

Plate 16 Residuals of the fit to catch-at-age data for Northeast Arctic cod (years 1990–2004, ages 3–11) found using the separable model.

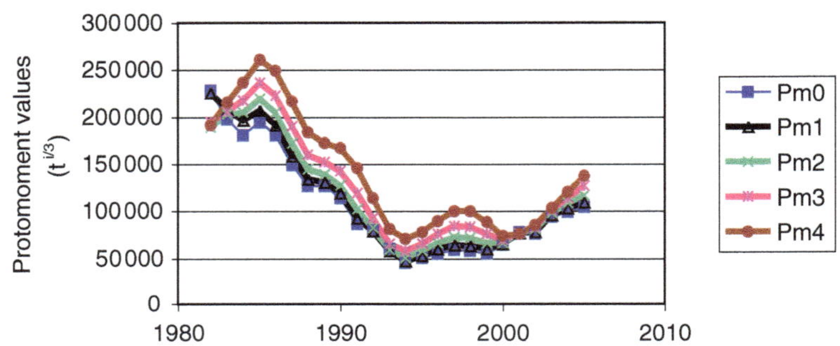

Plate 17 Protomoments (i = 0 to 4) of the Southern Newfoundland (3Ps) cod fitted to past data and projected on the assumption of a 15 K TAC from 2003 to 2008.

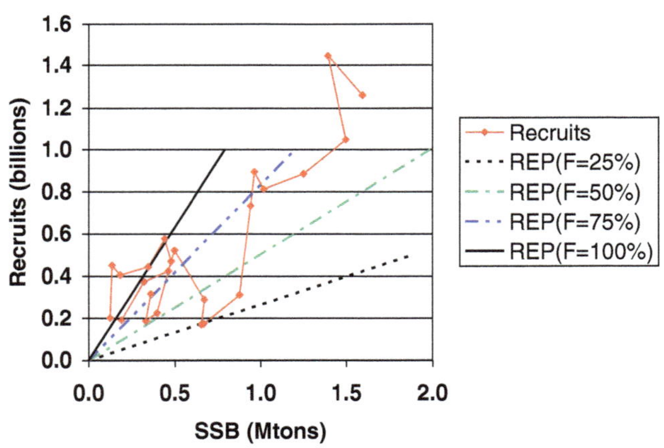

Plate 18 Recruitment plotted against spawning stock biomass for the Northern cod of Newfoundland (year classes 1962–1983). The replacement lines (REP%) are drawn at multiples of 100%, 75%, 50% and 25% of the average fishing mortality rate between 1962 and 1989. This fishing mortality rate averaged 0.47 between ages 5 and 10.

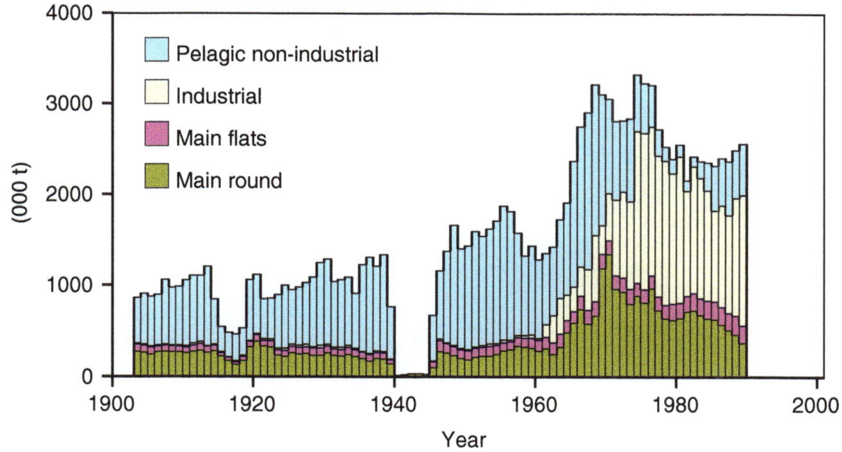

Plate 19 Trends of major components of North Sea catch 1902–1989.

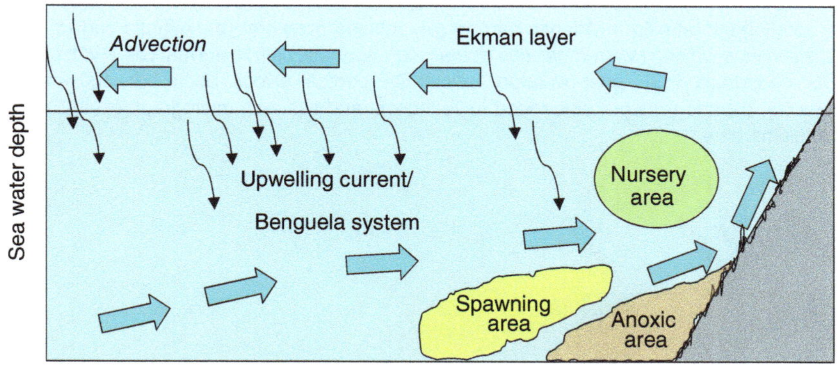

Plate 20 Fine-tuning of egg and larval buoyancy in the Benguela upwelling system as observed or indicated for several fish species, Cape hake (*Merluccius capensis*), Cape horse mackerel (*Trachurus trachurus capensis*), sardine (*Sardinops sagax*) and anchovy (*Engraulis capensis*). To avoid advective loss in the Ekman layer (offshore surface transport/wind-induced mixed layer about 20 m in depth) created by upwelling water masses (large arrows) and to reach favourable nursery grounds nearshore (high plankton production area), the eggs and larvae largely settle out of the Ekman layer (small arrows) and are transported back towards land. For eggs, the reason seems to be a combination of buoyancy and wind-induced mixing, while for larvae there seems also to be an active movement involved, especially for larger larvae. The figure specially refers to Cape hake, which spawn at 150–400 m depth close to anoxic bottom water. Other species such as sardine spawn just below the mixed layer. Thus, both mean and standard deviation in buoyancy seem to be uniquely adapted for each single species depending on spawning location, egg and larval characteristics. Note that salinity was observed to be slightly higher at lower water depths. Source: Sundby *et al.* (1999), Sundby *et al.* (2001), Stenevik *et al.* (2001, 2003); Svein Sundby, IMR (personal communication).

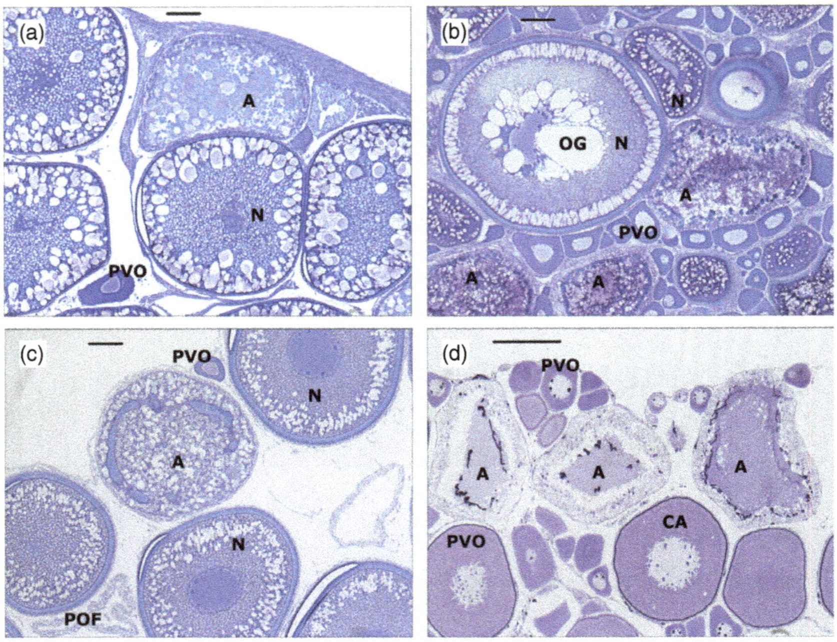

Plate 21 Vitellogenic atresia in Atlantic herring (a), Atlantic mackerel (b), Atlantic cod (c) and cortical alveoli atresia in blue whiting (*Micromesistius poutassou*). A: atretic cell; N: normal cell; PVO: previtellogenic oocyte; OG: oil globule; POF: post-ovulatory follicle; CA: cortical alveoli. Scale bar: 100 μm. Tissues in micrographs (a), (b) and (c) were embedded in Technovit, and tissue in micrograph (d) in Epon. Staining was with toluidine blue.

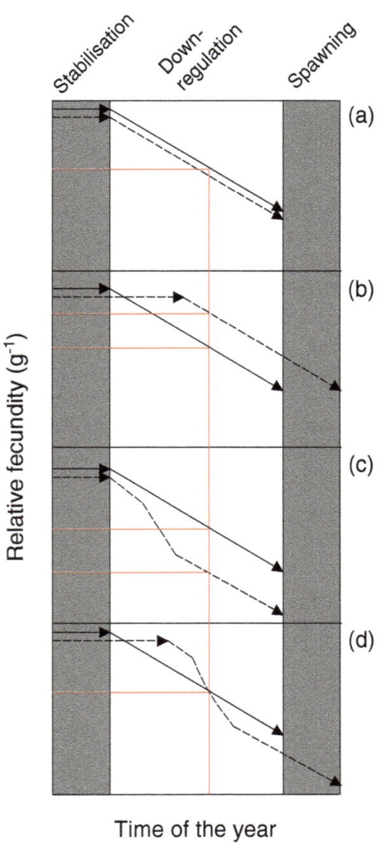

Plate 22 Different scenarios for how sampling at a specific point of time in the year may influence the understanding of relative potential fecundity (RF_P) in two contrasting groups (populations) of fish, depending on the temporal changes in the degree of downregulation. Vertical and horizontal thin (red) line: sampling time and corresponding RF_P; thick (black) line: reference (control) group; dashed line: response group, i.e., after a change in environmental or experimental conditions. (a) Same degree of downregulation and spawning time; correct conclusion; (b) Delayed spawning time in response group but equal degree of downregulation; overestimated RF_P in response group; (c) Same spawning time but larger degree of downregulation in response group: correct conclusion; (d) Delayed spawning time and larger degree of downregulation in response group: type II error (accepted false null hypothesis). Note for (c) and (d) that the response group was assumed to undertake a steep decline in RF_P during a specific part of the maturation cycle; that is, they showed a clear atretic window (Kurita *et al.* 2003).

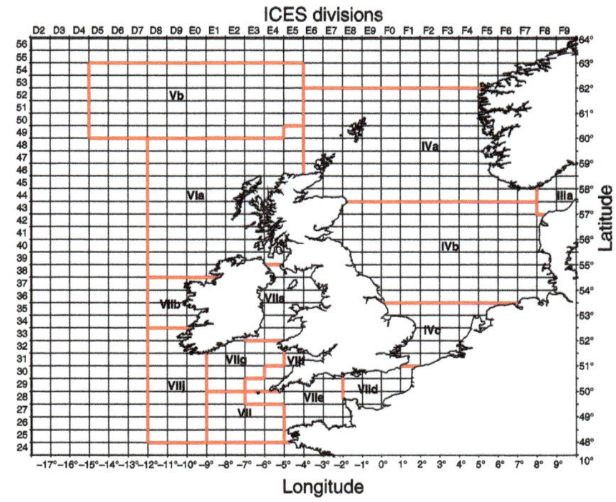

Plate 23 Sea region covered by the population dynamics model of European cod stocks.

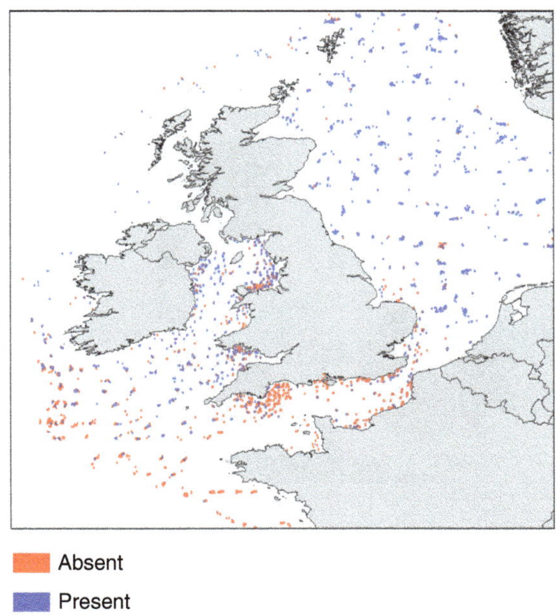

Plate 24 Cod distribution prediction map.

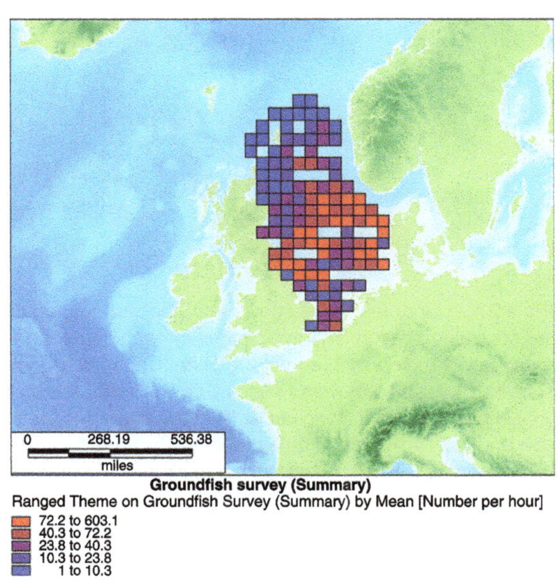

Plate 25 North Sea groundfish survey map showing catch-per-unit-effort (kg h^{-1}).

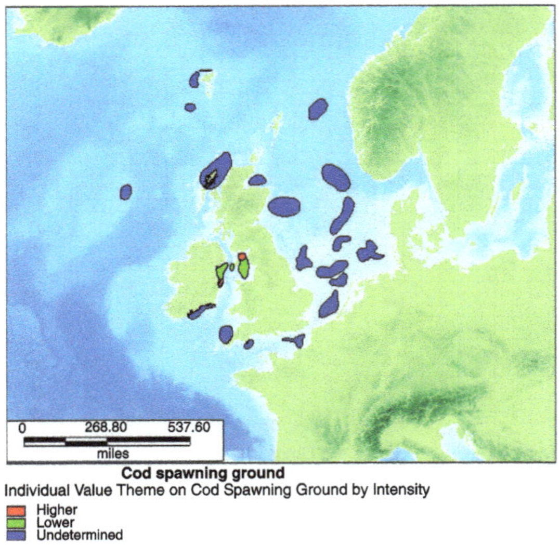

Plate 26 Cod spawning ground distribution map.

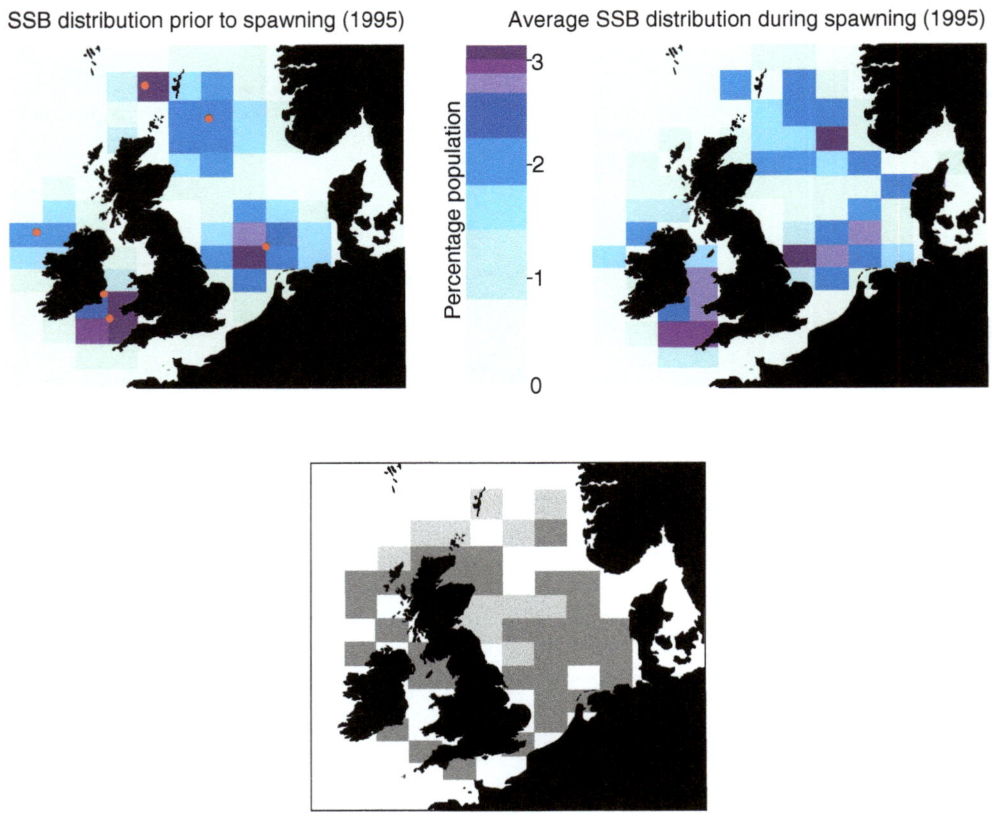

Plate 27 Upper left frame shows the predicted aggregated distribution of spawners immediately prior to the spawning season. The boundaries of each gathering region are shown in black, with the centre of the aggregation region marked with a circle. Upper right frame shows the predicted average distribution of spawners over the spawning season. Lower frame shows the observed distribution of spawning fish. Dark grey shows cells containing currently active spawning areas and light grey shows historical spawning areas now unused. Derived from data presented by Wright *et al.* (2003), Heath *et al.* (2003) and C.J. Fox (personal communication); cited in Andrews *et al.* (2006).

Chapter 8

Applied Fisheries Reproductive Biology: Contribution of Individual Reproductive Potential to Recruitment and Fisheries Management

Olav S. Kjesbu

8.1 Introduction

This chapter considers reproductive information used or having the potential to be used in the assessment of fish stocks in order to improve fisheries management advice (applied fisheries reproductive biology, AFRB). The focus of this chapter is on stock reproductive potential (SRP) (Trippel 1999), viewed at the cellular and individual level. The subsequent scaling from individual to population level is dealt with in Chapter 11, as is a more in-depth examination of basic reproductive biology in Chapter 2. The central concepts covered within this chapter are reproductive strategies and tactics expressed as various reproductive traits of interest. Special focus is paid to fecundity regulation, including methodological aspects and recent advancements, bearing in mind that numerical considerations are the unit of interest in the present setting. Trippel *et al.* (1997) reviewed parental effects of relevance for assessment in response to the great interest in these topics during the 1980s and 1990s, especially in terms of maternal effects, though paternal effects were also considered to some degree. These authors presented a comprehensive list of variables/factors relating to these effects on stock dynamics. Thus, a lack of any effect for one particular factor should not lead automatically to a general conclusion that parental effects are nonexistent in the species under study. One of the strongest positive maternal effects (i.e., life history trait patterns related to fish size, condition or age) is relative fecundity (see definition below). At present, the focus of AFRB studies seems to be less on egg and larval quality aspects and more on fecundity and spawning time aspects. This situation is reflected in this chapter. Dealing with these latter issues seems no less complicated, being increasingly noticeable as one starts to know more.

The number of peer-reviewed articles within fish reproductive biology has increased markedly over the past 25 years, from a few dozen to a couple of hundred 'hits' on the internet

Fish Reproductive Biology: Implications for Assessment and Management, Second Edition.
Edited by Tore Jakobsen, Michael J. Fogarty, Bernard A. Megrey and Erlend Moksness.

per year (Lowerre-Barbieri *et al.* 2011). The amount of related reviews is also rapidly building. As will be apparent below, conceptual developments in this field of research, as in others, go hand in hand with methodological improvements. This is easily seen when studying the various consulted reviews (and traditional papers) from back in time until the present day. Nikolskii (1969), Bagenal (1978), Murua *et al.* (2003), Kjesbu *et al.* (2003), Kamler (2005) and Rijnsdorp & Witthames (2005) have all provided overviews of the growth in information regarding marine fish reproductive biology which will, as far as possible, not be repeated here. Since the earlier production (2008) and subsequent release (2009) of this chapter, several newer reviews have appeared, both in books or anthologies and journals. These reviews range from being more conceptually (Green 2008, Witthames & Marshall 2008, Lowerre-Barbieri 2009, Wright & Trippel 2009, Lowerre-Barbieri *et al.* 2011, Bernal *et al.* 2012, Ganias 2013, McBride *et al.* 2013) to more methodologically oriented (Witthames *et al.* 2009, Armstrong & Witthames 2012, Wright 2013). The books edited by Rocha *et al.* (2008) and Cole (2010), addressing respectively fish reproductive physiology and conservation (including reproductive diversity), expand the insight in these matters further. This list of reviews is in no way meant to be complete, as there are also more specific ones addressing for instance spermatogenesis. In this chapter, an attempt is made to outline the present status of knowledge in AFRB and simultaneously point to where further research should be undertaken. The listing of detailed information on reproductive traits for various species was avoided, but when found necessary commercially exploited temperate, oviparous species were selected as examples due to their data richness. Nevertheless, in some situations weak data suggested poor quality conclusions and in such cases the information was labelled as indicative only. As will be seen below, the chapter brings together many facets of AFRB, which hopefully will be helpful for the reader in the planning and completion of new studies.

8.2 Reproductive styles of major commercial species

8.2.1 Sources of information

Several handbooks and articles in animal/fish reproductive biology have catalogued and described the rich variety of the different reproductive styles found in chondrichthyes (cartilaginous fish: chimaeras, sharks and skates) and osteichthyes (bony fish). However, although a very valuable conceptual framework, the applied fisheries reproductive biologist typically uses a more limited or specific scientific vocabulary, defined and agreed upon in research consortia (the foremost example being California Cooperative Oceanic Fisheries Investigations: http://www.calcofi.org/) and working groups (such as ICES Working Group on Mackerel and Horse Mackerel Egg Surveys: http://www.ices.dk/). Examples of this terminology are given in Murua & Saborido-Rey (2003), Murua *et al.* (2003) and Lowerre-Barbieri *et al.* (2011). However, we still have the problem that important concepts are presented by many different names, although meaning exactly the same. Hence, the standardisation of frequently used reproductive terms has been an issue over the last years (Lowerre-Barbieri *et al.* 2011), especially regarding maturity keys (Brown-Peterson *et al.* 2011, ICES 2012). In consulting the more general literature, the book of Wootton (1998) *Ecology of Teleost Fishes* still gives an excellent introduction to various reproductive strategies found in fish. Likewise, research scientists working within the field of fish reproductive physiology have over the years produced a substantial number of publications on gametogenesis, endocrine regulation and influences of environmental cues; compare for example Taranger *et al.* (2010), Zohar *et al.* (2010) and Levavi-Sivan *et al.* (2010). It is apparent that the rapid implementation of a range of molecular techniques has considerably increased insight into fundamental processes impossible to grasp by earlier traditional tools, for example in egg quality studies (Skjærven *et al.* 2013).

8.2.2 *Reproductive strategies and tactics*

Reproductive styles and reproductive strategies are often used as synonyms (Table 8.1). The latter term is subdivided into reproductive tactics in line with definitions used in the military (Wootton 1984), where strategy refers to the planning of operations and tactics to the *de facto* execution of plans. According to Wootton (1998), the reproductive strategy of a species summarises the different phenotypic expressions of reproductive traits (e.g., fecundity or egg size) over the full range of environmental situations, while reproductive tactics refer to which specific reproductive traits are expressed in a specific environmental situation, that is, the level of phenotypic plasticity (Green 2008). Logically, life-history strategy is a more general concept than reproductive strategy, as the former is defined as '... the coordinated evolution of all the life history traits together' (Stearns 1992). The use of the term reproductive strategy should therefore have some limitations, especially when discussing offspring development. A useful upper borderline might be when the larvae switch from endogenous (parental-derived) to exogenous feed resources.

Some reproductive strategies appear to be common across several species in response to specific environmental situations, such as seen in up-welling zones (Figure 8.1; Plate 20). In an attempt to clarify the meaning of reproductive strategies versus reproductive tactics, the Atlantic cod (*Gadus morhua*) was chosen as an example consisting of several stocks spread throughout the North Atlantic, including also low-salinity environments such as the Baltic Sea (Sundby 2000). This species has adopted a strategy of production of multiple batches of pelagic eggs (Kjesbu 1989), but where marine cod eggs are significantly smaller in diameter than brackish cod eggs (Nissling & Westin 1991, Kjesbu *et al.* 1992) (Figure 8.2). On the other hand, the latter eggs show an extremely high water content and a very low specific gravity (Thorsen *et al.* 1996) (Figure 8.2); they are clearly able to be kept afloat above unhealthy bottom water conditions (Nissling & Westin 1991). This egg characteristic of high buoyancy is again related to significantly more of the yolk being hydrolysed into free amino acids during final maturation/major water uptake (Thorsen *et al.* 1996). In other words, a fundamental physiological difference exists between marine and brackish water cod stocks in these respects. Thus, adopted reproductive tactics have links with the concept of stock separation, in particular for geographically or ecologically isolated populations with anomalous genetics (Johannesson & André 2006). This view is, however, complicated by the formal definition of reproductive strategy (Table 8.1), where all the individuals displaying the various reproductive tactics should belong to the same gene pool, defined as all the genes/alleles in an *interbreeding* population (Lawrence 2000). It seems unlikely that the Baltic cod stock interbreeds to a noticeable extent with marine cod stocks due to the very special conditions existing in the whole Baltic Sea distribution area (Nissling & Westin 1991). The simple solution seems to be to discuss reproductive tactics within a given species regardless of the level of interbreeding, as is frequently done in the large number of articles found on these subjects in general databases.

Table 8.1 Key definitions within applied fisheries reproductive biology.

Term	Definition	References	Examples of expressions or variables used, and various types of comments
Atresia	Resorption of oocytes	See Hunter & Macewicz (1985a) and historic references therein	Comment: refers normally to oocytes containing yolk only (i.e. vitellogenic oocytes), dealt with more specifically in Table 8.3
Capital breeder	An organism that uses stored energy for reproduction	Stearns (1992)	Comment: examples include individuals undertaking long spawning migration without feeding, or over-wintering species
Evolutionary trade-off	A change in one trait in a population that increases fitness but which is linked to a change in another trait that decreases fitness	Stearns (1992)	Egg size vs fecundity; fecundity vs off-spring survival; current vs future reproduction; age at maturity vs life-time growth rate
Fecundity	Number of oocytes or eggs produced in an individual ovary at a given point in the maturation cycle, most often just before spawning starts	Many sources but Nikolskii (1969) and Bagenal (1978) give early reviews, and Murua *et al.* (2003) a recent review	A high number of specific definitions exist dealt with more specifically in Table 8.3
Gonadosomatic index	Weight of gonad in relation to measures of body size	Many sources but West (1990) gives a useful review	
Income breeder	An organism that uses energy acquired during the spawning period rather than stored energy for reproduction	Stearns (1992)	Comment: typical for tropical fish
Maturity phase	Subdivison of a maturity stage into more detailed, gradual developments	Used in articles produced in the former Soviet Union, e.g. Shirokova (1977)	
Physiological trade-off	Allocation decisions between two or more processes that compete directly with one another for limited resources within a single individual	Stearns (1992)	Reproductive effort vs parental growth and survival
Portion of eggs spawned	Fraction of total number of eggs spawned by an individual at time of sampling	Kjesbu *et al.* (1990)	Comment: normally reported as a ratio less than 1 or in percentage
Post-ovulatory follicles	The remaining follicle layers after ovulation (release) of a female sex cell	Hunter & Macewicz (1985b)	Comment: specially used to estimate spawning fraction
Reproductive effort	Cost of reproduction within a season including both reproductive investment and cost of reproductive behaviour	Stearns (1992), Wootton (1998)	Comment: stable-isotope signatures techniques (Gauthier *et al.* 2003) and light manipulation (to block maturation) (Kjesbu & Witthames 2007) might give useful estimates

(Continued)

Table 8.1 *(Continued)*

Term	Definition	References	Examples of expressions or variables used, and various types of comments
Reproductive investment	The amount of mass or energy used to produce gonads within a current season	In a high number of publications such as Karlsen *et al.* (1995), Nash *et al.* (2000), Rijnsdorp *et al.* (2005)	Gonadosomatic index; fecundity; fecundity multiplied with egg size; loss in body weight during spawning
Reproductive strategy (reproductive style)	Combinations of reproductive traits characteristics of individuals belonging to the same gene pool/species	Wootton (1998), Murua & Saborido-Rey (2003)	Comment: see definitions in Chapter 2 of oviparity; ovoviviparity; viviparity; semelparity; and iteroparity. Total (single) spawners; batch spawners
Reproductive tactic	Variation in expression of reproductive traits within a species in response to environmental changes	Wootton (1998)	
Reproductive trait	Distinct reproductive, phenotypic character, which may be either genetic or environmentally determined, or both	Lawrence (2000)	Time of spawning; egg size; egg buoyancy; fecundity; batch size; spawning frequency; length of spawning season; egg incubation time; intrinsic egg mortality rate; age and size at maturity
Skipped spawning	Omission of annual spawning in a sexually mature, iteroparous individual	See review of Rideout *et al.* (2005)	Mass atresia, arrested or aborted stage-specific gonad development
Somatic weight	Gonad-free body weight	Multiple sources	
Spawning fidelity	The degree of limitation of a species to a particular spawning habitat	Lawrence (2000), cf., also Wright *et al.* (2006)	Comment: spawning fidelity = return of spawning individuals to the same spawning ground every year
Spawning fraction	The proportion of mature females spawning per day	See Hunter & Macewicz (2003) and review of Murua *et al.* (2003)	Comment: normally reported as a ratio less than 1 or in percentage
Spawning frequency	Average time between successive spawnings of females in a population but presented in frequency terms	Hunter & Macewicz (1985b)	Comment: normally reported as 'once per X days' and is deduced from the inverse of spawning fraction
Spawning (batch) interval	Time between successive spawnings of an individual	Eaton & Farley (1974)	Comment: normally measured in number of days or hours
Spawning period	Length of spawning of an individual fish within a season	Hickling & Rutenberg (1936)	Comment: normally measured in number of days
Spawning season (spawning duration)	Time in the year when the population produce eggs/offspring	Multiple sources	Comment: often set to be synonymous to spawning period
Surplus energy	The energy left after the cost of metabolism has been met	Stearns (1992), Wootton (1998)	Surplus energy = growth + reproduction
Surplus production	The sum of somatic growth and reproductive investment	Rijnsdorp (1990)	

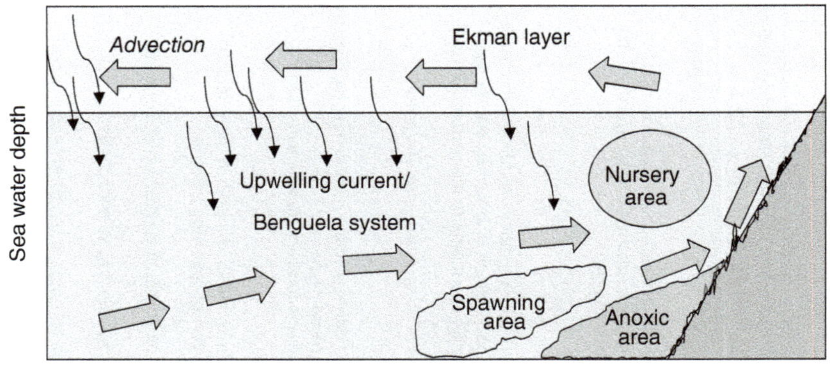

Figure 8.1 Fine-tuning of egg and larval buoyancy in the Benguela upwelling system as observed or indicated for several fish species, Cape hake (*Merluccius capensis*), Cape horse mackerel (*Trachurus trachurus capensis*), sardine (*Sardinops sagax*) and anchovy (*Engraulis capensis*). To avoid advective loss in the Ekman layer (offshore surface transport/wind-induced mixed layer about 20 m in depth) created by upwelling water masses (large arrows) and to reach favourable nursery grounds nearshore (high plankton production area), the eggs and larvae largely settle out of the Ekman layer (small arrows) and are transported back towards land. For eggs, the reason seems to be a combination of buoyancy and wind-induced mixing, while for larvae there seems also to be an active movement involved, especially for larger larvae. The figure specially refers to Cape hake, which spawn at 150–400 m depth close to anoxic bottom water. Other species such as sardine spawn just below the mixed layer. Thus, both mean and standard deviation in buoyancy seem to be uniquely adapted for each single species depending on spawning location, egg and larval characteristics. Note that salinity was observed to be slightly higher at lower water depths. For a color version of this figure, please see Plate 20 in the color plate section. Source: Sundby *et al.* (1999), Sundby *et al.* (2001), Stenevik *et al.* (2001, 2003); Svein Sundby, IMR (personal communication).

8.2.3 Operational definitions

Any description of reproductive styles and methodological issues involved requires a toolbox of definitions (Table 8.1). The importance of the definitions lies in that any wrong use may lead to wrong conclusions. In line with the argument above, only those definitions found to be commonly used among research scientists working within AFRB were selected – that is, on either Egg Production Methods (EPM) or on SRP (it might be discussed if parts of aquaculture should be associated with AFRB, but this is not normally done). The so-called 'Paulik diagram' is sometimes used as a reference framework in recruitment studies (see Rothschild 1986, and references therein). In practice, Quadrant 2 of the Paulik diagram (Egg Production or Spawning Module) may include results from both fields of research, the only operational difference being that EPM goes from the total number of eggs (in some cases also from larvae, Quadrant 3) to SSB (spawning stock biomass), while SRP goes in the opposite direction from SSB to total number of eggs (and thereafter eventually to subsequent life stages). The presented definitions in Table 8.1 are a mixture of general and specific ones. A few are further specified and extended below. In some circumstances it is impossible to locate the original source of the particular definition but note that several appear for the first time in articles on life-history theory defining the academic reason for their introduction.

The distinction between evolutionary and physiological trade-off is normally not operational as both are lumped together as simply 'trade-offs'. According to Stearns (1992), the trade-off between growth and reproduction is the best-documented one among species. Trade-offs between fecundity and egg size have been seen in fish species producing a single batch of

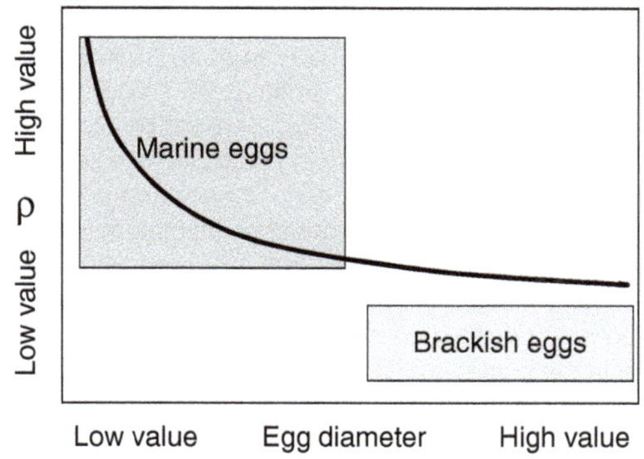

Figure 8.2 An illustration of egg specific gravity (ρ) of marine and brackish Atlantic cod eggs in relation to egg diameter. The curve (solid line) refers to modelled change in ρ of marine eggs with egg diameter, provided that all other intrinsic components influencing ρ are kept constant (i.e., internal egg specific gravity including specific gravity of ovoplasm, and specific gravity and amount of eggshell (chorion) material). It is clear that marine eggs do not reach the low specific gravity of brackish eggs by simply growing larger in egg size; this points to more fundamental underlying changes being in operation (see main text). Source: Kjesbu *et al.* (1992).

eggs per season. In the case of Atlantic herring (*Clupea harengus*), this pattern was found by comparing data across stocks (Parrish & Saville 1965, van Damme *et al.* 2009). Jonsson & Jonsson (1999) also found a significant, negative relationship (30% of the variance being explained) between egg size and relative fecundity, after excluding for body size effects, in brown trout (*Salmo trutta*) caught in Norwegian rivers. This phenomenon, if present, should be held separate from the general increase in reproductive investment with age, size and spawning experience noted in long-lived fish (Nash *et al.* 2000) resulting from both bigger eggs and/or a higher relative fecundity, as recently summarised for various species by Hixon *et al.* (2014) addressing the implication of BOFFFFs (big old fat fecund female fish) in stock conservation schemes.

The term 'reproductive effort' is frequently used incorrectly in the literature; an example is reporting of gonadosomatic index (GSI) as 'reproductive effort'. This concept includes the whole package of reproductive costs and consists therefore both of reproductive behaviour (spawning migration and courtship behaviour) and reproductive investment. In other words, it is an expression that is extremely difficult to assess. For obvious reasons due to the overall purpose of their investigations, applied fisheries reproductive biologists focus on reproductive investment instead of reproductive effort, but the cost of reproductive behaviour is central in several life-history modelling studies (Jørgensen & Fiksen 2006, Jørgensen *et al.* 2006), though direct field documentations are still limited (Mittelbach *et al.* 2014). Within a species, and especially within a stock, fluctuations in fecundity might be considered as an expression of variation in reproductive investment, although fecundity data alone is not enough when contrasting reproductive investment in different species (Table 8.2). As seen from this simple comparison between Atlantic cod, herring and salmon (*Salmo salar*), the three species show highly different relative fecundities, but comparable reproductive investments when egg size is taken into account. In more detail, the data indicate that salmon shows the highest reproductive investment and reproductive effort (based on post-spawning mortality) but, nevertheless,

Table 8.2 Overview of reproductive strategies of Atlantic cod, herring and salmon. *Sources*: Atlantic cod (*Gadus morhua*): Kjesbu *et al.* (1996a, 1998), Thorsen *et al.* (1996); Atlantic herring (*Clupea harengus*): Høie *et al.* (1999), Óskarsson *et al.* (2002), Kurita *et al.* (2003); Atlantic salmon (*Salmo salar*): Kittelsen (1986), Thorsen (1989), Jonsson *et al.* (1996), Christiansen & Torrissen (1997), Tom Hansen, Matre Research Station, IMR, Norway (personal communication). *Note*: The data refer to North East Atlantic stocks, more specifically North East Arctic cod, Norwegian spring-spawning herring and 'Eastern type'/Norwegian salmon. The estimation was undertaken on mature females of small to large size (but not extremely small or extremely large): cod: 2000–12 000 g; herring: 200–400 g; salmon: 2000–12 000 g. Relative fecundity is the number of maturing (vitellogenic) oocytes per gram whole wet body weight. The same denominator was used in the estimation of reproductive investment and individual egg dry weight was multiplied with fecundity in the numerator. The notation 'single' egg batch does not imply that all eggs are shed at exactly the same time, but that all oocytes to be spawned develop in uniformity inside the ovary.

	Species		
Term	Atlantic cod *Gadus morhua*	Atlantic herring *Clupea harengus*	Atlantic salmon *Salmo salar*
Fecundity	800 000-6 000 000	30 000-80 000	3 000-16 500
Relative fecundity (g^{-1})	380-480	150-200	1-2
Oocyte recruitment	Determinate	Determinate	Determinate
Number of egg batches during spawning period	Multiple	Single	Single
Egg size (mm)	1.2-1.5	1.4-1.6	5-6
Stored energy and location	Moderate, liver	Extreme, flesh and viscera	Extreme, flesh and viscera
Reproductive investment (mg egg dry x g wet body weight $^{-1}$)	30-65	45-70	50-85
Type of eggs	Pelagic, marine	Demersal, marine	Demersal, fresh water
Post-spawning mortality	Low	Low	Moderate-high

the difference in investments as such was lower than hypothesised based on the high mortality seen for spent salmon. Populations of sole (*Solea solea*) in the North Sea and Bristol Channel demonstrate similar levels of investments, as shown in Table 8.2 (35 cm female sole: 45–80 mg egg dry weight/g wet body weight) (Rijnsdorp & Witthames 2005). This range in investment for sole was found to be independent of the general trend of an increased fecundity and a decreased egg size from Portuguese waters towards the North Sea (Rijnsdorp & Witthames 2005), which lends support to the existence of some type of trade-off between egg size and fecundity, at least with change in geographical location (here latitude). However, as all these estimates on reproductive investment refer to pre-spawning fish only, they should be treated with caution, since atresia (see below) during spawning might result in a significant reduction in reproductive output, especially in species such as sole and cod, both of which are multiple batch spawners. Pre-spawning values for single batch spawners such as herring and salmon should, however, in principle, be correct. There is evidence that populations living in more variable environments (e.g., northerly populations) not only live longer and mature later but also show lower relative investments (see Heibo *et al.* 2005, and references therein). This should be seen in light of general life history theory; compare the quotation of Leggett & Carscadden (1978) on American shad (*Alosa sapidissima*): '… intraspecific variation in reproductive characteristics represents a fine tuning of life history to long-term features of the environment by natural selection'.

In the estimation of SRP, atresia (McMillan 2007) and skipped spawning (Rideout *et al.* 2005) have recently received increased attention; at least atresia was earlier only considered in depth in EPM (Hunter & Macewicz 1985a, Witthames 2003), while further extensive use of post-ovulatory follicles (POFs) (Saborido-Rey & Junquera 1998) as a marker of past spawning in histological slides has facilitated quantification of the fraction of 'skippers' at the stock level, found to be significant in long-migratory gadoids (see Skjæraasen *et al.* 2012 and references therein). A potentially emerging topic is *apoptosis* (programmed cell death) (Tilly 1996), especially in terms of male sperm production (Schulz & Miura 2002, Kristoffersen 2005). However, apoptosis might also be important for the initiation of atresia (Tilly 1996) and thereby the downregulation of fecundity (see below). Thomé *et al.* (2012) concluded in their immunohistochemical study on the freshwater teleost *Prochilodus argenteus*, that oogonial proliferation was increased and oogonial apoptosis decreased under favourable environmental conditions, and vice versa when the environmental conditions were unfavourable.

8.2.4 *Underlying definitions*

To complete the list of frequently used definitions, a few more basic concepts, which apparently need better clarification, are detailed in this subsection. Emphasis is placed on the definition and use of maturity stage classification systems, the concept of oogenesis, the formal distinction between oocytes and eggs, and when in the reproductive cycle to report egg size.

8.2.4.1 Applications in maturity studies

Biologists in the former Soviet Union have produced a large number of definitions within the field of fish maturation dynamics; see for instance references in Sorokin (1957) and Nikolskii (1969). Depending on the research question asked, a particularly useful point is the separation of each maturity stage into different 'phases'. Shirokova (1977) used this principle and introduced nine phases describing in a most detailed manner the histological changes taking place in the Atlantic cod ovary during development from an immature to an early maturing stage. More specifically, Shirokova employed morphological changes taking place in the cytoplasm and nucleus of oocytes as criteria. There is evidence that the 'synthesis apparatus' (RNA concentrations), located in the cytoplasm, represented by Balbiani bodies (Wallace & Selman 1981), or the analogous circumnuclear ring (CNR) (Woodhead & Woodhead 1965, Kjesbu & Kryvi 1989, Tomkiewicz *et al.* 2003, Tobin *et al.* 2010), appears before cortical alveoli formation and vitellogenesis (Shirokova 1977). Subsequently, Holdway & Beamish (1985), Kjesbu *et al.* (2011) and McPherson & Kjesbu (2012) adopted Shirokova's principles, and categorised the specific movement of the CNR from the centre to the periphery of the cytoplasm. This subdivision into phases appears to be necessary when the specific aim is to classify individuals as either sexually immature or sexually mature in cases when the samples are taken well outside the spawning season. At least for cod, a peripheral CNR (and subsequent oocytic development) documents sexual maturity (Kjesbu *et al.* 2011). This should not be understood as being a general recommendation to start gonad sampling programmes far from the spawning season, due to other problems such as 'immatures' which actually would have become sexually mature individuals, if the survey had been delayed during the year. More interestingly, the results of these highly detailed microscopic studies strongly indicated that the solstice (early CNR formation) and equinox (cortical alveoli formation) are the first- and second-decision windows for oocyte development in both cod (Kjesbu *et al.* 2011) and herring (McPherson & Kjesbu 2012), in line with the physiological threshold theory of Rowe & Thorpe (1990).

As outlined by West (1990), fish maturity can be determined by many different methods, such as visual (macroscopic) staging, gonadosomatic index and sex cell size. A major problem with the visual staging method (which is the most common) is the introduction and definition of far too many stages. It should be realised that the naked human eye has shortcomings both in terms of colour interpretation (which varies between people) and in actually noticing the presence of small particles such as cells. Thus, most calibration studies reach very logical answers: there are problems with separating immature fish from early maturing fish, and separating spent fish from early maturing fish. For these reasons a review panel concluded that four macroscopic stages would normally be sufficient: immature (completely translucent); maturing; spawning; and 'other' (early maturing/spent). However, it was also felt that '... some grading systems are so inaccurate and imprecise that it is a waste of time to validate them' (Kjesbu *et al.* 2003). Nonetheless, macroscopic staging systems are here to stay because of their low cost and quickness (Hunter & Macewicz 2003). One obvious problem with the present situation is the undertaking of too many personal judgements when deciding upon the stage for each fish. For this reason, it would likely be better to report characteristics only and allow the experts or specific computer programs (macros) to formulate conclusions about stage at a later time. This suggestion is based on the following statement of Hunter & Macewicz (2003): 'Using the presence or absence of yolked oocytes, presence and absence of hydrated (hyaline) oocytes, one can accomplish all the primary functions of anatomical grading without the misleading complexities or ephemeral characters of the older systems'.

Special studies should be dealt with separately, including the consideration of other methods. In studies of spawning season or time, egg surveys should preferably be used, or cellular processes studied inside the gonad. Examples of the latter include the presence of oocytes in final maturation and/or hydrated/ovulated oocytes, both clearly pointing to spawning fish (West 1990). However, the implementation of image analysis has very much opened up possibilities for the rapid measurement of oocytes (Thorsen & Kjesbu 2001); the leading oocyte size can be used to predict spawning time (West 1990, Kjesbu 1994), and oocyte size standard deviation to estimate the stage of spawning (Kjesbu *et al.* 1990). The use of males to estimate the length of the spawning season is met with the classic problem that they are 'always ready'; males are running both before and after the females have stopped spawning (Dahle *et al.* 2003). The individual male may, however, not be producing sperm of high quality throughout the whole season (Wootton 1998, Babiak *et al.* 2006).

In this respect, a distinction should be made between academically related research questions and specific needs within statutory stock monitoring programs (as part of tactical fisheries management). Here, the recently agreed terminology of ICES (ICES 2012) has turned around the previous meaning of 'phase' and 'stage'; phase being said to refer to the classification of the whole gonad while stage referring to classification of the sex cells, for example the whole ovary being in the 'developing phase' and its most developed oocytes in the 'Vtg 3 stage' (Vtg = vitellogenic) (Brown-Peterson *et al.* 2011). Logically, in this new system the old Russian 'phase' should be 'sub-stage'. However, this implementation is somewhat challenging because the old meaning of 'stage' appears imprinted in many types of adopted schemes. Accordingly, the 'spawning-capable phase' of Brown-Peterson *et al.* (2011) was replaced by ICES with the more traditional term 'spawning' (ICES 2012), although otherwise ICES largely adopted the Brown-Peterson *et al.* (2011) scale. These discussions about denotations should not impede the important standardisation studies being conducted as maturity misclassification and misinterpretation are still common, affecting stock status judgements (ICES 2012). Taken together the 2012 expert recommendation of ICES is that the common scale should consist of six 'codes' (but with an option to use subcodes to handle species with special gonad developmental characteristics): 'immature'; 'developing'; 'spawning'; 'regressing/regenerating'; 'omitted spawning'; and 'abnormal' (ICES 2012).

As an additional note, the earlier frequently used term 'previtellogenic oocytes' is apparently losing momentum, and is being replaced by 'primary oocytes', while the term 'vitellogenic oocytes' is still popular, though 'secondary oocytes' is often used synonymously. The longlasting debate on whether cortical alveoli oocytes should be considered as primary or secondary oocytes is still ongoing, but for several reasons this type of oocyte should normally be placed in the last category (see Brown-Peterson *et al.* 2011 and references therein).

8.2.4.2 Mixed definitions of oogenesis

One term that often creates confusion is oogenesis, which has a double meaning:

(1) The development/growth of female gametes from oogonia to eggs, that is, the counterpart to spermatogenesis, and thereby jointly termed gametogenesis (Gilbert 1997).
(2) The transfer/multiplication of oogonia into oocytes (Selman & Wallace 1989).

The latter is synonymous with oogonial proliferation (Selman & Wallace 1989). On consulting the medical literature several more definitions emerge, such as oocytogenesis, which corresponds to point (1) above. Although, today, the definition as given in point (1) seems to be generally adopted, it is occasionally replaced with the related expression folliculogenesis – that is, the intimate development of follicle layers and a true oocyte to form a follicle (Grier & Lo Nostro 2000). However, confusion regarding terminology are also seen here (Grier 2012); folliculogenesis is completed when the follicle has all the defined surrounding layers in place which occurs during the first part of primary growth (Grier 2012), and in this case oogenesis has a much broader meaning than folliculogenesis. More relevant to AFRB, existing information indicates a cyclic production of oogonia at spawning or after spawning in iteroparous species (Barr 1963, plaice (*Pleuronectes platessa*); Greer Walker *et al.* 1994, Atlantic mackerel (*Scomber scombrus*)). This should be held separate from those patterns seen in semelparous species, where oogonial proliferation is restricted to early life stages (Bromage & Cumaranatunga 1988). Consulting studies on iteroparous freshwater fish, further examples of a cyclic production of oogonia during the regressing/regenerating phase (i.e., in post-spawning individuals) have also been presented (Wildner *et al.* 2013). However, the actual quantification of oogonia still appears not to be in place methodologically (as opposed to the situation for larger primary oocytes; see below). These fundamental studies are important in the sense that they form a platform of understanding with regards to when in the year the large 'pool' of primary oocytes would be expected to be 'refilled', and thereby likely the sensitive period to environmental stressors (see also Lowerre-Barbieri *et al.* 2011).

8.2.4.3 Reporting of ultimate female sex cell size

In Table 8.1, the elementary definitions of an oocyte and an egg were excluded, yet surprisingly these concepts are sometimes mixed even today. This is not trivial, at least not in cases when oocyte size is wrongly used as a substitute for egg size. An oocyte (female sex cells in meiosis) is growing to a great extent in size (both in terms of diameter and absolute yolk content) throughout development, whereas an egg (at second meiotic division) does not, provided that the eggshell hardening is completed (Wallace & Selman 1981, Tyler *et al.* 2000). It should be noted also that parts of the incorporation of yolk in developing oocytes can be extremely fast, as seen in Atlantic cod where up to 20% of the egg dry weight is due to uptake during oocyte maturation between successive egg batches (Kjesbu *et al.* 1996b); this is in line with earlier findings for *Fundulus heteroclitus* (Wallace & Selman 1985). The possibility that oocyte maturation may include a rapid sequestration of vitellogenin (at least in its first part) is still largely ignored (cf. the above-mentioned common procedure of setting secondary growth equal to vitellogenesis). In practice, the egg concept can be applied from the point of ovulation (Wallace & Selman 1981) or – even more easily – at spawning for several marine fish, which occurs only a few hours after ovulation (McEvoy 1984, turbot (*Scophthalmus maximus*); Kjesbu *et al.* 1990, cod). This is not the case for salmonids (Lahnsteiner 2000, rainbow trout (*Oncorhynchus mykiss*) or some clupeid family members (Ware & Tanasichuk 1989, Carolsfeld *et al.* 1996: Pacific herring (*Clupea harengus pallasi*)), where eggs might be kept in the ovarian lumen for several days until the appropriate environmental signal appears. This phenomenon is designated as the 'ripe holding period' (Carolsfeld *et al.* 1996) or just 'holding' (Ware & Tanasichuk 1989).

Formally speaking, measurements of oocyte size should be reported as follicle size as the thin surrounding follicle layer (consisting of two main layers/cell types separated by a basement membrane: theca (external) cells and granulosa/follicle (internal) cells, cf. handbooks in developmental biology) is included in the estimate. Although occasionally mentioned as a minor point (especially at workshops), such a replacement in terminology may introduce some confusion as there exists a lower threshold diameter value where the follicle concept no longer is valid; that is, the oogonia and the very smallest oocytes are not yet trapped in a follicle layer (Grier & Lo Nostro 2000). Nevertheless, an internet search with the combination of words 'fish' and 'follicles' revealed that 'vitellogenic follicles' is becoming increasingly popular in use, and thereby might gradually replace the term 'vitellogenic (or secondary) oocytes'.

8.3 Fecundity regulation

8.3.1 Background information

In this section, attention is focused on the principles and concepts of importance to consider when presenting individual fecundity results. Bagenal & Braum (1978), Hunter *et al.* (1992), Murua *et al.* (2003), Witthames *et al.* (2009), Armstrong & Witthames (2012) and Ganias (2013) have each provided complementary information on a range of issues of relevance to the estimation and understanding of fecundity data. Here, special emphasis is placed on the suggestion that fecundity regulation should be seen in relation to underlying processes that operate during the whole maturation cycle and affect the presented results. These processes are exemplified in the following paragraphs.

Fecundity information is gathered for many reasons:

- Inclusion in life-history tables, for example in calculations of the intrinsic rate of increase in the Euler–Lotka equation (Stearns 1992, Denney *et al.* 2002), or in reaction norms studies (Rijnsdorp *et al.* 2005).
- To be included along with sex ratio and field production of eggs to indicate spawning stock biomass, that is, in EPM.
- To be included in recruitment analyses, which is commonly achieved by replacing SSB with total egg production (TEP; see Chapter 11).

The theory and practise of EPM was developed during the early 1980s and is still a key method for biomass estimation of a high number of stocks worldwide, especially in warmer waters for small pelagics (Somarakis *et al.* 2004), for species difficult to target acoustically (e.g., mackerels; Anon. 2006a), or for other scientific reasons such as change in spatial and temporal distribution (Bernal *et al.* 2012). EPM estimates are in several cases used as tuning indices in various assessment models (ICES 2006b, Bernal *et al.* 2012). The 1980s was also a period with high creativity in terms of fecundity method development (Hunter & Goldberg 1980, DeMartini & Fountain 1981, Hunter *et al.* 1985). Several reports on advancements in analysis techniques and the regulation of fecundity were produced in subsequent years, including those by Witthames & Greer Walker (1987), Alheit (1988), Emerson *et al.* (1990), Kjesbu *et al.* (1990) and Witthames & Greer Walker (1995). In some respects, the different 'schools' approached fecundity estimations in indeterminate species in different ways (see definitions below). Over the past few years, fecundity estimations have again attracted increased attention, however, especially with reference to environmental changes and reproductive success (Lowerre-Barbieri *et al.* 2011), metapopulation studies (e.g., Silva *et al.* 2013, but see also Dickey-Collas *et al.* 2010), reproductive traits in depleted stocks (Hutchings *et al.* 2012), as well as continued interest in EPM (Bernal *et al.* 2012). Many reports on fecundity are rather similar in nature, and are of less interest outside research groups investigating particular species/assessments. On consulting these reports, it has also become clear that the earlier principles and concepts of fecundity regulation in several situations should be complemented with new versions.

8.3.2 How to obtain new insight

Logically, new insight into fecundity regulation can be achieved in many ways, but in particular by the implementation of advanced methodologies (image analysis, modern stereology and immunohistochemistry), advanced experimental studies, and broad-scale sampling programmes in the field. This is particularly true if combinations of these approaches are brought together when establishing and testing hypotheses. Each of these approaches is briefly discussed in the following paragraphs.

The 'auto-diametric method' (often with no hyphen) was developed to provide model studies with extensive fecundity data within realistic budget frames (Thorsen & Kjesbu 2001). Today, this digital method, following its correct calibration (see Alonso-Fernández *et al.* 2009), not only permits rapid estimations of the potential fecundity (see Table 8.3) of a large number of individuals in the laboratory (on the order of 100 specimens per day), but simultaneously provides detailed oocyte size frequency distributions. The latter point is essential to understand where the female had resided during the maturation cycle (West 1990, Murua & Motos 2006,

Table 8.3 Central expressions used in fecundity work.

Term	Definition	References	Comments
Alpha atresia	Oocytes that contain yolk and undergo resorption	See Hunter & Macewicz (1985a) and references therein	Comment: the typical stage examined for atresia, but also the beta (degenerating follicle cells) and gamma (flocculent granulosa cells) stage might be considered
Atretic duration (atretic turnover rate)	The time it takes to break down an atretic oocyte	See Witthames (2003) and references therein and Kjesbu *et al.* (1991)	Comment: normally reported in days
Atretic intensity (relative intensity of atresia)	Number of atretic cells divided by the sum of number of normal and atretic cells	See Witthames (2003) and references therein and Kjesbu *et al.* (1998)	Comment: in a few cases the ratio is atretic/normal cells, or excluding fish with no atresia
Potential fecundity	Number of developing oocytes present at a certain time during the maturation cycle	Multiple sources	Comment: if not otherwise stated refers to pre-spawning fish
Prevalence	Fraction of the sampled population with the reported reproductive character (e.g. atresia)	Witthames (2003)	Comment: normally reported as a ratio less than 1, or in percentage
Relative fecundity	Number of developing oocytes divided by fish body weight	Multiple sources	Comment: normally expressed as either (1) relative potential fecundity: potential fecundity divided by whole body weight, or (2) somatic relative potential fecundity: potential fecundity divided by ovary-free body weight
Residual fecundity	Number of developing oocytes still remaining in the ovary at the time of observation during the spawning period	Kjesbu *et al.* (1998) and others	Comment: used along with stage of spawning (see Table 8.1) or in detailed atresia or batch fecundity studies
Realised fecundity	Number of eggs spawned	Multiple sources	Comment: used to be named as actual or true fecundity (Macer 1974)

Thorsen *et al.* 2006), and also to handle the challenges related to various populations or subcomponents of the stock demonstrating different maturation patterns that influence when and how to correctly report fecundity (Witthames *et al.* 1995, Kjesbu *et al.* 1998, Nash *et al.* 2000, Yoneda & Wright 2004, Silva *et al.* 2013). Other related methods include the use of scanners (Friedland *et al.* 2005). Despite these positive aspects, various methodological problems persist with the automated estimation of fecundity in indeterminate spawners (Murua *et al.* 2003). This emphasises once more that the 'fecundity pattern' (synonymous with 'fecundity type', see Table 8.3) largely defines which fecundity method should be used. Ganias (2013) has provided an in-depth update on the progress made in fecundity methods over the past years, notably since the review of Witthames *et al.* (2009).

In principle, advances in stereology and histology have largely been made by adopting or adjusting methods originally developed within medical research, or related fields of biology. These include the disector method (Andersen 2003, Murua *et al.* 2003), staining protocols (Witthames 2003), and the systematic uniform random sampling (SURS) principle, using the physical fractionator (Korta 2010, Bucholtz *et al.* 2013). However, as teleost oocytes are well-defined structures (they are often almost spherical in form) this allows the use of physical packing density theory (Kurita & Kjesbu 2009). When combined with simple stereology to estimate relevant total volume fractions (cf. Delesse principle), this approach has provided detailed insight into numerical oocyte recruitment patterns in a series of species (see Schismenou *et al.* 2012 and references therein).

Within immunohistochemistry, the 5-bromo-2-deoxyuridine (BrdU) technique seems useful for detecting proliferating cells (including oogonia) (Méndez *et al.* 2003). Not surprisingly, both this method and the following method have 'pros' and 'cons' (Eva Andersson, IMR, Norway, personal communication). The main advantage of the BrdU technique seems to involve tracking studies – that is, when studying cell migration and development. The main problems here are the high costs and possible animal welfare issues (e.g., the injection of chemicals) when used on advanced living specimens such as broodstock. The other relevant method, the (anti-) histone technique (Nowak & Corces 2004) seems more ideal in many respects as it is cheaper, can be applied directly on Bouin-fixed tissue (slides), and is more specific in that it reports only cells in S phase, but it is not feasible for tracking studies.

Recent strong developments in marine aquaculture have contributed to the establishment of many well-equipped research stations which, among other purposes, also are suited to conduct recruitment-related studies on adults and their early life stages. However, few sites have been built for these types of study alone. The realisation that factors such as temperature, light regime, water current speed and feeding protocols now can be computer-controlled and added together in advanced combinations has opened a wide range of new possibilities, especially when studying species that require specific environmental situations to thrive, including those needed for correct gonad development. For many species, hormone injections/implants (Agulleiro *et al.* 2006) or stripping procedures (McEvoy 1984, Norberg *et al.* 1991) are required to gain access to the eggs. The fact that these species do not spawn naturally when in captivity may have many reasons, including too-small tanks (in terms of volume and depth), incorrect environmental conditions (including light and sound/noise) and a biased sex ratio (Mangor-Jensen & Holm 2005).

In the case of field studies, a major obstacle has been capacity limitations in the laboratory to work-up samples for fecundity, and also to obtain representative samples. Most fecundity analyses deal with a few tens to a few hundreds of individuals. If successfully implemented,

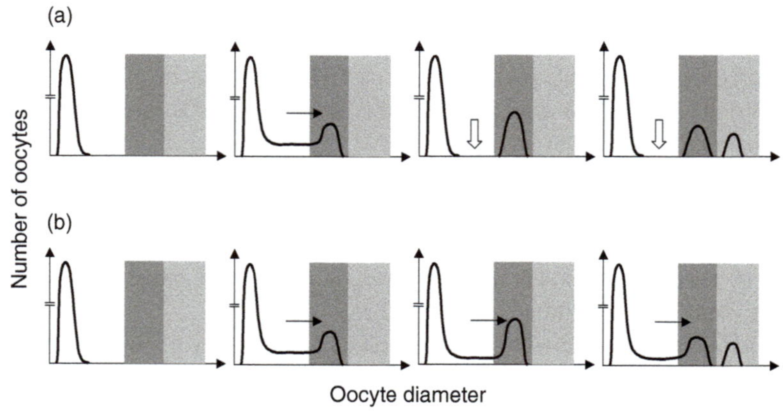

Figure 8.3 The principles of determinate (a) and indeterminate (b) oocyte recruitment (horizontal arrow) as used in applied fisheries reproductive biology. The dark grey box refers to the vitellogenic period, while the light grey box is final maturation/ovulation and egg batch formation. The number of previtellogenic oocytes (peak to left) is extremely high, which explains the broken y-axis scale. Vertical arrows show gap formation.

the use of image analysis methods (as noted above) should result in a significantly improved analysis capacity. As always, the fecundity field-sampling programme should be targeted to achieve a reasonable coefficient of variation (Hunter *et al.* 1992, ICES 2006a). Also, within a given set of institutional resources available, it is a much better strategy (in terms of precision) to take many small samples spread over the whole, relevant geographical area than a few, large ones at certain locations within the same area (Pennington & Vølstad 1994, Murua & Motos 2006, ICES 2006a). Taken together, future studies are expected to present a much more complex picture on fecundity variation, including handling the underlying problem with spawning migrating individuals, which might very quickly alter the demography of the stock unit under observation during the cruise.

8.3.3 Determinancy and indeterminacy classifications, and implications for individual fecundity estimation

8.3.3.1 Definitions of determinancy and indeterminacy

The concept of determinate and indeterminate spawners/spawning style was introduced as a result of studies undertaken by Hickling & Rutenberg (1936) and Yamamoto (1956), followed by popularisation by Hunter *et al.* (1985). This popularisation was later regretted due to the initiation of too many unproductive discussions, and the fact that all fish show an indeterminate style at some point in the maturation cycle (Hunter & Macewicz 2003). Nevertheless, these two concepts are probably here to stay, as they form the backbone on which fecundity methods are selected and used on either prespawning and/or spawning fish.

A determinate style means that there is no new (*de novo*) recruitment of oocytes to the vitellogenic mode(s) during spawning, while an indeterminate style implies that previtellogenic oocytes continuously take up yolk; that is, they turn into vitellogenic oocytes and recruit to the vitellogenic mode(s) during the active period of reproduction including at spawning (Figure 8.3). The distinction between the two styles could simply and easily be formulated based on the development (or not) of a gap (hiatus) between the previtellogenic and vitellogenic mode(s) in the oocyte size frequency distribution; a determinate fish shows such a gap, while an indeterminate fish does not. However, recent modelling studies have indicated that this may be too simplistic (Ganias *et al.* 2015). The gap found between vitellogenic oocytes and hydrated/ovulated oocytes in spawning individuals is held outside this definition. But, there are also other criteria/'lines of evidence' (Hunter *et al.* 1989, Greer Walker *et al.* 1994, Murua & Motos 2006) which basically are a consequence of the presence or absence of such a gap. For example, determinate spawners are expected to show a reduction in number but an increase in the size of vitellogenic oocytes during spawning. Introductory simulations focusing on the 'cannot make it in time' concept (cf. Line of Evidence 7 in Greer Walker *et al.* 1994) point to the length of the oocyte growth period (OGP) in relation to the length of the spawning period (SP) being central. In other words, species with a slow OGP, for instance due to a low ambient temperature (reflected as OGP/SP > 1) should opt for a determinate reproductive style, whereas OGP < 1 reflects indeterminate spawners (Ganias *et al.* 2015). Nevertheless, judgements of these lines of evidence are not always straightforward due to, for example, the appearance of size-specific atresia and seasonal changes in oocyte final maturation size (Greer Walker *et al.* 1994). Although likely very much an oversimplification, temperate, capital/partly capital breeders seem to be more inclined to be determinate spawners (e.g., Atlantic herring and cod), while tropical or subtemperate income breeders tend to be indeterminate spawners (e.g., hake *Merluccius* spp. and anchovy *Engraulis* spp.). In other words, an indeterminate style is expected to be more common among species showing a long spawning period or spawning throughout the whole year, while a determinate style should reflect shorter spawning periods. However, there are also trends related to the time of spawning in the year, with the determinate style being more typical among winter spawners and the indeterminate style among summer spawners (Rijnsdorp & Witthames 2005).

8.3.3.2 Determinancy and indeterminacy classification problems

It is normally assumed that a species is either a determinate or indeterminate spawner. However, this might be a too-strict approach, as reports for the determinate sole in the North Sea have indicated that individuals from some areas tend to approach the indeterminate style (Witthames & Greer Walker 1995). Although, in essence, the reason for this variation is presently unknown, it is thought to be related to differences in food productivity or exploration patterns (P.R. Witthames, Lowestoft, UK, personal communication). An additional common problem in experiments on cod reproduction is the overproduction of vitellogenic oocytes (Kjesbu *et al.* 1996a), which might relate to a longer indeterminate phase of oocyte recruitment in response to the stable feeding protocol normally established in tanks. This view is supported by the level of atresia in these animals (which are principally thought to be determinate spawners; Kjesbu *et al.* 1990), which increases towards the end of spawning (Kjesbu *et al.* 1991); this pattern is typically seen in indeterminate spawners (Wallace & Selman 1981, Greer Walker *et al.* 1994, Murua & Motos 2006). This end-of-spawning removal of any surplus production of vitellogenic oocytes is often mentioned as 'mopping up' (Wallace & Selman 1981). Another conflicting example is that of Atlantic/Western mackerel, which does not show any clear gap development prior to spawning, and thus in principle is an indeterminate fish. However, studies on oocyte growth rate have indicated that newly recruited vitellogenic oocytes during spawning will not have sufficient time to reach ovulation before the spawning season is completed (Greer Walker *et al.* 1994) (cf. Ganias *et al.* 2015).

These fundamental questions on oocyte recruitment styles can today be detailed at the cellular level by using appropriate staining protocols combined with image analysis measurements, or by using the above-described oocyte packing density (OPD) theory, or the Disector method (Ganias *et al.* 2014). In the latter case, if aiming to quantity the smallest oocytes or oogonia, the preferred practice is to use the Optical Disector (Myers *et al.* 2004), in which the embedded cells of interests are counted using focusing (confocal microscopy) techniques. This approach should be held separate from the Physical Disector, which includes the actual production of pair-wise histological slides. In the present framework, the Optical Disector is thought to be particularly suited for the estimation of oogonia, as these are generally too small (<10–20 μm in diameter; Bromage & Cumaranatunga 1988, Wallace *et al.* 1988) to be effectively numbered by the Physical Disector, which requires that the distance between successive sections are one-quarter to one-third of the smallest cells of interest (see citations in Andersen 2003, Murua *et al.* 2003). Nevertheless, Greer Walker *et al.* (1994), in their studies on Atlantic mackerel, presented convincing information on the seasonal production of previtellogenic oocytes by using former, specially calibrated, stereometric methods.

8.3.3.3 *Individual fecundity estimation*

Logically, the estimation of fecundity in an indeterminate fish is much more complicated than in a determinate fish. In the last case, the number of vitellogenic oocytes can simply be counted prior to spawning (i.e., potential fecundity), but any subsequent atresia (Table 8.3, Figure 8.4; Plate 21) must be subtracted to achieve a realised annual fecundity. For indeterminate species, batch fecundity is normally reported instead. In the following text, attention is focused on the estimation of annual realised fecundity (Table 8.3), as this is the ultimate target in many studies of this type.

The estimation of realised fecundity, F_R, can be based on the following three techniques (methods) and associated equations:

- *Total batch production method*

$$F_R = \sum F_B \quad (1)$$

where ΣF_B is the summed batch fecundity, that is, the total number of eggs spawned in all batches by the female during the length of the spawning period. To give accurate and precise data, this approach requires individual tracking of naturally spawning females in experimental set-ups (Hislop *et al.* 1978, Kjesbu 1989, Murua *et al.* 2003). This might be very difficult to achieve, depending on the species under study (see above). For pelagic egg producers, care should be taken to collect all eggs, as poor-quality eggs tend to settle out in the tanks (Kjesbu *et al.* 1996a).

- *Batch fecundity method*

$$F_R = N_{B,\,\text{estimated}} \times F_{B,\,\text{average}} \quad (2)$$

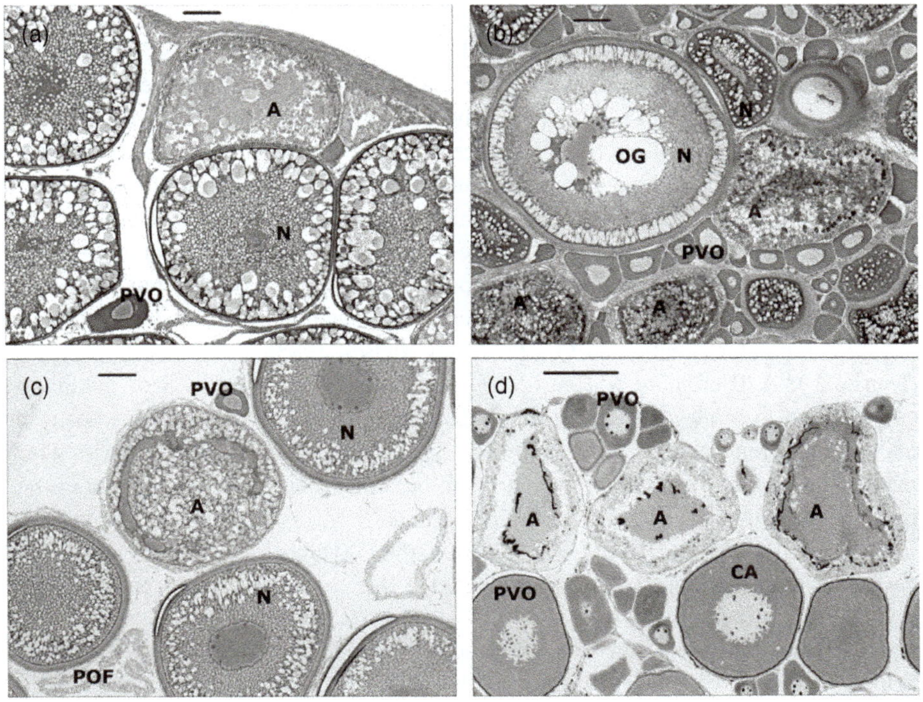

Figure 8.4 Vitellogenic atresia in Atlantic herring (a), Atlantic mackerel (b), Atlantic cod (c) and cortical alveoli atresia in blue whiting (*Micromesistius poutassou*). A: atretic cell; N: normal cell; PVO: previtellogenic oocyte; OG: oil globule; POF: post-ovulatory follicle; CA: cortical alveoli. Scale bar: 100 μm. Tissues in micrographs (a), (b) and (c) were embedded in Technovit, and tissue in micrograph (d) in Epon. Staining was with toluidine blue. For a color version of this figure, please see Plate 21 in the color plate section.

where $N_{B,\ estimated}$ is the estimated number of batches shed by a representative female, and $F_{B,\ average}$ is the average batch fecundity of the population under study. Thus, this approach can be used on either field samples or groups of fish in tanks. $N_{B,\ estimated}$ can be given from indications of the length of spawning period (SP) divided by the length of spawning (batch) interval. SP has been found by comparing the time (in days) between 50% of the females being very close to start spawning (showing oocytes in final maturation or hydrated oocytes) to 50% being in a spent/early recovering condition (see Karlou-Riga & Economidis 1996, Armstrong *et al.* 2001 and references therein). The mean batch interval and spawning frequency, the latter deduced from the inverse of spawning fraction (S) (Table 8.1), very much reflect the same type of information but the terms should be held separately; for example, a spawning fraction of 0.2 gives a spawning frequency of 'once every 5 days' or a mean batch interval of 5 days. From this, $N_{B,\ estimated} = SP \times S$. Normally, the prevalence of 1-day old post-ovulatory follicles is used as a criterion/tool to estimate S by adopting results achieved on northern anchovy (*Engraulis mordax*) (Hunter & Goldberg 1980). This is an integral part of EPM and will not be dealt with here (however, see recent applications by Murua *et al.* 2006). For more details the interested reader is also encouraged to consult the recent, free-accessible expert manual of Ganias *et al.* (2014).

For some indeterminate spawners, SP is in reality non-existent; spawning takes place more or less throughout the whole year. This implies that individual egg production cannot be adequately assessed. However, in such cases S can be estimated on a monthly basis in order to obtain a population daily fecundity curve (Murua *et al.* 2006). Interestingly, these authors found a pronounced change in S for European hake (*Merluccius merluccius*) throughout the year, equivalent to a change in length of mean batch interval from 5 to 12 days. Ceased-spawning females were excluded from the calculations. The time between successive batches is well known to be temperature-dependent (Hunter & Macewicz 1985b), and the reported maximum 2 °C difference reported by Murua *et al.* (2006) should, at least in principle, have an important influence on S, assuming the temperature coefficient Q_{10} (Schmidt-Nielsen 1983) to be 2 in the calculations (Kjesbu 1989). Claramunt & Roa (2001) also reported cyclicity in S of South American pilchard (*Sardinops sagax*) from studies on variations in the gonadosomatic index of 'hydrated' females. F_B, and thereby $F_{B,\ average}$, are based on counting oocytes in final maturation (nucleus migration) or hydration, but not ovulated eggs, which might be lost from the ovary during capture and handling (Hunter & Macewicz 2003). Thus, this method is often labelled as the 'Hydrated Oocyte Method' (Hunter *et al.* 1985). Ovaries with very new post-ovulatory follicles are excluded due to likely partial spawning (Murua *et al.* 2006).

Arguments in favour of the Batch Fecundity Method are many (Hunter *et al.* 1985, see below). However, the presentation of F_B is met by two problems often not mentioned. First, there is clear evidence that batch fecundity changes during the spawning period as a dome-shaped curve (Alheit 1988, Kjesbu 1989). Second, oocytes stay in final maturation/hydration only for a short period of time between successive batches, which means that only a restricted part of the fish catch can be used for batch fecundity studies (Kjesbu *et al.* 2003, Ganias *et al.* 2014). More extensive sampling during the spawning season, combined with precise information on when in the day the fish normally spawn, can be factors to solve the last problem. The first problem is more complicated to handle and might actually not even be required depending of the scientific aim, for example in the Daily Egg Production Method, which is normally based on a single egg survey and therefore, at least in theory, needs only a single daily batch production estimate (and the associated level of uncertainty). In the present context, however, aiming at the individual annual production, temporal changes in batch size should be considered, also because different size or age classes often spawn at different times in the season (Alheit 1988, Ware & Tanasichuk 1989). If the number of 'hydrated' ovaries is not sufficient due to the constraints mentioned above, an alternative way would be to analyse the vitellogenic oocyte size frequency distribution (mean, standard deviation, kurtosis and skewness) to estimate stage of spawning/portion of eggs spawned (Kjesbu *et al.* 1990) and thereby to reconstruct a general individual egg production pattern for the different classes of the spawning stock. Although, in practice, this might be impossible to pursue for those species, all likely being indeterminate, in which there may not necessarily be any such clear change in oocyte size distribution with time (Murua & Motos 2006: European hake). This observation is in marked contrast to the large changes noted in these respects for determinate spawners (Kjesbu *et al.* 1990). Thus, other measures reflecting trends in energy allocation patterns, such as in gonadosomatic index and condition indices (Rinchard & Kestemont 2003, Murua *et al.* 2006, de Oliveira *et al.* 2006), or changes in vitellogenin and sex hormone profiles (Johnson &

Casillas 1991, Methven *et al.* 1992), might be much more useful to pursue in some cases, or would yield valuable additional information.

- *Annual fecundity method*

$$F_R = F_P - F_A \tag{3}$$

$$F_A = F_P \times A_{RI} \times D/A_{T,\alpha-\text{stage}}, \tag{4}$$

where F_P is potential fecundity, F_A atretic fecundity (number of vitellogenic oocytes lost due to resorption), A_{RI} relative intensity of atresia (Table 8.3), D is the length of time (in days) from the point of sampling until the spawning period is fully completed, and $A_{T,\alpha-\text{stage}}$ is the atretic turnover rate (duration) (in days) for the α–stage at a particular water temperature (T) (Table 8.3). These sets of equations were established very much within the framework of the Annual Egg Production Method during the 1990s (see Armstrong *et al.* 2001, Witthames 2003, ICES 2006a), and also within a recruitment framework (Kjesbu *et al.* 1991, Ma *et al.* 1998). However, the underlying principles behind the outlined reduction in oocyte number through atresia resemble very much those seen in age-specific mortality equations frequently used in population dynamics. This was recognised by Kurita *et al.* (2003). By rearranging rearranging their equation (5b), used to model the length of the atretic turnover rate, the realised fecundity can be given as:

$$F_R = F_P \times (1 - A_{RI})^{D/AT,\alpha-\text{stage}}. \tag{5}$$

Importantly, equation (5) is more conservative than the combined use of equations (3) and (4), which means that the modelled decline in egg production of an individual falls notably less with an increase in A_{RI}.

Each of the independent variables in the previously mentioned equations can be set with reasonable accuracy and precision but, unfortunately, this positive situation applies to few species only (op. cit; Óskarsson *et al.* 2002, Kurita *et al.* 2003). Apparently this holds true even today – that is, two decades since the pioneer works started (Ganias *et al.* 2014). The value of D is typically given from previously established stock-specific oocyte growth curves. Note that $A_{T,\alpha-\text{stage}}$ is reported to be only in the order of one to two weeks (depending on species, temperature and methodological approach), which means that even a low A_{RI} may have a pronounced effect on F_R, at least if D (more specifically, the ‘atretic window’) is long. The main drawback with equations (3)–(5) is that they can only be used on determinate spawners or semi-indeterminate spawners; that is, in cases with no or insignificant de novo oocyte recruitment, to give meaningful fecundity results. Nevertheless, in the case of semi-indeterminate spawners, parts of the discussions in the literature have been too straightforward in light of the fact that equations (2)–(5) are all associated with important advantages and disadvantages (Table 8.4). On the other hand, energetic and numerical considerations have undoubtedly demonstrated for northern anchovy that major *de novo* oocyte recruitment is taking place during the season in this species (Hunter & Leong 1981). Thus, there are clear examples where equations (3)–(5) would not work. A central concept when applying equations (3)–(5) is ‘threshold value’, or in full, ‘lower size threshold of developing oocytes’ (Greer Walker *et al.* 1987, 1994, Witthames 2003). This value is a compromise between the observed maximum previtellogenic oocyte diameter in spent fish and the observed minimum cortical alveoli oocyte diameter for fish starting to mature. Consequently, all developing oocytes above the threshold value are counted to obtain F_P (or, in a few cases, residual fecundity; see Table 8.1).

Following the presentation of these equations it seems necessary to point out that the dominating number of fecundity publications still end with F_P; very few progress to F_R. A noticeable exception to this is reports on F_P in single batch spawners based on samples taken just prior to spawning (e.g. Tanasichuk & Ware 1987); F_P can in such situations safely be set equal to F_R, as any effect of A_{RI} as presented in equations (4)–(5) should be minimal or zero.

- *Additional comments*

  This section has focused on fecundity at the individual level. Naturally, any scaling to population level needs multiplication with the number of sexually mature females, and also other demographic considerations (see Chapter 11). Currently, equation (4) in this type of study (i.e., at the population level) is also multiplied with prevalence (the proportion of observed females with atresia) (Table 8.3) (Witthames 2003), while F_P, F_A and F_R in EPM studies are divided by body weight to obtain relative fecundities. Equations (3)–(4) should in future studies be replaced with equation (5), at least at high levels of atresia due to an underestimation of realised fecundity in traditional approaches.

  Several of the published reports which were read in writing this paragraph, but were not necessarily cited in the text, appeared to contain misnomers or conceptual errors. For instance, serial batch spawners were in one study consequently set equal to indeterminate spawners. This is not synonymous, as many determinate spawners are also serial (multiple) batch spawners. Apparently, many authors believe that the counting of number of peaks in an oocyte size frequency distribution gives the number of batches to be spawned by the female. But this is not the case for at least several gadoids, which may show only one vitellogenic mode but might spawn up to 20 batches (Kjesbu *et al.* 1996a) – that is, a number similar in size to what is observed in the highly indeterminate northern anchovy (Hunter & Leong 1981).

8.3.4 New or refreshed concepts of fecundity regulation

As seen above, fecundity estimates refer traditionally to a part of the maturation cycle, normally vitellogenic (potential) fecundity (see equation (3)), or final maturation/hydration (batch) fecundity combined with spawning fraction (equation (2)). Another way to look at this cycle – that is, instead of limiting the study to certain type of cells – would be to study the temporal, numerical production of female sex cells throughout the whole length of the maturation cycle (Figure 8.5). In the following, the primary focus is on changes in F_P due to the process of 'downregulation', that is, the underlying, gradual loss of developing oocytes, along with other factors thought to be important for fecundity regulation within a given year. The term downregulation is taken from medicine, where it refers normally to a reduction in the number of receptors at the cell surface (www.medterms.com). The noted extreme reduction of oogonia and oocytes as seen in the human ovary – that is, from a maximum of about seven million cells in the seven-month foster child to about 400 realised ovulations during the life of the woman – is considered the classical demonstration (see Gilbert 1997). Within fish biology the concept of downregulation was apparently for the first time used by Óskarsson *et al.* (2002) based on

Table 8.4 Pro and contra of the Batch Fecundity Method and the Annual Fecundity Method, with special reference to recruitment-related studies. F_P = potential fecundity; F_R = realised fecundity; F_B = batch fecundity; TEP = total egg production of the stock; DEPM = Daily Egg Production Method (Parker 1980); AEPM = Annual Egg Production Method (Armstrong *et al.* 2001); DFRM = Daily Fecundity Reduction Method (Lo *et al.* 1992).

Reproductive style	Advantage	Disadvantage
Batch fecundity method (Equation 2)	Inexpensive method, which is used in many laboratories worldwide	Degeneration rate of post-ovulatory follicles and thereby spawning fraction formulae are often based on results from one species (*E. mordax*) only
	Automated methods for counting hydrated oocytes are progressing	Experimental results show that both F_B and number of batches spawned vary between females, i.e., F_B alone does not reflect reproductive investment
	Applicable on all species with egg batch production	Spawning fraction varies seasonally
	Used as direct input in DEPM	Experimental results and partly field studies show that batch fecundity changes (as a dome-shaped curve) throughout the spawning period
		Difficult to get high number of samples due to short duration of the hydrated phase
		Few historic F_B data; the method was developed in the 1980s
		Body size or condition indices often correlate rather poorly with F_B
		Difficult to scale from F_B to TEP
Annual fecundity method (Equation 3)	F_P estimation is in principle straightforward and automated methods are in place	Not useful for clearly indeterminate spawners
	In many studies body size and condition explain a significant part of the variation in F_P	Expensive, complex routes, including atresia counting, to get F_R
	Some species have important historic F_P data	Full parameterisation of Equation 3 only available for few species
	F_P and F_R data are easy to use when scaling up to TEP	Limited F_R data exist
	F_R data are included in AEPM and DFRM	

the results of Kurita *et al.* (2000). As noted from the patterns in fecundity, Figure 8.5 concentrates on determinate spawners, that is, fish that show a definite window of *de novo* oocyte recruitment in the year. This is simply because of the existence of relevant published information. Also, in indeterminate spawners the phenomenon of downregulation, as defined, might be difficult to quantify or, if present, it might have an insignificant effect on F_P, being counterbalanced by *de novo* production. However, oocyte losses as such are well known to occur also in this category of fish, for example during the aforementioned 'mopping-up process' or at unfavourable environmental conditions (Hunter & Macewicz 1985a).

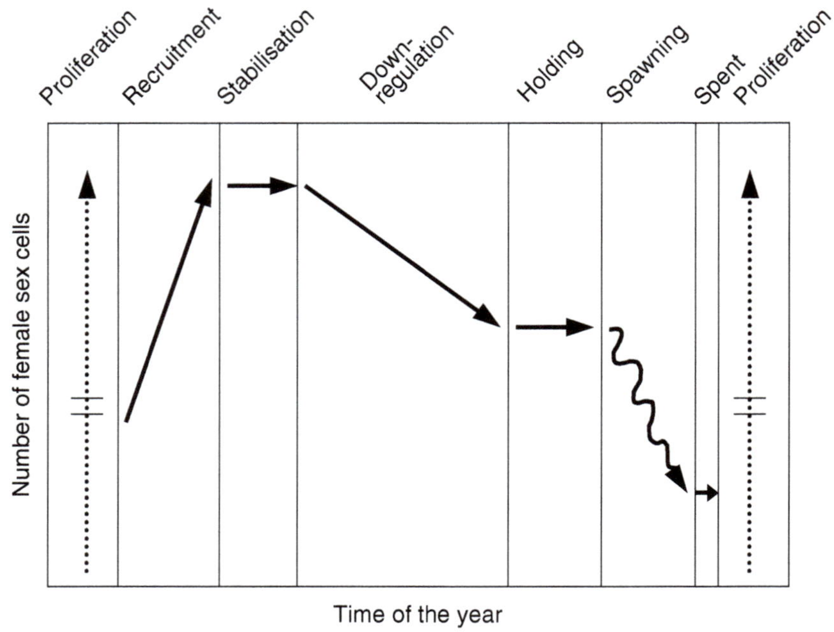

Figure 8.5 Generalised variation in numerical production and loss of sex cells in determinate, iteroparous fish ovary during a full maturation cycle. The production (proliferation) of oogonia is presently a 'black box' (note the dashed line) but is assumed to be particularly intense in the post-spawning period. Oocyte recruitment (from previtellogenic oocytes up to vitellogenic oocytes) is successively followed by the stabilisation period, the downregulation process, a short period of holding of ovulated oocytes in the ovary, spawning, the spent stage and, finally, likely a new cycle of proliferation. This figure was composed based on information given by Barr (1963), Kjesbu *et al.* (1991), Carolsfeld *et al.* (1996) and Kurita *et al.* (2003).

8.3.4.1 Presence of downregulation

During the 1950s, Vladykov (1956) recommended – based on observations of a reduction in oocyte size numbers throughout the maturation cycle in brook trout (*Salvelinus fontinalis*) – that any thorough fecundity analysis should be undertaken on females in the same maturity phase. A similar type of reduction in fecundity was noted for rainbow trout, and attributed to atresia (Bromage & Cumaranatunga 1988). The importance of these findings has now been implemented in a series of reports over the past decade or so, changing the understanding of how to adequately present fecundity figures – see for instance Witthames *et al.* (2009), considering both seasonal changes in flatfish and groundfish fecundities, and Kennedy *et al.* (2009), studying the complex reproductive cycle of Greenland halibut (*Reinhardtius hippoglossoides*) by this type of approach. In the model of Thorsen *et al.* (2006), fecundity in Atlantic cod was for the first time fully standardised to a certain mean oocyte diameter (700 μm), corresponding to an advanced phase of vitellogenesis. Other standardisation techniques, before the establishment of image analysis, have been used to construct fecundity time series based on maturing ovaries collected at the same geographical location in the same week of the year (Kjesbu *et al.* 1998). However, to deal with the problem of downregulation, this type of sampling scheme requires that: (i) the maturity phase is the same each year, that is, a stable spawning season; and (ii) there is no change in the spawning grounds. The first point might partly be the case

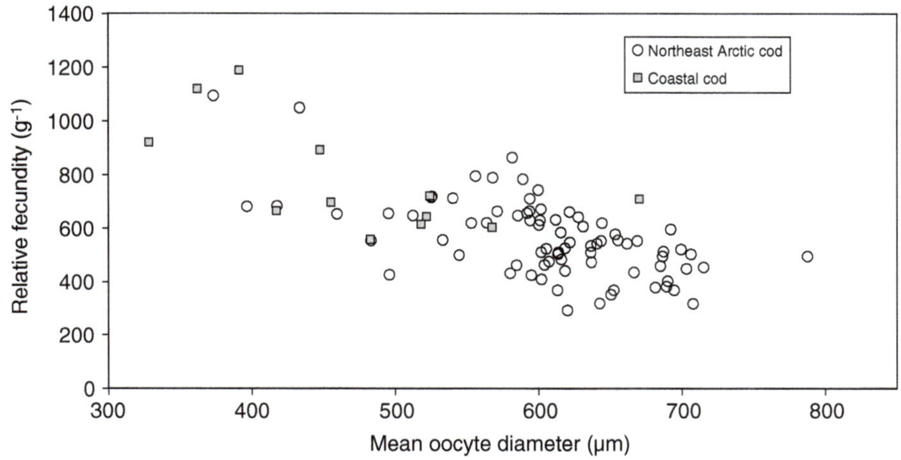

Figure 8.6 Relative potential fecundity in Atlantic cod versus mean oocyte diameter. The fish were sampled off Andenes, Northern Norway on 13–14 February and 8–9 March 2006, and classified from otolith readings to be either North East Arctic cod or Norwegian Coastal cod.

for North East Arctic cod (Pedersen 1984), but certainly not for Baltic cod, which has shown a large delay in spawning time (Wieland *et al.* 2000). Also, other species such as Pacific herring (Ware & Tanasichuk 1989) demonstrated a clear annual variability in spawning time. Different size classes may spawn at different times, complicating such a sampling approach even more (Ware & Tanasichuk 1989, Óskarsson *et al.* 2002). In terms of spatial aspects, multi-decadal temperature changes have been shown to influence the location of spawning grounds (Stenevik & Sundby 2007, Sundby & Nakken 2008), making a geographically fixed sampling location problematic. Thus, there is evidence to recommend the establishment of standardised protocols in F_P studies to handle the problem of maturity-phase differences, as illustrated in the next paragraph.

Kurita *et al.* (2003) estimated the fall in relative somatic potential fecundity ($RF_{P,S}$) (Table 8.3) of a 32-cm Norwegian spring-spawning herring to be about 60% from early to late vitellogenesis, although this figure is likely to be an overestimation due to the limited number of samples and examples of low-condition individuals accelerating the process of downregulation (Kennedy *et al.* 2011). The estimates presented for North East Arctic cod also indicate that the fecundity might fall significantly (e.g. in Thorsen *et al.* 2006; Figure 8.6), though the two studies are difficult to compare directly as the latter used whole body weight in the formula ($RF_{P,W}$) instead of somatic weight and a more narrow vitellogenic oocyte range, exemplifying another level of standardisation needed across studies and species. However, as for the herring, size-specific effects could not explain the observed fall in relative fecundity. Neither species take food during the main part of vitellogenesis (Table 8.2; see also citations in Kjesbu *et al.* 1991, Kurita *et al.* 2003) and consequently the whole body weight would be expected to fall as maturation advanced, driving the relative fecundity estimate upwards if the oocyte number is assumed to be held constant. This implies that the observed reduction in relative potential fecundity must have a clear baseline of reality, indirectly supporting the original view of Bromage & Cumaranatunga (1988) that atresia is the underlying driving force (cf. equations (3)–(5)).

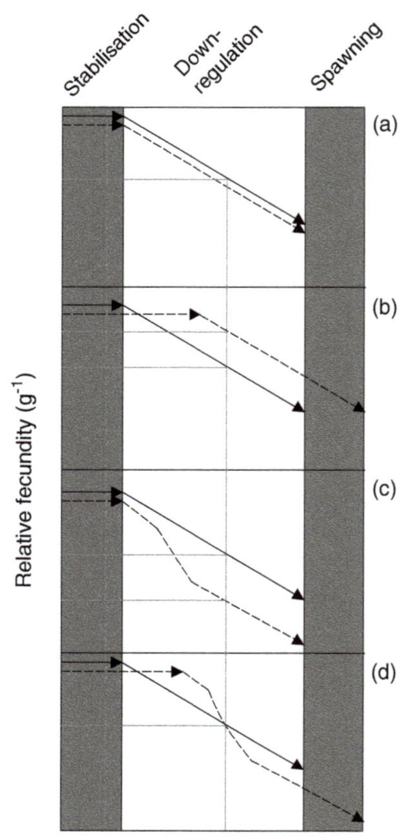

Figure 8.7 Different scenarios for how sampling at a specific point of time in the year may influence the understanding of relative potential fecundity (RF_P) in two contrasting groups (populations) of fish, depending on the temporal changes in the degree of downregulation. Vertical and horizontal thin (red) line: sampling time and corresponding RF_P; thick (black) line: reference (control) group; dashed line: response group, i.e., after a change in environmental or experimental conditions. (a) Same degree of downregulation and spawning time; correct conclusion; (b) Delayed spawning time in response group but equal degree of downregulation; overestimated RF_P in response group; (c) Same spawning time but larger degree of downregulation in response group: correct conclusion; (d) Delayed spawning time and larger degree of downregulation in response group: type II error (accepted false null hypothesis). Note for (c) and (d) that the response group was assumed to undertake a steep decline in RF_P during a specific part of the maturation cycle; that is, they showed a clear atretic window (Kurita *et al.* 2003). For a color version of this figure, please see Plate 22 in the color plate section.

8.3.4.2 Exclusion of downregulation: types of error likely to be made

In cases where no attention is paid to the phenomenon of downregulation when contrasting different series of fecundity data, such as among years, stocks or experimental groups, it might still be concluded correctly depending on how and when this downregulation takes place (Figure 8.7; Plate 22). However, as seen in Figure 8.7, there are situations when misinterpretations are likely to be made; four scenarios were selected leading to two correct and two incorrect conclusions. Other scenarios could have included different stabilisation plateaus (initial

RF_P), different temporal patterns for downregulation (different 'atretic windows'), or another shape than linear (e.g., negative exponential) for the downregulation curve. However, the four panels produced should suffice as examples to make the meaning clear: the risk of being wrong very much increases if spawning time differences exist among and within the materials analysed. Maturity-phase differences may appear for several reasons (e.g., due to temperature or pollution effects), but here they are limited to two: (i) size-specific effects; and (ii) condition-driven effects.

8.3.4.3 Size-specific effects and downregulation

As seen in several places in the fish literature, small and large females often differ in several reproductive traits. Only those being relevant in the present context are listed:

(1) Larger prespawning females have generally a higher RF_P than smaller females. For instance, a 120-cm North East Arctic cod female shows 30–80% higher $RF_{P,W}$ values than its 50-cm counterpart (Kjesbu *et al.* 1998). Analyses on Icelandic cod gave similar answers (Marteinsdottir & Begg 2002). However, such a result is not always found, especially if the material examined shows a narrow range in fish length (naturally, or, as frequently seen, for over-fished populations).
(2) Larger females (and most often also larger males) allocate a larger fraction of surplus energy to reproduction than smaller individuals (e.g., Bromley 2003) which, logically, is interconnected with the previous point. As a consequence, the common phenomenon of asymptotic growth observed at large body sizes is, physiologically speaking, thought to be due to increased reproductive investment and also increased total maintenance costs (cf. equations in Jørgensen & Fiksen 2006).
(3) Larger females often spawn earlier in the season (e.g., Ware & Tanasichuk 1989, Wright & Gibb 2005), which again is related to allometry and temperature sensitivity (Kjesbu *et al.* 2010).
(4) Oocyte growth rate seems higher in older females (Ramsay & Witthames 1996).
(5) Older females may start maturing earlier in the year than younger ones (Ramsay & Witthames 1996).
(6) The larger eggs from larger females (Trippel *et al.* 1997) are associated with larger oocyte size at final maturation (Kjesbu *et al.* 1996b).

Based on points (1)–(6), together with the previous description of downregulation, it is obvious that some considerations are needed to properly contrast different size or age categories of fish in terms of F_P or RF_P. First, a given oocyte size (e.g., as measured in the autodiametric method) for large and small females might reflect a different maturity phase, but note that similar-sized oocytes of Atlantic herring have been shown to have identical microstructure (including yolk content) (Óskarsson *et al.* 2002), including also when contrasting autumn and winter spawners (van Damme *et al.* 2009). Second, final maturation is well known to be a very quick process. Thus, the inclusion of an oocyte size correction factor or a more detailed examination of length of oocyte growth curves should be considered to handle this associated likely problem in fecundity estimation. However, the main issue is that standardisation protocols are implemented and that, eventually, some refinements are made later when the level of knowledge has increased. At present it is not fully clear how RF_P actually changes with maturity phase, in particular due to a lack of initial RF_P values represented by the height of the stabilisation plateau (see Figure 8.5), in combination with variation in fish size except, as mentioned, that RF_P are generally higher for larger females, at least towards spawning. With special reference to point (2), care should be taken to standardise by fish size in energy allocation studies, for example when plotting expressions of fecundity versus expressions of body growth rate, as in Rijnsdorp *et al.* (2005) who rather unexpectedly (see below) could find no relationship between F_P and growth (in cm per year) for 40-cm North Sea plaice. As these samples were taken between December and February, although at the spawning grounds (Rijnsdorp 1990), a part of the explanation might be differences in the level of F_P downregulation. Nevertheless, this lack of relationship in plaice is supported by ideas presented in the well-known Surplus Production Allocation Model of Rijnsdorp (1990), which in short says that RF_P is constant above a certain energetic threshold level. Thus, after this point is reached a relatively larger fraction of the surplus energy is directed towards body growth – that is, being prioritised and less to reproduction. This result is in agreement with recent experimental findings of Yoneda & Wright (2005), though a few other possible mechanisms might also be operating, as addressed below.

8.3.4.4 Stock-specific effects and downregulation

Several of the points addressed in the previous paragraph should also be consulted when contrasting different stocks of the same species. For example, does the entrance to final maturation occur at the same oocyte size across stocks? Or is the degree of downregulation related to final egg size (trade-offs)? These questions, related to reproductive tactics, are just on their way to being formulated today among fish reproductive biologists, as exemplified by the studies of van Damme *et al.* (2009) who documented a similar rate of downregulation in autumn and winter spawners of North Sea herring but ongoing for a longer period of time in the last category, as reflected by a comparatively lower fecundity but bigger eggs.

8.3.4.5 Condition-driven effects and downregulation

As summarised by Yoneda & Wright (2005), there is conflicting evidence whether energetic status (feeding ration) influences the time of start of spawning in the year, or not. The reasons for these differences in results – that is, from finding no effect at all (Karlsen *et al.* 1995) to seeing delays for poor-condition females (Springate *et al.* 1985, Kjesbu 1994, Yoneda & Wright 2005, Dunia González 2006) – seem complicated but include experimental length, initial energetic status, feed ration size, and the use of repeat or recruit spawners. Note that the study of Karlsen *et al.* (1995) is the only one that applied fluctuating feeding schemes: starvation weeks were mixed with full ration weeks. Based on reports from Springate *et al.* (1985), Kjesbu (1994) and Dunia González (2006), a delay on the order of two weeks is to be expected under poor feeding situations. Thus, a tactile response that is considered rather small but is expected to be so in light of the generally accepted Critical Period or Match–Mismatch Hypothesis (see Chapter 3). Using equation (3) and setting $A_{T,\alpha\text{–stage}} = 8$ days (Kurita *et al.* 2003), only very high A_{RI} values would be influential in a two-week period: 2% reduction in F_P for $A_{RI} = 0.01$, 17% reduction in F_P for $A_{RI} = 0.10$, and 70% reduction in F_P for $A_{RI} = 0.50$. At least in Atlantic cod and herring, a major drop in F_P over this time span seems unlikely, but cannot be fully excluded in extreme cases (Witthames *et al.* 2000, Óskarsson *et al.* 2002). Nonetheless, a failure to account for this general condition-driven effect on oocyte number related to changes in the time of start of spawning would likely not impact any major conclusions drawn on F_P or RF_p differences because of the above-mentioned likely small variation in the duration of this effect.

8.3.5 Condition effects

Generally, condition effects within the present framework of fish reproductive biology can be classified into three types that influence: (i) the timing of sexual maturity; (ii) reproductive investment; and (iii) life-history decisions. All three have 'lag effects', which means that any effect may not be detected in the current data set but is a consequence of a previous situation or weak underlying trends over time. These influences might have occurred a long time ago, including at the juvenile stage. Further information on these aspects, along with comments on points (i) and (iii), is provided by Saborido-Rey & Kjesbu (2005), Jørgensen & Fiksen (2006), and Jørgensen *et al.* (2006). Temporal changes in maturation schedules (maturity ogives) are an integral part of total egg production dynamics, and are dealt with in Chapter 11. Point (ii) can be split into condition effects on fecundity and condition effects on propagule (egg and larval) size and quality. A large number of reports have been made on the link between fecundity and condition, both from the field and the laboratory. Expressions of condition are either based on chemistry or morphology, or combinations of both, resulting in various indices. There are no simple answers as to which condition indices are the best to select and use; the approach to be taken depends very much on the question asked, whether the study operates at the individual or population level, and whether size effects are present, or not. A few issues of special interest in light of the concept of fecundity regulation are detailed in the following subsections.

8.3.5.1 Point of reference

Field studies often show no or only small changes in relative investment in reproduction, in agreement with the Surplus Production Allocation Model described in Section 8.3.4.3. The term energetic 'threshold level' used in this model is here renamed as 'reproductive investment satiation level' (in short: 'satiation level') to keep it separate from the above-mentioned, frequently used 'threshold value' that refers to the minimum size of developing oocytes. In contrast to the study of Rijnsdorp (1990) on North Sea plaice, Óskarsson *et al.* (2002) presented information for Norwegian spring-spawning herring that pointed towards a substantial variation in reproductive investment standardised by fish size: the RF_P in low-condition years was reported to be only 25% of the figures seen in high-condition years. Rijnsdorp (1990) also stated that RF_P might happen to be very low or even zero (skipped spawners) in special cases, such as captive females not taking food. Further elaborations on the topic of skipping are provided above. Several other studies have also shown important fluctuations in RF_P values over time, though typically not so large; for example, the lowest RF_P was 71% of the highest RF_P for North East Arctic cod (Kjesbu *et al.* 1998). This factor of potential increase of 1.4 (100/71) was significantly less than the corresponding factor of 2.5 apparently seen in another gadoid, the Norway pout (*Trisopterus esmarkii*) (Bagenal 1978). The key question, however, is whether there is a general upper satiation level or not, as defined in the Surplus Production

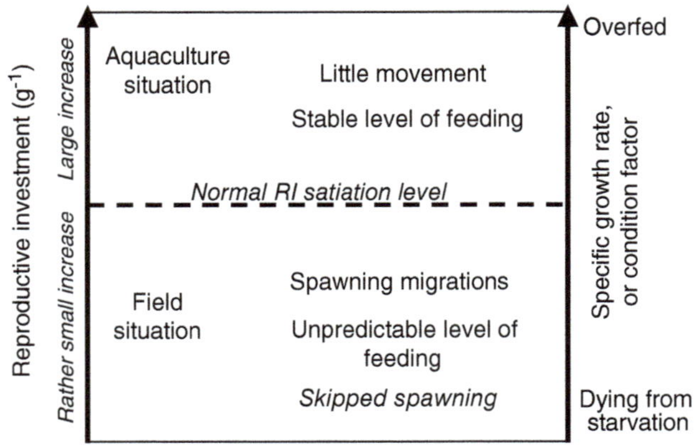

Figure 8.8 Maximum dynamic range in reproductive investment, specific growth rate and condition factor as observed when combining data from both the field and experimental/aquaculture studies in the analysis. RI = reproductive investment. Information on reproductive investment and condition was taken from Kjesbu (1989), skipped spawning from Rideout *et al.* (2005), and growth rate from Dutil & Brander (2003).

Allocation Model. This seems not to be so: on broadening the analyses to also include females in very good to extremely good condition (as seen in several aquaculture studies) the satiation level, as defined, disappeared; both somatic and reproductive investment continued to increase in parallel, finally reaching very high levels (Figure 8.8). Thus, there is a scope for a much higher reproductive investment than often believed. This again leads to the importance of establishing fecundity time series that correctly cover the whole dynamic range in environmental conditions. Unfortunately, however, many studies are based on rather limited data sources.

8.3.5.2 Cyclicity and feedback mechanisms

As this subject might be approached in various ways, this analysis was for simplicity limited to iteroparous fish with an annual reproduction, partially or fully based on accumulated energy reserves (capital breeders). Logically, there is a much tighter, instantaneous link between reproductive investment and food (energy) supply in income breeders, which use energy acquired during the spawning period rather than stored energy for reproduction (Stearns 1992, van Damme 2013). Stated in another way, Fulton's condition factor (Bagenal & Tesch 1978), or other types of condition index (Bolger & Connolly 1989), are likely much more meaningful predictors of egg production in capital breeders than in income breeders. Moreover, there is a saying in life history theory that 'current reproduction influences future reproduction' (Stearns 1992), as exemplified for fish by the well-known aquaria experiments of Reznik & Yang (1993). In the present context, however, it is of more interest to rephrase this to 'current reproduction is a function of previous investments' to first understand the noted fecundity variation in the analysed material and from this hindcast then, possibly, do any subsequent forecast.

Bearing in mind that there are feedback mechanisms operating, fecundity estimates should in principle not jump unrestrained and completely unpredictable from one year (season) to the next, but rather have links to the above-described overlying changes in the environment and the

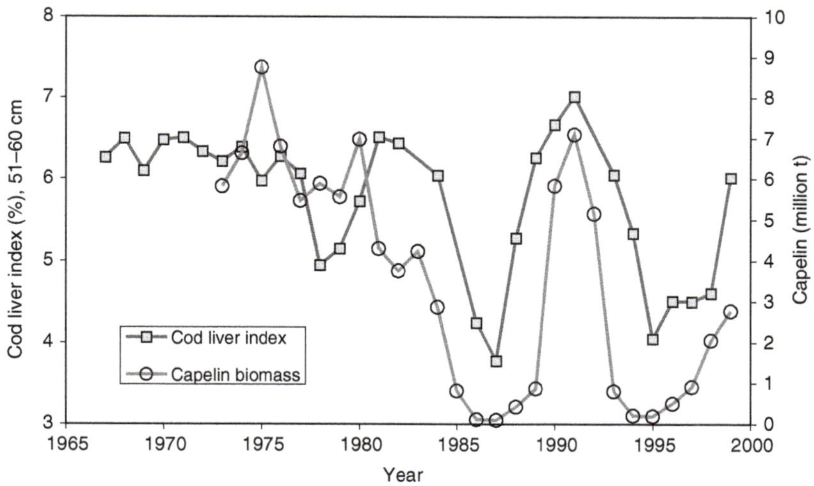

Figure 8.9 Time-series analysis of the relationship between liver index (100 × liver wet weight/whole wet body weight) of 51–60-cm North East Arctic cod and amount of Barents Sea capelin (*Mallotus villosus*), the main food item, in the Barents Sea ecosystem. Capelin is a short-lived species showing large variations in abundance due to natural fluctuations and a major fishery. Based on the data of Marshall *et al.* (1998).

underlying changes in the energy of the individual body. In first dealing with annual changes in measured fish condition, several long-term studies have shown cyclical changes. The foremost example here is the classical study of Hjort (1914); note Hjort's Figure 122 on variations in total body weight, liver and roe weight of spawning North East Arctic cod for the period 1880–1910 (liver). These data have recently been supplemented with additional archived liver data and modern statistics facilitating the construction of one of the longest biological time series existing, covering more than 150 years (Kjesbu *et al.* 2014). Similar messages of cyclic changes in energy allocation have been given by Holst (1996), Marshall *et al.* (1998, 2006) and Sandeman *et al.* (2008) (Figure 8.9). These five citations were selected from the literature based on the fact that they provide insight into the most comprehensive time series available for any fish, Norwegian spring-spawning herring and North East Arctic cod. In line with these results on condition (including also weight-at-length), published reports also give reason to believe that RF_P or size-specific F_P vary or tend to vary cyclically, depending on the species studied and the length of the time series established (Kjesbu *et al.* 1998, Rideout & Morgan 2007). Perhaps, more interesting, however, is that several sources of fecundity data indicate the existence of a negative feedback following poor-condition years (Hjort 1914, Kjesbu *et al.* 1998, Óskarsson *et al.* 2002), when the measured roe size or fecundity appear to be lower than expected from the corresponding measured condition. This type of consideration, although in full agreement with the previous statement from life-history theory, is in its early days (at least for fish species), although it might significantly improve our predictable capacities in terms of fecundity variation.

8.3.5.3 When to report condition in the maturation cycle

As implied above, the fecundity of an individual female is normally regressed on its corresponding body size or condition index/chemistry. Thus, the analysis is undertaken on x−y data

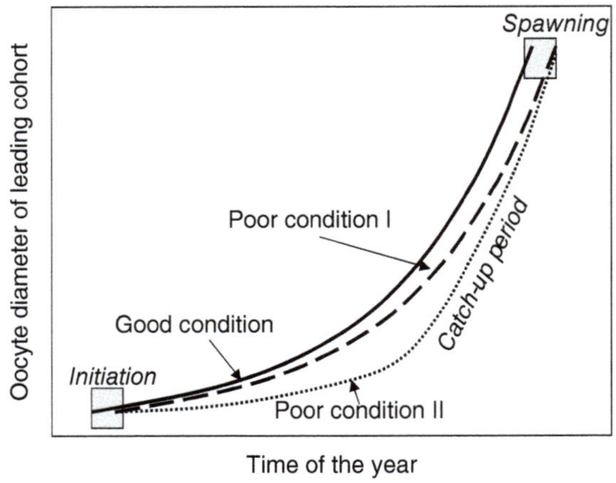

Figure 8.10 Schematic change in oocyte growth curves throughout one complete maturation cycle, from initiation of oocyte development until spawning of eggs for fish in good or poor condition. It is assumed that the delay in initiation of oocyte development for the poor-condition fish is reflected in a similar delay in start of spawning (grey boxes). The poor-condition situation is divided into: (I) oocyte growth curve as for good-condition fish but displaced to the right (later in the year) in accordance with the delay in spawning time; and (II) fish that are assumingly in very poor condition showing hardly any oocyte growth in the first part of the maturation cycle but then fast oocyte growth marked as 'Catch-up period'. Based on the results of Dunia González (2006).

sets matching, logically, exactly in time. Although giving often a high coefficient of determination (r^2), it might be questioned if such an approach is always the most relevant, biologically speaking, based on the following findings. In the experimental work of Tyler *et al.* (1994) on fecundity regulation and egg size determination in rainbow trout, it was made clear that unilateral ovariectomy (removal of one of the ovarian lobes) may result in compensatory effects, depending on when in the maturity cycle the surgery is performed. Mid-vitellogenesis appeared to be a particularly sensitive period for determining the level of fecundity, as operations performed at that time of the year resulted in an increased fecundity in the remaining lobe, whereas other females operated on at the same time showed no change in fecundity but instead an increase in egg size (all data were related to those for unoperated controls). A follow-up pilot study on cod indicated very much the same mechanism (Andersen *et al.* 2000). In the present context, however, a study in cod performed by Skjæraasen *et al.* (2006) was of particular interest, and is addressed more carefully in the following text. However, subsequent extensive experimental studies on the planktonic feeding Atlantic herring failed to show any such sensitive periods during early or late vitellogenesis, in contrast to earlier, published studies on ground- and flatfish (see Kennedy *et al.* 2010 and references therein). This indicated a more flexible *modus operandi* (MO) to food shortage in pelagic, shoaling categories of temperate teleosts, exemplified by the 'catch-up' MO outlined below.

The timing of the determination of fecundity versus the above-mentioned covariates should likely be coupled to oocyte growth patterns. As stated, poor-condition fish in temperate waters are expected to show delays in spawning time on the order of a couple of weeks. This might be explained by a similar temporal displacement of the whole standard oocyte growth curve or, as found by Dunia González (2006), for fish in very poor condition, slow oocyte growth initially followed by a fast 'catch-up' period (Figure 8.10). The latter pattern might, if not properly

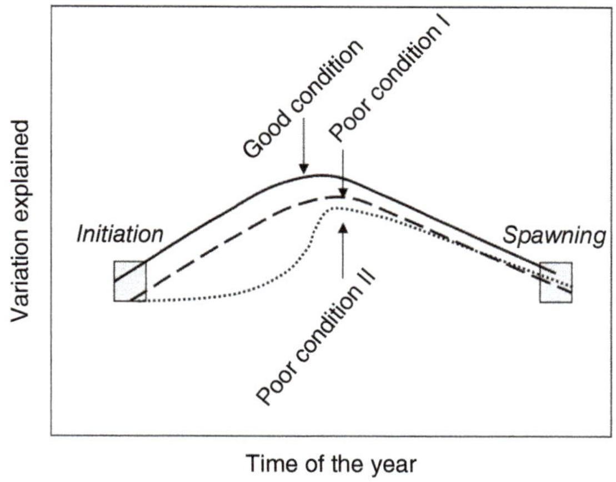

Figure 8.11 Assumed variation in explanatory power (r^2) of potential fecundity (F_P) as observed in prespawning females versus their condition (e.g., weight-at-length) at previous time points in the same maturation cycle. For background information, consult Figure 8.10. The results indicate that the largest variation (arrows) can be explained midway in the maturation cycle. Also, that it is more difficult to predict F_P in poor-condition fish than in good-condition fish, cf. the influence of atresia (see main text). Note that maximum r^2 is generally delayed in poor-condition fish. With reference to Figure 8.10, it can be deduced that r^2 starts to increase much later in the cycle for poor-condition II females than for poor-condition I females. Based on the data of Dunia González (2006) and Skjæraasen *et al.* (2006).

accounted for, result in an underestimation of the proportion of maturing fish in the stock in poor-condition years. In the report by Skjæraasen *et al.* (2006), the final (prespawning) F_P was related to body size or condition as measured during previous parts of the same maturation cycle by tracking individuals kept in tanks over time. (The same analytical model technique was also used in the herring studies of Dunia González (2006) and Kennedy *et al.* (2010).) Day length (short and normal day) was used to experimentally trigger oocyte development at different times in the year. Taken together, there was evidence to support that F_P shows the closest relationship (r^2) with condition measures early in vitellogenesis, a result which was very much in agreement with the principles outlined by Tyler *et al.* (1994) and Andersen *et al.* (2000). By adding in oocyte growth patterns as a part of this picture, the time for maximum prediction of F_P in the year should change, at least to some extent, depending on actual fish condition in that particular year (Figure 8.11). In short, prespawning data should as far as possible be supplemented by similar type of data taken earlier in vitellogenesis, in order to be able to predict fecundity in the best possible way. This statement is in line with the findings of Skjæraasen *et al.* (2006), although the modulating effect of condition variation on the timing of maximum r^2 was not considered as such in that particular study.

8.3.6 Predictors of fecundity

Previously published principles for the analysis of fecundity data (e.g., Bagenal & Braum 1978) are, with a few minor exceptions, still valid today. Consequently, these will not be studied in depth, but will be supplemented with other factors that might be useful to bear in mind. First, whole body weight is generally a significant better predictor of fecundity than total length, while age is the poorest of the three factors. This is logical, as total length is much

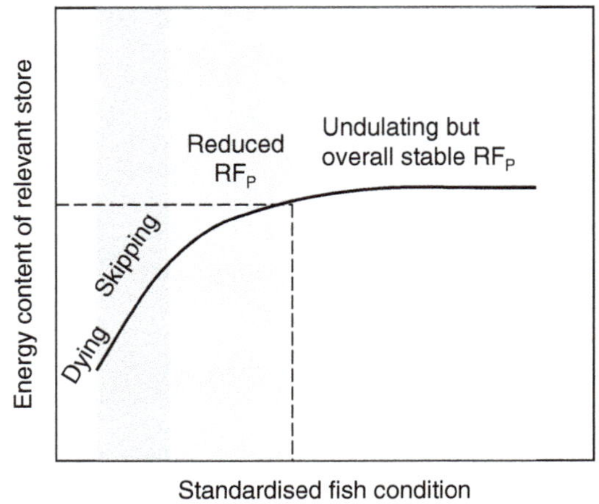

Figure 8.12 Energy content of relevant store (e.g., liver in 'lean fish' and muscle in 'fat fish') versus standardised fish condition (e.g., weight-at-length by length class and time of the year). The solid curve was based on data of Lambert & Dutil (1997), with special reference to their Figure 7 on muscle content versus condition factor, and liver energy content versus hepatosomatic (liver) index. Relative potential fecundity (RF_P) is set to start falling when the fish condition falls below a corresponding critical energy point in the relevant store (Lambert & Dutil 1997, Marshall *et al.* 1998, 2006, Kjesbu *et al.* 2014) (dashed line). RF_P above this point is supposed to be generally stable but undulating due to various feedback mechanisms operating (see main text). 'Skipping' refers to the omission of spawning in sexually mature, iteroparous individuals (see Table 8.1).

more stable ('resilient', cf. Skjæraasen *et al.* 2006) than whole body weight during the length of the maturation cycle, while fish of the same age might be highly different in total length and whole body weight. Clearly, whole body weight should also reflect energy reserves more realistically than the other two variables. As several studies have shown that there may be large variations in fecundity between years, one route forward might be to create a so-called 'year-independent' or 'general fecundity' formula to avoid the need to sample fecundity ovaries on a yearly basis. A first step is the establishment of year-specific multiple regressions with two or more independent variables. However, the main problem with general fecundity formulae is that they tend to overestimate fecundity for fish in poor condition (cf. the equations in Table IV in Thorsen *et al.* 2006), which clearly should be an issue to consider when dealing with stocks at risk. The reason for this inability to pick up points differing greatly from the others should be seen in light of the intrinsic range in the model data consulted (see above), and also that the energy content of different body compartments changes with overall body condition, and will not be constant. This 'unit-of energy' approach might be very useful, especially when referring to relevant energy stores, and RF_P instead of F_P (Figure 8.12). Note that there seems to be a critical energy point below which RF_P starts to fall. For Atlantic cod, this refers to an average liver index (100 × liver weight/whole body weight) of about 6–7% (Lambert & Dutil 1997, Marshall *et al.* 1998, Kjesbu *et al.* 2014).

Taken together, it might be useful to decompose the general fecundity formula into two – that is, above or below this critical point – in accordance with techniques used in growth studies with the establishment of so-called 'stanza' (Bagenal & Tesch 1978) or, more commonly, segmented equations. There are, however, two caveats. First, analyses of the energy content of an organ instead of its total energy (energy content × organ size) might mask or weaken the signal as the first parameter may change less, numerically speaking, than the last parameter (Kjesbu *et al.* 1991: seasonal water content of muscle versus seasonal total muscle mass). Second, the selected energy store may not be the relevant one for reproduction as such but rather for the condition of the fish (i.e., an indirect approach). Note that in this respect, protein should generally be considered as the dominant chemical compound used for reproduction, instead of fat. This is seen by examining the chemical composition of the ovary and the eggs, which are dominated by vitellogenin (yolk) consisting on the order of 70–80% protein on a dry weight basis (Plack *et al.* 1971, Tyler *et al.* 2000). However, this should not be interpreted as fat being unimportant for reproduction, which clearly is not the case for many reasons (Tyler *et al.* 2000). Nevertheless, the statement of Bradford (1993), when studying Atlantic herring, that '... the analyses indicate that storage lipids sustain routine and active metabolism while storage proteins are utilised to develop gonads' seems to be a useful conclusion, but is one which should be tested in specific cases, at least when considering other than data-rich species from boreal-temperate waters, which are the point of reference for this type of view.

8.4 Concluding remarks

As is apparent from this review, there are several topics within applied fisheries reproductive biology (AFRB) which require special attention in further studies. The complexity in the estimation of the above-mentioned reproductive traits and investments are made clear as far as possible, although some parts are indicative only. It might, however, be argued that going from SSB to Total Egg Production (TEP) is nothing else than an upscaling exercise – that is, it gives direct proportionality and thereby is of no additional use. This should be compared with the frequent use of SSB in assessment, which either directly or indirectly is assumed to reflect SRP (Marshall *et al.* 2006). As seen in Chapter 2 and in this chapter, this is incorrect and SSB alone might provide a too-optimistic view of the stock situation (as detailed in Chapter 11). However, the use of SSB is to a large extent inevitable because of a lack of more detailed information in the historical time series available. As the main centre of attention fisheries, assessment and management in the present context is to define a SSB reference point, which likely assures a good level of recruitment, and research scientists working at the cellular and individual level might find that their work is labelled as having a too-detailed focus. On the other hand, the quality of raw or processed data is central to all science, along with the principle that the sum of a high number of minor errors might build up to be a serious fault. Within the field of fish reproductive biology of today, various types of modern statistical models are often stressed as being the appropriate tools, and there is no reason to question this. Likewise, the discussion on different condition factors, including whether they are length-dependent or not, have been useful. However, if essential knowledge of fish reproductive biology is missing this does not result in any major advancement in this branch of science as such. As described earlier, many research groups investigating EPM demonstrate advanced skills in AFRB. A central theme in this chapter has been that advancements in knowledge often result from combining both field and experimental studies. However, the quality of experimental results is very much dependent on the quality of the experimental facilities.

At present, there is a strong momentum within recruitment-related studies to gain new insight into AFRB, and the present book is just one example of this newfound focus. The recent establishment of several international consortia and study groups, as well as significant method advancements (including automated procedures), together with the rapid appearance of many new articles and reviews, suggest that our knowledge will be largely improved in the near future.

8.5 Acknowledgements

I am most grateful to Yutaka Kurita, Hilario Murua and Richard Nash for giving me valuable feedback on the first-edition version of this chapter. A special appreciation goes to Olav Dragesund, Per Solemdal and Michael Greer Walker for introducing me to applied fisheries reproductive biology in the beginning of my career, and to Anders Thorsen, Merete Fonn, Peter R. Witthames and Fran Saberido-Rey for being my scientific fellow players for many years. The Institute of Marine Research and The Research Council of Norway (project no. 200497: *Adaptive Management of Living Marine Resources by Integrating Different Data Sources and Key Ecological Processes*) are thanked for supporting the production of this updated chapter version.

References

Agulleiro, M.J., Anguis, V., Canavate, J.P., Martinez-Rodriguez, G., Mylonas, C.C. & Cerda, J. (2006) Induction of spawning of captive-reared Senegal sole (*Solea senegalensis*) using different administration methods for gonadotropin-releasing hormone agonist. *Aquaculture*, **257**, 511–24.

Alheit, J. (1988) Reproductive biology of sprat (*Sprattus sprattus*): factors determining annual egg production. *Journal du Conseil International pour l'Exploration de la Mer*, **44**, 162–8.

Alonso-Fernández, A., Vallejo, A.C., Saborido-Rey, F., Murua, H. & Trippel, E.A. (2009) Fecundity estimation of Atlantic cod (*Gadus morhua*) and haddock (*Melanogrammus aeglefinus*) of Georges Bank: Application of the autodiametric method. *Fisheries Research*, **99**, 47–54.

Andersen, T.E. (2003) Unbiased stereological estimation of cell numbers and volume fractions: the disector and the principles of point counting. In: O.S. Kjesbu, J.R. Hunter & P.R. Witthames (Eds) *Report of the Working Group on Modern Approaches to Assess Maturity and Fecundity of Warm- and Cold-Water Fish and Squids*. pp. 11–18. Fisken og Havet, no. 12–2003.

Andersen, T.E., Witthames, P.R., Kjesbu, O.S., Bromley, P. & Tyler, C.R. (2000) Development of a surgery protocol for unilateral ovariectomy (ULO) on cod (*Gadus morhua*). In: B. Norberg, O.S. Kjesbu, G.L. Taranger, E. Andersson & S.O. Stefansson (Eds) *Proceedings of the 6th International Symposium on the Reproductive Physiology of Fish. Institute of Marine Research and University of Bergen, 4–9 July 1999*. pp. 304. John Grieg AS, Bergen.

Anon. (2006a) *Report of the Working Group on Mackerel and Horse Mackerel Egg Surveys* (WGMEGS). ICES CM 2006/LRC:09 Ref. RMC.

Armstrong, M.J. & Witthames, P.R. (2012) Developments in understanding of fecundity of fish stocks in relation to egg production methods for estimating spawning stock biomass. *Fisheries Research*, **117**, 35–47.

Armstrong, M.J., Connolly, P., Nash, R.D.M., Pawson, M.G., Alesworth, E., Coulahan, P.J., Dickey-Collas, M., Milligan, S.P., O'Neill, M.F., Witthames, P.R. & Woolner, L. (2001) An application of the annual egg production method to estimate the spawning biomass of cod (*Gadus morhua* L.), plaice (*Pleuronectes platessa* L.) and sole (*Solea solea* L.) in the Irish Sea. *ICES Journal of Marine Science*, **58**, 183–203.

Babiak, I., Ottesen, O., Rudolfsen, G. & Johnsen, S. (2006) Quantitative characteristics of Atlantic halibut, *Hippoglossus hippoglossus* L., semen throughout the reproductive season. *Theriogenology*, **65**, 1587–604.

Bagenal, T.B. (1978). Aspects of fish fecundity. In: S.D. Gerking (Ed.) *Ecology of Freshwater Fish Production.* pp. 75–101. Blackwell Scientific Publications, Oxford.

Bagenal, T.B. & Braum, E. (1978) Eggs and early life history. In: T. Bagenal (Ed.) *Methods for Assessment of Fish Production in Fresh Waters.* pp. 165–201. IBP Handbook No 3. Blackwell Scientific Publications, London.

Bagenal, T.B. & Tesch, F.W. (1978) Age and growth. In: T. Bagenal (Ed.) *Methods for Assessment of Fish Production in Fresh Waters.* pp. 101–36. IBP Handbook No. 3, Blackwell Scientific Publications, London.

Barr, W.A. (1963) The endocrine control of the sexual cycle in the plaice, *Pleuronectes platessa* (L.). I. Cyclical changes in the normal ovary. *General and Comparative Endocrinology*, **3**, 197–204.

Bernal, M., Somarakis, S., Witthames, P.R., van Damme, C.J.G., Uriarte, A., Lo, N.C.H. & Dickey-Collas, M. (2012) Egg production methods in marine fisheries: An introduction. *Fisheries Research*, **117**, 1–5.

Bolger, T. & Connolly, P.L. (1989) The selection of suitable indices for the measurement and analysis of fish condition, *Journal of Fish Biology*, **34**, 171–82.

Bradford, R.G. (1993) Differential utilization of storage lipids and storage proteins by Northwest Atlantic herring (*Clupea harengus harengus*). *Journal of Fish Biology*, **43**, 811–24.

Bromage, N. & Cumaranatunga, R. (1988) Egg production in the rainbow trout. In: J.F. Muir & R.J. Roberts (Eds) *Recent Advances in Aquaculture.* pp. 63–138. Croom Helm, London.

Bromley, P.J. (2003) The use of market sampling to generate maturity ogives and to investigate growth, sexual dimorphism and reproductive strategy in central and south-western North Sea sole (*Solea solea* L.). *ICES Journal of Marine Science*, **60**, 52–65.

Brown-Peterson, N.J., Wyanski, D.M., Saborido-Rey, F., Macewicz, B.J. & Lowerre-Barbieri, S.K. (2011) A standardized terminology for describing reproductive development in fishes. *Marine and Coastal Fisheries: Dynamics, Management, and Ecosystem Science*, **3**, 52–70.

Bucholtz, R.H., Tomkiewicz, J., Nyengaard, J.R. & Andersen, J.B. (2013) Oogenesis, fecundity and condition of Baltic herring (*Clupea harengus* L.): A stereological study. *Fisheries Research*, **145**, 100–13.

Carolsfeld, J., Scott, A.P., Collins, P.M. & Sherwood, N.M. (1996) Reproductive steroids during maturation in a primitive teleost, the Pacific herring (*Clupea harengus pallasi*). *General and Comparative Endocrinology*, **103**, 331–48.

Christiansen, R. & Torrissen, O.J. (1997) Effects of dietary astaxanthin supplementation on fertilization and egg survival in Atlantic salmon (*Salmo salar* L). *Aquaculture*, **153**, 51–62.

Claramunt, G. & Roa, R. (2001) An indirect approach of estimating spawning fraction as applied to *Sardinops sagax* from northern Chile. *Scientia Marina*, **65**, 87–94.

Cole, K.S. (Ed.) (2010) *Reproduction and Sexuality in Marine Fishes. Patterns and Processes.* University of California Press, Oakland, California.

Dahle, R., Taranger, G.L., Karlsen, Ø., Kjesbu, O.S. & Norberg, B. (2003) Gonadal development and associated changes in liver size and sexual steroids during the reproductive cycle of captive male and female Atlantic cod (*Gadus morhua* L.). *Comparative Biochemistry and Physiology Part A*, **136**, 641–53.

de Oliveira, J.A.A., Roel, B.A. & Dickey-Collas, M. (2006) Investigating the use of proxies for fecundity to improve management advice for western horse mackerel *Trachurus trachurus. ICES Journal of Marine Science*, **63**, 25–35.

DeMartini, E.E. & Fountain, R.K. (1981) Ovarian cycling frequency and batch fecundity in the queenfish, *Seriphus politus*: attributes representative of serial spawning fishes. *Fisheries Bulletin*, **79**, 547–60.

Denney, N.H., Jennings, S. & Reynolds, J.D. (2002) Life-history correlates of maximum population growth rates in marine fishes. *Proceedings of the Royal Society of London B*, **269**, 2229–37.

Dickey-Collas, M., Nash, R.D.M., Brunel, T., van Damme, C.J.G., Marshall, C.T., Payne, M.R., Corten, A., Geffen, A.J., Peck, M.A., Hatfield, E.M.C., Hintzen, N.T., Enberg, K., Kell, L.T. & Simmonds, E.J. (2010) Lessons learned from stock collapse and recovery of North Sea herring: a review. *ICES Journal of Marine Science*, **67**, 1875–86.

Dunia González, B. (2006) *Regulatory Reproductive Mechanisms and Somatic Growth in Atlantic Herring (Clupea harengus L.).* M. Phil. thesis, University of Bergen.

Dutil, J.D. & Brander, K. (2003) Comparing productivity of North Atlantic cod (*Gadus morhua*) stocks and limits to growth production. *Fisheries Oceanography*, **12**, 502–12.

Eaton, R.C. & Farley, R.D. (1974) Spawning cycle and egg production of zebrafish, *Brachydanio rerio*, in the laboratory. *Copeia*, **1**, 195–204.

Emerson, L.S., Greer Walker, M. & Witthames, P.R. (1990) A stereological method for estimating fish fecundity. *Journal of Fish Biology*, **36**, 721–30.

Friedland, K.D., Ama-Abasi, D., Manning, M., Clarke, L., Kligys, G. & Chambers, R.C. (2005) Automated egg counting and sizing from scanned images: rapid sample processing and large data volumes for fecundity estimates. *Journal of Sea Research*, **54**, 307–16.

Ganias, K. (2013) Determining the indeterminate: Evolving concepts and methods on the assessment of the fecundity pattern of fishes. *Fisheries Research*, **138**, 23–30.

Ganias, K., Murua, H, Claramunt, G., Dom'inguez-Petit, R., Gonçalves, P., Juanes, F., Kennedy, J., Klibansky, N., Korta, M., Kurita, Y., Lowerre-Barbieri, S., Macchi, G., Matsuyama, M., Medina, A., Nunes, C., Plaza, G., Rideout, R., Somarakis, S., Thorsen, A., Uriarte, A. & Yoneda, M. (2014) Egg production. In: R. Domínguez-Petit, H. Murua, F. Saborido-Rey & E. Trippel (Eds) *Handbook of Applied Fisheries Reproductive Biology for Stock Assessment and Management.* pp. 1–109. Vigo, Spain. http://hdl.handle.net/10261/87768.

Ganias, K., Lowerre-Barbieri, S.K. & Cooper, W. (2015) Understanding the determinate–indeterminate fecundity dichotomy in fish populations using a temperature dependent oocyte growth model. *Journal of Sea Research*, **96**, 1–10.

Gauthier, G., Bety, J. & Hobson, K.A. (2003) Are greater snow geese capital breeders? New evidence from a stable-isotope model. *Ecology*, **84**, 3250–64.

Gilbert, S.F. (1997) *Developmental Biology*. 5th edn. Sinauer Associates, Inc., Sunderland, Massachusetts.

Green, B.S. (2008) Maternal effects in fish populations. *Advances in Marine Biology*, **54**, 1–105.

Greer Walker, M., Witthames, P.R. & Bautista de los Santos, I. (1994) Is the fecundity of the Atlantic mackerel (*Scomber scombrus*: Scombridae) determinate? *Sarsia*, **79**, 13–26.

Greer Walker, M., Witthames, P., Emerson, L. & Walsh, M. (1987) Estimation of fecundity in the western mackerel stock, 1986. ICES C. M. 1987 /H:41.

Grier, H.J. (2012) Development of the follicle complex and oocyte staging in red drum, *Sciaenops ocellatus* Linnaeus, 1776 (Perciformes, Sciaenidae). *Journal of Morphology*, **273**, 801–29.

Grier, H.J. & Lo Nostro, F. (2000) The teleost germinal epithelium: a unifying concept. In: B. Norberg, O.S. Kjesbu, G.L. Taranger, E. Andersson & S.O. Stefansson (Eds) *Proceedings of the 6th International Symposium on the Reproductive Physiology of Fish. Institute of Marine Research and University of Bergen, 4–9 July 1999.* pp. 233–36. John Grieg AS, Bergen.

Heibo, E., Magnhagen, C. & Vøllestad, L.A. (2005) Latitudinal variation in life-history traits in Eurasian perch. *Ecology*, **86**, 3377–86.

Hickling, C.F. & Rutenberg, E. (1936) The ovary as an indicator of the spawning period in fishes. *Journal of the Marine Biological Association of the United Kingdom*, **21**, 311–16.

Hislop, J.R.G., Robb, A.P. & Gauld, J.A. (1978) Observations on effects of feeding level on growth and reproduction in haddock, *Melanogrammus aeglefinus* (L.) in captivity. *Journal of Fish Biology*, **13**, 85–98.

Hixon, M.A., Johnson, D.W. & Sogard, S.M. (2014) BOFFFFs: on the importance of conserving old-growth age structure in fishery populations. *ICES Journal of Marine Science*, **71**, 2171–85.

Hjort, J. (1914) Fluctuations in the great fisheries of northern Europe viewed in light of biological research. *Rapports et Procès-verbaux des Réunions, Conseil International pour l'Exploration de la Mer*, **20**, 1–228.

Holdway, D.A. & Beamish, F.W.H. (1985) The effect of growth rate, size, and season on oocyte development and maturity of Atlantic cod (*Gadus morhua* L.). *Journal of Experimental Marine Biology and Ecology*, **85**, 3–19.

Holst, J.C. (1996) *Long Term Trends in the Growth and Recruitment Pattern of the Norwegian Spring-Spawning Herring (*Clupea harengus Linnaeus *1758)* Dr. Scient. Thesis, University of Bergen.

Hunter, J.R. & Goldberg, S. (1980) Spawning incidence and batch fecundity in northern anchovy, *Engraulis mordax. Fisheries Bulletin*, **77**, 641–52.

Hunter, J.R. & Leong, R. (1981) The spawning energetics of female northern anchovy, *Engraulis mordax. Fisheries Bulletin*, **79**, 215–30.

Hunter, J.R., Lo, N.C.H. & Leong, R.J.H. (1985) Batch fecundity in multiple spawning fishes. In: R. Lasker (Ed.) *An Egg Production Method for Estimating Spawning Biomass of Pelagic Fish: Application to the Northern Anchovy,* Engraulis mordax. pp. 67–77. NOAA Technical Report. NMFS, 36, US Department of Commerce.

Hunter, J.R. & Macewicz, B.J. (1985a) Rates of atresia in the ovary of captive and wild northern anchovy, *Engraulis mordax. Fisheries Bulletin*, **83**, 119–36.

Hunter, J.R. & Macewicz, B.J. (1985b) Measurement of spawning frequency in multiple spawning fishes. In: R. Lasker (Ed.) *An Egg Production Method for Estimating Spawning Biomass of Pelagic Fish: Application to the Northern Anchovy,* Engraulis mordax. pp. 79–94. NOAA Technical Report. NMFS, 36, US Department of Commerce.

Hunter, J.R. & Macewicz, B.J. (2003) Improving the accuracy and precision of reproductive information used in fisheries. In: O.S. Kjesbu, J.R. Hunter & P.R. Witthames (Eds) *Report of the Working Group on Modern Approaches to Assess Maturity and Fecundity of Warm- and Cold-Water Fish and Squids.* pp. 57–68. Fisken og Havet., no. 12–2003.

Hunter, J.R., Macewicz, B.J. & Kimbrell, C.A (1989) Fecundity and other aspects of the reproduction of sablefish *Anoplopoma fimbria* in central Californian waters. *California Cooperative Oceanic Fisheries Investigative Report*, **30**, 61–72.

Hunter, J.R., Macewicz, B.J., Lo, N.C. & Kimbrell, C.A. (1992) Fecundity, spawning, and maturity of female Dover sole *Microstomus pacificus,* with an evaluation of assumptions and precision. *Fisheries Bulletin*, **90**, 101–28.

Hutchings, J.A., Butchart, S.H.M., Collen, B., Schwartz, M.K. & Waples, R.S. (2012) Red flags: correlates of impaired species recovery. *Trends in Ecology and Evolution*, **27**, 542–6.

Høie, H., Folkvord, A. & Johannessen, A. (1999) Maternal, paternal and temperature effects on otolith size of young herring (*Clupea harengus* L.) larvae. *Journal of Experimental Marine Biology and Ecology*, **234**, 167–84.

ICES (2006a) *Report of the Working Group on Mackerel and Horse Mackerel Egg Surveys* (WGMEGS). ICES CM 2006/LRC:09 Ref. RMC.

ICES (2006b) *Report of the Working Group on the Assessment of Mackerel, Horse Mackerel, Sardine and Anchovy* (WGMHSA). ICES CM 2006/ACFM:36.

ICES (2012) Report of the Workshop for Maturity Staging Chairs (WKMATCH). ICES CM 2012/ACOM:58 Ref. PGCCDBS.

Johannesson, K. & André, C. (2006) Life on the margin: genetic isolation and diversity loss in a peripheral marine ecosystem, the Baltic Sea. *Molecular Biology*, **15**, 2013–29.

Johnson, L.L. & Casillas, E. (1991) The use of plasma parameters to predict ovarian maturation stage in English sole *Parophrys-vetulus* Girard. *Journal of Experimental Marine Biology and Ecology*, **151**, 257–70.

Jonsson, N. & Jonsson, B. (1999) Trade-off between egg mass and egg number in brown trout. *Journal of Fish Biology*, **55**, 767–83.

Jonsson, N., Jonsson, B. & Fleming, I.A. (1996) Does early growth cause a phenotypically plastic response in egg production of Atlantic salmon? *Functional Ecology*, **10**, 89–96.

Jørgensen, C., Ernande, B., Fiksen, Ø. & Dieckmann, U. (2006) The logic of skipped spawning in fish. *Canadian Journal of Fisheries and Aquatic Sciences*, **63**, 200–11.

Jørgensen, C. & Fiksen, Ø. (2006) State-dependent energy allocation in cod (*Gadus morhua*). *Canadian Journal of Fisheries and Aquatic Sciences*, **63**, 186–99.

Kamler, E. (2005) Parent–egg–progeny relationships in teleost fishes: an energetics perspective. *Reviews in Fish Biology and Fisheries*, **15**, 399–421.

Karlou-Riga, C. & Economidis, P.S. (1996) Ovarian atretic rates and sexual maturity of European horse mackerel, *Trachurus trachurus* (L.), in the Saronikos Gulf (Greece). *Fisheries Bulletin*, **94**, 66–76.

Karlsen, Ø., Holm, J.C. & Kjesbu, O.S. (1995) Effects of periodic starvation on reproductive investment in first-time spawning Atlantic cod (*Gadus morhua* L.). *Aquaculture*, **133**, 159–70.

Kennedy, J., Gundersen, A.C. & Boje, J. (2009) When to count your eggs: Is fecundity in Greenland halibut (*Reinhardtius hippoglossoides* W.) down-regulated? *Fisheries Research*, **100**, 260–65.

Kennedy, J., Nash, R.D.M., Slotte, A. & Kjesbu, O.S. (2011) The role of fecundity regulation and abortive maturation in the reproductive strategy of Norwegian spring-spawning herring (*Clupea harengus*). *Marine Biology*, **158**, 1287–99.

Kennedy, J., Skjæraasen, J.E., Nash, R.D.M., Thorsen, A., Slotte, A., Hansen, T. & Kjesbu, O.S. (2010) Do capital breeders like Atlantic herring (*Clupea harengus*) exhibit sensitive periods of nutritional control on ovary development and fecundity regulation? *Canadian Journal of Fisheries and Aquatic Sciences*, **67**, 16–27.

Kittelsen, A. (1986) Settefiskanlegg. In: T. Gjedrem (Ed.) *Fiskeoppdrett med Framtid*. pp. 75–113. A/S Landbruksforlaget, Oslo (in Norwegian).

Kjesbu, O.S. (1989) The spawning activity of cod, *Gadus morhua* L. *Journal of Fish Biology*, **34**, 195–206.

Kjesbu, O.S. (1994) Time of start of spawning in Atlantic cod (*Gadus morhua*) females in relation to vitellogenic oocyte diameter, temperature, fish length and condition. *Journal of Fish Biology*, **45**, 719–35.

Kjesbu, O.S., Hunter, J.R. & Witthames, P.R. (Eds) (2003) *Report of the Working Group on Modern Approaches to Assess Maturity and Fecundity of Warm- and Cold-Water Fish and Squids*. Fisken og Havet, no. 12–2003.

Kjesbu, O.S., Klungsøyr, J., Kryvi, H., Witthames, P.R. & Greer Walker, M. (1991) Fecundity, atresia, and egg size of captive Atlantic cod (*Gadus morhua*) in relation to proximate body composition. *Canadian Journal of Fisheries and Aquatic Sciences*, **48**, 2333–43.

Kjesbu, O.S. & Kryvi, H. (1989) Oogenesis in cod, *Gadus morhua* L., studied by light and electron microscopy. *Journal of Fish Biology*, **34**, 735–46.

Kjesbu, O.S., Kryvi, H. & Norberg, B. (1996b) Oocyte size and structure in relation to blood plasma steroid hormones in individually monitored, spawning Atlantic cod. *Journal of Fish Biology*, **49**, 1197–215.

Kjesbu, O.S., Kryvi, H., Sundby, S. & Solemdal, P. (1992) Buoyancy variations in eggs of Atlantic cod (*Gadus morhua* L.) in relation to chorion thickness and egg size: theory and observations. *Journal of Fish Biology*, **41**, 581–99.

Kjesbu, O.S., Opdal, A.F, Korsbrekke, K., Devine, J.A. & Skjæraasen, J.E. (2014) Making use of Johan Hjort's unknown legacy: reconstruction of a 150-year coastal time series on Northeast Arctic cod (*Gadus morhua*) liver data reveals long-term trends in energy allocation patterns. *ICES Journal of Marine Science*, **71**, 2053–63.

Kjesbu, O.S., Righton, D., Krüger-Johnsen, M., Thorsen, A., Michalsen, K., Fonn, M. & Witthames, P.R. (2010) Thermal dynamics of ovarian maturation in Atlantic cod (*Gadus morhua*). *Canadian Journal of Fisheries and Aquatic Sciences*, **67**, 605–25.

Kjesbu, O.S., Solemdal, P., Bratland, P. & Fonn, M. (1996a) Variation in annual egg production in individual captive Atlantic cod (*Gadus morhua*). *Canadian Journal of Fisheries and Aquatic Sciences*, **53**, 610–20.

Kjesbu, O.S., Thorsen, A. & Fonn, M. (2011) Quantification of primary and secondary oocyte production in Atlantic cod (*Gadus morhua*) by simple oocyte packing density theory. *Marine and Coastal Fisheries: Dynamics, Management, and Ecosystem Science*, **3**, 92–105.

Kjesbu, O.S. & Witthames, P.R. (2007) Evolutionary pressure on reproductive strategies in flatfish and groundfish: relevant concepts and methodological advancements. *Journal of Sea Research*, **58**, 23–34.

Kjesbu, O.S., Witthames, P.R., Solemdal, P. & Greer Walker, M. (1990) Ovulatory rhythm and a method to determinate the stage of spawning in Atlantic cod (*Gadus morhua*). *Canadian Journal of Fisheries and Aquatic Sciences*, **47**, 1185–93.

Kjesbu, O.S., Witthames, P.R., Solemdal, P. & Greer Walker, M. (1998) Temporal variations in the fecundity of Arcto-Norwegian cod (*Gadus morhua*) in response to natural changes in food and temperature. *Journal of Sea Research*, **40**, 303–21.

Korta, M. (2010) *New Methodologies Applied to Quantify the Dynamics of the Ovary in Indeterminate Fecundity Species*. Ph.D. thesis, University of the Basque Country.

Kristoffersen, C. (2005). *A Histomorphometric Study of Testis of Atlantic Cod (*Gadus morhua *L.) Exposed to Continuous Light of Various Intensities*. Cand. Scient. Thesis, University of Bergen.

Kurita, Y. & Kjesbu, O.S. (2009) Fecundity estimation by oocyte packing density formulae in determinate and indeterminate spawners: Theoretical considerations and applications. *Journal of Sea Research*, **61**, 188–96.

Kurita, Y., Meier, S. & Kjesbu, O.S. (2003) Oocyte growth and fecundity regulation by atresia of Atlantic herring (*Clupea harengus*) in relation to body condition throughout the maturation cycle. *Journal of Sea Research*, **49**, 203–19.

Kurita, Y., Thorsen, A., Fonn, M., Svardal, A. & Kjesbu, O.S. (2000) Oocyte growth and fecundity regulation of Atlantic herring (*Clupea harengus*) in relation to declining body reserves during overwintering. In: B. Norberg, O.S. Kjesbu, G.L. Taranger, E. Andersson & S.O. Stefansson (Eds) *Proceedings of the 6th International Symposium on the Reproductive Physiology of Fish. Institute of Marine Research and University of Bergen, 4–9 July 1999*. pp. 85–7. John Grieg AS, Bergen.

Lahnsteiner, F. (2000) Morphological, physiological and biochemical parameters characterizing the over-ripening of rainbow trout eggs. *Fish Physiology and Biochemistry*, **23**, 107–18.

Lambert, Y. & Dutil, J.-D. (1997) Can simple condition indices be used to monitor and quantify seasonal changes in the energy reserves of Atlantic cod (*Gadus morhua*)? *Canadian Journal of Fisheries and Aquatic Sciences*, **54** (Suppl. 1), 104–12.

Lawrence, E. (2000) *Henderson's Dictionary of Biological Terms*, 12th edn. Pearson Education Limited, Harlow.

Leggett, W.C. & Carscadden, J.E. (1978) Latitudinal variation in reproductive characteristics of American shad (*Alosa sapidissima*): evidence for population specific life history strategies in fish. *Journal of the Fisheries Research Board of Canada*, **35**, 1469–78.

Levavi-Sivan, B., Bogerd, J., Mañanós, E.L., Gómez, A. & Lareyre, J.J. (2010) Perspectives on fish gonadotropins and their receptors. *General and Comparative Endocrinology*, **165**, 412–37.

Lo, N.C.H., Hunter, J.R., Moser, H.G., Smith, P.E. & Methot, R.D. (1992) The daily fecundity reduction method: a new procedure for estimating adult fish biomass. *ICES Journal of Marine Science*, **49**, 209–15.

Lowerre-Barbieri, S.K. (2009) Reproduction in relation to conservation and exploitation of marine fishes. In: B.G.M. Jamieson (Ed.) *Reproductive Biology and Phylogeny of Fishes (Agnathans and Bony Fishes)*. pp. 371–94. Science Publishers, Enfield, New Hampshire.

Lowerre-Barbieri, S.K., Brown-Peterson, N.J., Murua, H., Tomkiewicz, J., Wyanski, D.M. & Saborido-Rey, F. (2011) Emerging issues and methodological advances in fisheries reproductive biology. *Marine and Coastal Fisheries: Dynamics, Management, and Ecosystem Science*, **3**, 32–51.

Ma, Y., Kjesbu, O.S. & Jørgensen, T. (1998) Effects of ration on the maturation and fecundity in captive Atlantic herring (*Clupea harengus*). *Canadian Journal of Fisheries and Aquatic Sciences*, **55**, 900–8.

Macer, C.T. (1974) The reproductive biology of the horse mackerel *Trachurus trachurus* (L.) in the North Sea and English Channel. *Journal of Fish Biology*, **6**, 415–38.

Mangor-Jensen, A. & Holm, J.C. (2005) *Internet Manual for the Culture of Atlantic Halibut*. Institute of Marine Research, Norway. http://kveitemanualen.imr.no/

Marshall, C.T., Kjesbu, O.S., Yaragina, N.A., Solemdal, P. & Ulltang, Ø. (1998) Is spawner biomass a sensitive measure of the reproductive and recruitment potential of Northeast Arctic cod? *Canadian Journal of Fisheries and Aquatic Sciences*, **55**, 1766–83.

Marshall, C.T., Needle, C.L., Thorsen, A., Kjesbu, O.S. & Yaragina, N.A. (2006) Systematic bias in estimates of reproductive potential of an Atlantic cod (*Gadus morhua*) stock: implications for stock–recruit theory and management. *Canadian Journal of Fisheries and Aquatic Sciences*, **63**, 980–94.

Marteinsdottir, G. & Begg, G.A. (2002) Essential relationships incorporating the influence of age, size and condition on variables required for estimation of reproductive potential in Atlantic cod *Gadus morhua*. *Marine Ecology Progress Series*, **235**, 235–56.

McBride, R.S., Somarakis, S., Fitzhugh, G.R., Albert, A., Yaragina, N.A., Wuenschell, M.J., Alonso-Fernández, A. & Basilone, G. (2013) Energy acquisition and allocation to egg production in relation to fish reproductive strategies. *Fish and Fisheries* (online version). pp. 1–35. http://onlinelibrary.wiley.com/doi/10.1111/faf.12043/abstract

McEvoy, L.-A. (1984) Ovulatory rhythms and over-ripening of eggs in cultivated turbot, *Scophthalmus maximus* L. *Journal of Fish Biology*, **24**, 437–48.

McMillan, D.B. (2007) *Fish Histology: Female Reproductive Systems*. Springer, Dordrecht.

McPherson, L.R. & Kjesbu, O.S. (2012) Emergence of an oocytic circumnuclear ring in response to increasing day length in Atlantic herring (*Clupea harengus*). *Marine Biology*, **159**, 341–53.

Méndez, M.C., Ramirez, M., Varela, A.R., Chavez, B. & Pedernera, E. (2003) Follicle-stimulating hormone increases cell proliferation in the ovary and the testis of the chick embryo. *General and Comparative Endocrinology*, **133**, 181–8.

Methven, D.A., Crim, L.W., Norberg, B., Brown, J.A., Goff, G.P. & Huse, I. (1992) Seasonal reproduction and plasma levels of sex steroids and vitellogenin in Atlantic halibut (*Hippoglossus hippoglossus*). *Canadian Journal of Fisheries and Aquatic Sciences*, **49**, 754–9.

Mittelbach, G.G., Ballew, N.G. & Kjelvik, M.K. (2014) Fish behavioral types and their ecological consequences. *Canadian Journal of Fisheries and Aquatic Sciences*, **71**, 927–44.

Murua, H., Kraus, G., Saborido-Rey, F., Witthames, P.R., Thorsen, A. & Junquera, S. (2003) Procedures to estimate fecundity of marine fish species in relation to their reproductive strategy. In: E.A. Trippel (Ed.) *Reproductive Potential of Fish Populations of the North Atlantic. Journal of Northwest Atlantic Fishery Science*, **33**, 33–54.

Murua, H., Lucio, P., Santurtún, M. & Motos, L. (2006) Seasonal variation in egg production and batch fecundity of European hake *Merluccius merluccius* (L.) in the Bay of Biscay. *Journal of Fish Biology*, **69**, 1304–16.

Murua, H. & Motos, L. (2006) Reproductive strategy and spawning activity of the European hake *Merluccius merluccius* (L.) in the Bay of Biscay. *Journal of Fish Biology*, **69**, 1288–303.

Murua, H. & Saborido-Rey, F. (2003) Female reproductive strategies of marine fish species of the North Atlantic. In: E.A. Trippel (Ed.) *Reproductive Potential of Fish Populations of the North Atlantic. Journal of Northwest Atlantic Fishery Science*, **33**, 23–31.

Myers, M., Britt, K.L., Wreford, N.G.M., Ebling, F.J.P. & Kerr, J.B. (2004) Methods for quantifying follicular numbers within the mouse ovary. *Reproduction*, **127**, 569–80.

Nash, R.D.M., Witthames, P.R., Pawson, M. & Alesworth, E. (2000) Regional variability in the dynamics of reproduction and growth of Irish Sea plaice, *Pleuronectes platessa* L. *Journal of Sea Research*, **44**, 55–64.

Nikolskii, G.V. (1969) *Theory of Fish Population Dynamics as the Biological Background for Rational Exploitation and Management of Fishery Resources*. Oliver and Boyd, Edinburgh.

Nissling, A. & Westin, L. (1991). Egg buoyancy of Baltic cod (*Gadus morhua*) and its implications for cod stock fluctuations in the Baltic. *Marine Biology*, **111**, 33–5.

Norberg, B., Valkner, V., Huse, J., Karlsen, I. & Leröy Grung, G. (1991) Ovulatory rhythms and egg viability in the Atlantic halibut (*Hippoglossus hippoglossus*). *Aquaculture*, **97**, 365–71.

Nowak, S.J. & Corces, V.G. (2004) Phosphorylation of histone H3: a balancing act between chromosome condensation and transcriptional activation. *Trends in Genetics*, **20**, 214–20.

Óskarsson, G.J., Kjesbu, O.S. & Slotte, A. (2002) Predictions of realised fecundity and spawning time in Norwegian spring-spawning herring (*Clupea harengus*). *Journal of Sea Research*, **48**, 59–79.

Parker, K. (1980) A direct method for estimating northern anchovy, *Engraulis mordax*, spawning biomass. *Fisheries Bulletin*, **78**, 541–4.

Parrish, B.B. & Saville, A. (1965) The biology of the north-east Atlantic herring populations. *Oceanography and Marine Biology Annual Review*, **3**, 323–73.

Pedersen, T. (1984) Variation of peak spawning of Arcto-Norwegian cod (*Gadus morhua* L.) during the time period 1929–1982 based on indices estimated from fishery statistics. In: E. Dahl, D.S. Danielssen, E. Moksness & P. Solemdal (Eds) *The Propagation of Cod* Gadus morhua L. *An International Symposium, Arendal, 14–17 June 1983*. pp. 301–16. *Flødevigen rapportser*. 1, 1984. Institute of Marine Research, Flødevigen Biological Station, Arendal.

Pennington, M. & Vølstad, J.H. (1994) Assessing the effect of intra-haul correlation and variable-density on estimates of population characteristics from marine surveys. *Biometrics*, **50**, 725–32.

Plack, P.A., Pritchard, D.J. & Fraser, N.W. (1971) Egg proteins in cod serum. *Biochemical Journal*, **121**, 847–56.

Ramsay, K. & Witthames, P. (1996) Using oocyte size to assess seasonal ovarian development in *Solea solea* (L.). *Journal of Sea Research*, **36**, 275–83.

Reznick, D. & Yang, A.P. (1993) The influence of fluctuating resources on life history: patterns of allocation and plasticity in female guppies. *Ecology*, **74**, 2011–19.

Rideout, R.M. & Morgan, M.J. (2007) Major changes in fecundity and the effect on population egg production for three species of north-west Atlantic flatfishes. *Journal of Fish Biology*, **70**, 1759–79.

Rideout, R.M., Rose, G.A. & Burton, M.P.M. (2005) Skipped spawning in female iteroparous fishes. *Fish and Fisheries*, **6**, 50–72.

Rijnsdorp, A.D. (1990) The mechanism of energy allocation over reproduction and somatic growth in female North Sea plaice, *Pleuronectes platessa* L. *Netherlands Journal of Sea Research*, **25**, 279–90.

Rijnsdorp, A.D., Grift, R.E. & Kraak, S.B.M. (2005) Fisheries-induced adaptive change in reproductive investment in North Sea plaice (*Pleuronectes platessa*)? *Canadian Journal of Fisheries and Aquatic Sciences*, **62**, 833–43.

Rijnsdorp, A.D. & Witthames, P.R. (2005) Ecology of reproduction. In: R.N. Gibson (Ed.) *Flatfishes. Biology and Exploitation*. pp. 68–93. Fish and Aquatic Resources Series 9. Blackwell Science, Oxford.

Rinchard, J. & Kestemont, P. (2003) Liver changes related to oocyte growth in roach, a single spawner fish, and in bleak and white bream, two multiple spawner fish. *International Review of Hydrobiology*, **88**, 68–76.

Rocha, M.J., Arukwe, A. & Kapoor, B.G. (Eds) (2008) *Fish Reproduction*. Science Publisher, Enfield, New Hampshire. 632 pp.

Rothschild, B.J. (1986) *Dynamics of Marine Fish Populations*. Harvard University Press, Cambridge, Massachusetts.

Rowe, D.K. & Thorpe, J.E. (1990) Suppression of maturation in male Atlantic salmon (*Salmo salar* L.) parr by reduction in feeding and growth during spring months. *Aquaculture*, **86**, 291–313.

Saborido-Rey, F. & Junquera, S. (1998) Histological assessment of variations in sexual maturity of cod (*Gadus morhua* L.) at the Flemish Cap (north-west Atlantic). *ICES Journal of Marine Science*, **55**, 515–21.

Saborido-Rey, F., & Kjesbu, O.S. (2005) *Growth and Maturation Dynamics*. CSIC, Vigo, Spain. pp. 1–26. http://hdl.handle.net/10261/47150

Sandeman, L.R., Yaragina, N.A. & Marshall, C.T. (2008) Factors contributing to inter- and intra-annual variation in condition of cod *Gadus morhua* in the Barents Sea. *Journal of Animal Ecology*, **77**, 725–34.

Schismenou, E., Somarakis, S., Thorsen, A. & Kjesbu, O.S. (2012) Dynamics of de novo vitellogenesis in fish with indeterminate fecundity: an application of oocyte packing density theory to European anchovy, *Engraulis encrasicolus*. *Marine Biology*, **159**, 757–68.

Schmidt-Nielsen, K. (1983) *Animal Physiology: Adaptation and Environment*, 3rd edn. Cambridge University Press, Cambridge.

Schulz, R.W. & Miura, T. (2002) Spermatogenesis and its endocrine regulation. *Fish Physiology and Biochemistry*, **26**, 43–56.

Selman, K. & Wallace, R.A. (1989) Cellular aspects of oocyte growth in teleosts. *Zoological Science*, **6**, 211–31.

Shirokova, M.Y. (1977) Peculiarities of the sexual maturation of females of the Baltic cod, *Gadus morhua callarias*. *Journal of Ichthyology*, **17**, 574–81.

Silva, F.F.G., Slotte, A., Johannessen, A., Kennedy, J. & Kjesbu, O.S. (2013) Strategies for partition between body growth and reproductive investment in migratory and stationary populations of spring-spawning Atlantic herring (*Clupea harengus* L.). *Fisheries Research*, **138**, 71–9.

Skjæraasen, J.E., Nash, R.D.M., Korsbrekke, K., Fonn, M., Nilsen, T., Kennedy, J., Nedreaas, K.H., Thorsen, A., Witthames, P.R., Geffen, A.J., Høie, H. & Kjesbu, O.S. (2012) Frequent skipped spawning in the world's largest cod population. *Proceedings of the National Academy of Sciences of the United States of America*, **109**, 8995–99.

Skjæraasen, J.E., Nilsen, T. & Kjesbu, O.S. (2006) Timing and determination of potential fecundity in Atlantic cod (*Gadus morhua*). *Canadian Journal of Fisheries and Aquatic Sciences*, **63**, 310–20.

Skjærven, K.H., Penglase, S., Olsvik, P.A. & Hamre, K. (2013) Redox regulation in Atlantic cod (*Gadus morhua*) embryos developing under normal and heat-stressed conditions. *Free Radical Biology and Medicine*, **57**, 29–38.

Somarakis, S., Palomera, I., Garcia, A., Quintanilla, L., Koutsikopoulos, C., Uriarte, A. & Motos, L. (2004) Daily egg production of anchovy in European waters. *ICES Journal of Marine Science*, **61**, 944–58.

Sorokin, V.P. (1957) The oogenesis and reproduction cycle of the cod (*Gadus morhua* Linn.). *Trudy Pinro*, **10**, 125–44 (in Russian; English translation No. 72F49 Ministry for Agriculture, Fisheries and Food, UK, 1961).

Springate, J.R.C., Bromage, N.R. & Cumaranatunga, P.R.T. (1985) The effects of different ration on fecundity and egg quality in the rainbow trout (*Salmo gairdneri*). In: C.B. Cowey, A.M. Mackie & J.G. Bell (Eds) *Nutrition and Feeding in Fish*. pp. 371–93. Academic Press, London.

Stearns, S.C. (1992) *The Evolution of Life Histories*. Oxford University Press, Oxford.

Stenevik, E.K., Skogen, M., Sundby, S. & Boyer, D. (2003) The effect of vertical and horizontal distribution on retention of sardine (*Sardinops sagax*) larvae in the Northern Benguela – observations and modelling. *Fisheries Oceanography*, **12**, 185–200.

Stenevik, E.K. & Sundby, S. (2007) Impacts of climate change on commercial fish stocks in Norwegian waters. *Marine Policy*, **31**, 19–31.

Stenevik, E.K., Sundby, S. & Cloete, R. (2001) Influence of buoyancy and vertical distribution of sardine *Sardinops sagax* eggs and larvae on their transport in the Northern Benguela ecosystem. In: A.I.L. Payne, S.C. Pillar & R.J.M. Crawford (Eds) *A Decade of Namibian Fisheries Science. South African Journal of Marine Science*, **23**, 85–97.

Sundby, S. (2000) Recruitment of Atlantic cod stocks in relation to temperature and advection of copepod populations. *Sarsia*, **85**, 277–98.

Sundby, S. & Nakken, O. (2008) Spatial shifts in spawning habitats of Arcto-Norwegian cod related to multidecadal climate oscillations and climate change. *ICES Journal of Marine Science*, **65**, 953–62.

Sundby, S., Boyd, A.J., Hutchings, L., O'Toole, M.J., Thorisson, K. & Thorsen, A. (2001) Interactions between Cape hake spawning and the circulation in the Northern Benguela upwelling ecosystem. In: A.I.L. Payne, S.C. Pillar & R.J.M. Crawford (Eds) *A Decade of Namibian Fisheries Science. South African Journal of Marine Science*, **23**, 317–36.

Sundby, S., Skogen, M. & Kjesbu, O. (1999) Lysing i namibiske farvann – en økologisk tilpasning til et oppstrømningssystem. In: J. Aure (Ed.) *Havets Miljø*. pp. 68–75. Fisken og Havet, Særnummer 2 (in Norwegian).

Tanasichuk, R.W. & Ware, D.M. (1987) Influence of interannual variations in winter sea temperature on fecundity and egg size in Pacific herring (*Clupea harengus pallasi*). *Canadian Journal of Fisheries and Aquatic Sciences*, **44**, 1485–95.

Taranger, G.L., Carrillo, M., Schulz, R.W., Fontaine, P., Zanuy, S., Felip, A., Weltzien, F.A., Dufour, S., Karlsen, Ø., Norberg, B., Andersson, E. & Hansen, T. (2010) Control of puberty in farmed fish. *General and Comparative Endocrinology*, **165**, 483–515.

Thomé, R.G., Domingos, F.F.T., Santos, H.B., Martinelli, P.M., Sato, Y., Rizzo, E. & Bazzoli, N. (2012) Apoptosis, cell proliferation and vitellogenesis during the folliculogenesis and follicular growth in teleost fish. *Tissue Cell*, **44**, 54–62.

Thorsen, A. (1989) *Development of Atlantic Salmon Eggs: I. Oxygen Consumption, Ion Content, Survival and Growth of Atlantic Salmon (*Salmo salar L.*) Eggs at High Temperatures. II. Oxygen Availability and Hatching Mechanism of Atlantic Salmon (*Salmo salar L.*) Embryos.* Cand. Scient. Thesis, University of Bergen.

Thorsen, A. & Kjesbu, O.S. (2001) A rapid method for estimation of oocyte size and potential fecundity in Atlantic cod using a computer-aided particle analysis system. *Journal of Sea Research*, **46**, 295–308.

Thorsen, A., Kjesbu, O.S., Fyhn, H.J. & Solemdal, P. (1996) Physiological mechanisms of buoyancy in eggs from brackish water cod. *Journal of Fish Biology*, **48**, 457–77.

Thorsen, A., Marshall, C.T. & Kjesbu, O.S. (2006) Comparison of various potential fecundity models for north-east Arctic cod *Gadus morhua* L. using oocyte diameter as a standardizing factor. *Journal of Fish Biology*, **69**, 1709–30.

Tilly, J.L. (1996) Apoptosis and ovarian function. *Reviews of Reproduction*, **1**, 162–72.

Tobin, D., Wright, P.J. & O'Sullivan, M. (2010) Timing of the maturation transition in haddock *Melanogrammus aeglefinus. Journal of Fish Biology*, **77**, 1252–67.

Tomkiewicz, J., Tyberg, L. & Jespersen, Å. (2003) Micro- and macroscopic characteristics to stage gonadal maturation of female Baltic cod. *Journal of Fish Biology*, **62**, 253–75.

Trippel, E.A. (1999) Estimation of stock reproductive potential: history and challenges for Canadian Atlantic gadoid stock assessment. In: J. Morgan, J. Burnett & E. Aro (Eds) *Variations in Maturation, Growth, Condition and Spawning Stock Biomass Production in Groundfish. Journal of Northwest Atlantic Fishery Science*, **25**, 61–81.

Trippel, E.A., Kjesbu, O.S. & Solemdal, P. (1997) Effects of adult age and size structure on reproductive output in marine fishes. In: R.C. Chambers & E.A. Trippel (Eds) *Early Life History and Recruitment in Fish Populations*. pp. 31–62. Chapman & Hall, London.

Tyler, C.R., Nagler, J.J., Pottinger, T.G. & Turner, M.A. (1994) Effects of unilateral ovariectomy on recruitment and growth of follicles in the rainbow trout, *Oncorhynchus mykiss. Fish Physiology and Biochemistry*, **13**, 309–16.

Tyler, C.R., Santos, E.M. & Prat, F. (2000) Unscrambling the egg – cellular biochemical molecular and endocrine advances in oogenesis. In: B. Norberg, O.S. Kjesbu, G.L. Taranger, E. Andersson & S.O. Stefansson (Eds) *Proceedings of the 6th International Symposium on the Reproductive Physiology of Fish. Institute of Marine Research and University of Bergen, 4–9 July 1999.* pp. 273–80. John Grieg AS, Bergen.

van Damme, C.J.G. (2013) *Don't Put All Your Eggs in One Basket!* Ph.D. Thesis, Wageningen UR.

van Damme, C.J.G., Dickey-Collas, M., Rijnsdorp, A.D. & Kjesbu, O.S. (2009) Fecundity, atresia, and spawning strategies of Atlantic herring (*Clupea harengus*). *Canadian Journal of Fisheries and Aquatic Sciences*, **66**, 2130–41.

Vladykov, V.D. (1956) Fecundity of wild speckled trout (*Salvelinus fontinalis*) in Quebec Lakes. *Journal of the Fisheries Research Board of Canada*, **13**, 799–841.

Wallace, R.A. & Selman, K. (1981) Cellular and dynamic aspects of oocyte growth in teleosts. *American Zoologist*, **21**, 325–43.

Wallace, R.A. & Selman, K. (1985) Major protein changes during vitellogenesis and maturation of *Fundulus* oocytes. *Developmental Biology*, **110**, 492–8.

Wallace, R.A., Selman, K., Greeley, M.S., Jr, Begovac, P.C., Lin, Y.-W.P., McPherson, R. & Petrino, T.R. (1988) Current status of oocyte growth. In: D.R. Idler, L.W. Crim & J.M. Walsh (Eds) *Proceedings of the Third International Symposium on the Reproductive Physiology of Fish. St John's, Newfoundland, Canada, 2–7 August 1987.* pp. 167–77. Marine Sciences Research Laboratory, St John's, Newfoundland, Canada.

Ware, D.M. & Tanasichuk, R.W. (1989) Biological basis of maturation and spawning waves in Pacific herring (*Clupea harengus pallasi*). *Canadian Journal of Fisheries and Aquatic Sciences*, **46**, 1776–84.

West, G. (1990) Methods of assessing ovarian development in fishes: a review. *Australian Journal of Marine and Freshwater Research*, **41**, 199–222.

Wieland, K., Jarre-Teichmann, A. & Horbowa, K. (2000) Changes in the timing of spawning of Baltic cod: possible causes and implications for recruitment. *ICES Journal of Marine Science*, **57**, 452–64.

Wildner, D.D., Grier, H. & Quagio-Grassiotto, I. (2013) Female germ cell renewal during the annual reproductive cycle in Ostariophysians fish. *Theriogenology*, **79**, 709–24.

Witthames, P.R. (2003) Methods to assess maturity and realised fecundity illustrated by studies on Dover sole *Solea solea*. In: O.S. Kjesbu, J.R. Hunter & P.R. Witthames (Eds) *Report of the Working Group on Modern Approaches to Assess Maturity and Fecundity of Warm- and Cold-Water Fish and Squids*. pp. 125–37. Fisken og Havet, no. 12–2003.

Witthames, P.R., Andersen, T.E. & Kjesbu, O.S. (2000) The application of tank experiments to the study of reproductive potential using *Gadus morhua* as a test model. ICES C.M. 2000 R:09.

Witthames, P.R. & Greer Walker, M. (1987) An automated method for counting and sizing fish eggs. *Journal of Fish Biology*, **30**, 225–35.

Witthames, P.R. & Greer Walker, M. (1995) Determinancy of fecundity and oocyte atresia in sole (*Solea solea*) from the Channel, the North Sea and the Irish Sea. *Aquatic Living Resources*, **8**, 91–109.

Witthames, P.R., Greer Walker, M., Dinis, M.T. & Whiting, C.L. (1995) The geographical variation in the potential annual fecundity of Dover sole *Solea solea* (L.) from European shelf waters during 1991. *Netherlands Journal of Sea Research*, **34**, 45–58.

Witthames, P.R. & Marshall, C.T. (2008) The importance of reproductive dynamics in fish stock assessment. In: A. Payne, J. Cotter & T. Potter (Eds) *Advances in Fisheries Science: 50 Years on from Beverton and Holt*. pp. 306–24. Wiley-Blackwell, Chichester, U.K.

Witthames, P.R., Thorsen, A., Murua, H., Saborido-Rey, F., Greenwood, L.N., Dominguez, R., Korta, M. & Kjesbu, O.S. (2009) Advances in methods for determining fecundity: application of the new methods to some marine fishes. *Fisheries Bulletin*, **107**, 148–64.

Woodhead, A.D. & Woodhead, P.M.J. (1965) Seasonal changes in the physiology of the Barents Sea cod, *Gadus morhua* L., in relation to its environment. I. Endocrine changes particularly affecting migration and maturation. *ICNAF Special Publication*, **6**, 691–715.

Wootton, R.J. (1984) Strategies and tactics in fish reproduction. In: G.W. Potts & R.J. Wootton (Eds) *Fish Reproduction: Strategies and Tactics*. pp. 1–12. Academic Press, London.

Wootton, R.J. (1998). *Ecology of Teleost Fishes*, 2nd edn. Fish and Fisheries Series **24**, Kluwer Academic Publishers, Dordrecht.

Wright, P.J. (2013) Methodological challenges to examining the causes of variation in stock reproductive potential. *Fisheries Research*, **138**, 14–22.

Wright, P.J., Galley, E., Gibb, I.M. & Neat, F.C. (2006). Fidelity of adult cod to spawning grounds in Scottish waters. *Fisheries Research*, **77**, 148–58.

Wright, P.J. & Gibb, F.M. (2005) Selection for birth date in North Sea haddock and its relation to maternal age. *Journal of Animal Ecology*, **74**, 303–12.

Wright, P.J. & Trippel, E.A. (2009) Fishery-induced demographic changes in the timing of spawning: consequences for reproductive success. *Fish and Fisheries*, **10**, 283–304.

Yamamoto, K. (1956) Studies on the formation of fish eggs. I. Annual cycle in the development of ovarian eggs in the flounder, *Liopsetta obscura*. *Journal of the Faculty of Science, Hokkaido University Series VI*, **12**, 362–73.

Yoneda, M. & Wright, P.J. (2004) Temporal and spatial variation in reproductive investment of Atlantic cod *Gadus morhua* in the northern North Sea and Scottish west coast. *Marine Ecology Progress Series*, **276**, 237–48.

Yoneda, M. & Wright, P.J. (2005) Effects of varying temperature and food availability on growth and reproduction in first-time spawning female Atlantic cod. *Journal of Fish Biology*, **67**, 1225–41.

Zohar, Y., Muñoz-Cueto, J.A., Elizur, A., Kah, O. (2010) Neuroendocrinology of reproduction in teleost fish. *General and Comparative Endocrinology*, **165**, 438–55.

Part III

Incorporation of Reproductive Biology and Recruitment Considerations into Management Advice and Strategies

Chapter 9

Current Paradigms and Forms of Advice

Kevern L. Cochrane

9.1 Introduction

This chapter is an updated version of that provided under the same title in the first edition of this book, published in 2009. The fundamental principles, arguments, and many of the examples presented in the earlier version of the chapter remain valid in 2014. As a result, much of this second edition remains unchanged from the first, but where important new information and insights or interpretations have been found, they have been included. The most substantial change in this version has been the inclusion of climate change as an important consideration in the provision of advice in fisheries assessment and management. With hindsight, climate change had already been widely recognized by 2009 as an important driver in a number of marine areas and fisheries, frequently having implications for reproductive dynamics of populations, and should have been included in the first edition. There could be no excuses for not including it now.

The description of fisheries management, taken from FAO (1997), provided in the 2009 version of this chapter remains valid:

> "The integrated process of information gathering, analysis, planning, consultation, decision-making, allocation of resources and formulation and implementation, with enforcement as necessary, of regulations or rules which govern fisheries activities in order to ensure the continued productivity of the resources and accomplishment of other fisheries objectives."

Central to this circuitous and possibly confusing description is that it refers to a process with multiple objectives, only one of which is to ensure the continued productivity of the resource. The multiplicity of objectives is an important, but often forgotten, principle. All too frequently for the fisheries biologist and ecologist, achieving the biological objectives of fisheries is considered to be synonymous with fisheries management. Hilborn & Walters (1992) caution against this view and draw attention to the difference between assessing the biological potential of a resource and the decision-making process. They wrote, "Once the stock assessment is complete, choice remains."

Fish Reproductive Biology: Implications for Assessment and Management, Second Edition.
Edited by Tore Jakobsen, Michael J. Fogarty, Bernard A. Megrey and Erlend Moksness.

Cochrane (2000) referred to four primary realms in fisheries: biological, ecological, social, and economic considerations. Those four realms encompass the objectives of continued productivity of the resource and "other fisheries objectives" referred to in the FAO definition above. While the priority given to different goals will vary from case to case and over time, sustaining productivity is essential for achievement of any of the other goals in the longer term. Partly as a consequence of common awareness of this simple reality, continued productivity has been given a primacy in modern global policy, reflected in calls for sustainability in, for example, the FAO Code of Conduct for Sustainable Fisheries and the Plan of Implementation from the World Summit on Sustainable Development held in Johannesburg, South Africa in 2002, and reaffirmed at the United Nations Conference on Sustainable Development in 2012 (UNCSD 2012). These and other fisheries instruments are consistent with the United Nations Convention on the Law of the Sea of 10 December 1982, in particular Article 61 where it is stated that "The coastal state… shall ensure through proper conservation and management measures that the maintenance of the living resources in the exclusive economic zone is not endangered by over-exploitation." This chapter focuses on the role of management in attempting to ensure sustainable and productive use of the resource.

It is common knowledge that the coastal states of the world have not been very successful at living up to the responsibilities delegated to them in the Law of the Sea. FAO (2014) reported on the estimated status of fish stocks or species groups for which sufficient information was available for such an assessment. This analysis indicated that the percentage of stocks estimated as being fished at biologically sustainable levels had fallen from 90% in 1974 to 71.2% in 2011. This consisted of 61.3% that were being fully fished at that time and 9.9% that were being underfished. The remaining 28.8% were assessed as being overfished (FAO 2014). While there is considerable uncertainty about these estimates, as well as the status of the many stocks around the world for which insufficient information to determine status is available, these statistics indicate that the fishery resources of the world remain under considerable pressure and that precautionary management must be rigorously pursued, including restoration of those stocks that have been overfished and depleted.

There are many reasons for the failure, which include (Cochrane 2000, Cochrane & Doulman 2005):

- High levels of biological and ecological uncertainty in the information used for management.
- The common conflict between short-term economic and social objectives and the longer-term objectives of sustainability.
- Poorly defined objectives in fisheries.
- Institutional weaknesses, in particular the absence, or inappropriate systems, of user rights and the inadequate participation by stakeholders.
- The continued existence of open-access and quasi-open-access fisheries in many parts of the world.
- A tendency for political decision-makers to avoid taking decisions that would be unpopular in the short-term.
- Inadequate capacity in many national administrations.
- Inadequate fisheries MCS systems.

Of particular relevance to this chapter – and highlighted by Cochrane *et al.* (2011) as occurring within the fisheries system and therefore, in principle, the control of the fisheries administrations – are poor scientific information, including data on recruitment and the impacts of management decisions on future recruitment, poorly defined objectives, and – particularly but not exclusively – an inadequate capacity for management. The last of these typically includes an inadequate capacity for relevant scientific research and monitoring.

Lessons are being learned from the failures that have been experienced in fisheries, but it is also important to learn from the successes, and there have been many notable successes. Mace (2004) reviewed some of these and concluded that, contrary to widespread opinion within the scientific community, conventional, single-species models and concepts are not fundamentally flawed but that management recommendations based on them have only rarely been followed. This conclusion is consistent with those of a review on "Sustaining Marine Fisheries" undertaken under the auspices of the National Research Council of the United States of America, which recommended that substantial progress could be made towards achieving sustainable marine fisheries by the adoption of a conservative, risk-averse, single-species approach (NRC 1999). In a study of the trends in 166 fish stocks worldwide, Worm *et al.* (2009), concluded that there were increasing efforts to rebuild stocks, and that the objectives of fisheries and conservation were attainable through combining different management measures such as catch controls, gear regulations, and closed areas as required by the specifics of each case. These conventional management measures typically remain fundamental to sound fisheries management within the context of integrated management and ecosystem-based approaches.

This chapter considers, first, the conventional, single-species models and management approaches and the extent to which they take reproductive biology into account. It then considers some of the emerging issues that may need greater attention. In doing so, attention is focused on the biological and ecological goal of maintaining productivity of the resources. The reader should bear in mind that, in fisheries, such a goal can be achieved in different ways and that different options will, except in extreme cases of overexploitation, generally allow for the choices referred to by Hilborn & Walters (1992).

9.2 Early recognition

It would be difficult to pin down when those responsible for fisheries management first began to take reproductive processes and biology seriously but, given the close links of many cultures to pastoralism, it is likely that the importance of reproduction was recognized more or less simultaneously with the initial awareness of the need to manage fisheries for sustainability. In 1652, the Massachusetts Bay Colony prohibited fishing for cod, hake, haddock, pollock, and mackerel during the spawning season, and by 1877 many states in the USA were using gear regulations and fishing seasons to manage some fisheries – measures that addressed both recruitment and growth overfishing. The measures addressing recruitment overfishing were mainly aimed at ensuring that the age at first capture was sufficient to allow fish to have spawned at least once before being harvested (Ross 1997, p. 159). In 1880, Frank Buckland, a pioneer of fisheries research in Great Britain, wrote in his book *Natural History of British Fisheries*, "We also want to know the times and places of the spawning of sea fish. Where do the soles lay their eggs? When and how do the plaice, turbot, brill, halibut, &c., spawn?... Again, how are we to devise a mesh of net that shall let go the small soles and under-sized fry of other fish, keeping the marketable fish only, allowing the others to escape and grow" (Buckland 1880).

In 1883, at the Great International Fisheries Exhibition in London, Huxley made his frequently quoted optimistic assertions about fish productivity, motivated strongly by the apparently enormous reproductive potential of fish. On the same occasion, Sir Ray Lankester provided a more cautious perspective stating "... the thousands of apparently superfluous young produced by fishes are not really superfluous, but have a perfectly definite place in the complex interactions of the living beings within their area" (in Smith 1994). Lankester reasoned that sustainable fishing therefore required a means of compensating for the impact of fishing in removing potential producers of offspring. Reducing the abundance of predators, protecting young fish from predators and stocking with artificially spawned eggs or larvae were considered by him as possible means of achieving this. At much the same time, the Fishery Board of Scotland was established and, under T.W. Fulton, started research into various aspects of fish biology and ecology, including fertility and growth rates of fish (Smith 1994). In 1892, Ernest Holt reported to the Marine Biological Association in Plymouth that there had been a reduction in flatfish abundance, and proposed a size limit for plaice that was designed to reduce fishing on plaice nursery grounds in the North Sea (Nicolson 1979).

These ideas pre-date the onset of the Newtonian view of fisheries, based on fundamental population dynamics, which started to permeate fisheries science early in the 20th century. The earlier concepts showed an understanding of the inter-relationship between the dynamics of a population and its ecosystem that has never been lost but was widely side-lined as fisheries scientists and managers struggled to keep some control while technology and fishing effort soared after the industrial revolution, and particularly since World War II (Caddy & Cochrane 2001, Garcia & Cochrane 2009). Regrettably, but perhaps inevitably, with limitations in knowledge and in computational capacity, the links between fishery and ecosystem were lost in practice fairly early on in the battle against the impacts of technological growth of fishing. As population dynamicists grasped the mathematical principles of stock fluctuations, single species approaches took over. The importance of considering the reproductive biology, or at least reproductive behaviour, and recruitment in management was retained, however.

9.3 Single-species theory

Chapter 7 of this book provides a detailed account of the theory and practice of stock assessment. The briefer accounts in this chapter are intended to focus on the provision of advice, and also to set the background and context for consideration of areas where current approaches could be strengthened, with particular emphasis on the reproductive dynamics of exploited fish species.

Building on accumulated experiences and observations, the first theoretical fisheries model was the well-known catch equation proposed by Baranov in 1918, which explicitly included recruitment (N_0). Unfortunately, Baranov's proposal was not really understood until Ricker translated it from Russian more than 20 years later (Smith 1994, p. 308), and a paper published by E.S. Russell some 13 years after Baranov's equation was undoubtedly more influential in driving the direction of fisheries science and management for the next 50 years or more. Russell, who in 1922 became head of the Lowestoft Fisheries Laboratory, proposed the following equation for the biomass of a population:

$$S_2 = S_1 + (A + G) - (C + M) \quad (1)$$

where S_t is the biomass of the population at time t; A is the mass of new fish growing into the fishable component of the population (i.e., the recruits); G is the increase in mass of the new recruits plus other fish already of a fishable size; C is the mass caught; and M is the mass of fish that died naturally.

Within this equation, and Russell's interpretation of it, can be found the foundations of post-World War II fisheries management. In evaluating the implications of the equation, Russell recognized:

- The possibility of recruitment overfishing through eradication of the spawning stock.
- That an increase in fishing mortality, F, would lead to a decrease in the mean size of the catch.
- The theoretical existence of a constant maximum yield but the impracticality of obtaining that stable maximum because of the variability in recruitment (A).
- The relationship between the theoretical maximum and the age of first capture.

Graham's paper, making use of the logistic growth equation, followed in 1935 and Smith (1994) suggested that by the end of World War II there were three primary lines of research that he described as partial theories. These were the surplus production theory, the spawner recruit theory, and the yield per recruit theory.

9.3.1 Surplus production or biomass dynamic theory

Caddy (1999) suggested that the theory of the surplus production curve in fisheries arose in the context of the North Sea when Graham (1935), working from a study of human growth in the USA, developed a model based on biomass. This was subsequently popularized by Schaefer (1954). Mace (2001) suggested the roots also lay in the work of Russell and Hjort in the early 1930s.

Surplus production models are the simplest type of fisheries model (Hilborn & Walters 1992), and the Schaefer model is simply a logistic growth model. The equation for the Schaefer model is:

$$B_{t+1} = B_t + rB_t^* \left(1 - \frac{B_t}{K}\right) - C_t \tag{2}$$

where B_t is the biomass at time t; r is the intrinsic growth rate of the population; K is mean unexploited biomass; and C_t is catch in time t.

The essential features of a surplus production model are that, if unfished, a stock will tend to fluctuate around a level of biomass or numbers, the mean unexploited biomass (K). This can be considered as the mean environmental carrying capacity for that stock. If the biomass starts at, or is reduced to, a very low level and allowed to grow, it will grow to that unexploited level, K. Initially, the population growth rate will be the product of the biomass (B) and the intrinsic growth rate of the population (r), but as the population increases in size and biomass, density-dependent influences will reduce the effective population growth rate until it reaches a mean of 0 at K. In the Schaefer model, the maximum rate of population growth, commonly referred to as maximum sustainable yield (MSY), occurs at a biomass of $K/2$, but different model forms can be used to simulate maximum population growth at biomasses of a different fraction of K. The surplus production model therefore subsumes recruitment and reproductive processes into the product of the intrinsic growth rate, r, and the term in brackets estimating the extent of density-dependent influences.

Caddy (1999) reported that the surplus production, or biomass dynamic, theory of fish population dynamics was the source of the only fisheries reference point included in the United Nations Law of the Sea, where it is stated (Article 61.3) "Such (conservation and management) measures shall also be designed to maintain or restore populations of harvested species at levels which can produce the maximum sustainable yield… ".

Biomass dynamic models have been widely applied to tuna (e.g., International Commission for the Conservation of Atlantic Tunas; ICCAT) and a number of tropical and temperate fish stocks (Punt & Hilborn 1996). ICCAT, for example, makes use of production models, as well as age-structured models, for assessing the state of the yellowfin *Thunnus albacares* and bigeye *Thunnus obesus* stocks, and uses an age-structured production model to assess the status of the South Atlantic albacore *Thunnus alalunga* stock (ICCAT 2014). The most valuable fishery in South Africa – that for the two hake species *Merluccius capensis* and *M. paradoxus* – is assessed through an age-structured production model in which the two species are assessed as independent stocks (Rademeyer *et al.* 2008). It is managed primarily through total allowable catches (TACs) that are set through an operational management procedure designed to allow recovery of the deep-water hake stock to 20% of its estimated pre-exploitation level over a 20-year period (DAFF 2012).

The concept of a maximum sustainable yield has been criticized, most notably by P.A. Larkin in his legendary paper, "An epitaph for the concept of maximum sustainable yield" (Larkin 1977). Larkin's far-sighted paper addressed several biological limitations of the concept of MSY, if used as a target reference point. In relation to reproductive biology, he pointed out two primary problems with the use of MSY. The first problem is that when a population is fished down to a biomass close to B_{MSY} (i.e., the biomass on the surplus production curve at which the population is estimated to generate MSY on average), the surviving spawning population is likely to be dominated by young spawners. Larkin warned that this reduction in the number of spawning year classes would increase the vulnerability of the population to catastrophic events, and might also lead to a reduction in the quality of eggs produced (Larkin 1977). The issue of quality of reproductive products is specifically addressed later in this paper. It should be noted, however, that Larkin still recognized the potential value of MSY as a "valuable rough index of production potential", and his epitaph was qualified with the statement that the concept of MSY "will alone not be sufficient".

Since the publication of Larkin's paper, the use of MSY as a reference point has evolved, as well described by Mace (2001), who discussed a transition of the use of MSY from a fixed, annual target yield to the maximum average yield over time (typically implemented in terms of F_{MSY}, the fishing mortality rate that, on average, will yield MSY) and, as commonly used today, to a maximum limit to be avoided. Notwithstanding the findings of Hilborn (2010), which are discussed further later in this chapter, that a sustainable "pretty good yield" could be obtained from stock sizes between 20% and 50% of pristine stock abundance, the use of MSY as a limit reference point would go some way towards addressing the reproductive risks described by Larkin. Further, Mace (2001) pointed out that fishing below F_{MSY} will usually result in a relatively large gain in average stock biomass, which should give a proportionate gain in year-class structure, for a relatively small loss in average catch. Overall, therefore, while the concept and estimation of a surplus production model for a population aggregates all reproductive biology as a component of two parameters (r

and *K*), if used in a precautionary manner it can still be useful within a single-species framework.

9.3.2 *Stock recruit theory*

The underlying principle of stock recruit theory is that, at low population abundance, more eggs will be produced the greater the number of spawning adults and, following density-independent mortality, the greater the number of recruits that will enter the fishable biomass. As the biomass of the stock increases, density-dependent effects will become more pronounced and there will be a reduced per capita net production of recruits. Density-dependent mortality may be linked to, for example, the survival of eggs or pre-recruit life stages, competition for food or for space, or cannibalism (Gabriel & Mace 1999). The fundamental assumption of a stock–recruitment relationship is that if the abundance of spawners is reduced far enough, at some point the average recruitment produced by those spawners will be reduced. The two fundamental stock–recruitment models are those of Beverton & Holt (1957), which assumes that recruitment increases towards an asymptote with increasing spawner numbers or biomass, and of Ricker which, in contrast to the Beverton and Holt model, assumes declining recruitment above a certain spawner population size or biomass (e.g., Hilborn & Walters 1992).

The concept has underpinned management targets of many important fisheries. The Magnuson–Stevens Fishery Conservation and Management Act of the United States requires that conservation and management measures shall prevent overfishing. Mace & Sissenwine (1993) stated that, at that time, 60% of the fisheries in the USA defined overfishing in terms of minimum levels of spawner biomass per recruit which had been set, arbitrarily in their view, at between 20% and 35% of the spawner biomass per recruit that would occur in the absence of fishing. In addition, the majority of the more than 100 stocks from the Northeast Atlantic are assessed by the International Council for the Exploration of the Sea (ICES). ICES uses age-structured assessments that are related to limit reference points based on fishing mortality and spawning biomass (FAO 2005). The minimum spawning stock biomass limit reference point, B_{lim}, is based on available historical data and represents a spawner biomass below which there is a high risk that recruitment will be impaired and will, on average, be significantly lower than at higher spawner biomasses (ICES 2004).

In addition, precautionary reference points established by ICES have been determined for individual stocks, and are based on the need to avoid a low spawner biomass that could lead to reduced productivity. The precautionary limit on spawning-stock biomass, B_{pa}, is the estimated level for a given stock at and above which there is still a low probability that productivity will be impaired. In its management advice, ICES stresses that this biomass must be seen as a lower bound on biomass, not as a target (Piet & Rice 2004). The slope at the origin of a stock–recruitment relationship is also used as a limit reference point to reduce the risk of recruitment overfishing (Mace 2001). Yet another example of a spawning biomass limit reference point is found in the fishery for squid *Ilex argentinus* in the Southwest Atlantic. Management of this fishery was through effort control, with effort set each year so as to maintain the spawner biomass above a minimum, below which there is considered to be a high probability of low recruitment (Basson *et al.* 1996).

Even in the absence of good stock–recruitment data and a reasonably well-estimated stock–recruitment relationship, a spawner biomass threshold can still be used as a reference point in fisheries management. In the South African anchovy fishery, density-independent variability in recruitment is typically very high, and any estimated stock–recruitment relationship would have very high uncertainty. Underlying the management of the fishery during the 1990s and early 2000s was an empirically based assumption that recruitment varies randomly around a mean for any spawner biomass above 20% of the mean spawner biomass in the absence of exploitation (*K*), but will decline linearly on average, as the spawner biomass declines below 20% (Butterworth *et al.* 1997, Cochrane *et al.* 1998). Since 2004, the parameters of this stock–recruitment relationship have been estimated as part of the fit of the assessment model as a whole to the available data using a Bayesian estimation approach (de Moor & Butterworth 2009). The estimated stock–recruitment relationship is included in the operating model used to test and select decision rules for setting TACs within a management procedure.

As in the case of South African anchovy, a primary problem in the use of stock–recruitment relationships for fishery management is that the residual uncertainty in an estimated relationship is typically large (see e.g., de Moor & Butterworth 2009). This reflects not only observation, model, and estimation uncertainty, but also process uncertainty resulting from the many factors other than spawning biomass that can affect recruitment. The abundance of spawners is therefore only one measure of reproductive potential. Some of the other factors relate to the reproductive process; for example, Rätz & Lloret (2003) found that stocks of Atlantic cod *Gadus morhua* in poor condition were more likely to generate low recruitment when the spawner biomass was low than when the stocks were in good condition. Blanchard *et al.* (2003) estimated that a change in condition factor from 0.8 to 1.0 resulted in a doubling of fecundity-at-length for eastern Scotian Shelf haddock *Melanogrammus aeglefinus*. Those factors are likely to impact on total egg production, but would not necessarily be included in a standard stock–recruitment relationship.

There are also many external factors – both density-dependent and density-independent – that can influence the final year-class strength (Lasker 1985). In the case of small pelagic fish, for example, some of the other factors include cannibalism (Smith 1985, Valdés Szeinfeld & Cochrane 1992), unfavourable transport (Bakun & Parrish 1982, Shannon *et al.* 1992), and food availability on the spawning grounds (Cochrane & Hutchings 1995). Agnew *et al.* (2000) developed a model incorporating sea surface temperature and spawning that explained 77% of the variance in recruitment of *Loligo gahi* around the Falkland Islands, but a linear model with only sea surface temperature as the independent variable explained 66% of the variance. Statistically significant correlations between recruitment and the environment have also been found in North Atlantic albacore, south Pacific albacore and North Pacific albacore. In the case of North Atlantic albacore, it was demonstrated that spawning stock biomass and environmental effects can influence recruitment simultaneously, and that the density-dependent impacts are moderated by the environmental situation (Arregui *et al.* 2006).

Hilborn & Walters (1992) listed other examples where environmental effects on the stock–recruitment relationship have been demonstrated, but caution against the widespread use of this information, suggesting that it is nearly impossible to be sure that the identified correlation is not spurious, and also warning of the statistical implications of including too many variables in a statistical model. They also argued that the single largest problem in estimating stock–recruitment relationships is the difficulty of obtaining good estimates of the fundamental variables: spawner stock size and recruitment. Nevertheless, while those authors described determining the relationship between stock and recruitment as being the most difficult problem in assessment of fisheries, they also considered it the most important.

A recent study, also by Hilborn, re-emphasizes the importance of stock–recruitment relationships for productivity and resilience of the stocks. A meta-analysis using a simple age-structured model indicated that a sustainable yield of at least 80% of MSY could be obtained from a stock size between 20% and 50% of pristine stock abundance. The most important factor that determined the range was the steepness of the stock recruit curve (Hilborn 2010).

9.3.3 *Yield and biomass per recruit*

Thompson & Bell (1934) described the first dynamic pool model, and the concept was developed further by Beverton & Holt (1957) who formulated the concept of yield per recruit as a function of fishing mortality. The concept has had a profound effect on fisheries management. Per recruit models (also referred to as dynamic pool models) take into account the age structure of a stock, typically represented in the form of annual cohorts. They explicitly take into account the three processes of Russell's earlier model: recruitment (*A* in Russell's model), mortality (*C* and *M*), and growth (*G*). Age-specific fishing mortality rates can be included, for example in the simplified form of knife-edge recruitment. The inclusion of age at sexual maturity allows the estimation of spawner biomass in a biomass-per-recruit model.

A yield-per-recruit model used in isolation focuses on growth and mortality, and excludes the consideration of recruitment, and thereby reproductive biology. The well-known F_{max} reference point reflects only the maximization of the yield from one, or a constant number of, recruits. A second category of application of the yield-per-recruit concept to management has commonly been through use of the derived, but essentially *ad hoc*, reference point of $F_{0.1}$ (the fishing mortality rate at which the slope of the yield-per-recruit function is 10% of the slope at the origin). This target fishing mortality was, for example, the policy applied to the Canadian groundfish fishery (Parsons & Beckett 1997). As a yield-per-recruit reference point, $F_{0.1}$ does not take into account recruitment and is, in fact, purely arbitrary without any foundation in theory. However, it has been found to work in practice and has been shown by Deriso (1987, referred to in Hilborn & Walters 1992) not to reduce the abundance of spawners excessively.

Reproduction and recruitment can be brought into consideration in per-recruit approaches through the application of spawner biomass per recruit models, and a range of reference points have been applied or proposed based on a minimum acceptable spawner biomass, commonly measured as a proportion of unexploited biomass. Mace (2001) suggested that $F_{30\%}$ and $F_{40\%}$ (the fishing mortality that will lead to a reduction in spawner biomass, on average, to the indicated percentage of mean unexploited biomass) are proxies for F_{MSY}, while values of fishing mortality from $F_{10\%}$ to $F_{30\%}$ can be indicators of recruitment overfishing. Obviously, the specific threshold and its use – whether limit or target – will depend on the characteristics of the stock under consideration. Other common indicators include: P_m, the number or biomass of spawners needed to obtain maximum recruitment; P_r, the replacement spawner abundance which will generate recruitment equal to the parent stock; and $P_{s,}$ the number or biomass of spawners required to obtain maximum sustainable yield (Gabriel & Mace 1999).

Use of per-recruit models in management also allows an examination of the use of different age-structured exploitation patterns on a stock. For example, a per-recruit model can be used to consider a combination of total effort (as reflected in *F*) and gear or season regulation to determine the size at first capture in a fishery (Cochrane 2009). Per-recruit models do not usually include density-dependent or density-independent features of stock dynamics: the rates in a model typically vary by age but are not influenced by other factors (Gabriel & Mace 1999).

The application of yield-per-recruit models translates into two primary and complementary management approaches: (i) regulation of size at first capture as a means of avoiding growth overfishing; and (ii) the concept of F_{max}, the fishing mortality at which yield per recruit is maximized. In the mid-20th century, following the establishment of the International Commission for the Northwest Atlantic Fisheries and the Permanent Commission in the Northeast Atlantic, the regulation of North Atlantic groundfish was focused on controlling the size of fish that could be caught through restrictions on gear, especially on mesh size, limits on minimum sizes at capture, and closed area and season controls (Halliday & Pinhorn 2002). These regulations were underpinned by Beverton and Holt's yield-per-recruit equation and had the goal of maximizing yield, although within the USA and the Faroe Islands the use of a minimum size was motivated by a desire to protect the spawning stock. In addition to using surplus production models as a basis for scientific advice for management, ICCAT also uses yield-per-recruit methods as a basis for management advice. Yield-per-recruit analyses undertaken as part of an assessment of yellowfin tuna in 2003 indicated that an increase in effort would probably lead to a decrease in the yield-per-recruit, while a reduction in the fishing mortality on fish less than 3.2 kg could lead to an increase in yield-per-recruit and a more moderate increase in spawning biomass-per-recruit (ICCAT 2005). Yield and biomass per recruit reference points were applied in some data-poor groundfish fisheries on the Brazil–Guianas shelf as a basis for providing advice on controlling fishing effort (Hackett *et al.* 2000, Booth *et al.* 2000).

9.4 Management Measures

There is only a limited number of management tools available through which the impact of a fishery on a stock or resource can be controlled. These include gear restrictions, closed seasons, closed areas including marine protected areas (MPAs), input controls, and output controls. In practice, a combination of these options will commonly be applied to address a number of different operational objectives related to the resource and the ecosystem.

Regulations to reduce catches of small fish were introduced to England in 1605, and the relationship between mesh size and the size of fish that were caught was first described towards the end of the 19th century (Halliday & Pinhorn 2002). Since then, gear restrictions have become a widely used means of attempting to restrict the selectivity of fishing to preferred species and size classes within species, particularly as a means of attempting to avoid the capture of smaller fish so as to result in better recruitment to larger size classes. Bjordal (2009) described the ideal fishing gear as having the following characteristics: highly selective for the target species and sizes; effective, resulting in high catches of target fish at the lowest price; and producing catches of high quality.

The total selectivity of a fishing method results from a combination of the intrinsic selectivity characteristics of the gear and the manner in which the gear is operated. Although the "ideal gear" does not exist, different fishing gears have specific properties and specific contributions for sustainable fishing (Bjordal 2009). Bjordal therefore argued that understanding the properties, function and operation of the different fishing gear is essential for decision-making in fisheries management. Wileman *et al.* (1996, see their Table 3.1.4) provided an extensive list of factors that can affect the selectivity of gear for size, including, for example, mesh size and shape, twine characteristics, attachments to the codend, and the hauling procedure. In addition, the use of sorting devices, such as sorting grids and large mesh panels, can reduce selectivity for small fish.

Despite these possibilities, Halliday & Pinhorn (2002) concluded that gear regulation does not provide a reliable means of controlling size at capture, and pointed out that where a fishery is harvesting a multispecies community, gear selectivity will not be able to cater for the different sizes at age of the different species. In their review of management policies aimed at controlling the size at first capture for the North Atlantic groundfish fisheries, Halliday & Pinhorn concluded that attempting to manage fisheries by regulating the capture of small individuals is inadequate, and that it will need to be supplemented by regulation of overall fishing mortality rate.

Time and area closures can also provide a means of reducing undesired catch, such as limiting the catch of fish at specific life stages (Hall 2009). In a study on the economic aspects of fisheries in the OECD countries, a list was provided of 52 fisheries within 11 countries that used time and area closures (prohibition of fishing within a particular area during a specified time period). This total included examples where there had been complete moratoria on the fishery for a specified period. The study reported that, amongst other objectives, this measure was used as a means of avoiding capture of small sizes, including juveniles, by closing nursery areas. One example of this was closures of prawn and shrimp fisheries early in the season while the animals were still small (OECD 1997). The study concluded that, while the results were complicated, the measure can contribute to resource conservation but is not sufficient on its own.

There are a number of definitions for MPAs. FAO (2011) provides the broad definition of "any marine geographical area that is afforded greater protection than the surrounding waters for biodiversity conservation or fisheries management purposes will be considered an MPA". As a result of a combination of advocacy and greater awareness of the need for a broader approach to fisheries management, the use of MPAs in fisheries management is growing, although not always with the necessary scientific and social planning (Cochrane 2007). There are a number of examples of MPAs being used for a primary purpose of protecting vulnerable life stages. For example, an area in Japan where female zuwai or tanner crab *Chionoecetes opilio* aggregate to spawn has been closed (Gell & Roberts 2003), and MPAs of a total area of 17 000 km^2 were established by the USA on the Georges Bank in areas important to groundfish spawning and recruitment, in particular, to incorporate seasonal spawning grounds of haddock *Melanogrammus aeglefinus* and on the basis of the distribution of yellowtail flounder *Limanda ferrugineus* (Murawski *et al.* 2000).

A large number of reef fishes aggregate to spawn for short periods at specific localities. Russell (2001) related that 49 species of fish had been reported as forming spawning aggregations in the Great Barrier Reef Marine Park, and another 84 species that occurred in the park were known to form aggregations outside the park. Sadovy *et al.* (2003) proposed that, in addition to a number of urgently required management measures, the protection of spawning aggregation sites or of spawning fish during the spawning season was a serious requirement for sustainable use of humphead wrasse *Cheilinus undulatus*. Whether or not those areas should be designated as marine protected areas or as broader time area closures is open to differing interpretations, but the key issue is how effective they are.

Input controls regulate the effective effort that can be exerted in a fishery. Typically, they would include controls on the number and size of vessels that can be used in a fishery, and on the amount of time that vessels or fishing units can fish. Output controls regulate the amount of fish caught or, more accurately in many cases, landed. This can be in the form of a TAC for the fishery as a whole (which should be split into allocated quotas to avoid a race to fish), or bag limits on the number of fish that can be caught in a day. Commonly, input and output controls form the backbone of fisheries management, and are aimed at regulating the overall fishing mortality on a stock or resource as a whole. The fundamental biological goal underlying these measures is to maintain the biomass of the reproductively mature portion of the population sufficiently high so as not to jeopardize future recruitment. Gear regulations, closed areas, and closed seasons can be used to supplement input or output controls, to reduce the fishing mortality on particular size classes and species, or to divert it from selected areas to avoid undesirable mortality on sizes, species, or habitats.

Most effective modern fisheries use a combination of measures, determined and tuned in each case on the basis of stock assessments and other scientific advice, in order to achieve the agreed objectives (ecological, social, and economic). Management capacity, including the ability and efficiency of enforcing particular measures, should also be an important consideration in deciding on suitable management measures. As discussed earlier, notwithstanding the many problems being faced in fisheries around the world, these conventional approaches have led to a number of significant success stories (Mace 2004, Worm *et al.* 2009). Nevertheless, caution may be needed. Larkin (1977), a far-sighted prophet, concluded that the S in MSY cannot be forever and could not be for more than 50 to 100 years. He made that specific forecast in relation to the impact of inter-specific relationships on the productivity of a single species, but later in the same paper also linked it to reductions in genetic diversity. The next section in this chapter considers evidence and examples related to reproductive biology, and of where current management practices – heavily influenced by the three partial theories – may be taking too narrow a view.

9.5 Factors neglected or omitted in the three partial theories

9.5.1 Sex ratios and size-at-maturity

As sexually reproducing organisms, the presence of both sexes in proportions appropriate to breeding behaviour and characteristics is essential for the sustainability of fish stocks. The sexes can be assessed and modelled separately with both stock–recruit and per-recruit methods, but this is infrequently done in practice and the common assumption in stock assessments is that a single measure of spawner biomass provides an index of reproductive potential. In an equivalent search for statistical parsimony, but fed also by widespread unawareness, the assumption is also usually made that the eggs and larvae from all fish within a stock are of equal potential and value to the sustainability of the stock (Birkeland & Dayton 2005). Larkin (1977) drew attention to the danger of this approach and both of these assumptions are currently giving rise to concern and being challenged.

The sex ratio can be important and, particularly where there is sex-based selectivity in a fishery, minimum abundance thresholds may be required for one or both sexes instead of, or in addition to, thresholds on total abundance. Huchette *et al.* (2004) reported that in blacklip abalone *Haliotis rubra*, eggs with a larger cytoplasm were more likely to be fertilized when sperm densities were low, and that this could lead to an evolutionary reduction in the variability in size and energy reserves of oocytes, thereby reducing the ability of the population to cope with environmental variability. Huchette and colleagues suggested that insufficient male densities and skewed sex ratios could be a problem in a number of invertebrate populations, and proposed that an improved understanding of the reproductive behaviour and requirements of invertebrates – and abalone in particular, because it is frequently subjected to particularly high exploitation pressure – is a requirement for their sustainable use. The potential problem has also been described for hermaphrodite fishes which undergo a change in sex at a critical size, for example porgys (Sparidae) and groupers (Epinephalinae). Concerns have been expressed in some of these cases that an over-exploitation of males could lead to reduced fertilization rates and resulting depensation at low stock sizes (Jennings *et al.* 2001).

Evidence of differential survival rates for eggs and larvae is being accumulated for a growing number of fish species, and the importance of taking this into account in the management of such species is becoming increasingly apparent. Convincing evidence has been collected indicating that in species such as black rockfish *Sebastes melanops*, haddock *Melanogrammus aeglefinus*, and Atlantic cod *Gadus morhua*, larvae produced by older females grow faster and have a greater chance of survival than those from smaller individuals (Birkeland & Dayton 2005, Berkeley *et al.* 2004). In the case of black rockfish, older individuals have been found to produce larger larvae, each of which is provided with a reserve of energy-rich lipids that allows it to grow faster and to be less susceptible to starvation, thereby substantially increasing the probability of its survival. This enhanced survival multiplicatively increases the already higher contribution of larger females arising from their typically higher fecundity. Birkeland & Dayton (2005) suggested that increasing larval "quality" with maternal age is probably a particular feature of longer-lived species, and that it may not apply in shorter-lived, faster-growing species such as some tuna species and dolphinfish *Coryphaena hippurus*.

Two other mechanisms have been proposed by which longevity, in an unexploited population, could contribute to reduce the risk of poor recruitment in any given year (Berkeley *et al.* 2004). First, it has been observed in several studies that older marine teleosts commonly spawn before younger fish in the spawning season. Thus, it is argued, the presence of a healthy age structure (i.e., one that has not been excessively truncated by fishing mortality) provides a bet-hedging strategy against unfavourable conditions during different periods of the spawning season. Similarly, the same authors suggested that different age classes may spawn in different localities, which would result in an equivalent spatio-temporal bet-hedging strategy.

The implications of these investigations are clear. In fish stocks where one or more of the age-related factors applies (i.e., a higher fecundity or a higher survival rate of larvae in older females, spatial distribution by age class or temporal distribution of spawning by age class), fisheries management needs to ensure that the age composition of the fished population is maintained in a structure that does not lead to any of those properties being compromized beyond some minimum threshold. Maintenance of biomass at a level defined purely in terms of an MSY level may therefore not be sufficient to ensure that average recruitment is maintained. In support of this hypothesis, Marteinsdottir & Thorarrinsson (1998, cited in Berkeley *et al.* 2004) reported that strong year classes of Icelandic cod only occurred when the population had a broad age structure.

It has been suggested that there are three means by which fisheries management can help to ensure that the age structure of an exploited fish population retains adequate representation of older fish (Berkeley *et al.* 2004). These are to keep fishing mortality at a suitably low level, to implement "slot" size limits, and through the use of marine reserves. In the case of groundfish, the authors concluded that marine reserves would be the best option, and that this may also apply to other groundfish where the adults are essentially resident. More generally, the ideal approach – or, more likely, combination of approaches – will vary according to the particular life history of the species and nature of the fisheries in which it is captured.

9.5.2 Genetic considerations

While aquaculture and stock enhancement have given considerable attention to controlling and enhancing the genetic make-up of individual fish and invertebrates to optimize production, conventional fisheries management, premised upon the population dynamics concepts of the three partial theories, has typically failed to take finer genetic structure of populations into account. The potential risks of this omission have been revealed by more recent insights from genetic studies that have led to a better understanding of the genetic complexity of populations, including: the realization that, far from being essentially homogenous, marine populations exhibit substantial genetic structure. This genetic structure is associated with high levels of biocomplexity and adaptive differentiation that increases the resilience of populations to perturbation, including by fishing, and evidence that effective population sizes are several orders of magnitude smaller than previously thought, indicating that recruitment dynamics are considerably more complex than has previously been recognized (Hauser & Carvalho 2008).

Another important consequence of the, still prevailing, failure to take genetic structure into account in assessment and management has been the tendency to ignore the possibility of evolutionary consequences of fishing. The potential problems associated with this short-term view are now becoming more apparent. Some of the best-known work on the potential evolutionary impacts of selective fishing has come from experimental studies on the Atlantic silverside *Menidia menidia*. Conover & Munch (2002) used captive populations of the annual species and examined the effects of strongly size-selective fishing over four generations. These authors found that the yield in the experimental treatment involving selection of large individuals was initially higher than in the random and small-size treatments. However, by the fourth generation, the yield in the small-size harvest was double that of the large size harvest. The difference was ascribed to changes in the genetic structure related to mean growth rate (Conover & Munch 2002, Conover *et al.* 2005). The implications of these findings for fisheries have been moderated somewhat by further experimental studies which revealed that five generations of selective fishing did not reduce juvenile survival, and that there was still sufficient phenotypic variation to allow recovery of the exploited populations. Consequently, it was predicted that a full recovery would take about 12 generations (Conover *et al.* 2009). Although encouraging, 12 generations is a long time for slower-growing and maturing species, assuming that the result is applicable to a wider range of species than just the *Menidia menidia* used in these experiments.

Evidence from capture fisheries, however, has indicated that there can be a loss of genetic diversity from selective fishing. By using scales collected over a 50-year time interval, Hauser *et al.* (2002) were able to measure a significant decrease in the heterozygosity and mean number of alleles in a population of New Zealand snapper *Pagrus auratus* in Tasman Bay, from the time the fishery commenced in 1950 until 1998. These authors concluded that commercial fishing may have resulted in selective genetic changes to the population and in a reduced genetic diversity caused by genetic drift. In another example, the Northwest Atlantic cod

G. morhua, substantial historical declines in age and size at maturity, associated with a sustained over-exploitation and targeting of individuals that mature late and at larger sizes, are considered to have been genetic rather than phenotypic responses. The distinction is important because phenotypic responses are typically easily reversed when the driving force – fishing mortality in this case – is reduced or removed, whereas genetic recovery is likely to be much slower. Modeling studies performed by Hutchings (2005) indicated that the reductions in age-at-maturity and size-at-maturity would have led to reductions in the intrinsic rate of population growth, *r*. With a reduction in age-at-maturity from six to four years and a reduction in size-at-maturity, Hutchings estimated that *r* could decline by 29%.

As is usually the case in nature, the reality is probably even more complex than these studies have shown. An analysis of changes in the pike *Esox lucius* population in Lake Windermere, suggested that the changes in somatic growth and reproductive investment (defined in this study as gonad weight, egg number, and egg weight) during the period considered were driven by a combination of harvest selection and natural selection. It was concluded that the dominant driver of these two had changed during the time series, and that only during one short period, when the prey abundance had collapsed, did the two drivers operate in the same direction (Edeline *et al.* 2007).

9.5.3 Spatial considerations

The importance of taking the spatial distribution and characteristics of fish stocks and communities has been increasingly recognized in recent years, and is now the focus of many studies (e.g., Kruse *et al.* 2001). Failure to consider the spatial features of stocks and fisheries can lead to a wide variety of problems, including misinterpreting catch per unit of effort (cpue) trends with time, a failure to protect stocks at times of the year or life history stages when they are most vulnerable to capture, and ineffective placement of protected areas or closed areas.

In relation to reproductive biology, a spatial approach to management – for example, the use of temporary or permanent closed areas – has commonly been applied as a means to prevent fishing on particularly vulnerable life stages such as spawning aggregations or areas where juveniles aggregate (Hall 2009). As discussed earlier in this chapter, during recent years there has been particular concern about spawning aggregations of many reef species. Such aggregations present attractive fishing opportunities and have led to overfishing in many cases. Examples of fish which form spawning aggregations and that are or have been subjected to aggregation fisheries include the orange roughy *Hoplostethus atlanticus* off Australia, New Zealand and Namibia, the Nassau grouper *Ephinephelus striatus* in the western central Atlantic Ocean, and the emperor *Lethrinus nebulosus* in Egypt (Sadovy de Mitcheson *et al.* 2008, Sadovy & Domeier 2005). Information collected by the Society for the Conservation of Reef Fish Aggregations showed that, of the 60% of available records on spawning aggregations studied for which information on status was available, only a few were protected and there were indications of declining landings for 79% of these. Where there were management regulations in place, the most common were bans on sale during spawning seasons and time or area closures (Sadovy de Mitcheson *et al.* 2008). Sadovy & Domeier (2005) argued for the protection of spawning aggregations, pointing out that although the problem is widely recognized, only a few aggregations are managed or protected by MPAs. They recommended that seasonal closures and effort limitation could be used more frequently.

Spawning aggregations provide an obvious and extreme example of a need to regulate the spatial distribution of fishing mortality to protect reproduction, but there appear also to be more subtle examples. Berkeley *et al.* (2004) suggested that at least one of the reasons for the failure of fisheries management to prevent the depletion of a number of groundfish stocks on the west coast of the USA has been that it did not take into account the fine-scale spatial distribution of recruitment. They referred to the Hedgehog hypothesis that, in any one year, only a small fraction of spawners is responsible for most of the successful recruitment, and that success arises from matching the presence of spawners with those areas in the range that happen to have favourable conditions for spawning and larval survival in any given year. A number of studies had been undertaken, covering four different species of rockcod *Sebastes spp.*, that provided tests of the hypothesis, but with mixed results. However, these examples were sufficient for Berkeley *et al.* (2004) to conclude that the hypothesis may be true under some circumstances, and therefore that management should include an objective of maintaining spawning biomass over the full geographic range of the species.

Extending this argument, the improved understanding of the genetic structure of populations has important implications for the design of spatial controls, including MPAs. In planning MPAs, optimal size, position and spacing in networks is dependent on the connectivity features of populations and their population structure (Hauser & Carvalho 2008). However, the implications go beyond only spatial controls, and Hauser & Carvalho (2008) have argued that genetic principles and methods need to be taken into account in fisheries management in general.

9.5.4 Ecosystem approaches

The early awareness of the links between fisheries and their ecosystems, as briefly presented in Section 9.1, was largely lost in the population dynamics coup of the 20th century. Larkin (1977) described how "For almost 100 years, working from a European base, limnologists had been developing holistic schemes of trophic status… I vividly recall being proselytized by Bill Kennedy, a disciple of the new doctrine of population dynamics, about the futility of the old-fashioned limnological approaches… ". Notwithstanding the real progress and important contribution of the "new doctrine", the wheel is turning and consideration of the ecosystem has, again, been recognized as being an essential ingredient of management for sustainable use (e.g., FAO 2003). Much of the ecosystem approach is concerned about interactions between different species and between fisheries and non-target species. In general, while potentially very important, those considerations are frequently not directly linked to reproductive biology. However, the ecosystem approach recognizes the need to take into account not only the impact of fisheries on the ecosystem but also the impact of the ecosystem on fisheries. The ecosystem can – and does – impact on fisheries in many ways, including through direct impacts on the reproductive potential of target and otherwise directly impacted species.

One of the most obvious examples of ecosystem impacts on fisheries is that the reproductive biology of many fish species is linked to specific habitats, and any factor affecting that habitat can influence reproductive success. The destruction of critical habitats is a major problem in many parts of the world. In the western central Atlantic, for example, there are concerns about damage to coral reefs, seagrass beds, mangroves, coastal lagoons, and other habitats. The causes of the problems include pollution, disease, anchor damage, destructive fishing and tourism practices, hurricane damage, overloading by nutrients, and sediments from poor land-use (Cochrane 2005). These habitats play critical roles in the reproductive and recruitment processes of a many species that are important to fisheries and biodiversity conservation. Management needs to recognize this role and take steps to ensure that sufficient habitat is maintained to support the sustained productivity of fish resources. Without this, both r and K will certainly decline in affected species.

The importance of critical habitats for reproduction is well recognized in the FAO international guidelines for the management of deep-sea fisheries (FAO 2009). One of the criteria to be applied in identifying vulnerable marine ecosystems (VMEs) is whether an area provides habitats that are necessary for spawning, reproduction or specific life-history stages, such as nursery or rearing areas. The guidelines specify that states and regional fisheries management organizations or arrangements should take appropriate steps to prevent significant adverse impacts on VMEs, including the closing of areas if necessary (FAO 2009).

Taking a broader ecosystem approach can also lead to consideration of the impact of environmental changes on reproductive success of fished species, and the implications of this for fisheries. Many large fish populations undergo considerable medium- to long-term variability in abundance and productivity, driven by environmental variability (Bakun 1998, Maunder 1998). Many studies have attempted to understand the environmental processes involved, and to include them in management decisions and actions. In one case, the South African anchovy fishery where young-of-the year make up a substantial part of the catch, simulation studies indicated that, with the management procedure then being used, it would be possible to increase the average annual catch of anchovy by 48% for the same level of risk if it were possible to predict recruitment, with a coefficient of variation of 0.1, at the start of the fishing season (Cochrane & Starfield 1992).

The problems inherent in using environmental information in stock–recruitment relationships, as discussed earlier in this chapter, are clearly also relevant here. Those problems were the risk that an observed correlation may be spurious, and the statistical consequences of including too many variables in a statistical model (Hilborn & Walters 1992). Despite the clear economic and conservation gains that could be obtained from being able to forecast future recruitment, there are only a few examples in which environmental information as a predictor of, for example, recruitment or breeding success is used in practice in fisheries management and assessment (MacKenzie *et al.* 2008). These and related issues are considered in greater depth in Chapter 10 of this volume.

9.6 Climate change and variability

The widespread impacts of climate change and longer-term climate variability on fish and fisheries are now widely recognized. Two of the best-studied examples are the poleward shifts in the distribution of fish in response to increases in temperature over the past several decades in the North East Atlantic (e.g., Perry *et al.* 2005, Brander 2010, IPCC 2014a), and the similar trends being observed in South East Australia where the Eastern Australian Current has been extending further south than previously (Last *et al.* 2011, IPCC 2014a). There are many other examples around the world (IPCC 2014a,b,c). These changing temperatures have led to changes in the species composition of global catches, with catches of warm water species having increased at higher latitudes, while the relative proportion of subtropical species in catches from the tropics has decreased (Cheung *et al.* 2013).

The mechanisms through which climate change drives the abundance and distribution of fish have been well studied, and include effects on the recruitment, growth, mortality, abundance, and distribution of fish stocks (IPCC 2014b), of which the effects on recruitment are most relevant to this chapter. Based on a review of published information on fish in the North East Atlantic, Rijnsdorp *et al.* (2009) concluded that, while there is still uncertainty about the pathways of climate change impacts on fish, the evidence indicates that impacts on recruitment are a key factor. The mechanisms through which sea temperature rise leads to changes in recruitment include impacts on growth, adult reproductive performance, and reproductive success, changes in mortality of early life stages (through for example direct temperature effects, effects on matches and mismatches in food-web dynamics, and retention and transport dynamics), as well as at other life stages (Pörtner & Peck 2010). Pörtner & Peck (2010) have reported that the relationship between water temperature and recruitment is generally dome-shaped in both demersal and pelagic species, with an optimal recruitment range, above and below which recruitment success declines.

One example of these impacts is the strong link that has been identified between climate and recruitment in the Arcto-Norwegian cod *Gadus morhua* stock in the Barents Sea, with serious implications for fisheries management (Ottersen *et al.* 2006). The survival of individuals of this species through the early life stages is enhanced at higher temperatures, resulting in a variability in recruitment. The authors argued that the reduction in age and length composition of the spawning stock between the 1940s and 1990s had led to the cod population being less resilient to unfavourable climate conditions, which had increased the effect of climate on recruitment. Subsequently, Ottersen and colleagues suggested that this pattern is likely to be found also in other systems – a conclusion supported by other studies. In a review of the sensitivity of marine systems to climate change, Perry *et al.* (2010) noted that fishing tends to result in a loss of the older age classes, a reduction in area covered by a stock, a loss of sub-stocks, and the alteration of life histories. These consequences, together with a typical reduction in mean trophic level of communities, will in turn lead to a reduction in turnover times and result in heavily fished populations tracking climate variability more closely than would otherwise be the case, and also being more sensitive to climate change and variability.

The nature and direction of the impacts of climate varies from species to species and stock to stock, which means that changes in any particular area will result in differential shifts in the competitiveness of the different species, and therefore their relative abundance and productivity. An effect of this differential response is that species with a wider thermal tolerance, short generation times and high variability in phenotypes will be favoured by climate change (Pörtner & Farrell 2008, IPCC 2014c).

Climate change is therefore a further factor impacting reproductive dynamics and stock abundance, adding to the impacts of fisheries and other human-induced changes in habitat and water quality. In those areas where it has led to, or is projected to lead to, significant changes it adds an additional uncertainty for fisheries management. A key consideration for management – and hence for the provision of advice for management – in fisheries where climate change is taking place must be to ensure that the resilience of fish stocks, communities, and ecosystems is maintained and, where already impacted, is restored (Cochrane *et al.* 2009, Perry *et al.* 2010, IPCC 2014a). The maintenance of an adequate spawner biomass with sufficient age, genetic, and spatial structure to cope with environmental variability is a critical component of ensuring resilience. Maintaining and, where required, restoring stock and ecosystem resilience requires the implementation of an ecosystem approach in management, which in turn requires that reasonable flexibility is built into the management system and the fishery sector to allow for rapid response, as needed, to any changes driven by climate (Cochrane *et al.* 2009, IPCC 2014a). It has been proposed that, while the existing management measures remain relevant and necessary under climate change, they need to be applied within systems that strive to encourage resilient ecosystems and that explicitly incorporate consideration of uncertainties about future conditions and the effect of adaptation (Grafton 2010).

9.7 Conclusions and future directions

The conclusion of the first edition of this chapter, that fisheries management has made use of knowledge about the reproductive biology of fishes in different ways and at different levels of aggregation for centuries, has not changed. The three partial theories that have underpinned fisheries assessment and management in the post-industrial era of fisheries, as described by Smith (1994), remain central to resource management in nearly all fisheries, but it is clear from this updated chapter that even in the five years that have elapsed since the 2009 publication, the increased understanding of reproductive processes and complexities is changing perspectives in assessment and management, although practice is lagging behind.

The fundamental principle in fisheries resource management of ensuring that sufficient adult biomass remains in the population to ensure that recruitment is maintained remains paramount, as evidenced by the reference points derived from all of the partial theories and endorsed in international and national fisheries laws and regulations.

Notwithstanding the criticisms directed against conventional fisheries management, it has had notable successes and it is not too much of an exaggeration to state that, where conventional management has been seriously applied it has, with few exceptions, been more or less successful. It should certainly not be discarded and, despite revolutionary rhetoric and protest, no radically different alternatives, supported by rigorous theoretical or empirical evidence, have been proposed. The conclusion reached 15 years ago by the NRC (1999) panel that good progress could be made towards achieving sustainable marine fisheries by the adoption of a conservative, risk-averse, single-species approach should remain, in this author's view, the fall-back position in all but the most species-diverse situations. Even in highly species-diverse cases, some or other derivation of a conventional single-species approach, for example monitoring and reacting to trends in indicator species, appears to be the only reliable option at present.

The 2009 edition highlighted a number of cracks of varying size in the current approaches, even in the many cases of relatively successful conventional management. Scientists and managers have to operate with the human and financial resources available to them, including the best available information, but it is essential that they are aware of the cracks and the risks the advice could present. As a minimum, the performance of a management system must be monitored carefully against target and limit reference points, and management measures adjusted when necessary in an adaptive framework and with a precautionary approach. This would go a considerable way to minimizing the risks of unexpected developments and changes, even with imperfect models and incomplete understanding of why the management measures are not performing as initially expected. At the same time, the reasons for the deviations from expectation, which may be caused by cracks in the partial theories of stock dynamics, need to be investigated and further refinements to the management system introduced as the underlying cause or causes of problems are better identified. An approach that combines tactical adaptation with strategic learning becomes particularly important and achievable as fisheries management, at least in some cases, predominantly in developed countries, gives greater attention to long-term strategies, as suggested by O'Brien in Chapter 10 of this volume.

Some of the more important, or better-known, weaknesses of the current approaches have been described in this chapter. These include:

- A common failure to ensure that the sex ratio of the exploited fish population does not depart excessively from the unexploited sex ratio.
- Allowing the size and age structure of the population to be truncated to a point at which recruitment may be seriously reduced through excess reduction in the number or quality of reproductive outputs.
- Insufficient attention to the spatial distribution of a population and the extent to which sustained recruitment and population responses to environmental variability are dependent on the undisturbed spatial distribution.
- Insufficient knowledge of the genetic complexity of populations and the importance of maintaining adaptive diversity as a critical component of the resilience of populations.
- Related to the previous point, a lack of recognition of the potential consequences for reproductive output and success of the selective removal of productive individuals.
- The many potential or actual problems that arise if the interactions between a fished stock and the ecosystem in which it occurs are not taken into account. For example, in the context of reproductive biology, the maintenance of a critical habitat for different stages in the reproductive process, or ensuring the maintenance of adequate food organisms or sources for each life stage.
- Insufficient attention to the maintenance of resilient populations and ecosystems, and failures to adjust assessment and management measures in response to the often slow and difficult-to-discern impacts of climate change and longer-term variability on reproduction and recruitment.

To address these problems does not require a departure from a conventional approach, but does require the inclusion of additional targets and, frequently, additional management measures or refinement of existing measures. The hurdles to adding greater complexity are, of course, the same hurdles that have frequently created problems in the provision of high-quality scientific advice, even under the simplified partial theories and effective implementation of that advice. Those hurdles include inadequate human and financial capacity for monitoring, research and implementation (Cochrane & Doulman 2005). Where there are substantial capacity constraints, care needs to be taken to avoid creating new cracks on old ground by compromising conventional but essential assessment and management functions and tools through simply diverting capacity from those to the recently emerging considerations. Rather, recognition of the greater complexity in resource characteristics and dynamics requires greater attention to monitoring management performance and informed prioritization of scientific and management tasks to address shortcomings in management.

Depending on needs and priorities, actions to address the shortcomings in the current approaches could include additional research into the impacts of sex ratio and population size structure on recruitment as the basis for setting essential and realistic reference points for the structure of the populations. Similarly, if there is evidence that fishing or other human activities have significantly reduced or are reducing the distribution of a population, the risk that recruitment could have been or will be adversely affected should be recognized and precautionary measures put in place. This could entail the addition of spatial objectives, reference points, and management measures to existing approaches to halt or reverse the decline in distribution.

Management responses to the selective removal of reproductively productive individuals or of genetic subcomponents of a population may need to address one or more of at least three attributes depending on the biology and ecology of the fished population: (i) the maintenance of or restoration to a population size structure that is close to that of the unfished structure; (ii) maintaining the spatial structure of the population at different life stages; and (iii) avoiding selective fishing that could affect any differential timing of spawning in the population. These may not all be necessary in any particular case, but a precautionary approach could be to ensure that fishing mortality as a whole is kept appropriately low and well-distributed over time and space so as to distribute selectivity more or less evenly across the key genetic variables in the population. Clearly, in addition, extra caution will be required at critical times and places in the life history.

The consequences of failing to implement an ecosystem approach could include a loss of critical habitat from fishing or other human impacts, adverse impacts on food sources for different life stages or undesirable changes in predator, prey or competitor species. Further, blindly trying to manage a stock by regulating fishing mortality while ignoring impacts from outside the fishery, for example from climate change or other human impacts, is bound to fail.

The 2009 version of this chapter concluded that, while ecosystem issues are not inherently more important or urgent in managing fishing on the target species than the other neglected factors presented here, ecosystem factors may be the ones most likely to be overlooked in a short-sighted application of the partial theories. In the view of the author, this conclusion still applies. Further, the effective implementation of an ecosystem approach, which must include precautionary adaptive management based on the best-available scientific information, should lead to addressing – whether directly or indirectly – the other currently neglected factors of size structure and reproductive capacity, genetic diversity, spatial structure, and impacts of climate change.

References

Agnew, D.J., Hill, S. & Beddington, J.R. (2000) Predicting the recruitment strength of an annual squid stock: *Loligo gahi* around the Falkland Islands. *Canadian Journal of Fisheries and Aquatic Sciences*, **57**, 2479–87.

Arregui, I., Arrizabalaga, H., Kirby, D.S. & Martín-González, J.M. (2006) Stock-environment–recruitment models for North Atlantic albacore (*Thunnus alalunga*). *Fisheries Oceanography* **15**, 402–12.

Bakun, A. (1998) Ocean triads and radical interdecadal variation: bane and boon to scientific fisheries management. In T.J. Pitcher, P.J.B. Hart & D. Pauly (Eds) *Reinventing Fisheries Management.* (Proceedings of Symposium held in Vancouver, Canada, February 20–24, 1996). pp. 331–58. Chapman & Hall, London.

Bakun, A. & Parrish, R. (1982) Turbulence, transport and pelagic fish in the California and Peru current systems. *California Cooperative for Oceanic Fisheries Investigative Reporting*, **23**, 99–112.

Basson, M., Beddington, J.R., Crombie, J.A., Holden, S.J., Purchase, L.V., Tingley, G.A. (1996) Assessment and management techniques for migratory annual squid stocks: the *Illex argentinus* fishery in the Southwest Atlantic as an example. *Fisheries Research*, **28**, 3–27.

Berkeley, S.A., Hixon, M.A., Larson, R.J. & Love, M.S. (2004) Fisheries sustainability via protection of age structure and spatial distribution of fish populations. *Fisheries*, **29**, 23–32.

Beverton, R.J.H. & Holt, S.J. (1957) *On the dynamics of exploited fish populations.* Fisheries Investigations Series II, **19**. MAFF, London. 533 pp.

Birkeland, C. & Dayton, P.K. (2005) The importance in fishery management of leaving the big ones. *Trends in Ecology and Evolution*, **20**, 356–8.

Bjordal, Å. (2009) Regulation of Fishing Gears and Methods. In: K.L. Cochrane & S.M. Garcia (Eds) *A Fishery Manager's Guidebook, 2nd Edition*. pp. 367–95. FAO and Wiley-Blackwell Publishers.

Blanchard, J.L., Frank, K.T. & Simon, J.E. (2003) Effects of condition on fecundity and total egg production of eastern Scotian Shelf haddock (*Melanogrammus aeglefinus*). *Canadian Journal of Fisheries and Aquatic Sciences*, **60**, 321–32.

Booth, A., Cochrane, K.L., Hackett, A. & Shepherd, D. (2000) Application of a multispecies-multigear per recruit model that incorporates parameter variability to the shrimp and groundfish fisheries of Guyana. In: *Report of the Third Workshop on the Assessment of Shrimp and Groundfish Fisheries on the Brazil–Guianas Shelf.* FAO Fisheries Technical Report, **628**, 154–72.

Brander, K. (2010) Impacts of climate change on fisheries. *Journal of Marine Systems*, **79**, 389–402.

Buckland, F. (1880) Preface. In: *Natural history of British Fishes: Their Structure, Economic Uses, and Capture by Net and Rod.* Society for Promoting Christian Knowledge, London. http://home.planet.nl/~zoete004/buckl&.htm. Accessed 16 March 2005.

Butterworth, D.S., Cochrane, K.L. & de Oliveira, J.A.A. (1997) Management procedures: a better way to manage fisheries? The South African experience. In: E.K. Pikitch, D.D. Huppert & M.P. Sissenwine (Eds) *Global Trends: Fisheries Management* (Proceedings of the Symposium held in Seattle, Washington, 14–16 June, 1994). pp. 83–90. American Fisheries Society Symposium **20**, Bethesda, Maryland.

Caddy, J.F. (1999) Fisheries management in the twenty-first century: will new paradigms apply? *Reviews in Fish Biology and Fisheries*, **9**, 1–43.

Caddy, J.F. & Cochrane, K.L. (2001) A review of fisheries management past and present and some future perspectives for the third millennium. *Ocean & Coastal Management*, **44**, 653–82.

Cheung, W.W.L., Watson, R. & Pauly, D. (2013) Signature of ocean warming in global fisheries catches. *Nature*, **497**, 365–8.

Cochrane, K.L. (2000) Reconciling sustainability, economic efficiency and equity in fisheries: the one that got away? *Fish and Fisheries*, **1**, 3–21.

Cochrane, K.L. (2005) Western Central Atlantic, FAO Statistical Area 31. In: *Review of the state of world marine fishery resources.* FAO Fisheries Technical Paper, **457**, 31–42.

Cochrane, K.L. (2007) Marine protected areas as management measures: tools or toys? In: M.H. Nordquist, R. Long, T.H. Heidar & J.N. Moore (Eds) *Law, Science & Ocean Management.* pp. 701–37. Proceedings of the 30th Virginia Law of the Sea Conference, Dublin, Ireland, 12–14 July 2006. Martinus Nijhoff.

Cochrane, K.L. (2009) The use of scientific information. In: K.L. Cochrane & S.M. Garcia (Eds) *A Fishery Manager's Guidebook, 2nd Edition*. pp. 336–69. FAO and Wiley-Blackwell Publishers.

Cochrane, K.L., Butterworth, D.S., de Oliveira, J.A.A. & Roel, B.A. (1998) Management procedures in a fishery based on highly variable stocks and with conflicting objectives: experiences in the South African pelagic fishery. *Reviews in Fish Biology and Fisheries*, **8**, 177–214.

Cochrane, K.L., De Young, C., Soto, D. & Bahri, T. (Eds) (2009) Climate change implications for fisheries and aquaculture. Overview of current scientific knowledge. *FAO Fisheries and Aquaculture Technical Paper*, **530**. FAO, Rome. 212 pp.

Cochrane, K.L. & Doulman, D. (2005) The rising tide of fisheries instruments and the struggle to keep afloat. *Transactions of the Royal Society Series B*, **360** (1453), 77–94.

Cochrane, K.L., Emerson, W. & Willmann, R. (2011) Sustainable fisheries: the importance of the bigger picture. In: W.W. Taylor, A.J. Lynch & M.G. Schechter (Eds) *Sustainable fisheries: multi-level approaches to a global problem.* pp. 3–19. American Fisheries Society, Bethesda Maryland.

Cochrane, K.L. & Hutchings, L. (1995) A structured approach to using biological and environmental parameters to forecast anchovy recruitment. *Fisheries Oceanography*, **4**, 102–27.

Cochrane, K.L. & Starfield, A.M. (1992) The potential use of predictions of recruitment success in the management of the South African anchovy resource. In: A.I.L. Payne, K.H. Brink, K.H. Mann

& R. Hilborn (Eds) *Benguela Trophic Functioning. South African Journal of Marine Science*, **12**, 891–902.

Conover, D.O., Munch, S.B. & Arnott, S.A. (2009) Reversal of evolutionary downsizing caused by selective harvest of large fish. *Proceedings of the Royal Society B: Biological Sciences*, **276**, 2015–20.

Conover, D.O., Arnott, S.A., Walsh, M.R. & Munch S.B. (2005) Darwinian fishery science: lessons from the Atlantic silverside (*Menidia menidia*). *Canadian Journal of Fisheries and Aquatic Sciences*, **62**, 730–7.

Conover, D.O. & Munch, S.B. (2002) Sustaining fisheries yields over evolutionary time scales. *Science*, **297**, 94–6.

DAFF (2012) Status of the South African marine fishery resources. Department of Agriculture, Forestry and Fisheries, Cape Town. 75 pp.

de Moor, C.L. & Butterworth, D.S. (2009) The 2004 re-assessment of the South African sardine and anchovy populations to take account of revisions to earlier data and recent record abundances. *African Journal of Marine Science*, **31**, 333–48.

Deriso, R.B. (1987) Optimal $F_{0.1}$ criteria and their relationship to maximum sustainable yield. *Canadian Journal of Fisheries and Aquatic Sciences*, **44** (Suppl. 2), 339–48.

Edeline, E., Carlson, S.M., Stige, L.C., Winfield, I.J., Fletcher, J.M., James, J.B., Haugen, T.O., Vøllestad, L.A. & Stenseth, N.C. (2007) Trait changes in a harvested population are driven by a dynamic tug-of-war between natural and harvest selection. *Proceedings of the National Academy of Sciences of the United States of America*, **104**, 15799–804.

FAO (1997) Fisheries management. *FAO Technical Guidelines for Responsible Fisheries*. No. 4. FAO, Rome. 82 pp.

FAO (2003) The ecosystem approach to fisheries. *FAO Technical Guidelines for Responsible Fisheries*. No. 4, Supplement 2. FAO, Rome.

FAO (2005) Review of the state of world marine fisheries resources. *FAO Fisheries Technical Paper*, **457**, FAO, Rome. 235 pp.

FAO (2009) International guidelines for the management of deep-sea fisheries in the high seas. FAO, Rome. 73 pp.

FAO (2011) Fisheries management. 4. Marine protected areas and fisheries. *FAO Technical Guidelines for Responsible Fisheries*. No. **4**, Supplement 4. FAO, Rome. 198 pp.

FAO (2014) The State of World Fisheries and Aquaculture 2014. Rome. 223 pp.

Gabriel, W.L. & Mace, P.M. (1999) *A Review of Biological Reference Points in the Context of the Precautionary Approach*. Proceedings, 5th NMFS National Stock Assessment Workshop. 1999. NOAA Technical Memorandum. NMFS–F/SPO–40. 12 pp.

Garcia, S.M. & Cochrane, K.L. (2009) From past management to future governance: a perspective view. In K.L. Cochrane & S.M. Garcia (Eds) *A Fishery Manager's Guidebook, 2nd Edition*. pp 447–72. FAO and Wiley-Blackwell Publishers.

Gell, F.R. & Roberts, C.M. (2003) Benefits beyond boundaries: the fishery effects of marine reserves. *Trends in Ecology and Evolution*, **18**, 448–54.

Grafton, R.Q. (2010) Adaptation to climate change in marine capture fisheries. *Marine Policy*, **34**, 606–15.

GrahamM. (1935) Modern theory of exploiting a fishery and application to North Sea trawling. *Conseil International pour Exploration de la Mer*, **10**, 264–74.

Hackett, A., Cochrane, K.L. & Booth, A. (2000) Assessment of fisheries in Guyana for bangamary (*Macrodon ancylodon*) and butterfish (*Nebris microps*). In: *Report of the Third Workshop on the Assessment of Shrimp and Groundfish Fisheries on the Brazil–Guianas Shelf*. FAO Fisheries Technical Report, **628**, 110–23.

Hall, S. (2009) Area and time restrictions. In: K.L. Cochrane & S.M. Garcia (Eds) *A Fishery Manager's Guidebook, 2nd Edition*. pp 196–219. FAO and Wiley-Blackwell Publishers.

Halliday, R.G. & Pinhorn, A.T. (2002) A review of the scientific and technical basis for policies on the capture of small fish in North Atlantic groundfish fisheries. *Fisheries Research*, **57**, 211–22.

Hauser, L., Adcock, G.J., Smith, P.J., Bernal Ramirez, J.H. & Carvalho. G.R. (2002) Loss of microsatellite diversity and low effective population size in an overexploited population of New Zealand snapper (*Pagrus auratus*). *Proceedings of the National Academy of Science of the United States of America*, **99**, 11742–7.

Hauser, L. & Carvalho, G.R. (2008) Paradigm shifts in marine fisheries genetics: ugly hypotheses slain by beautiful facts. *Fish and Fisheries* **9**, 333–62.

Hilborn, R. (2010) Pretty good yield and exploited fishes. *Marine Policy*, **34**, 193–6.

Hilborn, R. & Walters, C.J. (1992) *Quantitative Fisheries Stock Assessment. Choice, Dynamics and Uncertainty.* Chapman & Hall, New York. 570 pp.

Huchette, S.M.H., Soulard, J.P., Koh, C.S. & Day, R.W. (2004) Maternal variability in the blacklip abalone, *Haliotis rubra* leach (mollusca: Gastropoda): effect of egg size on fertilisation success. *Aquaculture*, **231**, 181–95.

Hutchings, J.A. (2005) Life history consequences of overexploitation to population recovery in Northwest Atlantic cod (*Gadus morhua*). *Canadian Journal of Fisheries and Aquatic Sciences*, **62**, 824–32.

ICES (2004) Report of the ICES Advisory Committee on Fishery Management and Advisory Committee on Ecosystems, 2004. pp. 1–4. http://www.ices.dk/products/icesadvice/Book1Part1.pdf. Last accessed 22 March 2005.

ICCAT (2005) Stock assessments. http://www.iccat.es/assess.htm. Last accessed 21 May 2005.

ICCAT (2014) Species summary reports obtained from 'Stock assessments', http://www.iccat.es/assess.htm. Last accessed 14 August 2014.

IPCC (2014a) Chapter 7. Food Security and Food Production Systems. Fifth Assessment Report, Intergovernmental Panel on Climate Change. http://ipcc-wg2.gov/AR5/images/uploads/WGIIAR5-Chap7_FGDall.pdf. Last accessed 14 August 2014.

IPCC (2014b) Chapter 30. The Ocean. Fifth Assessment Report, Intergovernmental Panel on Climate Change. http://ipcc-wg2.gov/AR5/images/uploads/WGIIAR5-Chap30_FGDall.pdf. Last accessed 14 August 2014.

IPCC (2014c) Chapter 6. Ocean Systems. Fifth Assessment Report, Intergovernmental Panel on Climate Change. http://ipcc-wg2.gov/AR5/images/uploads/WGIIAR5-Chap6_FGDall.pdf. Last accessed 14 August 2014.

Jennings, S., Kaiser, M.J. & Reynolds, J.D. (2001) *Marine Fisheries Ecology*. Blackwell Science, Oxford. 417 pp.

Kruse, G.H., Bez, N., Booth, A., Dorn, M.W., Hills, S., Lipcius, R.N., Pelletier, D., Roy, C., Smith, S.J. & Witherell, D. (Eds) (2001) *Spatial processes and management of marine populations*. University of Alaska Sea Grant, AK–SG–01–02, Fairbanks. 720 pp.

Larkin, P.A. (1977) An epitaph for the concept of maximum sustainable yield. *Transactions of the American Fisheries Society*, **106**, 1–11.

Lasker, R. (1985) What limits clupeoid production? *Canadian Journal of Fisheries and Aquatic Sciences*, **42** (Suppl. 1), 31–8.

Last, P., White, W., Gledhill, D., Hobday, A., Brown, R., Edgar, G. & Pecl, G. (2011) Long term shifts in abundance and distribution of a temperate fish fauna: a response to climate change and fishing practices. *Global Ecology and Biogeography*, **20**, 58–72.

Mace, P.M. (2001) A new role for MSY in single-species and ecosystem approaches to fisheries stock assessment and management. *Fish and Fisheries*, **2**, 2–32.

Mace, P.M. (2004) In defence of fisheries scientists, single species models and other scapegoats: confronting the real problems. *Marine Ecology Progress Series*, **274**, 269–303.

Mace, P.M. & Sissenwine, M.P. (1993) How much spawning per recruit is enough? In: S.J. Smith, J.J. Hunt, & D. Rivard (Eds) *Risk evaluation and biological reference points for fisheries management*. pp. 101–118. *Canadian Special Publication in Fisheries and Aquatic Sciences*, **120**.

MacKenzie, B.R., Horbowy, J. & Köster, F.W. (2008) Incorporating environmental variability in stock assessment: predicting recruitment, spawner biomass, and landings of sprat (*Sprattus sprattus*) in the Baltic Sea. *Canadian Journal of Fisheries and Aquatic Sciences*, **65**, 1334–441.

Marteinsdottir, G. & Thorarinsson, K. (1998) Improving the stock-recruitment relationship in Icelandic cod (*Gadus morhua*) by including age diversity of spawners. *Canadian Journal of Fisheries and Aquatic Sciences*, **55**, 1372–7.

Maunder, M.N. (1998) Problems with an environmental-based recruitment index: examples from a New Zealand snapper assessment (*Pagrus auratus*). In: F. Funk, T.J. QuinnII, J. Heifetz, J.N. Ianelli, J.E. Powers, J.F. Schweigert, P.J. Sullivan & C.-I. Zhang (Eds) *Fishery Assessment Models*. University of Alaska Sea Grant, AK–SG–98–01, Fairbanks, pp. 679–92.

Murawski, S.A., Brown, R., Lai, H.L., Rago, P.J. & Hendrickson, L. (2000) Large-scale closed areas as a fisheries management tool in temperate marine systems: the Georges Bank experience. *Bulletin of Marine Science*, **66**, 775–98.

Nicolson, J. (1997) *Food from the Sea*. Cassell, London. 205 pp.

NRC (1999) *Sustaining marine fisheries*. National Academy Press. Washington D.C. 164 pp.

OECD (1997) *Towards Sustainable Fisheries. Economic Aspects of The Management of Living Marine Resources*. OECD, Paris. 268 pp.

Ottersen, G., Hjermann, D.O. & Stenseth, N.C. (2006) Changes in spawning stock structure strengthen the link between climate and recruitment in a heavily fished cod (*Gadus morhua*) stock. *Fisheries Oceanography*, **15** (3), 230–43.

Parsons, L.S. & Beckett, J.S. (1997) Fisheries management in Canada: the case of Atlantic groundfish. In: E.K. Pikitch, D.D. Huppert & M.P. Sissenwine (Eds) *Global Trends: Fisheries Management*. (Proceedings of the Symposium held in Seattle, Washington, 14–16 June, 1994). pp. 73–9. American Fisheries Society Symposium **20**, Bethesda, Maryland.

Perry, A.L., Low, P.J., Ellis, J.R. & Reynolds, J.D. (2005) Climate change and distribution shifts in marine fishes. *Science*, **308**, 1912–15.

Perry, R.I., Cury, P., Brander, K., Jennings, S., Möllmann, C. & Planque, B. (2010) Sensitivity of marine systems to climate and fishing: concepts, issues and management responses. *Journal of Marine Systems*, **79** (3–4), 427–35.

Piet, G.J. & Rice, J.C. (2004) Performance of precautionary reference points in providing management advice on North Sea fish stocks. *ICES Journal of Marine Science*, **61**, 1305–12.

Pörtner, H.-O. & Farrell, A.P. (2008) Ecology: Physiology and climate change. *Science*, **322** (5902), 690–2.

Pörtner, H.-O. & Peck, M.A. (2010) Climate change effects on fishes and fisheries: towards a cause-and-effect understanding. *Journal of Fish Biology*, **77** (8), 1745–79.

Punt, A.E. & Hilborn, R. (1996) Biomass dynamic models. User's manual. *FAO Computer Information Series,* **10**. FAO, Rome. 62 pp.

Rademeyer, R.A., Butterworth, D.S. & Plagányi, É. E. (2008) Assessment of the South African hake resource taking its two-species nature into account. *African Journal of Marine Science*, **30**, 263–90.

Rätz, H.-J. & Lloret, J. (2003) Variations in fish condition between Atlantic cod (*Gadus morhua*) stocks, the effect on their productivity and management implications. *Fisheries Research*, **60**, 369–80.

Rijnsdorp, A.D., Peck, M.A., Engelhard, G.H., Möllmann, C. & Pinnegar, J. K. (2009) Resolving the effect of climate change on fish populations. *ICES Journal of Marine Science*, **66**, 1570–83.

Ross, M.R. (1997) *Fisheries Conservation and Management*. Prentice-Hall, Upper Saddle River, New Jersey. 374 pp.

Russell, M. (2001) *Spawning aggregations of reef fishes on the Great Barrier Reef: implications for management*. Fishery Issues Group, Great Barrier Reef Marine Park Authority, Queensland, Australia.

Sadovy, Y. & Domeier., M. (2005) Are aggregation-fisheries sustainable? Reef fish fisheries as a case study. *Coral Reefs*, **24**, 254–62.

Sadovy, Y., Kulbicki, M., Labrosse, P., Letourneur, Y., Lokani, P. & Donaldson, T.J. (2003) The humphead wrasse, *Cheilinus undulatus*: synopsis of a threatened and poorly known giant coral reef fish. *Review of Fish Biology and Fisheries*, **13**, 327–64.

Sadovy de Mitcheson, Y., Cornish, A., Domeier, M., Colin, P.L, Russell, M. & Lindeman, K.C. (2008) A global baseline for spawning aggregations of reef fishes. *Conservation Biology*, **22**, 1233–44.

Schaefer M.B. (1954) Some aspects of the dynamics of populations important to the management of the commercial marine fisheries. *Inter-American Tropical Tuna Commission Bulletin*, **1**, 25–56.

Shannon, L.V., Crawford, R.J.M., Pollock, D.E., Hutchings, L., Boyd, A.J., Taunton-Clark, J., Badenhorst, A., Melville-Smith, R., Augustyn, C.J., Cochrane, K.L., Hampton, I., Nelson, G., Japp, D.W. & Tarr, R.J.Q. (1992) The 1980s – a decade of change in the Benguela ecosystem. *South African Journal of Marine Science*, **12**, 271–96.

Smith, P.E. (1985) Year-class strength and survival of 0-group clupeoids. *Canadian Journal of Fisheries and Aquatic Sciences*, **42** (Suppl. 1), 69–82.

Smith T.D. (1994) *Scaling fisheries*. Cambridge University Press, Cambridge. 392 pp.

Thompson, W.F. & Bell, F.H. (1934) Biological statistics of the Pacific halibut fishery. 2. Effects of changes in intensity upon total yield and yield per unit of gear. *Report of the International Fisheries Commission*, **8**. 49 pp.

UNCSD (2012) The Future We Want: Outcome document adopted at Rio+20. United Nations Conference on Sustainable Development, Rio de Janeiro, Brazil, June 2012. http://www.uncsd2012.org/content/documents/727The%20Future%20We%20Want%2019%20June%201230pm.pdf. Last accessed 9 August 2014.

Valdés Szeinfeld, E.S. & Cochrane, K.L. (1992) The potential effects of cannibalism and intraguild predation on anchovy recruitment and clupeoid fluctuations. In: A.I.L. Payne, K.H. Brink, K.H. Mann & R. Hilborn (Eds) *Benguela Trophic Functioning. South African Journal of Marine Science.* **12**, 695–702.

Wileman, D.A., Ferro, R.S.T., Fonteyne R. & Millar, R.B. (1996) Manual of methods of measuring the selectivity of towed fishing gears. *ICES Cooperative Research Report*, **215**. 126 pp.

Worm, B., Hilborn, R., Baum, J. *et al.* (2009) Rebuilding global fisheries. *Science*, **325**, 578–85.

Chapter 10
Management: New Approaches to Old Problems

Carl M. O'Brien

10.1 Introduction

10.1.1 Background and context

In the earlier chapters of this book, the authors have argued that a full understanding of reproductive dynamics is critical for assessing the full impacts of harvesting on fish populations and in devising appropriate management strategies. Attempts to ascertain limits to exploitation and defining optimal harvesting strategies have typically been based on proxies of reproductive potential of stocks – most notably simple measures of the biomass of the adult population.

It has been argued that there is a need for measures of the actual reproductive capacity and output of the population, and that an understanding of these factors requires a detailed understanding of reproduction biology, behaviour and demographic characteristics of the population to determine an adequate measure of reproductive capacity – the production of viable eggs. In addition, there is a need to understand the factors that affect the survival through the early life stages before recruitment to the fishery. The confluence of factors affecting egg condition and environmental effects on survival are critical in this regard.

Accounting for these factors in management will place renewed emphasis on demographic and other characteristics of the stock. Attention to the age and size structure of the population, sex ratios, and other pertinent biological factors will lead to new ways of measuring the reproductive capacity and replace simpler measures such as the spawning stock biomass (SSB). Management tools specifically to address these issues will require a shift from simple considerations of a total allowable catch (TAC) to measures that are designed both to limit the catch and to control its demographic composition (e.g., Mangi *et al.* 2013). This may entail consideration of strategies such as marine protected areas to protect segments of the population, the development of more selective fishing gears (e.g., Colloca *et al.* 2013) and a whole suite of biological and technical measures (e.g., Little *et al.* 2014). Consideration of factors such as preserving multiple reproductive opportunities for individual females will become increasingly important.

Fish Reproductive Biology: Implications for Assessment and Management, Second Edition.
Edited by Tore Jakobsen, Michael J. Fogarty, Bernard A. Megrey and Erlend Moksness.

Furthermore, one can anticipate a shift towards increased emphasis on long-term management strategies and their evaluation (e.g., de Oliveira *et al.* 2008) from the current focus on short- and, in some cases, medium-term management. Critical to the success of such a shift will be the need for a proper understanding of the relation between harvest limits and true population processes (Pastoors *et al.* 2007). A full understanding of the stock–recruitment relationship will be essential in this endeavour and this has been discussed by other authors in this book.

The sentiments in these opening paragraphs of this chapter remain as appropriate and relevant today in this second edition, as at the time of publication of the first edition. The new Common Fisheries Policy (CFP) basic regulation entered into force on 1st January 2014. The new regulation (EU No. 1380/2013) makes major changes to the way that Europe's fisheries are managed, with the aim of managing fish stocks sustainably within a healthy marine environment. One of the most significant changes relates to the progressive introduction of a landing obligation, which prohibits the discarding of fish; necessitating the need to limit catch rather than landings *per se*. Quoting from Article 2(2) of the reformed CFP:

> *The CFP shall apply the precautionary approach to fisheries management, and shall aim to ensure that exploitation of living marine biological resources restores and maintains populations of harvested species above levels which can produce the maximum sustainable yield.*
>
> *In order to reach the objective of progressively restoring and maintaining populations of fish stocks above biomass levels capable of producing maximum sustainable yield, the maximum sustainable yield exploitation rate shall be achieved by 2015 where possible and, on a progressive, incremental basis at the latest by 2020 for all stocks.*

The target dates of either 2015 or 2020 relate solely to ensuring that the fishing mortality rate itself is set at the appropriate level. However, this supports the objective of progressively restoring and maintaining populations of fish stocks above biomass levels capable of producing maximum sustainable yield (MSY). Once again, in the short-term at least, the stock–recruitment relationship will remain fundamental to the future successful achievement of this objective.

10.1.2 Unravelling the strands

There is often considerable variation in recruitment for a given level of spawning stock biomass. This variation is frequently attributed to environmental effects on survival at particular life-history stages. However, there is increasing evidence that the age, size, and spatial structure of the spawning stock and the physiological condition of spawners can influence the number of surviving recruits.

The task of understanding and unravelling these strands is crucial and can best be considered as being divided into four components (O'Brien & Little 2006):

(1) Comprehension of processes and prediction of recruitment: approaches that aim at understanding the underlying processes that contribute to the recruitment process and, hence, have the potential for a future prediction of recruitment. This necessitates a consideration of biological processes, physical processes and environmental drivers (e.g., Andrews *et al.* 2006).

(2) Simulation of recruitment patterns: approaches that generate ensemble recruitment appropriate for simulation purposes. These may not *per se* be good for predicting recruitment, but may allow evaluation of recruitment uncertainty within fishery assessment and management models (O'Brien 1999a,b).
(3) Fishery management: incorporation of prediction models and simulation models into short-term predictions and medium-term projections of stock status. Quantification of appropriate levels of uncertainty and risk (e.g., Basson 1999, Roel *et al.* 2004).
(4) Single species versus multispecies: how much is recruitment variability attributable to multispecies effects such as predation/competition rather than to environmental drivers (e.g., Bogstad *et al.* 1994, Bogstad & Mehl 1997).

10.1.3 Marine systems

Marine systems are complex, and care is needed not to focus too narrowly on a single aspect or species. Too narrow a focus could mean that crucial links or factors are missed. Mechanisms may also be flexible and change, depending on a whole suite of parameters. A number of these aspects have been discussed in the earlier chapters of this book by other authors, and it is important in this chapter to ensure as wide a vision as possible.

Case studies that provide examples of how accounting for the effect of environmental variability has had an impact on the perception of the status of a stock through assessment, and subsequent fisheries management, could be instructive but by and large those studies that have been published are limited in their scope and applicability. O'Brien & Little (2006) documented three area-based case studies in the Baltic Sea, Barents Sea and North Sea, but these are far from complete and merely illustrate the difficulty of the task. The authors provided a brief summary of process studies in the three selected areas; touching on the potential for using the results from such studies in management advice.

Although it is not possible to control factors such as sea temperature and wind directly, changes in the environment clearly influence recruitment and the future size of fish stocks. A better understanding of the relationships between environment, parental stock and recruitment should contribute to the design of robust management for commercial fisheries. Research aimed at addressing these issues continues unabated.

Simulation models can play an important role in identifying whether and where benefits to management are most likely to accrue and where it would be best to focus attention in terms of other (e.g., process) studies.

10.1.4 Remainder of this chapter

The majority of this chapter is devoted to the presentation of a single well-developed and documented case study (Andrews *et al.* 2006), a population model of cod stocks in the waters around the United Kingdom (UK), incorporating the accumulated knowledge of the biology of the species and the environmental influences on recruitment and survival, in order to improve the basis for defining sustainable harvesting strategies. The example deals with the incorporation of reproductive biology; namely, those biological aspects on reproduction highlighted in the synopsis of this book; and recruitment considerations into management advice and strategies.

The case study is presented to encourage and motivate the development of similar studies in the future, but first a few words must be noted on a number of related pertinent issues.

10.2 Biological knowledge: Modelling, assessment, projections and management

10.2.1 Multispecies recruitment issues

The recruitment of any fish species depends in large part on mortality during the pelagic phase of its early life. Mortality will be determined not only by the physical environment but also potentially by the abundance of other creatures – predators and prey, for example. The principal mechanism is likely to be predation pressure from older larvae, juveniles, or adults of various species, including commercially important fish; other possibilities include changes in prey availability and competition. If environmental conditions change in the future, the abundance or distribution of predators and prey may also change. It is not obvious that levels of predation in the future will be the same as they have been historically, so it is not obvious that historically derived relationships between environment and recruitment will persist. Therefore, it is important to remember possible multispecies effects when considering environmental influences.

10.2.1.1 Barents Sea

Management of multispecies resources calls for considerations on the interaction between species caused by predation and food competition, so that changing fishing mortality on one stock may influence the production of other stocks (prey or competitors).

Barents Sea cod is a well-known example. Good recruitment requires warm temperatures, so a warming trend might suggest an increased recruitment in the future. However, good recruitment may demand a low abundance of key predatory species, so if the warming trend were also to lead to increased abundance of those key predatory species, the overall effect might be to decrease cod recruitment.

Furthermore, it is believed that, to a large extent, the state of the Barents Sea ecosystem will be revealed through the state of the stocks of Northeast Arctic cod, Barents Sea capelin, and Norwegian spring-spawning herring (Hamre 1994). Both, total fish production in the Norwegian–Barents Sea area (including Norwegian coastal waters) and other aspects of the total ecosystem are believed to be closely linked to the development of these stocks. Cod prey on capelin, herring and young cod (Bogstad & Mehl 1997), while herring is an important predator on capelin larvae (Huse & Toresen 2000). Cod growth is also affected by prey (especially capelin) abundance (Mehl & Sunnanå 1991). All three stocks show large variations in recruitment. Multispecies interactions are now routinely included in the single-species assessment and management of the capelin, cod and haddock stocks in the Barents Sea.

10.2.1.2 Interactions

Predation among larvae is less well studied, but is also potentially important. In the North Sea, stock–recruitment fits can be greatly improved by including other species' SSBs as covariates (via the exponent of a Ricker model); this can represent either adult predation or intralarval predation. Unfortunately, there are many models for possible interaction, and results from multispecies forecasts depend critically on which model is selected. Unless the range of likely interactions can be narrowed, there will be little benefit in using complex multispecies models for projection. The same may apply to projections incorporating environmental links to single-species recruitment.

To narrow the range of likely interaction, one needs information on processes. There are potentially many sources of biological/physical information that could be useful, including diet, spawning times and places, and stock sizes. One way to start would be to use hydrographic models to predict likely overlaps – or to eliminate unlikely overlaps – in larval/adult distribution, at sizes when predation is likely. Note that it is not necessary to understand in detail every stage of the population dynamics; once a plausible set of interactions has been established, empirical data analysis (multispecies stock–recruitment relationships) can be used to parametrise the relationship. However, information on which interactions are plausible and how strong they might be is critical.

10.2.2 Assessing the effects of incorporating process information in assessments and projections

Routine stock assessments are sometimes criticised for failing to include biological information, particularly in relation to reproductive parameters (Trippel 1999). There are undoubtedly many cases where this is a valid criticism. However, although it seems intuitive that the inclusion of additional biological knowledge should improve the assessments, as with all assumptions made in the stock assessment process, it is appropriate to verify that this assumption is correct.

First, to take a hypothetical example, a fishery might be closed, based on the estimated SSB of the stock falling below some pre-set level. If a more biologically detailed measure of the stock's reproductive potential were used instead of simple SSB, then it is possible that the decision to close the fishery may have been reached at a different time. However, it is likely that, for subsequent management of the fishery, the two approaches could give different pictures of the stock. If the stock were rebuilt to some threshold SSB, it is likely that this would still represent a rather low reproductive potential owing to the relatively high contribution of recently recruited year classes. Hence, decisions based on reproductive potential would tend to be more conservative and hence to delay the reopening of the fishery until the actual reproductive potential had improved substantially (Scott *et al.* 1999).

While there is clear value in using a reproductive index rather than SSB in a management context, as in the above example, the situation is less clear-cut when the same information is used in a stock–recruitment relationship. The use of an index of stock reproductive potential in place of SSB on the horizontal axis of a stock–recruitment plot may result in an improvement in the coefficient of determination of the model fit. While this represents a statistical improvement in the assessment, it may not have a significant impact on the results of the assessment and projections. It is the latter which needs to be assessed to determine the contribution made by the additional biological information.

A stock–recruitment model within a routine stock assessment is used primarily to predict recruitment in stock projections. Generally, these are used to estimate the uncertainty associated with stock forecasts under different levels of fishing mortality (F) in medium-term projections. Hence, an appropriate test for the effectiveness of additional biological information might be the extent to which it results in reduced uncertainty in this context. The approach developed by Patterson *et al.* (2000) to test medium-term projection methods may be useful in this context.

An analogous problem occurs when several different models are available and criteria are needed to judge whether the choice of model will affect the management advice. An example of this is the stock–recruitment relationship for the east Baltic cod stock. Here, Sparholt (1996), Jarre-Teichmann *et al.* (2000) and Köster *et al.* (1999) have developed stock–recruitment models with varying degrees of complexity and which reflect environmental influence to differing extents. The stock–recruitment model used during stock assessments is different from all these (see, for example, ICES 2000). Clearly, in this case it is desirable to have some criteria by which the best model can be selected from among these competing candidates. Again, thorough testing as performed by Patterson *et al.* (2000) may be appropriate.

10.2.3 Modelling environmental change on fish stocks

It is now acknowledged that changes in the environment do affect fish population dynamics. This is demonstrated both by the spatial scales over which recruitment appears synchronised for several stocks (Myers *et al.* 1995, 1997), and by apparent correlations between recruitment and environmental variables (Ottersen & Sundby 1995, Myers 1998, Planque & Frédou 1999, Fox *et al.* 2000b). A significant problem with advancing this work is that variables such as temperature are correlated with many other environmental variables that also affect fish at the individual and, by inference, at the population level. For example, sea surface temperatures in the Northeast Atlantic during the first six months of the year are correlated with atmospheric conditions (depicted by the so-called NAO, denoting the North Atlantic Oscillation) and, thus, with windstress and average direction. Windstress, in turn, can affect primary productivity through water column processes, while varying levels of turbulence have been linked with changes in the feeding success of fish larvae (Alcaraz 1997, Dower *et al.* 1997, Fiksen *et al.* 1998). Environment also affects the growth of adult fish, both directly and through links with prey availability (Brown *et al.* 1989, Brander 1995, Dutil *et al.* 1999, Ottersen & Loeng 2000, Purchase & Brown 2001). For mature fish, feeding conditions affect condition and the amount of energy that can be put into reproduction. Combined with the status of the fish as first-time or repeat spawners, this could potentially affect subsequent recruitment success (Marshall *et al.* 1998, Marshall & Frank 1999). Changes in environmental variables such as temperature can also directly affect fish distribution and may lead to changes in spatial overlap with prey (Sparholt *et al.* 1991, deYoung & Rose 1993, Heessen 1993, Swain & Wade 1993, Ottersen *et al.* 1998).

Models developed to explore such relationships vary from simple statistical approaches through single-species population dynamics models to complex coupled hydrodynamic–biological models. To date, the last category has tended to focus on the biology of early life-history stages (Hermann *et al.* 1996, Leising & Franks 1999). Their complexity and subsequent computer resource requirements currently prevent their application over multiple years, which is a requirement for extended time-series population dynamics modelling (Hermann *et al.* 2001). Similarly, it can prove prohibitive to undertake multiple model runs for sensitivity testing, although applications of engineering simulation theory can help in this respect (Megrey & Hinckley 2001). Extended time-series population dynamics modelling has thus tended to use simpler, single-species population dynamics models, but representations of at least some biological processes can be included. Usually, the problem is knowing how much biological realism to include to generate realistic results. The most useful aspect of such models is that they allow exploration of the sensitivity of the population dynamics to changes in rates and processes at different life-history stages over several decades of simulated time (Quinlan & Crowder 1999). Modern computers allow large numbers of model runs to be made, producing the scope to explore the effects of stochastic variability. The challenge for the future appears to be in producing models capable of bridging these extremes.

O'Brien & Little (2006) presented some results from a simple, single-species population dynamics model for cod which has been used to explore how North Sea cod population dynamics might be affected by sea temperature (Clark *et al.* 2003). The model was used to assess the relative impacts of temperature on stock dynamics through effects on recruitment (early life-history stages) and adult growth. The working question addressed was whether reduced recruitment linked with increased sea temperatures can be compensated for by increases in adult growth rates? One assumption of the model used was that condition (i.e., the relation between length and weight) is constant. In a further examination, survey-based and market sampling data available for North Sea cod and plaice were examined for changes in adult condition factors and a simple exploratory analysis was conducted to assess whether changes in such factors could be modelled using simple environmental indices as covariates.

10.2.4 Fishery management considerations

Changes in the environment – physical, physiological and ecological factors – have major effects on fish population dynamics. Fishery managers by-and-large cannot control the physical environment; however, they can influence the nature and intensity of fishing practices. The challenge for fishery management is to devise and test strategies for controlling fishing that meet conservation and socioeconomic objectives as far as possible.

10.2.4.1 Management strategy

A management strategy has scientific as well as political aspects; the scientific aspects include the assessment strategy. Some aspects, such as technical measures, are not particularly affected by environmental factors. However, choosing an assessment strategy may well entail decisions on a number of issues to which the environment is potentially relevant:

- Appropriate stock structure.
- Assessment technique.
- Fitting of stock–recruitment relationship.
- Single-species reference point calculations.
- Medium-term projection.
- Method for advising on TACs (linked to reference points).

What are the scientific criteria for deciding how to deal with the above? A good management/assessment strategy must cope with the possible effect of the environment on fish populations and the inherent uncertainty about these effects. This does not mean that successful strategies always have to be complicated. For some stocks, it will be worth using a more complex model; for others, it will not. Although optimal strategies are nice, it is more important to be robust. A good robust management strategy should not perform disastrously (or disastrously worse than any other single strategy) under a range of plausible environmental relationships, population dynamics, and future conditions.

In order to decide whether and how to include environmental information, one needs a way to check robustness and general performance of different assessment strategies, under a range of plausible future scenarios. Even a simple strategy needs to be tested under realistically complex simulation models to guarantee its robustness. In such simulations, it is neither necessary nor even desirable to have a unique and definite idea about how the environment affects recruitment; it is more important to be realistic about the range of possible effects that might occur. Of course, as more becomes known about environmental links to recruitment and about the likely future state of the environment conditions, it becomes possible to exclude certain scenarios. In principle, this allows the tuning of a management strategy to deliver higher yields, more stable catches, and so on, while keeping conservation risks low. Strategy testing is a key area where scientists and managers should try to take account of all available information about environmental effects on recruitment.

The task of relating recruitment to the environment is difficult. However, there is a long history of trying to identify links, many without a precise understanding of underlying processes.

10.2.4.2 Potential benefits and drawbacks

There are several ways in which the incorporation of environmental factors (*E*) into stock assessment can inform or improve management. Although environmental factors can affect the population in many different ways (e.g., growth, spatial distribution), the main focus here is on recruitment (*R*), which is a major influence on the population dynamics of most fish stocks.

In terms of historical data, the incorporation of *E* into a stock–recruitment model could lead to improved estimates of stock–recruitment parameters, which should provide a better basis for setting biological reference points. If *E* is predictable, the stock–recruitment model could be used with predictions of *E* to obtain better short-term predictions of recruitment. This is likely to be particularly valuable for short-lived (or overexploited) stocks that are managed by TAC because, in such cases, recruitment may contribute a large proportion to predicted catches. If TACs were to be set for more than one year ahead, further benefits are likely. For example, if the catch is dominated by 3-year-olds, then a good prediction of recruits (at age 0) in the current year would affect the catch prediction three years ahead.

If multi-annual TACs were to be considered, advice would have to be provided for a longer period, based on predictions further in the future than is currently the case. In such a situation, the incorporation of an environmental factor could lead to predictions that are wrong by far more than predictions based on mean recruitment values. This is partly because both *E* and *R* need to be predicted, and also because the incorporation of *E* gives predictions of *R* further away from the mean. This could have a severe effect on the stock, particularly as the effects are cumulative (see, for example, O'Brien & Little 2006).

Medium- to long-term predictions of *E* are clearly difficult. Nonetheless, if it is assumed that the basic characteristics of the *E*-series would persist, then this can be used in medium-term projections. If the *E*-series is simply random, results from runs including and excluding

E would be essentially identical (O'Brien & Little 2006). However, if there is a great deal of structure (e.g., cycles, autocorrelation) in the *E*-series, then the variance of, say, projected SSB could be different if *E* is included in projections.

The incorporation of *E* may lead to an impression of increased certainty, and this could lead to a tendency to sail closer to the wind (e.g., harvesting at higher *F*s), implying more risky harvesting strategies. This could have serious implications for the stock if predictions turn out to be poor or if the relationship with *E* breaks down.

One of the main benefits for medium-term projections, or management strategy evaluations, is likely to occur in situations where there appears to be a distinct change in the environment (e.g., warmer temperatures in the most recent decade in the North Sea). One can evaluate what would happen if the environment returns to historical patterns or persists at recent levels, and base management advice on this information. Of course, this type of 'what-if' or scenario modelling can be done even if we do not know what is causing the observed changes in recruitment (see, for example, Needle *et al.* 2003, Kell *et al.* 2005, Horwood *et al.* 2006).

There are also potential benefits if hydrographic models can inform us about spatial aspects of recruitment. For example, if larvae from a spawning area (A) drift to another area (B) to settle, this process may be highly dependent on certain hydrographic conditions at the right time of year. If recruitment fails in area B in a given year, this may have nothing to do with harvesting at area B, but could simply stem from hydrographic conditions. This is an example where the effect of *E* may be more like an on-off switch rather than a continuous positive or negative influence. This type of information can be particularly useful when considering issues such as closed areas or stock structure/distribution.

There is also the possibility of making Type I or Type II errors (cf., O'Brien & Little 2006), and this would have implications for management if *E* is included. Sometimes, Type I and II errors can be accounted for by increasing time-series length, by performing meta-analyses, by increasing the accuracy of the variables measured, by correcting for autocorrelation, or by testing a priori hypotheses. Even when Type I and II errors are accounted for, the link detected remains only a statistical link until a mechanistic understanding can be suggested and tested. There are a number of situations in which strong correlations can be driven by non-causal links (e.g., if the environmental index is correlated with the true environmental forcing).

10.2.4.3 When to incorporate environmental factors

Evidence of a clear relationship between *E* and population dynamics may point to the inclusion of *E*, though the strength of the effect and potential benefits to management should also be explored. In some cases, evidence may be relatively indirect, coming from a range of studies in different geographical areas, and contain a high level of common sense plausibility.

Progress in the inclusion of *E*-series in recruitment prediction has been hampered by purely correlative studies that easily generate spurious correlations. A good correlation between some *E*-series and a biological variable (e.g., recruitment) is not enough on its own to justify inclusion into assessment. The least requirements are plausible hypotheses for mechanisms, while the best is some evidence for the hypotheses. There is also a need for some confidence that the detected relationship will persist. These issues are best addressed by studies aimed at identifying one or more likely mechanisms. An increased understanding of the possible reason for observing a strong correlation will strengthen trust that there really is a link and, therefore, increase confidence to incorporate *E*. An understanding of the mechanism also provides information on how/where in a functional relationship the *E*-term should enter as a covariate.

In addition to the strength and nature of a possible link between *E* and a biological process, the nature and characteristics of the environmental factor is very important.

10.2.4.4 Ways to incorporate environmental factors

The potential benefits to management (or improved assessment/prediction) are closely related to the way in which environmental factors are incorporated. In principle, an *E*-series can be used as a tuning index in an assessment, although possible nonlinearities may be a problem. If a direct index of the variable being estimated is available, for example larval abundance index to estimate recruitment, this may be preferred. Although both direct and indirect indices can be used together, there may be technical details, such as relative weighting of the indices that still need to be resolved.

An index of some environmental conditions can also be used in a two-step approach. Here the idea is to run an analytical stock assessment without the environmental index in Step 1. In Step 2, results (e.g., estimates of recruitment) and the environmental index are calibrated to allow the index to be used in a predictive way. Again, there may be technical details that need to be further elaborated, and some issues would have to be addressed on a case-by-case basis. Environmental series can be incorporated into stock–recruitment (S–R) models as covariates. If the mechanism is not known, tractable formulations of S–R models with *E*-terms will tend to be fitted. Different formulations, however, may have different implications, and it may not be possible to distinguish between models based on goodness-of-fit. The recruitment dynamics at low SSB, which is usually outside the range of the data, could be strongly affected.

Improvements to a S–R model fit where *E* is included may suggest changes in biological reference points (in terms of SSB and/or *F*). Here, however, a set of reference points (e.g., F_{crash} and MBAL (minimum biologically acceptable level)) would be associated with a given level of the environmental factor. This is because there is no longer a single S–R curve, but rather a surface, that is, a different curve for each value of *E*. Given that reference points should not change from year to year, and that it may be impossible to predict future *E* values, exactly how reference points should be adjusted for *E* still requires careful thought and further work.

In situations where *E* appears to have changed in level in the recent past (e.g., the Irish Sea and North Sea examples discussed by O'Brien & Little 2006), there may also be a need to re-evaluate current reference points to ensure that they are compatible with the current and the assumed near-future environmental conditions. The key here is the assumption about future environmental conditions.

Assumptions about the future environmental dynamics can be incorporated in medium-term projections. Clearly, it is not essential to predict *E* exactly into the medium-term projection in order to explore the likely future dynamics of the stock. Instead, a (statistical) distribution of future *E* is assumed. The projections can be done using a what-if approach to see whether different assumptions about the future distribution of *E* make a difference to results, or to identify worst-case scenarios.

The incorporation of *E* into management strategy evaluation can also play an important role. These studies can help to identify the likely benefits gained from incorporating *E* into assessment, for example in terms of stock conservation or yield. One can also explore the implications of:

- Incorporating E into assessment when there really is no persistent link.
- Ignoring E when there really is a link between E and recruitment, for example.

10.3 Applications and investigations

Published examples of complete studies that deal fully with the incorporation of reproductive biology and recruitment considerations into fisheries management advice and strategies are still few in number.

O'Brien & Little (2006) presented a limited number of area-based case studies that illustrate a sequential introduction of biological processes into estimates of stock biomass. Recruitment was viewed from both a biological and a stock assessment perspective, while potential impacts of spawning characteristics on reference points for fishery management were reviewed and discussed in the context of the current precautionary approach to fishery management. However, by and large, those studies that have been published and reviewed are limited in their scope and applicability.

Hence, this section is devoted to the presentation, and discussion, of a single well-developed case study (Andrews *et al.* 2006). It is presented to encourage and motivate the development of similar studies in the future, and to re-emphasise the need not to have a too-narrow focus when dealing with the incorporation of reproductive biology and recruitment considerations into management advice and strategies.

In the example, a spatially explicit physiologically structured demographic model for cod in the waters surrounding the UK is presented. The method of Gurney *et al.* (2001) has been used to combine spatial and life-history effects efficiently. The model recognises four distinct movement and life-history regimes for fish during their life span. The model includes spatially heterogeneous development rates and carrying capacities determined from data on the region. To my knowledge and that of my coworkers, this is the first developed model with such spatial and biological coverage which is capable of parameter fitting and hypothesis testing. Various models of movement for adult fish are examined, leading to the conclusion that the movement of settled individuals is key to understanding the spatial distributions of cod in the waters around the UK. In order to replicate these spatial patterns it is necessary to assume that adult cod make intentional migrations to spawning grounds. Additionally, it will be shown that assuming advective movement of settled stages of cod provides the closest match to spatial survey data. The best-fit model is then used to investigate the potential effects of long-term application of a series of regional fishing closure policies.

10.3.1 Case study

10.3.1.1 Introduction

Cod stocks around the UK have supported valuable fisheries for many years. They are major components of whitefish fisheries in the North Sea, Irish Sea and the West of Scotland, as well as smaller fisheries in the Celtic Sea and Channel (Defra 2004, 2005, 2006). During recent decades most of these cod stocks have declined, and recent advice from the International Council for the Exploration of the Sea (ICES) has highlighted severe problems with a number of them, with the possibility of collapse in some cases (ICES 2005, 2006). There is little doubt that excessive fishing pressure is a significant factor in stock decline, but there are also indications that environmental effects may have contributed to the demise of these stocks. There is scientific evidence linking higher sea temperatures to poorer cod recruitment, for example.

The stocks of cod (*Gadus morhua* (L.)) across the entire geographic range in the North Atlantic were subjected to intense fishing pressure during the 20th century, and many suffered well-documented collapses such as the north-western cod stock off Newfoundland (cf., Shelton & Lilly 2000, Rose 2004, Shelton 2005). Closer to home in the UK, the stock of North Sea cod is under continued pressure because of over-fishing, and is also threatened by a decline in the production of young cod that has paralleled the warming of the North Sea over recent decades (O'Brien *et al.* 2000). Taking account of the dependence of recruitment on stock size alone, it has been proposed (Cook *et al.* 1997) that the collapse of this stock may be imminent, as the SSB is currently at a low level (O'Brien *et al.* 2000, ICES 2005, 2006). The combination of exploitation with the recent changes in North Sea temperature, a low SSB and a stock dominated by young immature individuals, means that fishery managers must continue to take precautionary management measures (Longhurst 1998). In order to give the mature stock a chance to rebuild, fishing mortality rates need to be reduced to at least the precautionary levels advised by ICES. Other cod stocks of economic importance to the UK occupying the Irish and Celtic Seas, and Western Channel are under similar pressures to those of North Sea cod.

The case study presented in this section is focused on a model which encompasses all of these regions, as well as the waters further north including Scottish, North-Icelandic and South-Icelandic waters (see Figure 10.1; Plate 23). This is necessary in order to allow for the potential movement of fish into, and out of, the target waters and allow the potential for greater biological realism to be incorporated into the modelling approaches. The case study illustrates the depth of analyses required and data needs for such modelling approaches, and involves detailed analyses of historical data, together with computer-based modelling and simulation techniques.

Current ICES practice is to advise on medium-term exploitation strategies that are predicted to maintain each stock within safe biological limits. These limits are defined for each stock in terms of an upper threshold of fishing mortality and a lower threshold of spawning stock biomass, beyond which the precedent of historical data suggests that the risk of collapse becomes unacceptable. However, this approach assumes that historically based relationships between spawning stock and recruitment will continue to apply in the future, and the underlying models largely fail to take account of the now extensive knowledge of the biology and ecology of many exploited species. There are good reasons to suppose that greater realism in the models could significantly alter both the predictions and the perception of safe biological limits, but this is by no means certain. The approach adopted within this case study is to develop a series of linked stage-structured models capable of representing the various spawning populations of cod in northern European waters extending from the English Channel to the northern North Sea and including the West of Scotland, Irish Sea and Celtic Sea. The models explicitly represent the dynamics of pre-recruitment stages, as well as the survival of immature and adult fish, and incorporate environmental influences on growth and survival.

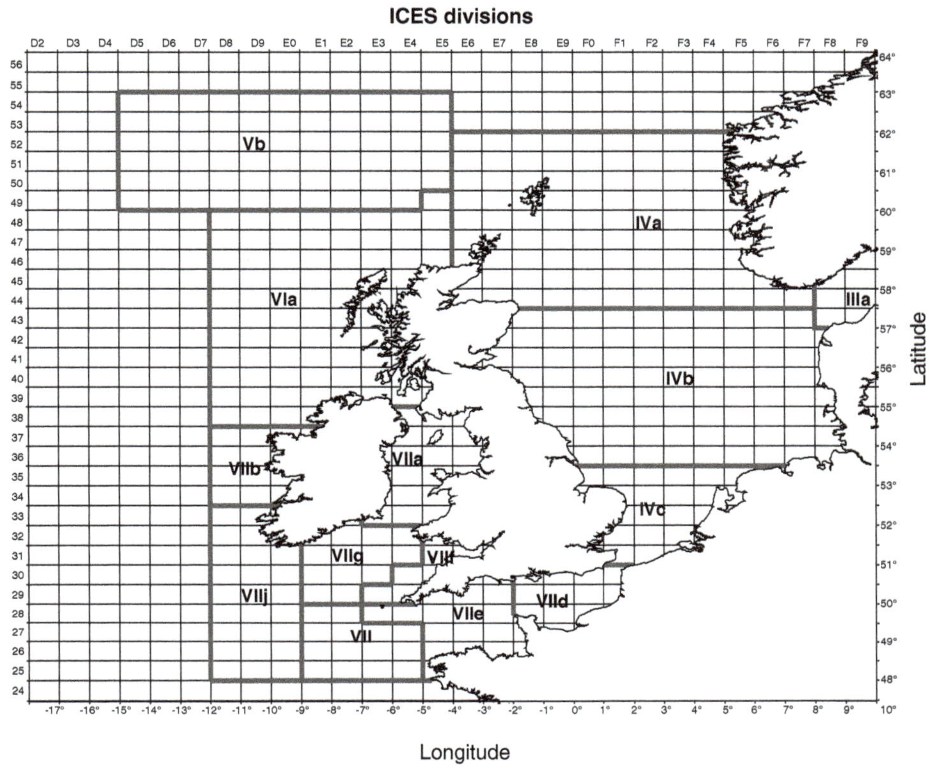

Figure 10.1 Sea region covered by the population dynamics model of European cod stocks. For a colour version of this figure, please see Plate 23 in the colour plate section.

The models developed are capable of representing the various spawning populations of cod in the northern European waters extending from the English Channel to the northern North Sea and including the West of Scotland, Irish Sea and Celtic Sea. The models have been used to investigate the sensitivity of current stock forecasts to the inclusion of greater biological realism and to possible environmental influences. Model development has been supported by a series of data analyses designed to feed data into the models.

Cod are still amongst the most valuable of the groundfish species exploited by the UK fishing fleets. Any conservation measures introduced to combat the low levels of European cod stocks will have significant economic impact on the UK fishing industry. It is important to the four UK devolved administrations, Defra (Department for Environment, Food and Rural Affairs), Marine Scotland (formerly, SEERAD - Scottish Executive Environment and Rural Affairs Department), DARDNI (Department of Agriculture and Rural Development—Northern Ireland) and the Welsh Government, that the scientific basis for the assessments and projections of the state of these fisheries is as scientifically credible as possible.

10.3.1.2 Modular structure

The research presented in this case study was undertaken in a modular fashion with component modules and sub-modules clearly defined. The first of these – Data assembly (module 1; Section 10.3.2.1) – incorporates the assimilation and collation of spatially resolved cod catch and survey data, the growth of cod eggs and larvae, time series of biological data on juvenile and adult cod (age and size at maturity, size at age), multi-species virtual population analysis (MSVPA) output to caricature natural mortality, flow-fields and tagging data, spatially and temporally resolved temperature data, reproductive biology, and settlement ecology (settlement areas, timing of settlement, densities of juvenile fish at settlement). New sources of information on biological processes (module 2; Section 10.3.2.2) include a study of mixing rates between discrete spawning units and the growth of adult cod. Population modelling (module 3; Section 10.3.2.3) and Implementation, testing and analysis (module 4; Section 10.3.2.4) consist of the simulation of the population dynamics of cod, sensitivity analysis of the model to structural, spatial and parameter variability, together with the analysis of the consequences of changes in exploitation for cod in the North Sea.

The work conducted within these four modules and the results obtained are presented and discussed in the remainder of this chapter. Full details may be found by consulting Andrews *et al.* (2006).

10.3.2 Population dynamics models of European cod stocks

10.3.2.1 Data assembly

The purpose of the model being developed within this case study is to provide a national capability to simulate the population dynamics of cod throughout the period 1960 to the present day. The research team's main concern was to determine whether the environmental links in the population model were capable of simulating, in general terms, the exceptional period of recruitment associated with the cold winter conditions in UK waters during the 1960s, and the subsequent fading of recruitment up to the present day.

Spatially resolved catch and survey data for cod covering the period 1960 to the present day

The population model required data on cod landing statistics resolved by age, size, time of year and stock. Much of this data was previously assembled and compiled by the Northwest Atlantic Fisheries Organisation (NAFO) into tables (NAFO Working Group on Reproductive Potential 2003) listing relevant reference sources. This has, however, only been compiled for the North Sea. Hence, historical catch and survey data from the other stock areas (including the Irish Sea and English Channel) were collated from the participating fisheries laboratories (Cefas, FRS and DARD) covering the period 1960 until the present day. The data were resolved temporally and spatially to the finest possible scale (by ICES rectangle) and modelled appropriately (see, for example, Figure 10.2; Plate 24).

Additionally, ICES holds records of catch and survey data from a multitude of areas, and over a variety of time periods, including the history of North Sea cod stocks landings and fishing mortality data (ICES 2001) (see, for example, Figure 10.3; Plate 25), internationally combined market data and estimated catch numbers-at-age (O'Brien *et al.* 2001), and average cod catch number-at-age from the 3rd quarter International Bottom Trawl Survey (IBTS) (ICES 2003).

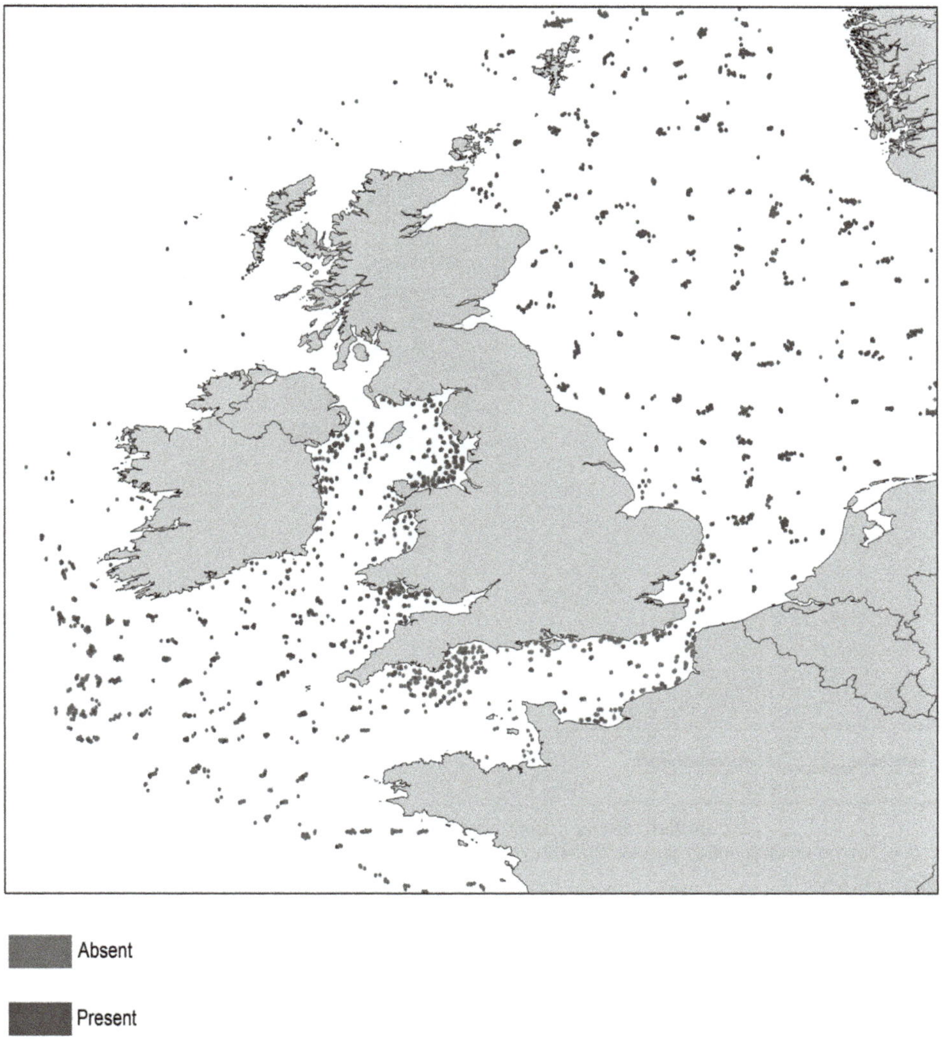

Figure 10.2 Cod distribution prediction map. For a colour version of this figure, please see Plate 24 in the colour plate section.

For much of the 20th century, three documents have normally been available from every RV cruise, irrespective of its duration or purpose: a cruise report; a scientist's logbook; and vessel station sheets. In addition, if large numbers of fish were measured, the lengths were recorded in a fish measurement logbook, and plankton tows were recorded in plankton logbooks. Hydrographic data were recorded separately and archived. In addition to these formal cruise documents, there are substantial quantities of published and grey literature which describe the methods used during some of these surveys, and provide analyses and interpretations of results. All relevant archived material has been searched during this project in order to provide as much information as possible for each survey.

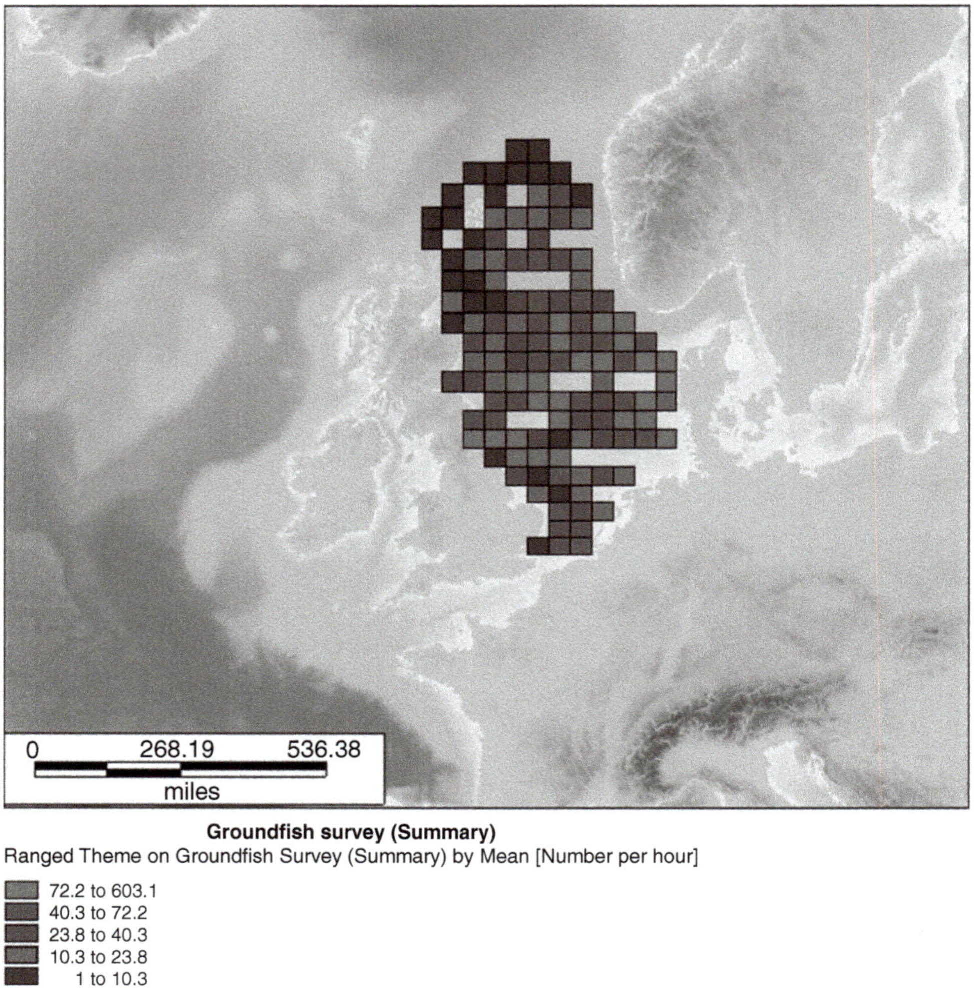

Figure 10.3 North Sea groundfish survey map showing catch-per-unit-effort (kg h^{-1}). For a colour version of this figure, please see Plate 25 in the colour plate section.

Growth of cod eggs and larvae

The temperature dependence of cod egg development and mortality rates are well-documented in the published scientific literature. Pepin (1991) derived a general relationship between egg mortality and temperature, though cod were not included in this compilation project. Models of cod larval growth have been produced by Campana & Hurley (1989), based on collections from the Gulf of Maine. These studies both suggest that temperature is the primary factor affecting growth. The development and growth of cod eggs and larvae must be dynamically represented in the population model. Temperature-dependent egg development models for cod have been determined based upon studies of fish from the North Sea.

Figure 10.4 (Plate 26) indicates the regions around the UK that have historically been spawning grounds for cod (cf., Daan 1978). The data used in the map are taken from that collated

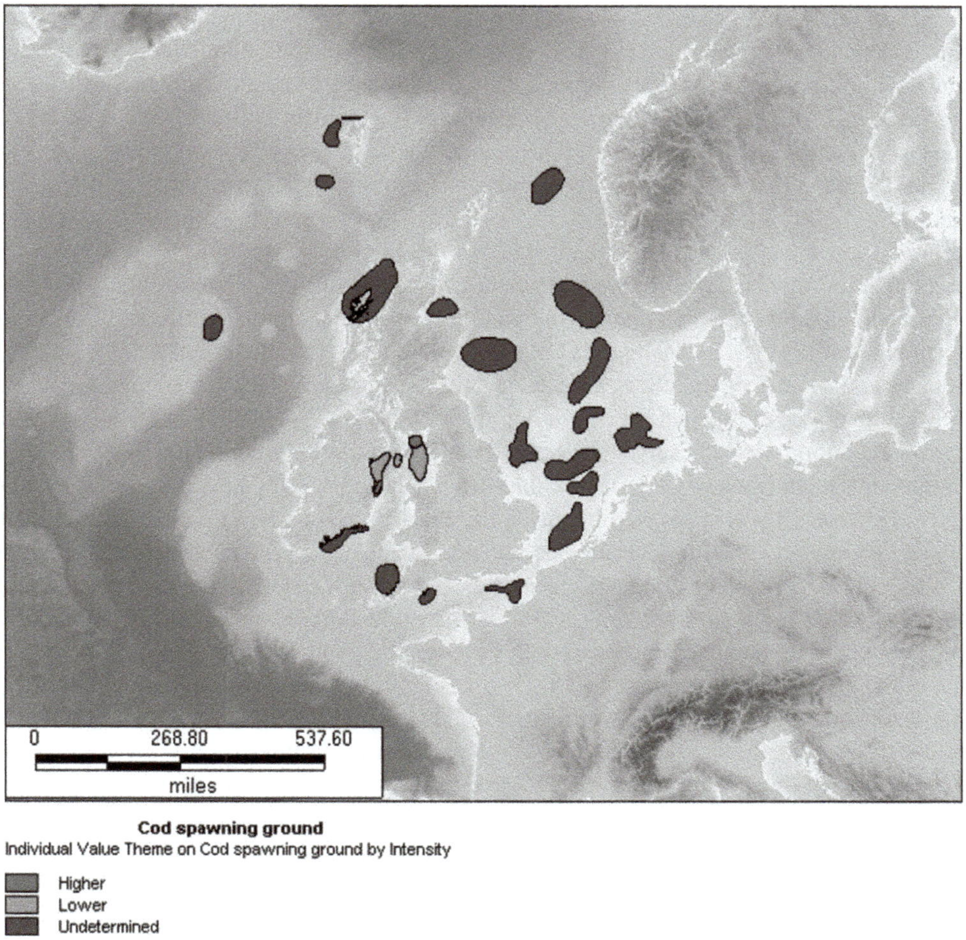

Figure 10.4 Cod spawning ground distribution map. For a colour version of this figure, please see Plate 26 in the colour plate section.

by Cefas, Lowestoft. Recently, further information regarding spawning grounds in the North Sea has been documented by the ICES Planning Group on North Sea Cod and Plaice Egg Surveys in the North Sea [PGEGGS] (ICES 2003). Spawning takes place from the beginning of January through to April, perhaps being slightly later with increasing latitude, at least around the British Isles. It occurs offshore in waters of salinity 34–35 PSU (Riley & Parnell 1984).

Campana & Hurley (1989) presented empirical relationships between size, age and accumulated temperature history for individual cod and haddock larvae in the Northwest Atlantic. The case study re-parameterised the relationship for cod larvae using archive otolith microstructure data. This involved back-calculating the temperature history of sampled cod larvae using particle tracking and hydrodynamic models. The derived data were used iteratively to determine the best-fit parameters of the growth relationship based on the body length and age of the sampled larvae. The Campana & Hurley equations were reformulated to describe the rate of change in stage development as a function of temperature.

Page & Frank (1989) have shown that the time between deposition and hatching of cod eggs is dependent on water temperature, so the case study classified individuals by the currently achieved proportion of development between release and hatch – an index which runs from zero to 1 – seeking an appropriate temperature development rate (Andrews *et al.* 2006). In order to obtain parameter values for the model the case study has exploited the work of Daan (1978) and Fox *et al.* (2000a), who indicate the similarity between the eggs of cod and haddock.

The development and growth of cod eggs and larvae must be dynamically represented in the population model. Temperature-dependent egg development models for cod have been determined based on fish from the North Sea. This model may apply to other areas, although recent studies (EU AIR3-CT94-2263: Development of stock assessment methodologies; egg production estimates of Irish Sea demersal stocks) suggest that there may be regional differences. With this as the subject for continuing work in the Irish Sea (MF0426: Fish egg development and mortality studies in the Irish Sea, and EU Studies Programme DGXIV 98/090), it was anticipated that area specific temperature-dependent egg development rates would be generated. However, this was not possible given the problem in accurate egg identification highlighted by Fox *et al.* (2000a). The latter study showed that, on the basis of size and the presence of oil globules, early-stage cod eggs cannot be distinguished readily from those of haddock (*Melanogrammus aeglefinus*) or large eggs of whiting (*Merlangius merlangus*); therefore generating specific temperature-dependent egg development rates for cod would be subject to chances of egg mis-identification.

After encountering the reliability problems in cod egg identification via historically reported methods, new reliable means of identification were necessary. Using an ICES PGEGGS survey as a platform of opportunity, some efforts were made to test the reliability of cod egg identification via a newly developed genetic approach, during the RV CORYSTES 01/04 North Sea research cruise. Eggs collected during the ichthyoplankton surveys in the North Sea were pre-sorted for genetic probing by the University of East Anglia (UEA). Additional eggs of cod and haddock from aquarium rearing were inserted into the samples as blind controls; that is, UEA were not aware of which samples were from the field (North Sea samples) and which are known (aquarium-reared). The preliminary results showed a correct identification of 95% of the control cod eggs, and 97% of the control haddock eggs.

Time series of biological data on juvenile and adult cod (age and size at maturity, size at age)

Data needed to be assembled on age and size at maturity, and size at age of cod from the various populations. A subset of these data were analysed to derive the parameters of the functions describing the rate of change of development stage index with time, and its dependence on environmental variables (in particular temperature) and the rate of maturation as a function of development stage. An independent subset of the data was used for comparison with the results from the model simulation of the changes in population structure over specific periods of time. Data on these biological parameters was partially completed as part of an ICES Workshop on Gadoid Stocks in the North Sea during the 1960s and 1970s. These data were aggregated at the management unit level, and were disaggregated as far as possible. This was necessary to better estimate the environmental temperature experienced by the populations during the lifetime of sampled year classes, and the relationship between development rate and temperature was then parameterised. This was also undertaken for the Irish Sea.

Fertility was modelled using ICES data; namely, the proportion maturity-at-age for cod was combined with a fecundity estimate. Assumptions were made that female fish spawn continuously between February and April, that eggs are released in the cell that is currently occupied by the female concerned, and that 50% of fish are female (i.e., a 1:1 sex ratio) (Andrews *et al.* 2006). To determine the fecundity of females in each length class the case study calculated the equivalent biomass from the global weight–length relationship, and then used the allometric relationship between weight and length (Marteinsdottir & Begg 2002, Andrews *et al.* 2006).

The ICES estimates of proportion mature at age for the whole North Sea were fitted using a Weibull distribution, and then transformed using an age–length key into a length-based distribution using a growth curve fit to all North Sea data (Andrews *et al.* 2006). The corresponding maturity at age for other regions is a fairly reasonable estimate of the ICES proportions, when back-converted using growth rates fitted to the other regions. Andrews *et al.* (2006) presented a figure that shows the implied equivalent maturity-at-age curves for individuals growing at the fastest (comparable to south Irish Sea cod) and slowest rates (comparable to central-eastern North Sea cod).

Recent trends in maturity at age in Irish Sea cod have been investigated using samples collected during Northern Ireland groundfish surveys (NIGFS) carried out annually from March 1992 to 2002 (Armstrong *et al.* 2004). Almost all one-year-old cod were immature, and all three-year-olds were mature. The proportions of two-year-old mature cod have fluctuated, with a minimum mean of 42% in 1995 to 89% in 2000.

Biological time-series data is available in various forms and from a number of different sources. Weight- and length-at-age data for individual North Sea cod is available from the English third-quarter groundfish survey; a historical comparison of the changes of length–maturity ogives is given in Oosthuizen & Daan (1974) extending from the 1890s to the 1970s. This shows a shift in L_{50} from approximately 75 cm to just above 50 cm in length at maturity over this long-term period. North Sea cod maturity data is collected during ICES IBTS, while ICES maturity ogives used in VIIe-k cod assessments are based on data collected during the UK Westerly Groundfish Surveys.

In terms of assessing reproductive potential, it appears that age diversity is a useful tool for describing major changes in the population age structure that will impact on reproduction (Marteinsdottir & Thorarinsson 1998, Secor 2000). For North Sea cod the correlation between age diversity and recruitment was only significant when the variable proportion mature in the stock was used to estimate age diversity. Time-invariant maturity ogives give an incorrect perception of age diversity, as well as other aspects of the stock dynamics. The age diversity of North Sea cod has shown a long-term decline since 1963. Since 1993, age diversity has been below the long-term mean value. During the 1990s, fishing pressure was very high on the North Sea cod stocks and there has been an accompanying decline in SSBs.

MSVPA output to caricature natural mortality

The implementation of natural mortality as a variable, rather than a constant term, was a critical feature of the population model. The main variable feature of natural mortality that we represented was the predation on cod by other piscivorous fish in the ecosystem. This predation changed with the species and size composition of the fish assemblage, which affected the overall level of harvesting from the ecosystem, as well as other environmental factors. The main aim was to caricature the relationship between ecosystem level harvesting rate and the natural mortality of cod development stages.

The relationship between total ecosystem catch rate and stage-specific predation mortality of cod was parameterised using results from the North Sea MSVPA, which simulates the consumption of fish by other fish in the system. As a MSVPA is not available for the Irish or Celtic Sea, assumed natural mortality was used as in the single species Virtual Population Analysis (VPA). However, when the multispecies estimates for the North Sea and Skagerrak were applied to the stock assessments of Irish Sea cod, it illustrated the sensitivity of the current Irish Sea stock assessments to potential changes in natural mortality-at-age.

Mortality was modelled using ICES data. The STEREO Project (Stock Effects on Recruitment relationships, EU FAIR CT98-4122) fit for natural mortality (based on North Sea data) was used for all regions, having been altered to depend on length. This was done, as the North Sea is the only region to have such data based on age, rather than being consistent across all age groups. To calculate the daily mortality for eggs (0.232 d^{-1}) we used the relationship between egg size and mortality given by Rijnsdorp & Jaworski (1990), with an assumed egg diameter of 1.4 mm (Andrews *et al.* 2006). For settled individuals in both juvenile and adult segments, it was assumed that the background annual mortality rate is a universal function of length, as discussed in detail for the North Sea by Heath *et al.* (2003), and which it was assumed applies equally to the other assessment areas (Andrews *et al.* 2006).

Individuals who are large enough to settle but cannot do so because the cell they occupy had no suitable ground, are subject to the mortality rate appropriate to a settled fish of the same size multiplied by a factor that increases exponentially with length (Andrews *et al.* 2006). Given the limited amount of data that exists for mortality rates of larvae and pelagic juveniles, this model assumes that the pelagic mortality rate is independent of length, but shows spatiotemporal variation driven by local average temperature and time (Andrews *et al.* 2006).

The instantaneous natural mortality of North Sea cod obtained from traditional methods (i.e., age-frequency data or mark–recapture methods) by Beverton & Holt (1959) and Ware (1980) gives a rate of 0.2. However, Myers & Doyle (1983) calculated the instantaneous estimate of natural mortality to be 0.22, under the assumption that post-maturational growth is due to a concave egg conversion function.

Flow-fields and tagging data

To generate temperature fields for driving growth and survival in the particle tracking model, a mixed-layer model was set up for three different regions of interest: (i) Scottish waters; (ii) North-Icelandic waters; and (iii) South-Icelandic waters. Although these regions are outside our target areas for cod stocks, as noted in Section 10.1 these waters are still encompassed by the model to ensure that movements into and out of the North Sea and coastal Scottish waters are accounted for. The model was converted into a user-friendly version. Initial and forcing data were collected which encompass meteorological data and initial temperature and salinity data. The model coding, initial data and forcing data were transferred to the partners responsible for implementing the particle tracking.

Particle tracking methodology (following the methodology of Gurney *et al.* 2001) to be used in the computation of mixing coefficients required the use of flow-fields generated by three-dimensional hydrodynamic modelling. Flow-field data at 14 km resolution are already available; these were calculated using a statistical characterisation (SNAC; Logemann *et al.* 2004) of output from the Hamburg Shelf Ocean Model (HAMSOM) (Harms *et al.* 2000) and bathymetry from NOAA's National Marine Fisheries Service (Andrews *et al.* 2006).

Archived tagging data on cod was available at participating laboratories. ICES quarterly IBTS data on the distribution of cod year classes, together with ancillary data such as depth and temperature, were assembled. A review of available tagging data describing the movement and migrations between Celtic and Irish Sea is reported in Fox *et al.* (2005). Recent tagging has been described in Connolly & Officer (2001), where 2220 cod were tagged between 1997 and 2000; it was found that although there was some movement of cod between the Irish Sea and Celtic Sea, the component of the Irish Sea cod in the Celtic Sea was low and that no cod tagged in the Celtic Sea were recovered in the Irish Sea. The results from thousands of tagged cod being released in the North Sea showed that there is very little interchange of cod between the North Sea and West of Scotland, but that there is much more exchange between the Eastern Channel (VIId) and the Skagerrak (IIIa) (Cefas Tagged Fish Database). It is for this reason that ICES assess the stock for Sub-area IV, and Divisions IIIa (Skagerrak) and VIId combined.

Spatially and temporally resolved temperature data

Sea surface and bottom temperature time series for the European shelf seas are available in raw form from ICES for the period from 1900 to the present day. The data from 1955 to 1995 have previously been analysed, aggregated to monthly temporal resolution and ICES statistical rectangle spatial resolution for the North Sea, and filling routines developed to deal with missing data. During this project, the spatial coverage was extended to include the waters west of the UK. Data were derived from ICES and COADS (Comprehensive Ocean Atmosphere Dataset, National Center for Atmospheric Research, Boulder, Colorado, USA) databases, from satellite data (recent years) and from a heating stirring model of the Irish Sea developed by Cefas. These were the main environmental inputs to the population model.

The physical environment of the model consists of temperature fields taken from the statistical model reported by Heath *et al.* (2003) and Andrews *et al.* (2006). Interannual changes in North Sea cod recruitment have been related to changes in sea surface temperature (Brander 1997, Planque & Fox 1998, Planque & Frédou 1999, O'Brien *et al.* 2000).

Historically, strong year classes have been associated with lower-than-average temperatures during the first half of the year. However, weak year classes have also occurred during cold years (such as 1986 and 1987), but only when the SSB was low (O'Brien *et al.* 2000). Since 1988, mean sea temperatures have been higher than during the previous three decades, correlating with the current low level of SSB and the proposal (Cook *et al.* 1997) that the collapse of this stock may be imminent (O'Brien *et al.* 2000).

Reproductive biology

The effects of fish age and size on the realised fecundity (potential fecundity minus atretic resorption of unshed oocytes) and timing of spawning in cod stocks around the UK are unknown. Cod from the Irish Sea and the North Sea show radically different growth rates, and both are different from those found at Iceland and Norway, where most previous such studies have been conducted. Data from Iceland and Norway cannot simply be assimilated for use in the context of UK–European cod stocks. Similarly, recent studies in Norway have shown large interannual variations in realised fecundity of cod in relation to temperature and the abundance of the main prey of adult cod in the region (capelin). Cod in European waters have a more diverse diet, so a strong relationship to particular prey species will probably be more difficult to discern, but temperature effects may be strongly reflected.

A considerable amount of data on fecundity of cod for the Irish Sea existed prior to this study, based on studies conducted in 1995 (AIR3-CT94-2263) and 2000 (EU Studies Programme DGXIV 98/090). Historical data also existed for North Sea cod.

Fish size is the most important factor influencing potential fecundity. Whilst length is the measure most generally used, weight has been found to explain a higher proportion of the variation in some cases (Kjesbu *et al.* 1998, McIntyre & Hutchings 2003). As food intake affects the amount of energy available for allocation into the gonads, instantaneous indices of energy reserves may also explain some of the residual variation around potential fecundity–size relationships.

The most relevant index of energy available will depend on the main mode of lipid storage in the fish. In cod, the energy content of the liver increases exponentially in relation to the increase in the lipid accumulation (Lambert & Dutil 1997). As such, liver index in gadoids could be a useful measurement for prediction of fecundity (Marshall *et al.* 1999).

During the Annual Egg Production Method for Stock Assessment in 1995 and 2000, Armstrong *et al.* (2001) and Armstrong (2002) reported on the intensively studied reproductive biology of cod. This was summarised by Fox *et al.* (2005) – the spawning duration of individual cod in 1995 was around 40 days with low levels of atresia, mainly confined to spent fish towards the end of the spawning season. Estimated realised fecundity was 854 eggs g^{-1} female^{-1}. The spawning duration of individual cod in 2000 was around 45 days. Levels of atresia were highest at the beginning of the spawning season (22% of samples, $n = 22$) but then declined sharply. Over the whole spawning season, atresia resulted in an estimated loss of potential fecundity of only 2%, resulting in a realised fecundity of 1046 eggs g^{-1} female^{-1}.

Ovarian atresia seems mainly to be initiated under low nutritional condition prior to spawning (Kjesbu *et al.* 1998, Witthames *et al.* 2000), implying that atresia is a regulating factor between the energy resource accumulation during the feeding period prior to spawning and the number of oocytes likely to complete maturation successfully during spawning. There is good evidence that, on some occasions and in some species including cod (Hardardottir *et al.* 2003), potential fecundity does not closely approximate to realised fecundity, due to high levels of atresia during the spawning season. Studies with Icelandic cod (Hardardottir *et al.* 2003) have found no relationship with size. Higher intensities of ovarian atresia are seen prior to spawning in females with a low hepatosomatic index (HIS) (Rideout *et al.* 2000).

Settlement ecology (settlement areas, timing of settlement, densities of juvenile fish at settlement)

Data were required on settlement areas, timing of settlement and densities of juvenile fish at settlement. In addition, information on prey levels available to settling fish was required. It is known that cod probably require a cryptic habitat into which to settle, and that this may be a mechanism to avoid predators (Gregory & Anderson 1997, Bromley & Kell 1999). It is therefore possible that successful settlement could be limited to relatively small areas of suitable habitat (ICES 2003).

The IBTS data were examined and used in this module. The data were mapped using presence/absence generalised additive models (GAMs) using an estimation procedure analogous to that employed by Fox *et al.* (2000a) for the modelling of ichthyoplankton survey data (see Figure 10.2; Plate 24).

Settlement of fish was dependent on a derived carrying capacity (Andrews *et al.* 2006), where each square gives the number of fish that could be supported using the area in the square between 25 and 200 m, with each fish needing 2 m^2. This was converted to a biomass, by comparing the numbers for the four regions covered (North Sea, Irish Sea, West of Scotland, Celtic Sea), with the maximum total biomass recorded by ICES. With the carrying capacity having been set, a fish is allowed to settle in a square if the capacity of the square has not already been surpassed. Any fish that are unable to settle will experience higher mortality and try to settle at each subsequent growth update of the model. If any square surpasses its carrying capacity due to adult movement, a higher level of mortality is imposed on the statistical rectangle.

One might regard the settlement habitat as being defined simply on the basis of depth, and hypothesise further that the area-specific carrying capacity might vary between regions and linearly with temperature and time, and optimise the parameter values to maximise the model fit to observations of SSB (Andrews *et al.* 2006). During the settlement phase, several density-dependent and -independent processes may determine juvenile mortality and subsequent recruitment to the exploited population. In addition, settling and newly settled juveniles may perform migratory movements, a process that would influence the mixing coefficients between population units (this is discussed further below).

10.3.2.2 New sources of information on biological processes

Mixing rates between discrete spawning units

The concept of mixing between sub-stocks encompasses two types of process. Initially, the planktonic stages are dispersed and mixed by advection and diffusion from the spawning areas. Subsequently, the fish perform active migrations according to some environmental and/or innate homing cues. A general representation of these processes was incorporated into the population model to characterise the mixing of individuals from each of the modelled population units.

The task of this sub-module was to investigate and quantify, by a separate modelling exercise, the extent and stage dependence of mixing between populations. The planktonic part of the mixing process was investigated using a particle tracking approach. The outcome was a statistical relationship between stage-specific mixing coefficient for each pair of population units and the NAO index (or other suitable climatic/environmental indices), for implementation in the population model. The active migratory mixing of individuals from each population unit was derived from statistical modelling of field data, such as the ICES quarterly IBTS in the North Sea, and the use of an individual-based model where migratory rules were applied and compared to the outcome of random migratory movements.

The population model works with two different types of updates: movement and life history. Movement updates take place every seven days and are done using transition matrices (corrected for diffusion), calculated by a tracking program using SNAC currents as input. Diffusion is set in the tracking process to be just large enough to be sure that movement, even if it is slow, will be registered.

There are several biological processes that must be considered by the model. Within this project, these are processes separated into growth, settlement, fertility, natural mortality and fishing mortality. These are all updated separately from the movement process. The updates are made until another update would require going over the weekly movement interval. At this point all squares then undergo the movement mixing process, which is possible as the same length classes are used in each square in the model.

Growth of adult cod

The growth of fish is mediated both by temperature and food availability, and an assumption of constant growth is therefore not acceptable. The population dynamics models required data on how post-recruitment growth has varied over space and time. A general temperature-dependent growth model for post-juvenile cod from several stocks has been developed by Brander (1995); however, this was based on stock averages and data on individual variability are required for population modelling. Such data may be generated by back-calculation from historical collections of otoliths.

This case study has presented a spatial model for cod in the North Sea that uses the length of a fish as the defining characteristic. To do this, the population is tracked by storing the number of millions of fish in each length class in each spatial square of a simulation grid. Each model square runs on an individual timer that depends on the growth rate for the square. Hence, at regular intervals all fish are moved up the growth ladder, with mortality, spawning, and any settlement being calculated at each time interval. Within each week there are a minimum of three such updates, with the fastest-growing squares having approximately six such updates.

Growth for each square was fitted using age-length key data for each ICES statistical rectangle. A von Bertalanffy growth curve was fitted using the data, with minimum size set to 0, and maximum size set to 197, as this gave the best fit over the whole region. It should be noted that it is necessary to have these values the same for every square in order to ensure that length classes can be consistent throughout the domain. The growth rate for each statistical rectangle was the fit, and these fits were then compared with annual bottom temperature across the region as well as longitude. This gave a linear fit with adjusted r^2 of 0.7155, and was used to calculate the growth rate for statistical rectangles for which there was no age–length key data.

10.3.2.3 Population modelling

The basic model unit was a developmental stage-structured representation of a cod population with a closed life cycle, explicitly representing spawning and all development stages from eggs to the oldest adults. The model represented growth and maturation as functions of environmental conditions and population density, and imposed variable mortality rates as functions of stage, population density, and environment. The developmental index (stage) formed the basis for the biological structure of the model, and was a composite of age and size, and transformation rules were provided to convert between stage, age and size for comparison between model results and observations. The basic population model was constructed from a set of differential equations describing the development, mortality and reproduction processes.

The cod life cycle begins in the spawning season (February to April; Brander 1994) with the release of eggs which float at or near the surface, drifting with the prevailing current (Andrews *et al.* 2006). The temperature dependence of cod egg development is well documented in the literature. After surviving incubation, the eggs hatch into larvae, which feed and grow into juvenile fish (Brander 1994, Brander & Mohn 2004). At this stage feeding begins, the larvae move deeper and begin to actively swim (Heath *et al.* 2003). However, given their small size (<3.5 cm), the mid-water currents to which they are exposed must still be the principal determinant of long-term movement (Andrews *et al.* 2006).

Pelagic juveniles whose length exceeds 3.5 cm move down the water column to seek a region where they can begin demersal life, where lower current speeds and increasing body size combine to imply that long-term relocation is likely to be largely facultative (see Brander 1994, Robichaud & Rose 2004). Settled juveniles have lower natural mortality than pelagic juveniles and grow rapidly, becoming vulnerable to commercial fishing. Surviving individuals mature into reproductively active adults, who continue to grow and spawn yearly until death, with large females being considerably more fecund than small ones (Marteinsdottir & Begg 2002).

Where no independent data exists for parameters, they were determined using numerical optimisation. The unknown parameters include the 12 region-dependent parameters that define the relationship between temperature, time and the mortality rate for larvae and unsettled juveniles. This also includes the 12 region-dependent parameters that determine the relationship between the same predictor variables and cell carrying capacity. In addition, for the models with aggregated spawning, it is further assumed that there are six areas to which individuals that are about to spawn aggregate. The position and size of these areas are also regarded as unknowns.

Values for the set of unknown parameters were obtained by fitting to a test dataset comprising smoothed trends in total SSB for each assessment region over the period 1970–2000, and a decadal average cell-resolution spatial distribution of SSB for the period 1990–2000.

The first element of the test dataset is the total SSB for each of the four assessment regions: North Sea, West Coast, Celtic Sea and Irish Sea. The second element of the test dataset is the spatial distribution of SSB. For a full explanation of the numerical optimisation used and the test dataset, refer to Andrews *et al.* (2006) where full mathematical content can be found in Appendix B.

The case study then differentiates between the following seven hypotheses about the movements of settled individuals:

(1) No pelagic transport or adult movement.
(2) Unaggregated spawning, no adult movement.
(3) Unaggregated spawning, adults diffuse.
(4) Unaggregated spawning, adults advect and diffuse with local current.
(5) Aggregated spawning, no adult movement.
(6) Aggregated spawning, adults diffuse.
(7) Aggregated spawning, adults advect and diffuse with local current.

Table 10.1 Minimised errors for model variants. Spatial errors are proportional squared errors of the average catch averaged over the cells by the number of hours of surveying in each cell. Trend errors are proportional squared errors of the average spawning stock biomass in each region, averaged over the number of data points. The total error is the sum of 80% of the spatial error and 20% of the trend error. All errors ×100 (after Andrews *et al.* 2006).

Spawning	Movement	Spatial error	Trend error	Total error
No transport		141	1	113
Unaggregated	1 None	73.1	0.99	58.7
	2 Diffusive	76.2	1	61.1
	3 Adv. + diff	58.2	1	46.7
Aggregated	1 None	38.5	1	31.0
	2 Diffusive	47.4	1	38.1
	3 Adv. + diff	28.6	0.99	23.1

To do this the model must be fitted to the historical dataset and a comparison made of the best attainable fits.

Table 10.1 shows the overall quality of the best fit obtainable from these models. The variants with unaggregated spawning, while fitting the trend to the stipulated accuracy have spatial errors which are more than twice that of the best performing variant. By comparison, all model variants with aggregated spawning show a low spatial error, with predicted relative distributions with many of the key features of that which is observed.

It was concluded that a spatial demographic model is essential for understanding the population dynamics of cod in the waters around the UK, that aggregated spawning is essential to qualitatively correctly predict the spatial distribution of spawning stock, and that long-term advection movement of settled individuals is primarily a result of net transport by local bottom currents.

The test data-set was then fitted to the model in order to determine which configuration provided the best fit. Each region was treated individually and the best-fit values for parameters that determine the spatial distribution of carrying capacity and larval and juvenile mortality were ascertained.

The best-fit model proposes that, immediately prior to the spawning season, the adults who will spawn that year are gathered into six relatively small regions. Each individual goes to the nearest aggregation centre. Immediately after being gathered into the distribution (shown in the left-hand frame of Figure 10.5; Plate 27), the adults begin to move again in conformance with the normal movement rules. Eggs are thus deposited over the region through which the spawners move during the spawning season. The average density of adult individuals over the spawning season is shown in the right-hand frame of Figure 10.5 (Plate 27). The comparison between the best-fit model and observations (lower frame of Figure 10.5; Plate 27) shows a good qualitative agreement over most of the domain.

10.3.2.4 Implementation, testing and analysis

Simulation of the population dynamics of cod

The assembled population models were tested as to whether they were capable of simulating the population responses to events such as those seen historically and refined. The results have been presented in Section 10.3.2.3 above, as it is not possible to separate the modelling from

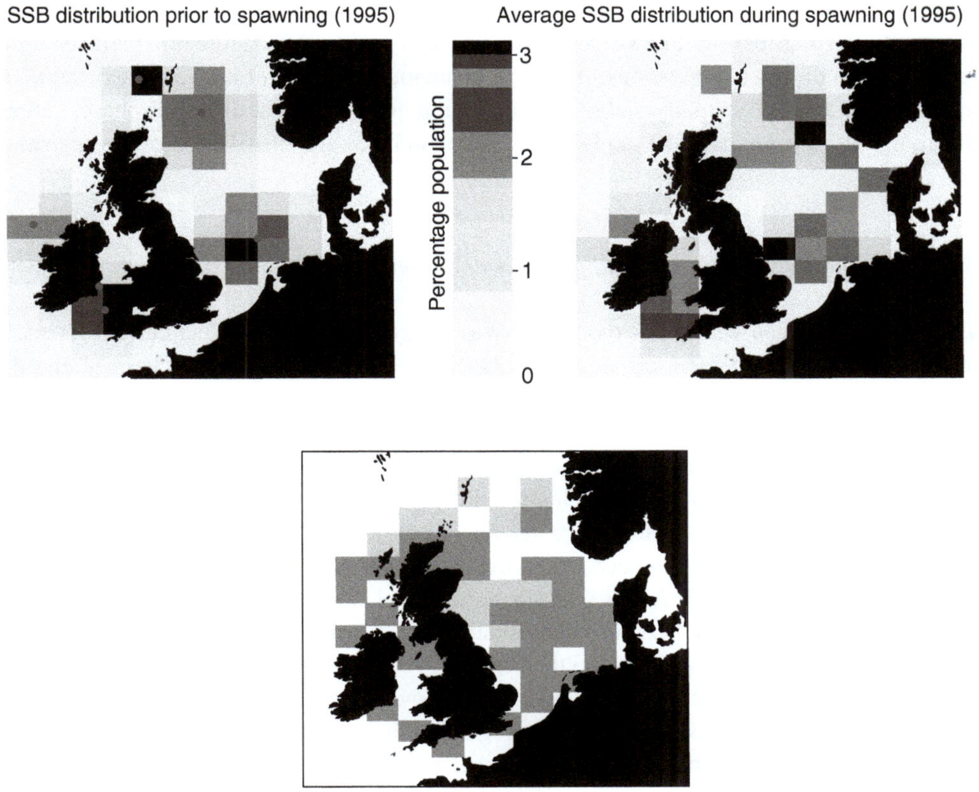

Figure 10.5 Upper left frame shows the predicted aggregated distribution of spawners immediately prior to the spawning season. The boundaries of each gathering region are shown in black, with the centre of the aggregation region marked with a circle. Upper right frame shows the predicted average distribution of spawners over the spawning season. Lower frame shows the observed distribution of spawning fish. Dark grey shows cells containing currently active spawning areas and light grey shows historical spawning areas now unused. For a colour version of this figure, please see Plate 27 in the colour plate section. Derived from data presented by Wright *et al.* (2003), Heath *et al.* (2003) and C.J. Fox (personal communication); cited in Andrews *et al.* (2006).

its implementation and testing. Complete details may be found in the report of Andrews *et al.* (2006).

Sensitivity analysis of the model to structural, spatial and parameter variability

The sensitivity of the outputs from the population models to variability in the structure and parameters were assessed using computer-intensive simulation techniques. The models were used to inform the assessment process by conducting an analysis of the sensitivity of predictions of the future state of stocks to the structure and parameterisation of the models.

The main sensitivity analysis of the model focused on the structural uncertainty surrounding the migration behaviours of juvenile and adult cod. For this, the other major areas of uncertainty (pelagic mortality and the specific carrying capacity of the environment) were treated as free fitting, and each structural variant of the model was optimised to the established levels, trends and distributions of SSB. The analysis clearly showed that structural variants of the model, which do not recognise two modes of behaviour in the adult fish: an active seasonal migration to a set of spatially stable spawning sites followed by a dispersal phase, cannot under any fitting circumstances explain both the abundance and distribution of the spawning stock.

Analysis of the consequences of changes in exploitation for cod in the North Sea

The work undertaken within this case study was always intended to reflect current issues of relevance to the management of cod stocks. Over recent years, spatial management and the identification of sea areas to be closed to fishing have become ever more important. The models developed in this case study by Andrews *et al.* (2006) were adapted and applied by Darby *et al.* (2006) in a Defra-funded project entitled: 'Investigations into closed area management of the North Sea cod', in order to identify potential future options for the management of fisheries. The investigations were restricted to the sea area of the North Sea, but there is no reason to suppose that the modelling approach would not extend into other sea regions.

Spatial management in the North Sea. The basic model developed by Andrews *et al.* (2006) was extended to include the best current estimate of the distribution of fishing effort, and then refitted to the test dataset to emphasise the period 1990–2000. A set of hydrodynamic driving functions was generated for the period 2004–2020 by cyclic repetitions of the years 1994–2004 inclusive, and this produced temperature driving functions by assuming that the spatiotemporal temperature distribution remained at that observed in 1999. The model was then run under five scenarios:

- Scenario 1: Fishing effort and distribution remains constant.
- Scenario 2: The EU Commission's 2001 closure during the months of February, March and April remains in force.
- Scenario 3: The closure of the area estimated by the STECF (2003) meeting to be that from which 60% of the international catches of cod were reported to be taken in 2002 remains in force permanently.
- Scenario 4: The North-east English coastal fishery is closed permanently.
- Scenario 5: All fishing effort is removed from the North Sea.

For each scenario (except scenarios 1 and 5), two variants were examined. In the first, variant (a) it was assumed that fishing effort from the closed areas is redistributed over the remaining open area in proportion to existing catch-per-unit-effort (CPUE), while in the second variant (b) it was assumed that displaced effort is removed from the fishery.

Without any change in regulatory policy, the continuation of recent trends leads to the SSB being halved by 2020. By extreme contrast, the entire removal of fishing effort from 2001 onwards results in the recovery of the SSB to its carrying capacity by 2015. All of the closure policies except scenario 4 produce significant improvements, but only those in which effort is removed, rather than redistributed, result in serious progress being made in moving SSB towards a significant fraction of the carrying capacity.

With no change in regulatory policy, the North Sea spawning stock appears to have moved south and to be primarily concentrated in an area between the central southern English coast and Skagerrak. Interestingly, the two closure policies which produce significant population enhancement do so by reinforcing this population concentration, which one may note does not lie to any significant degree within the areas from which fishing effort is excluded. This emphasises the need for spatial models in designing and evaluating possible closure policies.

In the intervening years since the publication of the first edition, much progress has been made towards the integration of genetics, biology, behaviour and modelling into the design and study of management measures (e.g., Heath *et al.* 2013). Some of the messages, however, are not new but the integration of different disciplines into single studies is becoming more frequent and can only be for the good.

10.4 Conclusions

The relationship between spawning stock and recruitment is fundamental to the scientific approach to fisheries management. Environmental factors can influence how a stock might be managed, yet there continues to be a real concern within the fisheries science community regarding the low rate of uptake of potential environmental indicators in stock assessment and management advice. It is still unclear what variety of environmental information should be used, and whether it would leave advice open to criticism for using speculative information. However, a debate is long overdue, and it is hoped that this chapter, along with the earlier chapters in this book, will motivate such discussions.

10.4.1 Incorporation of biological realism into assessment and management models

A few published studies have dealt fully with the incorporation of reproductive biology and recruitment into fisheries management advice. The studies of O'Brien & Little (2006) have been cited within this chapter, and the research of Andrews *et al.* (2006) has been presented and discussed.

The key findings from these studies, and of this chapter, may best be summarised as follows:

- Marine systems are complex and care is needed not to focus too narrowly on a single aspect or species. A too-narrow focus could mean that crucial links or factors are missed. Mechanisms may also be flexible and change, depending on a whole suite of parameters.
- Considerations of environmental factors can make a difference in how one might manage a stock. Simulation models can play an important role in identifying whether and where benefits to management are most likely to accrue, and where it would be best to focus attention in terms of other (e.g., process) studies.
- The convincing incorporation of environmental factors involves a great deal of work, particularly in terms of fieldwork where mechanisms are being explored and in terms of long-term observations. Long-term studies are crucial to the success of this type of work. Progress should be viewed as an iterative process of improvements, and most benefits from such studies are likely to accrue after several years rather than a single year or less.

The low rate of uptake of potential environmental indicators in stock assessment, as well as the incorporation of biological realism into assessment and management models, remain a concern. The move towards an ecosystem-based management of marine resources should redress the balance and the results of simulation studies should be used to guide future biological studies.

There is an increasing body of work (e.g., Parma 2000) showing that very simple management strategies or harvest control rules can perform well. The key issue is that proposed strategies be tested for robustness and performance, and this issue is addressed in the next section.

10.4.2 Management strategies: Should reference points and environmental factors be linked?

One question regarding fisheries management in the context of productivity that exhibits cycles or regime shifts is whether to (and how to) change biological reference points. This question is considered by Basson (1999), who was motivated by the observation that sea surface temperatures (SSTs) in the North Sea have been above the long-term average for most of the past decade. There is increasing evidence of a link between cod recruitment and SST (Planque & Frédou 1999). Furthermore, in the North Sea, for example, above average temperatures appear to impair recruitment (O'Brien *et al.* 2000). Therefore, it is relevant to ask whether current reference points would remain appropriate if above average temperatures were to persist.

Theoretical considerations and simulation studies were used by Basson (1999) to explore potential problems with, and the merits of, adjusting reference points according to an environmental factor. Theoretical considerations highlighted the following:

- The Model choice is important (e.g., Ricker or Beverton–Holt formulation). How the environmental effect is incorporated into the formulation.
- When analysing real stock–recruitment data it is usually impossible to distinguish statistically between different candidate models.
- Ignoring the environmental effect could lead to F being too high if there is a trend or pattern in the environmental series.
- There are potential conceptual difficulties with a B_{MSY}-based threshold because it suggests a lower threshold when recruitment is low than when recruitment is average or high.

The simulation studies used a relatively simple age-based projection model. A distinction was made between the stock–recruitment model used to generate the underlying recruitment series (called the correct model), and that used for calculation of reference points, F_{MSY} in particular. This allowed a performance evaluation of a management strategy based on a wrong model (i.e., a model differing from the so-called correct model). An imaginary SST series with a distinct regime shift to above-average temperatures was used to drive recruitment in scenarios where recruitment was linked to SST. It is important to note that the simulations did not attempt to predict year-to-year recruitment, did not include any management inaccuracies, and did not incorporate assessment bias or error. Therefore, it is neither a reflection of the existing modus operandi for advising on TACs, for example, nor is it a full-scale management evaluation. The main aim of this particular exercise was to explore the implications of model uncertainty and environmentally driven recruitment for four different management strategies.

Some of the main conclusions drawn from the simulation model are as follows. If the target harvest rate (F_{MSY}) is adjusted as a response to changes in SST, then:

- The model formulation used for the adjustments is important.
- There is a need to predict SST, and if a recent average is used, a lag in response can develop.
- Adjustment of F, particularly upward adjustments, has effort implications.
- Model performance is poor if recruitment is not really affected by SST.

When a harvest control rule which does not respond to changes in SST, but that reduces F linearly between F_{MSY} and 0 for $0.5\ B_{MSY} < SSB < B_{MSY}$, is used, then:

- Performance is good, even if the model choice is wrong (and recruitment is affected by SST).
- There is no need to predict SST.
- Adjustment in F still has effort implications, but there is an upper limit at F_{MSY}.
- Performance is good if recruitment is not affected by SST.

One of the main reasons that the harvest control rule performs well under these scenarios is because it maintains SSB at a higher level than the strategies that adjust the reference points (to lower B_{MSY}) as SST increases. The harvest control rule and the adjusted reference point strategies all lead to increased variability in catches compared with a fixed F_{MSY} strategy (as might be expected). The average catches, however, are very similar under the different scenarios when the model choice is correct.

Basson's paper concludes that one approach to management of a system that may be subjected to substantial changes in productivity (e.g., regime shifts or cycles) is to design robust, smooth harvest strategies. Such strategies may not even incorporate a factor such as SST directly, but it is crucial that robustness tests are performed for a wide range of hypotheses about, or scenarios for, the environmental driver.

Many details and issues are not considered in the simulation study, and need to be explored in a case-specific context. Issues such as assessment error and bias and management inaccuracies (e.g., catches not being equal to the TAC) have already been mentioned. Other issues include testing with a wide range of environmental scenarios; for example, not only should regime shifts that lead to reduced recruitment be considered, but also those that lead to increased recruitment.

10.5 Future directions

It has been clearly stated throughout this book that its principal objective is to present an argument for stronger linkages between basic and applied research on fisheries recruitment, assessment and management of exploited fish stocks. This is right and proper of a scientific text but is somewhat limited in scope, and it would be remiss not to mention a number of emerging wider issues. These still remain as appropriate and relevant today in this second edition, as at the time of publication of the first edition.

10.5.1 Climate change

Climate change could have serious scientific and economic repercussions for fisheries (Klyashtorin 2001, Sumaila *et al.* 2011), but the implications are not clear (Cook & Heath 2005) even though there is a need to consider mitigation and adaptation policies to minimize impacts on fisheries. During the past decades, fishing pressure coupled with warming has put pressure on North Sea cod, for example, catches of which are mainly comprised of young immature fish (O'Brien *et al.* 2000).

Temperature is frequently used as an environmental variable in recruitment models (Cardinale *et al.* 2004) because temperature is recognised as affecting fish biology (Brander 1995). Temperature affects fish directly and indirectly through ecological and physiological mechanisms (Drinkwater 2005, Ottersen *et al.* 2005).

The marine ecosystem is full of complex interactions that are not fully understood. There is evidence of fish changing their latitude, depth, and boundaries with climate change (Perry *et al.* 2005). Consequently, there is ambiguity about how these changes will alter predator–prey relationships and habitat availability (Perry *et al.* 2005). The long-term changes in the abundance and distribution of stocks will depend on complex predator–prey interactions, habitat availability, and extent of climate variability (Perry *et al.* 2005). A greater knowledge of fish biology is required to understand exactly how climate change may act.

The combination of fishing and climate change are placing pressures on fish stocks (McFarlane *et al.* 2000, Schiermeier 2004, Cook & Heath 2005), and precautionary action needs to be taken to help rebuild stocks (O'Brien *et al.* 2000). Current rates of fishing mortality are not sustainable and may need to be cut to ensure the recovery of depleted stocks (Cook & Heath 2005). If fishing mortality were reduced to half of the current levels, biomass precautionary reference points could possibly be reached (Kell *et al.* 2005). Precautionary action needs to be taken by fishery managers as a North Sea fishery of young immature fish has emerged from an amalgamation of changes in North Sea temperature, low SSB, and levels of exploitation (Longhurst 1998). The collapse of fish stocks may be exacerbated more by warming than by fishing alone (Clark *et al.* 2003).

General circulation models (GCMs) predict that global sea surface temperatures will increase as a result of climate change, though the extent of the warming is uncertain (Drinkwater 2005). The HadCM3, a coupled Atmosphere-Ocean General Circulation Model (AOGCM) predicts that by 2040, North Sea surface temperature will have risen by 1 °C (Gordon *et al.* 2000, Pope *et al.* 2000). The regional climate model HadRM3 predicts that mean annual North Sea surface temperatures will increase by 0.5–1.0 °C, 1.0–2.5 °C, and 1.5–4.0 °C by 2020, 2050, and 2080, respectively, based on different emission scenarios (Hulme *et al.* 2002).

Much uncertainty surrounds the relationship that temperature shares with fish (Cardinale *et al.* 2004), because the reaction of individual species to climate change is ambiguous (Perry *et al.* 2005). Some species may benefit, while others become less productive as a consequence of higher sea temperatures (Cook & Heath 2005). Species that spawn in shallow, low-salinity waters may be more affected than those species which spawn in deeper, hydrographically stable water (Rose 2005). Cod may respond unexpectedly to climate change, as the Earth and oceans experience temperatures not previously witnessed (Kell *et al.* 2005). Sustained temperature changes may cause stocks to become distressed, causing them to diminish (Drinkwater 2005).

Environmental data are built into stock assessments to account for climate change (Planque *et al.* 2003). When a climate effect is found, temperature is associated with recruitment, which suggests that it should be incorporated into modelling (Cardinale *et al.* 2004). Although incorporating climate change into stock–recruitment models helps to improve stock assessment predictions, the process is extremely complex and has a high degree of uncertainty (Cardinale *et al.* 2004).

Temperature shares a complex relationship with recruitment (Cook & Heath 2005). The Ricker stock–recruitment model is frequently used to examine the effect of temperature on recruitment (Cook & Heath 2005, Kell *et al.* 2005). Temperature affects recruitment through juvenile survival (Planque & Frédou 1999, Clark *et al.* 2003) and habitat availability (Fromentin *et al.* 2001). The impact of climate change may depend on whether juvenile survival or carrying capacity are affected by temperature (Kell *et al.* 2005).

The considerable variation in the strength and sign of the relationship that species share with temperature consequently makes it difficult for fisheries managers to produce reliable stock predictions (Cook & Heath 2005). The benefits associated with the incorporation of environmental parameters into stock–recruitment models will depend on whether they can be well predicted, and if a strong relationship exists between the environmental variable and recruitment (Basson 1999). Stock assessment advice may need to be adjusted frequently to incorporate the uncertainty of climate change. Annual revisions to stock assessment advice are believed to be the most efficient strategy, but are also thought to be the most expensive (Roel *et al.* 2004). Scientists face a difficult task in making reliable predictions in the face of future climate change because it is challenging to produce accurate stock forecasts in the longer term (Rothschild *et al.* 2005). Modelling capabilities are increasing rapidly, and with the development of regional climate models (RCMs) scientists will be able to bring more certainty to recruitment predictions, which will be particularly important to fishery managers and policy makers.

10.5.2 Integration of fisheries and environmental sciences under a common theme

Traditionally, fisheries science and environmental science have been largely separated, with fisheries being centred logically on the status of populations and on the wider ecosystem consequences of the activity of fishing on living resources, and environmental science being primarily centred on impacts arising from the use and exploitation of non-living resources. With the increasing requirement for ecosystem-based science and advice, there is a need for scientists to adapt to the challenge that this poses and to integrate fisheries and environmental sciences under a common theme which might be referred to as Marine Resource Management (MRM).

The greatest challenges to the management of off-shore marine living resources arise from fishing activities, whereas coastal, estuarine and freshwater resources are more likely to be impacted by an array of additional or alternative environmental perturbations, both anthropogenic and natural.

A science of MRM will need to be focused on understanding and assessing the impacts of human activities on the marine environment in order to better support, and inform, both policy and regulatory decisions for sustainable development. In the short term, this will necessitate further development of the two traditional sciences dealing with:

- *Impacts on fishery resources:* Scientists are continually updating and developing their suite of tools in order to study the status of marine resources, and to provide better predictions of the likely effects of fishery management decisions on both fish stocks and fishers.
- *Impacts on other human activities:* This covers the assessment of a range of other activities potentially affecting both living and non-living resources, including navigational dredging, disposal of dredged material, marine constructions (e.g., wind farms, coastal protection/sea defences, harbour works and outfalls), the extraction of marine aggregates (sand and gravels) from the seabed and the extraction of oil and gas from beneath the seabed.

Naturally, these will include the further development of innovative tools and techniques, improvements to the methodologies employed in the evaluation of marine and freshwater resources and impacts arising from their use or exploitation, and objective approaches to enhance sustainability and protect these resources.

The main element of European environmental legislation of relevance to fisheries management is the Marine Strategy Framework Directive (MSFD). This includes a requirement to achieve what is known as *Good Environmental Status* (GES) by 2020. GES covers a wide range of elements, not all of which are impacted by fishing, but it does include Descriptors related to the abundance of each stock as well as their place in the food-web. The introduction of the landing obligation under the reformed CFP, previously mentioned in Section 10.1.1, will be important for determining how fishing vessels respond to fishery management actions as TACs become more limiting on their activity than was previously the case when discarding was allowed.

In the longer term, there will be a requirement to focus on understanding and assessing the impacts of human activities on the marine environment in order to better support, and inform, both policy and regulatory decisions for sustainable development.

New scientific challenges should emerge which will necessitate the further integration of fisheries and environmental sciences (e.g., Morris *et al.* 2011). With a move towards ecosystem-based management of marine resources, two immediate challenges can be identified and others will be identified in due course.

Challenge 1: Integrating evidence to support the management of human activities

The development of the European Marine Strategy Directive (MSD) has focused attention on the overarching goal of promoting sustainable use of the seas and the conservation of marine ecosystems. To achieve sustainable development will require a more integrated approach to the management of human activities (Fletcher 2007). Such development of a more coherent, streamlined, and integrated framework for the management of human activities will necessitate further supporting the development of Marine Spatial Planning (MSP). Since the publication of the first edition of this book, progress has been made, for example, within European Member States towards identifying and implementing a range of marine protected areas so ensuring that key habitats are protected from damaging activities; delivering benefits both in terms of conservation and the associated wider ecosystem services.

Integrated management requires integrated scientific advice on the level of risk, and on the expected impact of proposed new developments given the degree of pressure and impact from existing human activities. Some of these activities, such as oil and gas development, aggregate extraction and wind farm development, are managed through the issue of licences that apply over relatively small spatial scales. Fishing management regulations are, by contrast, set at much larger spatial scales. Spatial discrepancies and the difficulty in obtaining accurate geo-referenced data describing pressures caused by each of these activities acts as a serious impediment to the development of integrated assessments of risk and impact.

Providing integrated advice on current and expected levels of impact will thus require the development of new risk assessment (RA) systems with predictive capabilities. These RA systems will need to be spatially structured and capable of assimilating multi-source geo-referenced data that describe (and, where necessary, predict) both human pressures and ecosystem state. The system would use information on the relationships between pressure and state to allow assessments of impact from new developments given existing pressures. Impacts caused by pressure from multiple human activities would need to be considered in order to allow cumulative effects assessments. Having developed such a system, spatial conflicts could be more easily identified and management decisions taken based on an integrated assessment of the level of risk posed by new developments.

Making progress with MSP at an early stage will provide robust tools to address important policy questions related to the spatial impacts of multiple human activities, and the design of management responses that are justifiable and acceptable to society.

One of the most significant developments, since the first edition of this book was published, has been the movement towards a sustainable exploitation of fish stocks using data-limited methods where previously assessments were not undertaken. In 2012, ICES implemented the Data-Limited Stocks (DLS) framework to provide quantitative advice, not only for fish stocks for which there is sufficient data to conduct full analytical assessments, but also for those fish stocks considered data-limited. At the time, this represented a six-fold increase in quantitative advice provided for data-limited stocks (ICES 2012). The framework relies on the principle that available information should be used, and that advice should follow a precautionary approach and be based on the same principles as applied to data-rich stocks for which analytical assessments are available. The framework therefore calls for the determination of exploitation relative to F_{MSY} and the consideration of stock trend, and provides a variety of methods that can be applied to facilitate this – the choice of method will depend on the information and data available for a given stock. However, to date, a large proportion of data-limited stocks (which mainly consisted of those that have unreliable catch data) have utilised a single method that relies on recent survey trends; simulation testing has shown that although this method stabilises SSB in the short-term (for around five years), its performance starts to deteriorate in the longer term because it does not include a target (ICES 2013). It is critically important to understand how well the candidate suite of methods performs in recovering stock status. Rosenberg *et al.* (2014) developed a simulation framework for testing the performance of data-limited methods under factorial combinations of a number of factors: initial depletion; harvest dynamics; time-series length; life-history type; recruitment variability; environmental forcing on recruitment; and observation errors in catches. Investigations into the interactive effects of the design variables are facilitated in a statistically rigorous factorial design. Much scientific work remains to be done and some is on-going (e.g., le Quesne *et al.* 2013). The future looks good for both the development of these methods and the delivery of sustainability. Sustainable fisheries management is a prerequisite for providing food security, as well as conserving the biodiversity of the world's oceans. Fisheries policy in Europe is moving towards a more integrated ecosystem approach to fisheries management (EAFM) that takes account of multiple drivers, encompassing ecological, economic and social, with humans as part of the ecosystem rather than an external actor on it.

Challenge 2: Linking biophysical processes with food-web and human activity dynamics for the evaluation of management options

Demand for scientific advice on the ecosystem approach to fisheries management will require predictive tools capable of exploring the impacts of environmentally driven influences and management actions on the sustainability of resources and dependent industry. It requires models linking biophysical processes with food-web and fisheries dynamics.

Scientists have been working towards the development of coupled three-dimensional hydrodynamic–ecosystem models composed of two dynamic models; examples are the Princeton Ocean Model (POM) (Blumberg & Mellor 1978), which provides the physics, and the European Regional Seas Ecosystem Model (ERSEM) (Baretta *et al.* 1995), which describes the biogeochemical processes. From a fisheries perspective, such coupled models will be needed for what-if scenarios to investigate management strategies including reductions in trawled areas and reductions in fishing effort, together with a more detailed suite of scenarios including both spatial and temporal closures.

Due to the different disciplinary approaches and needs of customers, the existing models of biophysical processes and fisheries interactions are disconnected. There are complementarities between the models that can be exploited for developing the tools needed to deliver more holistic advice. Simply stated, the outputs of biophysical models can provide the inputs for models of food-web and fisheries interactions.

Biophysical and fisheries models still need to come together to promote a more comprehensive understanding of the ecosystem. Although such models reach some way beyond our current capability, it can be anticipated that within the next five years they will be needed to provide scientific support to management on a routine basis.

References

Alcaraz, M. (1997) Copepods under turbulence: grazing, behaviour and metabolic rates. *Scientia Marina*, **61**, 177–95.

Andrews, J.M., Gurney, W.S.C., Heath, M.R., Gallego, A., O'Brien, C.M., Darby, C. & Tyldesley, G. (2006) Modelling the spatial demography of Atlantic cod (*Gadus morhua*) on the European continental shelf. *Canadian Journal of Fisheries and Aquatic Sciences*, **63**, 1027–48.

Armstrong, M.J. (2002) Development and validation of egg-production based biomass estimates, using cod and plaice stocks in the Irish Sea. Commission of the European Communities, Directorate General for Fisheries Contract 98/090. 256 pp.

Armstrong, M.J., Connolly, P., Nash, R.D.M., Pawson, M.G., Alesworth, E., Coulahan, P.J., Dickey-Collas, M., Milligan, S.P., O'Neill, M.F., Witthames, P.R. & Woolner, L. (2001) An application of the annual egg production method to estimate the spawning biomass of cod (*Gadus morhua* L.), plaice (*Pleuronectes platessa* L.) and sole (*Solea solea* L.) in the Irish Sea. *ICES Journal of Marine Science*, **58**, 183–203.

Armstrong, M.J., Gerritsen, H.D., Allen, M., McCurdy, W.J. & Peel, J.A.D. (2004) Variability in maturity and growth in a heavily exploited stock: cod (*Gadus morhua* L.) in the Irish Sea. *ICES Journal of Marine Science*, **61**, 98–112.

Baretta, J.W., Ebenhoh, W. & Ruardij, P. (1995) The European Regional Seas Ecosystem Model – a complex marine ecosystem model. *Netherlands Journal of Sea Research*, **33**, 233–46.

Basson, M. (1999) The importance of environmental factors in the design of management procedures. *ICES Journal of Marine Science*, **56**, 933–42.

Beverton, R.J.H. & Holt, S.J. (1959) A review of the lifespans and mortality rates of fish in nature, and their relation to growth and other physiological characteristics. *Ciba Foundation Colloquium on Ageing*, **5**, 142–80.

Blumberg, A.F. & Mellor, G.L. (1978) A coastal ocean numerical model. In: J. Sunderman & K. Holtz (Eds) *Proceedings of the International Symposium on Mathematical Modelling of Estuarine Physics*. pp. 203–14. Springer-Verlag, Hamburg.

Bogstad, B., Lilly, G.R., Mehl, S., Pálsson, Ó.K. & Stefánsson, G. (1994) Cannibalism and year-class strength in Atlantic cod (*Gadus morhua* L.) in Arcto-boreal ecosystems (Barents Sea, Iceland, and eastern Newfoundland). *ICES Marine Science Symposia*, **198**, 576–99.

Bogstad, B. & Mehl, S. (1997) Interactions between cod (*Gadus morhua* L.) and its prey species in the Barents Sea. In: *Forage Fishes in Marine Ecosystems*. Proceedings of the International Symposium on the Role of Forage Fishes in Marine Ecosystems. pp. 591–615. Alaska Sea Grant College Program Report No. 97–01. University of Alaska, Fairbanks.

Brander, K.M. (1994) The location and timing of cod spawning around the British Isles. *ICES Journal of Marine Science*, **51**, 71–89.

Brander, K.M. (1995) The effect of temperature on growth of Atlantic cod (*Gadus morhua* L.) *ICES Journal of Marine Science*, **52**, 1–10.

Brander, K. (1997) Effects of climate change on cod (*Gadus morhua*) stocks. In: C.M. Wood & D.G. McDonald (Eds) *Global Warming: Implications for Freshwater and Marine Fish*. pp. 255–78. Cambridge University Press, Cambridge.

Brander, K. & Mohn, R. (2004) Effect of the North Atlantic Oscillation on recruitment of Atlantic cod (*Gadus morhua*). *Canadian Journal of Fisheries and Aquatic Sciences*, **61**, 1558–64.

Bromley, P.J. & Kell, L.T. (1999) Vertical migration and spatial distribution of pelagic 0-group gadoids (cod, haddock, whiting and Norway pout) prior to and during settlement in the North Sea. *Acta Adriatica*, **40**, 7–17.

Brown, J.A., Pepin, P., Methven, D.A. & Somerton, D.C. (1989) The feeding, growth and behaviour of juvenile cod, *Gadus morhua* L., in cold environments. *Journal of Fish Biology*, **35**, 373–80.

Campana, S.E. & Hurley, P.C.F. (1989) An age- and temperature-mediated growth model for cod (*Gadus morhua*) and haddock (*Melanogrammus aeglefinus*) larvae in the Gulf of Maine. *Canadian Journal of Fisheries and Aquatic Sciences*, **46**, 603–13.

Cardinale, M., Hjelm, J. & Casini, M. (2004) Effect of climate on recruitment success of clupeid stocks in the North Atlantic. ICES CM Document 2004/K:01.

Clark, R.A., Fox, C.J., Viner, D. & Livermore, M. (2003) North sea cod and climate change – modelling the effects of temperature on population dynamics. *Global Change Biology*, **9**, 1669–80.

Colloca, F., Cardinale, M., Maynou, F., Giannoulaki, M., Scarcella, G., Jenko, K., Bellido, J. M., *et al.* (2013) Rebuilding Mediterranean fisheries: a new paradigm for ecological sustainability. *Fish and Fisheries*, **14**, 89–109.

Connolly, P. & Officer, R. (2001) The use of tagging data in the formulation of the Irish Sea Cod recovery plan. ICES CM Document 2001/O:05.

Cook, R.M. & Heath, M.R. (2005) The implications of warming climate for the management of North Sea demersal fisheries. *ICES Journal of Marine Science*, **62**, 1322–6.

Cook, R., Sinclair, A. & Stefansson, G. (1997) Potential collapse of North Sea cod stocks. *Nature*, **385**, 521–2.

Daan, N. (1978) Changes in cod stocks and cod fisheries in the North Sea. *Rapports et Procés-verbaux des Réunions, Conseil Permanent International pour l'Exploration de la Mer*, **172**, 39–57.

Darby, C., Hutton, T., Andrews, J., Gurney, W.S.C., Beveridge, D. & Hiddinck, J.G. (2006) Investigations into closed area management of the North Sea cod. Cefas Contract Report C2465. 110 pp.

Defra (2004) *United Kingdom Sea Fisheries Statistics 2003*. Department for Environment, Food and Rural Affairs, London. 145 pp.

Defra (2005) *United Kingdom Sea Fisheries Statistics 2004*. Department for Environment, Food and Rural Affairs, London. 149 pp.

Defra (2006) *United Kingdom Sea Fisheries Statistics 2005*. Department for Environment, Food and Rural Affairs, London. 151 pp.

de Oliveira, J.A.A., Kell, L.T., Punt, A.E., Roel, B.A. & Butterworth, D.S. (2008) Managing without best predictions: the Management Strategy Evaluation framework. In: A.I.L. Payne, A.J.R. Cotter & T. Potter (Eds) *Advances in Fisheries Science: 50 years on from Beverton and Holt*. pp. 104–134. Blackwell Publishing, Oxford.

deYoung, B. & Rose, G.A. (1993) On recruitment and distribution of Atlantic cod (*Gadus morhua* L.) off Newfoundland. *Canadian Journal of Fisheries and Aquatic Sciences*, **50**, 2729–41.

Dower, J.F., Miller, T.J. & Leggett, W.C. (1997) The role of microscale turbulence in the feeding ecology of larval fish. *Advances in Marine Biology*, **31**, 170–220.

Drinkwater, K.F. (2005) The response of Atlantic cod (*Gadus morhua*) to future climate change. *ICES Journal of Marine Science*, **62**, 1327–37.

Dutil, J.-D., Castonguay, M., Gilbert, D. & Gascon, D. (1999) Growth, condition, and environmental relationships in Atlantic cod (*Gadus morhua* L.) in the northern Gulf of St Lawrence and implications for management strategies in the Northwest Atlantic. *Canadian Journal of Fisheries and Aquatic Sciences*, **56**, 1818–31.

Fiksen, Ø., Utne, A.C.W., Aksnes, D.L., Eiane, K., Helvik, J.V. & Sundby, S. (1998) Modelling the influence of light, turbulence and ontogeny on ingestion rates in larval cod and herring. *Fisheries Oceanography*, **7**, 355–63.

Fletcher, S. (2007) Converting science to policy through stakeholder involvement: an analysis of the European Marine Strategy Directive. *Marine Pollution Bulletin*, **54**, 1881–6.

Fox, C.J, Armstrong, M. & Blanchard, J.L. (2005) Irish Sea (ICES Division VIIa). In: *Spawning and Life History Information for North Atlantic Cod Stocks*. pp. 62–75. ICES Cooperative Research Report, No. 274. 154 pp.

Fox, C.J., O'Brien, C.M., Dickey-Collas, M. & Nash, R.D.M. (2000a) Patterns in the spawning of cod (*Gadus morhua* L.), sole (*Solea solea* L.) and plaice (*Pleuronectes platessa* L.) in the Irish Sea as determined by generalized additive modeling. *Fisheries Oceanography*, **9**, 33–49.

Fox, C.J., Planque, B. & Darby, C.D. (2000b) Synchrony in the recruitment time-series of plaice (*Pleuronectes platessa* L.) around the United Kingdom and the influence of sea temperature. *Journal of Sea Research*, **44**, 159–68.

Fromentin, J.-M., Myers, R.A., Bjørnstad, O.N., Stenseth, N.C., Gjøsæter, J. & Christie, H. (2001) Effects of density-dependent and stochastic processes on the regulation of cod populations. *Ecology*, **82**, 829–36.

Gordon, C., Cooper, C., Senior, C.A., Banks, H., Gregory, J.M., Johns, T.C., Mitchell, J.F.B. & Wood, R.A. (2000) The simulation of SST, sea ice extents and ocean heat transports in a version of the Hadley Centre coupled model without flux adjustments. *Climate Dynamics*, **16**, 147–68.

Gregory, R.S. & Anderson, J.T. (1997) Substrate selection and use of protective cover by juvenile Atlantic cod *Gadus morhua* in inshore waters of Newfoundland. *Marine Ecology Progress Series*, **146**, 9–20.

Gurney, W.S.C., Speirs, D.C., Wood, S.N., Clarke, E.D. & Heath, M.R. (2001) Simulating spatially and physiologically structured populations. *Journal of Animal Ecology*, **70**, 881–94.

Hamre, J. (1994) Biodiversity and exploitation of the main fish stocks in the Norwegian–Barents Sea ecosystem. *Biodiversity and Conservation*, **3**, 473–92.

Hardardottir, K., Kjesbu, O.S. & Marteinsdottir, G. (2003) Atresia in Icelandic cod (*Gadus morhua* L.) prior to and during spawning. *Fisken og Havet*, **12**, 51–5.

Harms, I., Heath, M., Bryant, A., Backhaus, J. & Hainbucher, D. (2000) Modelling the north-east Atlantic circulation – Implications for the spring invasion of shelf regions by *Calanus finmarchicus*. *ICES Journal of Marine Science*, **57**, 1694–707.

Heath, M.R., Culling, M.A., Crozier, W.W., Fox, C.J., Gurney, W.S.C., Hutchinson, W.F., Nielsen, E.E., O'Sullivan, M., Preedy, K.F., Righton, D.A., Speirs, D.C., Taylor, M.I., Wright, P.J. & Carvalho, G.R. (2013) Combination of genetics and spatial modeling highlights the sensitivity of cod (*Gadus morhua*) population diversity in the North Sea to distributions of fishing. *ICES Journal of Marine Science*, **71**, 794–807.

Heath, M., MacKenzie, B., Ådlandsvik, B., Backhaus, J., Begg, G., Drysdale, A., Gallego, A., Gibb, F., Gibb, I., Harms, I., Hedger, R., Kjesbu, O., Logemann, K., Marteinsdottir, G., McKenzie, E., Michalsen, K., Nielsen, E., Scott, B., Strugnell, G., Thorsen, A., Visser, A., Wehde, H. & Wright, P. (2003) *An Operational Model of the Effect of Stock Structure and Spatio-temporal Factors on Recruitment*. Final Report of the EU-STEREO Project FAIR-CT98-4122. Fisheries Research Services Contract Report 10/03.

Heessen, H.J.L. (1993) The distribution of cod (*Gadus morhua* L) in the North Sea. *NAFO Scientific Council Studies*, **18**, 59–65.

Hermann, A.J., Hinckley, S., Megrey, B.A. & Napp, J.M. (2001) Applied and theoretical considerations for constructing spatially explicit individual-based models of marine larval fish that include multiple trophic levels. *ICES Journal of Marine Science*, **58**, 1030–41.

Hermann, A.J., Rugen, W.C., Stabeno, P.J. & Bond, N.A. (1996) Physical transport of young pollock larvae (*Theragra chalcogramma* L.) near Shelikof Strait as inferred from a hydrodynamic model. *Fisheries Oceanography*, **5**, 58–70.

Horwood, J., O'Brien, C. & Darby, C. (2006) North Sea cod recovery? *ICES Journal of Marine Science*, **63**, 961–8.

Hulme, M., Jenkins, G.J., Lu, X., Turnpenny, J.R., Mitchell, T.D., Jones, R.G., Lowe, J., Murphy, J.M., Hassell, D., Boorman, P., McDonald, R. & Hill, S. (2002) *Climate Change Scenarios for the United Kingdom: The UKCIP02 Scientific Report*. Tyndall Centre for Climate Change Research, School of Environmental Sciences, University of East Anglia, Norwich, UK. 120 pp.

Huse, G. & Toresen, R. (2000) Juvenile herring prey on Barents Sea capelin larvae. *Sarsia*, **85**, 385–91.

ICES (2000) Report of the Baltic Fisheries Assessment Working Group, Copenhagen, 10–19 April 2000. ICES Document CM 2000/ACFM:14.

ICES (2001) Report of the workshop on gadoid stocks in the North Sea during the 1960s and 1970s. ICES Cooperative Research Report, No. 244. 55 pp.

ICES (2003) Report of the Planning Group on North Sea Cod and Plaice Egg Surveys in the North Sea. Ijmuiden, The Netherlands, 24–26 June 2003. ICES CM Document 2003/G:06.

ICES (2005) Report of the ICES Advisory Committee on Fishery Management, Advisory Committee on the Marine Environment and Advisory Committee on Ecosystems 2005. ICES Advice. Volumes 1–11. 418 pp.

ICES (2006) Report of the ICES Advisory Committee on Fishery Management, Advisory Committee on the Marine Environment and Advisory Committee on Ecosystems 2006. ICES Advice. Volumes 1–10. 310 pp.

ICES (2012) Report of the Workshop to Finalize the ICES Data-limited Stock (DLS) Methodologies Documentation in an Operational Form for the 2013 Advice Season and to make Recommendations on Target Categories for Data-limited Stocks (WKLIFE II), 20–22 November 2012, Copenhagen, Denmark. ICES CM 2012/ACOM:79. 46 pp.

ICES (2013) Report of the Workshop on the Development of Quantitative Assessment Methodologies based on LIFE-history traits, exploitation characteristics, and other key parameters for Data-limited Stocks (WKLIFE III), 28 October–1 November 2013, Copenhagen, Denmark. ICES CM 2013/ACOM:35. 98 pp.

Jarre-Teichmann, A., Wieland, K., MacKenzie, B.R., Hinrichsen, H.H., Plikshs, M. & Aro, E. (2000) Stock–recruitment relationships for cod (*Gadus morhua* callarias L.) in the central Baltic Sea incorporating environmental variability. *Archive of Fishery and Marine Research*, **48**, 97–123.

Kell, L.T., Pilling, G.M. & O'Brien, C.M. (2005) Implications of climate change for the management of North Sea cod (*Gadus morhua*). *ICES Journal of Marine Science*, **62**, 1483–91.

Kjesbu, O.S., Witthames, P.R., Solemdal, P. & Greer Walker, M. (1998) Temporal variations in the fecundity of Arcto-Norwegian cod (*Gadus morhua*) in response to natural changes in food and temperature. *Journal of Sea Research*, **40**, 303–21.

Klyashtorin, L.B. (2001) Climate change and long-term fluctuations of commercial catches – the possibility of forecasting. Federal Institute for Fisheries and Oceanography, Moscow. FAO Fisheries Technical Paper 410. 86pp.

Köster, F.W., Hinrichsen, H.-H., Schnack, D., St John, M.A., MacKenzie, B., Tomkiewicz, J. & Plikshs, M. (1999) Stock–recruitment relationships of Baltic cod incorporating environmental variability and spatial heterogeneity. ICES Document CM 1999/Y:26.

Lambert, Y. & Dutil, J.D. (1997) Condition and energy reserves of Atlantic cod (*Gadus morhua*) during the collapse of the northern Gulf of St Lawrence stock. *Canadian Journal of Fisheries and Aquatic Sciences*, **54**, 2388–400.

Leising, A.W. & Franks, P.J.S. (1999) Larval Atlantic cod (*Gadus morhua*) and haddock (*Melanogrammus aeglefinus*) growth on Georges Bank: a model with temperature, prey size, and turbulence forcing. *Canadian Journal of Fisheries and Aquatic Sciences*, **56**, 25–36.

le Quesne, W., Brown, M., de Oliveira, J. & O'Brien, C. (2013) Data-deficient fisheries in EU waters. Directorate-General for Internal Policies, Policy Department B: Structural and Cohesion Policies, Fisheries, Study IP/B/PECH/IC/2012-118.

Little, A.S., Needle, C.L., Hilborn, R., Holland, D.S. & Marshall, T. (2014) Real-time spatial management approaches to reduce bycatch and discards: experiences from Europe and the United States. *Fish and Fisheries* (doi: 10.1111/faf.12080).

Logemann, K., Backhaus, J. & Harms, I. (2004) SNAC: a statistical emulator of the north-east Atlantic circulation. *Ocean Modelling*, **7**, 97–110.

Longhurst, A. (1998) Cod: perhaps if we all stood back a bit? *Fisheries Research*, **38**, 101–8.

Mangi, S.C., Dolder, P.J., Catchpole, T.L., Rodmell, D. & de Rozarieux (2013) Approaches to fully documented fisheries: practical issues and stakeholder perceptions. *Fish and Fisheries* (doi: 10.1111/faf.12065).

Marshall, C.T. & Frank, K.T. (1999) The effect of interannual variation in growth and condition on haddock recruitment. *Canadian Journal of Fisheries and Aquatic Sciences*, **56**, 347–55.

Marshall, C.T., Kjesbu, O.S., Yaragina, N.A., Solemdal, P. & Ulltang, Ø. (1998) Is spawner biomass a sensitive measure of the reproduction and recruitment potential of Northeast Arctic cod? *Canadian Journal of Fisheries and Aquatic Sciences*, **55**, 1766–83.

Marshall, C.T., Yaragina, N.A., Lambert, Y. & Kjesbu, O.S. (1999) Total lipid energy as a proxy for total egg production by fish stocks. *Nature*, **402**, 288–90.

Marteinsdottir, G. & Begg, G. (2002) Essential relationships incorporating the influence of age, size and condition on variables required for estimation of reproductive potential in Atlantic cod *Gadus morhua* stocks. *Marine Ecology Progress Series*, **235**, 23–56.

Marteinsdottir, G. & Thorarinsson, K. (1998) Improving the stock–recruitment relationship in Icelandic cod (*Gadus morhua* L.) by including age diversity of spawners. *Canadian Journal of Fisheries and Aquatic Sciences*, **55**, 1372–7.

McFarlane, G.A., King, J.R. & Beamish, R.J. (2000) Have there been recent changes in climate? Ask the fish. *Progress in Oceanography*, **47**, 147–69.

McIntyre, T.M. & Hutchings, J.A. (2003) Small-scale temporal and spatial variation in Atlantic cod (*Gadus morhua*) life history. *Canadian Journal of Fisheries and Aquatic Sciences*, **60**, 1111–21.

Megrey, B.A. & Hinckley, S. (2001) Effect of turbulence on feeding of larval fishes: a sensitivity analysis using an individual-based model. *Journal of Marine Research*, **58**, 1015–29.

Mehl, S. & Sunnanå, K. (1991) Changes in growth of Northeast Arctic cod in relation to food consumption in 1984–1988. *ICES Marine Science Symposia*, **193**, 109–12.

Morris, D., O'Brien, C. & Larcombe, P. (2011) Actually achieving marine sustainability demands a radical re-think in approach, not 'more of the same'. *Marine Pollution Bulletin*, **62**, 1053–7.

Myers, R.A. (1998) When do environment–recruitment correlations work? *Reviews in Fish Biology and Fisheries*, **8**, 285–305.

Myers, R.A. & Doyle, R.W. (1983) Predicting natural mortality rates and reproduction–mortality trade-offs from fish life history data. *Canadian Journal of Fisheries and Aquatic Sciences*, **40**, 612–20.

Myers, R.A., Mertz, G. & Barrowman, N.J. (1995) Spatial scales of variability in cod recruitment in the North Atlantic. *Canadian Journal of Fisheries and Aquatic Sciences*, **52**, 1849–62.

Myers, R.A., Mertz, G. & Bridson, J. (1997) Spatial scales of interannual recruitment variations of marine, anadromous, and freshwater fish. *Canadian Journal of Fisheries and Aquatic Sciences*, **54**, 1400–17.

NAFO Working Group on Reproductive Potential (2003) Atlantic Cod in ICES Area IV. Scientific Council Studies, **37**, 85–91.

Needle, C.L., O'Brien, C.M., Darby, C.D. & Smith, M.T. (2003) Incorporating time-series structure in medium-term recruitment projections. *Scientia Marina*, **67** (Suppl. 1), 201–9.

O'Brien, C.M. (1999a) An approach to stock–recruitment modelling based upon GLMs, HGLMs, and DLMs. ICES Document CM 1999/T:01.

O'Brien, C.M. (1999b) Time-series models in fish recruitment – a journey from classical statistics to dynamic models and Bayesian forecasting. ICES Document CM 1999/T:02.

O'Brien, C.M., Darby, C.D., Rackham, B.D., Maxwell, D.L., Degel, H., Flatman, S., Mathewson, M., Pastoors, M.A., Simmonds, E.J. & Vinther, M. (2001) The precision of international market sampling for North Sea cod (*Gadus morhua* L.) and its influence on stock assessment. ICES CM 2001/P:14.

O'Brien, C.M., Fox, C.J., Planque, B. & Casey, J. (2000) Climate variability and North Sea cod. *Nature*, **404**, 142.

O'Brien, C.M. & Little, A.S. (2006) Incorporation of Process Information into Stock–Recruitment Models. ICES Cooperative Research Report, No. 282. 152 pp.

Oosthuizen, E. & Daan, N. (1974) Egg fecundity and maturity of North Sea cod, *Gadus morhua*. *Netherlands Journal of Sea Research*, **8**, 378–97.

Ottersen, G., Alheit, J., Drinkwater, K., Frieland, K., Hagen, E. & Stenseth, N.C. (2005) The responses of fish populations to ocean climate fluctuations. In: N.C. Stenseth, G. Ottersen, J.W. Hurrel & A. Belgrano (Eds) *Marine Ecosystems and Climate Variation: The North Atlantic – A Comparative Perspective*. pp.73–94. Oxford University Press, Oxford.

Ottersen, G. & Loeng, H. (2000) Covariability in early growth and year-class strength of Barents Sea cod, haddock, and herring: the environmental link. *ICES Journal of Marine Science*, **57**, 339–48.

Ottersen, G., Michalsen, K. & Nakken, O. (1998) Ambient temperature and distribution of northeast Arctic cod. *ICES Journal of Marine Science*, **55**, 67–85.

Ottersen, G.S. & Sundby, S. (1995) Effects of temperature, wind and spawning-stock biomass on recruitment of Arcto-Norwegian cod. *Fisheries Oceanography*, **4**, 278–92.

Page, F.H. & Frank, K.T. (1989) Spawning time and egg stage duration in Northwest Atlantic haddock (*Melanogrammus aeglefinus*) stocks with emphasis on Georges and Browns Bank. *Canadian Journal of Fisheries and Aquatic Sciences*, **46**, 68–81.

Parma, A.M. (2000) *In search of robust harvest rules in the face of uncertain assessments and decadal changes in productivity*. Presented at the Symposium on Targets, Reference Points, and the Burden of Proof in Fishery Management, 31 October–2 November 2000, MOTE Marine Laboratory, Sarasota, FL.

Pastoors, M.A., Poos, J.J., Kraak, S.B.M. & Machiels, M.A.M. (2007) Validating management simulation models and implications for communicating results to stakeholders. *ICES Journal of Marine Science*, **64**, 818–24.

Patterson, K.R., Cook, R.M., Darby, C.D., Gavaris, S., Mesnil, B., Punt, A.E., Restrepo, V.R., Skagen, D.W., Stefánsson, G. & Smith, M. (2000) Validating three methods for making probability statements in fisheries forecasts. ICES Document CM 2000/V:06.

Pepin, P. (1991) Effect of temperature and size on development, mortality and survival rates of the pelagic early life history stages of marine fish. *Canadian Journal of Fisheries and Aquatic Sciences*, **48**, 503–18.

Perry, A.L., Low, P.J., Ellis, J.R. & Reynolds, J.D. (2005) Climate change and distribution shifts in marine fishes. *Science*, **308**, 1912–15.

Planque, B. & Fox, C.J. (1998) Interannual variability in temperature and the recruitment of Irish Sea cod. *Marine Ecology Progress Series*, **172**, 101–5.

Planque, B., Fox, C.L., Saunders, A. & Rockett, P. (2003) On the prediction of short-term changes in the recruitment of North Sea cod (*Gadus morhua* L.) using statistical temperature forecasts. *Scientia Marina*, **67** (Suppl.1), 211–18.

Planque, B. & Frédou, T. (1999) Temperature and the recruitment of Atlantic cod (*Gadus morhua*). *Canadian Journal of Fisheries and Aquatic Sciences*, **56**, 2069–77.

Pope, V.D., Gallani, M.L., Rowntree, P.R. & Stratton, R.A. (2000) The impact of new physical parameterizations in the Hadley Centre climate model: HadAM3. *Climate Dynamics*, **16**, 123–46.

Purchase, C.F. & Brown, J.A. (2001) Stock-specific changes in growth rates, food conversion efficiencies, and energy allocation in response to temperature change in juvenile Atlantic cod. *Journal of Fish Biology*, **58**, 36–52.

Quinlan, J.A. & Crowder, L.B. (1999) Searching for sensitivity in the life history of Atlantic menhaden: inferences from a matrix model. *Fisheries Oceanography*, **8**, 124–33.

Rideout, R.M., Burton, M.P.M. & Rose, G.A. (2000) Observations on mass atresia and skipped spawning in northern Atlantic cod, from Smith Sound, Newfoundland. *Journal of Fish Biology*, **57**, 1429–40.

Rijnsdorp, A. & Jaworski, A. (1990) Size-selective mortality in plaice and cod eggs – a new method in the study of egg mortality. *Journal du Conseil International pour l'Exploration de la Mer*, **47**, 256–63.

Riley, J.D. & Parnell, W.G. (1984) The distribution of young cod. In: E. Dahl, D.S. Danielssen, E. Moksness & P. Solemdal (Eds) *The Propagation of Cod* Gadus morhua L. pp. 563–80. Flødevigen Rapportser, 1, Institute of Marine Research, Arendal, Norway.

Robichaud, D. & Rose, G.A. (2004) Migratory behaviour and range in Atlantic cod: inference from a century of tagging. *Fish and Fisheries*, **5**, 185–214.

Roel, B.A., O'Brien, C.M. & Basson, M. (2004) Management options for the Blackwater herring, a local spring-spawning stock in the Thames Estuary. *ICES Journal of Marine Science*, **61**, 297–307.

Rose, G. (2004) Reconciling over-fishing and climate change with stock dynamics of Atlantic cod (*Gadus morhua*) over 500 years. *Canadian Journal of Fisheries and Aquatic Sciences*, **61**, 1553–7.

Rose, G.A. (2005) On distributional responses of North Atlantic fish to climate change. *ICES Journal of Marine Science*, **62**, 1360–74.

Rosenberg, A.A., Fogarty, M.J., Cooper, A.B., Dickey-Collas, M., Fulton, E.A., Gutiérrez, N.L., Hyde, K.J.W., Kleisner, K.M., Kristiansen, T., Longo, C., Minte-Vera, C., Minto, C., Mosqueira, I., Chato Osio, G., Ovando, D., Selig, E.R., Thorson, J.T. & Ye, Y. (2014) Developing new approaches to global stock status assessment and fishery production potential of the seas. FAO Fisheries and Aquaculture Circular No. 1086. FAO, Rome.

Rothschild, B.J., ChangshengChen & Greg Lough, R. (2005) Managing fish stocks under climate uncertainty. *ICES Journal of Marine Science*, **62**, 1531–41.

Schiermeier, Q. (2004) Climate findings let fishermen off the hook. *Nature*, **428**, 4.

Scott, B., Marteinsdottir, G. & Wright, P. (1999) The potential effects of maternal factors on spawning stock–recruitment relationships under varying fisheries pressure. *Canadian Journal of Fisheries and Aquatic Sciences*, **56**, 1882–90.

Secor, D.H. (2000) Spawning in the nick of time? Effect of adult demographics on spawning behaviour and recruitment of Chesapeake Bay striped bass. *ICES Journal of Marine Science*, **57**, 403–11.

Shelton, P.A. (2005) Did over-reliance on commercial catch rate data precipitate the collapse of northern cod? *ICES Journal of Marine Science*, **62**, 1139–49.

Shelton, P.A. & Lilly, G.R. (2000) Interpreting the collapse of the northern cod stock from catch and survey data. *Canadian Journal of Fisheries and Aquatic Sciences*, **57**, 2230–9.

Sparholt, H. (1996) Causal correlation between recruitment and spawning stock size of central Baltic cod? *ICES Journal of Marine Science*, **53**, 771–9.

Sparholt, H., Aro, E. & Modin, J. (1991) The spatial distribution of cod (*Gadus morhua* L.) in the Baltic Sea. *Dana*, **9**, 45–56.

STECF (2003) Report of meeting on cod assessment and technical measures, 28 April–7 May 2003, Brussels. Edited by H.-J. Rätz.

Sumaila, U.R., Cheung, W.W.L., Lam, V.W.Y., Pauly, D. & Herrick, S. (2011) Climate change impacts on the biophysics and economics of world fisheries. *Nature Climate Change*, **1**, 449–56.

Swain, D.P. & Wade, E.J. (1993) Density-dependent geographic distribution of Atlantic cod (*Gadus morhua* L.) in the southern Gulf of St Lawrence. *Canadian Journal of Fisheries and Aquatic Sciences*, **50**, 725–33.

Trippel, E.A. (1999) Estimation of stock reproductive potential: history and challenges for Canadian Atlantic gadoid stock assessments. *Journal of Northwest Atlantic Fishery Science*, **25**, 61–81.

Ware, D.M. (1980) Bioenergetics of stock and recruitment. *Canadian Journal of Fisheries and Aquatic Sciences*, **37**, 1012–24.

Witthames, P.R., Greenwood, L. & Lyons, B. (2000) Ovarian atresia in *Sole solea* (L.) In: B. Norberg, O.S. Kjesbu, G.L. Taranger, E. Andersson & S.O. Stefansson (Eds) *Proceedings of the Sixth International Symposium on the Reproductive Physiology of Fish*. Institute of Marine Research, Arendal, Norway.

Wright, P., Gibb, F., Gibb, I., Heath, M. & McLay, H. (2003) North Sea Cod Spawning Grounds. Fisheries Research Services Internal Report 17 03.

Chapter 11

Implementing Information on Stock Reproductive Potential in Fisheries Management: The Motivation, Challenges and Opportunities

C. Tara Marshall

11.1 Introduction

Simplifying assumptions are unavoidable when estimating what is essentially unknowable: the absolute number of fish belonging to a marine fish stock. Through ubiquitous use, these assumptions become ingrained in our understanding of population dynamics to the point that they are rarely scrutinised. The necessity of simplifying assumptions should not, however, pre-empt regular scrutiny, particularly when new knowledge casts doubt on their validity. For example, the assumption that natural mortality is constant and equal to 0.2 is widely applied in stock assessment despite its obvious fallacy. Using empirical data for a closed fishery to estimate natural mortality (M) for cod in the southern Gulf of St Lawrence, Sinclair (2001) suggested that M for this stock was temporally variable and frequently closer to 0.4. Although some stocks adjust the value of M to reflect higher rates of mortality in specific time periods the majority assume M is constant and equal to 0.2.

A simplifying assumption underlying the stock/recruit (S/R) relationship is that spawning stock biomass (SSB) is proportional to total egg production (TEP). In the original formulations of the S/R relationship (Ricker 1954, Beverton & Holt 1957), recruitment was recognised to be functionally dependent on TEP by the stock. However, both studies introduced the assumption of direct proportionality. For example, Beverton & Holt (1957) formulated TEP as:

$$\text{TEP} = s\chi\text{SSB} \tag{1}$$

where s is the proportion of females and χ is the relative fecundity (no. of eggs per unit body weight). The assumption of proportionality thus depends on both s and χ being constant through time. Even a cursory analysis would suggest that these two constancy assumptions are invalid. Dimorphic growth is common among fishes (Parker 1992, Henderson *et al.*

Fish Reproductive Biology: Implications for Assessment and Management, Second Edition.
Edited by Tore Jakobsen, Michael J. Fogarty, Bernard A. Megrey and Erlend Moksness.

2003, Purchase *et al.* 2005, Schutz *et al.* 2006). Differential maturation and morality schedules for males and females will, in turn, contribute to variable sex ratios with age, particularly for long-lived species. In cases where males exhibit higher growth and/or earlier maturation (e.g. Hunt 1996), sex ratios often become progressively female-biased with increasing age or length (Ajiad *et al.* 1999, Marshall *et al.* 2006). Consequently, the loss of large spawners that often results from high fishing mortality will disproportionately reduce the number of females in the spawning stock and reduce s of the stock. Demographic shifts towards smaller spawners can also reduce the χ of the stock if the relative fecundity of individuals is positively correlated with length (Kjesbu *et al.* 1998, Marshall *et al.* 1998, Oskarsson & Taggart 2005). Not surprisingly, an empirical test of the two constancy assumptions found that, over a 55-year time series, both s and χ exhibited considerable interannual variation that was systematic with trends in the length composition of the stock such that both parameters decreased as the stock became progressively more dominated by smaller-sized spawners (Marshall *et al.* 2006). Interannual variability in condition will also contribute to variation in χ of the stock given that individual fecundity is influenced by condition in several stocks (Yoneda & Wright 2004, Marshall *et al.* 2006).

The use of SSB as a proxy for TEP in fisheries management has come under increasing challenge (Marshall *et al.* 1998, Cardinale & Arrhenius 2000, Berkeley *et al.* 2004, Spencer *et al.* 2007). At the same time longevity overfishing, resulting from the selective removal of large proportions of old fish which are disproportionately more important for stock reproductive potential, is of concern for management because of the possibility that it results in decreased reproductive potential (Beamish *et al.* 2006, Hsieh *et al.* 2006, Trippel *et al.* 1997). Despite the growing realisation that essential features of population dynamics are being overlooked by using relatively coarse measures of stock reproductive potential, there are currently no examples where an index other than SSB is used to formulate management advice for many important groundfish, flatfish and pelagic species (with the exception of stocks assessed using egg production methods which are detailed in Chapter 8). The overall aim of this chapter is to summarise why, when and how fisheries management should incorporate improved knowledge about life-history traits relevant to reproduction. This aim will be achieved by: summarising arguments that have been used in the past to justify continued use of SSB; reviewing the evidence that suggests that SSB is not accurately representing the dynamic range in stock reproductive potential; summarising the criteria by which the alternative indices of reproductive potential should be evaluated; illustrating how alternative indices of reproductive potential have been estimated for several cod stocks; indicating how these estimates can be incorporated into conventional management protocols; and lastly, highlighting several novel directions for future research.

11.2 Justifying the use of spawning stock biomass to represent stock reproductive potential

To paraphrase an old adage, old biases die hard in fisheries science. In spite of growing evidence that SSB is a biased estimator of stock reproductive potential, the use of SSB by management is a difficult habit to break. Some of the justifications for this practice are reviewed below.

11.2.1 Lack of requisite data on fecundity

Almost certainly, the lack of fecundity data in the 1950s necessitated the original assumption by Beverton, Holt and Ricker that SSB is directly proportional to TEP. Since then, large-scale sampling programmes and research into reproductive biology of commercial species have increased the availability of empirical data that can be used to quantify stock reproductive potential. Sex is routinely recorded in fisheries databases (Tomkiewicz *et al.* 2003); consequently there is no need to continue to assume that *s* is constant. For many stocks historical data have been used to develop a time series for sex ratios and maturity ogives that are specific to females (e.g. Tomkiewicz *et al.* 1997, Morgan & Brattey 2005, Marshall *et al.* 2006). This allows female-only SSB (FSB) to be estimated for the full assessment time period. This simple change would, in itself, lead to a more sensitive index of reproductive potential given that in many exploited stocks the spawning stock has become progressively more male-biased through the loss of large, old spawners that are disproportionately female (Marshall *et al.* 2006). Fecundity data remain in scarce supply (Tomkiewicz *et al.* 2003). Even for stocks having multi-year data on fecundity (e.g. Thorsen *et al.* 2006), the number of years for which fecundity has been estimated is small compared with the number of years represented in the stock assessment.

The availability of fecundity data can reasonably be expected to increase in the future. The development of automated procedures for fecundity estimation (Thorsen & Kjesbu 2001, Friedland *et al.* 2005) has substantially reduced the amount of time required to estimate fecundity. Consequently, a larger number of government laboratories are routinely determining fecundity for commercial fish stocks. Over time this will give a better understanding of the magnitude of interannual variation in relative fecundity and help identify the environmental factors that contribute to this interannual variation, e.g. condition and temperature. This knowledge allows statistical models to be developed which hindcast fecundity for years that lack fecundity data (e.g. Kraus *et al.* 2002, Marshall *et al.* 2006, Thorsen *et al.* 2006, Rideout & Morgan 2007). Once time series of age-specific fecundity become available for data-rich stocks then it becomes difficult to justify the continued use of SSB as a proxy for TEP.

11.2.2 Universality of spawning stock biomass

The continued use of SSB could be justified by the desire to apply generic stock assessment methodology thereby maintaining consistency among stocks. Whether consistency is a desirable goal for stocks spanning a gradient from data-rich to data-poor is questionable. Indeed, it has been argued that a completely different suite of management protocols are required for data-poor stocks (Kelly & Codling 2006). In the case of data-rich stocks maintaining consistency requires overlooking relevant scientific knowledge and data in the interest of conforming to the standards set by data-poor stocks. Such a non-scientific approach is not easily defended.

In reality, there are already considerable differences between stocks in how they represent reproductive potential. For *K*-selected species experiencing high rates of exploitation, e.g. elasmobranchs and sharks, it intuitively makes sense to estimate FSB and accommodate reproductive rates that are low and age-dependent. In the northwest Atlantic, FSB is being used to estimate biological reference points for spiny dogfish *(Squalus acanthias),* and fecundity data, in terms of number of pups produced per female at length, has been incorporated in population projections (Northeast Fisheries Science Center 2003). Because small females produce smaller pups that may have a lower survival rate, the population projections account for differential survival rates (Northeast Fisheries Science Center 2003).

While it makes intuitive sense to define and estimate stock reproductive potential as precisely as possible for *K*-selected species, similar reasoning has not been applied to highly fecund *r*-selected species. This could stem from an innate expectation that the impact of fecundity variation on population dynamics is negligible when individual fecundity ranges from tens of thousands to millions. This viewpoint has a long history. Kurlansky (1997) quotes early fish biologists, J. Smith Homans Sr & Jr: 'Cod—It is amazingly prolific. Leewenhoek counted 9 384 000 eggs in a cod-fish of middling size—a number that will baffle all the efforts of man to exterminate'. It is clearly evident that even prolific species such as cod can be reduced to the point of stock extinction. It is possible that we do not yet fully understand the role that stock reproductive potential plays in predisposing individual stocks to collapse because we are not accurately quantifying it.

11.3 Why should fisheries management change how it estimates reproductive potential?

The preceding section established that the scientific justification for using SSB as an index of stock reproductive potential partly stems from a lack of fecundity data and the desire for consistency across stocks. This section will examine the drivers for change from a variety of perspectives, ranging from simple conceptual arguments and progressing through to more empirically-based reasons.

11.3.1 Conceptual advantages

There are several conceptual advantages to replacing SSB with TEP. Because SSB is a proxy for TEP, TEP is the better measure of reproductive potential by definition. TEP incorporates a larger suite of factors that, in combination, determine stock reproductive potential (abundance, maturation rates, sex ratios, condition and fecundity). SSB represents a smaller subset of these factors (usually abundance, maturation rates and weight) and thus could be considered to be less dynamic than TEP because it incorporates fewer factors. This basic logic could justify switching to TEP provided that the additional terms included in TEP do not introduce greater uncertainty into the estimate of stock reproductive potential (see Section 11.3.2). Using TEP would also be advantageous because it is defined as the potential number of recruits at time equal zero and is measured in the units of absolute abundance. Therefore, it allows mortality to be calculated from time zero. Having an estimate of starting abundance of the cohort is useful for a range of ecological modelling applications (see Section 11.6). Finally, TEP is an informative axis to include in the well-known Paulik diagram (Nash 1998) which tracks the changes in abundance throughout the life cycle.

11.3.2 Uncertainty of the estimates

It would be a strong disincentive to use an alternative index of reproductive potential, e.g. TEP, if the estimation error was substantially larger than for SSB. Although it is not possible to quantify the degree of error in estimates of SSB or TEP in absolute terms, it is possible to assess errors qualitatively. Different sources of uncertainty in the metrics used by fisheries management can be categorised into several different types (Kell *et al.* 2005). The following types of errors are relevant to assessing the relative degree of uncertainty in different estimates of stock reproductive potential:

(1) *Model error:* related to the ability of the model structure to capture the true complexity of the quantity of interest.
(2) *Observation error:* sampling error and measurement error.
(3) *Estimation error:* arising when estimating parameters of the models used to estimate the quantity of interest.

To assess the differences between SSB and TEP in model error it is necessary to compare the component parts of the equations used to estimate them. The formula for estimating SSB in a given year from age-based information is:

$$SSB = \sum_{a} n_a \times m_a \times w_a \tag{2}$$

where a indexes age and n_a, m_a and w_a are the numbers-, proportion mature- and weight-at-age, respectively. By comparison, TEP in a given year can be estimated as:

$$TEP = \sum_{a} n_a \times s_a \times m_{a|f} \times e_a \tag{3}$$

where s_a is the sex ratio at age, $m_a|_f$ is the proportion of females that are mature at age and e_a is the number of eggs produced by a mature females of age a. Clearly, the model error for TEP is less than that for SSB because it incorporates a higher degree of biological realism into the model. This would not necessarily be true if simpler formulations of TEP are used. For example, the formulation could also be used for TEP:

$$TEP = \sum_{a} n_a \times 0.5 \times m_a \times w_a \times re_a \tag{4}$$

where 0.5 converts n_a to the number of females and re_a is the number of eggs produced by mature female of a given age per unit weight (i.e. χ expressed at the individual level for specific age classes). In cases where re_a is assumed to be constant across age classes and time then there is no difference between Equation (2) and (4) other than a scaling difference; both would have the same degree of model error. In cases where re_a is estimated on an annual basis then Equation (4) would have less model error than Equation (2) because a valid component of the stock reproductive potential had been incorporated into the model formulation.

When comparing the degree of observation error in Equations (2) and (3) it is reasonable to assume that the observation error for $m_a|_f$ is comparable in magnitude to m_a. For many species determining gender is essentially error-free. Consequently, the s_a term can reasonably be considered as having a negligible degree of observation error when the databases used to estimate s_a are of sufficient size. The problem of assessing whether Equation (3) has more observation error than Equation (2) therefore reduces to comparing the relative magnitude of observation errors in w_a and e_a. Analytically, individual weight is estimated gravimetrically with a high degree of precision. However, individual fecundity can also be estimated with a high degree of precision using automated methods (Thorsen & Kjesbu 2001). In both w_a and e_a the major contributor to observation error would be biases resulting from unrepresentative sampling strategies (too few samples, restricted size range, restricted spatial sampling). There are potentially sources of bias in both e_a and $w_{a.}$ Individual e_a is obviously a property of mature females sampled close to the start of spawning. This restricts the measurement of e_a for a particular gender and maturity stage making it easier to define a 'representative' sample. It is considerably more difficult to define a representative sample for w_a because individual weight increases throughout the year, exhibits regional variability and varies between immature and mature individuals or between sexes.

Estimation error is another aspect of uncertainty that is relevant to estimating stock reproductive potential. As fecundity data are only available for a limited number of years, values of e_a must be hindcast for the full time period in order to establish a time series of TEP (e.g. Marshall *et al.* 2006). The accuracy of this hindcasting model depends on the data set that it is conditioned on and whether the data set captures the full range of dynamic behaviour in $e_{a.}$ If fecundity is only estimated in years when individuals were consistently in good or poor physiological condition or for a sample of small females then it will be inappropriate to apply the general model derived from that data set to a long time period during which physiological condition of individuals comprising the spawning stock varies considerably or the size composition included a high proportion of large size classes. Even though hindcasting models introduce estimation error into $e_{a,}$ it should be recognised that estimates of w_a can also be poorly estimated. For some stocks, w_a is derived from mean length-at-age by applying a fixed weight/length relationship. This practice arose in an era when it was technically difficult to measure individual weights at sea due to the motion of the ship. Estimates of w_a obtained this way will have a high degree of estimation error when weight-at-length varies interannually.

For any individual stock, the qualitative arguments above amount to assessing whether the reduction in model error that is achieved by using TEP outweighs the possible increases in observation or estimation errors in TEP. A simulation study by De Oliveira *et al.* (2006) using data for an indeterminate spawner (western horse mackerel) concluded that when basic knowledge and data describing reproductive potential of the stock are unknown, i.e. model and observation errors are large, proxies for TEP will degrade management advice compared with assuming that reproductive output (expressed relative to SSB) is constant. In the absence of a rigorous quantitative analysis of the actual magnitudes of the errors associated with SSB and TEP the magnitude of uncertainty in TEP should not necessarily be assumed to be greater than that for SSB, particularly given known sources of systematic bias in SSB (Marshall *et al.* 2006).

11.3.3 Accounting explicitly for maternal effects on reproductive potential

There is growing literature describing maternal effects on different aspects of population dynamics including fecundity, viability of eggs and larvae, the temporal and spatial structure of spawning and possibly recruitment. This section provides a brief overview of a selection of these studies and indicates, in general terms, how these maternal effects could be incorporated in fisheries management.

11.3.3.1 Maternal effects on potential and realised fecundity

A large number of studies have shown that individual fecundity is dependent on maternal characteristics. These include the following:

- Relative fecundity (χ) varies interannually by 90% for Northeast Arctic cod (Kjesbu *et al.* 1998) and by 33% for Baltic cod (Kraus *et al.* 2000).
- Interannual variation in spawner condition affects the age- or size-specific fecundity for many fish species (Hislop *et al.* 1978, DeMartini 1991, Koslow *et al.* 1995, Kjesbu *et al.* 1998, Ma *et al.* 1998).
- Spawners in poor condition can skip spawning seasons altogether (Burton *et al.* 1997, Marshall *et al.* 1998, Rideout *et al.* 2005, Rideout & Rose 2006, Rideout *et al.* 2006).
- The size, age and condition of female cod affects the size of eggs and larvae (Kjesbu 1989, Chambers & Waiwood 1996, Marteinsdottir & Steinarsson 1998).
- Fecundity and total egg dry weight were significantly lower in poor-condition female cod (Lambert & Dutil 2000).
- Fecundity of herring was positively correlated with size and condition (Oskarsson *et al.* 2002)
- Atresia (the resorption of developing oocytes) is higher in poor condition females (Kjesbu *et al.* 1991, Ma *et al.* 1998, Oskarsson *et al.* 2002, Kurita *et al.* 2003).
- Relative fecundity of cod in the North Sea is influenced by nutritional status, as measured by liver condition index (Yoneda & Wright 2004).
- Female energy reserves 3 to 4 months before spawning impact relative fecundity of cod (Skjaeraasen *et al.* 2006).

Condition effects on fecundity have not been found in all stocks (McIntyre & Hutchings 2003, Koops *et al.* 2004). One possible explanation is that stocks growing in highly productive ecosystems would be unlikely to be in poor physiological condition when fecundity samples are taken. However, for stocks experiencing large amplitude fluctuations in feeding conditions the evidence that condition impacts various aspects of reproductive potential, including condition, is convincing. For these stocks, it is imperative that the index used to represent stock reproductive potential is responsive to the effects of condition. Fecundity models that are used to reconstruct a time series of e_a often include a term for condition (Marshall *et al.* 2006). This makes the corresponding estimate of TEP more sensitive to environmental variation and reduces model error. Many current measures of SSB are relatively insensitive to the effect that interannual variation in condition has on stock reproductive potential. This is especially true for stocks for which SSB is estimated by converting mean length-at-age to mean weight-at-age using a fixed weight/length relationship. In this case estimates of SSB are entirely insensitive to the effect of environment on reproductive potential.

11.3.3.2 Maternal effects on viability of eggs and larvae

The growing literature on this topic was recently reviewed by Lambert *et al.* (2003). However, some examples of maternal effects on egg and larval viability include:

- Eggs from recruit (first-time) spawners have a lower hatching success compared with second- or third-time spawners (Solemdal *et al.* 1995).

- Larvae from large eggs have been shown to have higher yolk reserves (Trippel 1998), greater feeding success (Knutsen & Tilseth 1985), a higher probability of developing functional swimbladders (Marteinsdottir & Steinarsson 1998) and higher rates of swimming activity and growth (Solemdal *et al.* 1992).
- Positive correlations have also been detected between female condition and larval feeding success and larval specific growth rates (Marteinsdottir & Steinarsson 1998).
- Larval growth and survival are positively correlated with egg size in Baltic cod (Nissling *et al.* 1998).
- Larval production per gram of maternal body weight was one order of magnitude higher for repeat-spawning cod relative to recruit spawners (Trippel 1998).
- Larvae produced by older female black rockfish grow more than three times as fast and survive starvation for longer periods compared with larvae produced by younger females (Berkeley *et al.* 2004).
- Larvae produced by large female striped bass had faster growth-in-length rates compared with larvae produced to small females. However, mean survival rates were unrelated to maternal size (Monteleone & Houde 1990).

It is therefore relevant to fisheries management to distinguish between potential TEP and realised TEP. The former term is equivalent to the number of oocytes in the ripening ovaries of the spawning stock, whereas the latter term corresponds to the number of eggs that are spawned by the stock and proceed through fertilisation to hatching, i.e. total hatched egg production. To incorporate this distinction into estimates of stock reproductive potential requires a model describing how the observed maternal characteristics (age, length, condition) relate to hatching success. At the present time there are relatively few data sets which could be used to develop a model that can be applied to the stock level. Further research is required before information can be incorporated into fisheries management in any detailed way. Simulation studies (e.g. Scott *et al.* 1999, 2006, Murawski *et al.* 2001, O'Farrell & Botsford 2006) are an alternative until the models are developed and validated.

11.3.3.3 Maternal effects on the temporal and spatial structure of spawning

The start time of spawning can show considerable interannual variation and the duration of the spawning season of many temperate fish species extends over several weeks to months. These aspects of spawning can contribute to variability in recruitment by influencing the synchrony between offspring production and favourable environmental conditions. There can be selective advantages to spreading reproduction over time to ensure that not all offspring experience suboptimal conditions for growth and survival (Mertz & Myers 1994). Similarly, spawning location can contribute to recruitment success through interactions with environmental factors. Studies which have shown maternal effects on timing and location of spawning include the following:

- In most batch-spawning species the older/larger females begin spawning earlier in the season and continue for longer than younger and smaller females (DeMartini & Fountain 1981, Parrish *et al.* 1986, Buckley *et al.* 1991, Hutchings & Myers 1993, Marteinsdottir & Bjornsson 1999).

- Single batch-spawning species may also have a population spawning season of variable duration through age-related differences in the timing of spawning (Lambert 1990).
- For all species that exhibit an age- or size-related spawning the removal of older members of a stock will shorten the population spawning season and change the peak time of egg production (Trippel *et al.* 1997).
- In simulations incorporating environment conditions, age structure and the age-specific spatial distribution of spawners, differential survival to the end of the pelagic stage was observed between haddock eggs spawned in different areas (Heath & Gallego 1998). The study concluded that the progressive loss of older age classes can reduce the maximum recruitment rate purely due to spatial factors.
- Estimates of spawner biomass that incorporate spatial variability in age and size at maturity of spawners are more temporally variable than those estimated by the VPA (Rochet 2000).
- There is considerable spatial variation in the size and age of spawners among adjacent geographical locations within the region generally referred to as the main spawning ground of cod (Marteinsdottir *et al.* 2000). Variations among subregions in egg production must be considered in relation to environmental differences.

Simulation analysis showed that fisheries-induced shifts in size composition towards smaller spawners result in a spawning season that is 4 months shorter and a 2 week shift in the date of peak spawning (Scott *et al.* 2006). Unfortunately, it is difficult to validate the predictions of simulation models because high resolution data describing the structure and location of the stock at the time of spawning are lacking for most stocks.

11.3.3.4 Proxies for maternal effects on recruitment

Because many stocks lack requisite data (principally fecundity) for estimating TEP, proxies for reproductive potential have been considered (Lambert *et al.* 2003). There are several studies which have constructed proxies for stock reproductive potential and tested for a significant relationship with recruitment. These studies, which suggest the maternal effect on recruitment may not be properly accounted for using SSB, include the following:

- There is a higher probability of above average recruitment when age diversity is high for Norwegian spring-spawning herring (Lambert 1987, 1990) and Icelandic cod (Marteinsdottir & Thorarinsson 1998).
- High fishing mortality reduces the number of old/large spawners (Trippel *et al.* 1997). Old/large spawners contribute disproportionately to the number of successful recruits (Scott *et al.* 1999). Consequently, the reproductive potential of heavily exploited stocks may have been reduced to a greater degree than is indicated by decreases in spawner biomass.
- Positive associations between condition and recruitment have been observed for both demersal (Marshall & Frank 1999, Marshall *et al.* 1999) and pelagic (Boyd *et al.* 1998) stocks.
- For Georges Bank haddock the fit of the S/R relationship was improved if estimates of spawning stock biomass excluded the first-time spawners (Wigley 1999).
- Older repeat-spawning cod contribute the largest amount of eggs having the highest survival probability (Cardinale & Arrhenius 2000).

- For Baltic cod, replacing spawner biomass with total viable egg production has improved the S/R relationship (Jarre-Teichmann *et al.* 2000).
- For Northeast Arctic cod the total lipid energy in the livers of mature females showed a positive, linear correlation with recruitment over a 50-year time period (Marshall *et al.* 2000).
- Recruitment of Baltic cod is positively correlated to the fraction of eggs produced by old females (Vallin & Nissling 2000).
- Proxies representing age composition and the proportion of first-time spawners did not consistently explain recruitment variation in three Atlantic cod stocks and an American plaice stock (Morgan *et al.* 2007).

Proxies are useful means of identifying factors that may be important to recruitment and misrepresented using SSB as an index of reproduction. As proxies, they are indirect measures of stock reproductive potential having an inherent degree of uncertainty. The failure of a proxy to explain a higher proportion of variation in recruitment compared with SSB could simply indicate that the proxy was of low value. Many proxies are responding to a single aspect of stock dynamics (e.g. condition, age diversity) and not the full suite of factors that, in combination, determine stock reproductive potential (abundance, maturation rates, sex ratios, condition, fecundity). Consequently, they should be applied with suitable caution and interpreted with appropriate caveats. Alternatively, proxies could be evaluated by examining their relationship with residuals from the S/R relationship (Morgan *et al.* 2007).

11.3.4 Reformulating the stock/recruit relationship

The strongest justification for switching to an alternative index of reproductive potential would be that it explains a higher proportion of recruitment variability. There are too few stocks which have reconstructed the S/R relationship using well-founded estimates of TEP to be able to draw a general consensus on this point. Incorporating increasing levels of biological detail into the alternative estimates of reproductive potential resulted in incremental improvements in the amount of variation explained in the S/R relationship for the Baltic cod stock (Köster *et al.* 2001). This was not necessarily the case for Northeast Arctic cod (Marshall *et al.* 2006) or three cod stocks on the Northwest Atlantic (Morgan & Brattey 2005). The customary explanation for the weakness of the S/R relationship is that there is a high degree of interannual variability in mortality during early life-history stages. Given the importance of mortality during early life stages for recruitment it would be inappropriate to require the alternative indices of reproductive potential to explain a substantially higher degree of recruitment variability than SSB (notwithstanding that they do so for Baltic cod). They should not, however, explain substantially less recruitment variation. For NA cod the r^2 values obtained using two alternative indices of reproductive potential (SSB and TEP) were slightly lower those obtained using SSB (Marshall *et al.* 2006). The small reduction in explanatory power that would accompany replacing SSB with either of these two alternatives should be judged against the benefit of correcting for a systematic source of bias in the use of SSB.

11.4 How to estimate alternative indices of stock reproductive potential

There are a small but growing number of studies which have estimated TEP for the time period used by management to formulate the S/R relationship. These different studies use a variety of approaches depending on the type and quality of data available for each stock. A key consideration is the availability of body size information over the full time period. For long-lived species many reproductive traits, including sex ratio, maturity and fecundity, are influenced by body size as well as age. Weight-at-age is routinely reported because it is used to estimate SSB. In some cases weight-at-age is derived from length-at-age using a fixed weight/length relationship and thus is insensitive to variation in weight-at-length (i.e. condition). The impact of year-to-year variation in growth and condition on stock reproductive potential can be accounted for if length-at-age and weight-at-length can be estimated empirically for the assessed time period. If this information is not readily available for the full time period, then TEP can be estimated by multiplying weight-at-age by relative fecundity (Equation 4). This approach circumvents the problem of not having length-at-age information. However, it minimises the contrast between SSB and TEP because weight-at-age term is common to both (see Equation (2) and Equation (4)). This method of estimating stock reproductive potential cannot distinguish between short, heavy spawners and long, lean spawners having the same weight-at-age. Despite the ubiquity of length information, it is sometimes difficult to compile information suitable for describing the aggregate stock, particularly when the stock is sampled by different countries having separate monitoring systems and databases. For ICES stocks, it is not standard practice to include time series of length-at-age in assessment reports. This constrains many length-based research initiatives, e.g. growth modelling.

Three studies which have estimated the reproductive potential of five different cod stocks are summarised briefly in the next section. Full details can be obtained by consulting the original publications. It is important to consider that although the different approaches used in these studies have differing degrees of biological resolution, this is also true of the varying formulations of SSB that are used for different stocks (e.g. knife-edge maturity, time invariant weight/length relationships). Of particular interest is how each of the stocks reconstructed historical variation in individual fecundity given that only a limited number of years were sampled.

11.4.1 Three approaches to estimating TEP

The TEP by the Baltic cod stock has been estimated between 1976 and 1999 by estimating FSB and then multiplying this value by a year-specific estimate of relative fecundity (Kraus *et al.* 2002). Determinations of annual average relative fecundity were made for 10 years and these observed values were used to develop a predictive equation for estimating relative fecundity that could be applied to the entire time series. In the predictive model relative fecundity was positively correlated with an index of clupeid prey availability expressed relative to cod abundance. This is equivalent to incorporating a condition effect on fecundity because condition reflects feeding success. It should be noted that by using one value of relative fecundity, this approach assumes that young, small cod have the same relative fecundity as old, large cod. The study concluded that TEP was a superior estimate of stock reproductive potential than SSB because the correlation with survey-based estimates of realised egg production was higher (Kraus *et al.* 2002). Highly resolved estimates of realised egg production from ichthyoplankton surveys allowed environmental effects on larval production to be identified (Köster *et al.* 2003). Although TEP explains a higher degree of recruitment variation, SSB continues to be used as the basis of management advice for the stock (ICES ACFM 2006).

The study by Morgan & Brattey (2005) estimated TEP for three cod stocks in the Northwest Atlantic over extended time periods (1963–2002 for Northern cod, 1960–2002 for southern Grand Bank, and southern Newfoundland cod). For comparative purposes, maturity was es-timated using both a coarse approach (knife edge maturity at either age 7 or age 6) and a detailed approach (cohort-specific maturity ogives estimated for females only). Historical fe-cundity data were used to develop a length-based fecundity model for each stock. For each stock the same fecundity model was used for all years. Thus, fecundity variation was driven solely by variation in length-at-age and the potential effects of interannual variation in condition on fecundity were not incorporated into the estimates of TEP. Unlike the Baltic cod example, there was no clear evidence of an improvement in the S/R relationship. However, the study concluded that different formulations of stock reproductive potential gave different perceptions of stock productivity, as measured by the relative rate of recruitment and in the spawning stock produced per recruit.

The long time series of biological information that is available for the Northeast Arctic cod stock was used by Marshall *et al.* (2006) to estimate both FSB and TEP over the time period 1946–2001. In this study, age/length keys were used to convert numbers-at-age from the assessment to numbers-at-length (n_l) and the following length-based formulation of Equation (3) was used to estimate TEP:

$$\mathrm{TEP} = \sum_{l} n_l \times s_l \times m_{l|f} \times e_l \tag{5}$$

where s_l is the sex ratio at length, $m_l|_f$ is the proportion of females that are mature at length and e_l is the number of eggs produced by a mature female of a given length. The fecundity term was estimated (in millions) as:

$$e_l = \exp(-15.090 + 3.595(\ln(L)) + 1.578(\ln(Kn))) \tag{6}$$

where L is individual length (cm) and Kn is the corresponding length-specific value of relative condition. For each year and length Kn was estimated by the ratio of the 'observed' weight-at-length for that year to the 'long-term' weight-at-length. In this way, interannual variation in condition was incorporated into the estimate of fecundity. Similar to the cod stocks in the Northwest Atlantic (Morgan & Brattey 2005), the relationship between TEP and recruitment was as variable as the relationship between SSB and recruitment. However, in recent years (1980–2001) TEP fell below the threshold reference level more frequently than SSB fell below its corresponding threshold reference level. This suggests that using SSB as a measure of stock reproductive potential could lead to overly optimistic assessments of stock status in some years.

11.4.2 Refining estimates of TEP

As mentioned earlier (Section 11.3.3.2), the number of eggs contained in the ovary prior to or at the start of the spawning does not correspond to the number of fertilised eggs or the number of eggs that survive to the hatching stage. Accordingly, Kraus *et al.* (2002) estimated both potential TEP (using abundance and reproductive characteristics of the spawning females) and realised TEP (using egg production estimates derived from ichthyoplankton surveys). As ichthyoplankton surveys are not routinely conducted for the majority of commercial fish stocks, it is impossible to estimate realised TEP in the field. A limited number of experimental studies have identified maternal traits that impact egg and early larval survival. This knowledge can potentially be adapted for implementation at the stock level provided that the maternal trait that impacts survival has been monitored consistently throughout the time period used by management. The possible refinements that can be made to estimates of potential TEP to *presumably* increase the degree of correlation with realised TEP (presumably, because it is not possible to estimate realised TEP in most cases) are identified below.

11.4.2.1 Correcting for other maternal effects on egg and larval quality

In species which produce larvae that derive nutrition from their yolk-sac, there are three discrete periods of elevated endogenous mortality: during the incubation period, during hatching and at final yolk absorption. Kamler (2005) distinguishes between embryonic survival, which includes the first two sources of mortality, and starvation mortality, which includes the latter source of mortality. Embryonic survival is determined by a combination of parental and environmental factors (Lambert *et al.* 2003). It has been suggested that the maternal factors contributing to embryonic survival include maternal age (Pedersen *et al.* 1997), spawner experience (recruit vs repeat spawner; Solemdal *et al.* 1995), and maternal condition (Marteinsdottir & Steinarsson 1998, Laine & Rajasilta 1999). Provision of energy reserves in the yolk sac also reflects maternal condition (Gagliano & McCormick 2006). Poorly provisioned larvae have higher starvation mortality (Berkeley *et al.* 2004). These 'parent-egg-progeny' relationships (Kamler 2005) potentially offer a mechanism for pro-rating estimates of TEP according to measurable traits of the spawning stock that are most likely to cause the realised TEP to be substantially lower than the potential TEP.

To date, there has not been a parent-egg-progeny relationship developed specifically for application to the stock level. For any given species (or stock) the number of experimental studies which have investigated parental effects on embryonic survival or starvation mortality is limited because of the logistical difficulties of designing experiments. An exception is the pond study of Blom *et al.* (1994) which compared survival and growth of two different strains of Atlantic cod from release as eggs or larvae through to 120 days post-hatching. This study did not find between-strain differences in growth and mortality rates that were consistent in both of the 2 years of the experiment. An interim step has been to explore the possible consequences of maternal effects on realised TEP through simulations. Empirical relationships taken from the Marteinsdottir & Steinarsson (1998) study for Icelandic cod were used by Scott *et al.* (1999) to incorporate maternal effects into an estimate of individual fecundity corrected for egg viability. These corrected values of individual fecundity were then used to estimate realised TEP for a generic cod stock. Murawski *et al.* (2001) incorporated maternal effects into stock dynamics for Georges Bank cod by pro-rating the TEP according to spawning experience to estimate the number of hatched eggs. Hatching success was modelled semi-qualitatively as increasing from 13% to 62% to 100% from the first to second to third spawning seasons, respectively. This estimate of realised TEP was then corrected for the proportion of larvae surviving using the relationship from Marteinsdottir & Steinarsson (1998) for Icelandic cod to derive the total number of viable larvae produced by the stock. Scott *et al.* (2006) incorporated spawner experience and atresia (pre-spawning and seasonal) into estimates of realised TEP. To incorporate spawner experience the relative viability of eggs produced by recruit and repeat spawners was treated simply as 0.5 or 1, respectively. Pre-spawning atresia was assumed to be an exponentially decreasing function of spawner condition based on an empirical study by Hardardottir (2001) for Icelandic cod. Similarly, the empirical relationships developed for black rockfish *(Sebastes melanops)* by Berkeley *et al.* (2004) were used to derive a larval mortality function dependent on maternal age (O'Farrell & Botsford 2006). The effects of age-specific fishing mortality were then simulated by estimating the relative effects on total lifetime egg or larval production (O'Farrell & Botsford 2006). Harvest rate proxies for black rockfish became more conservative as the index of reproductive potential used to construct the S/R relationship changed from SSB to one based on the production of viable larvae (Spencer *et al.* 2007).

Simulation studies, such as those summarised above, raise awareness about the consequences of an age structure truncated through overfishing (Birkeland & Dayton 2005). This awareness will presumably benefit conservation efforts. However, it is important to recognise that the em-pirical basis of the assumptions used in these simulations is often very limited. It is frequently assumed that recruit spawners produce offspring having a lower viability than those produced by repeat spawners (Murawski *et al.* 2001). The empirical basis of this assumption is studies by Solemdal *et al.* (1995) and Trippel (1998), both of which used relatively few spawners. A direct test of the assumption that repeat spawners had higher quality offspring found no difference between recruit and repeat spawners when female spawners were of approximately the same size and condition (Clemmeson *et al.* 2003). Due to the paucity of data, simulation studies often 'borrow' information from one stock and apply it to another stock. For example, the Marteinsdottir & Steinarsson (1998) study for Icelandic cod is the basis of simulations for Georges Bank cod (Murawski *et al.* 2001). There are also examples of general insights from one species being applied to another species. For example, a simulation study by Beamish *et al.* (2006) on the population dynamics of sablefish acknowledged that there were no appropriate biological data for the species and therefore based their assumptions about differential larval mortality rates on the results of the study by Berkeley *et al.* (2004) on black rockfish. Caution is clearly warranted when attempting to generalise from one species or stock to another (Marteinsdottir & Begg 2002). Assessing whether fisheries management can be improved by using more sensitive indices of reproductive potential will depend on evidence from studies having a firm empirical basis and a minimum of assumptions.

11.4.2.2 Correcting for variable egg production in time and space

Larger females often spawn earlier in the season (Ware & Tanasichuk 1990, Wright & Gibb 2005). It is also possible that the timing of spawning is influenced by maternal condition (Kjesbu 1994). The dependency of start time of spawning and duration of spawning on maternal attributes provides another mechanism for linking observable parental traits to the survival of offspring because survival will be determined by the degree to which the production of first-feeding larvae coincides with favourable environmental conditions (Kjesbu *et al.* 1996). Simulations have once again been used to explore the implications of variable population structure and timing of reproduction on reproductive output under a variety of fishing scenarios (Scott *et al.* 2006). Unusually, there is field-based support suggesting the importance of this mechanism for population dynamics. An analysis of the birth-date distribution of juvenile haddock in the North Sea (Wright & Gibb 2005) indicated that survivorship was consistently low when age 2 females contributed the most to TEP. These studies suggest that for batch-spawning fish, information about the intra-annual timing of egg production could be used to develop more precise estimates of realised TEP. For stocks exhibiting distinct spatial gradients in growth and (or) condition it could also be important to estimate egg production by the stock on a spatially disaggregated basis (Heath & Gallego 1998, Begg & Marteinsdottir 2003).

11.5 Implementation in management

The previous section illustrated basic approaches to estimating potential TEP and identified refinements that could be made to adjust the estimates so that they presumably correlate more closely to realised TEP. Admittedly, there is a need for further research to increase the empirical support for key assumptions regarding parental effects on offspring production and viability such that accurate (or at least defensible) estimates of realised TEP are possible. Recognising that information is not currently being used in management of data-rich stocks such as Baltic cod and Northeast Arctic cod we must consider the nature of the impediments.

11.5.1 Current situation

A review of stocks spanning an information gradient showed that SSB continues to be the index of stock reproductive potential used by management even when practicable alternatives exist for data-rich stocks (Marshall *et al.* 2003). At the institutional level the motivation for change would have to become considerably stronger to overcome the innate caution when making changes having economic consequences. It could be argued that the protocols used by fisheries management do not adequately reflect the improvements to basic knowledge being generated through research programmes on a range of species (most notably cod). A parallel can be made with the increase in knowledge related to the possibility of genetic effects on long-term trends in maturation rates. Conventional fisheries management has not yet developed a mechanism for taking genetic effects into account when assessing current and future stock productivity.

11.5.2 Implementation of information on stock reproductive potential in the precautionary approach

A central tenet in the application of the precautionary approach in fisheries management is the use of reference points (Mace 1994). The International Council for the Exploration of the Sea (ICES) states that: 'in order for stocks and fisheries exploiting them to be within safe biological limits, there should be a high probability that (1) the spawning stock biomass is above the threshold where recruitment is impaired' (ICES ACFM 2003). Management advice for the upcoming fishing year is formulated according to the probability of staying above this threshold by a pre-specified margin of error. Within the ICES, piecewise linear regression (Barrowman & Myers 2000) is often used to identify objectively a changepoint for highly indeterminate S/R relationships. The same analytical approach can be used to identify threshold values of TEP (e.g. Marshall *et al.* 2006) to establish whether the stock is within safe biological limits. This would have several conceptual advantages. If estimates of TEP incorporated condition effects on fecundity then the biological reference points defined for TEP would account for the impact of interannual variation in condition on stock productivity (Ratz & Lloret 2003). This would be more consistent with the precautionary approach in the sense that it would avoid the scenario whereby SSB was above the threshold level when TEP was below the threshold level. Estimates of TEP that incorporate well-resolved fecundity data would also account for differences between old/large fish and young/small fish in terms of their relative fecundity. Reference points for the number of eggs-per-recruit are used in the management of lobsters (Gendron 2005), thus accounting for observed sex ratios and proportions of ovigerous females.

11.5.3 Impediments to implementation

For many stocks, researchers undertake research and stock assessment scientists do annual assessments. This division of labour creates a barrier to the exchange of information between the academic and applied branches of fisheries science. The best opportunity for integrating more detailed information on reproductive potential into management is when there is a high degree of communication between scientists undertaking the annual assessment of stock status with scientists engaged in research related to quantifying stock reproductive potential. The cod stocks off the coast of Newfoundland and Labrador are notable in this respect, as in some cases individuals have responsibilities for both assessment (Department of Fisheries and Oceans 2006) and research (Morgan & Brattey 2005). In the past, a dedicated symposium on Variations in Maturation, Growth, Condition and Spawning Stock Biomass Production in Groundfish (9–11 September 1998, Lisbon, Portugal) was attended by both researchers and assessment scientists, in part because it preceded the annual meeting of the North Atlantic Fisheries Organisation (NAFO) annual Scientific Council meeting. This symposium provided a valuable opportunity for dialogue between researchers and applied management regarding the possibilities for incorporating information. As a consequence of this meeting, the NAFO Working Group on Reproductive Potential (Chair: E.A. Trippel) was founded. The terms of reference of the working group (Table 11.1) have included terms that specifically focus on the implementation of information in management. To achieve this goal an effective two-way dialogue between assessment scientists and biologists is required.

11.6 Future directions

Considerable progress has been made towards quantifying the magnitude of maternal effects on stock reproductive potential. As outlined above, the knowledge is amenable to incorporation into management for several data-rich stocks, particularly cod. The utility of knowledge about reproductive potential is not limited to implementation in conventional management protocols. Indeed, having highly resolved population-level estimates of TEP is a unique

Table 11.1 Past and current terms of reference for the North Atlantic Fisheries Organisation (NAFO) Working Group on Reproductive Potential (Chair: E.A. Trippel).

Period	Term of reference	Products to date
2000–2002	ToR 1: Explore and review availability of information and existing data on reproductive potential by areas and species	Morgan *et al.* 2003, Tomkiewicz *et al.* 2003
	ToR 2: Explore possibilities to develop standard internationally coordinated research protocols to estimate egg and larval production	Murua & Saborido-Rey 2003, Murua *et al.* 2003, Thorsen *et al.* 2003, Lambert & Thorsen 2003, Trippel 2003
	ToR 3: Explore and evaluate alternative methods to estimate reproductive potential annually or part of routine in monitoring and sampling schemes (such as HSI)	Lambert *et al.* 2003
	ToR 4: Review possibilities to develop methods and opportunities to estimate stock reproductive potential for assessment and management	Marshall *et al.* 2003, O'Brien *et al.* 2003
2003–2007	ToR 1: Complete inventory of available data in standardised format on reproductive potential for fish stocks of the North Atlantic and Baltic Sea	
	ToR 2: Explore the use of correlation analysis to estimate the reproductive potential of fish stocks having limited data availability	
	ToR 3: Model the interannual and interstock variability in size-dependent fecundity for stocks having multi-year estimates	
	ToR 4: Explore how the current use of biological reference points and medium-term projections can be adapted to include new information on reproductive potential	Morgan & Brattey 2005, Marshall *et al.* 2006
	ToR 5: Explore the consequences of fishery-induced changes in the timing and location of spawning to reproductive success	
	ToR 6: Provide recommendations for the collection of required data in existing research surveys, sentinel fisheries and captive fish experiments that are required to improve annual estimates of reproductive potential for stocks varying in data availability	
	ToR 7: Explore the effects of the environment on Stock Reproductive Potential and how these relate of ToRs 2, 3 and 4	
2008–2010	ToR 1: Explore and conduct evaluation of underlying assumptions of protocols used to estimate total realised egg production of selected marine species and stocks	
	ToR 2: Explore and investigate the potential effects of changes in water temperature and food supply on reproductive success in selected marine species and stocks	
	ToR 3: Undertake appraisal of methods to improve fish stock assessments and fishery management advice that incorporate new biological data for highly exploited and closed fisheries	

resource in ecological research. Several directions for future research are described briefly below.

11.6.1 Delayed effects on reproductive potential

Delayed or intergenerational effects arise when the experience of the parent influences the reproduction and survival of their offspring and possibly later descendents (Metcalfe & Mon-aghan 2001). Intergenerational effects on reproduction have been detected for species ranging from insects (Plaistow *et al.* 2006) to finches (Nagiub *et al.* 2006) to humans (Lummaa 2003, Wells 2003). These studies suggest that early developmental stress can have long-lasting ef-fects affecting the reproductive success of future generations. Poor parental provisioning in the egg or juvenile stages could have effects on long-term traits (Metcalfe & Monaghan 2001). Consequently, conditions experienced in the past could exert an influence on reproductive traits including maturation rate (Morita & Fukuwaka 2006). This area of research generally requires longitudinal studies which are not always feasible for long-lived species. The impact of delayed effects on reproductive potential would presumably be modulated for iteroparous fish species where the spawning stock is composed of several cohorts. Nevertheless, it is worth considering the implications of this emerging research area for fisheries management, especially in the case of fast-growing species.

11.6.2 Life tables

Classical population ecology suggests that the intrinsic rate of population increase (r)isakey determinant of the resiliency of individual populations (Holling 1973, Sibly & Hone 2002). The resiliency of fish stocks to fishing is partly determined by *r* (Musick 1999). The scale of *r* is straightforward: populations that are potentially increasing have positive values of *r*, whereas negative values of *r* indicate a decreasing population. Negative values of *r* have been implicated in the collapse of the Northern cod stock (Hutchings & Myers 1994). The implications of *r* for sustainable harvesting levels have been explored for *K*-selected species, notably elasmobranch species (Cailliet 1992, Beerkircher *et al.* 2003). Direct methods for estimating *r* (Caswell 2001) are based on solving the Euler-Lotka equation:

$$\sum_{x=\alpha}^{w} l_x e^{-rx} m_x = 1.0 \tag{7}$$

where m_x is the fecundity at age x (equating to the number of female offspring produced per female at age x), l_x is the probability of survival to age x, a is the age at first maturity and w is the maximum reproductive age. Thus, r can be interpreted as the number of female offspring produced per female per unit time (usually one year) at low population densities. Given this definition, it is a natural parameter to estimate when highly resolved estimates of fecundity are available. The value of r is also appealing because it represents the reproductive potential of the individual and thus complements stock-level metrics such as TEP which incorporate abundance.

11.6.3 Implications for recovery strategies

Age (and size) structures that are truncated by fishing, potentially have decreased reproductive potential and resiliency to adverse environmental conditions (Secor 2000, Hsieh *et al.* 2006). Restoring a 'normal' age structure has therefore been identified as a factor in successful stock rebuilding strategies (Caddy & Agnew 2004). However, determining the time scales that it will take for stock rebuilding requires a forward projection of future stock sizes and reproductive rates. Katsukawa *et al.* (2002) used an index of long-term (or lifetime) stock reproductive potential as an alternative to SSB. The index is based on Fisher's reproductive value (Fisher 1930) for an age *i* individual (RV_{i}) which is estimated as:

$$\mathrm{RV_i} = \sum_{x=i}^{w} e^{-r(x-i)} \times m_x \times l_x \tag{8}$$

where *r* is the intrinsic rate of population increase (obtained by solving Equation 7) and $e^{-(x-i)}$ is the discount rate of the egg value for an individual of age *i*. The long-term stock reproductive potential (LSRP) is then estimated as:

$$\mathrm{LSRP} = \sum_{x=i}^{w} \mathrm{N_i} \times \mathrm{RV_i} \tag{9}$$

where N_i is the abundance of age *i* individuals. Expressing the population as a matrix model it can be shown that the future stock level is proportional to the initial LSRP (Katsukawa *et al.* 2002). This approach to describing population dynamics gives weight to immature individuals which have all of their reproductive potential in the future but are not considered in the current assessment year using SSB. A variation on this approach has been used for elasmobranchs (Galluci *et al.* 2006).

11.6.4 Disentangling reproductive potential and harvestable biomass

The only biomass index used for formulating management advice regarding quotas is SSB, which is assumed to represent the reproductive potential of the stock and therefore correspond to what the fishery will see in future. However, SSB does not always reflect the potentially harvestable biomass which corresponds to what the fishery sees now. Most commercial fisheries are size-selective and do not discriminate between mature and immature individuals. Stocks having low growth and maturation rates typically become available to the fishery at sizes that are considerably smaller than the size at which they attain maturity. This creates a large discrepancy between harvestable biomass and SSB such that immature fish can comprise a substantial fraction of the commercial catch. The proportion of the landings that are immature fish is not routinely tracked by management, even though high rates of exploitation of immature fish will decrease the probability that an individual fish will reproduce at least once. Management measures to account for the reproduction foregone by harvesting have been proposed (Caddy & Seijo 2002). Switching from an open access fishery to an individual transferable quota for Alaskan sablefish eliminated the race for fish, thereby improving catching efficiency and decreasing the harvesting rate of immature fish (Sigler & Lunsford 2001).

11.7 Conclusion

The precautionary approach dictates that risk management measures, such as biological reference points, should be implemented when scientific uncertainty is large. Several data-rich cod stocks have clearly shown that SSB is an insensitive index of stock reproductive potential. This conclusion is supported by experimental studies documenting maternal effects on egg and larval quantity and quality in a variety of commercial fish species. As a result, the assumptions underlying the use of the S/R relationship are invalid. Despite having isolated and quantified a source of uncertainty this knowledge is not incorporated into management and the risk management measures remain unchanged. This is not consistent with the precautionary approach. Many scientists have encouraged fisheries management to give greater consideration of the effect changing life-history traits has on population dynamics (Solemdal 1997, Conover & Munch 2002, Young *et al.* 2006), particularly reproductive potential (Marshall *et al.* 1998, 2006, Marteinsdottir & Begg 2002, Palumbi 2004, Beamish *et al.* 2006). Implementing change would indicate that fisheries science is applying contemporary research results to move forward from the basic protocols established 50 years ago.

Acknowledgements

Colleagues on the NAFO WG on Reproductive Potential are thanked for many years of enjoyable collaborations. O.S. Kjesbu, M.J. Morgan and E.A. Trippel provided helpful comments on earlier drafts.

References

Ajiad, A., Jakobsen, T. & Nakken, O. (1999) Sexual difference in maturation of Northeast Arctic cod. *Journal of Northwest Atlantic Fisheries Science*, **25**, 1–15.

Barrowman, N.J. & Myers, R.A. (2000) Still more spawner-recruitment curves: the hockey stick and its generalizations. *Canadian Journal of Fisheries and Aquatic Science*, **57**, 665–76.

Beamish, R.J., McFarlane, G.A. & Benson, A. (2006) Longevity overfishing. *Progress in Oceanography*, **68**, 289–302.

Beerkircher, L., Shivji, M. & Cortes, E. (2003) A Monte Carlo demographic analysis of the silky shark *(Carcharhinus falciformis):* implications of gear selectivity. *Fisheries Bulletin*, **101**, 168–74.

Begg, G.A. & Marteinsdottir, G. (2003) Spatial partitioning of relative fishing mortality and spawning stock biomass of Icelandic cod. *Fisheries Research*, **59**, 343–62.

Berkeley, S.A., Chapman, C. & Sogard, S.M. (2004) Maternal age as a determinant of larval growth and survival in a marine fish, *Sebastes melanops. Ecology*, **85**, 1258–64.

Beverton, R.J.H. & Holt, S.J. (1957) *On the Dynamics of Exploited Fish Populations*. Fisheries Investigation Series II, **19.** MAFF, London.

Birkeland, C. & Dayton, P.K. (2005) The importance in fishery management of leaving the big ones. *Trends in Ecology and Evolution*, **20**, 356–8.

Blom, G., Svåsand, T., Jφrstad, K.E., Otterå, H., Paulsen, O.I. & Holm, J.C. (1994) Comparative survival and growth of two strains of Atlantic cod *(Gadus morhua)* through early life stages in a marine pond. *Canadian Journal of Fisheries and Aquatic Science*, **51**, 1012–23.

Boyd, A.J., Shannon, L.J., Schulein, F.H. & Taunton-Clark, J. (1998) Food, transport and anchovy recruitment in the southern Benguela upwelling system of South Africa. In: M.-H. Durand, P. Cury, R. Mendelssohn, C. Roy, A. Bakun & D. Pauly (Eds) *Global Versus Local Changes in Upwelling Systems*. pp.195–209. Editions de l'Orstom, Paris.

Buckley, L.J., Smigielski, A.S., Halavik, T.A., Caldarone, E.M., Burns, B.R. & Laurence, G.C. (1991) Winter flounder *Pseudopleuronectes americanus* reproductive success. II Effects of spawning time and female size on size, composition and viability of eggs and larvae. *Marine Ecology Progress Series*, **74**, 125–35.

Burton, M.P.M., Penney, R.M. & Biddiscombe, S. (1997) Time course of gametogenesis in Northwest Atlantic cod *(Gadus morhua)*. *Canadian Journal of Fisheries and Aquatic Science*, **54**(Suppl. 1), 122–31.

Caddy, J.F. & Agnew, D.J. (2004) An overview of recent global experience with recovery plans for depleted marine resources and suggested guidelines for recovery planning. *Reviews in Fish Biology and Fisheries*, **14**, 43–112.

Caddy, J.F. & Seijo, J.C. (2002) Reproductive contributions foregone with harvesting: a conceptual framework. *Fisheries Research*, **59**, 17–30.

Caillet, G.M. (1992) Demography of the central Californian population of the leopard shark *(Triakis semifasciata)*. *Marine and Freshwater Research*, **43**, 183–93.

Cardinale, M. & Arrhenius, F. (2000) The relationship between stock and recruitment: are the assumptions valid? *Marine Ecology Progress Series*, **196**, 305–9.

Caswell, H. (2001) *Matrix Population Models: Construction, Analysis and Interpretation*. 2nd edn. Sinauer Associates, Sunderland, MA.

Chambers, R.C. & Waiwood, K.G. (1996) Maternal and seasonal differences in egg sizes and spawning characteristics of captive Atlantic cod, *Gadus morhua*. *Canadian Journal of Fisheries and Aquatic Science*, **53**, 1986–2003.

Clemmesen, C., Buhler, V., Carvalho, G., Case, R., Evans, G., Hauser, L., Hutchinson, W.F., Kjesbu, O.S., Mempel, H., Moksness, E., Otteraa, H., Paulsen, H., Thorsen, A. & Svaasand, T. (2003) Variability in condition and growth of Atlantic cod larvae and juveniles reared in mesocosms: environmental and maternal effects. *Journal ofFish Biology*, **62**, 706–23.

Conover, D.O. & Munch, S.B. (2002) Sustaining fisheries yields over evolutionary time scales. *Science*, **297**, 94–6.

DeMartini, E.E. (1991) Annual variations in fecundity, egg size, and the gonadal and somatic conditions of Queenfish *Seriphus politus* (Sciaenidae). *Fisheries Bulletin*, **89**, 9–18.

DeMartini, E.E. & Fountain, R.K. (1981) Ovarian cycling frequency and batch fecundity in the queenfish, *Seriphus politus:* attributes representative of serial spawning fishes. *Fisheries Bulletin*, **79**, 547–60.

De Oliveira, J.A.A., Roel, B.A. & Dickey-Collas, M. (2006) Investigating the use of proxies for fecundity to improve management advice for western horse mackerel *Tracherus trachurus*. *ICES Journal of Marine Science*, **63**, 25–35.

Department of Fisheries and Oceans (2006) Stock assessment on Subdivision 3Ps cod. DFO Canadian Science Advisory Secretariat Science Advisory Report 2006/040.

Fisher, R.A. (1930) *The Genetical Theory of Natural Selection*. Oxford University Press, Oxford.

Friedland, K.D., Ama-Abasi, D., Manning, M., Clarke, L., Kligys, G. & ChambersR.C. (2005) Automated egg counting and sizing from scanned images: rapid sample processing and large data volumes for fecundity estimates. *Journal of Sea Research*, **54**, 307–16.

Gagliano, M. & McCormick, M.I. (2006) Maternal condition influences phenotypic selection on offspring. *Journal of Animal Ecology*, **76**, 174–82.

Galluci, V.F., Taylor, I.G. & Erzini, K. (2006) Conservation and management of exploited shark popula-tions based on reproductive value. *Canadian Journal of Fisheries and Aquatic Science*, **63**, 931–42.

Gendron, L. (2005) Impact of minimum legal size increases on egg-per-recruit production, size structure, and ovigerous females in the American lobster *(Homarus americanus)* population off the Magdalen Islands (Quebec, Canada): a case study. *New Zealand Journal of Marine and Freshwater Research*, **39**, 661–74.

Hardardottir, K. (2001) Relationship between atresia, fish size and condition in Icelandic cod *(Gadus morhua* L.). MSc thesis. Department of Fisheries and Marine Biology, University of Bergen, December 2001. 52 pp.

Heath, M.R. & Gallego, A. (1998) Biophysical modelling of the early life stages of haddock in the North Sea. *Fisheries Oceanography*, **7**, 110–25.

Henderson, B.A., Collins, N., Morgan, G.E. & Vaillancourt, A. (2003) Sexual size dimorphism of walleye *(Stizostedion vitreum vitreum). Canadian Journal of Fisheries and Aquatic Science*, **60**, 1345–52.

Hislop, J.R.G., Robb, A.P. & Gauld, J.A. (1978) Observations on effects of feeding level on growth and reproduction in haddock, *Melanogrammus aeglefinus* (L.) in captivity. *Journal of Fisheries Biology*, **13**, 85–98.

Holling, C.S. (1973) Resilience and stability of ecological systems. *Annual Review of Ecology and Systematics*, **4**, 1–24.

Hsieh, C., Riess, C.S., Hunter, J.R., Beddington, J.R., May, R.M. & Sugihara, G. (2006) Fishing elevates variability in the abundance of exploited species. *Nature*, **443**, 859–62.

Hunt, J.J. (1996) Rates of sexual maturation of Atlantic cod in NAFO Division 5Ze and commercial fishery implications. *Journal of Northwest Atlantic Fisheries Science*, **18**, 61–75.

Hutchings, J.A. & Myers, R.A. (1993) Effect of age on the seasonality of maturation and spawning of Atlantic cod, *Gadus morhua,* in the Northwest Atlantic. *Canadian Journal of Fisheries and Aquatic Science*, **50**, 2468–74.

Hutchings, J.A. & Myers, R.A. (1994) What can be learned from the collapse of a renewable resource? Atlantic cod, *Gadus morhua,* of Newfoundland and Labrador. *Canadian Journal of Fisheries and Aquatic Science*, **51**, 2126–46.

International Council for the Exploration of the Sea Advisory Committee on Fishery Management (2003) Report of the Study Group on biological reference points for Northeast Arctic cod. ICES CM 2003/ACFM:11.

International Council for the Exploration of the Sea Advisory Committee on Fishery Management (2006) Report of the Baltic Fisheries Assessment Working Group (WGBFAS). ICES SM 2006/ACFM:24.

Jarre-Teichmann, A., Weiland, K., MacKenzie, B.R., Hinrichsen, H.-H., Plikshs, M. & Aro, E. (2000) Stock-recruitment relationships for cod *(Gadus morhua callerias* L.) in the central Baltic Sea incorporating environmental variability. *Archive of Fisheries and Marine Research*, **48**, 97–123.

Kamler, E. (2005) Parent-egg-progeny relationships in teleost fishes: an energetics perspective. *Reviews in Fish Biology and Fisheries*, **15**, 399–421.

Katsukawa, T., Matsuda, H. & Matsumiya, Y. (2002) Population reproductive potential: evaluation of long-term stock productivity. *Fisheries Science*, **68**, 1106–12.

Kell, L.T., Pilling, G.M., Kirkwood, G.P., Pastoors, M., Mesnil, B., Korsbrekke, K., Abaunza, P., Aps, R., Biseau, A., Kunzlik, P., Needle, C., Roel, B.A. & Ulrich-Rescan, C. (2005) An evaluation of the implicit management procedure used for some ICES roundfish stocks. *ICES Journal of Marine Science*, **62**, 750–9.

Kelly, C.J. & Codling, E.A. (2006) 'Cheap and dirty' fisheries science and management in the North Atlantic. *Fisheries Research*, **79**, 233–8.

Kjesbu, O.S. (1989) The spawning activity of cod, *Gadus morhua* L. *Journal of Fish Biology*, **34**,195–206.

Kjesbu, O.S. (1994) Time of start of spawning in Atlantic cod *(Gadus morhua)* females in relation to vitellogenic oocyte diameter, temperature, fish length and condition. *Journal of Fish Biology*, **45**, 719–35.

Kjesbu, O.S., Klungsφyr, J., Kryvi, H., Witthames, P.R. & Greer Walker, M. (1991) Fecundity, atresia, and egg size of captive Atlantic cod *(Gadus morhua)* in relation to proximate body composition. *Canadian Journal of Fisheries and Aquatic Science*, **48**, 2333–43.

Kjesbu, O.S., Solemdal, P., Bratland, P. & Fonn, M. (1996) Variation in annual egg production in individual captive Atlantic cod *(Gadus morhua). Canadian Journal of Fisheries and Aquatic Science*, **53**, 610–20.

Kjesbu, O.S., Witthames, P.R., Solemdal, P. & Greer Walker, M. (1998) Temporal variations in the fecundity of Arcto-Norwegian cod *(Gadus morhua)* in response to natural changes in food and temperature. *Journal of Sea Research*, **40**, 303–21.

Knutsen, G. M. & Tilseth, S. (1985) Growth, development, and feeding success of Atlantic cod larvae *Gadus morhua* related to egg size. *Transactions of the American Fisheries Society*, **114**, 507–11.

Koops, M.A., Hutchings, J.A. & McIntyre, T.M. (2004) Testing hypotheses about fecundity, body size and maternal condition in fishes. *Fish and Fisheries*, **5**, 120–30.

Koslow, J. A., Bell, J., Virtue, P. & Smith, D.C. (1995) Fecundity and its variability in orange roughy: effects of population density, condition, egg size, and senescence. *Journal ofFish Biology*, **47**,1063–80.

Köster, F., Hinrichsen, H.-H., Schnack, D., St John, M., MacKenzie, B.R., Tomkiewicz, J., Mollman, C., Kraus, G., Plikshs, M., Makarchouk, A. & Aro, E. (2003) Recruitment of Baltic cod and sprat stocks: identification of critical life stages and incorporation of environmental variability into stock-recruitment relationships. *Scientia Marina*, **67**(Suppl. 1), 129–54.

Köster, F.W., Hinrichsen, H.-H., St John, M.A., Schnack, D., MacKenzie, B.R., Tomkiewicz, J. & Plikshs, M. (2001) Developing Baltic cod recruitment models. II. Incorporation of environmental variability and species interaction. *Canadian Journal of Fisheries and Aquatic Science*, **58**, 1534–66.

Kraus, G., Müller, A., Trella, K. & Köster, F. (2000) Fecundity of Baltic cod: temporal and spatial variation. *Journal of Fish Biology*, **56**, 1327–41.

Kraus, G., Tomkiewicz, J. & Köster, F. (2002) Egg production of Baltic cod *(Gadus morhua)* in relation to variable sex ratio, maturity, and fecundity. *Canadian Journal of Fisheries and Aquatic Science*, **59** 1908–20.

Kurita Y., Meier S. & Kjesbu O.S. (2003) Oocyte growth and fecundity regulation by atresia of Atlantic herring *(Clupea harengus)* in relation to body condition throughout the maturation cycle. *Journal of Sea Research*, **49**, 203–19.

Kurlansky, M. (1997) *Cod: A Biography of the Fish that Changed the World.* Vintage, Denmark.

Laine, P. & Rajasilta, M. (1999) The hatching success of Baltic herring eggs and its relation to female condition. *Journal of Experimental Marine Biology and Ecology*, **237**, 61–73.

Lambert, T.C. (1987) Duration and intensity of spawning in herring *Clupea harengus* as related to the age structure of the mature population. *Marine Ecology Progress Series*, **39**, 209–20.

Lambert, T.C. (1990) The effect of population structure on recruitment in herring. *Journal du Conseil International pour L'Exploracion de la Mer*, **47**, 249–55.

Lambert, Y. & Dutil, J.-D. (2000) Energetic consequences of reproduction in Atlantic cod *(Gadus morhua)* in relation to spawning level of somatic energy reserves. *Canadian Journal of Fisheries and Aquatic Science*, **57**, 815–25.

Lambert, Y. & Thorsen, A. (2003) Integration of captive and wild studies to estimate egg and larval production of fish stocks. *Journal of Northwest Atlantic Fisheries Science*, **33**, 71–9.

Lambert, Y., Yaragina, N.A., Kraus, G., Marteinsdottir, G. & Wright, P.J. (2003) Using environmental and biological indices as proxies for egg and larval production of marine fish. *Journal of Northwest Atlantic Fisheries Science*, **33**, 115–59.

Lummaa, V. (2003) Early developmental conditions and reproductive success in humans: downstream effects of prenatal famine, birthweight, and timing of birth. *American Journal of Human Biology*, **15**, 370–9.

Ma, Y., Kjesbu, O.S. & Jϕrgensen, T. (1998) Effects of ration on the maturation and fecundity in captive Atlantic herring *(Clupea harengus). Canadian Journal of Fisheries and Aquatic Science*, **55**, 900–8.

Mace, P.M. (1994) Relationships between common biological reference points used as thresholds and targets of fisheries management strategies. *Canadian Journal of Fisheries and Aquatic Science*, **51**, 110–22.

Marshall, C.T. & Frank, K.T. (1999) The effect of interannual variation in growth and condition on haddock recruitment. *Canadian Journal of Fisheries and Aquatic Science*, **56**, 347–55.

Marshall, C.T., Kjesbu, O.S., Yaragina, N.A., Solemdal, P. & Ulltang, ϕ (1998) Is spawner biomass a sensitive measure of the reproduction and recruitment potential of Northeast Arctic cod? *Canadian Journal of Fisheries and Aquatic Science*, **55**, 1766–83.

Marshall, C.T., Needle, C.L., Thorsen, A., Kjesbu, O.S. & Yaragina, N.A. (2006) Systematic bias in estimates of reproductive potential of an Atlantic cod *(Gadus morhua)* stock: implications for stock-recruit theory and management. *Canadian Journal of Fisheries and Aquatic Science*, **63**, 980–94.

Marshall, C.T., O'Brien, L., Tomkiewicz, J., Marteinsdottir, G., Morgan, M.J., Saborido-Rey, F., Köster, F.W., Blanchard, J.L., Secor, D.H., Kraus, F., Wright, P., Mukhina, N.V. & Bjornsson, H. (2003) Developing alternative indices of reproductive potential for use in fisheries management: case studies for stocks spanning an information gradient. *Journal of Northwest Atlantic Fisheries Science*, **33**, 161–90.

Marshall, C.T., Yaragina, N.A., Adlandsvik, B. & Dolgov, A.V. (2000) Reconstructing the stock/recruit relationship for Northeast Arctic cod using a bioenergetic index of reproductive potential. *Canadian Journal of Fisheries and Aquatic Science*, **57**, 2433–42.

Marshall, C.T., Yaragina, N.A., Lambert, Y. & Kjesbu, O.S. (1999) Total lipid energy as a proxy for total egg production by fish stocks. *Nature*, **402**, 288–90.

Marteinsdottir, G. & Begg, G.A. (2002) Essential relationships incorporating the influence of age, size and condition on variables required for estimation of reproductive potential in Atlantic cod Gadus morhua. *Marine Ecology Progress Series*, **235**, 235–56.

Marteinsdottir, G. & Bjornsson, H. (1999) Time and duration of spawning cod in Icelandic waters. ICES CM 1999/Y:34.

Marteinsdottir, G., Gudmundsdottir, A., Thorsteinsson, V. & Stefanson, G. (2000) Spatial variation in abundance, size composition and viable egg production of spawning cod *(Gadus morhua* L.) in Icelandic waters. *ICES Journal of Marine Science*, **57**, 824–30.

Marteinsdottir, G. & Steinarsson, A. (1998) Maternal influence on the size and viability of Iceland cod *(Gadus morhua L.)* eggs and larvae. *Journal of Fisheries Biology*, **52**, 1241–58.

Marteinsdottir, G. & Thorarinsson, K. (1998) Improving the stock-recruitment relationship in Icelandic cod *(Gadus morhua* L.) by including age diversity of spawners. *Canadian Journal of Fisheries and Aquatic Science*, **55**, 1372–7.

McIntyre, T.M. & Hutchings, J.A. (2003) Small-scale temporal and spatial variation in Atlantic cod *(Gadus morhua)* life history. *Canadian Journal of Fisheries and Aquatic Science*, **60**, 1111–21.

Mertz, G. & Myers, R.A. (1994) Match/mismatch predictions of spawning duration versus recruitment variability. *Fisheries Oceanography*, **3/4**, 236–45.

Metcalfe, N.B. & Monaghan, P. (2001) Compensation for a bad start: grow now, pay later? *Trends in Ecology and Evolution*, **16**, 254–60.

Monteleone, D.M. & Houde, E.D. (1990) Influence of maternal size on survival and growth of striped bass *Morone saxatalis* Walbaum eggs and larvae. *Journal of Experimental Marine Biology and Ecology*, **140**, 1–11.

Morgan, M.J. & Brattey, J. (2005) Effect of changes in reproductive potential on perceived productivity of three Northwest Atlantic cod *(Gadus morhua)* stocks. *ICES Journal of Marine Science*, **62**, 65–74.

Morgan, M.J., Burnett, J., Tomkiewicz, J. & Saborido-Rey, F. (2003) The availability of data for estimating reproductive potential for selected stocks in the North Atlantic. *Journal of the Northwest Atlantic Fisheries Society*, **37**, 1–3.

Morgan, M.J., Shelton, P.A. & Brattey, J. (2007) Age composition of the spawning stock does not always influence recruitment. *Journal of the Northwest Atlantic Fisheries Society*, **38**, 1–12.

Morita, K. & Fukuwaka, M. (2006) Does size matter most? The effect of growth history on probabilistic reaction norm for salmon maturation. *Evolution*, **60**, 1516–21.

Murawski, S.A., Rago, P.J. & Trippel, E.A. (2001) Impacts of demographic variation in spawning char-acteristics on reference points for fishery management. *ICES Journal of Marine Science*, **58**, 1002–14.

Murua, H., Kraus, G., Saborido-Rey, F., Witthames, P.R., Thorsen, A. & Junquera, S. (2003) Procedures to estimate fecundity of wild collected marine fish in relation to fish reproductive strategy. *Journal of Northwest Atlantic Fisheries Science*, **33**, 33–54.

Murua, H. & Saborido-Rey, F. (2003) Female reproductive strategies of marine fish species of the North Atlantic. *Journal ofNorthwest Atlantic Fisheries Science*, **33**, 23–31.

Musick, J.A. (1999) Criteria to define extinction risk in marine fishes—The American Fisheries Society initiative. *Fisheries*, **24**, 6–14.

Naguib, M., Nemitz, A. & Gil, D. (2006) Maternal developmental stress reduces reproductive success of female offspring in zebra finches. *Proceedings of the Royal Society B*, **273**, 1901–5.

Nash, R.D.M. (1998) Exploring the population dynamics of Irish Sea plaice, *Pleuronectes platessa* L., through the use of Paulik diagrams. *Journal of Sea Research*, **40**, 1–18.

Nissling, A., Larsson, R., Vallin, L. & Frohlund, K. (1998) Assessment of egg and larval viability in cod, *Gadus morhua:* methods and results from an experimental study. *Fisheries Research*, **38**, 169–86.

Northeast Fisheries Science Center (2003) Report of the 36th Northeast Regional Stock Assessment Workshop (37th SAW): advisory report. Northeast Fisheries Science Center Reference Document 03–17.

O'Brien, L., Rago, P.J., Berrien, P. & Lough, R.G. (2003) Incorporating early-life history parameters in the estimation of the stock-recruit relationship of Georges Bank Atlantic cod *(Gadus morhua). Journal of Northwest Atlantic Fisheries Science*, **33** 191–205.

O'Farrell, M.R. & Botsford, L.W. (2006) The fisheries management implications of maternal-age-dependent larval survival. *Canadian Journal of Fisheries and Aquatic Science*, **63**, 2249–58.

Óskarsson, G.J., Kjesbu, O.S. & Slotte, A. (2002) Predictions of realised fecundity and spawning time in Norwegian spring-spawning herring *(Clupea harengus). Journal of Sea Research*, **48**, 59–79.

Óskarsson, G.J. & Taggart, C.T. (2005) Fecundity variation in Icelandic summer-spawning herring and implications for reproductive potential. *ICES Journal of Marine Science*, **63**, 493–503.

Palumbi, S.R. (2004) Why mothers matter? *Nature*, **430**, 621–2.

Parker, G.A. (1992) The evolution of sexual size dimorphism in fish. *Journal of Fish Biology*, **41**(Suppl. B), 1–20.

Parrish, R.H., Mallicoate, D.L. & Klingbeil, R.A. (1986) Age dependent fecundity, number of spawning per year, sex ratio, and maturation stages in the northern anchovy, *Engraulis mordax. Fisheries Bulletin*, **84**, 503–17.

Pedersen, G.I., Gerup, J., Nilssen, L., Larsen, J.R. & Schneider, R. (1997) Body burdens of lipophilic xenobiotics and reproductive success in relation to low temperature and maternal nutritional condition. ICES CM Document 1997/U:10. 22 pp.

Plaistow, S.J., Lapsley, C.T. & Benton, T.G. (2006) Context-dependent intergenerational effects: the interaction between past and present environments and its effect on population dynamics. *American Nature*, **167**, 206–15.

Purchase, C.F., Collins, N.C., Morgan, G.E. & Shuter, B.J. (2005) Sex-specific covariation among life-history traits of yellow perch *(Percaflavescens). Evolutionary Ecology Research*, **7**, 549–66.

Ratz, H.-J. & Lloret, J. (2003) Variation in condition between Atlantic cod *(Gadus morhua)* stocks, the effect on their productivity and management implications. *Fisheries Research*, **60**, 369–80.

Ricker, W.E. (1954) Stock and recruitment. *Journal of the Fisheries Research Board of Canada*, **11**, 559–623.

Rideout, R.M. & Morgan, M.J. (2007) Major changes in fecundity and the effect on population egg production for three species of north-west Atlantic flatfishes. *Journal of Fisheries Biology*, **70**, 1759–79.

Rideout, R.M., Morgan, M.J. & Lilly, G.R. (2006) Variation in the frequency of skipped spawning in Atlantic cod *(Gadus morhua)* off Newfoundland and Labrador. *ICES Journal of Marine Science*, **63**, 1101–10.

Rideout, R.M. & Rose, G.A. (2006) Suppression of reproduction in Atlantic cod, *Gadus morhua. Marine Ecology Progress Series*, **320**, 267–77.

Rideout R.M., Rose G.A. & Burton, M.P.M. (2005) Skipped spawning in female iteroparous fishes. *Fish and Fisheries*, **6**, 50–72.

Rochet, M.-J. (2000) Spatial and temporal patterns in age and size at maturity and spawning stock biomass of North Sea gadoids. ICES CM 2000/N:26.

Schutz, D., Parker, G.A., Taborsky, M. & Sato, T. (2006) An optimality approach to male and female body sizes in an extremely size-dimorphic cichlid fish. *Evolutionary Ecology Research*, **8**, 1393–408.

Scott, B.E., Marteinsdottir, B., Begg, G.A., Wright, P.J. & Kjesbu, O.S. (2006) Effects of population size/age strucuture, condition and temporal dynamics of spawning on reproductive output in Atlantic cod *(Gadus morhua). Ecological Modelling*, **191**, 383–415.

Scott, B., Marteinsdottir, G. & Wright, P. (1999) The potential effects of maternal factors on spawning stock-recruitment relationships under varying fisheries pressure. *Canadian Journal of Fisheries and Aquatic Science*, **56**, 1882–90.

Secor, D.H. (2000) Spawning in the nick of time? Effect of adult demographics on spawning behaviour and recruitment in Chesapeake Bay striped bass. *ICES Journal of Marine Science*, **57**, 403–11.

Sibley, R.M. & Hone, J. (2002) Population growth rate and its determinants: an overview. *Philosophical Transactions of the Royal Society ofLondon B*, **357**, 1153–70.

Sigler, M.F. & Lunsford, C.R. (2001) Effects of individual quotas on catching efficiency and spawning potential in the Alaska sablefish fishery. *Canadian Journal of Fisheries and Aquatic Science*, **58**, 1300–12.

Sinclair, A.F. (2001) Natural mortality of cod *(Gadus morhua)* in the Southern Gulf of St Lawrence. *ICES Journal of Marine Science*, **58**, 1–10.

Skjaeraasen, J.E., Nilsen, T. & Kjesbu, O.S. (2006) Timing and determination of potential fecundity in Atlantic cod *(Gadus morhua). Canadian Journal of Fisheries and Aquatic Science*, **63**, 310–20.

Solemdal, P. (1997) Maternal effects—a link between the past and the future. *Journal of Sea Research*, **37**, 213–27.

Solemdal, P., Bergh, ф, Finn, R.N., Fyhn, H.J., Grahl-Nielsen, O., Homme, O., Kjesbu, O.S., Kjørsvik, E., Opstad, I. & Skiftesvik, A.B. (1992) The effects of maternal status of Arcto-Norwegian cod on egg quality and vitality of early larvae. II. Preliminary results of the experiment in 1992. ICES CM 1992/G:79.

Solemdal, P., Kjesbu, O.S. & Fonn, M. (1995) Egg mortality in recruit- and repeat-spawning cod—an experimental study. ICES CM 1995/G:35.

Spencer, P.D., Hanselman, D. & Dorn, M. (2007) The effect of maternal age of spawning on estimation of F_{msy} for Alaskan Pacific ocean perch. In: J. Heifetz, J. DiCosimo, A.J. Gharrett, M.S. Love, V.M. O'Connell & R.D. Stanley (Eds) *Biology, Assessment, and Management of North Pacific Rockfishes*. pp. 513–33. Alaska Sea Grant, University of Alaska, Fairbanks.

Thorsen, A. & Kjesbu, O.S. (2001) A rapid method for estimation of oocyte size and potential fecundity in Atlantic cod using a computer-aided particle analysis system. *Journal of Sea Research*, **46**, 295–308.

Thorsen, A., Marshall, C.T. & Kjesbu, O.S. (2006) Comparison of various potential fecundity models for north-east Arctic cod *Gadus morhua* L. using oocyte diameter as a standardizing factor. *Journal of Fish Biology*, **69**, 1709–30.

Thorsen, A. Trippel, E.A. & Lambert, Y. (2003) Experimental methods to monitor the production and quality of eggs of captive marine fish. *Journal of Northwest Atlantic Fisheries Science*, **33**, 55–70.

Trippel, E.A. (1998) Egg size and viability and seasonal offspring production of young Atlantic cod. *Transactions of the American Fisheries Society*, **127**, 339–59.

Trippel, E.A. (2003) Estimation of male reproductive success of marine fishes. *Journal of Northwest Atlantic Fisheries Science*, **33**, 81–113.

Trippel, E.A., Kjesbu, O.S. & Solemdal, P. (1997) Effects of adult age and size structure on reproductive output in marine fishes. In: R. C. Chambers & E.A. Trippel (eds) *Early Life History and Recruitment in Fish Populations.* pp. 63–102. Fish and Fisheries Series 21. Chapman & Hall, London.

Tomkiewicz, J., Eriksson. M., Baranova, T., Feldman, V. & Muller, H. (1997) Maturity ogives and sex ratios for Baltic cod: establishment of a database and time series. ICES CM Document No. 1997/CC:20. 22 pp.

Tomkiewicz, J., Morgan, M.J., Burnett, J. & Saborido-Rey, F. (2003) Available information for estimating reproductive potential of northwest Atlantic groundfish stocks. *Journal of Northwest Atlantic Fisheries Science*, **33**, 1–21.

Vallin, L. & Nissling, A. (2000) Maternal effects on egg size and egg buoyancy of Baltic cod, *Gadus morhua.* Implications for stock structure effects on recruitment. *Fisheries Research*, **49**, 21–37.

Ware, D.M. & Tanasichuk, R.W. (1990) A method for forecasting the timing of ripeness in Pacific herring *(Clupea harengus pallasi)* females. *Canadian Journal of Fisheries and Aquatic Science*, **47**, 2375–9.

Wells, J.C.K. (2003) The thrifty phenotype hypothesis: thrifty offspring or thrifty mother? *Journal of Theoretical Biology*, **221**, 1143–61.

Wigley, S.E. (1999) Effects of first time spawners on stock-recruitment relationships for two groundfish species. *Journal of Northwest Atlantic Fisheries Science*, **25**, 215–18.

Wright, P.J. & Gibb, F.M. (2005) Selection for birth date in North Sea haddock and its relation to maternal age. *Journal of Animal Ecology*, **74**, 303–12.

Yoneda, M. & Wright, P.J. (2004) Temporal and spatial variation in reproductive investment of Atlantic cod *Gadus morhua* in the northern North Sea and Scottish west coast. *Marine Ecology Progress Series*, **276**, 237–48.

Young J.L., Bornik Z.B., Marcotte M.L., Charlie K.N., Wagner G.N., Hinch S.G. & Cooke S.J. (2006) Integrating physiology and life history to improve fisheries management and conservation. *Fish and Fisheries*, **7**, 262–83.

Species Index

Fish Reproductive Biology: Implications for Assessment and Management, Second Edition.
Edited by Tore Jakobsen, Michael J. Fogarty, Bernard A. Megrey and Erlend Moksness.

Subject Index

Fish Reproductive Biology: Implications for Assessment and Management, Second Edition.
Edited by Tore Jakobsen, Michael J. Fogarty, Bernard A. Megrey and Erlend Moksness.

www.ingramcontent.com/pod-product-compliance
Ingram Content Group UK Ltd.
Pitfield, Milton Keynes, MK11 3LW, UK
UKHW050844101026
15624UKWH00018B/576